Reliability and Availability Engineering

Do you need to know what technique to use to evaluate the reliability of an engineered system? This self-contained guide provides comprehensive coverage of all the analytical and modeling techniques currently in use, from classical non-state and state space approaches, to newer and more advanced methods such as binary decision diagrams, dynamic fault trees, Bayesian belief networks, stochastic Petri nets, non-homogeneous Markov chains, semi-Markov processes, and phase type expansions. You will quickly understand the relative pros and cons of each technique, as well as how to combine different models together to address complex, real-world modeling scenarios. Numerous examples, case studies, and problems provided throughout help you put knowledge into practice, and a solutions manual and Powerpoint slides for instructors accompany the book online.

This is the ideal self-study guide for students, researchers, and practitioners in engineering and computer science.

Kishor S. Trivedi is a Professor of Electrical and Computer Engineering at Duke University, and a Life Fellow of the IEEE and a recipient of the IEEE Computer Society's Technical Achievement Award.

Andrea Bobbio is a Professor of Computer Science in the Dipartimento di Scienze e Innovazione Tecnologica at the Università degli Studi del Piemonte Orientale, Italy, and a senior member of IEEE.

T0320637

Reliability and Availability Engineering
Modeling, Analysis, and Applications

KISHOR S. TRIVEDI
Duke University, North Carolina

ANDREA BOBBIO
Università degli Studi del Piemonte Orientale

CAMBRIDGE
UNIVERSITY PRESS

CAMBRIDGE
UNIVERSITY PRESS

University Printing House, Cambridge CB2 8BS, United Kingdom

One Liberty Plaza, 20th Floor, New York, NY 10006, USA

477 Williamstown Road, Port Melbourne, VIC 3207, Australia

4843/24, 2nd Floor, Ansari Road, Daryaganj, Delhi – 110002, India

79 Anson Road, #06-04/06, Singapore 079906

Cambridge University Press is part of the University of Cambridge.

It furthers the University's mission by disseminating knowledge in the pursuit of education, learning, and research at the highest international levels of excellence.

www.cambridge.org
Information on this title: www.cambridge.org/9781107099500

© Cambridge University Press 2017

This publication is in copyright. Subject to statutory exception
and to the provisions of relevant collective licensing agreements,
no reproduction of any part may take place without the written
permission of Cambridge University Press.

First published 2017

A catalogue record for this publication is available from the British Library

Library of Congress Cataloguing in Publication data
Names: Trivedi, Kishor Shridharbhai, 1946– author. |
Bobbio, Andrea (Professor of computer science), author.
Title: Reliability and availability engineering : modeling, analysis, and
applications / Kishor Trivedi, Duke University, North Carolina,
Andrea Bobbio, Università degli Studi del Piemonte Orientale.
Description: New York, NY, USA : Cambridge University Press, 2017. |
Includes bibliographical references and index.
Identifiers: LCCN 2017003111 | ISBN 9781107099500 (hardback)
Subjects: LCSH: Reliability (Engineering) | Systems availability.
Classification: LCC TA169.T76 2017 | DDC 620/.00452–dc23
LC record available at https://lccn.loc.gov/2017003111

ISBN 978-1-107-09950-0 Hardback

Cambridge University Press has no responsibility for the persistence or accuracy
of URLs for external or third-party internet websites referred to in this publication,
and does not guarantee that any content on such websites is, or will remain,
accurate or appropriate.

Contents

Preface   *page* xiv

Part I   Introduction   1

1   Dependability   3
    1.1   Definition   3
    1.2   Dependability Measures and Metrics   4
    1.3   Examples of System Dependability Evaluation   7
        1.3.1   Pure Reliability Evaluation   7
        1.3.2   Safety Analysis of Critical Systems   8
        1.3.3   Availability and Maintainability Evaluation   8
        1.3.4   Software Dependability   8
        1.3.5   Service-Oriented Dependability   9
        1.3.6   Task-Oriented Dependability   9
    1.4   Predictive Dependability Assessment   10
    1.5   Further Reading   11
    References   11

2   Dependability Evaluation   15
    2.1   Quantitative Evaluation of Dependability   15
        2.1.1   Measurement-Based Evaluation   16
        2.1.2   Model-Based Evaluation   17
        2.1.3   Interplay between Measurement and Modeling   19
    2.2   The Modeling Process   20
        2.2.1   Studying/Understanding the System Being Modeled   21
        2.2.2   Development of a Conceptual Model   23
        2.2.3   Translation into an Operational Computerized Model   24
        2.2.4   Parametrization of the Operational Model   24
        2.2.5   Solution of the Operational Model and Presentation of the Results   25
        2.2.6   Verification, Validation and Improvement of the Model   25
        2.2.7   Use of the Model   26
    2.3   Modeling Formalisms   27
    2.4   Model Verification and Validation   29
        2.4.1   Model Verification   29
        2.4.2   Model Validation   31

2.5		Errors in Models	32
	2.5.1	Errors in the Construction of the Conceptual Model	32
	2.5.2	Errors in the Construction of the Operational Computerized Model	32
	2.5.3	Errors in Model Solution	33
2.6		The Probabilistic Approach	33
2.7		Statistical Dependence	34
2.8		Further Reading	36
		References	36

3   Dependability Metrics Defined on a Single Unit   41

3.1		Non-Repairable Unit	41
3.2		Common Probability Distribution Functions	46
	3.2.1	Exponential Distribution	46
	3.2.2	Shifted Exponential Distribution	50
	3.2.3	Weibull Distribution	50
	3.2.4	Normal and Lognormal Distributions	52
	3.2.5	Phase-Type Distribution	54
	3.2.6	Gamma Distribution	61
	3.2.7	Log-Logistic Distribution	61
	3.2.8	Bernoulli Distribution	62
	3.2.9	Poisson Distribution	63
	3.2.10	Mass at Origin	63
	3.2.11	Defective Distribution	64
3.3		Minimum and Maximum of Random Variables	66
	3.3.1	The Cdf of the Minimum of Random Variables	66
	3.3.2	The Cdf of the Maximum of Random Variables	68
3.4		Epistemic Uncertainty Propagation	69
	3.4.1	Distribution for Rate Parameter of an Exponential Distribution	70
	3.4.2	Reliability Distribution	70
3.5		Repairable Unit	72
	3.5.1	Measures for Repairable Systems	73
	3.5.2	Renewal Processes	74
	3.5.3	The Exponential Case: Poisson Process	77
	3.5.4	Modified Renewal Process	81
	3.5.5	Availability Analysis: Alternating Renewal Process	81
	3.5.6	Availability Analysis: State Transition Diagram	87
	3.5.7	Sensitivity of Steady-State Availability	91
	3.5.8	Scaled Sensitivity of Steady-State Availability	92
	3.5.9	Cumulative Downtime Distribution	93
3.6		Interval Reliability	94
3.7		Task-Oriented Measures	96
	3.7.1	No Failure and No Repair	96
	3.7.2	Failure and Repair	96
	3.7.3	Failure and Constrained Repair	98

3.8	Improving Dependability	99
3.9	Further Reading	99
References		99

Part II Non-State-Space (Combinatorial) Models   103

4 Reliability Block Diagram   105

4.1	Series Systems		105
	4.1.1	Reliability of Series Systems	106
	4.1.2	Importance (Sensitivity) Analysis of Series Systems	110
	4.1.3	The Parts Count Method	111
	4.1.4	Other Measures for Series Systems	112
4.2	Systems with Redundancy		115
	4.2.1	Parallel Redundancy	115
	4.2.2	Reliability of a Parallel System	116
	4.2.3	Parallel System: The Exponential Case	118
	4.2.4	Importance (Sensitivity) Analysis of Parallel Systems	118
	4.2.5	Parallel Systems: Availability	119
	4.2.6	Series-Parallel Systems	119
	4.2.7	Importance (Sensitivity) Analysis of Series-Parallel Systems	125
	4.2.8	System Redundancy vs. Component Redundancy	126
	4.2.9	Is Redundancy Always Useful?	131
4.3	k-out-of-n Majority Voting Systems		133
	4.3.1	The Exponential Case	136
	4.3.2	Application of k-out-of-n Redundancy	137
	4.3.3	The Consecutive k-out-of-n System	140
4.4	Factoring for Non-Series-Parallel Systems		141
4.5	Non-Identical k-out-of-n System		143
	4.5.1	k-out-of-n System with Two Groups of Components	144
4.6	Further Reading		147
References			148

5 Network Reliability   150

5.1	Networks and Graphs		151
5.2	Binary Probabilistic Networks		151
	5.2.1	Basic Properties of a Binary Decision Diagram	152
	5.2.2	Network Reliability Based on Minpath Analysis	154
	5.2.3	Network Reliability Based on Mincut Analysis	159
	5.2.4	Network Reliability Based on Factoring	162
	5.2.5	Graph-Visiting Algorithm	163
	5.2.6	Deriving Minpaths and Mincuts from BDD	167
5.3	Binary Probabilistic Weighted Networks		173
	5.3.1	Definition of BPWN	174
	5.3.2	Weight as Cost	175
	5.3.3	Weight as Capacity	178

5.4	Multi-State Networks		183
	5.4.1	Weighted Multi-State Networks	185
	5.4.2	Multi-Valued Decision Diagrams	185
	5.4.3	Basic Operations for MDD Manipulation	186
	5.4.4	Algorithmic Implementation and Probability Evaluation	189
5.5	Limitation of the Exact Algorithm		192
5.6	Computing Upper and Lower Bounds of Reliability		193
5.7	Further Reading		196
References			196

6	**Fault Tree Analysis**		201
6.1	Motivation for and Application of FTA		201
6.2	Construction of the Fault Tree		203
	6.2.1	The OR Logic Gate	203
	6.2.2	The AND Logic Gate	205
	6.2.3	Fault Trees with OR and AND Gates	206
	6.2.4	Fault Trees with Repeated Events	206
	6.2.5	The k-out-of-n Node	207
6.3	Qualitative Analysis of a Fault Tree		208
	6.3.1	Logical Expression through Minimal Cut Sets	208
	6.3.2	Logical Expression through Graph-Visiting Algorithm	212
	6.3.3	Qualitative Analysis of FTRE	213
	6.3.4	Fault Tree and Success Tree	216
	6.3.5	Structural Importance	217
6.4	Quantitative Analysis		219
	6.4.1	Probability of Basic Events	219
	6.4.2	Quantitative Analysis of an FT without Repeated Events	221
	6.4.3	Quantitative Analysis of an FT with Repeated Events	226
6.5	Modularization		228
6.6	Importance Measures		229
	6.6.1	Birnbaum Importance Index	230
	6.6.2	Criticality Importance Index	231
	6.6.3	Vesely–Fussell Importance Index	232
6.7	Case Studies		234
6.8	Attack and Defense Tree		243
	6.8.1	Countermeasures	243
	6.8.2	Weighted Attack Tree (WAT)	244
6.9	Multi-State Fault Tree		247
6.10	Mapping Fault Trees into Bayesian Networks		250
	6.10.1	Bayesian Network Definition	250
	6.10.2	Mapping an Algorithm from FT to BN	252
	6.10.3	Probabilistic Gates: Common Cause Failures	255
	6.10.4	Noisy Gates	256
	6.10.5	Multi-State Variables	257

		6.10.6	Sequentially Dependent Failures	259
		6.10.7	Dependability Analysis through BN Inference	261
	6.11	Further Reading		264
	6.12	Useful Properties of Boolean Algebra		265
	References			266

7	**State Enumeration**			**271**
	7.1	The State Space		271
		7.1.1	Characterization of System States: Truth Table and Structure Function	274
		7.1.2	Structural Importance and Frontier States	275
		7.1.3	Boolean Expression of the Structure Function	275
	7.2	The Failure Process Defined on the State Space		278
		7.2.1	Dependability Measures Defined on the State Space	279
	7.3	System Reliability with Independent Components		280
	7.4	Repairable Systems with Independent Components		284
	References			285

8	**Dynamic Redundancy**			**286**
	8.1	Cold Standby Case		286
	8.2	Warm Standby		289
	8.3	Hot Standby and k-out-of-n		291
	8.4	Imperfect Fault Coverage		292
	8.5	Epistemic Uncertainty Propagation		295
		8.5.1	Cold Standby with Identical Components	296
		8.5.2	Warm Standby	298
		8.5.3	Hot Standby and k-out-of-n	299
	References			300

| **Part III** | **State-Space Models with Exponential Distributions** | | | **303** |

9	**Continuous-Time Markov Chain: Availability Models**			**305**
	9.1	Introduction		305
		9.1.1	Chapman–Kolmogorov Equations	306
		9.1.2	The Infinitesimal Generator Matrix	308
		9.1.3	Kolmogorov Differential Equation	309
		9.1.4	Distribution of the Sojourn Time in a Given State	312
	9.2	Classification of States and Stationary Distribution		313
		9.2.1	Irreducible Markov Chain	313
		9.2.2	Expected State Occupancy	315
	9.3	Dependability Models Defined on a CTMC		317
		9.3.1	Expected Uptime and Expected Downtime	318

9.4   Markov Reward Models   329
       9.4.1   MRM with Reward Rates   330
       9.4.2   MRM with Impulse Reward   332
9.5   Availability Measures Defined on an MRM   333
       9.5.1   Instantaneous, Steady-State, and Interval Availability   333
       9.5.2   Expected Uptime and Expected Downtime   333
       9.5.3   Expected Number of Transitions   334
       9.5.4   Expected Number of Visits   334
       9.5.5   Expected Number of System Failures/Repairs   335
       9.5.6   Equivalent Failure and Repair Rate   337
       9.5.7   Defects per Million   340
9.6   Case Study: IBM Blade Server System   342
9.7   Parametric Sensitivity Analysis   349
       9.7.1   Parametric Sensitivity Analysis for a CTMC   351
9.8   Numerical Methods for Steady-State Analysis of Markov Models   352
       9.8.1   Power Method   352
       9.8.2   Successive Over-Relaxation   353
9.9   Further Reading   353
References   354

10   **Continuous-Time Markov Chain: Reliability Models**   357
10.1   Continuous-Time Markov Chain Reliability Models   357
       10.1.1   Convolution Integration Method for Transient
                Probabilities   363
       10.1.2   Solution with Laplace Transforms   364
       10.1.3   Other Dependability Measures Defined on a CTMC   373
10.2   Continuous-Time Markov Chains with Absorbing States   374
       10.2.1   Single Absorbing State: Cdf of Time to Absorption   375
       10.2.2   Single Absorbing State: Expected Time to Absorption   376
       10.2.3   Single Absorbing State: Moments of Time to Absorption   381
       10.2.4   Multiple Absorbing States   388
       10.2.5   Expected First-Passage Time   397
10.3   Continuous-Time Markov Chains with Self-Loops   405
       10.3.1   Uniformization of a CTMC   409
10.4   Transient Solution Methods   412
       10.4.1   Fully Symbolic and Semi-Symbolic Methods   412
       10.4.2   Transient Solution via Series Expansion   413
       10.4.3   Transient Solution via Uniformization (Jensen's method) 415
       10.4.4   ODE-Based Solution Methods   416
10.5   Further Reading   419
References   419

11   **Continuous-Time Markov Chain: Queuing Systems**   423
11.1   Continuous-Time Markov Chain Performance Models   424

11.2	The Birth–Death Process	424
11.3	The Single Queue	427
	11.3.1 $M/M/1$ Queue	428
11.4	$M/M/m$: Single Queue with m Servers	433
	11.4.1 $M/M/\infty$: An Infinite Number of Servers	435
11.5	$M/M/1/K$: Finite Storage	435
	11.5.1 $M/M/m/m$ Queue	438
11.6	Closed $M/M/1$ Queue	441
11.7	Queues with Breakdown	442
11.8	Further Reading	450
	References	451

12	**Petri Nets**	**453**
12.1	Introduction	453
12.2	From Petri Nets to Stochastic Reward Nets	455
	12.2.1 Petri Nets	455
	12.2.2 Structural Extensions to Petri Nets	458
	12.2.3 Stochastic Petri Nets	460
	12.2.4 Generalized Stochastic Petri Nets	463
12.3	Stochastic Reward Nets	469
12.4	Computing the Dependability and Performance Measures	476
12.5	Further Reading	482
	References	482

Part IV	**State-Space Models with Non-Exponential Distributions**	**487**

13	**Non-Homogeneous Continuous-Time Markov Chains**	**489**
13.1	Introduction	489
	13.1.1 Kolmogorov Differential Equation	489
13.2	Illustrative Examples	490
13.3	Piecewise Constant Approximation	494
13.4	Queuing Examples	498
13.5	Reliability Growth Examples	501
13.6	Numerical Solution Methods for NHCTMCs	503
	13.6.1 Ordinary Differential Equation Method	503
	13.6.2 Uniformization Method	504
13.7	Further Reading	505
	References	505

14	**Semi-Markov and Markov Regenerative Models**	**509**
14.1	Introduction	509
14.2	Steady-State Solution	510

14.3   Semi-Markov Processes with Absorbing States   528
       14.3.1   Mean Time to Absorption   528
       14.3.2   Variance of the Time to Absorption   535
       14.3.3   Probability of Absorption in SMPs with Multiple Absorbing
                States   538
14.4   Transient Solution   539
14.5   Markov Regenerative Process   542
       14.5.1   Transient and Steady-State Analysis of an MRGP   543
14.6   Further Reading   546
References   547

15   Phase-Type Expansion   551
15.1   Introduction   551
       15.1.1   Definition   552
15.2   Properties of the PH Distribution   556
15.3   Phase-Type Distributions in System Modeling   558
15.4   Task Completion Time under a PH Work Requirement   563
15.5   Phase-Type Distribution in Queuing Models   566
15.6   Comparison of the Modeling Power of Different Model Types   569
15.7   Further Reading   570
References   570

Part V   Multi-Level Models   575

16   Hierarchical Models   577
16.1   Introduction   577
16.2   Hierarchical Modeling   578
       16.2.1   Import Graph   579
16.3   Availability Models   580
16.4   Reliability Models   586
       16.4.1   Phased Mission Systems   590
       16.4.2   Behavioral Decomposition   592
16.5   Dynamic Fault Tree   599
16.6   Performance Models   604
16.7   Performability Models   608
16.8   Survivability Models   619
16.9   Further Reading   624
References   624

17   Fixed-Point Iteration   631
17.1   Introduction   631
       17.1.1   Revisiting the Import Graph   632
17.2   Availability Models   632
17.3   Reliability Models   635

17.4 Performance Models   636
17.5 Further Reading   640
References   640

Part VI   Case Studies   643

18   Modeling Real-Life Systems   645
18.1 Availability Models   645
18.2 Reliability Models   669
18.3 Combined Performance and Reliability Models   675
18.4 Further Reading   689
References   690

Author Index   695
Subject Index   704

Preface

The increasing dependence on technological infrastructure, especially in advanced economies, is not without its perils. Even short outages of the infrastructure can have drastic consequences, ranging from economic losses all the way to loss of human life in the worst case. It is thus imperative that our complex and critical infrastructure, including the likes of the Internet, transportation, energy supply, communications, health, financial services, and commerce, be designed and implemented with the assurance of a certain level of dependability. While the aforementioned scenarios are typical, how can their dependability be guaranteed *a priori* with a certain level of confidence? This makes it imperative that suitable techniques that enable us to assess, predict, verify, and validate the levels of reliability, availability, safety, and maintainability of these products, services, and infrastructures be developed. Now there are many books that provide some insights into the methods of dependability assurance, but none of them provides a comprehensive set of techniques in the context of real-life applications. The plethora of analytical techniques available for the evaluation of reliability and availability often leaves a practitioner in a quandary, unable to figure out the most suitable approach to analytically evaluate the system under investigation.

The primary aim of this book is to comprehensively consider in detail all the techniques together, showing the similarities and differences and pointing out the pros and cons of each approach. Furthermore, the book not only covers classical techniques like reliability block diagrams, fault trees, network reliability, and Markov models, but also more advanced techniques like non-exponential models and new approaches and analysis techniques, like the use of binary decision diagrams, dynamic fault trees, Bayesian belief networks, and stochastic Petri nets. Furthermore, the book particularly addresses multi-level modeling for the analysis of large systems, combining different modeling formalisms. Another unique feature of the book is that there are three chapters exclusively devoted to dealing with non-exponential distributions. A major strength of the book is the large number of solved examples, a significant number of real case studies and many problems that can be assigned as homework. Some of the examples are covered repeatedly in different chapters to show the application of different techniques to the same basic example. We are planning a solution manual and a set of PowerPoint slides for instructors using this book as a text. The software packages SHARPE and SPNP can be obtained by contacting the first author. Three key topics are not covered, except in a few examples, due to size limitations: statistical techniques, discrete-event simulation, and optimization.

Over forty years of the authors' experience in developing techniques and software applications, and real-life experience in consultancy and teaching, is distilled in this book. The book can be used for a two-semester course on reliability engineering; it can also be used by practicing engineers, as it is comprehensive and contains many real-life case studies; and it can be used as a reference by researchers in reliability engineering.

We are particularly indebted to Dr. Jogesh Muppala of Hong Kong University of Science and Technology for his invaluable assistance in developing this book. We would like to thank the following friends, colleagues, and students for helping with proofreading, developing examples, drawing figures, and in many other ways: Javier Alonso Lopez, Arthur Carran, Xiaolin Chang, Ester Ciancamerla, Daniele Codetta Raiteri, Amita Devaraj, Salvatore di Stefano, Flavio Frattini, Rafael Fricks, Rahul Ghosh, Siqian Gong, Andras Horváth, Dong Seong Kim, Jang Se Lee, Xiaodan Li, Jaeshik Lim, Dian Liu, Fumio Machida, Jose M. Martinez, Rivalino Matias, Jr., Michele Minichino, Kesari Mishra, Subroto Mondal, Gaorong Ning, Riccardo Pinciroli, Agapios Platis, Luigi Portinale, Yu Qiao, Fangyun Qin, Kun Qiu, Ricardo Rodriguez, Arpan Roy, Pablo Sartor, Cun Shi, Harish Sukhwani, Roberta Terruggia, Ishan Upadhyay, Alfredo Verna, Haoqin Wang, Nan Wang, Ruofan Xia, Liudong Xing, Beibei Yin, Meng Lai Yin, Xiaoyan Yin, Jing Zhao, Yijing Zhao, Zhibin Zhao, Zheng Zheng.

IIT Gandhinagar and its Director Sudhir Jain were excellent hosts during our sabbaticals, when the major initiative on this book took place. Our parent institutions, Duke University ECE Department and DiSit Università del Piemonte Orientale, are much appreciated. The patience shown by the publisher and their representatives is worthy of emulation. We would like to thank our families for their understanding during several periods of intense work on the book. This book is dedicated to our grandchildren: Ishaan, Kairav, Avi, Maia, Andrea, Lea, and Alice.

Part I

Introduction

The human race, in its endeavor to improve its life and well-being, has built around itself a marvellous technological framework, culminating in the recent information technology revolution. The increasing dependence on this technological infrastructure, especially in advanced economies, is not without its perils. Even short outages of the infrastructure can have drastic consequences, ranging from economic loss to loss of human life. It is thus imperative that this complex and critical infrastructure, including the likes of the Internet, transportation, energy supply, manufacturing, communications, health, financial services, and commerce, be designed and implemented with the assurance of a certain level of *dependability*.

Consider the following scenarios:

- A cloud service provider wants to assure its customer that the maximum service outage within a calendar year will be less than five minutes with a guarantee of 99.999%.
- A hard disk manufacturer wishes to provide a guaranteed mean time to hard disk failure of at least 1500 days.
- A telecommunications company needs to ensure that the number of calls lost per million calls is below a preset threshold, say 25 calls.

While the aforementioned scenarios are typical, how can the conditions be guaranteed *a priori* with a certain level of confidence? This makes it imperative that suitable techniques that enable us to assess, predict, verify, and validate the levels of reliability and availability of these products, services, and infrastructures be developed. Assurance techniques that rely solely on checklists and qualitative arguments regarding implemented features that enhance dependability will no longer suffice. A rigorous mathematical framework that enables us to provide these assurances is essential. This book is an attempt at providing one such framework.

The book is divided into six parts: Introduction, Non-State-Space Methods, State-Space Methods with Exponential Distributions, State-Space Methods with Non-Exponential Distributions, Multi-Level Models, and Case Studies. In Part I, we introduce the need for and metrics of dependability in Chapter 1, discuss evaluation methods in Chapter 2 and provide a thorough treatment of the reliability and availability of a non-redundant system in Chapter 3.

Since redundancy is used extensively to achieve high reliability and availability, the rest of the book develops methods of evaluation of systems with redundancy. Part II shows the use of so-called non-state-space methods that avoid the construction, storage, and solution of the underlying state space, leading to a highly efficient method of evaluation. But the simplicity and efficiency come with the (unrealistic) assumption of independence in the failure/repair behavior of the components and subsystems. Taking various kinds of dependency among system components into account requires the use of state-space methods. The most common state-space method is based on homogeneous continuous time Markov chains. These are described in Part III. Markov chains suffer from two major afflictions: the assumption that all inter-event times are exponentially distributed, and exponential growth in the state space. Part IV shows how to relax the assumption of exponential distribution, and Part V shows how to deal with the state-space explosion problem. Finally, Part VI covers a number of real-life case studies from the authors' own research and experiences.

1 Dependability

Terms like reliability, availability, maintainability, safety, and dependability are often used loosely in colloquial conversation without relying on their precise definition and representation. However, in scientific literature it is important to define these terms precisely so that they can be described, represented, and evaluated without any loss of precision. This chapter is an attempt to present this latter view in a comprehensive manner and set the stage for the detailed exposition that follows in the subsequent chapters.

1.1 Definition

The scientific use of the terms dependability, availability, reliability, maintainability, and safety were first defined rigorously by Laprie [1]. The terminology was further refined and comprehensive definitions relating to dependability were given in a landmark paper by Avizienis *et al.* [2].

Dependability is treated as an umbrella concept that encompasses attributes such as reliability, availability, safety, integrity, and maintainability. Security-related concepts such as confidentiality, availability, and integrity were also brought under the overall framework in this paper. Further elaboration on these ideas has been suggested elsewhere, including [3]. Let us now examine the term *dependability* in some detail.

The IEC international vocabulary [4] defines the dependability (of an item) as the "ability to meet success criteria, under given conditions of use and maintenance." In the information technology area [5], the meaning of the word "dependability" has been thoroughly investigated by the International Federation for Information Processing (IFIP) working group WG10.4 (on dependable and fault-tolerant computing), and a recent definition issued by this group appears in [2]:

The dependability of a computer system is the ability to deliver a service that can justifiably be trusted. The service delivered by a system is its behavior as it is perceived by its user(s); a user is another system (physical, human) that interacts with the former at the service interface.

A 1988 survey of several definitions of computer-based system dependability resulted in the following summary [6]:

Dependability of a computer system may be defined as the justifiable confidence the manufacturer has that it will perform specified actions or deliver specified results in a trustworthy and timely manner.

In this context a system can be a single component, a module, a subsystem of a complex system, such as a power or chemical or nuclear plant, a computer system, a software system such as a web server, a data center, a telephone network, smart grid, cloud, an avionic traffic control system or an aircraft flight control system.

Dependability is the result of a large number of physical, technological, and structural characteristics of the system, in addition to the impact of the environmental and operating (application) conditions: the materials with which the elementary components are built, the assembly techniques, the operating conditions including the workload being processed and the thermal conditions, the interaction with the human operator, the technical assistance and maintenance policies. Understanding, modeling, and analyzing the dependability of a system with the aim of pinpointing potential weaknesses and improving the capability to operate correctly requires the harmonious combination of different disciplines, from materials science to probability and statistics, from manufacturing engineering to man–machine interaction and production organization.

The degree to which a system is able to provide the expected operation or service for which it is designed needs to be quantitatively assessed by defining proper measurable quantities. The quantitative assessment of system dependability thus becomes essential in system design, planning, implementation, validation, manufacturing, and field operation. A number of requirements, methods, and techniques have been established and standardized to quantitatively evaluate the capability of a system for providing the desired operation.

Along the various phases of the design and manufacturing process of a system there is a need to predict with the lowest possible degree of uncertainty: how the system will behave in operation during its entire useful life; how malfunctions or failures will appear during operation and with what frequency; how long the duration of the outages will be and what resources are needed to maintain the system in a correct operating state. The quantitative evaluation of these attributes plays a critical role in assessing the effectiveness of design alternatives, the choice of appropriate materials or parts, and in determining the success of a product or system.

In order to quantitatively assess and assure the dependability attributes of a system, different measures are introduced and a set of modeling and analysis techniques have been developed to derive and evaluate the measures. The following sections examine these in further detail.

1.2   Dependability Measures and Metrics

Dependability, as defined earlier, is a qualitative property, an aptitude of the system to conform to design specifications and user expectations. However, in order to establish a quantitative theory of dependability, specific quantities that characterize different aspects or attributes of dependability have to be formally defined so that they can be evaluated through unambiguous mathematical techniques.

IFIP WG10.4 views dependability issues from three aspects: threats, attributes, and means. The taxonomy of the precise terms and their relationship is delineated in the

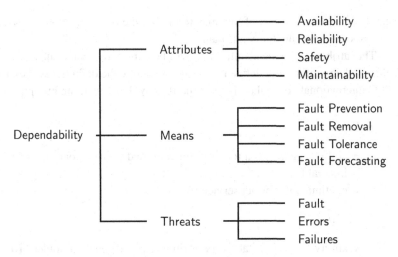

Figure 1.1   IFIP WG10.4 dependability tree.

dependability tree as shown in Figure 1.1 [2, 5]. The dependability tree has subsequently been extended to include security as well [2].

The threats to dependability can be viewed in terms of faults, errors, and failures. Faults are adjudged causes of errors and failures. A fault, when exercised, may produce an internal error. Errors, either singly or on accumulation, may give rise to a failure. Failure of a subsystem in turn becomes a fault at the system level. Faults for electronic hardware have been further classified into permanent, transient, and intermittent [7]. For software, faults were recently classified into "Bohrbugs," (non-aging-related) "Mandelbugs," and aging-related bugs [8], and for networks errors have been classified into single-bit, multiple-bit, and correlated errors in [9]. Failures have been classified into omission failures, value (or content) failures, and timing failures [10]. Timing failures are also known as performance failures or dynamic failures. Failures are also classified by their severity [11] or criticality [12]. For instance, failures in safety-critical and life-critical systems can be classified into safe and unsafe failures. For capturing security metrics, failures have been classified into those compromising confidentiality, those compromising integrity and those leading to a lack of access (that is, unavailability) [13, 14]. For further discussion on this topic, see [2].

The means to achieving and assuring dependability include fault prevention (or fault avoidance), which consists of carefully designing a system with a minimal number of faults, and fault removal, which is the process of finding and fixing bugs during testing or during operation. Fault tolerance constitutes a set of techniques that allow the system as a whole to continue to function in spite of component or subsystem failures. The use of redundancy is an essential part of fault tolerance. Using more components than required (massive redundancy), repeating an operation (time redundancy) or using more bits than required (information redundancy) are commonly used techniques. Furthermore, the management of redundancy including detection, location, reconfiguration and recovery is an essential aspect of fault tolerance. Finally, fault forecasting consists of methods to predict the occurrence of faults/errors/failures or associated dependability attributes.

The attributes of dependability that will be considered in the subsequent chapters of this book are informally defined next, by combining the definitions taken from both the IEC international vocabulary [4] and the IFIP WG10.4 working group [2].

Reliability

- ability to perform a required function under given conditions for a given time interval [4];
- continuity of correct service [2].

Availability

- ability to be in a state to perform a required function, under given conditions, at a given instant of time, or after enough time has elapsed, assuming that the required external resources are provided [4];
- readiness for correct service [2].

Maintainability

- ability to be retained in, or restored to, a state in which it can perform a required function, under given conditions of use and maintenance [4];
- ability to undergo modifications and repairs [2].

A fundamental difference between reliability and availability is that reliability refers to failure-free operation during an interval, while availability refers to failure-free operation at a given instant of time, usually the time when a device or system is accessed to provide a required function or service. Reliability is a measure that characterizes the failure process of an item, while availability combines the failure process with the restoration or repair process and looks at the probability that at a given time instant the item is operational independently of the number of failure/repair cycles already undergone by the item.

Each one of the above attributes can be characterized by different metrics that will be formally defined in Chapter 3. In the probabilistic approach to the quantitative dependability assessment, the above attributes (reliability, availability, maintainability) will be computed as the probability of occurrence of specific events.

- For computing the attribute *reliability*, we refer to the event that the system is operating continuously without failures in a given interval under specified operating and environmental conditions. The *reliability* is the probability that this event occurs.
- For computing the attribute *availability*, we refer to the event that the system is operating at a given point in time independently of the number of failures (and repairs) already incurred by the system. The *availability* is the probability that this event occurs.

- For computing the attribute *maintainability*, we refer to the event that when the system is not operational due to failures, it will be recovered to an operating condition. The *maintainability* is the probability that this event occurs.

The traditional dependability metrics defined above take a *system-oriented* perspective [2]. *Service-oriented* dependability metrics as defined by Tortorella [15], on the other hand, take a user-oriented perspective. Tortorella [15] classifies the figures of merit for service dependability into three attributes:

Service accessibility: the ability to initiate the service request when desired.
Service continuity: the successful continuation of a successfully initiated service request.
Service release: the successful completion of an initiated service request.

Bauer and Adams [16] state that since most services are reliable, it is more convenient to focus on the much smaller number of unreliable service events or service defects. These service defects are conveniently normalized as the number of calls or customer demands not served per million attempts, referred to as defects per million (DPM) [16, 17].

1.3 Examples of System Dependability Evaluation

So far, we have examined several conceptual and theoretical aspects of dependability. We present a brief overview of several examples of dependability evaluation in this section. The motivation behind presenting these examples is threefold:

- provide concrete evidence of the applicability of the techniques that will be encountered in the subsequent chapters of this book;
- illustrate how the dependability techniques are employed in the evaluation of specific systems;
- showcase different real-life case studies that span the whole gamut of dependability evaluation, ranging from pure reliability evaluation to systems showing attributes of availability, maintainability, and beyond.

1.3.1 Pure Reliability Evaluation

Mission-critical systems like aircraft and spacecraft flight control, nuclear reactor control systems and telecommunication systems are characterized by stringent requirements imposed by government regulatory bodies on the probability of catastrophic failure. As an example, the US Federal Aviation Administration (FAA) mandates that the catastrophic failure probability of aircraft should be below 10^{-9}/flight-hour [18, 19]. These are classic examples of systems requiring pure reliability evaluation. Ramesh *et al.* [20] consider the reliability evaluation of an aircraft in a flight operation time management scenario. They develop analytical and numerical methods for probabilistic risk analysis using fault trees, Markov chains, and stochastic Petri nets. Hjelmgren *et al.* [21] carry out the reliability analysis of a fault-tolerant Full Authority Digital Electronic Control system (FADEC) used for control of an aircraft gas turbine engine on an aircraft equipped with a single engine. Two redundancy options, including a two-channel hot standby and a three-channel triple modular redundancy (TMR) system that is reconfigured into a two-channel hot standby after an error has been detected and located, are evaluated. Markovian models are used to assess the probability of failure for the system. The operational framework for fault detection, identification and recovery in autonomous spacecraft has been studied in [22] using dynamic fault trees and Bayesian networks.

1.3.2   Safety Analysis of Critical Systems

In safety-critical systems, like chemical, nuclear or power plants, or intensive care units for emergency rooms or operating theatres in a hospital, the occurrence of a system failure may entail catastrophic consequences for human life or the environment. The interest in the analysis is to evaluate the probability that such a catastrophic condition may occur, and the time of the first occurrence of such a catastrophic condition.

1.3.3   Availability and Maintainability Evaluation

High-availability systems, such as the IBM BladeCenter®, are designed to provide commercial services such as e-commerce, financial, stock trading, and telephone communication services [23]. Blade servers are widely adopted because of their modular design, with industry-standard racks accommodating multiple servers together with shared power, cooling, and other services within the rack chassis. Availability requirements for such systems are in the region of 0.999 99 ("five nines"), with annual mean system downtime requirements below six minutes. Smith *et al.* [23] present a detailed availability analysis of an IBM BladeCenter comprising up to 14 separate blade servers contained within a single chassis. They construct a comprehensive two-level hierarchical model with a higher-level fault tree model of the system with the underlying subsystems and components being modeled as lower-level Markov chains. Specific dependability metrics for this model include the system availability and the downtime in minutes/year. In addition, they conduct sensitivity analysis of the system, including the contribution of each component to a single blade downtime.

1.3.4   Software Dependability

Software *aging* [24] is now a well-recognized phenomenon, resulting from degradation in the software state or the execution environment. This degradation is due to such causes as memory bloating and leaking, unreleased file locks, data corruption, storage space fragmentation and accumulation of roundoff errors. The net result is performance degradation of the software, possibly resulting in a crash/hang failure. Software aging-related bugs are included in the software fault classification (Bohrbugs or Mandelbugs) by [24]. To counteract the effects of aging-related problems, software *rejuvenation* is often employed, which entails occasionally stopping the software, cleaning up its state and the environment and restarting the software. The primary motivation behind this example [24] is to derive optimal rejuvenation schedules to maximize availability or minimize downtime cost. Toward this goal, the authors first construct a semi-Markov reward model based on workload and resource usage data collected from a Unix environment. The model is then solved to obtain estimated times to exhaustion for each resource. The results from the semi-Markov reward model are then fed into a higher-level semi-Markov availability model that accounts for failure followed by reactive recovery, as well as proactive recovery, to determine the optimal rejuvenation schedule.

1.3.5   Service-Oriented Dependability

Service-oriented environments like telecommunication systems are better evaluated using service-oriented dependability metrics. In this example, we focus on an environment that supports Voice over IP (VoIP) functionality using the session initiation protocol (SIP), an application-layer control protocol for creating, modifying, and terminating sessions, including Internet telephone calls, multimedia distribution, and multimedia teleconferences. A typical SIP-based communication involves interaction between two parties: the user agent client (UAC) and the user agent server (UAS). To establish a SIP call session, signaling messages are exchanged with the mediation of an application server. The application server, such as the IBM's WebSphere Application Server (WAS) [25] or BEA's WebLogic Server [26], is implemented as a software process and provides rich SIP functionality for the users. This example [17, 27] considers that the SIP application installed on the application server is a back-to-back user agent (B2BUA), which acts as a proxy for SIP messages in VoIP call sessions [28]. Any failure of the application server will result in lost calls, either newly arriving calls due to blocking, or in-progress calls due to cutoff. A blocked call event is defined as a call that was prevented from being successfully established due to failures. A cutoff call event occurs when a stable call is terminated prior to either party going on-hook. In this example, we are primarily interested in computing DPM, a commonly used service (un)reliability measure for telecommunication systems. DPM accounts for all types of call losses, including blocked and cutoff calls.

1.3.6   Task-Oriented Dependability

A task, which requires a specified amount of work to be executed, is processed by a system that may change its computational power in time due to failures and repairs or to variations in the performance level. A task-oriented view of such a system recognizes the fact that the task completion time is affected by changes in the system computational power. For example, the occurrence of a fault during the execution of a task may cause the task to be dropped, or to be preempted and then resumed at a later system recovery, or to be preempted and then restarted from the beginning when the system is up again. Analysis of the distribution of the task completion time under different interruption and recovery policies when the computing system changes its mode of operation randomly can be found in [29–31].

1.4　Predictive Dependability Assessment

Quantitative dependability analysis is conveniently applied in the earliest phases of a design process with the goal of evaluating the conformity to specifications and of comparing different design alternatives. Such methods have also been used in later phases such as verification or operational phases. Some areas where predictive dependability techniques are invaluable, not only from a technical but also from a social and economic point of view, and to determine the success of a project, are:

Risk assessment and safety analysis: The risk level associated with a technological activity is related to the probability that some malfunction appears in a part of a system and the consequences that this malfunction may have on the system as a whole, on the workers or persons working in the proximity of the malfunctioning system or on the surrounding environment [32]. Dependability theory offers a framework to evaluate the probability that a malfunction may appear in the system, thus providing a necessary input in the construction of a quantitative risk assessment [33]. Quantitative evaluation of safety-critical systems is also the objective of international electronic standards [34].

Design and contract specifications: In the design specifications of complex or safety-critical systems, dependability clauses are often included. We have already seen examples in Sections 1.3.1 and 1.3.3. These clauses may be in the form of a guaranteed availability, or in the form of an upper bound on the total expected downtime in a given period (e.g., one year), or in the form of probability of correct operation in a given mission time. The fulfillment of the dependability clauses requires the capability of quantitatively predicting the required measures from the earliest phases of system design.

Technical assistance and maintenance: The cost of the technical assistance and maintenance services depends on the expected number of requests that will be issued in a given period of time to keep or to restore the system in proper operating conditions. The number of requests depends on the expected number of failures, on their type, and severity, and on the complexity of the actions that are needed to recover to operating condition. Planning the technical assistance service requires an *a priori* estimation of the number of times the assistance will be needed. Furthermore, the knowledge of the behavior of the system in time may allow one to predict the expected cost of system outages in a given time interval of operation and to infer the logistics of the maintenance services – number of spares for each component or subsystem, scheduling of repair crews, location of repair facilities, and so forth. Warranty periods can also be seen as an aspect of guaranteed dependability, since, in this case, the manufacturers assume the cost of the potential malfunctions and resultant downtime.

Optimal preventive maintenance: Scheduling of preventive maintenance can be based on dependability quantification. This has been applied both for hardware preventive maintenance [35–37] and software preventive maintenance [24, 38].

Lifecycle cost: The cost of a system, considered over its complete lifecycle, may be considered as composed of several parts: an initial acquisition cost known as *CapX*, and the operational, labor, and a deferred maintenance cost distributed over the life of the system, collectively known as *OpX*. These cost components are often in conflict; a highly dependable system tends to have a higher initial acquisition cost but a reduced deferred cost. Dependability prediction techniques may help in finding an optimal tradeoff between these cost factors, by balancing the amount to invest in the initial system dependability to reduce maintenance, operational, and labor costs distributed over the useful life. Formal cost minimization techniques can be applied to solve this problem [39].

Market competitiveness: Dependability is often a key ingredient in product differentiation and valorization leading to the commercial success of a product.

Summary

In conclusion, we have presented in this chapter formal definitions of various dependability-related terms: availability, reliability, maintainability, and safety. We have briefly referred to various measures of dependability. Several case studies to illustrate the practical applicability of the techniques were briefly presented. This chapter sets the stage for the reader to start with a clear understanding of the terminology and expectations for the rest of the chapters.

1.5 Further Reading

Any journey into understanding dependability begins with [1, 2], where a comprehensive overview of the dependability terminology was presented. Another source of information is the IEC international vocabulary [4]. Two edited volumes by K. B. Misra provide excellent source material on all the topics discussed here [40, 41]. We caution the reader, though, that the terminology we use is somewhat different from that used in [41]. For instance, compare, and contrast our Figure 1.1 with Figure 1.2 in [41]. For software reliability engineering, a good source is [42]; for safe software in critical applications, [43]. More recent discussions on software failures and their mitigation can be found in [8, 24, 44]. Tortorella [15] presents detailed definitions of service-oriented dependability metrics. Computing of service-oriented metrics is discussed in [45]. The related field of risk assessment is covered in [33, 46].

References

[1] J.-C. Laprie, "Dependable computing and fault-tolerance," in *Proc. 15th Int. Symp. on Fault-Tolerant Computing (FTCS-15)*, 1985, pp. 2–11.

[2] A. Avizienis, J. Laprie, B. Randell, and C. Landwehr, "Basic concepts and taxonomy of dependable and secure computing," *IEEE Transactions on Dependable and Secure Computing*, vol. 1, no. 1, pp. 11–33, 2004.

[3] K. S. Trivedi, D. S. Kim, A. Roy, and D. Medhi, "Dependability and security models," in *Proc. 7th Int. Workshop on Design of Reliable Communication Networks (DRCN 2009)*, 2009, pp. 11–20.

[4] IEC 60050, *International Electrotechnical Vocabulary: Chapter 191: Dependability and Quality of Service*. IEC Standard No. 60050-191, 2nd edn., 2001.

[5] A. Avizienis and J.-C. Laprie, "Dependable computing: From concepts to design diversity," *Proceedings of the IEEE*, vol. 74, no. 5, pp. 629–638, 1986.

[6] B. Parhami, "From defects to failures: A view of dependable computing," *SIGARCH Comput. Archit. News*, vol. 16, no. 4, pp. 157–168, Sep. 1988.

[7] D. P. Siewiorek and R. S. Swarz, *Reliable Computer Systems: Design and Evaluation*, 3rd edn. A K Peters/CRC Press, 1998.

[8] M. Grottke and K. Trivedi, "Fighting bugs: Remove, retry, replicate, and rejuvenate," *Computer*, vol. 40, no. 2, pp. 107–109, 2007.

[9] D. Chen, S. Garg, and K. S. Trivedi, "Network survivability performance evaluation: A quantitative approach with applications in wireless ad-hoc networks," in *Proc. 5th ACM Int. Workshop on Modeling Analysis and Simulation of Wireless and Mobile Systems*, 2002, pp. 61–68.

[10] F. Cristian, "Understanding fault-tolerant distributed systems," *Commun. ACM*, vol. 34, no. 2, pp. 56–78, Feb. 1991.

[11] M. Grottke, A. Nikora, and K. Trivedi, "An empirical investigation of fault types in space mission system software," in *IEEE/IFIP Int. Conference on Dependable Systems and Networks (DSN)*, 2010, pp. 447–456.

[12] J. Musa, "Software reliability-engineered testing," *Computer*, vol. 29, no. 11, pp. 61–68, 1996.

[13] B. B. Madan, K. Goševa-Popstojanova, K. Vaidyanathan, and K. S. Trivedi, "A method for modeling and quantifying the security attributes of intrusion tolerant systems," *Performance Evaluation*, vol. 56, pp. 167–186, 2004.

[14] F. B. Schneider, "Blueprint for a science of cybersecurity," *The Next Wave*, vol. 19, no. 2, pp. 47–57, 2012.

[15] M. Tortorella, "Service reliability theory and engineering I: Foundations," *Quality Technology & Quantitative Management*, vol. 2, no. 1, pp. 1–16, 2005.

[16] E. Bauer and R. Adams, *Reliability and Availability of Cloud Computing*. Wiley-IEEE Press, 2012.

[17] K. S. Trivedi, D. Wang, and J. Hunt, "Computing the number of calls dropped due to failures," in *Proc. IEEE 21st Int. Symp. Software Reliability Engineering (ISSRE)*, 2010, pp. 11–20.

[18] FAA, *Certification Maintenance Requirements*, Advisory Circular num 25-19, Nov. 28, 1994.

[19] FAA, *System Design and Analysis*, Advisory Circular/Advisory Material Joint 25.1309, May 24, 1996.

[20] A. Ramesh, D. Twigg, U. Sandadi, and T. Sharma, "Reliability analysis of systems with operation-time management," *IEEE Transactions on Reliability*, vol. 51, no. 1, pp. 39–48, 2002.

[21] K. Hjelmgren, S. Svensson, and O. Hannius, "Reliability analysis of a single-engine aircraft FADEC," in *Proc. Annual Reliability and Maintainability Symposium*, 1998, pp. 401–407.

[22] A. Bobbio, D. Codetta-Raiteri, L. Portinale, A. Guiotto, and Y. Yushtein, "A unified modelling and operational framework for fault detection, identification, and recovery in autonomous spacecrafts," in *Theory and Application of Multi-Formalism Modeling*, eds. M. Gribaudo and M. Iacono. IGI-Global, 2014, ch. 11, pp. 239–258.

[23] W. E. Smith, K. S. Trivedi, L. Tomek, and J. Ackaret, "Availability analysis of blade server systems," *IBM Systems Journal*, vol. 47, no. 4, pp. 621–640, 2008.

[24] K. Vaidyanathan and K. S. Trivedi, "A comprehensive model for software rejuvenation," *IEEE Transactions Dependable and Secure Computing*, vol. 2, no. 2, pp. 124–137, Apr. 2005.

[25] IBM's WebSphere Application Server (WAS). [Online]. Available: www.ibm.com/software/ webservers/appserv/was/

[26] BEA's Web Logic Server. [Online]. Available: www.bea.com/content/products/weblogic/ server/

[27] S. Mondal, X. Yin, J. Muppala, J. Alonso Lopez, and K. S. Trivedi, "Defects per million computation in service-oriented environments," *IEEE Transactions on Services Computing*, vol. 8, no. 1, pp. 32–46, Jan. 2015.

[28] Back-To-Back User Agent (B2BUA) SIP Servers Powering Next Generation Networks – A Functional and Architectural Look At Back-To-Back User Agent (B2BUA) SIP Servers. RADVISION – an Avaya Company. [Online]. Available: www.radvision.com

[29] V. Kulkarni, V. Nicola, and K. Trivedi, "The completion time of a job on a multi-mode system," *Advances in Applied Probability*, vol. 19, pp. 932–954, 1987.

[30] P. Chimento and K. Trivedi, "The completion time of programs on processors subject to failure and repair," *IEEE Transactions on Computers*, vol. 42, pp. 1184–1194, 1993.

[31] A. Bobbio and M. Telek, "Task completion time in degradable systems," in *Performability Modelling: Techniques and Tools*, eds. B. R. Haverkort, R. Marie, G. Rubino, and K. S. Trivedi. Wiley, 2001, ch. 7, pp. 139–161.

[32] E. Zio, "Challenges in the vulnerability and risk analysis of critical infrastructures," *Reliability Engineering & System Safety*, vol. 152, pp. 137–150, 2016.

[33] E. Henley and H. Kumamoto, *Reliability Engineering and Risk Assessment*. Prentice Hall, 1981.

[34] IEC 61508, *Functional Safety of Electrical/Electronic/Programmable Electronic Safety-Related Systems*. IEC Standard No. 61508, 2011.

[35] D. G. Nguyen and D. N. P. Murthy, "Optimal preventive maintenance policies for repairable systems," *Operations Research*, vol. 29, no. 6, pp. 1181–1194, 1981.

[36] D. Chen and K. Trivedi, "Analysis of periodic preventive maintenance with general system failure distribution," in *Proc. Pacific Rim Int. Symp. on Dependable Computing (PRDC)*, 2001, pp. 103–107.

[37] D. Chen and K. S. Trivedi, "Closed-form analytical results for condition-based maintenance," *Reliability Engineering & System Safety*, vol. 76, no. 1, pp. 43–51, 2002.

[38] Y. Huang, C. M. R. Kintala, N. Kolettis, and N. D. Fultoni, "Software rejuvenation: Analysis, module and applications," in *Proc. Int. Symp. on Fault-Tolerant Computing (FTCS)*, 1995, pp. 381–390.

[39] R. Pietrantuono, S. Russo, and K. Trivedi, "Software reliability and testing time allocation: An architecture-based approach," *IEEE Transactions on Software Engineering*, vol. 36, no. 3, pp. 323–337, 2010.

[40] K. B. Misra, *New Trends in System Reliability Evaluation*. Elsevier, 1993.

[41] K. B. Misra, *Handbook of Performability Engineering*. Springer-Verlag, 2008.

[42] M. R. Lyu, *Software Fault Tolerance*. John Wiley & Sons, 1995.

[43] N. Leveson, *Safeware: System Safety and Computers*. Addison-Wesley, 1995.

[44] M. Grottke, D. S. Kim, R. K. Mansharamani, M. K. Nambiar, R. Natella, and K. S. Trivedi, "Recovery from software failures caused by Mandelbugs," *IEEE Trans. Reliability*, vol. 65, no. 1, pp. 70–87, 2016.

[45] D. Wang and K. S. Trivedi, "Modeling user-perceived reliability based on user behavior graphs," *International Journal of Reliability, Quality and Safety Engineering*, vol. 16, no. 04, pp. 303–329, 2009.

[46] H. Kumamoto and E. Henley, *Probabilistic Risk Assessment and Management for Engineers and Scientists*. IEEE Press, 1996.

2 Dependability Evaluation

It is abundantly clear from the discussion in the previous chapter that dependability is an important requirement for any system. According to the IEC international vocabulary [1], "dependability is achieved by due attention to failure prediction, prevention, and mitigation." Dependability evaluation and analysis is devoted to investigating the possible manifestations of failures (when and how they can occur), their impact on the system as a whole or on its parts, and the way to prevent and mitigate them. Various techniques available to evaluate the dependability attributes of a system are reviewed in this chapter.

2.1 Quantitative Evaluation of Dependability

The information about failures and their mitigation may be obtained from two distinct sources: *measurements* and *models*. Measurements involve collecting data from the observation of real physical systems; models, on the other hand, involve construction of an idealized representation of the system and subsequent evaluation of the formal model to evaluate qualitative information and to compute quantitative information. Note that measurements are often summarized in models such as regression models, but we shall refer to such models as "blackbox" models. In contrast, the models we shall explore in this book can be called "whitebox" models or structural models that start with information about measurable parts of the system and their interactions to create a formal mathematical model that can be used to reason about the overall system behavior.

The main methods of evaluation are summarized in Figure 2.1. The two methods, measurements and models, are not in conflict with each other. Rather, they should be integrated in a *prediction-to-validation* cycle. Models provide predictive information that is then validated through life testing experiments or field data as the system is put into operation. Conversely, experimental results and measurements provide the necessary statistical base for estimating the parameters that are fed into the models. Models in turn can be used to help guide measurement experiments. Quite often, during system design, the system as a whole is not available for measurements. However, subsystem- or component-level measurements may be available. They can be used as inputs to a system model that is put together based on the knowledge of how the subsystems interact with each other. Thus, a judicious combination of measurements

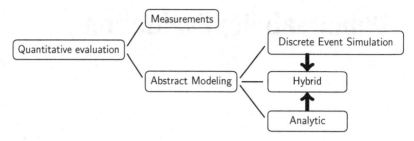

Figure 2.1   Quantitative evaluation taxonomy.

and models are used together in practice. The models can thus be seen as a way to derive the behavior of a large system out of individually testable components or subsystems [2].

2.1.1   Measurement-Based Evaluation

Measurement-based evaluation involves gathering information about the occurrences of failures, their causes and severity, their effect on the downtime duration, etc., from the observation of systems, either during normal operation or under specifically designed and controlled test conditions. Depending on the specific conditions under which the observations are collected, measurement-based evaluation can be further subdivided into field measurements and test measurements. In both cases, statistical inference techniques are needed to analyze the collected data to make inferences regarding the nature and parameters of various distributions [3–7]. Further, regression analysis [8], analysis of variance [9] and design of experiments [10, 11] have been utilized for analyzing measurement data.

Field Measurements
This approach involves gathering measurements from real components or systems under normal operational conditions. In this age where big data and data analytics have acquired great importance, failure data analytics is bound to gain prominence. A recent analysis of NASA satellite problem reports is an interesting example here [12, 13]. Such field measurements can be used via appropriate statistical analysis to feed models, and even for online control to improve system dependability [14]. Nevertheless, data confidentiality and paucity of information related to individual failure occurrences are some of the important challenges.

   This approach has the drawback that the observations cannot be easily extrapolated to different systems or in different operational or environmental conditions. Furthermore, collecting field data requires that the system under observation has been built and put into operation for a sufficiently long duration such that the amount of collected data is statistically significant. Often it may be necessary to have preliminary information in the early phases of the design, manufacturing, and field installation process, when the system is not yet working and field data is not available. In this case, physical observations at the component or subsystem level may be utilized as inputs to an overall system model. The methods of probability theory enable us to derive the behavior of combinations of interacting components to form a system, or combinations of multiple interacting systems to form a system of systems.

Test Measurements

This approach involves the design of specific procedures to be carried out to ascertain and measure dependability-related attributes of components or subsystems. In particular, life tests are conducted to estimate the useful life of an item under controlled stress and environmental or operational conditions. The test may be conducted in emulated field conditions or under accelerated test conditions. According to the IEC international vocabulary [1], an accelerated test is a "test in which the applied stress level is chosen to exceed specified operational conditions in order to shorten the time duration required to observe the stress response of the item." Acceleration may also be achieved by subjecting the parts under test to more severe environmental conditions (temperature, humidity, vibration, corrosive atmosphere, etc.) or operational conditions (e.g., stress cycles, temperature cycles, etc.). These tests may be repeated periodically to ascertain that new batches of the same parts maintain the desired properties. Statistical techniques for the design of experiments and accelerated life-testing are needed besides those mentioned earlier [15]. These techniques have recently also been applied to software systems [16, 17].

Apart from test measurements to acquire failure data, tests are often carried out to estimate parameters of recovery processes such as detection, reconfiguration, reboot, and related steps [18, 19]. Such experiments are known as fault/error injection experiments and a large body of literature and tools exists on this topic [20, 21].

A subtype of test measurements is prototype measurements. Prior to mass production, it is customary to produce and test in controlled operation a few copies of the component or system with the aim of gathering information on its quality and dependability. In this case, observations are made on a final system, but under controlled conditions.

2.1.2   Model-Based Evaluation

Measurement-based evaluation, as discussed in the previous section, is feasible only when the real system or its prototype is in existence. However, in many circumstances, the system may not even be available for carrying out the measurements. For example, in the early stage of the system lifecycle, only a conceptual design may be in existence. In such circumstances, the only alternative is to resort to a deductive model-based evaluation. Furthermore, model-based evaluation is in general faster and less expensive compared to measurement-based evaluation, but at the cost of fidelity.

A *model* is a *mathematical abstraction* of the system aimed at providing a simplified conceptualized representation of the behavior of the system. Model-based evaluation requires two steps: a modeling phase and a solution phase.

- In the *modeling phase*, the modeler constructs a simplified view of reality by abstracting the behavior details from the system under examination; this simplified view must preserve the main features of the system that the model is intended to analyze. The abstraction of the physical system should then be described in a formal language by choosing an appropriate modeling formalism. The nature of the problem under analysis guides the choice of a specific modeling formalism, but also vice versa – the modeling formalism imposes restrictions on the details that can be included in the conceptual system model.

- In the *solution phase*, the model of the system is solved by using appropriate mathematical, numerical or simulative approaches to compute the measures of interest. The solution techniques that can be applied are driven by the choice of the modeling formalism.

We refer to the ability of a formalism to correctly incorporate details of the real system as its *modeling power*, and the ability of a formalism to algorithmically determine the properties of the models as its *decision power* [22]. There is a tradeoff between the two properties; the higher the modeling power of a formalism, the lower its decision power. The success of any modeling formalism depends on the balance between its modeling power and its decision power. Since a model is a simplified representation of the real system, the trustworthiness of the computed results depends on the ability of the modeler to incorporate into the model the most relevant features that influence the system behavior. Greater accuracy and trustworthiness of a model comes at a greater cost in terms of the time and effort needed to obtain its solution. The skill of a modeler is in determining an appropriate balance between the modeling power of the formalism selected for constructing the model, and the effort needed to obtain its solution.

Solution techniques for modeling formalisms can be divided into two broad categories: *analytical* and *simulative* solutions. Sometimes, the same modeling formalism lends itself to both analytical and simulative solution approaches. In this case, the modeling phase is common, but the solution phase differs based on the solution approach used. It is also possible to combine formalisms from the two categories into a hybrid modeling formalism.

Analytical Solution

Sometimes it is possible to develop mathematical equations (or relations) depicting the dynamic behavior of the model. These relations then need to be solved using the available techniques of mathematical calculus combined with numerical analysis. According to the solution procedure that can be applied to the model, we can obtain a *closed-form* solution or only a *numerical* solution.

- *Closed-form* solutions are more informative since they describe the behavior of the system as a function of the unknown variables, thus facilitating parameterized evaluations and sensitivity studies. The values of the required quantities can be obtained by substituting the proper numerical values for the parameters. Usually, closed-form solutions are possible for problems of small or moderate size with simple or regular structure.

- *Numerical* solutions can be the only feasible alternative when the model does not possess a closed-form solution. In this case, a specific numerical value should be assigned to the parameters appearing in the model, and the result thus obtained is a value representing a single point in the solution space. Numerical solutions usually require more memory and computational time, and their application may be limited either by the memory or the computational time available. Nevertheless, with the advent of powerful computers and the availability of powerful solution packages, this avenue is increasingly feasible even for very large models.

Simulative Solution

Simulative solutions are based on the generation of a possible pattern of the evolution (*sample path*) of the modeled system by means of a computer program that does not solve equations, but reconstructs or mimics the behavior of the system in time. During the reconstruction of a possible evolution of the system, some significant outputs can be recorded for subsequent analysis. The generation of a single pattern is usually called a *simulation experiment* or *simulation run*. Repetition of a statistically relevant number of simulation experiments allows us to estimate the values of the recorded quantities with their associated statistical properties and uncertainties (*confidence intervals*). Simulative solutions are used for models whose governing equations cannot be easily written, or for solving models for which a numerical solution becomes infeasible because of excessive storage or computational power requirements [23].

2.1.3 Interplay between Measurement and Modeling

A complete dependability evaluation effort should utilize all possible information sources, and should then be conceived based on the *prediction-to-validation* cycle. Central to this effort is the organization of an information system as in Figure 2.2.

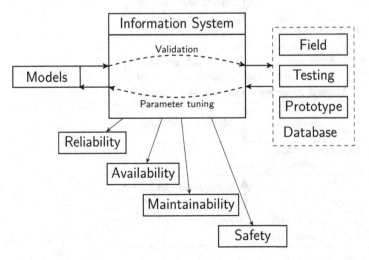

Figure 2.2 Information system for the dependability assessment cycle.

Data is gathered from available physical sources (field, tests, prototypes) and from the items under observation (components, parts, systems) and fed into a database. Data is manipulated in the database to provide statistical estimates of useful quantities. On the other side, the evaluation of abstract mathematical models requires that specific values are assigned to parameters; these values are conveniently inferred from physical observations.

Hence, on the one hand models provide a means to guide the design of a system by predicting its dependability attributes; on the other hand, models need to be instantiated with data. In the absence of specific information, data can be obtained from similar systems or components or extrapolated from field data, accelerated tests or simply previous human experience or expert judgment. When the system is built, the gathered data may be used to validate the model and provide new information for more accurate parameter instantiation, or for the prediction of future releases or the design of new or similar systems. The final goal of the dependability assessment cycle is to quantitatively evaluate the dependability attributes such as *reliability*, *availability*, and *maintainability* that were formally introduced in Section 1.2.

2.2   The Modeling Process

The modeling process, from the physical system to the final quantitative results, is a complex process that requires the implementation of many steps (Figure 2.3).

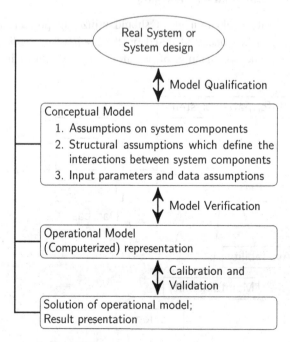

Figure 2.3   The overall modeling process.

1. Studying/understanding the system being modeled.
2. Development of a conceptual model.
3. Translation into an operational computerized model.
4. Parametrization of the operational model.
5. Solution of the operational model and presentation of the results.
6. Verification, validation, and improvement of the model.
7. Use of the model for dependability evaluation.

2.2.1 Studying/Understanding the System Being Modeled

Before embarking on any modeling activity it is essential to become intimately familiar with the structure and functioning of the system being modeled. This step requires the reliability specialist to interact closely with the team of system specialists that work directly with the system. The team should typically include hardware and software architects and designers, operating and assistance engineers, maintenance specialists, and managers devoted to the supervision of the system operation. Visual aids such as flow charts, block diagrams, fault trees, state charts, and timing diagrams are very useful in communicating with the team in this phase. The information that must be available at the end of the familiarization process includes the physical and functional structure of the system. Different failure modes or deviations of each block/component from its correct mode of operation need to be identified. Recovery actions following each type of failure need to be understood and identified. Interactions among system components, the environmental conditions and the system–human interaction need to be understood. The modeler then needs to filter out the unimportant details and abstract the system behavior to a sufficient level of detail to facilitate the development of an appropriate model.

This phase can take advantage of the qualitative systematic techniques for failure identification and analysis such as *failure mode and effect analysis (FMEA)* and *hazard operability analysis (HAZOP)*. In addition to the above information on component failure modes and their effects, information about the *fault/error handling (FEH)* and subsequent maintenance procedures must also be identified.

Failure Mode and Effect Analysis

FMEA is a systematic approach for identifying all possible failures in a design, manufacturing or assembly process, or a product or service [24–27]. The procedure starts by identifying failures or malfunctions in any part of the system that may cause an undesirable effect or a deviation of the system and jeopardize its operation. FMEA is usually supported by worksheets that help the analyst to identify potential failures and their consequences on the system operation. FMECA (failure mode effects and criticality analysis) is an extension to the FMEA approach that adds to each identified failure a criticality index that is needed in a subsequent risk assessment procedure.

Hazard Operability Analysis

HAZOP is a systematic technique aimed at identifying potential hazards and operability problems in a system. HAZOP assumes that potential risk to system operation is caused by deviations of influential parameters that affect the design or normal operating conditions [28]. Identification of such deviations is facilitated by using ad hoc spreadsheets and sets of guide words. A HAZOP study attempts to identify previously unconsidered failure modes by suggesting hypothetical faults for review and, where necessary, this is followed by the suggestion of means of overcoming identified hazards. The derivation of a fault tree from an operability analysis has been illustrated in [29]. HAZOP is conveniently applied to manufacturing processes, equipment, and facilities in the chemical, nuclear, electrical power, medical, and pharmaceutical industries for evaluating process safety hazards. The HAZOP approach has also been proposed for software products [30].

Coverage Factor: Fault/Error Handling

FEH refers to the set of all mechanisms and procedures that come into play upon the occurrence of component, subsystem or system failure. The failure may be *covered*, so the system continues to operate, or the failure may cause a catastrophic failure [31, 32]. This is sometimes known as FDIR (fault detection, isolation, and repair), and it involves stages of detection, isolation/location, switching out/in, and such other steps to bring the system back to operation (in a possibly degraded mode). A number of papers [18, 33] and corresponding software packages [34] elaborate further on this topic. In some cases the FEH may be condensed into a single parameter, called the *coverage factor*, that gives the probability that the exit from the FDIR is toward an operational condition.

The capabilities of a Bayesian belief network have also been exploited in designing an FDIR mechanism for autonomous systems [35]. In particular, issues like partial observability, uncertain system evolution, and system–environment interaction, as well as the prediction and mitigation of imminent failures, can be addressed in the framework of Bayesian networks.

Maintenance

There are two types of maintenance in common use: reactive (or unscheduled) and proactive (or scheduled or preventive). In the former case, repair or replacement and related maintenance is carried out after the occurrence of a failure, while in the latter case, maintenance is carried out before the occurrence of a failure, thus preventing or postponing failures. In the case of reactive repair, failure notification time, field service travel time, diagnostic time and the actual repair time and any subsequent cleanup action such as data reconstruction time needs to be considered [36, 37]. The number of spare parts and repair crews and their locations need to be determined. Further, the spare parts may be found to be not functioning (dead on arrival) and the repair may be imperfect. During the actual repair, the system as a whole may be down (cold swap), or it can continue functioning (hot swap) [36, 37]. It may sometimes be beneficial to defer component repair [37, 38].

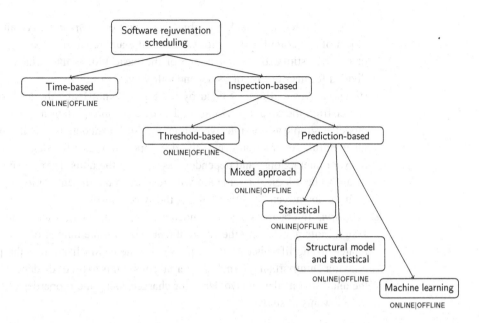

Figure 2.4   Software rejuvenation scheduling.

In the case of preventive maintenance, the scheduling method, and granularity of maintenance action need to be determined. Preventive maintenance as applied to software systems has been studied under the umbrella of software rejuvenation. Software rejuvenation scheduling methods include time-based and inspection-based techniques [39], as illustrated in Figure 2.4.

2.2.2   Development of a Conceptual Model

The development of a conceptual model is a purely mental activity that involves more art than science and requires that the complexity of the physical system be reduced and controlled, keeping in mind a possible operational formalization of the model. The development of a conceptual model proceeds along three activities: *decomposition*, *abstraction*, and *idealization*.

> *Decomposition:* Decomposition involves dividing the system into parts. Each part can in turn be a large subsystem, an assembly of several components or a single component. The partitioning of the system is driven primarily by two main guidelines:
>
> - The parts should be physically identifiable units and could possibly be loosely interacting.
> - Each part can be seen as a black box from the dependability point of view, in the sense that a distribution of the time to failure (and repair) of the part is known or can be inferred from similar parts and previous data or can be obtained through testing and statistical analysis.

We will refer to a black box object as a *unit* or *component*. A unit can be an object of any complexity but for which failure and repair distributions are known or can be estimated. A unit is one of the basic blocks into which a system is divided for the purpose of its dependability analysis.

Abstraction: Abstraction is guided by the type of knowledge the analyst wants to gather from the model. It is intended to discard unimportant details to make the model as simple as possible, while retaining sufficient details to make the analysis meaningful. Abstraction drives the decomposition and the identification of the relevant interactions (and dependencies) among the units. Here the interaction is meant to be a logical interaction with respect to the dependability: how the unit malfunction affects the other units or the system level.

Idealization: Interactions and dependencies are both internal and external to the system: interaction with the environment and with the human operator are often related to the life history of the physical system. Idealization is the process of relaxing unimportant internal or external constraints to focus on dependencies that the analyst considers unavoidable for characterizing the properties of the system he/she wants to study.

Part of the conceptualization step involves the definition of measures that are intended to be evaluated at the end of the solution step. The three elements – the list of components, their interactions/dependencies and the final output measures – can help guide the choice of a proper operational model that provides a formal language to translate the conceptual model.

2.2.3   Translation into an Operational Computerized Model

The conceptualization of the model is carried out looking at a possible formalization of the model into a computerized operational model and measures to be evaluated. The choice of the operational model is thus oriented by the nature of the conceptualization in the previous steps. Several operational model types are available for this purpose, and will be described in detail in the subsequent chapters. A brief summary of the modeling formalisms is presented in Section 2.3.

As a first classification we can distinguish the following categories of model types:

- *Monolithic* models use a single formalism for developing a monolithic model.
- *Multi-level* models; for a complex system, it is often necessary to use a divide and conquer strategy and develop a multi-level model. We may use the same formalism in each level (homogeneous case) or use multiple formalisms (heterogeneous case).

2.2.4   Parametrization of the Operational Model

Assignment of numerical values or distribution functions to parameters is a delicate and challenging problem. Parameter values are often unknown and difficult to derive by measurement or estimation. They may depend on external conditions like environment, application, logistics, maintenance policies and management. Setting up experimental campaigns for parameter estimation requires effort and time. Parameter estimation and testing may not even be feasible in the early stages of design for new products. The lack of real and appropriate data often hinders the use of quantitative modeling.

Data can be categorized into different classes (e.g., hardware component time to failure or time to repair, software failure times, recovery stage delays, repair times, imperfect coverage for various recovery/repair stages) for which specific testing and estimation techniques must be utilized [19].

Parameters can be known with *uncertainties* [40], and various techniques have been proposed in the literature to propagate the uncertainties in dependability models from the input parameters to the output measures [41–43].

2.2.5   Solution of the Operational Model and Presentation of the Results

The choice of the solution method is partly determined by the operational model, partly by the experience of the modeler and partly by the availability of software packages. Two broad categories of solution types can be envisaged: *transient* and *steady state*, and two main families of methods, *analytic* and *simulative*, as summarized in Figure 2.5. We note that some software packages offer both analytic and simulative solutions (SPNP [44, 45] or Möbius [46] are possible examples).

Analytic transient solution techniques may be further categorized into fully symbolic closed-form solutions (by hand or by symbolic software tools such as Mathematica [47]), semi-symbolic (or semi-numerical) transient solutions where the parameters are numerical values and the time variable t is symbolic, as supported by the SHARPE software package [48], or numerical where at each computation only the value of one single point in the solution space is computed. Steady-state analytic solution techniques can be categorized as fully symbolic (or closed-form) solutions or numerical solutions.

2.2.6   Verification, Validation and Improvement of the Model

Before embarking on a detailed evaluation of the dependability characteristics of the system under study, it is important to confirm that the above steps have yielded the most appropriate model for the system. In particular, the primary question to be addressed at

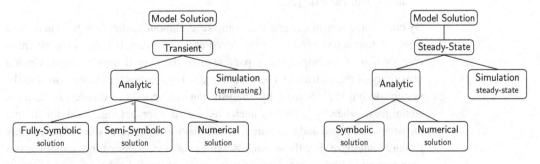

Figure 2.5   Solution techniques for computerized operational models.

this stage is: "Does the computerized model faithfully reflect the behavior of the real system?" This involves the following four issues to be addressed:

- *Conceptual model qualification.* This issue represents the process of assessing the degree to which a conceptual model is an accurate representation of the real world from the perspective of the model's intended applications.
- *Model verification.* This step is the process of determining that a computerized software implementation correctly represents the conceptual model and that the model is solved correctly.
- *Model validation.* Validation is the process of substantiating that the computerized model behaves with satisfactory accuracy consistent with the study objectives.
- *Data validation.* Finally, data validation is the process of confirming that the data used is accurate, complete, unbiased and appropriate in its original and transformed forms.

A detailed discussion of model verification and validation can be found in Section 2.4.

2.2.7   Use of the Model

The ultimate goal of the modeling process is to enable the detailed evaluation of the system dependability characteristics. The last step in the modeling process is aimed at facilitating this. A well-developed, verified and validated model is very amenable to the study of different variations and tradeoffs for system design alternatives. In particular, the model will enable the users to carry out evaluations including:

Parametric sensitivity analysis: This analysis is focused on evaluating a system whose output behavior can be expressed as a function of an input parameter θ [49, 50]. By using sensitivity analysis, we pinpoint the particular factor(s)/component(s) that are the most influential in affecting the output behavior of the system. The sensitivity of an output metric to the input parameter θ may have two trends:

- *Negative* sensitivity: if the value of the parameter θ increases, the measure of interest will decrease.
- *Positive* sensitivity: if the value of the parameter θ increases, the measure of interest will increase [51].

By conducting sensitivity analysis using several input parameters, we can deduce which parameter(s) need to be improved at design time. It reflects the potential contribution of a component or parameter to the overall improvement. Greater sensitivity of the metrics to a parameter indicates avenues for improvement in the system design [52]. Sensitivity analysis can be carried out through brute-force techniques, whereby each parameter is varied over a range, and the output measures computed and compared. While this is easy to carry out, it is extremely resource intensive. For those models, where the derivatives of the output metrics with respect to various input parameters can be computed either in closed form or numerically, the process helps identify and rank the parameters according to their importance.

Optimization: Optimization studies with respect to one or more parameters should be feasible depending on the flexibility in the model. Different kinds of optimization, including static vs. dynamic, constrained vs. unconstrained, integer vs. continuous, could be carried out.

Propagation of parametric (epistemic) uncertainty: The various parameters used in the model have uncertainties associated with them due to lack of knowledge or to the statistical estimation procedures. This parametric uncertainty is known as *epistemic* uncertainty [53], in contrast with the inherent uncertainty in the stochastic model that is known as *aleatory* uncertainty [54]. The uncertainties in the parameters could be propagated through the model to the output measures [41, 43, 55].

2.3   Modeling Formalisms

Several modeling formalisms are in common use in the field of dependability evaluation. The selection of the most suitable formalism for modeling the dependability of a specific system is often a contentious point among the modelers. Given below is a (non-exhaustive) list of important points for consideration when selecting an appropriate modeling formalism:

- the structure and behavior of the system being modeled;
- the ease of representation of the system's features and properties in the chosen modeling formalism;
- the set of dependability attributes that are of interest in the modeling endeavor.

The temptation for a modeler is to use the formalism that he/she is most familiar with, or a formalism for which a modeling toolkit is readily available. It is important to be more discriminating in picking the right formalism based on the considerations given above rather than convenience. Furthermore, multiple formalisms can be used for the same model as a method of model verification.

Modeling formalisms can be broadly classified into three types:

Non-state-space (or combinatorial) models: Solutions of these models can be determined relatively efficiently with the assumptions that the components are statistically independent. They provide high analytical tractability since the underlying state space need not be generated, but the modeling power is low as dependence is not allowed. We devote one chapter each for the three main non-state-space formalisms: reliability block diagrams (Chapter 4), reliability graphs or networks (Chapter 5), and fault trees (Chapter 6). The chapters on state enumeration (Chapter 7) and dynamic redundancy (Chapter 8) can be viewed as a bridge between non-state-space and state-space formalisms.

State-space models: These models can account for statistical as well as state and time dependencies and hence possess high modeling power but less analytical tractability as they suffer from the curse of dimensionality. State-space methods are broadly classified into Markovian (homogeneous and non-homogeneous) and non-Markovian methods. Three chapters are devoted to continuous-time homogeneous Markov chains (CTMC; Chapters 9, 10, and 11) and one devoted to stochastic Petri nets (Chapter 12), which enable automated generation of large Markov chains. In order to remove the restriction of an exponential distribution imposed by homogeneous continuous-time Markov chains, three chapters are devoted to non-homogeneous continuous-time Markov chains (Chapter 13), semi-Markov and Markov regenerative models (Chapter 14), and the method of phase-type expansions (Chapter 15).

Multi-level models: These combine the modeling power of state-space models with the efficiency of non-state-space models. Such multi-level models will be explored in two forms: hierarchical models (Chapter 16) and fixed-point iterative models (Chapter 17). The final chapter in this book will describe real-life case studies that naturally use such multi-level models (Chapter 18).

Figure 2.6 summarizes the different modeling formalisms in use. The subsequent chapters in this book examine each of these formalisms in great detail, including their advantages and disadvantages.

Malhotra *et al.* [56] developed a power hierarchy among the modeling formalisms according to their modeling power and conciseness of model specification. The power hierarchy is depicted in Figure 2.7. Among the non-state-space model types, they show that fault trees with repeated events (FTRE) are the most powerful in terms of the kinds of interactions among various system components that can be modeled. Reliability graphs (RG) are less powerful than fault trees with repeated events, but

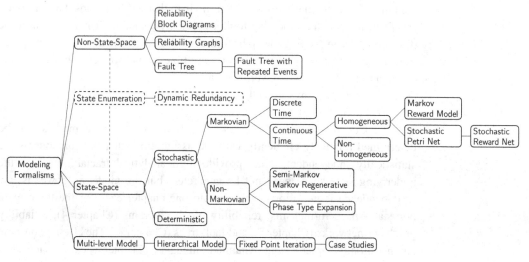

Figure 2.6   Modeling formalisms.

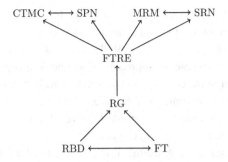

Figure 2.7 Power hierarchy of modeling formalisms.

more powerful than (series parallel) reliability block diagrams (RBD) and fault trees without repeated events (FT). Among the Markovian (state-space) model types, they consider homogeneous continuous-time Markov chains, (Markovian) stochastic Petri nets (SPN), Markov reward models (MRM) and stochastic reward nets (SRN). These are more powerful than non-state-space model types in that they can capture dependencies such as a repair facility shared between system components, standby redundancy, and common mode failures. Modeling formalisms that allow non-exponential distributions are not considered in the power hierarchy of Figure 2.7.

2.4 Model Verification and Validation

The ultimate aim of any modeling approach is to faithfully reflect the behavior of the real system to enable the accurate estimation of the system dependability characteristics. The modeling process detailed in Section 2.2 involves a step for verification and validation of the model. The verification process is aimed at answering the question: "Has the conceptual model been correctly implemented and solved?" The validation process is aimed at answering the question: "Does the computerized model faithfully reflect the behavior of the real system?" In the following subsections we give a brief overview of the techniques used for verification and validation in system modeling.

2.4.1 Model Verification

Verification refers to the process of determining that a computerized software implementation correctly represents the conceptual model and that the model is solved correctly. There are not many formal techniques to support the verification process, but a few guidelines can be followed to carry out the process.

> *Checking the model.* A simple yet straightforward approach is to let the software program be checked and exercised by an independent team, different from the modeling team. One of the important steps is to prepare detailed documentation that can help third parties to check the model implementation.

Checking the logical flow. Another technique is to check for logical flows in the construction of the program and in the input parameters, and to check whether the output results are reasonable or compatible with the physical system.

Multi-version modeling. Another useful approach is to check the possibility of implementing the model in different ways using different techniques or different software packages, at least in some specific special cases, and checking the correctness of the results obtained. An example is to verify analytic models via comparison with simulation results and vice versa.

Input/output graphics and animation. A useful support consists of a GUI interface for input and output and the possible animation of the model execution. If feasible, this enables the team to check whether the system behavior is reflected appropriately in the model. As an example, an SPN model can be examined by looking at its reachability graph. Similarly, job flows in a queuing network can be checked by animation.

Special cases. If a model is solved analytically/numerically, then we can check if special simplified cases can be solved in closed form and compared.

Alternative approaches:

- Alternate analytical approaches to modeling: Sometimes models can be constructed using several different formalisms and the results compared; for example, for the same problem, CTMC, SRN, queuing networks, hierarchical and fixed-point iterative models could be developed and cross-checked.
- Alternative solution methods: For a given formalism, several different solution approaches may be feasible, enabling verification by comparison. As an example, a fault tree could be solved using binary decision diagrams (BDD), sum of disjoint products (SDP) or factoring. Similarly, a CTMC could be solved in steady state using either power or successive over-relaxation (SOR) methods, and transient solutions using uniformization, ordinary differential equations or semi-numerical (transform) approaches. Some of these approaches may not be scalable and hence not used for large-scale models due to solution complexity, but may be quite feasible for smaller models.
- Alternative software packages: A given formalism could be solved using several different software packages, like Mathematica, Matlab, Möbius, SHARPE or SPNP, enabling us to cross-check the results.
- If a model can only be solved using simulation, we can verify if a special case can be solved analytically to compare and cross-check with the simulation results.
- If simulation is the only choice, then alternative solution techniques for the simulation like importance sampling, RESTART or regenerative simulation can be used. Also, simulations can be done using alternative simulation methods/packages like Csim, Matlab, ns-3, OPNET, OMNET, Simula, SPNP, and DEMOS, to cross-check the results.

2.4.2   Model Validation

There are no formal procedures to support the validation process. Naylor and Finger [57] outlined three steps for systematically carrying out validation:

1. Face validation: Discussion with experts about the conceptual model and its translation into the operational model is essential. This should result in an improved model with high fidelity.
2. Input–output validation: We need to confirm that the input data used is accurate, complete, unbiased, and appropriate with respect to the physical system. Similarly, the output results obtained from the model should be compared with those from measurements on the real system.
3. Validation of model assumptions: All assumptions should be clearly identified and documented. We should then either prove theoretically that the assumption is correct or do statistical hypothesis testing to validate each assumption. Rejection of a hypothesis regarding a model assumption based on measurement data should lead to reconsidering the assumption and improving the model.

The final goal of the model validation process is to substantiate evidence that the computerized model behaves with satisfactory accuracy and is consistent with the behavior of the system under study.

The verification and validation (V&V) process can be viewed as an iterative process, as shown in Figure 2.8. The results of the V&V process help in checking the results against the physical system and may suggest a need to update the conceptual model and its operational translation. The iterative process may be carried on until the V&V step does not suggest any further improvement. At this point the model is ready to be used.

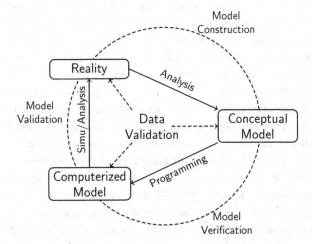

Figure 2.8   The cycle of the iterative verification and validation process.

2.5   Errors in Models

No modeling exercise is completely free of errors. Indeed, the endeavor is to keep the errors under control as far as possible. The V&V process is specifically intended to unveil these errors and find ways to remove them or mitigate their effects when removal is not possible. Several sources of error may compromise the correctness of a model output. The typical sources of error in a model are classified below based on the stage at which they are introduced. Furthermore, strategies for dealing with the errors are also delineated.

2.5.1   Errors in the Construction of the Conceptual Model

As detailed earlier, the construction of the conceptual model from understanding the behavior of the real system involves the selection of those features that should be included in the model and discarding the unimportant features. In this process, several key errors may creep in. The decomposition, aggregation, and idealization process may result in including details that may have little impact on the system dependability attributes, and inadvertently neglecting those features that may have significant impact. For instance, not explicitly considering the power subsystem in the model. Similarly, neglecting to include in the conceptual model important failure modes or including failure modes that are not significant can also introduce errors. The errors in this step could be mitigated by having detailed discussions with experts and conducting face validation of the model.

2.5.2   Errors in the Construction of the Operational Computerized Model

Even after getting the conceptual model right, translating it into the operational computerized model may introduce more errors. The selection of the appropriate formalism for constructing the operational model is important. Misinterpretation of the conceptualized model in the construction of the operational model can easily introduce errors. For example, if the formalism chosen is a CTMC, then missing or extra states, or missing or extra transitions, will introduce errors. Missing elements or designing wrong elements in the operational model or further ignoring details that may be relevant in the analysis may cause problems. Ignoring dependencies and common modes of failure in the conceptual model, either due to the inability to represent them appropriately in the chosen modeling formalism or to reduce the model complexity, may cause inaccuracies in the results [58, 59]. It is important to verify the assumptions through hypothesis testing. Simplifying assumptions may be made in order to fit to the chosen modeling formalism, for example assuming exponential distributions for various times to failure, times to repair, etc. [60]. If the real data shows significant deviation from the assumption, we should be prepared to modify the model. We can use goodness of fit tests and improve the model in case of poor fit with data collected from real (sub)systems. Sometimes the choice of modeling formalism may be driven by the modeler's familiarity with a specific formalism or the availability of tools that permit the analysis of the models constructed using a particular formalism. This may prove to be the ultimate undoing of the whole modeling exercise.

Errors in the estimation of the parameters used in the models may be the source of the problem. This can be mitigated to some extent through using parametric sensitivity analysis [50, 51, 61], bounds analysis [41, 62], and uncertainty propagation through models [55].

2.5.3 Errors in Model Solution

Analytical models, especially those based on a state-space approach, often face significant problems during the solution phase due to largeness and stiffness in the models. To mitigate the problems due to largeness and stiffness it is sometimes necessary to make simplifying approximations by resorting to decomposition techniques [63–65], state truncation [66–68], and fixed-point iteration [69, 70]. These techniques may introduce approximation errors. Furthermore, even in the absence of such approximations, numerical errors are unavoidable in any model solution approach and may derive from many causes: discretization of continuous variables, series truncation errors or roundoff errors. Numerical analysis texts suggest numerous methods to reduce the incidence of such errors.

Simulative solutions are also afflicted by sampling errors due to a limited sample size. The extent of the sampling error is reflected in the confidence interval.

2.6 The Probabilistic Approach

For mechanical, electrical or electronic components, the physical mechanisms that lead the item to failure are in general very complex and depend on a variety of factors: physical, chemical, mechanical, environmental, human. The specific influence of each factor and their mutual interactions that eventually lead to a failure are difficult to identify and to quantify. A discipline known as *reliability physics* or *failure physics* [71] is devoted to identifying the basic mechanisms that are responsible for the failure of such items. But due to the complexity of the failure mechanisms, the time at which a failure will appear in an item cannot be described by deterministic laws. We are then led to consider the time to failure of an item as a random variable. By this we mean that the most complete information that we can obtain about the failure event of an item is a measure of the probability that the event will occur by a given time or in a given interval. This directly leads to the distribution of the time to failure random variable (the cumulative distribution function, Cdf) and its complementary function known as the reliability function (the survival function, Sf). A similar argument can also be invoked for modeling by means of random variables the time to failure of a software system [12, 13].

For systems that are repairable and can be recovered to a functioning state after a failure, a measure that will be considered in the models is the duration of the downtime or the time that will elapse to restore a failed item into an operating state that we can identify as the *repair time*. In most cases, the duration of the downtime cannot be predicted deterministically since it depends on many variable factors such as the specific mechanism that has produced the failure, the availability of spare parts, and the availability and skill of the maintenance person. Thus, even in this case, we are led to consider the repair time as a random variable.

The above considerations are the basis for the probabilistic approach to dependability analysis that is the theme of this book.

2.7   Statistical Dependence

In Section 2.3, modeling formalisms were broadly classified based on whether component behaviors are considered *statistically independent* or *statistically dependent*. Statistical independence means that the failure (and repair) process of each component is not influenced by the presence of other components or by the way in which the component is linked to the other components through the system configuration. Therefore, statistical independence means that each component behaves as if it were in isolation. Statistical independence is assumed as a basic hypothesis throughout Part II of the book, while in subsequent parts of the book statistical dependence is captured by the models. The independence assumption has the consequence that by increasing the level of redundancy in the system one can reach any desired level of dependability. In practice, unfortunately, this is not the case. The statistical dependencies prevent the system from reaching all theoretically possible levels of dependability [72].

The hypothesis of statistical independence is often less realistic in the repair behavior of the components than in the failure behavior. In fact, deferred maintenance policies, prioritized maintenance lists and the limitation on the number of repair crews are all factors that introduce dependencies. The most common types of phenomena that introduce dependencies in the failure process are the following [59].

> *Load dependencies:* The failure of some components in a redundant configuration may increase the load on the components still functioning and modify their failure behavior.
>
> *Functional dependencies:* These dependencies are caused by the logic of operation of the system. A typical case occurs when a component is switched on upon the failure of other components (standby redundancy), thus modifying its status (from dormant to operational) as a function of the up or down condition of other components.
>
> *Human error:* The failure of a human operator to carry out a specific task or to perform the correct action may cause dependencies among components in a system.
>
> *Coincident fault dependencies:* Many of the reliability/availability computations are done assuming that the probability that two components fail simultaneously in a small interval is very small and can be neglected. However, this is not always the

case. A frequent case is the one in which the failure of different components is caused by a common phenomenon or a common cause.

Common cause failures: The IEC Standard 61508-4 [73] defines a *common cause failure* (CCF) as "the result of one or more events, causing concurrent failures of two or more separate channels in a multiple channel system, leading to system failure."

CCFs are failure mechanisms that can either be internal or external to the system and that act simultaneously on all (or a large number of) components, thus rendering ineffective the possible presence of redundancy. CCF acts on all the redundant lines in the same way. CCFs are a major concern in many safety-critical applications [74, 75]. The most typical CCFs are those due to external events like earthquakes, flooding, hurricanes, storms, and lightning. If lightning hits an item of electrical equipment it may destroy all the levels of redundancy at the same time. Some CCFs may be partially external, as in the outage of an electrical power supply; it is partially external since it means that the power supply was not considered part of the system and was not included in the model. There also exist subtle internal causes, sometimes related to weaknesses in the design, that can manifest themselves when triggered by the same event (e.g., redundant equipment that is poorly protected against rain, and must operate in the open air). A further category of CCFs is software bugs. If a piece of code contains a bug, and is replicated over all the equipment of the same class, all of them are exposed to the same bug even in the presence of redundancy.

Some studies (e.g., [76]) have estimated that the percentage of failures due to CCFs in complex plants can be of the order of 20% in the case that no special care is taken in the design, construction, and testing phase for dealing with CCF. This percentage may decrease to 2% or to 0.2% when specific mechanisms and procedures are designed to deal with CCFs [74]. However, universal and effective protection against CCFs does not exist, and the main technique to be followed is the use of *diversity* [77]. Diversity here refers to the physical location of objects (geographical diversity) and design diversity in their technology, software components, and operational procedures.

Summary

In this chapter we have reviewed the three main approaches to quantitative evaluation of dependability characteristics of systems: measurement, modeling, and simulation. We have reviewed the modeling process and the various stages of modeling and evaluation of systems. We have briefly reviewed various modeling formalisms that will be examined in more detail in the subsequent chapters. We have presented a power hierarchy among the modeling formalisms, and introduced modeling verification and validation, and potential sources of errors in models.

2.8   Further Reading

Several prominent books on reliability engineering, like Barlow and Proschan [78, 79], Shooman [80], Leemis [6], Birolini [81], Tobias and Trindale [7], Elsayed [82] and Kapur and Pecht [83], are good sources of information on various aspects of reliability evaluation, modeling, and engineering. Naylor and Finger [57] provide discussion on model verification and validation.

References

[1] IEC 60050, *International Electrotechnical Vocabulary: Chapter 191: Dependability and Quality of Service*. IEC Standard No. 60050-191, 2nd edn., 2001.

[2] S. C. Lee and D. M. Gregg, "From art to science: A vision for the future of information assurance," *Johns Hopkins APL Technical Digest*, vol. 26, pp. 334–342, 2005.

[3] H. Cramer, *Mathematical Methods of Statistics*. Princeton University Press, 1945.

[4] D. Kececioglu, *Reliability and Life Testing Handbook*. Available from the author, Vols. I and II, 1994.

[5] J. F. Lawless, *Statistical Models and Methods for Lifetime Data*. 2nd edn. Wiley, 2002.

[6] L. Leemis, *Reliability: Probabilistic Models and Statistical Methods*. 2nd edn. Lightning Source, 2009.

[7] P. A. Tobias and D. Trindade, *Applied Reliability*, 3rd edn. Chapman and Hall, CRC, 2011.

[8] K. Trivedi, *Probability and Statistics with Reliability, Queueing and Computer Science Applications*, 2nd edn. John Wiley & Sons, 2001.

[9] G. Iversen and H. Norpoth, *Analysis of Variance*, 2nd edn. Sage, 1987.

[10] D. Montgomery, *Design and Analysis of Experiments*, 8th edn. John Wiley & Sons, 2012.

[11] G. E. P. Box, G. M. Jenkins, and G. C. Reinsel, *Time Series Analysis: Forecasting and Control*, 4th edn. Wiley, 2008.

[12] J. Alonso, M. Grottke, A. Nikora, and K. S. Trivedi, "The nature of the times to flight software failure during space missions," in *Proc. IEEE Int. Symp. on Software Reliability Engineering (ISSRE)*, 2012.

[13] M. Grottke, A. Nikora, and K. Trivedi, "An empirical investigation of fault types in space mission system software," in *IEEE/IFIP Int. Conf. on Dependable Systems and Networks (DSN)*, 2010, pp. 447–456.

[14] T. Dohi, K. Goševa-Popstojanova, and K. Trivedi, "Estimating software rejuvenation schedules in high-assurance systems," *The Computer Journal*, vol. 44, no. 6, pp. 473–485, 2001.

[15] W. B. Nelson, *Accelerated Testing: Statistical Models, Test Plans, and Data Analysis*, 1st edn. Wiley-Interscience, 2004.

[16] R. Matias, P. Barbetta, K. Trivedi, and P. Filho, "Accelerated degradation tests applied to software aging experiments," *IEEE Transactions on Reliability*, vol. 59, no. 1, pp. 102–114, 2010.

[17] J. Zhao, Y. Wang, G. Ning, K. S. Trivedi, R. Matias, and K. Cai, "A comprehensive approach to optimal software rejuvenation," *Performance Evaluation*, vol. 70, no. 11, pp. 917–933, 2013.

[18] J. Bechta Dugan and K. Trivedi, "Coverage modeling for dependability analysis of fault-tolerant systems," *IEEE Transactions on Computers*, vol. 38, no. 6, pp. 775–787, Jun. 1989.

[19] K. S. Trivedi, D. Wang, J. Hunt, A. Rindos, W. E. Smith, and B. Vashaw, "Availability modeling of SIP protocol on IBM© WebSphere©," in *Proc. Pacific Rim Int. Symp. on Dependable Computing (PRDC)*, 2008, pp. 323–330.

[20] M. Hsueh, T. Tsai, and R. Iyer, "Fault injection techniques and tools," *IEEE Computer*, vol. 30, no. 4, pp. 75–82, Apr. 1997.

[21] R. K. Iyer, N. Nakka, W. Gu, and Z. Kalbarczyk, "Fault injection," in *Encyclopedia of Software Engineering*, Nov 2010, ch. 29, pp. 287–299.

[22] J. Peterson, *Petri Net Theory and the Modeling of Systems*. Prentice Hall, 1981.

[23] M. Ouyang, "Review on modeling and simulation of interdependent critical infrastructure systems," *Reliability Engineering and System Safety*, vol. 121, pp. 43–60, 2014.

[24] The International Marine Contractors Association, *Guidance on Failure Modes & Effects Analyses (FMEAs)*. IMCA M 166, 2002.

[25] R. McDermott, R. Mikulak, and M. Beauregard, *The Basics of FMEA*, 2nd edn. Productivity Press, 2008.

[26] IEC 60812, *Analysis Techniques for System Reliability – Procedure for Failure Mode and Effects Analysis (FMEA)*. IEC Standard No. 60812, 2nd edn., 2006.

[27] C. Carlson, *Effective FMEAs: Achieving Safe, Reliable, and Economical Products and Processes using Failure Mode and Effects Analysis*. John Wiley & Sons, 2012.

[28] ICH Expert Working Group, *Qualitative Risk Management – Q9*. ICH Harmonised Tripartite Guideline, 2005.

[29] N. Piccinini and I. Ciarambino, "Operability analysis devoted to the development of logic trees," *Reliability Engineering and System Safety*, vol. 55, pp. 227–241, 1997.

[30] J. McDermid and D. Pumfrey, "A development of hazard analysis to aid software design," in *Proc. 9th Ann. Conf. on Computer Assurance (COMPASS '94)*, 1994, pp. 17–25.

[31] T. Arnold, "The concept of coverage and its effect on the reliability model of a repairable system," *IEEE Transaction on Computers*, vol. C-22, pp. 251–254, 1973.

[32] H. Amer and E. McCluskey, "Calculation of coverage parameters," *IEEE Transactions on Reliability*, vol. R-36, pp. 194–198, 1987.

[33] S. Amari, A. Myers, A. Rauzy, and K. S. Trivedi, "Imperfect coverage models: Status and trends," in *Handbook of Performability Engineering*, ed. K. B. Misra. Springer, 2008, pp. 321–348.

[34] S. J. Bavuso, J. Bechta Dugan, K. Trivedi, E. M. Rothmann, and W. E. Smith, "Analysis of typical fault-tolerant architectures using HARP," *IEEE Transactions on Reliability*, vol. R-36, no. 2, pp. 176–185, Jun. 1987.

[35] D. Codetta-Raiteri and L. Portinale, "Dynamic Bayesian networks for fault detection, identification, and recovery in autonomous spacecraft," *IEEE Transactions on Systems, Man, and Cybernetics: Systems*, vol. 45, no. 1, pp. 13–24, Jan. 2015.

[36] W. E. Smith, K. S. Trivedi, L. Tomek, and J. Ackaret, "Availability analysis of blade server systems," *IBM Systems Journal*, vol. 47, no. 4, pp. 621–640, 2008.

[37] D. Tang and K. Trivedi, "Hierarchical computation of interval availability and related metrics," in *Proc. 2004 Int. Conf. on Dependable Systems and Networks (DSN)*, 2004, p. 693–698.

[38] H. de Meer, K. S. Trivedi, and M. Dal Cin, "Guarded repair of dependable systems," *Theor. Comput. Sci.*, vol. 128, no. 1–2, pp. 179–210, 1994.

[39] J. Alonso, R. Matias, E. Vicente, A. Maria, and K. S. Trivedi, "A comparative experimental study of software rejuvenation overhead," *Perform. Eval.*, vol. 70, no. 3, pp. 231–250, 2013.

[40] G. J. Klir, *Uncertainty and Information: Foundations of Generalized Information Theory.* Wiley-Interscience, 2005.

[41] P. Limbourg, *Dependability Modelling under Uncertainty.* Springer, 2008.

[42] D. Coit, "System reliability confidence intervals for complex systems with estimated component reliability," *IEEE Transactions on Reliability*, vol. 46, no. 4, pp. 487–493, Dec. 1997.

[43] K. Mishra and K. S. Trivedi, "Uncertainty propagation through software dependability models," in *Proc. IEEE Int. Symp. on Software Reliability Engineering (ISSRE)*, 2011, pp. 80–89.

[44] G. Ciardo, J. Muppala, and K. S. Trivedi, "SPNP: Stochastic Petri net package," in *Proc. Third Int. Workshop on Petri Nets and Performance Models*, 1989, pp. 142–151.

[45] C. Hirel, B. Tuffin, and K. S. Trivedi, "SPNP: Stochastic Petri Nets. Version 6," in *Int. Conf. on Computer Performance Evaluation: Modelling Techniques and Tools (TOOLS 2000)*, LNCS 1786, 2000, pp.354–357.

[46] D. Deavours, G. Clark, T. Courtney, D. Daly, S. Derisavi, J. Doyle, W. H. Sanders, and P. G. Webster, "The Möbius framework and its implementation," *IEEE Transactions on Software Engineering*, vol. 28, pp. 956–969, 2002.

[47] Wolfram Research Inc., Mathematica, 2010.

[48] R. Sahner, K. Trivedi, and A. Puliafito, *Performance and Reliability Analysis of Computer Systems: An Example-Based Approach Using the SHARPE Software Package.* Kluwer Academic Publishers, 1996.

[49] P. M. Frank, *Introduction to System Sensitivity Theory.* Academic Press, 1978, vol. 11.

[50] A. Bobbio and A. Premoli, "Fast algorithm for unavailability and sensitivity analysis of series-parallel systems," *IEEE Transactions on Reliability*, vol. R-31, pp. 359–361, 1982.

[51] R. Matos, P. Maciel, F. Machida, D. S. Kim, and K. S. Trivedi, "Sensitivity analysis of server virtualized system availability," *IEEE Transactions on Reliability*, vol. 61, no. 4, pp. 994–1006, Dec. 2012.

[52] N. Sato and K. S. Trivedi, "Stochastic modeling of composite web services for closed-form analysis of their performance and reliability bottlenecks," in *Proc. 5th Int. Conf. on Service-Oriented Computing*, eds. B. J. Krämer, K.-J. Lin, and P. Narasimhan. Springer, 2007, pp. 107–118.

[53] L. Swiler, T. Paez, and R. Mayes, "Epistemic uncertainty quantification tutorial," in *Proc. IMAC-XXVII*. Society for Experimental Mechanics Inc., 2009.

[54] A. D. Kiureghian and O. Ditlevsen, "Aleatory or epistemic? does it matter?" *Structural Safety*, vol. 31, no. 2, pp. 105–112, 2009.

[55] K. Mishra and K. S. Trivedi, "Closed-form approach for epistemic uncertainty propagation in analytic models," in *Stochastic Reliability and Maintenance Modeling*, vol. 9. Springer Series in Reliability Engineering, 2013, pp. 315–332.

[56] M. Malhotra and K. Trivedi, "Power hierarchy among dependability model types," *IEEE Transactions on Reliability*, vol. R-43, pp. 493–502, 1994.

[57] T. H. Naylor and J. M. Finger, "Verification of computer simulation models," *Management Science*, vol. 14, no. 2, pp. B-92–B-101, 1967.

[58] J. Muppala, M. Malhotra, and K. Trivedi, "Markov dependability models of complex systems: Analysis techniques," in *Reliability and Maintenance of Complex Systems*, ed. S. Özekici. Springer Verlag, 1996, pp. 442–486.

[59] R. Fricks and K. Trivedi, "Modeling failure dependencies in reliability analysis using stochastic Petri nets," in *Proc. European Simulation Multi-Conference (ESM '97)*, 1997.

[60] D. Wang, R. Fricks, and K. S. Trivedi, "Dealing with non-exponential distributions in dependability models," in *Performance Evaluation: Stories and Perspectives*, ed. G. Kotsis, 2003, pp. 273–302.

[61] J. T. Blake, A. L. Reibman, and K. S. Trivedi, "Sensitivity analysis of reliability and performability measures for multiprocessor systems," *SIGMETRICS Perform. Eval. Rev.*, vol. 16, no. 1, pp. 177–186, May 1988.

[62] A. V. Ramesh and K. S. Trivedi, "On the sensitivity of transient solutions of Markov models," *SIGMETRICS Perform. Eval. Rev.*, vol. 21, no. 1, pp. 122–134, Jun. 1993.

[63] P. J. Courtois, *Decomposability: Queueing and Computer System Applications*. Academic Press, 1977.

[64] A. Bobbio and K. S. Trivedi, "An aggregation technique for the transient analysis of stiff Markov chains," *IEEE Transactions on Computers*, vol. C-35, pp. 803–814, 1986.

[65] G. Ciardo and K. Trivedi, "A decomposition approach for stochastic reward net models," *Performance Evaluation*, vol. 18, pp. 37–59, 1993.

[66] R. R. Muntz, E. de Souza e Silva, and A. Goyal, "Bounding availability of repairable computer systems," *SIGMETRICS Perform. Eval. Rev.*, vol. 17, no. 1, pp. 29–38, Apr. 1989.

[67] J. K. Muppala, A. Sathaye, R. Howe, and K. S. Trivedi, "Hardware and software fault tolerance in parallel computing systems," in *Hardware and Software Fault Tolerance in Parallel Computing Systems*, ed. D. R. Avresky. Ellis Horwood, 1992, pp. 33–59.

[68] P. Brameret, A. Rauzy, and J. Roussel, "Automated generation of partial Markov chain from high level descriptions," *Reliability Engineering & System Safety*, vol. 139, pp. 179–187, 2015.

[69] V. Mainkar and K. S. Trivedi, "Sufficient conditions for existence of a fixed point in stochastic reward net-based iterative models," *IEEE Transactions on Software Engineering*, vol. 22, no. 9, pp. 640–653, Sep. 1996.

[70] R. Ghosh, K. S. Trivedi, V. Naik, and D. S. Kim, "End-to-end performability analysis for infrastructure-as-a-service cloud: An interacting stochastic models approach," in *Proc. IEEE Pacific Rim Int. Symp. on Dependable Computing (PRDC)*, 2010, pp. 125–132.

[71] IEEE Reliability Society, "Annual international reliability physics symposium." [Online]. Available: www.irps.org/

[72] K. Trivedi, A. Sathaye, O. Ibe, and R. Howe, "Should I add a processor?" in *Proc. 23rd Ann. Hawaii Int. Conf. on System Sciences HICSS-23*, 1990, pp. 214–221.

[73] IEC 61508, *Functional Safety of Electrical/Electronic/Programmable Electronic Safety-Related Systems*. IEC Standard No. 61508, 2011.

[74] A. Moslehl, D. M. Rasmuson, and F. M. Marshall, "Guidelines on modeling common-cause failures in probabilistic risk assessment," in *NUREG/CR-5485*. U.S. Nuclear Regulatory Commission, 1998, pp. 1–212.

[75] M. Stamatelatos and W. Vesely, *Fault Tree Handbook with Aerospace Applications*. NASA Office of Safety and Mission Assurance, 2002, vol. 1.1.

[76] B. Martin and R. Wright, "A practical method of common cause failure modelling," *Reliability Engineering*, vol. 19, no. 3, pp. 185–199, 1987.

[77] A. Avizienis and J.-C. Laprie, "Dependable computing: From concepts to design diversity," *Proceedings of the IEEE*, vol. 74, no. 5, pp. 629–638, 1986.

[78] R. Barlow and F. Proschan, *Mathematical Theory of Reliability*. John Wiley & Sons, 1965.

[79] R. Barlow and F. Proschan, *Statistical Theory of Reliability and Life Testing*. Holt, Rinehart and Winston, 1975.

[80] M. Shooman, *Probabilistic Reliability: An Engineering Approach*. McGraw Hill, 1968.

[81] A. Birolini, *Reliability Engineering: Theory and Practice*, 6th edn. Springer Verlag, 2010.

[82] E. A. Elsayed, *Reliability Engineering*, 2nd edn. John Wiley & Sons, 2012.

[83] K. Kapur and M. Pecht, *Reliability Engineering*. John Wiley & Sons, 2014.

3 Dependability Metrics Defined on a Single Unit

As we discussed in the previous chapters, a large system can be viewed as being composed of subsystems, each of which in turn can be further subdivided into smaller elements. Thus, the whole system forms a hierarchy of subsystems, which at the lowest level is composed of *components*. As an example, a hard disk can be viewed as a component in a storage subsystem, and a CPU can be viewed as a component of a blade server in a data center, which in turn forms part of a large-scale cloud computing system. In this chapter, we concentrate on modeling the dependability behavior of a single component, which we refer to as a *unit*.

Let A be a single unit that represents a component in a large system. By virtue of it being the smallest unit of the system, it can be considered as being in one of two exhaustive and mutually exclusive conditions (or states): *working* (or up) and *failed* (or down). A *failure* results in the unit moving from the up state to the down state. Based on the subsequent actions, two cases can be considered:

- The unit is *non-repairable*; once the unit reaches the failed state it remains there forever.
- The unit is *repairable*; once the unit reaches the failed state, a repair action is initiated to restore the unit to a working condition (the *up* state).

The rest of this chapter examines the mathematical representation of the unit's behavior and defines the proper metrics to characterize the unit in both the cases above.

3.1 Non-Repairable Unit

Since the unit is non-repairable, the only event of interest is the instant at which the unit experiences failure. Let the unit A be initially in the up state at time $t = 0$. The instant of time at which it will eventually fail (or enter the down state) is a continuous random variable (r.v.) $X \geq 0$, where X represents the time to failure (or the lifetime) of A. Since X is a random variable, its range is the set of all positive real numbers, including zero ($X \in \mathbb{R}^+ \cup \{0\}$).

The *cumulative distribution function* (Cdf) $F_X(t)$ of the time to failure X is defined as

$$F_X(t) = P\{X \leq t\}, \tag{3.1}$$

and represents the probability that unit A has already failed at or before time t; it is known as the *unreliability* of unit A at time t. From probability theory [1], we know that:

$$\begin{cases} F_X(0) \geq 0, \\ \lim_{t \to \infty} F_X(t) = 1, \\ F_X(t) \text{ is non-decreasing in } t. \end{cases} \qquad (3.2)$$

Typical behavior of a Cdf function $F_X(t)$ is depicted in Figure 3.1 (solid line). With reference to Figure 3.1, given a value $t = a$ and denoting by $q_a = F_X(a)$ the corresponding value of the Cdf, we have:

$$q_a = F_X(a) = P\{X \leq a\}.$$

Hence, q_a is the probability that the r.v. X is less than or equal to a, i.e., it is the probability associated with the event that unit A has failed by time $t = a$. Similarly, given two time instants $t = a$ and $t = b$ (with $a \leq b$), the probability that the r.v. X lies within the time interval $(a, b]$ gives the probability that the failure of unit A occurs in that interval. This probability is given by:

$$F_X(b) - F_X(a) = P\{a < X \leq b\}.$$

The *survival function* (Sf) of the r.v. X, denoted by $R_X(t)$, is defined as

$$R_X(t) = P\{X > t\} = 1 - F_X(t), \qquad (3.3)$$

and represents the probability that the r.v. X is greater than t, i.e., that the unit has been continuously working from time 0 to t, or, in other words, that its failure did not occur until t. Hence, the quantity $R_X(t)$ in (3.3) is the unit's *reliability*. Typical behavior of the reliability function $R_X(t)$ is shown in Figure 3.1 (dashed line), and $R_X(t)$ satisfies the

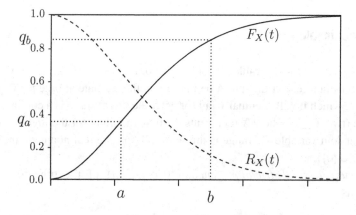

Figure 3.1   Typical behavior of the Cdf function $F_X(t)$ (solid line) and the reliability $R_X(t)$ (dashed line).

following conditions:

$$\begin{cases} R_X(0) = 1, \\ \lim_{t \to \infty} R_X(t) = 0, \\ R_X(t) \text{ is non-increasing in } t. \end{cases} \tag{3.4}$$

Given two time instants a and b (with $a \leq b$), the quantity

$$R_X(a) - R_X(b) = F_X(b) - F_X(a) = P\{a < X \leq b\}$$

gives us the probability that the unit lifetime lies within the interval $(a, b]$.

If the Cdf $F_X(t)$ has a derivative, the *probability density function* (pdf) $f_X(t)$ of the r.v. X is given by

$$f_X(t) = \frac{d F_X(t)}{dt} = -\frac{d R_X(t)}{dt}. \tag{3.5}$$

From the definition of the pdf and with reference to Figure 3.2, we have:

$$\int_a^b f_X(t)\, dt = P\{a < X \leq b\} = F_X(b) - F_X(a).$$

The *n*th *moment* of the r.v. X is defined as:

$$E[X^n] = \int_{-\infty}^{\infty} t^n \cdot f_X(t)\, dt. \tag{3.6}$$

The expected value is the moment of order one, and is usually referred to as the *mean time to failure* (MTTF). From Eq. (3.6), and taking into account that X is a non-negative random variable:

$$\text{MTTF} = E[X] = \int_0^{\infty} t \cdot f_X(t)\, dt = \int_0^{\infty} R_X(t)\, dt. \tag{3.7}$$

The right-most term in the above equation is obtained from the preceding one using integration by parts. The central moment of order n is defined as:

$$E[(X - E[X])^n] = \int_0^{\infty} (t - E[X])^n \cdot f_X(t)\, dt. \tag{3.8}$$

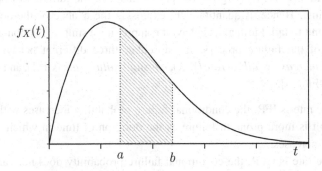

Figure 3.2 The probability density function $f_X(t)$.

The *variance* is the central moment of order two, and is given by:

$$\text{Var}[X] = \sigma_X^2 = E[(X - E[X])^2] = \int_0^\infty (t - E[X])^2 \cdot f_X(t)\, dt. \qquad (3.9)$$

For the variance, the following relation holds:

$$\text{Var}[X] = E[X^2] - (E[X])^2.$$

The squared *coefficient of variation* cv^2 of the r.v. X is defined as:

$$cv^2 = \frac{\text{Var}[X]}{(E[X])^2}.$$

The hazard rate $h_X(t)$ of an r.v. X is defined as the conditional probability that X takes a value in the interval $(t, t+dt]$, given that $X > t$. When X is the time to failure of a unit, the hazard rate is called the *failure rate* and $h_X(t)dt$ is the probability that the unit fails between t and $t+dt$ given that it was functioning up to time t [2].

Using conditional probabilities, we obtain:

$$h_X(t)dt = P\{t < X \le t+dt \,|\, X > t\} = \frac{P\{t < X \le t+dt, X > t\}}{P\{X > t\}}. \qquad (3.10)$$

From (3.5) and (3.3), Eq. (3.10) becomes:

$$h_X(t) = \frac{f_X(t)}{R_X(t)} = -\frac{1}{R_X(t)} \frac{dR_X(t)}{dt}. \qquad (3.11)$$

Inverting (3.11), we can express the reliability $R_X(t)$ as a function of the failure rate $h_X(t)$:

$$R_X(t) = \exp\left[-\int_0^t h_X(u)\, du\right] = e^{-\int_0^t h_X(u)\, du}. \qquad (3.12)$$

All the functions $F_X(t)$, $R_X(t)$, $f_X(t)$, and $h_X(t)$ are interrelated, such that given one of them, the remaining functions can easily be derived. This relationship is illustrated in Table 3.1. The choice of function to describe the random phenomenon is arbitrary and depends partly on the application domain and partly on the mathematical or graphical convenience of the representation. In reliability engineering the failure rate is often the chosen function since it has a direct physical meaning. Given a unit that is working at time t, the failure rate $h_X(t)$ gives the probability that the failure will occur in the next unit of time. Hence, it quantitatively expresses the chance or the aptitude of a still-working unit to fail. Further, its behavior can provide a simple (but non-exhaustive) classification of the failure process. In particular, three different behaviors can be distinguished: *increasing failure rate* (IFR), *constant failure rate* (CFR), and *decreasing failure rate* (DFR) [3].

- If the failure rate is IFR, the conditional failure probability increases with the unit's age. The unit is more prone to failure as the duration of time in which it has been operational increases.
- If the failure rate is CFR, the conditional failure probability does not vary with the unit's age.

Table 3.1 Relationship between $f(t)$, $F(t)$, $R(t)$, and $h(t)$.

	$f(t)$	$F(t)$	$R(t)$	$h(t)$
$f(t)$	1	$F'(t)$	$-R'(t)$	$h(t)e^{-\int_0^t h(u)du}$
$F(t)$	$\int_0^t f(u)du$	1	$1-R(t)$	$1-e^{-\int_0^t h(u)du}$
$R(t)$	$\int_t^\infty f(u)du$	$1-F(t)$	1	$e^{-\int_0^t h(u)du}$
$h(t)$	$\dfrac{f(t)}{\int_t^\infty f(u)du}$	$\dfrac{F'(t)}{1-F(t)}$	$-\dfrac{d}{dt}\ln R(t)$	1

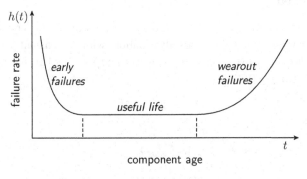

Figure 3.3 Bathtub-shaped curve for the failure rate.

- If the failure rate is DFR, the conditional failure probability decreases with the unit's age. The unit is less prone to failure as the duration of time in which it has been operational increases.

Some distributions have a monotonic behavior for the hazard rate and can be classified as IFR or CFR or DFR distributions. However, for many distributions the hazard rate is not monotonic and does not belong to any of the three classes listed above.

It has been experimentally observed that the failure rate for many but not all physical units follows the so-called *bathtub curve* (Figure 3.3). The curve is divided into three regions, as a function of the unit's age: an initial DFR region characterized by early failures or infant mortality, a CFR region characterized by a constant (or time-independent) failure rate and, finally, an IFR region characterized by wearout failures.

Early-life failures are assumed to be caused by a small deviant population, which shows defects that are not detected by the testing and quality control procedures. These defects may arise from poor materials, assembly, installation or other causes not covered by testing. These units fail in their initial operational life. The failure rate decreases up to the level of the "normal" population as the deviant units are exhausted.

The second period is characterized by an (almost) constant failure rate and is called the useful life region, since the unit is in operation at its lowest failure rate.

In the third period, the increasing failure rate denotes the appearance of possible aging phenomena due to chemical, physical or environmental factors such as wearout, corrosion, hardening or fatigue that irreversibly increase the proneness of the unit to fail.

3.2   Common Probability Distribution Functions

In this section, several commonly used distributions and their properties are presented.

3.2.1   Exponential Distribution

The exponential distribution is characterized by having a time-independent (constant) hazard rate; conversely, a continuous distribution with a constant hazard rate is exponential. If $h(t) = const = \lambda$, it is easily verified from (3.12) that the following expressions hold (for $t \geq 0$):

$$F(t) = 1 - e^{-\lambda t},$$

$$R(t) = e^{-\lambda t},$$

$$f(t) = \lambda \, e^{-\lambda t}, \qquad (3.13)$$

$$h(t) = \lambda.$$

Applying (3.7), the MTTF is given by:

$$E[X] = \text{MTTF} = \int_0^\infty t \cdot f(t) \, dt = \lambda \int_0^\infty t \cdot e^{-\lambda t} \, dt. \qquad (3.14)$$

The integral (3.14) is resolved using integration by parts:

$$E[X] = \text{MTTF} = \lambda \left[-\frac{t \cdot e^{-\lambda t}}{\lambda} \right]_0^\infty + \lambda \int_0^\infty \frac{e^{-\lambda t}}{\lambda} \cdot dt = \frac{1}{\lambda}. \qquad (3.15)$$

Equation (3.15) shows that an exponentially distributed r.v. X has an MTTF equal to the reciprocal of the failure rate: $\text{MTTF} = 1/\lambda$. This result has three important implications that favor the use of exponential distributions in reliability analysis:

- The constant failure rate assumes a precise physical meaning since it is equal to the reciprocal of the MTTF. We should note that this reciprocal relationship between failure rate and MTTF only applies to the exponential distribution.
- The exponential distribution is completely specified by a single parameter given by its expected value. Thus, in the rest of the book an exponential distribution with parameter λ will be denoted by $\text{EXP}(\lambda)$.
- The expected value is the simplest and most direct parameter to statistically estimate based on experimental observations.

It should be pointed out that the time to failure (TTF) is a random variable, while MTTF is its expected value – the distinction between these two should always be kept in mind. Furthermore, the units of TTF and MTTF are time units (e.g., hours or years), while that of the failure rate are in failures per unit of time.

The variance of the exponential distribution and the squared coefficient of variation are given by:

$$\text{Var}[X] = \int_0^\infty \left(t - \frac{1}{\lambda}\right)^2 \cdot f(t)\, dt = \frac{1}{\lambda^2},$$

$$cv^2 = \frac{\text{Var}[X]}{\text{MTTF}^2} = 1.$$

Typical behavior of the exponential distribution is shown in Figure 3.4 for a value of $\lambda = 2$.

The exponential distribution possesses the *memoryless* property. If X is an EXP(λ) r.v., the survival probability of the unit, given that it is alive after a generic time $t = a$, equals the survival probability at time $t = 0$. More formally, we can define the residual life of a unit after a units of time as a new random variable $Y = X - a$, which represents the remaining life of the unit conditioned on the fact that the unit was alive at time a (Figure 3.5).

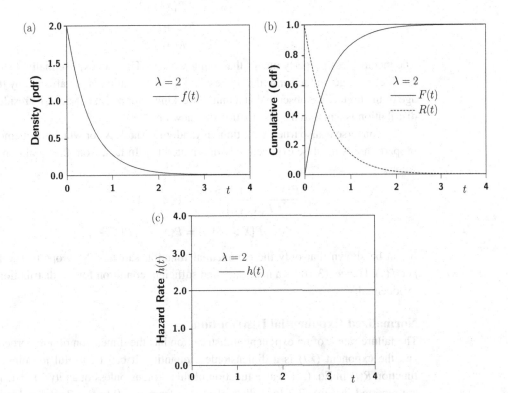

Figure 3.4 Exponential distribution ($\lambda = 2$). (a) Density $f(t)$; (b) Cdf $F(t)$, and reliability $R(t)$; (c) failure rate $h(t) = \lambda$.

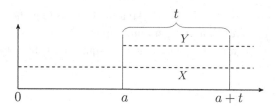

Figure 3.5   The residual life distribution.

The memoryless property implies that the distribution of the residual life Y is the same as the distribution of the entire life X [1]. With reference to Figure 3.5, the memoryless property can be formally expressed as:

$$P\{Y > t | X > a\} = P\{X > t\}. \tag{3.16}$$

Elaborating (3.16), we can write:

$$
\begin{aligned}
P\{Y > t | X > a\} &= P\{X > a+t | X > a\} \\
&= \frac{P\{X > a+t, X > a\}}{P\{X > a\}} \\
&= \frac{P\{X > a+t\}}{P\{X > a\}} \\
&= \frac{R(a+t)}{R(a)} = e^{-\lambda t}.
\end{aligned} \tag{3.17}
$$

The memoryless property means that a unit whose TTF is an exponentially distributed r.v. does not age, in the sense that its residual life distribution is not affected by the real age of the unit. If we observe that a unit is working after a units of time, its residual life distribution is equal to the distribution of a new unit.

The converse is also true. A continuous random variable X for which the memoryless property holds must be exponentially distributed [1]. In fact, from the condition (3.16) we get:

$$\frac{P\{X > a+t, X > a\}}{P\{X > a\}} = P\{X > t\}$$

$$P\{X > a+t\} = P\{X > a\} P\{X > t\}. \tag{3.18}$$

It can be shown that only the exponential function satisfies the property $f(x+y) = f(x)f(y)$. Hence, (3.16) is a necessary and sufficient condition for the distribution to be exponential.

Normalized Exponential Distribution

The failure rate λ of an exponential distribution has the dimension of a reciprocal time and the exponent (λt) is a dimensionless quantity. It can be useful to compute the function $R(t)$ in Eq. (3.13) as a function of this dimensionless quantity λt. The results are reported in Table 3.2 for values of $u = \lambda t$ from $u = 0$ to $u = 2$, and in Figure 3.6 from $u = 0$ to $u = 5$.

Table 3.2 The normalized exponential reliability function $R(u)$ as a function of the dimensionless parameter $u = \lambda t$.

$u = \lambda t$	$R(u)$	$u = \lambda t$	$R(u)$	$u = \lambda t$	$R(u)$	$u = \lambda t$	$R(u)$
0.00	1.000000	0.55	0.576950	1.00	0.367879	1.55	0.212248
0.05	0.951229	0.60	0.548812	1.10	0.332871	1.60	0.201897
0.15	0.860708	0.65	0.522046	1.15	0.316637	1.65	0.192050
0.20	0.818731	0.70	0.496585	1.20	0.301194	1.70	0.182684
0.25	0.778801	0.75	0.472367	1.25	0.286505	1.75	0.173774
0.30	0.740818	0.80	0.449329	1.30	0.272532	1.80	0.165299
0.35	0.704688	0.85	0.427415	1.35	0.259240	1.85	0.157237
0.40	0.670320	0.90	0.406570	1.40	0.246597	1.90	0.149569
0.45	0.637628	0.95	0.386741	1.45	0.234570	1.95	0.142274
0.50	0.606531			1.50	0.223130		

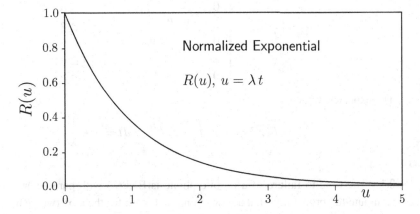

Figure 3.6 The normalized exponential reliability function.

The entry $\lambda t = 1$ in Table 3.2 corresponds to a value of $t = 1/\lambda = $ MTTF. The reliability value is $R(u) = e^{-1} = 0.3679$. This means that the probability of a unit with exponentially distributed lifetime surviving over a mission time equal to its MTTF is $R(1) = 0.3679$. In order to operate a unit with a high reliability, the mission time needs to be short compared to the MTTF. Looking at Table 3.2, in order to operate a unit with a reliability greater than $R(u) > 0.95$ we must have $\lambda t < 0.05$, which implies $t < 0.05/\lambda = 0.05\,$MTTF. The mission time should not exceed 5% of the MTTF.

In general, given a value of the unit reliability that we wish to achieve, we can compute the mission time that corresponds to the required reliability level. If a unit A is required to operate with a reliability $R(u) > r_A$, we can write:

$$R(u) = e^{-u} = e^{-\lambda t} = r_A \tag{3.19}$$

$$u = \lambda t = -\ln r_A$$

$$t = -\frac{1}{\lambda} \ln r_A = -\text{MTTF} \ln r_A.$$

Equation (3.20) shows that the value $-\ln r_A$ is the fraction of the MTTF for which the unit exhibits a reliability greater than r_A. To achieve a reliability $r_A = 0.99$, the unit must operate for a time less than the fraction $-\ln r_A = 0.01005$ of its MTTF.

3.2.2   Shifted Exponential Distribution

The exponential distribution introduced in Section 3.2.1 is meant to start at $t = 0$. In general, however, the distribution can start at any positive value $a \geq 0$. The corresponding distribution is called a *shifted exponential* distribution. The Cdf, survivor function, and density of a shifted exponential starting at $a \geq 0$ are given by:

$$
\begin{aligned}
F(t) &= \begin{cases} 1 - e^{-\lambda(t-a)} & t \geq a \\ 0 & t < a \end{cases} \\
R(t) &= \begin{cases} e^{-\lambda(t-a)} & t \geq a \\ 0 & t < a \end{cases} \\
f(t) &= \begin{cases} \lambda e^{-\lambda(t-a)} & t \geq a \\ 0 & t < a \end{cases} \\
h(t) &= \begin{cases} \lambda & t \geq a \\ 0 & t < a. \end{cases}
\end{aligned}
\tag{3.20}
$$

The expected value is:

$$
E[X] = a + \int_a^\infty e^{-\lambda(t-a)}\, dt = a + \frac{1}{\lambda}.
\tag{3.21}
$$

Problems

3.1 The time to failure of a mobile phone is EXP(λ) with parameter $\lambda = 1/6\,\mathrm{yr}^{-1}$. Compute the probability that a new phone will work for the next two years.

3.2 As in Problem 3.1, the time to failure of a mobile phone is EXP(λ) with parameter $\lambda = 1/6\,\mathrm{yr}^{-1}$. Compute the maximum time horizon in which the mobile phone can be utilized maintaining a reliability greater than 0.95 or 0.90.

3.3 The time to failure of a mobile phone is EXP(λ) with parameter $\lambda = 1/6\,\mathrm{yr}^{-1}$. If you buy a used phone that has been in use for two years, compute the probability that your mobile phone will work for three more years.

3.2.3   Weibull Distribution

The two-parameter Weibull distribution is defined for $t \geq 0$ and is characterized by the following expressions:

$$
\begin{aligned}
F(t) &= 1 - \exp\left[-(\eta t)^\beta\right], \\
R(t) &= \exp\left[-(\eta t)^\beta\right], \\
f(t) &= \beta\eta \cdot (\eta t)^{\beta-1} \cdot \exp\left[-(\eta t)^\beta\right], \\
h(t) &= \beta\eta \cdot (\eta t)^{\beta-1},
\end{aligned}
\tag{3.22}
$$

where $\eta > 0$ is called the *scale parameter* since it locates the position of the pdf on the time axis, and $\beta > 0$ is the *shape parameter* since it determines the shape of the pdf.

A Weibull distribution will be denoted throughout the book as $\text{WEIB}(\eta, \beta)$. Typical behaviors of the functions $f(t)$, $F(t)$, and $h(t)$ are plotted in Figure 3.7 for a fixed value of $\eta = 1$ and for $\beta = 0.5$, 1 and 3. In particular, looking at the expression of the failure rate in (3.22), we can distinguish three different types of behavior:

$$\beta < 1 \implies h(t) \quad \text{is a decreasing function of time;}$$
$$\beta = 1 \implies h(t) \quad \text{is constant;}$$
$$\beta > 1 \implies h(t) \quad \text{is an increasing function of time.}$$

The extensive use of the Weibull distribution in the interpretation and analysis of failure phenomena is mainly related to the fact that the shape of its failure rate curve depends on a single parameter and that the exponential distribution is a member of the Weibull family with $\beta = 1$.

If X is a Weibull-distributed r.v., its MTTF, variance, and squared coefficient of variation are given by:

$$E[X] = \text{MTTF} = \frac{1}{\eta} \Gamma\left(1 + \frac{1}{\beta}\right), \tag{3.23}$$

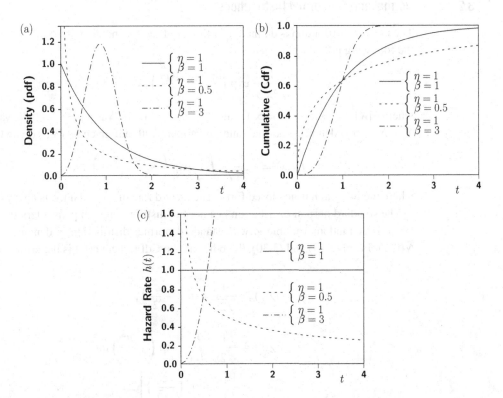

Figure 3.7 Weibull distribution for $\eta = 1$ and $\beta = 0.5, 1, 3$. (a) Density; (b) Cdf; (c) hazard rate.

$$\text{Var}[X] = \left(\frac{1}{\eta}\right)^2 \left[\Gamma\left(1+\frac{2}{\beta}\right) - \left[\Gamma\left(1+\frac{1}{\beta}\right)\right]^2\right], \tag{3.24}$$

$$cv^2 = \frac{\Gamma(1+2/\beta)}{[\Gamma(1+1/\beta)]^2} - 1, \tag{3.25}$$

where the gamma function $\Gamma(x)$ is defined as:

$$\Gamma(x) = \int_0^\infty u^{x-1} e^{-u} du. \tag{3.26}$$

If x is a positive integer, denoted by k, then:

$$\Gamma(k) = (k-1)!$$

An alternative form of the Weibull distribution function is written as [1]:

$$F(t) = 1 - \exp(-\alpha t^\beta), \tag{3.27}$$

$$h(t) = \alpha \beta t^{\beta-1},$$

where $\beta > 0$ is the shape parameter, as in (3.22), and $\alpha > 0$ is an alternative expression for the scale parameter. There is also a three-parameter Weibull distribution function [1]:

$$F(t) = 1 - \exp(-\alpha (t-\theta)^\beta). \tag{3.28}$$

3.2.4   Normal and Lognormal Distributions

The normal distribution is defined over the set of the real numbers $(-\infty, +\infty)$, and has the following pdf:

$$f_X(x) = \frac{1}{\sqrt{2\pi}\,\sigma} \exp\left(-\frac{(x-\mu)^2}{2\sigma^2}\right), \qquad -\infty < x < \infty, \tag{3.29}$$

where $E[X] = \mu$ $(-\infty < \mu < \infty)$ is the expected value and $\text{Var}[X] = \sigma^2 > 0$ the variance. $X \sim N(\mu,\sigma^2)$ denotes a normal random variable with pdf given by (3.29). The Cdf is:

$$F_X(x) = \frac{1}{\sqrt{2\pi}\,\sigma} \int_{-\infty}^x \exp\left(-\frac{(u-\mu)^2}{2\sigma^2}\right) du, \tag{3.30}$$

where the integral has no closed form. The hazard rate of a normal r.v. is always IFR.

The standard normal distribution is a normal distribution with parameters $\mu = 0$ and $\sigma^2 = 1$. A random variable Z with standard normal distribution is denoted by $Z \sim N(0,1)$. From (3.29) and (3.30), the pdf, and the Cdf (where erf() is the error function) are:

$$f_Z(z) = \frac{1}{\sqrt{2\pi}} \exp\left(-\frac{z^2}{2}\right),$$

$$\tag{3.31}$$

$$F_Z(z) = \frac{1}{\sqrt{2\pi}} \int_{-\infty}^z \exp\left(-\frac{u^2}{2}\right) du$$

$$= \frac{1}{2}\left[1 + \text{erf}\left(\frac{z}{\sqrt{2}}\right)\right].$$

The shape of the standard normal pdf, Cdf, and hazard rate, with $\mu = 0$ and $\sigma = 1$, is shown in Figure 3.8 as a solid line. For comparison, Figure 3.8 also shows the normal curves for $\mu = 0$ and two different values of the standard deviation, $\sigma = 0.5$ and $\sigma = 2$. For a particular value x of a normal random variable $X \sim N(\mu, \sigma^2)$, the corresponding value of the standardized normal random variable $Z \sim N(0, 1)$ is given by $z = (x - \mu)/\sigma$. Then, by applying this change of variables we obtain

$$F_X(x) = F_Z\left(\frac{x - \mu}{\sigma}\right),$$

so that a value of the Cdf of a random variable $X \sim N(\mu, \sigma^2)$ at x can be obtained from the corresponding value of the standardized normal variable $Z \sim N(0, 1)$ at $z = \frac{x - \mu}{\sigma}$.

The normal distribution is extremely important, mainly in statistical applications, since the central limit theorem states that under very general assumptions the mean of a sample of n independent identically distributed (i.i.d.) random variables is normally distributed in the limit $n \to \infty$. However, because the normal distribution is defined over the entire set of real numbers, from $-\infty$ to $+\infty$, its application is questionable for random variables that are defined only for non-negative real numbers. To properly account for this, a truncated normal density can be used [1].

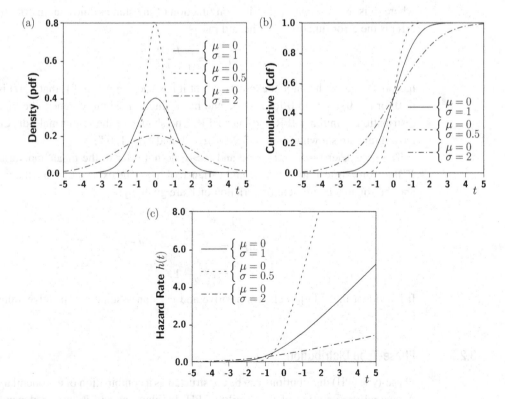

Figure 3.8 Normal distribution for $\mu = 0$ and $\sigma = 1, 0.5, 2$. (a) Density; (b) Cdf; (c) hazard rate.

Lognormal Distribution

The lognormal distribution is a distribution of a random variable whose logarithm is normal. If X is a normal random variable $X \sim N(\mu, \sigma^2)$, then the random variable $Y = e^X$ has a *lognormal* distribution with parameters μ and σ. Conversely, we can say that if Y has a lognormal distribution with parameters μ and σ, then $X = \ln Y$ is normally distributed, $X \sim N(\mu, \sigma^2)$. We indicate a lognormal random variable as $Y \sim LN(\mu, \sigma^2)$.

According to its definition, a lognormal random variable is defined only on the non-negative real axis ($Y \in \mathbb{R} > 0$). The pdf is given by:

$$f_Y(t) = \frac{1}{\sigma t \sqrt{2\pi}} \exp\left(-\frac{(\ln t - \mu)^2}{2\sigma^2}\right), \qquad t \geq 0. \tag{3.32}$$

The Cdf $F_Y(t)$ can be computed by observing that

$$P\{Y \leq a\} = P\{\ln Y \leq \ln a\},$$

and that $\ln Y \sim N(\mu, \sigma^2)$. Hence:

$$F_Y(t) = \Phi\left(\frac{\ln x - \mu}{\sigma}\right) \tag{3.33}$$

$$= \frac{1}{2}\left[1 + \operatorname{erf}\left(\frac{\ln t - \mu}{\sigma\sqrt{2}}\right)\right],$$

where Φ is the cumulative distribution function of the standard normal distribution and erf() is the error function. The hazard rate

$$h_Y(t) = \frac{f_Y(t)}{1 - F_Y(t)}$$

has no simple mathematical expression, but it has been shown in [4] that $h_Y(t)$ is equal to 0 for $t = 0$, tends to 0 for $t \to \infty$ and has a single maximum; hence, the lognormal distribution is initially IFR and later DFR. The shapes of the lognormal pdf, Cdf, and hazard rate are shown in Figure 3.9 for $\mu = 0$ and $\sigma = 1, 0.5, 2$.

The parameters $-\infty < \mu < \infty$ and $\sigma > 0$ do not refer to the mean and variance of Y, they refer instead to the mean and variance of $\ln Y$. The expected value, the variance, and the squared coefficient of variation of Y are given by:

$$E[Y] = e^{\mu + \sigma^2/2},$$

$$\operatorname{Var}[Y] = e^{2\mu + 2\sigma^2} - e^{2\mu + \sigma^2}, \tag{3.34}$$

$$cv^2 = e^{\sigma^2} - 1.$$

It is evident that $E[Y]$ is always positive and cv^2 can assume any positive value from $0 < cv^2 < \infty$.

3.2.5   Phase-Type Distribution

Phase-type (PH) distributions can be constructed as a combination of exponential stages. A general presentation of the family of PH distributions and its use in dependability

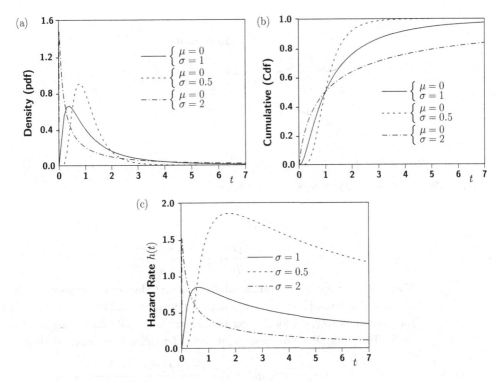

Figure 3.9 Lognormal distribution for $\mu = 0$ and $\sigma = 1, 0.5, 2$. (a) Density; (b) Cdf; (c) hazard rate.

modeling is deferred to Chapter 15. In this section, two subclasses of PH distributions that have a simple representation are explicitly examined.

Hyperexponential Distribution

The hyperexponential distribution is a special case of the PH distribution, obtained as a mixture of several exponential distributions. Its Cdf can be expressed as a weighted sum of the Cdfs of the individual exponential distributions. Mathematically and graphically, a hyperexponential distribution can be represented as the combination of k parallel exponential stages, where each stage is selected according to an initial (mixing) probability mass function. A hyperexponential distribution with two stages is depicted in Figure 3.10. Stage 1 is $\text{EXP}(\lambda_1)$ and it is selected with probability $q_1 = q \geq 0$; stage 2 is $\text{EXP}(\lambda_2)$ and it is selected with probability $q_2 = (1 - q) \geq 0$, so that $q_1 + q_2 = 1$. Both stages represent a working condition for the unit, while state 0 is the down state.

This two-stage hyperexponential random variable X has three parameters and will be denoted by $X \sim \text{HYPER}(\lambda_1, \lambda_2; q)$. The Cdf, Sf, pdf, and hazard rate are given by:

$$F(t) = q(1 - e^{-\lambda_1 t}) + (1 - q)(1 - e^{-\lambda_2 t}),$$

$$R(t) = q e^{-\lambda_1 t} + (1 - q) e^{-\lambda_2 t}, \tag{3.35}$$

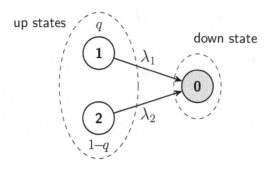

Figure 3.10   A two-stage hyperexponential distribution.

$$f(t) = q\lambda_1 e^{-\lambda_1 t} + (1-q)\lambda_2 e^{-\lambda_2 t},$$

$$h(t) = \frac{q\lambda_1 e^{-\lambda_1 t} + (1-q)\lambda_2 e^{-\lambda_2 t}}{q e^{-\lambda_1 t} + (1-q) e^{-\lambda_2 t}}.$$

If $\lambda_1 = \lambda_2 = \lambda$, the hyperexponential distribution reduces to an exponential distribution of rate λ. The hazard rate of the two-stage hyperexponential ranges from $q\lambda_1 + (1-q)\lambda_2$ for $t = 0$ to $\min\{\lambda_1, \lambda_2\}$ for $t \to \infty$ and is a monotonically decreasing function of time [3, 5]. To see the DFR property, compute the derivative with respect to time:

$$\frac{dh(t)}{dt} = -q(1-q)\frac{(\lambda_1 - \lambda_2)^2 e^{-(\lambda_1 + \lambda_2)t}}{(q e^{-\lambda_1 t} + (1-q)e^{-\lambda_2 t})^2} < 0. \tag{3.36}$$

The expected value, the variance, and the squared coefficient of variation are:

$$E[X] = \frac{q}{\lambda_1} + \frac{(1-q)}{\lambda_2},$$

$$\mathrm{Var}[X] = E[X^2] - (E[X])^2 = 2\left(\frac{q}{\lambda_1^2} + \frac{(1-q)}{\lambda_2^2}\right) - \left(\frac{q}{\lambda_1} + \frac{(1-q)}{\lambda_2}\right)^2,$$

$$cv^2 = \frac{2\left(\dfrac{q}{\lambda_1^2} + \dfrac{(1-q)}{\lambda_2^2}\right)}{\left(\dfrac{q}{\lambda_1} + \dfrac{(1-q)}{\lambda_2}\right)^2} - 1. \tag{3.37}$$

By substituting $\alpha = \lambda_1/\lambda_2$, the coefficient of variation can be written as:

$$cv^2 = \frac{2\left(\dfrac{q}{\alpha^2} + (1-q)\right)}{\left(\dfrac{q}{\alpha} + (1-q)\right)^2} - 1.$$

By letting $\alpha \to 0$ we obtain $cv^2 = 2/q - 1$. Since $0 < q < 1$, the coefficient of variation ranges from $cv^2 = 1$ for $q = 1$ to $cv^2 \to \infty$ as $q \to 0$. Figure 3.11 plots the density, Cdf, and hazard rate of three different two-stage hyperexponential distributions with expected values $E[X] = 1$ and parameters $(\lambda_1 = 1, \lambda_2 = 1, q = 0.5)$, $(\lambda_1 = 5, \lambda_2 = 0.555, q = 0.5)$ and $(\lambda_1 = 5, \lambda_2 = 0.918, q = 0.1)$. The hyperexponential distribution

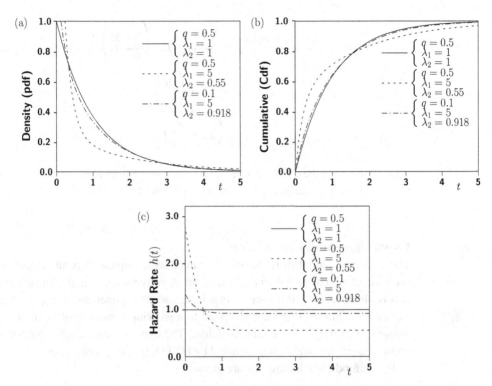

Figure 3.11 A two-stage hyperexponential distribution with $E[X] = 1$ and with three sets of parameters: $(\lambda_1 = 1, \lambda_2 = 1, q = 0.5)$, $(\lambda_1 = 5, \lambda_2 = 0.555, q = 0.5)$, $(\lambda_1 = 5, \lambda_2 = 0.918, q = 0.1)$. (a) Density; (b) Cdf; (c) hazard rate.

can be generalized to a mixture of k independent exponential distributions. We say that k is the *order* of the hyperexponential distribution. Let $\text{EXP}(\lambda_i)$ be the ith exponential distribution of the mixture $(i = 1, 2, \ldots, k)$, and let $q_i > 0$ and $\sum_{i=1}^{k} q_i = 1$.

The Cdf, pdf, and hazard rate are then given by:

$$F(t) = \sum_{i=1}^{k} q_i (1 - e^{-\lambda_i t}),$$

$$f(t) = \sum_{i=1}^{k} q_i \lambda_i e^{-\lambda_i t}, \tag{3.38}$$

$$h(t) = \frac{\sum_{i=1}^{k} q_i \lambda_i e^{-\lambda_i t}}{\sum_{i=1}^{k} q_i e^{-\lambda_i t}}.$$

It can be shown that the hazard rate is always DFR [1]. Furthermore:

$$E[X] = \sum_{i=1}^{k} \frac{q_i}{\lambda_i},$$

$$\text{Var}[X] = 2 \sum_{i=1}^{k} \left(\frac{q_i}{\lambda_i^2} \right) - \left(\sum_{i=1}^{k} \frac{q_i}{\lambda_i} \right)^2,$$

$$cv^2 = \frac{2 \sum_{i=1}^{k} \left(\frac{q_i}{\lambda_i^2} \right)}{\left(\sum_{i=1}^{k} \frac{q_i}{\lambda_i} \right)^2} - 1. \tag{3.39}$$

Also, in this case the coefficient of variation is greater than 1 unless all the λ_i are equal; then, the hyperexponential reduces to an exponential.

Hypoexponential Distribution

The hypoexponential distribution is obtained as a sequential combination of stages, each one of which is exponentially distributed. A hypoexponential distribution of order two is obtained by a sequence of two independent exponential stages, as depicted in Figure 3.12. Stages 1 and 2 correspond to a working condition of the unit, while the attainment of stage 3 represents its failure. The hypoexponential of order two depends on two parameters and will be denoted by $\text{HYPO}_2(\lambda_1, \lambda_2)$, with $\lambda_1, \lambda_2 > 0$.

The Cdf, pdf, and hazard rate are given by:

$$F(t) = 1 - \frac{\lambda_2}{\lambda_2 - \lambda_1} e^{-\lambda_1 t} + \frac{\lambda_1}{\lambda_2 - \lambda_1} e^{-\lambda_2 t},$$

$$f(t) = \frac{\lambda_1 \lambda_2}{\lambda_2 - \lambda_1} (e^{-\lambda_1 t} - e^{-\lambda_2 t}), \tag{3.40}$$

$$h(t) = \frac{\lambda_1 \lambda_2 (e^{-\lambda_1 t} - e^{-\lambda_2 t})}{\lambda_2 e^{-\lambda_1 t} - \lambda_1 e^{-\lambda_2 t}}.$$

The above equations are defined only for $\lambda_1 \neq \lambda_2$. The hazard rate ranges from $h(0) = 0$ for $t = 0$ to $h(t) \to \min\{\lambda_1, \lambda_2\}$ for $t \to \infty$, and it is monotonically increasing in time since its derivative is always positive. Indeed,

$$\frac{d\,h(t)}{d\,t} = \lambda_1 \lambda_2 \frac{(\lambda_1 - \lambda_2)^2 e^{-(\lambda_1 + \lambda_2)t}}{(\lambda_2 e^{-\lambda_1 t} - \lambda_1 e^{-\lambda_2 t})^2} > 0. \tag{3.41}$$

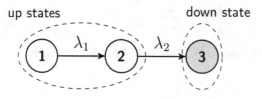

Figure 3.12   A two-stage hypoexponential distribution.

The expected value, variance, and coefficient of variation are given by:

$$E[X] = \frac{\lambda_1 \lambda_2}{\lambda_2 - \lambda_1} \int_0^\infty t(e^{-\lambda_1 t} - e^{-\lambda_2 t}) dt = \frac{\lambda_1 + \lambda_2}{\lambda_1 \lambda_2} = \frac{1}{\lambda_1} + \frac{1}{\lambda_2},$$

$$\text{Var}[X] = \frac{\lambda_1 \lambda_2}{\lambda_2 - \lambda_1} \int_0^\infty \left(t - \frac{\lambda_1 + \lambda_2}{\lambda_1 \lambda_2}\right)^2 (e^{-\lambda_1 t} - e^{-\lambda_2 t}) dt = \frac{1}{\lambda_1^2} + \frac{1}{\lambda_2^2}, \qquad (3.42)$$

$$cv^2 = \frac{\lambda_1^2 + \lambda_2^2}{(\lambda_1 + \lambda_2)^2}.$$

The expected value is the sum of the expected values of the two sequential exponential distributions, while the coefficient of variation is always less than one. Fixing λ_1, the cv is maximum for $\lambda_2 = 0$ and minimum for $\lambda_2 = \lambda_1$. Summarizing:

$$cv^2 = \begin{cases} 1 & \text{for} \quad \lambda_2 = 0; \quad \lambda_1 \text{ any value,} \\ 1/2 & \text{for} \quad \lambda_2 = \lambda_1. \end{cases} \qquad (3.43)$$

The case $\lambda_2 = 0$ reduces to an exponential distribution of rate λ_1; the case $\lambda_2 = \lambda_1$ is degenerate – the corresponding distribution is called the Erlang distribution and will be considered in the next subsection. As from (3.43), in a $\text{HYPO}_2(\lambda_1, \lambda_2)$, we have $1/2 \le cv^2 \le 1$.

The above results can be generalized to a hypoexponential distribution of order k generated as the sequence of k independent stages each one with an exponential distribution of rate λ_i, $i = 1, 2, \ldots, k$ (Figure 3.13). The states from 1 to k represent up states, and state $k + 1$ is the only down state.

The expected value, variance, and squared coefficient of variation are given by:

$$E[X] = \sum_{i=1}^k \frac{1}{\lambda_i}, \qquad \text{Var}[X] = \sum_{i=1}^k \frac{1}{\lambda_i^2}, \qquad cv^2 = \frac{\sum_{i=1}^k \frac{1}{\lambda_i^2}}{\left(\sum_{i=1}^k \frac{1}{\lambda_i}\right)^2}.$$

The hypoexponential distribution of any order is IFR with coefficient of variation $1/k \le cv^2 \le 1$. As the hypoexponential is an IFR distribution, it is often used to capture aging phenomena both in hardware and software [6], particularly for optimal preventive maintenance scheduling [7, 8].

Erlang Distribution

When all the k stages of a hypoexponential distribution have the same rate λ, the hypoexponential distribution is then known as an Erlang distribution of order k, denoted by $\text{ER}_k(\lambda)$. The Erlang distribution depends on two parameters: the order k and the

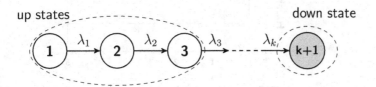

Figure 3.13 A k-stage hypoexponential distribution.

rate λ. The order is a positive integer ($k \in \mathbb{Z}^+ \geq 1$); the rate is a positive real number ($\lambda \in \mathbb{R} > 0$).

The Cdf, pdf, and hazard rate are given by:

$$F(t) = 1 - \sum_{i=0}^{k-1} \frac{(\lambda t)^i}{i!} e^{-\lambda t},$$

$$f(t) = \frac{\lambda^k t^{k-1}}{(k-1)!} e^{-\lambda t}, \tag{3.44}$$

$$h(t) = \frac{\lambda^k t^{k-1}}{(k-1)! \sum_{i=0}^{k-1} \frac{(\lambda t)^i}{i!}}.$$

Figure 3.14 compares three different Erlang distributions with the same expected value $E[X] = 1$: $ER_1(\lambda = 1)$ (coincident with EXP(1)), $ER_2(\lambda = 2)$, and $ER_5(\lambda = 5)$.

The hazard rate is an increasing function of time, so that the $ER_k(\lambda)$ distribution is always IFR. The expected value, variance, and squared coefficient of variation are given by:

$$E[X] = \frac{k}{\lambda}, \qquad \text{Var}[X] = \frac{k}{\lambda^2}, \qquad cv^2 = \frac{1}{k}. \tag{3.45}$$

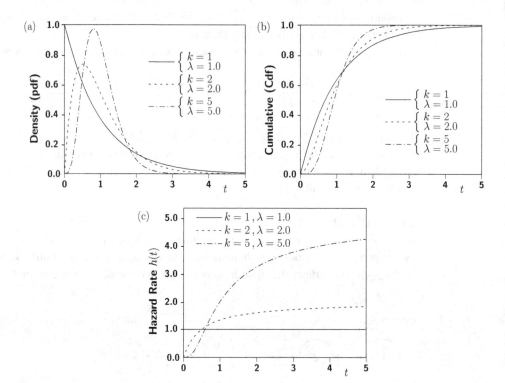

Figure 3.14   Erlang distribution with $E[X] = 1$ and three values of the parameters ($k = 1$, $\lambda = 1.0$; $k = 2$, $\lambda = 2.0$; $k = 5$, $\lambda = 5.0$). (a) Density; (b) Cdf; (c) hazard rate.

It has been shown in [9] that the coefficient of variation $cv = 1/\sqrt{k}$ of the Erlang distribution of order k is the minimum attainable value for the class of hypoexponential distributions of order k. For this reason, the Erlang distribution is sometimes used to approximate a deterministic variable [1, 10].

3.2.6 Gamma Distribution

The gamma distribution can be derived from the Erlang distribution when the order parameter k assumes a real positive value. Hence, the gamma distribution depends on two real positive parameters and will be denoted by $GAM(r, \lambda)$. The pdf is given by:

$$f(t) = \frac{\lambda^r t^{r-1} e^{-\lambda t}}{\Gamma(r)} \qquad t > 0, r > 0, \lambda > 0, \tag{3.46}$$

where the gamma function $\Gamma(r)$ is defined in (3.26). When the parameter r is an integer, the gamma distribution coincides with the Erlang distribution of order r, and for $r = 1$ with the exponential distribution. It can be shown that the $GAM(r, \lambda)$ is IFR for $r > 1$, DFR for $r < 1$, and CFR for $r = 1$. Thus it can model each behavior of the bathtub curve.

The expected value, variance, and coefficient of variation are given by:

$$E[X] = \frac{r}{\lambda}, \qquad \text{Var}[X] = \frac{r}{\lambda^2}, \qquad cv^2 = \frac{1}{r}. \tag{3.47}$$

The coefficient of variation versus r is:

$$\begin{cases} cv > 1 & \text{for} \quad r < 1, \\ cv = 1 & \text{for} \quad r = 1, \\ cv < 1 & \text{for} \quad r > 1. \end{cases} \tag{3.48}$$

3.2.7 Log-Logistic Distribution

The log-logistic distribution, denoted by $LL(\alpha, \beta)$, may present a hazard rate function with a maximum like the lognormal distribution, but it has the advantage over the latter that the density, Cdf, and hazard rate function can be expressed in closed form; it is therefore more practical for use in modeling and statistical estimation [11]. Log-logistic distributions have been used in reliability studies in hardware [12] as well as in software [13]. If X is a log-logistic-distributed random variable, the Cdf, density, and hazard rate functions have the following expressions:

$$F(t) = \frac{(\alpha t)^\beta}{1 + (\alpha t)^\beta},$$

$$f(t) = \frac{\alpha \beta (\alpha t)^{\beta-1}}{(1 + (\alpha t)^\beta)^2},$$

$$h(t) = \frac{\alpha \beta (\alpha t)^{\beta-1}}{1 + (\alpha t)^\beta}.$$

Examples of $LL(1, \beta)$ for a number of values of β are given in Figure 3.15.

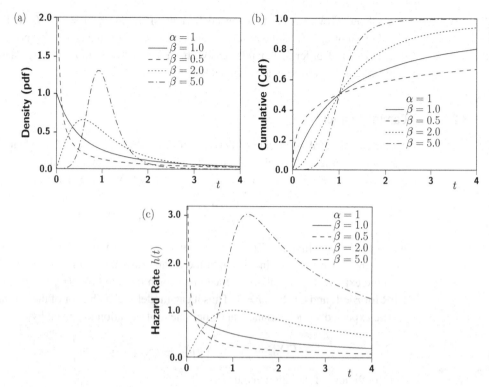

Figure 3.15   Log-logistic distribution with $\alpha = 1$ and $\beta = 0.5$, 1.0, 2.0, 5.0. (a) Density; (b) Cdf; (c) hazard rate.

By denoting $b = \pi/\beta$, the expected value is given by

$$E[X] = \frac{b}{\alpha \, \sin b}, \quad \beta > 1.$$

If $\beta \le 1$ the hazard rate is DFR, while if $\beta > 1$ the hazard rate function is non-monotonic, starting with IFR behavior at $t = 0$ and changing to DFR behavior as t increases.

3.2.8   Bernoulli Distribution

So far we have considered only continuous random variables. Quite often the need for discrete and mixed distributions arises. For instance, when experimentally assessing the reliability of a web service, we could make a large number, n, of requests, and if k requests are successful we will use k/n as a point estimate of the reliability of the web service [14]. So in this context reliability is simply a probability rather than a function of time. This view of reliability is common in the software testing community. Another instance of the use of such a notion of reliability occurs in "one-shot" systems such as missiles, rockets, and guns [15], or in switching devices, where we want to evaluate the probability of correct operation upon demand. To capture such scenarios, we consider a binary random variable X that takes values 0 (indicating failure) or

1 (indicating success). The probability mass function (pmf) of this discrete random variable is given by:

$$p_X(0) = P\{X = 0\} = q \geq 0,$$

$$p_X(1) = P\{X = 1\} = 1 - q = p \geq 0. \tag{3.49}$$

The Cdf is given by

$$F_X(x) = \begin{cases} 0 & x < 0, \\ q & 0 \leq x < 1, \\ 1 & x \geq 1. \end{cases} \tag{3.50}$$

The expected value, variance, and coefficient of variation are given by

$$E[X] = p, \qquad \text{Var}[X] = p(1-p), \qquad cv^2[X] = \frac{1-p}{p}. \tag{3.51}$$

3.2.9 Poisson Distribution

A discrete random variable X defined over the non-negative integers $(0, 1, \ldots)$ has a Poisson distribution of parameter λ if its pmf and Cdf are given by

$$P\{X = k\} = \frac{\lambda^k}{k!} e^{-\lambda}, \qquad P\{X \leq k\} = \sum_{i=0}^{k} \frac{\lambda^i}{i!} e^{-\lambda},$$

with expected value and variance given by:

$$E[X] = \lambda, \qquad \text{Var}[X] = \lambda.$$

As a consequence of the central limit theorem, a Poisson distribution of parameter λ tends to be distributed as a normal distribution of mean $\mu = \lambda$ and variance $\sigma^2 = \lambda$, as $\lambda \to \infty$. We can use this result to approximate Poisson probabilities using the normal distribution for large λ. Hence,

$$X \sim N(\lambda, \lambda) \qquad \text{and} \qquad \frac{X - \lambda}{\sqrt{\lambda}} \sim N(0, 1), \quad \lambda \to \infty.$$

3.2.10 Mass at Origin

Earlier, we assumed that the unit is operational at time $t = 0$. However, this may not always be the case. If we assume that the unit is defective to begin with, with probability q, then we can model its TTF random variable X as a modified Cdf with a mass at origin:

$$P\{X = 0\} = q,$$

$$P\{X \leq t\} = q + (1 - q)F_X(t). \tag{3.52}$$

If X is otherwise EXP(λ), the modified Cdf, the expected value, and variance are given by:

$$P\{X \leq t\} = q + (1 - q)(1 - e^{-\lambda t}) = 1 - (1 - q)e^{-\lambda t}, \tag{3.53}$$

$$E[X] = (1-q)\frac{1}{\lambda},$$

$$\text{Var}[X] = (1-q)^2 \frac{1}{\lambda^2}. \tag{3.54}$$

Note that this random variable is neither continuous nor discrete; it is a mixed random variable.

3.2.11   Defective Distribution

A random variable X is said to have a defective distribution if its Cdf is such that

$$\lim_{t\to\infty} F(t) = \alpha < 1,$$

where the value $(1-\alpha)$ is called the defect.

Example 3.1   *A Safety-Critical Switching Device*
A switching device S in a safety-critical system is used to switch off equipment when some critical conditions arise. The device S has two competing failure modes, short circuit, and open circuit, but only the open circuit failure is dangerous from the safety point of view since it prevents the controller from switching off the equipment. We say that the open circuit is a *fail-danger* failure mode while the short circuit is a *fail-safe* failure mode since it leads the system to a failed but safe state. In a safety study, only the open circuit failure mode is relevant. Assume that the time to short circuit failure is exponentially distributed with rate λ_c and the time to open circuit failure is exponentially distributed with rate λ_o. The failure process of the switch S can be represented as in Figure 3.16.

The time to reach the unsafe state open is given by:

$$F_o(t) = \frac{\lambda_o}{\lambda_o + \lambda_c} - \frac{\lambda_o}{\lambda_o + \lambda_c} e^{-(\lambda_o + \lambda_c)t}, \tag{3.55}$$

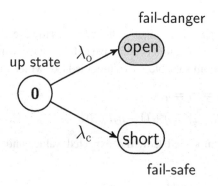

Figure 3.16   Switch failure with two failure modes.

$$\text{and} \qquad \lim_{t \to \infty} F_o(t) = \frac{\lambda_o}{\lambda_o + \lambda_c} < 1.$$

The defect $1 - \frac{\lambda_o}{\lambda_o + \lambda_c} = \frac{\lambda_c}{\lambda_o + \lambda_c}$ represents the probability of eventually reaching the failed but safe condition of short circuit.

Problems

3.4 The TTF of a component is distributed according to an $ER_2(\lambda)$. Compute the residual life of the component after a units of time. Compare the distribution of the residual life with the distribution of the life of a new component, and show that the $ER_2(\lambda)$ distribution represents wearout phenomena. Compute the MTTF of the life of a new component and the MTTF of the residual life.

3.5 In Problem 3.4, assume now the following numerical values: $\lambda = 0.2$ and $a = 10$. Plot the distribution of the TTF of a new component and the distribution of its residual life after a units of time, and compute the two MTTFs.

3.6 The TTF of a component is distributed according to a $HYPER_2(\lambda_1, \lambda_2; q)$. Compute the residual life of the component after a units of time. Compare the distribution of the residual life with the distribution of the life of a new component and show that the $HYPER_2(\lambda_1, \lambda_2; q)$ distribution represents early failure phenomena.

3.7 In Problem 3.6, assume now the following numerical values: $\lambda_1 = 0.2$, $q = 0.4$ and $a = 10$. First determine the value of λ_2 such that $MTTF = 10$ for the life of the new component. With the above values, plot the distribution of the TTF of a new component and the distribution of its residual life after a units of time, and compute the two MTTFs.

3.8 Return to Problem 3.3, but now assume that the time to failure of the mobile phone is Weibull distributed with scale parameter η and shape parameter β in such a way that the mean lifetime is six years, as in Problem 3.3. If you buy a used phone that has been in use for two years, compute the probability that your mobile phone will work for three more years. Consider three different values of β: 0.5, 1, and 2.

3.9 For the shifted Weibull distribution, write down formulas for its mean, variance, and the coefficient of variation.

3.10 Given two random variables X and Y with the same mean (i.e., $E[X] = E[Y] = 10000\,\text{hr}$), with X being exponentially distributed and Y being Weibull distributed, find the scale parameter of Y given the shape parameter $\beta = 0.5$, 1.5, 2.0. Then plot and hence compare the reliability functions for X and Y as functions of time.

3.11 In Example 3.1, suppose now that the TTF of the open circuit failure is an $ER_2(\lambda_o)$ and the TTF of the short circuit failure an $ER_2(\lambda_c)$. Find the Cdf of reaching the state open at time t and its defect.

3.12 Consider two items A and B. The TTF of item A is an $ER_2(\lambda)$ with $\lambda = 1/2000\,\text{f}\,\text{hr}^{-1}$. The TTF of item B is a $HYPER_2(\gamma_1, \gamma_2, q)$ with $\gamma_1 = 1/5000\,\text{f}\,\text{hr}^{-1}$, $\gamma_2 = 1/500\,\text{f}\,\text{hr}^{-1}$ and $q = 0.8$.

(a) Compare the MTTF of A and B.

(b) Compare the reliability of A and B at $t = 1000\,\text{hr}$.

(c) Given that both items A and B have survived for $a = 500\,\text{hr}$, compute the probability of finding them operating at $t = 1000\,\text{hr}$.

(d) Discuss the results.

3.3   Minimum and Maximum of Random Variables

Suppose we have n independent random variables $\{X_1, X_2, \ldots, X_n\}$ with Cdfs $F_1(t), F_2(t), \ldots, F_n(t)$, respectively. In many practical applications, the distribution of the minimum or the maximum of the given n random variables is needed.

3.3.1   The Cdf of the Minimum of Random Variables

Let us start with two independent r.v.s X_1 and X_2 with Cdfs $F_1(t)$ and $F_2(t)$, respectively. We introduce a new random variable Z defined as $Z = \min\{X_1, X_2\}$, and we wish to determine the distribution $F_Z(t)$ of Z. We observe that the event $Z > t$ occurs if and only if both the events $X_1 > t$ and $X_2 > t$ occur. Since the r.v.s are independent,

$$P\{Z > t\} = P\{X_1 > t\} P\{X_2 > t\},$$

and hence $\quad 1 - F_Z(t) = (1 - F_1(t))(1 - F_2(t)),$ $\qquad$ (3.56)

or, in terms of survival functions, $\quad S_Z(t) = S_1(t) S_2(t).$

The generalization to n independent r.v.s is straightforward. Given $Z = \min\{X_1, X_2, \ldots, X_n\}$ we have:

$$P\{Z > t\} = P\{X_1 > t\} P\{X_2 > t\}, \ldots, P\{X_n > t\},$$

and finally, $\quad F_Z(t) = 1 - \prod_{i=1}^{n} (1 - F_i(t)),$ $\qquad$ (3.57)

or, $\quad S_Z(t) = \prod_{i=1}^{n} S_i(t).$

Suppose now that we wish to determine the distribution of the minimum of n r.v.s when the minimum is a specific r.v., say X_i, i.e., $Z = \min\{X_1, X_2, \ldots, X_n\} = X_i$. Conditioning to $X_i = t$, we can write:

$$P\{Z > t | X_i = t\} = P\{X_1 > t, X_2 > t, \ldots, X_{i-1} > t, X_{i+1} > t, \ldots X_n > t | X_i = t\},$$

$$1 - F_{Z|X_i}(t) = \prod_{j=1; j \neq i}^{n} (1 - F_j(t)),$$

$$S_{Z|X_i}(t) = \prod_{j=1; j \neq i}^{n} S_j(t).$$

Unconditioning with respect to X_i using the continuous version of the theorem of total probability, we can finally write:

$$1 - F_{Z=X_i}(t) = \int_0^t \prod_{j=1; j\neq i}^{n} [1 - F_j(u)] f_i(u) \, du,$$

$$S_{Z=X_i}(t) = \int_0^t \prod_{j=1; j\neq i}^{n} S_j(u) f_i(u) \, du, \tag{3.58}$$

and the probability that the r.v. X_i is eventually the minimum becomes:

$$1 - F_{Z=X_i}(\infty) = \int_0^\infty \prod_{j=1; j\neq i}^{n} [1 - F_j(u)] f_i(u) \, du,$$

$$S_{Z=X_i}(\infty) = \int_0^\infty \prod_{j=1; j\neq i}^{n} S_j(u) f_i(u) \, du. \tag{3.59}$$

Example 3.2 *The Exponential Case*
We assume that all the random variables are exponentially distributed, $X_i \sim \text{EXP}(\lambda_i)$, $i = 1, 2, \ldots, n$. From (3.57) we have:

$$S_Z(t) = \prod_{i=1}^{n} S_i(t), \tag{3.60}$$

$$S_Z(t) = e^{-\lambda t} \quad \text{with } \lambda = \sum_{i=1}^{n} \lambda_i. \tag{3.61}$$

The minimum of n exponentially distributed random variables is also exponentially distributed, so that $Z \sim \text{EXP}(\lambda)$ with rate λ is equal to the sum of the rates of the n random variables.

From (3.58) we obtain:

$$S_{Z=X_i}(t) = \int_0^t \prod_{j=1; j\neq i}^{n} S_j(u) f_i(u) \, du$$

$$= \int_0^t e^{-\sum_{j=1; j\neq i}^{n} \lambda_j u} \lambda_i e^{-\lambda_i u} \, du$$

$$= \lambda_i \int_0^t e^{-\lambda u} \, du = \frac{\lambda_i}{\lambda} (1 - e^{-\lambda t}),$$

and the probability that X_i is eventually the minimum is:

$$S_{Z=X_i}(\infty) = \frac{\lambda_i}{\lambda}.$$

Example 3.3 *A Safety-Critical Switching Device*
Comparing Eq. (3.55) of Example 3.1 with the results of Example 3.2, we see that the probability that the fail-danger state is eventually reached is the minimum of the two competing failure modes.

3.3.2   The Cdf of the Maximum of Random Variables

To find the maximum among independent random variables we proceed in a similar way to Section 3.3.1. Suppose we have two independent r.v.s X_1 and X_2 with Cdfs $F_1(t)$ and $F_2(t)$, respectively. We introduce a new random variable Z defined as $Z = \max\{X_1, X_2\}$, and we wish to determine the distribution $F_Z(t)$ of Z. We observe that the event $Z \leq t$ occurs if and only if both the events $X_1 \leq t$ and $X_2 \leq t$ occur. Since the r.v.s are independent,

$$P\{Z \leq t\} = P\{X_1 \leq t\}P\{X_2 \leq t\},$$

and hence $F_Z(t) = F_1(t)F_2(t)$.

Given $Z = \max\{X_1, X_2, \ldots, X_n\}$, the generalization to n independent r.v.s leads to

$$P\{Z \leq t\} = P\{X_1 \leq t\}P\{X_2 \leq t\}, \cdots, P\{X_n \leq t\},$$

and finally, $F_Z(t) = \prod_{i=1}^{n} F_i(t)$.   (3.62)

The distribution of the maximum of n r.v.s when the maximum is a specific r.v., say X_i, i.e., $Z = \max\{X_1, X_2, \ldots, X_n\} = X_i$, is given by

$$F_{Z=X_i}(t) = \int_0^t \prod_{j=1; j \neq i}^{n} F_j(u) f_i(u)\, du,$$ (3.63)

and the probability that the r.v. X_i is eventually the maximum becomes

$$F_{Z=X_i}(\infty) = \int_0^\infty \prod_{j=1; j \neq i}^{n} F_j(u) f_i(u)\, du.$$ (3.64)

Example 3.4   *The Exponential Case*

We assume that all the random variables are exponentially distributed, $X_i \sim \text{EXP}(\lambda_i)$, $i = 1, 2, \ldots, n$. From (3.62) we have:

$$F_Z(t) = \prod_{i=1}^{n} F_i(t)$$ (3.65)

$$= \prod_{i=1}^{n} [1 - e^{-\lambda_i t}].$$ (3.66)

From (3.63) we obtain:

$$F_{Z=X_i}(t) = \int_0^t \prod_{j=1; j \neq i}^{n} F_j(u) f_i(u)\, du$$

$$= \int_0^t \prod_{j=1; j \neq i}^{n} [1 - e^{-\lambda_j u}] \lambda_i e^{-\lambda_i u}\, du.$$

Problems

3.13 A computer program is composed of three modules whose execution times are denoted by A, B, and C, and are exponentially distributed with rates α, β, and γ, respectively. Assuming the necessary independence assumptions, derive the Cdf and the expected value of the program execution time in the following cases:

(a) The three modules are executed in parallel and the program is considered completed as soon as one of the three modules has completed.
(b) The three modules are executed in parallel and the program is considered completed when all the three modules have completed.
(c) The three modules are executed in parallel and the program is considered completed only if B is the last one to complete.
(d) The three modules are executed in sequence.

3.14 In Problem 3.13, compute the Cdf of the execution time at $t = 0.005$ s and $t = 0.05$ s, and the expected value in the four different cases given the numerical values $\alpha = 10\,\text{s}^{-1}$, $\beta = 50\,\text{s}^{-1}$, and $\gamma = 20\,\text{s}^{-1}$.

3.4 Epistemic Uncertainty Propagation

The probability distributions introduced to quantify the failure of a unit depend on parameters to which a fixed value needs to be assigned. However, the input parameter values of any model have uncertainty associated with them as they are derived either from a finite number of observations (from lifetime-determining experiments or field data) or are based upon expert guesses. This uncertainty in model input parameter values (known as epistemic uncertainty) is not normally taken into account in reliability and availability calculations [16, 17]. The uncertainty in the model parameters may be expressed in the form of a distribution of parameter values themselves or in the form of bounds or a confidence interval for the parameter values, obtained from lifetime experiments or from manufacturer datasheets. The reliability value computed using fixed values of the model parameters can be considered to be conditional upon the parameter values used. To propagate the uncertainty of model input parameters to the model output, it needs to be unconditioned by applying the theorem of total probability for continuous random variables. The unconditioning takes the form of an integral that may be solved in closed form in simple cases or via numerical integration or using Monte Carlo sampling.

Let the input parameters of a reliability model be $(\Lambda_1, \Lambda_2, \ldots, \Lambda_l)$. The reliability $R(t)$ at time t can be viewed as a random variable (function) g of the l input parameters as $R(t) = g(\Lambda_1, \Lambda_2, \ldots, \Lambda_l)$. Due to the uncertainty associated with the model parameters, computing the reliability at specific parameter values can be seen as computing the conditional reliability $R(t|\Lambda_1 = \lambda_1, \Lambda_2 = \lambda_2, \ldots, \Lambda_l = \lambda_l)$ [denoted by $R(t|\bullet)$ in Eq. (3.67)].

Applying the theorem of total probability [1], this can be unconditioned to compute the distribution of reliability via the joint density $f_{\Lambda_1, \Lambda_2, \ldots, \Lambda_l}(\lambda_1, \lambda_2, \ldots, \lambda_l)$ of the input parameters (denoted by $f(\bullet)$):

$$F_{R(t)}(x) = \int \cdots \int I_{\{R(t|\bullet) \leq x\}} f(\bullet) d\lambda_1 \ldots d\lambda_l, \tag{3.67}$$

where $I_{\{Event\}}$ is the indicator variable of the event *Event*. The unconditional expected reliability at time t can be computed as:

$$E[R(t)] = \int \cdots \int R(t|\bullet) f(\bullet) d\lambda_1 \ldots d\lambda_l. \tag{3.68}$$

3.4.1   Distribution for Rate Parameter of an Exponential Distribution

Let X be an exponentially distributed random variable of rate Λ and let $f_\Lambda(\lambda)$ be the prior density of Λ. We observe a sample of k i.i.d. random realizations of X, say $X_1, X_2, \ldots, X_k$. The maximum likelihood estimator for Λ is proportional to the sum $S = \sum_{i=1}^{k} X_i$, which has a k-stage Erlang distribution $\mathrm{ER}_k(\lambda)$. Therefore, the conditional density of S given $\Lambda = \lambda$ is:

$$f_{S|\Lambda}(s|\lambda) = \frac{\lambda^k s^{k-1} e^{-\lambda s}}{(k-1)!}. \tag{3.69}$$

Then, applying the continuous version of Bayes' theorem to Eq. (3.69), the conditional density of Λ given $S = s$ becomes

$$\begin{aligned} f_{\Lambda|S}(\lambda|s) &= \frac{f_\Lambda(\lambda) f_{S|\Lambda}(s|\lambda)}{\int_0^\infty f_\Lambda(\lambda) f_{S|\Lambda}(s|\lambda) d\lambda} \\ &= \frac{f_\Lambda(\lambda) \dfrac{\lambda^k s^{k-1} e^{-\lambda s}}{(k-1)!}}{\int_0^\infty f_\Lambda(\lambda) \dfrac{\lambda^k s^{k-1} e^{-\lambda s}}{(k-1)!} d\lambda}. \end{aligned} \tag{3.70}$$

Using as an improper prior (due to Jeffreys) $f_\Lambda(\lambda) = s/\lambda$, Eq. (3.70) becomes an $\mathrm{ER}_k(s)$ pdf:

$$f_{\Lambda|S}(\lambda|s) = \frac{\lambda^{k-1} s^k e^{-\lambda s}}{(k-1)!}. \tag{3.71}$$

3.4.2   Reliability Distribution

The reliability $R(t)$ calculated at a fixed rate value λ can be thought of as conditioned on a specific value $\Lambda = \lambda$. If the rate value is estimated from a sample of k data points, applying the theorem of total probability, using Eqs. (3.67) and (3.71), the unconditional Cdf at time t can be computed as:

$$F_{R(t)}(x) = \int_0^\infty I_{\{R(t) \leq x\}} \frac{\lambda^{k-1} s^k e^{-\lambda s}}{(k-1)!} d\lambda. \tag{3.72}$$

The integral is non-zero only for values of λ for which $R(t) \leq x$. Since, by assumption, $R(t) = e^{-\lambda t}$, the condition stated by the indicator variable $I_{\{R(t) \leq x\}}$ is satisfied for $\lambda >$

$\lambda_a = -\ln x/t$. Further, the maximum likelihood estimator for the rate parameter λ with k observations is $\hat{\lambda} = k/s$, which can be written as $s = k/\hat{\lambda}$. From Eq. (3.72) we can write:

$$F_{R(t)}(x) = \int_{\lambda_a}^{\infty} \frac{\lambda^{k-1} s^k e^{-\lambda s}}{(k-1)!} d\lambda = 1 - \underbrace{\int_0^{\lambda_a} \frac{\lambda^{k-1} s^k e^{-\lambda s}}{(k-1)!} d\lambda}_{\text{Erlang Cdf}}$$

$$= \sum_{i=0}^{k-1} e^{-s\lambda_a} \frac{(s\lambda_a)^i}{i!}$$

$$= \sum_{i=0}^{k-1} e^{\frac{k\ln x}{\hat{\lambda}t}} \frac{\left(\frac{-k\ln x}{\hat{\lambda}t}\right)^i}{i!}. \tag{3.73}$$

At any time t the reliability $R(t)$ of the system will have the Cdf given by Eq. (3.73), and hence will have an expected value, a variance, and a confidence interval. In [17], a plot of $F_{R(t)}(x)$, computed at different values of t from (3.73) and with the number of observations $k = 10$, is presented. The plot shows that the Cdf of $R(t)$ shifts toward 0 as t increases, as expected.

The Erlang distribution for Λ in Eq. (3.71) has expected value $\mu = k/s = \hat{\lambda}$ and variance $\sigma = k/s^2 = \hat{\lambda}/k$ [Eq. (3.45)]. When $k \to \infty$, from the central limit theorem [1] we know that the Erlang distribution tends to a normal distribution with $\mu = k/s = \hat{\lambda}$ and $\sigma \to 0$. Hence, the density of Λ tends to a Dirac-delta function at $\hat{\lambda}$ and $R(t)$ tends to $R(t) \to e^{-\hat{\lambda}t}$. From Eq. (3.73),

$$\lim_{k \to \infty} F_{R(t)}(x) = u(x - e^{-\hat{\lambda}t}).$$

Expected Reliability

Given that the point estimate $\hat{\lambda}$ of λ was obtained from k observations, the unconditional expected reliability at time t can be evaluated from Eq. (3.73):

$$E[R(t)] = \int_0^{\infty} e^{-\lambda t} \cdot \underbrace{\frac{\lambda^{k-1} s^k e^{-\lambda s}}{(k-1)!}}_{\text{Erlang pdf}} d\lambda$$

$$= \left(\frac{s}{s+t}\right)^k = \left(\frac{1}{1+\hat{\lambda}t/k}\right)^k. \tag{3.74}$$

The above equation makes use of the expression $\hat{\lambda} = k/s$, that is, the maximum likelihood estimate (MLE) of λ. Using the identity $\lim_{h \to \infty}(1 + 1/h)^h = e$ [1], the limiting value of unconditional expectation of reliability at time t is:

$$\lim_{k \to \infty} E[R(t)] = \frac{1}{\lim_{k \to \infty}\left(1 + \frac{\hat{\lambda}t}{k}\right)^k} = e^{-\hat{\lambda}t}. \tag{3.75}$$

Figure 3.17 plots the expected value of $R(t)$ at $t = 5000$ as a function of k. The value of $\hat{\lambda}$ is chosen to be 5.7078×10^{-5} hr^{-1}, corresponding to an MTTF of 17 520 hr; this

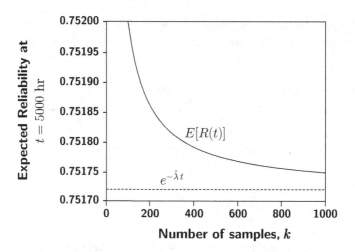

Figure 3.17   Expected reliability of a single-component system at $t = 5000$ hr.

was used for failure of software in [18]. It can be seen that $E[R(t)]$ tends to $e^{-\hat{\lambda}t}$ as k increases.

3.5   Repairable Unit

In Sections 3.1 and 3.2 we considered units that, upon failure, remain in the failed state forever, and we defined suitable metrics to characterize the behavior of such units. However, in practice many units may be restored to an operational state after a failure. We define this class of objects as *repairable*. The unit in this context may be a complete system, a subsystem or a component. The behavior in time of a repairable object will not only depend on its lifetime and on its failure mode but also on the way in which the object can be recovered and the duration of the recovery time. Hence, in the study of repairable systems we need to include the failure as well as the repair behavior of the object. A general framework to deal with extensive analysis of repairable systems will be delayed until the third part of the book on state-space models. In this section, we introduce the main measures to characterize a repairable system and how they can be quantitatively evaluated by means of some simplifying hypotheses.

In particular, we consider a unit whose behavior can be described by two exhaustive and mutually exclusive states, an up state and a down state. If the system is repairable, its dynamic behavior can be represented as an alternating of periods spent in the up state and the down state, as shown in Figure 3.18.

Transitions from level up to level down represent the passage from an operating condition to a non-operating condition, and hence represent the occurrence of a failure. Conversely, transitions from level down to level up represent the restoration of the unit to an operating condition, indicating the completion of a repair or recovery action.

The periods spent in the up state represent the successive uptimes, and, with reference to Figure 3.18, are denoted by the sequence of random variables $(X_1, X_2, \ldots, X_n, \ldots)$. Similarly, the periods spent in the down state represent the successive downtimes,

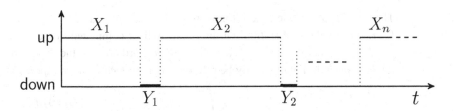

Figure 3.18 A sample realization of a repairable item.

which can also be considered as random variables denoted by $(Y_1, Y_2, \ldots, Y_n, \ldots)$ (Section 2.6). Hence, X_i denotes the elapsed time between the $(i-1)$th repair and the ith failure, while Y_i denotes the elapsed time between the ith failure and the ith repair. A quantitative analysis of the behavior of a repairable unit in time requires that the uptime random variables X_i and the downtime random variables Y_i are completely specified, i.e., their Cdfs are known. In general, the probability distribution of the ith uptime may differ from the probability distribution of the preceding or succeeding uptimes. Similarly, the probability distribution of the jth downtime may differ from the probability distribution of the preceding or succeeding downtimes. However, in order to develop analytically tractable models a simplifying assumption is usually made that all the successive uptimes X_i are i.i.d. with a common Cdf $F_X(t)$, and all the successive downtimes Y_j are i.i.d. with a common Cdf $G_Y(t)$. Hence, $F_X(t)$ represents the Cdf of each uptime and $G_Y(t)$ the Cdf of each downtime. This assumption is rather realistic when the restoration is achieved by replacing the failed unit with a new one that is statistically equivalent. The same assumption may be less realistic with other restoration policies; these will be considered in subsequent chapters of the book. Also, multiple types of failures and different types of recovery/repair will be considered in Part IV. By adopting these two assumptions, the alternating behavior of uptimes and downtimes depicted in Figure 3.18 is completely characterized by only two random variables, the uptime X, and the downtime Y. We refer to this assumption as the *as good as new* assumption, since after any restoration the behavior of the system is identical to the new one.

3.5.1 Measures for Repairable Systems

Maintainability

According to Section 1.2, the maintainability is the ability to return or to be restored to a good state after a failure. To quantify this concept, let Y be the random variable, defined over the positive reals ($Y \in \mathbb{R}^+ > 0$), that characterizes the i.i.d. downtimes of a repairable unit. Let $G_Y(t) = P\{Y \le t\}$ denote its Cdf. By definition, $G_Y(t)$ represents the probability that a unit whose restoration process starts at $t=0$ will be repaired before time t, and is a measure of the item's *maintainability*. The pdf and hazard rate are:

$$g_Y(t) = \frac{dG_Y(t)}{dt},$$

$$h_Y(t) = \frac{g_Y(t)}{1 - G_Y(t)}.$$

The hazard rate of the random downtime will be referred to as the *repair rate*, since $h_Y(t)dt$ gives the probability that the repair will be completed in the time interval t to $t + dt$, given that the system is down at time t. Furthermore, the expected value of the random variable Y will usually be referred to as the *mean time to repair* (MTTR):

$$E[Y] = \text{MTTR} = \int_0^\infty t g_Y(t) dt = \int_0^\infty (1 - G_Y(t)) dt.$$

Y can in principle have any distribution defined over the positive reals. We note that the time to repair (TTR) is a random variable, MTTR is its expected value (or its mean) and the reciprocal of MTTR is the repair rate if the distribution of TTR is exponential.

Availability

To characterize the behavior in time of a repairable unit that alternates between an operational and a non-operational state as shown in Figure 3.18, we introduce a new measure called *availability*, denoted by $A(t)$. According to Section 1.2, the availability depicts the readiness for correct service. In order to make this concept quantifiable, we define the availability as the probability that at time t the unit is in an operational state. The *unavailability* is the probability that the unit is in a non-operational state at time t:

$$A(t) = P\{\text{unit up at time } t\},$$

$$U(t) = P\{\text{unit down at time } t\}. \tag{3.76}$$

Since the up and down states are collectively exhaustive and mutually exclusive:

$$A(t) + U(t) = 1.$$

We recall that while the reliability indicates the probability that a system is continuously working up to time t, the availability gives the probability that the system is up at a given time epoch, independent of the number of breakdowns and restorations that the system has undergone before time t. Availability is thus the level of trustworthiness or readiness that we can place on a system that needs to be used at time t [19], and takes into account both the way in which the system fails and the way in which the system is restored. A highly available system is a system with a low probability of breaking down (high reliability) and a high probability of being quickly repaired (high maintainability).

The quantitative analysis of availability and related measures requires studying the dynamic evolution of the system up to time t, and an appropriate formalism for its evaluation is postponed to later chapters. Here, we introduce the analysis of a single repairable unit based on the description of the mission profile of the unit by means of sequences of i.i.d. random variables (renewal processes), and alternatively based on the representation of the system behavior by means of the state-space representation.

3.5.2   Renewal Processes

A *renewal process* [20] is generated by a sequence of i.i.d. random variables $(X_1, X_2, \ldots, X_n, \ldots)$ with Cdf $F(t)$, density $f(t)$ and expected value $E[X] = m_1$ (see Figure 3.19). The process so generated is called a *renewal process* since it represents

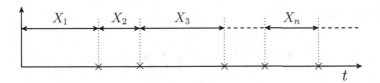

Figure 3.19 A renewal process.

the behavior of a unit that is replaced by an *as good as new* unit at each failure. For now, we assume that the repair time is negligible, that is, we assume instantaneous repair (we shall soon remove this restriction):

$$F(t) = P\{X \le t\}, \qquad f(t) = \frac{dF(t)}{dt}, \qquad E[X] = m_1.$$

We define the corresponding Laplace transforms (LT; denoted by the symbol $*$) as:

$$F^*(s) = \mathscr{L}[F(t)], \qquad f^*(s) = \mathscr{L}[f(t)]. \tag{3.77}$$

Let $S_k = \sum_{i=1}^{k} X_i$ denote the time of the occurrence of the kth event in the renewal process, and let $F_k(t)$ and $f_k(t)$ denote its Cdf and density, respectively. We can write:

$$F_k(t) = P\{S_k \le t\}, \qquad f_k(t) = \frac{dF_k(t)}{dt}. \tag{3.78}$$

Applying the convolution theorem for Laplace transforms, Eq. (3.78) can be written in the Laplace domain:

$$F_k^*(s) = \frac{1}{s}[f^*(s)]^k, \qquad f_k^*(s) = [f^*(s)]^k. \tag{3.79}$$

If $N(t)$ is the number of renewals in the interval $(0, t]$, then the following relation holds:

$$N(t) < k \quad \text{if and only if} \quad S_k > t, \tag{3.80}$$

from which:

$$P\{N(t) < k\} = P\{S_k > t\} = 1 - F_k(t). \tag{3.81}$$

From Eq. (3.81), we obtain:

$$P\{N(t) = k\} = P\{N(t) < k + 1\} - P\{N(t) < k\}$$
$$= F_k(t) - F_{k+1}(t). \tag{3.82}$$

We define the *renewal function* $H(t)$ as the expected number of renewals in the interval $(0, t]$. Then:

$$H(t) = E[N(t)] = \sum_{k=0}^{\infty} k P\{N(t) = k\}$$

$$= \sum_{k=0}^{\infty} k[F_k(t) - F_{k+1}(t)] = \sum_{k=1}^{\infty} F_k(t). \tag{3.83}$$

In Laplace transforms:

$$H^*(s) = \sum_{k=1}^{\infty} F_k^*(s) = \frac{1}{s} \sum_{k=1}^{\infty} f^{*k}(s)$$

$$= \frac{1}{s} \{f^*(s) + f^{*2}(s) + \cdots + f^{*k}(s) + \cdots\} \qquad (3.84)$$

$$= \frac{1}{s} \frac{f^*(s)}{1 - f^*(s)}$$

(here we assume that $f^*(s)$ satisfies appropriate convergence conditions). Manipulating the last expression in (3.84), we can write:

$$H^*(s) = \frac{f^*(s)}{s} + H^*(s) f^*(s). \qquad (3.85)$$

Equation (3.85) can be transformed back to the time domain to get:

$$H(t) = F(t) + \int_0^t H(t-u) \cdot f(u) \, du. \qquad (3.86)$$

Equation (3.86) is known as the *fundamental renewal equation*. We state without proof the following property:

$$\lim_{t \to \infty} \frac{H(t)}{t} = \frac{1}{E[X]}. \qquad (3.87)$$

Equation (3.87) states a rather intuitive result that the expected number of renewals in a given interval of time equals the length of the interval divided by the expected duration of a single renewal. The renewal density $h(t)$ is defined as the time derivative of the renewal function $H(t)$. By virtue of (3.83),

$$h(t) = \frac{dH(t)}{dt} = \sum_{k=1}^{\infty} f_k(t). \qquad (3.88)$$

The renewal density $h(t)$ multiplied by Δt may be interpreted as the probability of occurrence of one or more renewals in an interval of length Δt. Indeed, we can obtain the same result of (3.88) by the following argument:

$$h(t) = \lim_{\Delta t \to 0} \frac{P\{\text{one or more events occur in } (t, t + \Delta t]\}}{\Delta t}$$

$$= \sum_{k=1}^{\infty} \lim_{\Delta t \to 0} \frac{P\{k\text{th event occurs in } (t, t + \Delta t]\}}{\Delta t}$$

$$= \sum_{k=1}^{\infty} f_k(t). \qquad (3.89)$$

Taking the LT of (3.89), we get:

$$h^*(s) = \sum_{k=1}^{\infty} f_k^*(s)$$

$$= f^*(s) + [f^*(s)]^2 + \cdots + [f^*(s)]^k + \cdots$$

$$= \frac{f^*(s)}{1 - f^*(s)},$$

from which we obtain the *renewal equation* in the LT and time domains:

$$h^*(s) = f^*(s) + h^*(s) \cdot f^*(s),$$

$$h(t) = f(t) + \int_0^t h(t-u) \cdot f(u) \, du.$$

3.5.3 The Exponential Case: Poisson Process

A renewal process in which the generating random variable X is exponentially distributed is called a *Poisson process*. Assume that $X \sim \text{EXP}(\lambda)$; then:

$$F(t) = 1 - e^{-\lambda t} \qquad \Longrightarrow \qquad F^*(s) = \frac{\lambda}{s(s+\lambda)},$$

$$f(t) = \lambda e^{-\lambda t} \qquad \Longrightarrow \qquad f^*(s) = \frac{\lambda}{s+\lambda}. \tag{3.90}$$

The Cdf and density of the time up to the kth event S_k can be obtained from Eq. (3.79):

$$F_k^*(s) = \frac{\lambda^k}{s(s+\lambda)^k}, \qquad f_k^*(s) = \left(\frac{\lambda}{s+\lambda}\right)^k. \tag{3.91}$$

Transforming the LT density (3.91) for $k = 1, 2, \ldots$ back to the time domain we obtain

$$f_1(t) = \mathcal{L}^{-1}\left[\frac{\lambda}{s+\lambda}\right] = \lambda e^{-\lambda t},$$

$$f_2(t) = \mathcal{L}^{-1}\left[\frac{\lambda^2}{(s+\lambda)^2}\right] = \lambda^2 t e^{-\lambda t},$$

$$\vdots$$

$$f_k(t) = \mathcal{L}^{-1}\left[\frac{\lambda^k}{(s+\lambda)^k}\right] = \frac{\lambda (\lambda t)^{k-1}}{(k-1)!} e^{-\lambda t},$$

and for the Cdf:

$$F_k(t) = \int_0^t f_k(u) \, du = 1 - \sum_{i=0}^{k-1} \frac{(\lambda t)^i}{i!} e^{-\lambda t},$$

which is the Cdf of the k-stage Erlang random variable with parameter λ ($\text{ER}_k(\lambda)$).

Denote by $P_k(t)$ the probability of having k renewals in a time duration of length t; then, from Eq. (3.83):

$$P_k(t) = P\{N(t) = k\} = F_k(t) - F_{k+1}(t).$$

Taking Laplace transforms:

$$P_k^*(s) = \frac{\lambda^k}{s(s+\lambda)^k} - \frac{\lambda^{k+1}}{s(s+\lambda)^{k+1}} = \frac{\lambda^k}{(s+\lambda)^{k+1}}.$$

Inverting again to the time domain, we obtain the Poisson probability mass function (pmf):

$$P_k(t) = P\{N(t) = k\} = \frac{(\lambda t)^k}{k!} e^{-\lambda t}, \tag{3.92}$$

or, in a recursive form:

$$P_k(t) = P\{N(t) = k\} = \frac{(\lambda t)}{k} P_{k-1}(t). \tag{3.93}$$

Given a time t, the number of renewals by time t follows a Poisson distribution with parameter (λt) (Section 3.2.9).

The expected number of renewals $H(t)$ and the renewal density $h(t)$ in a Poisson process assume the form, respectively:

$$H(t) = E[N(t)] = \sum_{k=0}^{\infty} k P\{N(t) = k\} = \sum_{k=0}^{\infty} k P_k(t)$$

$$= e^{-\lambda t} \sum_{k=1}^{\infty} \frac{(\lambda t)^k}{(k-1)!}$$

$$= \lambda t \cdot e^{-\lambda t} \left(1 + \lambda t + \frac{(\lambda t)^2}{2!} + \cdots \right) = \lambda t,$$

$$h(t) = \lambda.$$

Example 3.5 The Poisson pmf $P_k(t)$ in (3.92) can be plotted as a function of k and of the dimensionless parameter λt, which also represents the expected number of renewals in $(0, t]$. Figure 3.20 displays the values of the Poisson pmf for $\lambda t = 1$ in tabular and graphical form.

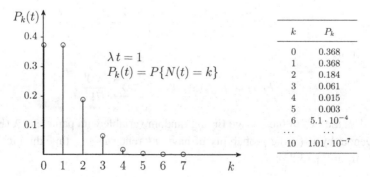

k	P_k
0	0.368
1	0.368
2	0.184
3	0.061
4	0.015
5	0.003
6	$5.1 \cdot 10^{-4}$
...	...
10	$1.01 \cdot 10^{-7}$

Figure 3.20 Poisson pmf for $\lambda t = 1$.

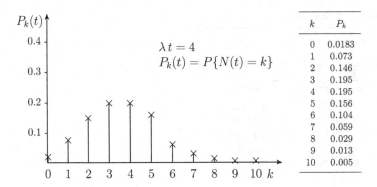

Figure 3.21   Poisson pmf for $\lambda t = 4$.

Figure 3.21 displays the values of the Poisson pmf for $\lambda t = 4$ in tabular and graphical form.

Buffer Design with a Poisson Process

Jobs arrive according to a Poisson process with rate λ and must be stored in a buffer during an interval of duration τ. The design problem consists in evaluating the integer number k of slots in the buffer so that the probability of refusing an incoming job in the interval of duration τ is less than a prescribed (small) risk α $(0 \leq \alpha \leq 1)$.

Let $N(\tau)$ be the number of arrivals in the interval of duration τ and let k be the number of slots to be determined. The design problem can be formulated as:

$$P\{N(\tau) > k\} \leq \alpha,$$
$$P\{N(\tau) \leq k\} > 1 - \alpha, \tag{3.94}$$
$$P\{N(\tau) \leq k\} = \sum_{j=0}^{k} \frac{(\lambda\tau)^j}{j!} e^{-\lambda\tau} > 1 - \alpha.$$

The value of k is the smallest integer that satisfies the above equation.

Example 3.6   A server receives input bits according to a Poisson process with rate $\lambda = 0.8\,\text{s}^{-1}$ and the server is scanned and depleted cyclically every $\tau = 5\,\text{s}$. We wish to determine the size of the buffer that must be attached to the server so that the risk of losing bits per cycle is less than $\alpha = 0.1$, or $\alpha' = 0.01$.

Denoting by k the size in bits of the buffer, we need to find the smallest integer k such that

$$P\{N(\tau) > k\} \leq \alpha,$$
$$P\{N(\tau) \leq k\} > 1 - \alpha.$$

To this end, we observe that the expected number of arrivals in a time span τ is $\lambda\tau = 4$. Using this value, we can compute from (3.92) the probability of observing exactly k arrivals in time τ and the probability of observing no more that k arrivals in time τ. These values are reported in Table 3.3 for increasing k.

Table 3.3 Sizing a buffer.

i	$P_i(\tau)$	$\sum_{j=0}^{k} P_j(\tau)$
	$\frac{(\lambda\tau)^i}{i!} e^{-\lambda\tau}$	$\sum_{j=0}^{k} \frac{(\lambda\tau)^j}{j!} e^{-\lambda\tau}$
0	0.01831	0.01831
1	0.07326	0.09158
2	0.14652	0.23810
3	0.19536	0.43347
4	0.19536	0.62883
5	0.15629	0.78513
6	0.10419	0.88932
7	0.05954	0.94886
8	0.02977	0.97863
9	0.01323	0.99187

Table 3.3 shows that at a risk level $\alpha = 0.1$ the buffer dimension should be greater than $k = 7$ bits, and at a risk level $\alpha' = 0.01$ it should be greater than $k = 9$ bits.

Example 3.7 *Spare Part Computation with Assigned Risk Level*
The spare part computation of a component with exponentially distributed lifetime can be formulated in a similar way [3]. Suppose that a component has a constant failure rate λ and it is replaced upon failure with a new component with the same failure rate. The successive replacements form a Poisson process of rate λ. The number of spare parts to be allocated over a mission time τ equals the number of renewals in the same timeslot. If we assign a risk level α to the probability of exhausting the spare parts, we can compute k according to (3.94).

As a numerical example, assume that we want to compute the number of spares to be allocated in a time $\tau = 2/\lambda$.

We build up Table 3.4, which shows that with a risk $\alpha = 0.1$ we must allocate more than $k = 4$ spares, and with a risk $\alpha = 0.01$ we must allocate more than $k = 6$ spares.

Table 3.4 Number of spare parts.

i	$P_i(\tau)$	$\sum_{j=0}^{k} P_j(\tau)$
	$\frac{(\lambda\tau)^i}{i!} e^{-\lambda\tau}$	$\sum_{j=0}^{k} \frac{(\lambda\tau)^j}{j!} e^{-\lambda\tau}$
0	0.13533	0.13533
1	0.27067	0.40600
2	0.27067	0.67667
3	0.18044	0.85712
4	0.09022	0.94734
5	0.03609	0.98343
6	0.01202	0.99546

3.5.4 Modified Renewal Process

A modified renewal process is a renewal process in which the distribution $F_1(t)$ of the first renewal time X_1 (see Figure 3.19) differs from the distribution $F(t)$ of all the subsequent renewals.

Let $S_k = \sum_{i=1}^{k} X_i$ denote, as before, the time of the occurrence of the kth event in the modified renewal process, and let $F_k(t)$ and $f_k(t)$ denote its Cdf and density, respectively. We can write in the LT domain:

$$F_k^*(s) = \frac{1}{s} f_1^*(s)[f^*(s)]^{k-1}, \qquad f_k^*(s) = f_1^*(s)[f^*(s)]^{k-1}. \qquad (3.95)$$

Accordingly, the expected number of renewals in the interval $(0, t]$ in the modified renewal process is given in Laplace transforms by:

$$H^*(s) = \sum_{k=1}^{\infty} F_k^*(s) = \frac{1}{s} f_1^*(s) \sum_{k=0}^{\infty} f^{*k}(s)$$

$$= \frac{1}{s} f_1^*(s)\{1 + f^*(s) + f^{*2}(s) + \cdots + f^{*k}(s) + \cdots\}$$

$$= \frac{1}{s} \frac{f_1^*(s)}{1 - f^*(s)}. \qquad (3.96)$$

Elaborating, we get:

$$H^*(s) = \frac{f_1^*(s)}{s} + H^*(s)f^*(s),$$

and, in the time domain $\quad H(t) = F_1(t) + \int_0^t H(t-u)f(u)\,du.$

The renewal density is

$$h^*(s) = sH^*(s) = \frac{f_1^*(s)}{1 - f^*(s)}, \qquad (3.97)$$

and, in the time domain $\quad h(t) = f_1(t) + \int_0^t h(t-u)f(u)\,du.$

3.5.5 Availability Analysis: Alternating Renewal Process

An *alternating renewal process* consists of an intertwined sequence of a random variable X of *Type I*, with Cdf $F(x)$ and density $f(x)$, followed by a random variable Y of *Type II*, with Cdf $G(x)$ and density $g(x)$. The successive random variables of Type I and Type II are i.i.d., and we assume that the process starts with probability 1 with a Type I variable. A possible sample path of an alternating renewal process is sketched in Figure 3.22.

In the figure, the symbol $\times$ denotes the occurrence of a Type I event, while the symbol o denotes the occurrence of a Type II event. If we assume that the Type I variables represent system failure and the Type II variables represent system repair, the alternating

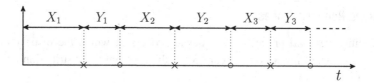

Figure 3.22   An alternating renewal process.

renewal process models the behavior of a repairable system that uses an *as good as new* repair policy, as exemplified in Figure 3.18.

If we consider the subsequence formed by the occurrence of the Type II events, representing the completion of a failure/repair cycle (sequence of o symbols), the process is an ordinary renewal process with inter-renewal time random variable ($Z = X + Y$). The expected value of the random variable Z is called the *mean time between failures* (MTBF) and satisfies the following relation:

$$\text{MTBF} = E[Z] = E[X] + E[Y] = \text{MTTF} + \text{MTTR}.$$

The expected number of Type II event occurrences in the interval $(0, t]$, $H_Y(t)$, satisfies the following Laplace transform equation [see (3.84)]:

$$H_Y^*(s) = \frac{f^*(s) \cdot g^*(s)}{s\{1 - f^*(s) \cdot g^*(s)\}}.$$

For the Type I event occurrences we have a modified renewal process. The LT of the expected number of Type I occurrences, $H_X^*(s)$, in the interval $(0, t]$ is obtained from (3.96):

$$H_X^*(s) = \frac{f^*(s)}{s\{1 - f^*(s) \cdot g^*(s)\}}.$$

The renewal density can be obtained from the expected number of renewals, applying (3.97). In particular,

$$h_X^*(s) = s H_X^*(s) = \frac{f^*(s)}{\{1 - f^*(s) \cdot g^*(s)\}}, \tag{3.98}$$

$$h_Y^*(s) = s H_Y^*(s) = \frac{f^*(s) \cdot g^*(s)}{\{1 - f^*(s) \cdot g^*(s)\}}, \tag{3.99}$$

where $h_X(t)\,dt$ can be interpreted as the probability of occurrence of a renewal of Type I (indicated by the symbol $\times$ in Figure 3.22) in a small interval of duration dt, and $h_Y(t)\,dt$ as the probability of occurrence of a renewal of Type II (indicated by the symbol o in Figure 3.22) in a small interval of duration dt.

With the above definitions, we can say that the instantaneous availability $A(t)$ is the probability that at time t a variable of Type I is active (the system is in an up state), while $U(t)$ is the probability that at time t a variable of Type II is active (the system is in a down state).

In order to compute the availability $A(t)$, we observe that the event that a Type I variable is active at time t is the union of two mutually exclusive events: either *event (i)*, the component never failed before t; or *event (ii)*, the last repair occurred at an instant of time $u < t$, and from the last repair the component has continued to function without failure up to time t.

Event (i): No Type I event occurs in $(0, t]$. The probability associated with this event is $1 - F(t)$.

Event (ii): A Type II event occurred in $(u, u + du]$ $(u < t)$, and no Type I events have occurred in the remaining interval $(u, t]$. The probability associated with this event is

$$\int_0^t h_Y(u)[1 - F(t - u)]\, du.$$

Since the two events are mutually exclusive, the instantaneous availability is given by:

$$A(t) = [1 - F(t)] + \int_0^t h_Y(u)[1 - F(t - u)]\, du. \tag{3.100}$$

By a similar argument, we can say that the probability that at time t a Type II variable is active (instantaneous unavailability) can be computed by the probability of the event that at a time $u < t$ a Type I event occurred (a failure occurred) and in the interval $(u, t]$ repair is not completed:

$$U(t) = \int_0^t h_X(u)[1 - G(t - u)]\, du. \tag{3.101}$$

Taking the Laplace transform of (3.100) and (3.101), we obtain:

$$A^*(s) = \frac{1 - f^*(s)}{s} + \frac{f^*(s) \cdot g^*(s)}{1 - f^*(s) \cdot g^*(s)} \frac{1 - f^*(s)}{s}$$

$$= \frac{1 - f^*(s)}{s\{1 - f^*(s) \cdot g^*(s)\}},$$

$$\tag{3.102}$$

$$U^*(s) = \frac{f^*(s)}{1 - f^*(s) \cdot g^*(s)} \frac{1 - g^*(s)}{s}$$

$$= \frac{f^*(s)(1 - g^*(s))}{s\{1 - f^*(s) \cdot g^*(s)\}}.$$

Rearranging (3.102), we obtain:

$$A^*(s) = H_Y^*(s) - H_X^*(s) + \frac{1}{s},$$

$$A(t) = H_Y(t) - H_X(t) + 1.$$

The Limiting Behavior of an Alternating Renewal Process

We are often interested in the state of the system after a sufficiently long period of time. To this end, we consider the limiting behavior of the alternating renewal process when $t \to \infty$. We define the *steady-state availability* A and the *steady-state unavailability* $U = 1 - A$ as the limiting values of $A(t)$ and $U(t) = 1 - A(t)$, respectively, as $t \to \infty$.

Assuming that the conditions for the application of the final value theorem for Laplace transforms are satisfied, we can write:

$$A = \lim_{t \to \infty} A(t) = \lim_{s \to 0} sA^*(s), \qquad (3.103)$$

$$U = \lim_{t \to \infty} U(t) = \lim_{s \to 0} s U^*(s).$$

In order to evaluate expressions (3.103) for small values of s, we can use a first-order approximation:

$$e^{-st} \simeq 1 - st,$$

hence,

$$f^*(s) = \int_0^\infty e^{-st} f(t) \, dt$$

$$\simeq \int_0^\infty f(t) \, dt - s \int_0^\infty t f(t) \, dt$$

$$\simeq 1 - s E[X]. \qquad (3.104)$$

Similarly,

$$g^*(s) \simeq 1 - s E[Y].$$

Applying the above results, Eq. (3.103) takes the form:

$$A = \lim_{s \to 0} \frac{1 - (1 - s E[X])}{1 - (1 - s E[X])(1 - s E[Y])} = \frac{E[X]}{E[X] + E[Y]}, \qquad (3.105)$$

$$U = \frac{E[Y]}{E[X] + E[Y]}.$$

Since $E[X] = \text{MTTF}$ is the expected value of the uptime and $E[Y] = \text{MTTR}$ the expected value of the downtime, the above results are usually written in the form:

$$A = \frac{\text{MTTF}}{\text{MTTF} + \text{MTTR}}, \qquad U = \frac{\text{MTTR}}{\text{MTTF} + \text{MTTR}}. \qquad (3.106)$$

The result in (3.106) states that the steady-state availability is the ratio of the expected duration of an up state (MTTF) and the expected duration of a complete failure/repair cycle (MTTF + MTTR). We note that the above equation is valid for any distribution of X and Y as long as they both have finite means. The main assumption was that of a single unit. For a system with redundancy the equation needs to be used carefully, replacing MTTF by MTTF_{eq} and MTTR by MTTR_{eq} as illustrated in Section 9.5.6 [1].

The Exponential Case

Assume that the Type I variable is exponentially distributed with rate λ and the Type II variable is exponentially distributed with rate μ, then:

$$f(t) = \lambda e^{-\lambda t}, \quad f^*(s) = \frac{\lambda}{s + \lambda},$$

$$g(t) = \mu e^{-\mu t}, \quad g^*(s) = \frac{\mu}{s + \mu}.$$

From (3.102),

$$A^*(s) = \frac{s + \mu}{s(s + \lambda + \mu)}$$

$$= \frac{\mu}{\lambda + \mu} \cdot \frac{1}{s} + \frac{\lambda}{\lambda + \mu} \cdot \frac{1}{s + \lambda + \mu}, \tag{3.107}$$

$$U^*(s) = \frac{\lambda}{s(s + \lambda + \mu)}.$$

Inverting (3.107), we finally obtain, in the time domain,

$$A(t) = \frac{\mu}{\lambda + \mu} + \frac{\lambda}{\lambda + \mu} e^{-(\lambda + \mu)t}, \tag{3.108}$$

$$U(t) = \frac{\lambda}{\lambda + \mu} - \frac{\lambda}{\lambda + \mu} e^{-(\lambda + \mu)t}. \tag{3.109}$$

(Note that for any t, $A(t) + U(t) = 1$.) From the first equation we get $A(0) = 1$, $U(0) = 0$, and in the limit the steady-state behavior becomes:

$$A = \lim_{t \to \infty} A(t) = \frac{\mu}{\lambda + \mu}, \quad U = \lim_{t \to \infty} U(t) = \frac{\lambda}{\lambda + \mu}. \tag{3.110}$$

Taking into account the relation between hazard rates and expected values for the exponential distribution, Eqs. (3.110) can be rewritten [see also Eq. (3.106)] as:

$$A = \lim_{t \to \infty} A(t) = \frac{\mu}{\lambda + \mu} = \frac{\text{MTTF}}{\text{MTTF} + \text{MTTR}}, \tag{3.111}$$

$$U = \lim_{t \to \infty} U(t) = \frac{\lambda}{\lambda + \mu} = \frac{\text{MTTR}}{\text{MTTF} + \text{MTTR}}.$$

The expression (3.108) can be directly obtained in the time domain starting from (3.100), which, in the exponential case, becomes:

$$A(t) = e^{-\lambda t} + \int_0^t h_Y(u) e^{-\lambda(t-u)} du, \tag{3.112}$$

where $h_Y(t)$ is obtained by inverting (3.99) to get:

$$h_Y(t) = \frac{\lambda \mu}{\lambda + \mu} \left(1 - e^{-(\lambda + \mu)t} \right). \tag{3.113}$$

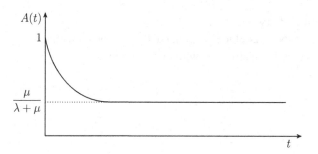

Figure 3.23   Availability $A(t)$ versus time.

Typical behavior of the availability versus time is shown in Figure 3.23. For small t, the availability has an exponential decay with rate $(\lambda + \mu)$, then tends to reach the steady-state value $A = \frac{\mu}{\lambda+\mu}$.

The Transient Part of $A(t)$

We now consider more closely the transient part of the function $A(t)$ in (3.108). Given a negative exponential term with rate, say, γ, we have that $e^{-\gamma t} \simeq 0.01$ for $t \simeq 4.5/\gamma$ (compare with Figure 3.6). This result means that after a time t approximately equal to $4.5/(\lambda + \mu)$ the transient term in (3.108) reduces to a value of $\simeq 1\%$ of the initial value and can be considered negligible. Since in a real system the mean time to failure is orders of magnitude larger than the mean time to repair, we have MTTF $\gg$ MTTR, and therefore $\lambda \ll \mu$. As a consequence, we can assume that the transient term becomes negligible, and hence the availability reaches its steady-state value after a very short time with respect to the mean time to failure.

Example 3.8   Given a system with MTTF $= 10^4$ hr and MTTR $= 10$ hr, we have $\lambda = 1 \cdot 10^{-4} \, \text{hr}^{-1}$ and $\mu = 1 \cdot 10^{-1} \, \text{hr}^{-1}$. With these failure and repair rate values, the transient term of $A(t)$ has negligible effect after a time of the order of $t = 45$ hr, which is very short with respect to a reasonable useful lifetime for the system.

The Steady-State Availability

We know from (3.106) that the availability approaches the ratio of the expected operational time and the expected time of a failure/restoration cycle, after a sufficiently long time has elapsed, independently of the distribution function of the operational and non-operational times. In the exponential case the steady-state availability assumes the form (3.111). We refer to the cycle time as the *mean time between failures*, for which MTBF $=$ MTTF $+$ MTTR. Under the assumption MTTF $\gg$ MTTR, MTBF and MTTF are nearly equal and hence are often used interchangeably in practice.

In system availability studies, it is customary to refer to the MTTF as the mean uptime with the symbol MUT, and to the MTTR as the mean downtime with the symbol MDT. The reason for this is that the downtime of a system can be due to many different causes, also external to the system itself, such as the outage of the electrical power, time devoted to various tests, hardware/software updates or preventive maintenance on the system, shortage of parts to be processed by the system in an assembly line, etc. In an availability study all these causes contribute to the system downtime and should be taken into account. The MDT includes the mean downtime not only due to a repair after a failure but due to the other causes mentioned above as well. With this notation, the steady-state availability is written as:

$$A = \frac{\text{MUT}}{\text{MUT} + \text{MDT}}, \qquad U = \frac{\text{MDT}}{\text{MUT} + \text{MDT}}. \tag{3.114}$$

Problems

3.15 Consider a unit with $HYPO_2(\lambda_1, \lambda_2)$ time to failure distribution and $EXP(\mu)$ time to repair distribution. Use Eq. (3.102) to write the Laplace domain version of the instantaneous availability, and thence invert using partial fraction expansion to get a time domain expression for $A(t)$.

3.16 In Problem 3.15, find the steady-state availability and the steady-state unavailability.

3.17 Consider a unit with $EXP(\lambda)$ time to failure distribution and $HYPO_2(\mu_1, \mu_2)$ time to repair distribution. Use Eq. (3.102) to write the Laplace domain version of the instantaneous availability, and thence invert to get a time domain expression for $A(t)$.

3.18 In Problem 3.17, find the steady-state availability and the steady-state unavailability.

3.19 As a variation of the above problems, compare the transient and steady-state availabilities of two units A and B:

  Unit A: TTF distribution $EXP(\lambda)$ and TTR distribution $ER_2(\gamma)$ such that MTTR $= 1/\mu$.
  Unit B: TTF distribution $ER_2(\beta)$ such that MTTF $= 1/\lambda$ and TTR distribution $EXP(\mu)$.

3.5.6   Availability Analysis: State Transition Diagram

Assume that the Type I variable is exponentially distributed with rate λ and the Type II variable is exponentially distributed with rate μ. The evolution in time of the system can be represented by means of the labeled state transition graph of Figure 3.24, where the state labeled W represents the working or up condition and the state labeled D represents the non-working or down condition.

The arc labeled λ represents the failure transition (passage form W to D), while the arc labeled μ represents the repair transition (from D to W). The study and solution of such state transition diagrams with constant transition rates will be carried out in Part III of the book. Here, we sketch a solution procedure for the simple case of Figure 3.24. Let $\pi_W(t)$ and $\pi_D(t)$ represent the probabilities of being in state W and D, respectively, at time t since the beginning of system operation. By definition, we can identify $\pi_W(t)$

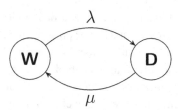

Figure 3.24   Two-state diagram of a repairable system.

and $\pi_D(t)$ as the system availability $A(t)$ and unavailability $U(t)$, respectively. We apply to the system a probability balance equation. The rate of change in the probability of a given state at time t equals the difference between the flow of the probability in and out of that state. The balance equation is in the form of a set of two linear differential equations with constant coefficients:

$$\begin{cases} \dfrac{d\pi_W(t)}{dt} = -\lambda\pi_W(t) + \mu\pi_D(t), \\ \dfrac{d\pi_D(t)}{dt} = \lambda\pi_W(t) - \mu\pi_D(t). \end{cases} \tag{3.115}$$

Assuming an initial condition $\pi_W(0) = 1$ (at time $t = 0$ the system is operational with probability 1), and solving Eq. (3.115), we obtain the same results expressed by Eqs. (3.108) and (3.109):

$$A(t) = \pi_W(t) = \frac{\mu}{\lambda + \mu} + \frac{\lambda}{\lambda + \mu} e^{-(\lambda+\mu)t}, \tag{3.116}$$

$$U(t) = \pi_D(t) = \frac{\lambda}{\lambda + \mu} - \frac{\lambda}{\lambda + \mu} e^{-(\lambda+\mu)t}.$$

(Note that for any t, $\pi_W(t) + \pi_D(t) = 1$.) From the first of Eq. (3.116) we get:

$$A(0) = 1, \qquad \lim_{t \to \infty} A(t) = A = \frac{\mu}{(\lambda + \mu)}. \tag{3.117}$$

In the above derivation, it is certainly possible to allow for the case that at $t = 0$, the unit is up with some probability, say $q \leq 1$ (and down with probability $1 - q$). The differential equations (3.115) can be solved to yield, in this case,

$$A(t) = \pi_W(t) = \frac{\mu}{(\lambda + \mu)} + \frac{\lambda q}{(\lambda + \mu)} e^{-(\lambda+\mu)t} + \frac{\mu(1-q)}{(\lambda + \mu)} e^{-(\lambda+\mu)t}. \tag{3.118}$$

Note that in the limit $t \to \infty$ the result is independent of q (i.e., is independent of the initial condition) and given by Eq. (3.117).

Expected Interval Availability

The expected interval availability is defined as the expected fraction of time that the system is in the up state during the interval $(0, t]$ [21]. It can be expressed as:

$$A_I(t) = \frac{1}{t} \int_0^t A(u)\, du.$$

In the case that both the failure times and repair times are exponentially distributed with rates λ and μ, respectively, we obtain from Eq. (3.116):

$$A_I(t) = \frac{\mu}{\lambda + \mu} + \frac{\lambda}{t(\lambda + \mu)^2} (1 - e^{-(\lambda + \mu)t}).$$

Hence, in the limit:

$$\lim_{t \to \infty} A_I(t) = \lim_{t \to \infty} A(t) = \frac{\mu}{\lambda + \mu}.$$

The steady-state value of the availability can also be interpreted as the expected fraction of time that the system is up in an interval of duration t, when enough time has elapsed to reach the steady-state condition.

Failure and Repair Frequencies

According to (3.105), in steady state the following relation holds:

$$\mathrm{MTBF} = \mathrm{MTTF} + \mathrm{MTTR}$$

(with $\mathrm{MTTF} = E[X]$ and $\mathrm{MTTR} = E[Y]$), independently of the particular distribution of X and Y. Hence, we can define:

- Failure frequency for the system:

$$\nu = \frac{1}{\mathrm{MTBF}} = \frac{1}{\mathrm{MTTF} + \mathrm{MTTR}}.$$

- Equivalent failure and repair rate:

$$\lambda_{eq} = \frac{1}{\mathrm{MTTF}}, \qquad \mu_{eq} = \frac{1}{\mathrm{MTTR}}.$$

Example 3.9 Though availability is a function of time, it is usually represented by a fraction or a percentage, by implicitly assuming that we are considering its steady-state value. If the steady-state availability of a system is $A = 0.95$, it means that after a sufficiently long time has elapsed to reach a steady-state condition, given a time period τ, the system is expected to be operational for a time equal to $A \times \tau$. Conversely, we expect the system to be non-operational for an expected total time $U \times \tau = (1 - A) \times \tau$. If τ is taken to be 8760 hr (hours in a year) then the annual expected total downtime in minutes is given by $D_I = (1 - A) \times 8760 \times 60 \min \mathrm{yr}^{-1}$.

A taxonomy for different classes of systems according to the (steady-state) availability together with the associated downtimes in minutes per year is given in [22], and is displayed in Table 3.5.

Example 3.10 Consider a production plant that has a rated productivity of W parts yr^{-1}. However, failures and repairs reduce the total number of parts that can really be produced by a quantity $W \cdot U$ parts yr^{-1}, where U is the system unavailability. Given the unavailability, we can compute the expected productivity loss due to the downtime periods.

Table 3.5 A taxonomy for system availability [22].

System Type	Downtime (min yr^{-1})	Availability (%)
Unmanaged	52560	90
Managed	5256	99
Well managed	526	99.9
Fault tolerant	53	99.99
High availability	5	99.999
Very high availability	0.5	99.9999
Ultra high availability	0.05	99.99999

A chemical plant is designed to produce $W = 2$ tons of ammonia per day with a non-stop production cycle ($W = 2$ tons d^{-1}). In previous years the plant has shown an unavailability of $U = 0.15$. We can expect that in the long run the real average productivity will be $(1 - U) \cdot W = 0.85 \cdot 2 = 1.7$ tons d^{-1}, with a loss of 0.3 tons d^{-1}.

Example 3.11 Knowledge of the unavailability can also be exploited at the design level. Suppose that the technical assistance service of a company has the charge of n similar objects (a fleet of commercial vehicles, or personal computers in a bank), and that the unavailability of each individual object is U. The number of items that are expected to be simultaneously out of service is therefore $n \cdot U$. If we assume, as in Example 3.9, $A = 0.95$, the mean number of items requiring repair will be equal to 5% of the total. This information is useful to correctly estimate the capacity of the assistance service (in terms of both manpower and spare parts).

Example 3.12 In a real system, the operational periods are generally very much longer than the non-operational ones, i.e., MUT $\gg$ MDT. Taking this inequality into account, Eq. (3.116) can be written in an approximate form as:

$$U \simeq \text{MDT/MUT}, \qquad A \simeq 1 - \text{MDT/MUT}. \qquad (3.119)$$

If the uptimes can be considered as exponentially distributed with rate λ and the downtimes exponentially distributed with rate μ, then $\mu \gg \lambda$, and

$$U \simeq \lambda/\mu, \qquad A \simeq 1 - \lambda/\mu. \qquad (3.120)$$

For mechanical items a reasonable failure rate is of the order of $\lambda = 10^{-4}$ f hr^{-1}, corresponding to an MTTF $= 10^4$ hr (remember, 1 yr $= 8760$ hr). Varying the repair rate μ from $\mu = 0.01$ r hr^{-1} (corresponding to an MTTR $= 100$ hr) to $\mu = 1.0$ r hr^{-1} (corresponding to an MTTR $= 1$ hr), the exact unavailability values (3.116) and their approximate values computed from (3.120) are compared in Table 3.6. The approximation error decreases as the unavailability decreases. Using (3.120) as a first-order computation thus provides a reasonable approximation.

Example 3.13 *System with Reliability and Availability Specification [23]*
Reliability and availability constraints can be combined in a system specification.

Table 3.6 Accuracy of the unavailability approximation
(3.120).

μ	Approximate value	Exact value	Error
0.01	0.01	0.009 900 9	1×10^{-4}
0.1	0.001	0.000 999 0	1×10^{-6}
1	0.0001	0.000 099 99	1×10^{-8}

Assume that a system is requested to operate with an availability greater than 0.999 but at the same time with a reliability greater than 0.9 over an interval of 1000 hr. The reliability specification means that we need to guarantee a probability of operating without failures over a period of 1000 hr.

The reliability specification is satisfied if:

$$R(t) = e^{-\lambda t} = 0.9 \text{ for } t = 1000 \text{hr};$$

$$\text{then,} \quad \lambda = -\ln(0.9)/1000 \leq 1.05 \cdot 10^{-4} \text{fhr}^{-1}.$$

To satisfy the availability specification it is required that:

$$\mu/(\lambda + \mu) = 0.999, \quad \text{hence} \quad \mu \geq 0.1052 \text{rhr}^{-1}.$$

The above expressions allow a designer to estimate the values of λ and μ that satisfy the required specifications.

3.5.7 Sensitivity of Steady-State Availability

Steady-state availability A depends on the two quantities MTTF and MTTR. To improve the steady-state availability one can either increase the MTTF or decrease the MTTR. Parametric sensitivity analysis (Section 2.2.7) is a method for determining the factors that are most influential on model results, and is performed by computing the partial derivatives of the measure of interest with respect to each input parameter. A positive sensitivity indicates that an increase of a parameter causes an increase in the measure of interest, and a negative sensitivity indicates that if a parameter value increases the measure decreases. The ranking of sensitivities (in absolute value) shows which parameter is more effective in modifying the measure of interest.

For the steady-state availability A, the sensitivities are:

$$S_F = \frac{\partial A}{\partial \text{MTTF}} = \frac{\text{MTTR}}{(\text{MTTF} + \text{MTTR})^2},$$

$$S_R = \frac{\partial A}{\partial \text{MTTR}} = -\frac{\text{MTTF}}{(\text{MTTF} + \text{MTTR})^2}. \tag{3.121}$$

Since S_F is always positive, A increases with an increasing MTTF, whereas since S_R is always negative, A decreases with an increasing MTTR, as expected. Now we compare the absolute values:

$$|S_R| > S_F$$

$$\frac{\text{MTTF}}{(\text{MTTF} + \text{MTTR})^2} > \frac{\text{MTTR}}{(\text{MTTF} + \text{MTTR})^2}.$$

Since in any practical case MTTF $\gg$ MTTR, it turns out that $|S_R| \gg S_F$. Thus, reducing the MTTR is much more effective than increasing the MTTF [24].

Example 3.14   Assuming the values of Example 3.12, with MTTF $= 10^4$ hr and MTTR varying from 1 to 100 hr, the results are reported in Table 3.7.

Table 3.7   Sensitivity of A vs. MTTF and MTTR.

MTTR (hr)	A	S_F	$|S_R|$	$0.5 \times$ MTTF %A	$0.5 \times$ MTTR %A
1	0.999 900 000	$9.998\,00 \times 10^{-9}$	0.999 900 010	0.000 033 3311	0.000 049 9975
10	0.999 000 999	$9.980\,03 \times 10^{-8}$	0.999 900 010	0.000 333 1110	0.000 499 7500
100	0.990 099 010	$9.802\,96 \times 10^{-7}$	0.999 000 999	0.003 311 2580	0.004 975 1240

The first column reports the value of MTTR, the second A, and the third and fourth S_F and $|S_R|$, respectively. Finally, we have computed the percentage increment of A by incrementing the MTTF by 50% in column five, and by decrementing the MTTR by 50% in column six.

3.5.8   Scaled Sensitivity of Steady-State Availability

The *relative* or *scaled* sensitivity (denoted by SS) is a dimensionless measure of parameter importance. For the steady-state availability it is computed using Eq. (3.121) as:

$$SS_F = \frac{\text{MTTF}}{\partial \text{MTTF}} \frac{\partial A}{A} = \frac{\text{MTTR}}{\text{MTTF} + \text{MTTR}},$$

$$SS_R = \frac{\text{MTTR}}{\partial \text{MTTR}} \frac{\partial A}{A} = -\frac{\text{MTTR}}{\text{MTTF} + \text{MTTR}}.$$

The scaled sensitivities turn out to be equal in absolute value for both MTTF and MTTR.

The results of these two sections may guide a designer toward a more cost-effective availability growth program. If the cost of improving the availability is proportional to the amount of increase (or decrease) in MTTF (MTTR), the unscaled sensitivities as computed in Section 3.5.7 imply that reducing the MTTR is more effective than increasing the MTTF. If, however, the costs are proportional to the relative changes in the values of MTTF and MTTR, the two actions have the same importance and the availability growth program can be based on different considerations.

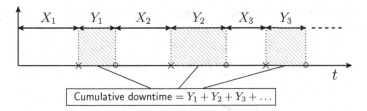

Figure 3.25 Cumulative downtime in an alternating renewal process.

3.5.9   Cumulative Downtime Distribution

Let the random variable $D(t)$ be the cumulative total time spent by the unit in the down state during the interval $(0,t]$ [25] (Figure 3.25). We denote by $F_D(t,x)$ the Cdf of $D(t)$ defined as:

$$F_D(t,x) = P\{D(t) \le x\}.$$

$F_D(t,x)$ depends on the distribution of the uptime intervals $F(t)$ and the distribution of downtime intervals $G(t)$, and on the time duration of observation t. If we assume that at the initial time 0 the system is already in a steady state, we obtain the cumulative downtime distribution in steady state.

Let $D_k = Y_1 + Y_2 \cdots + Y_k$ denote the accumulated downtime after k cycles, and let $D_k(t)$ be the accumulated downtime in $(0,t]$ after k cycles. Similarly, denote by U_k the accumulated uptime after k cycles and by $U_k(t)$ the accumulated uptime in $(0,t]$ after k cycles. Two cases are possible depending on whether the alternating renewal process starts at time $t = 0$ with a Type I variable or a Type II variable. The following has been proven in [26].

If the process starts with a Type I variable:

$$P\{D(t) \le x, k\,\text{cycles}\} = P\{D_k \le x, U_k \le t-x\} - P\{D_k \le x, U_{k+1} \le t-x\}$$

$$= P\{D_k \le x\}[P\{U_k \le t-x\} - P\{U_{k+1} \le t-x\}]$$

$$= [F_k(t-x) - F_{k+1}(t-x)]G_k(x).$$

The above relation makes use of the independence of the X and Y sequences; see also (3.82).

$$F_D^{(I)}(t,x) = \sum_{k=1}^{\infty} P\{D(t) \le x, k\,\text{cycles}\} = \sum_{k=1}^{\infty} [F_k(t-x) - F_{k+1}(t-x)]G_k(x).$$

If the process starts with a Type II variable:

$$P\{D(t) \le x, k\,\text{cycles}\} = P\{D_k \le x, U_{k-1} \le t-x\} - P\{D_k \le x, U_k \le t-x\}$$

$$= P\{D_k \le x\}[P\{U_{k-1} \le t-x\} - P\{U_k \le t-x\}]$$

$$= [F_{k-1}(t-x) - F_k(t-x)]G_k(x).$$

The above relation also makes use of the independence of the X and Y sequences; see also (3.82).

$$F_D^{(II)}(t,x) = \sum_{k=1}^{\infty} P\{D(t) \leq x, k\,\text{cycles}\}$$

$$= \sum_{k=1}^{\infty} [F_k(t-x) - F_{k+1}(t-x)]\,G_{k+1}(x).$$

Finally,

$$F_D(t,x) = P\{\text{Type I at } t = 0\}\,F_D^{(I)}(t,x) + P\{\text{Type II at } t = 0\}\,F_D^{(II)}(t,x). \tag{3.122}$$

For a system whose failure times are exponentially distributed with rate λ and whose repair times are exponentially distributed with rate μ, the steady-state cumulative downtime distribution can be calculated as follows. In steady state:

$$\begin{cases} P\{\text{Type I at } t = 0\} = \dfrac{\mu}{\lambda + \mu}, \\[2mm] P\{\text{Type II at } t = 0\} = \dfrac{\lambda}{\lambda + \mu}. \end{cases} \tag{3.123}$$

Furthermore:

$$[F_i(x) - F_{i+1}(x)] = \frac{(\lambda x)^i}{i!}\,e^{-\lambda x},$$

$$1 - G_k(x) = \sum_{i=0}^{k-1} \frac{(\mu x)^i}{i!}\,e^{-\mu x}.$$

Substituting the above expressions in (3.122), we obtain the final result:

$$1 - F_D(t,x) = \frac{\mu}{\lambda + \mu} \sum_{k=1}^{\infty} \frac{(\lambda(t-x))^k}{k!}\,e^{-\lambda(t-x)} \sum_{i=0}^{k-1} \frac{(\mu x)^i}{i!}\,e^{-\mu x}$$

$$+ \frac{\lambda}{\lambda + \mu} \sum_{k=1}^{\infty} \frac{(\lambda(t-x))^k}{k!}\,e^{-\lambda(t-x)} \sum_{i=0}^{k} \frac{(\mu x)^i}{i!}\,e^{-\mu x}. \tag{3.124}$$

3.6   Interval Reliability

Interval reliability was introduced by Barlow and Hunter in 1961 [27] for repairable systems. Given an interval $(t, t+\tau]$, the *interval reliability* is defined as the probability that the unit is up at time t and remains operational up to time $t+\tau$. Informally, we require the unit to be up when needed and as long as needed.

Assume the unit is described by an alternating renewal process of the type depicted in Figure 3.22. The interval reliability $IR(t,\tau)$ is the probability that a Type I variable is continuously active in the interval $(t, t+\tau]$, and this event can be decomposed into two mutually exclusive events: either *event (i)*, starting from time 0, the component never failed until time $t + \tau$; or *event (ii)*, the component did fail at some point prior to t, the last repair occurred at an instant $u < t$, and from the last repair the component has continued to function up to $t + \tau$.

Event (i): No Type I event occurs during the interval $(0, t + \tau]$. The probability associated with this event is $1 - F(t + \tau)$.

Event (ii): A Type II event occurred in the interval $(u, u + du]$ $(u < t)$, and no Type I events occur in the interval $(u, t + \tau]$. The probability associated with this event is

$$\int_0^t h_Y(u) [1 - F(t + \tau - u)] du.$$

Since the two events are exhaustive and mutually exclusive, the interval reliability is given by:

$$IR(t, \tau) = [1 - F(t + \tau)] + \int_0^t h_Y(u) [1 - F(t + \tau - u)] du. \qquad (3.125)$$

In the exponential case we get, from (3.125),

$$IR(t, \tau) = e^{-\lambda(t + \tau)} + \int_0^t h_Y(u) e^{-\lambda(t + \tau - u)} du$$

$$= e^{-\lambda \tau} \left\{ e^{-\lambda t} + \int_0^t h_Y(u) e^{-\lambda(t - u)} du \right\}. \qquad (3.126)$$

The term in braces coincides with the availability $A(t)$ given in Eq. (3.112), so that Eq. (3.126) finally becomes:

$$IR(t, \tau) = e^{-\lambda \tau} A(t). \qquad (3.127)$$

In other words, the system should be up at time t and not fail in the interval $(t, t + \tau]$. Thus, by taking appropriate limits we can get reliability, instantaneous availability and steady-state availability as special cases of the interval reliability.

Example 3.15 For systems such as missile control, we can use interval reliability as the probability of correct operation for the firing of a missile if the missile firing command is given at time t and it takes τ time units to carry out the firing action. Since prior to time t the system is dormant, we can assign a (dormant) failure rate λ_d for the interval $(0, t]$ that is less than the active failure rate λ_a applied for the interval $(t, t + \tau]$. In this case, the probability of correctly firing the missile is:

$$P_M(t, \tau) = e^{-\lambda_a \tau} \left[\frac{\mu}{\lambda_d + \mu} + \frac{\lambda_d}{\lambda_d + \mu} e^{-(\lambda_d + \mu)t} \right]. \qquad (3.128)$$

Example 3.16 Missile control system parts are often checked periodically, and may be repaired or replaced if found to be defective. Suppose that the total failure rate of those parts that are checked periodically is λ_c, while that of those parts that are not checked is λ_{nc}. Assume that the last check was done at time t_c, and that the firing command was given at time t_F. In this case, the probability of correctly firing the missile is:

$$P_M(t_c, t_F, \tau) = e^{-(\lambda_c + \lambda_{nc})\tau} e^{-\lambda_c(t_F - t_c)} e^{-\lambda_{nc} t_F}. \tag{3.129}$$

Problems

3.20   Generalize Eq. (3.129) to the Weibull time to failure distributions for both the checked and not-checked parts of the missile control system.

3.7   Task-Oriented Measures

The task-oriented point of view was introduced in Section 1.3.6. In this section, we concentrate on a task-oriented measure that refers to the probability of completing a task of a given duration on a unit (system) that alternates between up and down states. We assume that the time to failure and time to repair are exponentially distributed with rates λ and μ, respectively. Consider a task that needs τ units of time to be completed in the absence of failures. Let $T(\tau)$ be the r.v. representing the task completion time. We define two measures, the distribution of the task completion time $F_T(x, t, \tau)$ and the probability of completing the task (or task reliability) $R_T(t, \tau) = \lim_{x \to \infty} F_T(x, t, \tau)$, where x is the current time variable, t is the time at which the task starts to be performed and τ is the task time requirement on the server in absence of failures.

3.7.1   No Failure and No Repair

Assume that the task arrives at time t. At time $x = t + \tau$ the task completes with probability 1. The task completion time is $T_1(\tau) = t + \tau$, and the distribution of the task completion time and the task reliability are:

$$F_{T_1}(x, t, \tau) = u(x - t - \tau),$$

$$R_{T_1}(t, \tau) = 1,$$

where $u(x - t - \tau)$ is the unit step function located at $x = t + \tau$.

3.7.2   Failure and Repair

We can consider various cases depending on whether the task arrives when the unit is up (with probability 1) or at a generic time t. Furthermore, once the task starts execution, it may be interrupted by a server failure, in which case we consider three possibilities: (i) the task is dropped and the work done is lost (we refer to this policy as the *drop policy* (dp)); (ii) when the unit is repaired and is up again the task resumes from the point it was interrupted (we refer to this policy as *preemptive resume* (prs)); (iii) when the unit is up again the task is restarted from scratch (we refer to this policy as *preemptive repeat* (prt)) [28].

Task Arrives when Unit is Up
We assume that the task of length τ arrives at time $t = 0$ when the system is up with probability 1.

Drop policy: The task completes if there is an interval of length τ before system failure. The distribution of the task completion time and the task reliability are:

$$F_{dp}(x, \tau) = e^{-\lambda \tau} u(x - \tau),$$
$$R_{dp}(\tau) = e^{-\lambda \tau}. \tag{3.130}$$

The distribution $F_{dp}(x, \tau)$ is defective (Section 3.2.11), and the defect $(1 - e^{-\lambda \tau})$ equals the probability that a failure occurs before the task completes (or task unreliability).

Preemptive resume: The analysis of the distribution of the completion time is quite complex in this case, and can be found in [28, 29]. The task completion time distribution $F_{prs}(x, \tau)$ is obtained in terms of the Laplace–Stieltjes transform, for which no closed-form inversion is available. The inversion can be carried out numerically. The task reliability is:

$$R_{prs}(\tau) = \lim_{x \to \infty} F_{prs}(x, \tau) = 1.$$

As time x goes to infinity the task eventually completes.

Preemptive repeat: In this case also the distribution of the completion time $F_{prt}(x, \tau)$ can be obtained in the form of the Laplace–Stieltjes transform [29], for which no closed-form inversion is available. The task reliability is:

$$R_{prt}(\tau) = \lim_{x \to \infty} F_{prt}(x, \tau) = 1.$$

As time x goes to infinity the task eventually completes.

Task Arrives at a Generic Time t
Upon its arrival at time t, the task may find the unit operational or non-operational. The distribution of the task completion time in this case can be written using the theorem of total probability:

$$F_{\bullet}(x, t, \tau) = A(t) F_{\bullet}(x, \tau) + U(t) (F_{\bullet}(x, \tau) \star G(x)),$$

where the $\bullet$ stands for one of the three policies dp, prs or prt. The $\star$ symbol represents convolution, and indicates the fact that if the task arrives when the unit is down it has to wait until the termination of the current repair time to resume (under the prs policy) or to restart (under the prt policy) execution. In our case, $G(x) = 1 - e^{-\mu x}$ is the distribution of the residual repair time.

Drop policy:

$$F_{dp}(x, t, \tau) = A(t) e^{-\lambda \tau} u(x - \tau) + U(t) e^{-\lambda \tau} \mathcal{K}_{\{x \geq \tau\}}[1 - e^{-\mu(x - \tau)}],$$

$$R_{dp}(t, \tau) = \lim_{x \to \infty} F_{dp}(x, t, \tau) = e^{-\lambda \tau}.$$

Preemptive resume, preemptive repeat:

$$R_{prs}(t,\tau) = \lim_{x \to \infty} F_{prs}(x,t,\tau) = R_{prt}(t,\tau) = \lim_{x \to \infty} F_{prt}(x,t,\tau) = 1.$$

3.7.3   Failure and Constrained Repair

A task of duration τ should be completed on a unit that undergoes an alternating renewal process with uptimes and downtimes exponentially distributed with rates λ and μ, respectively. The unit can tolerate downtime of a limited duration, but fails if a single downtime exceeds a critical threshold θ [30]. This case occurs in many practical applications: (i) real-time systems where interruptions may hinder the correct operation of the system; (ii) safety systems where upon occurrence of an accident the protective system has a limited time to intervene before a crash; (iii) guaranteed interval availability [31], where contract specifications may contain clauses that fix the maximum tolerated downtime before the manufacturer incurs some penalty; (iv) unit degrades during downtime until the degradation level is no longer tolerated.

Assume that the unit starts from the up state with probability 1. We denote by Y the total accumulated uptime before unit failure. The Cdf of Y has been derived in [30] by resorting to Laplace–Stieltjes transform (LST) analysis. With constant failure and repair rates the closed-form inversion is possible and is given by:

$$F_Y(t,\theta) = 1 - e^{-\lambda t e^{-\mu \theta}}. \tag{3.131}$$

The distribution of the accumulated uptime is exponential with parameter $\lambda e^{-\mu \theta}$. Now, assume a task of length τ is initiated in an up state and that it is executed during the uptimes only. Upon failure, either a prs or prt policy can be considered.

Under a prs policy, the probability $R_{prs}^c(\tau,\theta)$ that the task completes before unit failure (task reliability) equals the probability that the accumulated uptime is greater than τ. Hence, from (3.131),

$$R_{prs}^c(\tau,\theta) = 1 - F_Y(\tau,\theta) = e^{-\lambda \tau e^{-\mu \theta}}.$$

When the downtime is not tolerated, $\theta \to 0$:

$$\lim_{\theta \to 0} R_{prs}^c(\tau,\theta) = e^{-\lambda \tau}.$$

The task must complete before the first failure, as in Eq. (3.130). When there is no constraint on the accumulated downtime, $\theta \to \infty$:

$$\lim_{\theta \to \infty} R_{prs}^c(\tau,\theta) = 1,$$

and the task eventually completes.

Under a prt policy, the task reliability has been derived in [30] to give:

$$R_{prt}^c(\tau,\theta) = \frac{1}{1 - (1 - e^{\lambda \tau})e^{-\mu \theta}}.$$

Again:

$$\lim_{\theta \to 0} R^c_{prt}(\tau,\theta) = e^{-\lambda \tau},$$

$$\lim_{\theta \to \infty} R^c_{prt}(\tau,\theta) = 1.$$

For a more extensive analysis of the task completion time problem on a single unit, with possible restrictions on the total accumulated downtime, the interested reader may consult [30, 32].

3.8 Improving Dependability

We have considered the reliability and availability of a single unit in this chapter. It is quite likely that the level of reliability and/or availability thus obtained is well below requirements. We briefly consider techniques to help increase reliability and availability that will be explored in the rest of the book. In order to improve reliability, we have three methods: fault avoidance (or fault prevention), fault removal, and fault tolerance. Fault avoidance amounts to building in high quality, and fault removal is based on extensive testing/debugging. It is also possible to avoid or postpone system failure by means of preventive maintenance provided the unit has an IFR behavior. The use of redundancy is the primary method used to achieve fault tolerance and subsequent high reliability in systems. We will be exploring different methods of predicting the reliability of systems employing fault tolerance as well as those employing preventive maintenance.

In order to improve system availability, we may use all the techniques described above to improve system reliability, but, in addition, we could also reduce the downtime due to failures. This could be achieved by employing escalated levels of recovery, as is done in telecommunication systems [33], or by reducing detection delay, restart delay, reboot delay, failover delay, logistics delay, etc. The actual repair time can be reduced by designing in a hot swap capability and improved provision for spare parts. We will explore a large number of techniques that will enable us to predict the availability of systems employing different types of redundancy and different methods of recovery after the occurrence of failures.

3.9 Further Reading

Readers interested in learning more on the various probability distributions can refer to Barlow and Proschan [3], Ross [34], Papoulis [2], and Trivedi [1]. Detailed discussions on stochastic processes can be found in Ross [35, 36].

References

[1] K. Trivedi, *Probability and Statistics with Reliability, Queueing and Computer Science Applications*, 2nd ed. John Wiley & Sons, 2001.

[2] A. Papoulis, *Probability, Random Variables and Stochastic Processes*. McGraw Hill, 1965.

[3] R. Barlow and F. Proschan, *Statistical Theory of Reliability and Life Testing*. Holt, Rinehart, and Winston, 1975.

[4] A. Sweet, "On the hazard rate of the lognormal distribution," *IEEE Transactions on Reliability*, vol. 39, pp. 325–328, 1990.

[5] J. Mi, "A new explanation of decreasing failure rate of a mixture of exponentials," *IEEE Transactions on Reliability*, vol. 47, pp. 460–462, 1998.

[6] Y. Huang, C. M. R. Kintala, N. Kolettis, and N. D. Fulton, "Software rejuvenation: Analysis, module and applications," in *Proc. Int. Symp. on Fault-Tolerant Computing (FTCS)*, 1995, pp. 381–390.

[7] A. Bobbio and A. Cumani, "A Markov approach to wear-out modelling," *Microelectronics and Reliability*, vol. 23, pp. 113–119, 1983.

[8] D. Chen and K. S. Trivedi, "Closed-form analytical results for condition-based maintenance," *Reliability Engineering & System Safety*, vol. 76, no. 1, pp. 43–51, 2002.

[9] D. Aldous and L. Shepp, "The least variable phase type distribution is Erlang," *Stochastic Models*, vol. 3, pp. 467–473, 1987.

[10] A. Bobbio and M. Telek, "Computational restrictions for SPN with generally distributed transition times," in *First European Dependable Computing Conference (EDCC-1)*, LNCS 852, eds. D. H. K. Echtle and D. Powell, Springer Verlag, 1994, pp. 131–148.

[11] N. Balakrishnan, H. Malik, and S. Puthenpura, "Best linear unbiased estimation of location and scale parameters of the log-logistic distribution," *Communications in Statistics – Theory and Methods*, vol. 16, no. 12, pp. 3477–3495, 1987.

[12] G. Rao and R. Kantam, "Estimation of reliability in multicomponent stress-strength model: Log-logistic distribution," *Electronic Journal of Applied Statistical Analysis*, vol. 3, pp. 75–84, 2010.

[13] S. Gokhale and K. Trivedi, "A time/structure based software reliability model," *Annals of Software Engineering*, vol. 8, no. 1-4, pp. 85–121, 1999.

[14] N. Sato and K. Trivedi, "Stochastic modeling of composite web services for closed-form analysis of their performance and reliability bottlenecks," in *Proc. 5th Int. Conf. on Service-Oriented Computing, ICSOC '07*. Springer-Verlag, 2007, pp. 107–118.

[15] J. L. Fleming, "Relcomp: A computer program for calculating system reliability and MTTF," *IEEE Transactions on Reliability*, vol. 20, pp. 102–107, 1971.

[16] L. Swiler, T. Paez, and R. Mayes, "Epistemic uncertainty quantification tutorial," in *Proc. IMAC-XXVII*. Society for Experimental Mechanics Inc., 2009.

[17] K. Mishra and K. S. Trivedi, "Closed-form approach for epistemic uncertainty propagation in analytic models," in *Stochastic Reliability and Maintenance Modeling*. Springer, 2013, vol. 9, pp. 315–332.

[18] W. E. Smith, K. S. Trivedi, L. Tomek, and J. Ackaret, "Availability analysis of blade server systems," *IBM Systems Journal*, vol. 47, no. 4, pp. 621–640, 2008.

[19] A. Avizienis, J. Laprie, B. Randell, and C. Landwehr, "Basic concepts and taxonomy of dependable and secure computing," *IEEE Transactions on Dependable and Secure Computing*, vol. 1, no. 1, pp. 11–33, 2004.

[20] D. R. Cox, *Renewal Theory*. Chapman & Hall, 1962.

[21] D. Tang and K. Trivedi, "Hierarchical computation of interval availability and related metrics," in *Proc. Int. Conf. on Dependable Systems and Networks (DSN)*, 2004, p. 693.

[22] J. Gray and D. Siewiorek, "High-availability computer systems," *IEEE Computer*, pp. 39–48, Sep. 1991.

[23] M. Shooman, *Probabilistic Reliability: An Engineering Approach*. McGraw Hill, 1968.

[24] G. Candea, A. Brown, A. Fox, and D. Patterson, "Recovery-oriented computing: Building multitier dependability," *Computer*, vol. 37, no. 11, pp. 60–67, Nov. 2004.

[25] R. Barlow and F. Proschan, *Mathematical Theory of Reliability*. John Wiley & Sons, 1965.

[26] L. Takacs, "On certain sojourn time problems in the theory of stochastic processes," *Acta Mathematica Academiae Scientiarum Hungaricae*, vol. 8, pp. 169–191, 1957.

[27] R. Barlow and L. Hunter, "Optimum preventive maintenance policies," *Operations Research*, vol. 8, pp. 90–100, 1961.

[28] V. Kulkarni, V. Nicola, and K. Trivedi, "The completion time of a job on a multi-mode system," *Advances in Applied Probability*, vol. 19, pp. 932–954, 1987.

[29] P. Chimento and K. Trivedi, "The completion time of programs on processors subject to failure and repair," *IEEE Transactions on Computers*, vol. 42, pp. 1184–1194, 1993.

[30] A. Goyal, V. Nicola, A. Tantawi, and K. Trivedi, "Reliability of systems with limited repair," *IEEE Transactions on Reliability*, vol. R-36, pp. 202–207, 1987.

[31] A. Goyal and A. Tantawi, "A measure of guaranteed availability and its numerical evaluation," *IEEE Transactions on Computers*, vol. C-37, pp. 25–32, 1988.

[32] V. Nicola, A. Bobbio, and K. Trivedi, "A unified performance reliability analysis of a system with a cumulative down time constraint," *Microelectronics and Reliability*, vol. 32, pp. 49–65, 1992.

[33] K. S. Trivedi, D. Wang, J. Hunt, A. Rindos, W. E. Smith, and B. Vashaw, "Availability modeling of SIP protocol on IBM© WebSphere©," in *Proc. Pacific Rim Int. Symp. on Dependable Computing (PRDC)*, 2008, pp. 323–330.

[34] S. M. Ross, *Introduction to Probability Models*. Academic Press, 2014.

[35] S. M. Ross, *Stochastic Processes*. John Wiley & Sons, 1996, vol. 2.

[36] S. M. Ross, *Applied Probability Models with Optimization Applications*. Courier Corporation, 2013.

Part II

Non-State-Space (Combinatorial) Models

Several traditional methods for the analysis of system dependability can be classified under the umbrella of *non-state-space* (sometimes called *combinatorial*) methods. Part II covers the following traditional techniques:

- reliability block diagram (Chapter 4)
- network reliability or reliability graph (Chapter 5)
- fault tree analysis (Chapter 6)
- state-space enumeration (Chapter 7)
- dynamic redundancy (Chapter 8).

The last two chapters, 7 and 8, can be seen as an introduction to the subsequent parts of the book, related to state-space models.

All the above techniques are characterized by two common fundamental assumptions:

Assumption 1: The system is decomposed into elementary blocks or units called components.

Assumption 2: Components of the system behave in a statistically independent manner.

A third major assumption is usually made about the behavior of each component:

Assumption 3: The components or units into which the system is divided have a binary behavior since they can be in only two mutually exclusive and collectively exhaustive conditions identified as working (or up) and failed (or down).

According to assumption 3, a generic component i can be associated with a Boolean (binary) variable that will be indicated by a letter, for instance X_i. Throughout the book we associate the binary value $X_i = 1$ to the up state and the binary value $X_i = 0$ to the down state. Alternatively, we indicate with X_i the component i in the up state and by the same overlined symbol $\overline{X}_i$ the component i in the down state. In a similar way, the whole system is also usually assumed to have a binary behavior, in the sense that the whole system can also only be in one of two mutually exclusive states, an up state or a down state.

Multi-state components and systems may be considered as well, by relaxing assumption 3. In the following chapters some specific sections will explicitly address the problem of modeling and handling multi-state systems of multi-state components.

Let E_i represent the event that component i is in the up state and $\overline{E}_i$ the event that the component i is in the down state. We indicate as $P\{E_i\}$ the probability of the event E_i, and by $P\{\overline{E}_i\}$ the probability of the event $\overline{E}_i$. Since the up state and the down state are exhaustive and mutually exclusive, we have $P\{E_i\} + P\{\overline{E}_i\} = 1$.

We can consider different measures associated with the events E_i and $\overline{E}_i$, depending on the nature of the component and of the problem at hand.

- If the components has a point mass probability of being in the up or down state, then $P\{E_i\}$ is the probability that the component is functioning and $P\{\overline{E}_i\}$ is the probability that the component is not functioning. Thus, the time to failure distribution is a time-independent Bernoulli distribution as is needed for one-shot systems.
- If the component is non-repairable and $E_i(t)$ is the event that component i has been continuously up in the interval $(0,t]$, $P\{E_i(t)\} = R_i(t)$ is the component reliability at time t and $P\{\overline{E}_i(t)\} = F_i(t) = 1 - R_i(t)$ is the component unreliability. To quantify these probabilities the distribution function of the component TTF is needed.
- If the component is repairable and $E_i(t)$ is the event that component i is up at time t, $P\{E_i(t)\} = A_i(t)$ is the component instantaneous availability at time t. $P\{\overline{E}_i(t)\} = 1 - A_i(t)$ is the component unavailability. To quantify these measures the distribution functions of the component TTF and TTR need to be known.
- If the component is repairable and $E_i(t)$ is the event that component i is up as $t \to \infty$, $P\{E_i(t)\} = A_i$ is the component's steady-state availability. To quantify these measures, the MTTF and the MTTR need to be known (Section 3.5.1).
- As defined in Section 3.6, given an interval $(t, t+\tau]$, $E_i(t+\tau)$ is the event that the unit is up at time t and remains operational up to time $t + \tau$. In this case, $P\{E_i(t+\tau)\} = IR(t,\tau)$ is the interval reliability. To quantify these measures the distribution functions of the component TTF and TTR need to be known.

From the above discussion, $P\{E_i\}$ or $P\{\overline{E}_i\}$ can assume different meanings depending on the definition of E_i. Assumption 2 of statistical independence required by all the techniques mentioned in Part II implies that the components behave independently and the dependability measure for the whole system can be deduced from the dependability measures of each component considered in isolation. Hence, only the probability measures related to any individual components is what is needed to study the dependability of a complex system. Dependence will be extensively studied from Part III onwards.

4   Reliability Block Diagram

The reliability block diagram (RBD) method is extensively described in the international standard IEC 1078, *Analysis Techniques for Dependability; Reliability Block Diagram Method* [1]:

> An RBD is a pictorial representation of a system's reliability performance. It shows the logical connection of (functioning) components needed for system success. The modelling techniques described are intended to be applied primarily to systems without repair and where the order in which failures occur does not matter. At any instant in time, an item is considered to be in only one of two possible states: operational or faulty.

Most engineers in practice tend to assume that the RBD as a method is applicable only for non-repairable systems as described in the IEC 1078 standard quoted above. We demonstrate that this restriction is unnecessary, since the dominant assumption is that of statistical independence. If the components are repairable and the failure and repair processes of all the components are statistically independent, the RBD method is just as applicable.

Once the system structure has been investigated and the failure modes established, the system must be partitioned into appropriate logical blocks for the purpose of carrying out the analysis. The partition into blocks is guided by the idea that to start the RBD method we should know the reliability (or availability) of each block. Then, the logical connections among the blocks must be defined, keeping in mind that the RBD structure does not necessarily correspond to the way the physical components are connected. The RBD method shows how to evaluate the reliability (or availability) of the whole system given the reliability of each block.

4.1   Series Systems

Components of a system are said to be connected in series if each and every one of them must be operational for the system to be operational, i.e., the failure of any one of its components causes the system to fail. The RBD of a series system with n components is depicted in Figure 4.1. If the failure (and repair) events of components in the system are mutually independent, the probability of the system being operational is given by the probability that all the components are operational.

Let $P_i = P\{E_i\}$ be the probability associated with event E_i that component i is functioning. Then, by the assumption of series connection, the probability $P\{S\}$ that

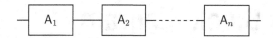

Figure 4.1   Series system with n components.

the system S is functioning is:

$$P\{S\} = P\{E_1 \cap E_2 \cap \cdots \cap E_n\}$$
$$= P\{E_1\}P\{E_2\} \cdots P\{E_n\}$$
$$= \prod_{i=1}^{n} P\{E_i\}. \tag{4.1}$$

Note that we have assumed that the events $E_1, E_2, \ldots, E_n$ are mutually independent. This assumption means that the failure (and repair) process of each component in the system is independent of the other components, and hence each component behaves as if it were in isolation.

4.1.1   Reliability of Series Systems

If E_i represents the event that component i has been continuously up in the interval $(0, t]$, $P\{E_i\} = R_i(t)$ is the component reliability at time t. In this case, denoting by $R_s(t)$ the series system reliability, i.e., the probability that the series system has been continuously up in the interval $(0, t]$, we have, from Eq. (4.1),

$$R_s(t) = \prod_{i=1}^{n} R_i(t). \tag{4.2}$$

The reliability of a series system of independent components is expressed in Eq. (4.2) by the simple *product law of reliabilities*. Since the reliability of each component is a positive number less than one, the product value is less than each term. It follows that the reliability of a series system is less than the reliability of each constituent component and, hence, is less than the reliability of its least reliable component. The product law of reliabilities in Eq. (4.2) also expresses the fact that the system reliability degrades with an increasing number of components. For example, if a system consists of five components in series, each having a reliability of $R_i(t) = 0.95$, then the system reliability is $R_s(t) = (0.95)^5 = 0.7738$. If the system complexity is increased to contain ten equal components, its reliability would reduce to $R_s(t) = (0.95)^{10} = 0.5987$. Adding complexity (components in series) always degrades the reliability. Conversely, if we have a reliability target of a series system of $R_s(t) = 0.95$, Table 4.1 shows the reliability that each individual component should have, as a function of the number of components in the system.

Table 4.1 Per-component reliability for series system reliability to be $R_s(t) = 0.95$.

Number of components	Component reliability
1	0.95000
2	0.97468
3	0.98305
4	0.98726
5	0.98979
10	0.99488

As an alternative derivation of (4.2), consider a series system of n independent components, where X_i denotes the lifetime of component i and X_s denotes the lifetime of the system. Then

$$X_s = \min\{X_1, X_2, \ldots, X_n\}.\tag{4.3}$$

With reference to Section 3.3.1, and applying (3.57), we get:

$$R_X(t) = \prod_{i=1}^{n} R_{X_i}(t) \le \min_i \{R_{X_i}(t)\}.\tag{4.4}$$

Then

$$E[X] = \int_0^\infty R_X(t)dt \le \min_i \left\{ \int_0^\infty R_{X_i}(t)dt \right\}$$

$$= \min_i \{E[X_i]\}$$

$$0 \le E[X] \le \min\{E[X_i]\},\tag{4.5}$$

which gives rise to the common remark that a series system is weaker than its weakest link.

Series System: The Exponential Case

If the TTF distribution of component i is $X_i \sim \text{EXP}(\lambda_i)$, the reliability of each component at time t is given by $R_i(t) = e^{-\lambda_i t}$. Equation (4.1) for the series system reliability, see also (3.60), becomes:

$$R_s(t) = e^{-\lambda_s t}, \text{ with } \lambda_s = \sum_{i=1}^{n} \lambda_i.\tag{4.6}$$

The series system of n components with time-independent failure rates behaves as a single component with time-independent failure rate given by the sum of the failure rates of the constituent components. From Eqs. (3.7) and (4.6), we see that the MTTF of a series system of time-independent failure rate components is given by:

$$\text{MTTF} = E[X] = \int_0^\infty R_s(t)dt = \frac{1}{\lambda_s} = \frac{1}{\sum_{i=1}^{n} \lambda_i}.\tag{4.7}$$

As already observed in Eq. (4.5), the MTTF of a series system is less than the MTTF of any of its components.

Example 4.1   *A Fluid Level Controller*

Consider a system used to maintain the fluid level in a tank at a constant value. The system is depicted in Figure 4.2 and is composed of four components (or blocks): a level sensor S, control logic L, a group motor/pump P, and a valve V. The components are connected in series since the correct operation of each component is needed to guarantee the correct functioning of the system. In order to compute the reliability of the system we need to know the reliability of its components. Often, the information that is available from a data bank or from specific life experiments is in the form of a (constant) failure rate, thus implicitly assuming an exponential failure time distribution.

Consulting a data bank, the following values were obtained (expressed in failures per hour or $f\,hr^{-1}$):

$$
\begin{aligned}
S &= \text{Level sensor} & \lambda_S &= 2 \times 10^{-6}\,f\,hr^{-1}, \\
L &= \text{Control logic} & \lambda_L &= 5 \times 10^{-6}\,f\,hr^{-1}, \\
P &= \text{Group motor/pump} & \lambda_P &= 2 \times 10^{-5}\,f\,hr^{-1}, \\
V &= \text{Valve} & \lambda_V &= 1 \times 10^{-5}\,f\,hr^{-1}.
\end{aligned}
$$

The reliability of each component after 1 year ($t = 8760\,hr$) of continuous operation is given by

$$R_S(t = 8760\,hr) = e^{-\lambda_S t} = 0.983,$$

$$R_L(t = 8760\,hr) = e^{-\lambda_L t} = 0.957,$$

$$R_P(t = 8760\,hr) = e^{-\lambda_P t} = 0.839,$$

$$R_V(t = 8760\,hr) = e^{-\lambda_V t} = 0.916,$$

and for the series system:

$$R_s(t = 8760\,hr) = e^{-\lambda_S t} e^{-\lambda_L t} e^{-\lambda_P t} e^{-\lambda_V t}$$

$$= 0.983 \cdot 0.957 \cdot 0.839 \cdot 0.916 = 0.723. \tag{4.8}$$

Alternatively, the result in Eq. (4.8) can be obtained by summing the failure rates of the constituent components as in Eq. (4.6):

$$\lambda_s = \lambda_S + \lambda_L + \lambda_P + \lambda_V = 3.7 \times 10^{-5}\,f\,hr^{-1},$$

$$R_s(t = 8760\,hr) = e^{-\lambda_s t} = 0.723.$$

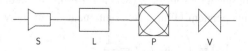

Figure 4.2   Series system for guaranteeing a constant level in a tank.

From (4.8) we can compute:

$$\text{MTTF}_s = \frac{1}{\lambda_S + \lambda_L + \lambda_P + \lambda_V} = \frac{1}{\lambda_s} = 27\,027\,\text{hr}.$$

If the components are independently repairable with constant repair rate μ_X ($X = $ S, L, P, V), the steady-state availability becomes:

$$A_{sys} = \frac{\mu_S}{\lambda_S + \mu_S} \frac{\mu_L}{\lambda_L + \mu_L} \frac{\mu_P}{\lambda_P + \mu_P} \frac{\mu_V}{\lambda_V + \mu_V}.$$

Components with Generally Distributed TTFs

Equation (4.2) can be applied to any system of n independent components with any lifetime distribution function. Suppose the TTF of the ith component is Weibull distributed (see Section 3.2.3) with the same value for the shape parameter β and different scale parameters η_i:

$$R_i(t) = e^{-\eta_i^\beta t^\beta} \text{ for } i = 1, 2, \ldots, n.$$

Then the system reliability $R(t)$ is given by

$$R(t) = \prod_{i=1}^{n} R_i(t) = e^{-t^\beta \left(\sum_{i=1}^{n} \eta_i^\beta \right)}.$$

Thus the system TTF is also Weibull distributed, with scale parameter $\eta = \left(\sum_{i=1}^{n} \eta_i^\beta \right)^{1/\beta}$ and shape parameter β.

Example 4.2 *A Fluid Level Controller: Weibull Distribution*

Suppose now that the component TTFs in Example 4.1 are Weibull distributed. We have:

$$R_S(t) = e^{-\eta_S^\beta_S t^{\beta_S}}, \ R_L(t) = e^{-\eta_L^{\beta_L} t^{\beta_L}}, \ R_P(t) = e^{-\eta_P^{\beta_P} t^{\beta_P}}, \ R_V(t) = e^{-\eta_V^{\beta_V} t^{\beta_V}},$$

and

$$R_s(t) = e^{-\eta_S^{\beta_S} t^{\beta_S}} e^{-\eta_L^{\beta_L} t^{\beta_L}} e^{-\eta_P^{\beta_P} t^{\beta_P}} e^{-\eta_V^{\beta_V} t^{\beta_V}}.$$

Example 4.3 As a concrete example [2], consider a liquid-cooling cartridge system that is used in enterprise-class servers designed by Sun Microsystems. The series system consists of a blower, a water pump, and a compressor. Table 4.2 gives the Weibull data for the three components.

Table 4.2 Weibull data for a server cooling system.

Component	L10 (hr)	Shape parameter (β)
Blower	70 000	3.0
Water pump	100 000	3.0
Compressor	100 000	3.0

L10 is the *10th percentile* of the component lifetime distribution, which is the time at which 10% of the components are expected to have failed or, equivalently, $R(L10) = 0.9$. The system reliability expression for the blower can be derived as

$$R_B(L10) = e^{-\eta_B^3 \cdot 70\,000^3} = 0.9,$$

which gives $\eta_B^3 = 3.0717 \times 10^{-16}$. Similarly, for the water pump and compressor we can find $\eta_W^3 = \eta_C^3 = 1.0536 \times 10^{-16}$. Hence, $\eta_B^3 + \eta_W^3 + \eta_C^3 = 5.1789 \times 10^{-16}$. The system reliability expression can now be written as

$$R(t) = e^{-5.1789 \cdot 10^{-16} t^3} = e^{-(8.0305 \cdot 10^{-6} t)^3},$$

with $\eta = 8.0305 \times 10^{-6}$ and $\beta = 3$. From Eq. (3.23) the MTTF becomes:

$$E[X] = \text{MTTF} = \frac{1}{\eta} \Gamma\left(1 + \frac{1}{\beta}\right) = \frac{1}{8.0305 \times 10^{-6}} \Gamma\left(1 + \frac{1}{3}\right) = 111\,197.66\,\text{hr}.$$

4.1.2   Importance (Sensitivity) Analysis of Series Systems

Importance analysis is the term usually adopted in dependability to indicate the sensitivity of a system measure on the variation of the component's features (see Section 2.2.7). Components, in general, do not each have the same importance in determining the reliability of the whole system. Importance analysis is aimed at quantifying the importance of a component in determining the reliability of a system by determining how the improvement of the reliability of a single component influences the reliability of the whole system. The importance of the *i*th component is computed as the derivative of the system reliability with respect to the component's reliability. This index is also called the *Birnbaum importance index* after [3]. The importance index is defined as $I_i = \partial R_s / \partial R_i$. From Eq. (4.1), we obtain [4]:

$$I_i = \frac{\partial R_s}{\partial R_i} = \frac{R_s}{R_i} = \prod_{j \neq i} R_j. \tag{4.9}$$

Note that the importance index of component *i* does not depend on its reliability R_i. The maximum gain in system reliability is obtained by improving the component with the highest Birnbaum index, which, from Eq. (4.9), is the one with the lowest reliability; hence, the optimal strategy consists of incrementing the reliability of the least reliable component (this strategy also tends to be optimal from the cost point of view: improving a less reliable component is usually less costly than improving a highly reliable component).

Example 4.4 *A Fluid Level Controller (continued)*

In the system of Example 4.1, assume that it is possible to improve the reliability either of the pressure sensor S or of the pump P by an absolute increment $\Delta R_i = 0.015$. If the increment is applied to the sensor, the system reliability becomes:

$$R_s^*(t = 8760\,\text{hr}) = (0.983 + 0.015) \cdot 0.957 \cdot 0.839 \cdot 0.916 = 0.734.$$

If the same increment is applied to the pump, the system reliability becomes:

$$R_s^{**}(t = 8760\,\text{hr}) = 0.983 \cdot 0.957 \cdot (0.839 + 0.015) \cdot 0.916 = 0.736.$$

In agreement with Eq. (4.9), the computation shows that $R_s^*(t) < R_s^{**}(t)$.

4.1.3   The Parts Count Method

This method refers to systems composed of a large number of components with constant failure rate and logically connected in series, in the sense that each component is needed for correct system operation. Sometimes, the large number of components can be grouped in families of identical parts. This is typical for the case of electronic equipment, where components belong to a restricted number of families and a unit may have various components belonging to the same family. Moreover, for electronic equipment there is experimental evidence that the constant failure rate is usually a satisfactory assumption.

Let n be the number of different component families identified in the system, n_i the number of components belonging to family i and λ_i the failure rate of each component in family i. Thus n_i is the *parts count* of components of family i, while the total number of components is $\sum_{i=1}^{n} n_i$.

The total failure rate λ_s of the system is then:

$$\lambda_s = \sum_{i=1}^{n} n_i \lambda_i, \tag{4.10}$$

from which the system reliability $R_s(t)$ can be calculated using Eq. (4.6).

This method, suggested for instance by reliability handbooks for electronic components like MIL-HDBK-217 [5], consists in counting how many components of each family are present in the system and applying formula (4.10).

Example 4.5   The parts count method is usually applied for computing the reliability of electronic equipment composed of boards with a large number of components. To apply the method a table is prepared for each board, according to the example of Table 4.3. The columns in Table 4.3 report:

- description and identification of the component family;
- number of components (n_i) of the same family i;
- failure rate (λ_i) of a component of family i;

Table 4.3 Structuring a table for the parts count method.

Component description	Number of components n_i	Individual failure rate λ_i	Group failure rate $n_i \cdot \lambda_i$	Incidence (%)
Film resistor	51	9.01×10^{-9}	4.59×10^{-7}	27.4
Wire-wound resistor	8	4.05×10^{-8}	3.24×10^{-7}	19.3
Electrolytic capacitor	3	1.05×10^{-7}	3.15×10^{-7}	18.8
Transistor Si NPN	2	1.06×10^{-7}	2.12×10^{-7}	12.7
Ceramic capacitor	6	2.33×10^{-8}	1.40×10^{-7}	8.3
Printed board	1	7.17×10^{-8}	7.17×10^{-8}	4.3
IC, linear, bipolar	2	3.29×10^{-8}	6.58×10^{-8}	3.9
Zener diode	2	1.01×10^{-8}	2.03×10^{-8}	1.2
IC, linear, bipolar	1	1.95×10^{-8}	1.95×10^{-8}	1.2
IC, linear, bipolar	2	1.87×10^{-8}	1.87×10^{-8}	1.1
IC, digital, CMOS (SSI)	1	1.49×10^{-8}	1.49×10^{-8}	0.9
Connectors	1	8.16×10^{-9}	8.16×10^{-9}	0.5
Si diode	6	1.04×10^{-9}	6.22×10^{-9}	0.4
Connectors	1	1.77×10^{-9}	1.77×10^{-9}	0.1
Total (λ_s)			1.68×10^{-6}	100

- failure rate ($n_i \cdot \lambda_i$) for the whole group of components of family i;
- the contribution (as a percentage) of components of family i to the failure rate of the whole board.

To help the analyst, the component groups can be listed in the table in decreasing order of importance (last column), thus highlighting the most critical groups. The sum of the values of the fourth column ($\lambda_s = 1.68 \times 10^{-6}$ in the case of Table 4.3) provides the failure rate of the whole board according to Eq. (4.10).

In the printed circuit board analyzed in Table 4.3, the group labeled "Film resistor," which has 51 components, contributes 27.4% to the failure rate of the whole board.

The parts count method can easily be computerized and automatically connected to a reliability database or a reliability handbook. From the component description, the failure rate values can be retrieved from the database to fill a form similar to Table 4.3. In this way, the procedure is accelerated and the possibility of making transcription errors is reduced or avoided.

4.1.4   Other Measures for Series Systems

If the events E_i in Eq. (4.1) have different meanings as illustrated in the introduction to Part II, we obtain different measures for the series system.

Series System: Availability

Assume that the components in a series system are repairable, and that their failure and repair processes are independent. Let E_i represent the event that the (repairable) component i is up at time t. Hence, $P\{E_i\} = A_i(t)$ is the component's instantaneous availability at time t. In this case, denoting by $A_s(t)$ the instantaneous availability of the series system, we have, from Eq. (4.1),

$$A_s(t) = \prod_{i=1}^{n} A_i(t).$$ (4.11)

In the limit as $t \to \infty$, $P\{E_i\} = A_i$ is the component steady-state availability, and we have, from (4.1),

$$A_s = \prod_{i=1}^{n} A_i,$$ (4.12)

where A_s is the steady-state availability of the system.

Example 4.6 *Availability of Hardware and Software Systems*
A usual approximation in systems that have both a hardware component and a software component is to consider the two components as separate blocks, as depicted in Figure 4.3.

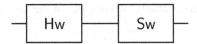

Figure 4.3 An RBD for a hardware and software system.

Assume that λ_h is the failure rate of the hardware component, λ_s the failure rate of the software component, μ_h the repair rate of the hardware component and μ_s the repair (presumably by reboot) rate of the software component. The system steady-state availability A_{RBD} is then:

$$A_{RBD} = \frac{\mu_h}{(\lambda_h + \mu_h)} \frac{\mu_s}{(\lambda_s + \mu_s)}.$$ (4.13)

We note that the independence assumption behind Eq. (4.13) implies that the software may incur cycles of failure/repair even if the hardware is down, and vice versa. Later, in Example 9.7, we will show how to remove this assumption.

Example 4.7 *Availability of the Cisco Router GSR 12000*
A Cisco router is modeled by the RBD shown in Figure 4.4 [6], where each block in the RBD represents a subsystem. Each of these subsystems will be separately modeled in Example 18.1 and their steady-state availabilities computed.

Figure 4.4 Reliability block diagram of Cisco 12000 GSR.

The system steady-state availability is computed by multiplying the individual subsystem availabilities.

Series System: Interval Reliability

In the case where each individual component in a series system is independently repairable, let E_i represent the event that the (repairable) component i is functioning at time t and then remains continuously up in the interval $(t, t+x]$. $P\{E_i\} = IR_i(t,x)$ is the component's interval reliability. In this case, denoting by $IR_s(t,x)$ the interval reliability of the series system, we have, from Eq. (4.1),

$$IR_s(t,x) = \prod_{i=1}^{n} IR_i(t,x).$$

(4.14)

Example 4.8 *Computer System: Interval Reliability*

For a job to be successfully executed on a computer system, the CPU, main memory, and cache all need to be functional. Let λ_C, λ_M and λ_h be their respective failure rates, and let μ_C, μ_M, and μ_h be their respective repair rates. Assuming independence of failure and repair events, the interval reliability of the CPU in the interval $(t+x]$ is written, from Eq. (3.127), as

$$IR_C(t,x) = \left[\frac{\mu_C}{(\lambda_C + \mu_C)} + \frac{\lambda_C}{(\lambda_C + \mu_C)} e^{-(\lambda_C + \mu_C)t} \right] e^{-\lambda_C x}.$$

Similar expressions hold for the main memory and the cache:

$$IR_M(t,x) = \left[\frac{\mu_M}{(\lambda_M + \mu_M)} + \frac{\lambda_M}{(\lambda_M + \mu_M)} e^{-(\lambda_M + \mu_M)t} \right] e^{-\lambda_M x},$$

$$IR_h(t,x) = \left[\frac{\mu_h}{(\lambda_h + \mu_h)} + \frac{\lambda_h}{(\lambda_h + \mu_h)} e^{-(\lambda_h + \mu_h)t} \right] e^{-\lambda_h x},$$

and hence we can write down the job's overall interval reliability using (4.14):

$$IR_s(t,x) = IR_C(t,x) IR_M(t,x) IR_h(t,x).$$

(4.15)

Example 4.9 *Web Service Request*

In the previous example, all the resources were simultaneously needed for the execution of a job. There are other situations where the overall job execution is divided into a sequence of phases so that the duration x for which the resource i is needed is resource dependent and the time t at which resource i is needed is also dependent on i.

Consider the processing of a web service request that needs the web server ($i = 1$) and the database server ($i = 2$) in sequence. The interval reliability for the webserver is then:

$$IR_1(t_1, x_1) = \left[\frac{\mu_1}{(\lambda_1 + \mu_1)} + \frac{\lambda_1}{(\lambda_1 + \mu_1)} e^{-(\lambda_1 + \mu_1)t_1} \right] e^{-\lambda_1 x_1}.$$

Once the execution on the web server is successful, at time $t_1 + x_1$, the execution on the database server can begin. Hence the interval reliability for the second phase of execution is:

$$IR_2(t_1 + x_1, x_2) = \left[\frac{\mu_2}{(\lambda_2 + \mu_2)} + \frac{\lambda_2}{(\lambda_2 + \mu_2)} e^{-(\lambda_2 + \mu_2)(t_1 + x_1)} \right] e^{-\lambda_2 x_2}.$$

The overall interval reliability of the web service request is then the product of the above two interval reliabilities:

$$IR_{\text{web request}}(t, x) = IR_1(t_1, x_1) IR_2(t_1 + x_1, x_2). \tag{4.16}$$

For a detailed study of task completion time with multiple execution steps, see [7].

4.2 Systems with Redundancy

When the reliability of a series system does not reach the design goals and improvement of the constituent components is not feasible (more reliable parts are not available or are too costly), it becomes necessary to act at the structure level and resort to fault-tolerant configurations by incrementing the number of available resources. A system configuration is said to be redundant when the occurrence of a component failure does not necessarily cause a system failure. Various redundant structures have been studied and applied in practice, and they will be illustrated in the following sections.

4.2.1 Parallel Redundancy

The conceptually simplest redundant configuration is parallel redundancy. This consists of a replication scheme with multiple units so that correct operation of at least one of the replicated units is sufficient to guarantee the correct operation of the system. Such parallel redundant systems are sometimes referred to as active/active or hot standby redundancy. Note that the replicated components can either be statistically identical or different. First, consider a system consisting of two independent components such that it will fail only if both components fail. The RBD of the system is shown in Figure 4.5(a).

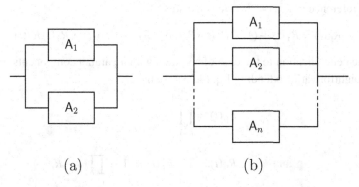

(a)　　　　　　　　(b)

Figure 4.5　Parallel systems: (a) with two components; (b) with n components.

Define, as usual, the event E_i = "the component i ($i = 1, 2$) is functioning properly," and the event $\overline{E_i}$ = "the component i has failed." Similarly, define the event E_p = "the parallel system is functioning properly," and the event $\overline{E_p}$ = "the parallel system has failed." To establish a relation between E_p and the E_i events, it is easier to consider the complementary events. Thus:

$$\overline{E_p} = \text{"the parallel system has failed"} = \overline{E_1} \cap \overline{E_2}. \tag{4.17}$$

Therefore, assuming independence of failure events,

$$P\{\overline{E_p}\} = P\{\overline{E_1} \cap \overline{E_2}\} = P\{\overline{E_1}\} P\{\overline{E_2}\}. \tag{4.18}$$

Applying De Morgan's rule to Eq. (4.17), we can also write:

$$E_p = E_1 \cup E_2 = E_1 \cup (\overline{E_1} \cap E_2), \tag{4.19}$$

where the last term in (4.19) is in the form of the union of disjoint terms (see also Section 6.12). Consequently:

$$P\{E_p\} = P\{E_1 \cup E_2\} = P\{E_1\} + P\{E_2\} - P\{E_1\} P\{E_2\}$$
$$= P\{E_1\} + P\{\overline{E_1} \cap E_2\} = P\{E_1\} + (1 - P\{E_1\}) P\{E_2\}. \tag{4.20}$$

4.2.2   Reliability of a Parallel System

If event E_i represents that component i has been continuously up in the interval $(0, t]$, then $P\{E_i\} = R_i(t)$ is the component reliability at time t and $P\{\overline{E_i}\} = F_i(t) = 1 - R_i(t)$ is the component unreliability. In this case, denoting by $R_p(t)$ ($F_p(t)$) the parallel system reliability (unreliability), we have, from Eq. (4.18),

$$F_p(t) = \prod_{i=1}^{2} F_i(t) = \prod_{i=1}^{2} (1 - R_i(t))$$

and
$$R_p(t) = 1 - F_p(t) \tag{4.21}$$
$$= 1 - (1 - R_1(t))(1 - R_2(t)) \tag{4.22}$$
$$= R_1(t) + R_2(t) - R_1(t) R_2(t).$$

With reference to Eq. (4.20), we can also write:

$$R_p(t) = R_1(t) + (1 - R_1(t)) R_2(t) = R_1(t) + R_2(t) - R_1(t) R_2(t). \tag{4.23}$$

The generalization to the case of a system with n parallel components, Figure 4.5(b), is straightforward. Extending Eq. (4.23), we have:

$$F_p(t) = \prod_{i=1}^{n} F_i(t)$$

and
$$R_p(t) = 1 - F_p(t) = 1 - \prod_{i=1}^{n} (1 - R_i(t)). \tag{4.24}$$

Table 4.4 Reliability of a parallel system with n components each with reliability 0.95.

Number of components, n	System reliability, $R_p(t)$	Reliability increment
1	0.950 000 000 0000	—
2	0.997 500 000 0000	5.000×10^{-2}
3	0.999 875 000 0000	2.381×10^{-3}
4	0.999 993 750 0000	1.188×10^{-4}
5	0.999 999 687 5000	5.938×10^{-6}
10	0.999 999 999 9999	1.855×10^{-12}

Thus, for a parallel system of independent components, we have the *product law of unreliabilities* dual to the product law of reliabilities for series systems. It follows that the reliability of a parallel system is greater than the reliability of any one of the constituent components and, hence, a parallel system is more reliable than its most reliable component.

If we increase the number of parallel components the system reliability increases, as shown in Table 4.4, where the first column indicates the number of statistically identical parallel components (each one with a reliability $R_i(t) = 0.95$), and the second column lists the parallel system reliability. The third column indicates the relative increment in the system reliability by augmenting the number of parallel components by one.

Looking at Table 4.4 one should be aware of a *law of diminishing returns*, according to which the rate of increase in reliability with each additional component decreases rapidly as n increases.

As an alternative derivation of Eq. (4.24), consider a parallel system of n independent components, where X_i denotes the lifetime of component i and X_p denotes the lifetime of the system. Then

$$X_p = \max\{X_1, X_2, \ldots, X_n\}. \tag{4.25}$$

From Section 3.3.2, we can write:

$$(1 - R_{X_p}(t)) = \prod_{i=1}^{n}(1 - R_{X_i}(t)), \tag{4.26}$$

and hence,

$$R_{X_p}(t) = 1 - \prod_{i=1}^{n}[1 - R_{X_i}(t)] \geq 1 - [1 - R_{X_i}(t)], \quad \text{for all } i, \tag{4.27}$$

which implies that the reliability of a parallel redundant system is larger than that of any of its components. Further,

$$E[X_p] = \int_0^\infty R_{X_p}(t)dt \geq \max_i \left\{ \int_0^\infty R_{X_i}(t)dt \right\}$$

$$= \max_i\{E[X_i]\}. \tag{4.28}$$

The mean lifetime of the parallel system is greater than the mean lifetime of any of its components.

4.2.3   Parallel System: The Exponential Case

If the TTF distribution of the ith component is EXP(λ_i), Eq. (4.24) becomes

$$R_{\mathrm{p}}(t) = 1 - \prod_{i=1}^{n}(1 - e^{-\lambda_i t}). \tag{4.29}$$

In the case of a parallel system with two independent components:

$$R_{\mathrm{p}}(t) = e^{-\lambda_1 t} + e^{-\lambda_2 t} - e^{-(\lambda_1 + \lambda_2)t},$$

$$\mathrm{MTTF} = \int_0^\infty R_{\mathrm{p}}(t)\, dt = \frac{1}{\lambda_1} + \frac{1}{\lambda_2} - \frac{1}{\lambda_1 + \lambda_2}. \tag{4.30}$$

If the two components are i.i.d. with distribution EXP(λ), the two equations in (4.30) become:

$$R_{\mathrm{p}}(t) = 2e^{-\lambda t} + e^{-2\lambda_2 t},$$

$$\mathrm{MTTF} = \frac{3}{2\lambda}.$$

Problems

4.1 Consider a parallel system with n statistically identical components with exponentially distributed lifetimes each with a failure rate λ. Write down an expression for the system reliability in this case. Next show that system MTTF is given by:

$$E[X_{\mathrm{p}}] = \frac{1}{\lambda}\sum_{i=1}^{n}\frac{1}{i} = \frac{H_n}{\lambda} \simeq \frac{\ln(n) + C}{\lambda}, \tag{4.31}$$

where $H_n = \sum_{i=1}^{n}\frac{1}{i}$ is the sum of the first n terms of the harmonic series, and $C\ (= 0.577)$ is Euler's constant.

4.2 Continuing with the previous problem, plot the expected life of a parallel system of independent components as a function of n. Assume that $\lambda = 0.00001\,\mathrm{f\,hr}^{-1}$.

4.3 Consider a three-component parallel redundant system where the time to failure distribution of the ith component is EXP(λ_i). Derive an expression for the system MTTF assuming independence.

4.2.4   Importance (Sensitivity) Analysis of Parallel Systems

In a parallel system, not all the components may have the same importance in determining the overall system reliability. We define the *importance index* I_i of component i as the sensitivity of the system reliability with respect to the reliability R_i of component i [4]:

$$I_i = \frac{\partial R_{\mathrm{p}}}{\partial R_i} = \frac{1 - R_{\mathrm{p}}}{1 - R_i}. \tag{4.32}$$

The highest importance is assigned to the most reliable component. Hence, in order to improve the reliability of a parallel system the optimal strategy consists in improving the most reliable component (note that this action may not be convenient from the point of view of the balance of cost vs. reliability).

Example 4.10 A system is composed of two components A and B in a parallel configuration with reliabilities $R_A = 0.50$ and $R_B = 0.85$, respectively. From Eq. (4.23),

$$R_p = 0.50 + 0.85 - (0.50 \cdot 0.85) = 0.925.$$

Assume that it is possible to acquire components with the reliability incremented by a value $\Delta R_i = 0.10$. If the improved component B is utilized, we have:

$$R_p^* = 0.50 + (0.85 + 0.10) - [0.50 \cdot (0.85 + 0.10)] = 0.975.$$

If, instead, the improved component A is utilized, we have:

$$R_p^{**} = (0.50 + 0.10) + 0.85 - [(0.50 + 0.10) \cdot 0.85] = 0.940.$$

We see that $R_p^* > R_p^{**}$. The reliability increment assigned to the more reliable component produces a higher increment in the overall system reliability.

4.2.5 Parallel Systems: Availability

As in Section 4.2.2, if each component in the parallel system fails independently and is repaired independently, the probability that a component is up at time t is the component availability $P\{E_i\} = A_i(t)$, and Eq. (4.24) applies to the component availabilities (unavailabilities).

4.2.6 Series-Parallel Systems

Equations (4.2) and (4.24) for the reliability computation of series and parallel systems can be used to compute the reliability of any series-parallel system [4]. The computational procedure consists of a progressive reduction of the system structure by replacing blocks of components in series with a single equivalent block with reliability given by Eq. (4.2), and blocks of components in parallel with a single equivalent block with reliability given by Eq. (4.24).

Example 4.11 Consider the series-parallel system of Figure 4.6(a). Let R_A, R_B, R_C, R_D, and R_E be the reliabilities of components A, B, C, D, and E, respectively.

At the first level of reduction, Figure 4.6(b), the parallel block formed by components A and B is replaced by a single equivalent block U with reliability $R_U = R_A + R_B - R_A R_B$, and the series block formed by components D and E is replaced by a single equivalent block V with reliability $R_V = R_D R_E$.

At the second level of reduction, the series connection between U and C is replaced by a single equivalent block W with reliability $R_W = R_U R_C$. Finally, at the last step,

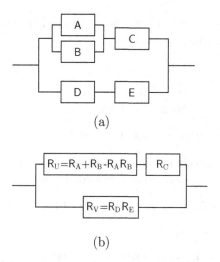

(a)

(b)

Figure 4.6 A series-parallel system with five components.

the parallel connection between V and W is reduced to obtain the system reliability $R_s = R_V + R_W - R_V R_W$. Following this procedure, we have

$$R_s = (R_A + R_B - R_A R_B) R_C + R_D R_E - (R_A + R_B - R_A R_B) R_C R_D R_E$$
$$= R_A R_C + R_B R_C + R_D R_E - R_A R_B R_C - R_A R_C R_D R_E -$$
$$- R_B R_C R_D R_E + R_A R_B R_C R_D R_E. \tag{4.33}$$

If we assume that all the components have the same reliability ($R_A = R_B = R_C = R_D = R_E = R$), Eq. (4.33) becomes:

$$R_s = 3R^2 - R^3 - 2R^4 + R^5. \tag{4.34}$$

Example 4.12 *A Fluid Level Controller with Redundancy*
Consider again the system of Example 4.1. If the reliability computed for the series system does not meet the required specification, a possible design alternative is to duplicate the chain as represented in Figure 4.7(a). In order to calculate the system

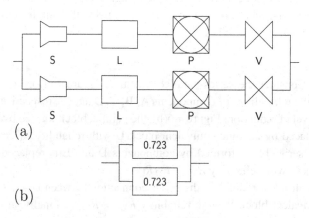

(a)

0.723

0.723

(b)

Figure 4.7 Parallel system with two level control lines.

reliability over one year (= 8760 hr) of continuous operation, it is possible to reduce each parallel line to a single equivalent component, using the results of Eq. (4.8) [Figure 4.7(b)]. Applying Eq. (4.23),

$$R_p(t = 8760\,\text{hr}) = 2 \cdot 0.723 - (0.723)^2 = 0.923.$$

The reliability improvement thus gained must be balanced against the increased cost of the employed redundancy.

A common case of series-parallel systems is that of n serial stages where stage i consists of n_i statistically identical components in parallel [8]. Let the reliability of each component at stage i be R_i. Assuming that all the components are independent, the system reliability R_{sp} can be computed from the formula

$$R_{sp} = \prod_{i=1}^{n} [1 - (1 - R_i)^{n_i}]. \tag{4.35}$$

Partial Parallel Redundancy
The system designer can decide the level of redundancy used for each component in a series-parallel system. That is, the values of n_i in Eq. (4.35) can be chosen to balance reliability and cost. It is thus possible to implement only partial redundant configurations, in which only selected components are replicated.

Example 4.13 *A Fluid Level Controller with Redundancy*
With reference to the level control system discussed in Examples 4.1 and 4.12, it is possible to realize a configuration in which only one component is replicated at a time. If only the sensor S is replicated we obtain the configuration shown in Figure 4.8(a). The standard series-parallel reduction in Figure 4.8(b) provides the reliability value $R_s^*(t = 8760\,\text{hr}) = 0.735$.

By replicating the pump component P (Figure 4.9), the following reliability value is obtained: $R_s^{**}(t = 8760\,\text{hr}) = 0.839$. Replicating component P is more effective than

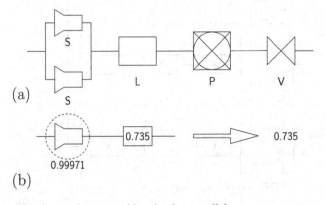

(a)

(b)

Figure 4.8   A level control system with redundant parallel sensor.

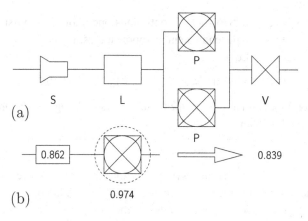

Figure 4.9   A level control with redundant parallel pump.

replicating S. This result could have been predicted from the discussion in Section 4.1.2, since the pump is the least reliable component of the chain while the sensor is the most reliable one.

Example 4.14 *Multi-Voltage Propulsion System of a High-Speed Train [9]*
A multi-voltage propulsion system designed for the Italian high-speed railway system [9] consists of three equivalent modules in a parallel redundant configuration. Each module is modeled as a series of four blocks (transformer T, filter F, inverter I, and motor M) and two parallel converters (C_1 and C_2) that feed the induction motors. The RBD of the system is given in Figure 4.10. In [9], constant failure rates are assumed, with the values reported in Table 4.5.

Denoting by $\lambda = \lambda_T + \lambda_F + \lambda_I + \lambda_M$ the sum of the failure rates of the series components, and by γ the failure rate of a single converter, the reliability of a single

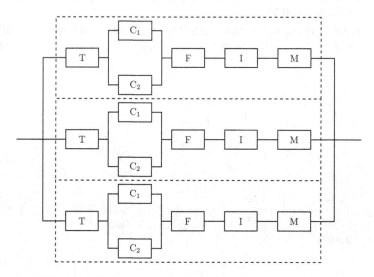

Figure 4.10   Reliability block diagram of a train propulsion system.

Table 4.5 Component failure rates for a
train propulsion system.

Component		Failure rate $(\mathrm{f\,hr^{-1}})$
Transformer	$\lambda_T =$	2.2×10^{-6}
Filter	$\lambda_F =$	0.4×10^{-6}
Inverter	$\lambda_I =$	3.8×10^{-5}
Motors	$\lambda_M =$	3.2×10^{-5}
Converter	$\gamma =$	2.8×10^{-5}

module i ($i = 1, 2, 3$) becomes

$$R_i(t) = R_T(t)R_F(t)R_I(t)R_M(t)(2R_C(t) - R_C^2(t)) = 2\,e^{-(\gamma+\lambda)t} - e^{-(2\gamma+\lambda)t}, \quad (4.36)$$

and the reliability of the system

$$R(t) = 1 - (1 - R_i^3(t)).$$

Example 4.15 *Base Repeater in a Cellular Communication System*
Consider a base repeater in a cellular communication system with two control channels
and three voice channels [2].

Assume that the system is up as long as at least one control channel and at least one
voice channel is functioning. The reliability block diagram for this problem is shown
in Figure 4.11. Let R_c be the reliability of a control channel and R_v the reliability of a
voice channel. The reliability of the system is then given by:

$$R = (1 - (1 - R_c)^2) \cdot (1 - (1 - R_v)^3) = (1 - F_c^2)(1 - F_v^3) = 1 - F_c^2 - F_v^3 + F_c^2 F_v^3.$$

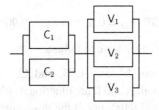

Figure 4.11   Reliability block diagram of a base repeater.

Example 4.16 *Storage Module for a Sun Microsystems High-Availability Platform*
Telecommunication systems require very high levels of availability. In the Sun
carrier-grade platform the storage system is composed of dual SCSI chains for internal
and external disks [10]. For the purpose of availability computation, the storage system
is modeled as the block diagram shown in Figure 4.12, where the SCSI block is split
into three parts. That part denoted SCSI$_S$, whose failure brings both disks down, is put
in series while the parts that only bring their own disk down, denoted SCSI$_1$ and SCSI$_2$,
are put in parallel. The drive blocks, Drive$_1$ and Drive$_2$, are modeled as a single block

Table 4.6 Failure and repair rate and steady-state availability for the blocks of Figure 4.12.

	SCSI$_S$	SCSI$_{1,2}$	Drive
Failure rate ($\lambda\,hr^{-1}$)	1.33542×10^{-7}	6.67710×10^{-8}	0.000001000
Repair rate ($\mu\,hr^{-1}$)	0.4	4	0.115384615
Steady-state availability	0.999999967	0.999999980	0.999991330

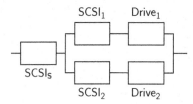

Figure 4.12   Reliability block diagram M_{SCSI} of the disk subsystem of a carrier-grade platform (from [10]).

in series with their SCSI. Since this model will be used later in this book, we name it M_{SCSI}.

The failure and repair rates for the different units are reported in the first two rows of Table 4.6 [10]. Applying Eq. (3.110), the steady-state availabilities for the different blocks are reported in the last row of Table 4.6. Then, the availability of the dual chain becomes:

$$A_{dual} = A_{SCSI_S}(2A_{SCSI_1}A_{Drive} - (A_{SCSI_1}A_{Drive})^2) = 0.999999967. \qquad (4.37)$$

Problems

4.4 Compute the reliability of an RBD with five serial stages of parallel components, with $n_1 = n_5 = 1$, $n_2 = n_4 = 2$ and $n_3 = 3$. Also,

$$R_A = 0.95,\ R_B = 0.80,\ R_C = 0.70,\ R_D = 0.85 \text{ and } R_E = 0.90.$$

4.5 For Example 4.14, compute the system mean time to failure.

4.6 A water supply system has three pumps labeled A, B, and C. The system is operational if either pump C is functional or both A and B are functional. Write down the expression for system reliability in terms of the reliability of the three pumps assuming independence. Next, assume that the times to failure are exponentially distributed with respective parameters λ_A, λ_B, and λ_C, and write down expressions for the system reliability and system MTTF.

4.7 For the water supply system of Problem 4.6, write down the expression for the system reliability assuming that the times to failure are Weibull distributed with respective scale parameters η_A, η_B, and η_C, and identical shape parameter β. Thence, write down an expression for the system MTTF.

4.8 For the water supply system of Problem 4.6, assume that the three pumps are repairable; the TTFs are ER$_2$ distributed with parameters λ_A, λ_B, and λ_C, respectively, and the TTRs are HYPER$_2$ with parameters $(\mu_{A1}, \mu_{A2}, q_A)$, $(\mu_{B1}, \mu_{B2}, q_B)$ and $(\mu_{C1}, \mu_{C2}, q_C)$, respectively. Find the steady-state availability of the system.

4.9 For the base repeater of Figure 4.11, assume now that the system is up as long as both control channels are up and all the three voice channels are up. Find the reliability of the system if R_c is the reliability of a single control channel and R_v the reliability of a single voice channel.

4.2.7 Importance (Sensitivity) Analysis of Series-Parallel Systems

The importance or sensitivity measure of a series-parallel system can be obtained during the series-parallel reduction procedure, avoiding any extra computation [4]. During a series reduction step the formula in (4.9) is applied, and during a parallel reduction step the formula in (4.32) is applied.

Example 4.17 With reference to the system of Figure 4.6, the derivative of the system reliability with respect to the various component reliabilities are calculated during the reduction procedure for computing the system reliability [4], and are given by:

$$I_A = \frac{\partial R_s}{\partial R_W} \frac{\partial R_W}{\partial R_U} \frac{\partial R_U}{\partial R_A} = \frac{1 - R_s}{1 - R_W} \frac{R_W}{R_U} \frac{1 - R_U}{1 - R_A},$$

$$I_B = \frac{\partial R_s}{\partial R_W} \frac{\partial R_W}{\partial R_U} \frac{\partial R_U}{\partial R_B} = \frac{1 - R_s}{1 - R_W} \frac{R_W}{R_U} \frac{1 - R_U}{1 - R_B},$$

$$I_C = \frac{\partial R_s}{\partial R_W} \frac{\partial R_W}{\partial R_C} = \frac{1 - R_s}{1 - R_W} \frac{R_W}{R_C},$$

$$I_D = \frac{\partial R_s}{\partial R_V} \frac{\partial R_V}{\partial R_D} = \frac{1 - R_s}{1 - R_V} \frac{R_V}{R_D},$$

$$I_E = \frac{\partial R_s}{\partial R_V} \frac{\partial R_W}{\partial R_E} = \frac{1 - R_s}{1 - R_V} \frac{R_V}{R_E}.$$

Problems

4.10 For the base repeater in Example 4.15, derive expressions of the importance of system reliability with respect to R_c and R_v.

4.11 Continuing with the base repeater of Example 4.15, now assume that the times to failure of each channel are exponentially distributed. Derive expressions of the sensitivity of the system reliability with respect to the respective failure rates of the control and voice channels, λ_c and λ_v.

4.12 For the base repeater in Example 4.15, derive expressions of the importance of system steady-state availability with respect to the respective steady-state availabilities A_c and A_v. Thence, derive expressions for the sensitivities of the system steady-state availability with respect to the respective failure and repair rates of the two types of channels: λ_c, λ_v, μ_c, and μ_v.

4.2.8   System Redundancy vs. Component Redundancy

We further explore the effectiveness of the system configuration with respect to the system reliability. Consider a system composed of two series components A and B [Figure 4.13(a)]. This configuration is denoted as configuration (a), and its reliability is denoted by R_a. According to Eq. (4.2), $R_a = R_A R_B$.

If we decide to improve the reliability of the system by applying redundancy using one single replica for each component, two solutions are possible. In Figure 4.13(b) the complete line is replicated (*system redundancy*), while in Figure 4.13(c) each component is individually replicated (*component redundancy*). By indicating with R_b the reliability of configuration (b) (system redundancy), and with R_c the reliability of configuration (c) (component redundancy), we obtain

$$R_a = R_A R_B,$$

$$R_b = 2R_A R_B - (R_A R_B)^2$$
$$= R_A R_B (2 - R_A R_B), \tag{4.38}$$

$$R_c = (2R_A - R_A^2)(2R_B - R_B^2)$$
$$= R_A R_B [4 - 2(R_A + R_B) + R_A R_B].$$

It is easy to see that $R_b > R_a$ and $R_c > R_a$; both redundant configurations are more reliable than the original system. Next we compare the two redundant configurations (b) and (c) [11]:

$$\frac{R_c}{R_b} = \frac{4 - 2(R_A + R_B) + R_A R_B}{2 - R_A R_B} = 1 + \frac{2(1 - R_A)(1 - R_B)}{2 - R_A R_B} > 1. \tag{4.39}$$

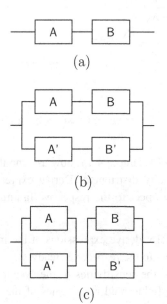

Figure 4.13   (a) The original system; (b) System redundancy; (c) Component redundancy.

Independently of the value of the reliability of each single block, configuration (c) (component-level redundancy) is always more reliable than configuration (b) (system-level redundancy). It should, however, be noted that configuration (c) is more complex than configuration (b) since, in order to exploit the redundancy, any component of type A needs to be possibly connected with any component of type B. This higher complexity might require additional control logic (not considered in the formulas) that might reduce the benefits indicated by the formulas. The effect of the additional complexity is further elaborated in Example 4.23.

The reason why configuration (c) is more reliable than configuration (b) can also be explained on a qualitative and intuitive basis, noticing that there are failure combinations of basic blocks, like $A'B$ and AB', that cause failure of configuration (b) but not of configuration (c).

Example 4.18   *A Fluid Level Controller with Redundancy*
Consider again the fluid level control system of Example 4.12 and identify as block A the series subsystem consisting of sensor and control logic, and as block B the series subsystem consisting of pump and valve. The component redundant configuration is represented in Figure 4.14. From Eq. (4.39), the reliability becomes $R_c(t = 8760\,\text{hr}) = 0.942$, which is greater than that calculated for the system redundant configuration examined in Example 4.12.

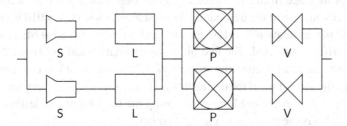

Figure 4.14   Component redundant configuration for the level control system.

Example 4.19   *Shuffle-Exchange Network*
An $N \times N$ shuffle-exchange network (SEN) has $N = 2^n$ inputs and N outputs and is formed by n stages, where each stage has $N/2$ switching elements (SEs) [12]. Assume that the stages are labeled from 1 to n and the SEs at each stage are labeled from 0 to $N/2 - 1$. An SE is a 2×2 circuit as shown in Figure 4.15 that can perform two operations. It can either transmit the input directly to the output (T) as in Figure 4.15(a), or it can exchange the inputs (X) as in Figure 4.15(b).

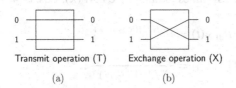

Transmit operation (T)   Exchange operation (X)

(a)   (b)

Figure 4.15   A 2×2 switching element.

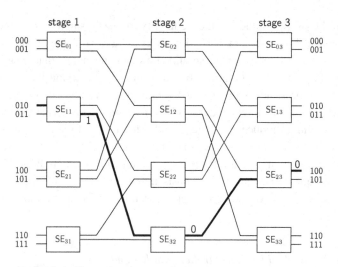

Figure 4.16   An 8×8 shuffle-exchange network.

An 8×8 SEN is shown in Figure 4.16. It has eight inputs, eight outputs, and three stages, each with four SEs. The network complexity, defined as the total number of SEs, is $(N/2)(\log_2 N)$; in the case of Figure 4.16, this is equal to 12. The SEN is a self-routing network. A message from any source to a given destination is routed according to the binary representation of the destination. For instance, if source $S = 010$ wants to send a message to the destination $D = 100$, the message is first presented at SE_{11}, to which the input $S = 010$ is connected. The first bit of the destination address $D = \underline{1}00$ is used by SE_{11} for routing and selecting the output link 1. The output link is connected to SE_{32}, which uses the second bit of the destination address to select the correct output $D = 1\underline{0}0$. That output is connected to SE_{23}, which, using the third bit of the destination address $D = 10\underline{0}$, delivers the message to the right output.

The SEN has a unique path from any source to any destination, and hence to be fully operational all the $(N/2)(\log_2 N)$ SEs should be operational. From the point of view of reliability, the SEN can be viewed as a series system of $(N/2)(\log_2 N)$ elements, and if $r_{SE}(t)$ denotes the reliability of a single SE, the reliability of the system becomes:

$$R_{SEN}(t) = [\,r_{SE}(t)\,]^{(N/2)(\log_2 N)},$$

$$MTTF_{SEN} = \int_0^\infty R_{SEN}(t)\,dt. \tag{4.40}$$

In the special case in which the SE times to failure are $EXP(\lambda)$, Eqs. (4.40) become:

$$R_{SEN}(t) = e^{-(N/2)(\log_2 N)\lambda t}, \tag{4.41}$$

$$MTTF_{SEN} = \frac{1}{(N/2)(\log_2 N)\lambda}. \tag{4.42}$$

Example 4.20 *Shuffle-Exchange Network Plus*

An SEN is a non-redundant network. A common way to add redundancy is to add an extra stage to obtain a so-called shuffle-exchange network plus (SEN+) [12], whose layout is shown in Figure 4.17. The number of SEs in an $N \times N$ SEN+ is $(N/2)(\log_2 N + 1)$.

The extra stage in the SEN+ is labeled stage 0. Adding the extra stage allows two possible paths of communication between any source–destination pair. The paths in the first and last stages are not disjoint, though in the intermediate stages they do traverse disjoint links. As an example, we follow the paths between the same source–destination pair used in Example 4.19, from $S = 010$ to $D = 100$. Starting from input $S = 010$, at SE_{10} two paths can be activated: one from the exit 0 and one from the exit 1. Then, from stage 1 to stage 3 the message follows the same routing rules as the SEN.

To compare the reliabilities of the SEN and SEN+ we study in detail the case of the 4×4 network whose layout is shown in Figure 4.18(a), for the SEN, and in Figure 4.18(b) for the SEN+.

For the 4×4 SEN we have, from Eq. (4.40),

$$R_{\text{SEN}_4}(t) = [r_{\text{SE}}(t)]^4. \tag{4.43}$$

For the 4×4 SEN+ we can observe that stage 0 and stage 2 are not redundant, while stage 1 can tolerate one failure without preventing an active connection between any

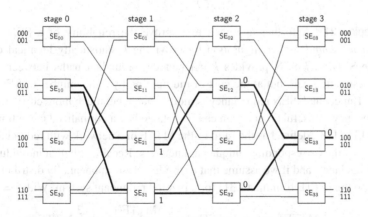

Figure 4.17 An 8×8 augmented shuffle-exchange network.

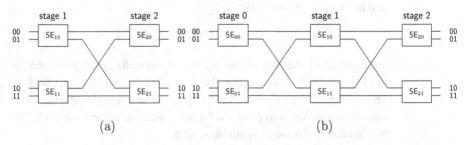

Figure 4.18 A 4×4 shuffle-exchange network. (a) SEN; (b) SEN+.

source–destination pair. Hence, from the point of view of the reliability evaluation a 4×4 SEN+ can be represented as the series-parallel system of Figure 4.19, whose reliability is:

$$R_{SEN+_4}(t) = [r_{SE}(t)]^4 [1 - (1 - r_{SE}(t))^2]. \qquad (4.44)$$

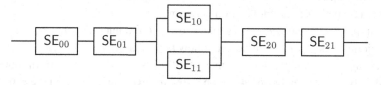

Figure 4.19   A series-parallel RBD equivalent to a 4×4 SEN+.

Comparing Eq. (4.43) with Eq. (4.44), it is clear that the 4×4 SEN is more reliable than the 4×4 SEN+. However, it has been shown in [12] that on increasing the dimension of the network, for $N \geq 8$ the $N \times N$ SEN+ has a higher reliability than the $N \times N$ SEN.

If the single SE TTF is EXP(λ), for the 4×4 case we have:

$$MTTF_{SEN_4} = \frac{1}{4\lambda},$$

$$MTTF_{SEN+_4} = \frac{2}{5\lambda} - \frac{1}{6\lambda} = \frac{7}{30\lambda} < MTTF_{SEN_4}.$$

Example 4.21   A further approach to improve the reliability of a SEN is to adopt a *system redundancy* configuration by providing k statistically identical copies of a single SEN. A k-SEN provides k independent redundant paths between any source and destination pair, but at the cost of multiplying by k the number of SEs of a single SEN. Hence the k-SEN redundancy is more reliable but much more costly than a SEN+ redundancy. A useful comparison can be obtained by a normalized mean time to failure (NMTTF) [13], defined as the ratio of the MTTF of a network with redundancy and the MTTF of the corresponding unique path network. Recalling that an individual SEN has no redundancy, and if we assume that each SE has an exponentially distributed TTF, we can use the MTTF formula (4.31) for a parallel redundant system. With $k = 2$ we have

$$MTTF_{2\text{-SEN}} = 2 MTTF_{SEN} - \frac{MTTF_{SEN}}{2} = \frac{3}{2} MTTF_{SEN},$$

and the normalized value is

$$NMTTF_{2\text{-SEN}} = \frac{3}{2},$$

independent of the size of the SEN. The normalized value for the SEN+ is studied in [13], where it is shown that the normalized value increases with N, and for $N \geq 16$ overcomes the value of NMTTF$_{2\text{-SEN}}$. For networks of size larger than 16, the reliability improvement achieved by using an extra stage is superior than that obtained by duplicating the unique-path basic SEN.

4.2.9   Is Redundancy Always Useful?

In a perfect world, if components are independent then the reliability increases with the degree of redundancy, as can be inferred from Eq. (4.24), and, notwithstanding the law of diminishing returns (see Table 4.4), it would be always possible to reach any value of reliability close to 1 simply by increasing the level of redundancy. However, there are two unavoidable effects that prevent this result from being realized: the first is that the system complexity increases with the level of redundancy and forces us to introduce new mechanisms to properly manage the redundancy. The second, and the more important one, is the presence of various effects that prevent components being statistically independent: a brief survey of the causes of dependence was reported in Section 2.7.

Example 4.22   Consider a component A with reliability R_A. In order to improve the reliability of the system we adopt the strategy of duplicating the component so as to have a parallel redundant system with two statistically identical components of type A. However, in order to properly operate the system a switch S is required to automatically assure that in case of failure of one of the components, the still-working component is correctly connected. But, the switch may also fail. We assume that the failure of one item does not affect the lifetime of the other one, and that a switch failure causes the system to fail. The question we wish to address is: "when is the redundant configuration more reliable than a single component?" The system with the switch is shown in Figure 4.20.

The reliability of the overall system is:

$$R = (2R_A - R_A^2)R_S.$$

This expression is to be compared with the reliability of a single unit. Thus, the parallel configuration is more convenient than a single unit if:

$$(2R_A - R_A^2)R_S > R_A, \quad \text{i.e.,} \quad R_S > \frac{R_A}{(2R_A - R_A^2)} = \frac{1}{2 - R_A}. \tag{4.45}$$

From (4.45), we observe that R_S should be, in any case, greater than 0.5 and $R_S > R_A$. The switch should be more reliable than the single component.

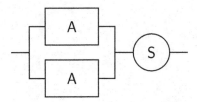

Figure 4.20   Parallel redundant configuration with a switch.

Example 4.23 A system is composed of two series components A and B. In order to improve the dependability of the overall system, a redundancy scheme may be applied according to system redundancy, Figure 4.21(a), or component redundancy, Figure 4.21(b). As already observed in Section 4.2.8, the higher complexity of the component redundancy configuration of Figure 4.21(b) requires additional logic to establish the correct connection between the still-working components. We identify the additional logic by means of the switch S between the two parallel blocks. The presence of the component S reduces the reliability of the scheme in Figure 4.21(b).

We assume that all the components' TTFs are exponentially distributed and we wish to compute an upper bound on the failure rate λ_S of the switch S such that scheme in Figure 4.21(b) is more reliable than the scheme in Figure 4.21(a) at a mission time $t = 10000\,\text{hr}$, when the failure rates of the components have the following values:

$$\lambda_A = \lambda_{A'} = 2.0 \times 10^{-5}\,\text{fhr}^{-1}, \qquad \lambda_B = \lambda_{B'} = 1.5 \times 10^{-6}\,\text{fhr}^{-1}.$$

The reliability of the configuration in Figure 4.21(a) is:

$$R_{\text{conf-a}}(t) = 2R_A(t)R_B(t) - (R_A(t)R_B(t))^2.$$

The reliability of the configuration in Figure 4.21(b) is:

$$R_{\text{conf-b}}(t) = (2R_A(t) - R_A(t)^2)R_S(t)(2R_B(t) - R_B(t)^2).$$

In order for the configuration 4.21(b) to have a reliability greater than configuration 4.21(a), the following relation must hold:

$$R_S(t) \geq \frac{2R_A(t)R_B(t) - (R_A(t)R_B(t))^2}{(2R_A(t) - R_A(t)^2)(2R_B(t) - R_B(t)^2)} = \frac{R_A(t)R_B(t)(2 - (R_A(t)R_B(t)))}{R_A(t)R_B(t)(2 - R_A(t))(2 - R_B(t))}$$

$$= \frac{2 - (R_A(t)R_B(t))}{(2 - R_A(t))(2 - R_B(t))}.$$

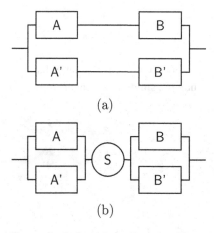

(a)

(b)

Figure 4.21   Comparison of system redundancy **(a)** vs. component redundancy with a switch **(b)**.

Taking into account that $R_A(t) = e^{-\lambda_A t}$ and $R_B(t) = e^{-\lambda_B t}$, and that the above inequality should be satisfied for $t = 10\,000\,\text{hr}$, we get:

$$R_S(t) \geq \frac{2 - e^{-(\lambda_A + \lambda_B)t}}{(2 - e^{-\lambda_A t})(2 - e^{-\lambda_B t})}$$

$$= \frac{2 - e^{-(0.2 + 0.015)}}{(2 - e^{-0.2})(2 - e^{-0.015})} = \frac{1.1935}{1.1988} = 0.9956.$$

Assuming $R_S(t) = e^{-\lambda_S t}$, we have

$$R_S(t = 10\,000) = e^{-\lambda_S \, 10\,000} = 0.9956$$

$$-\lambda_S \, 10\,000 = \ln 0.9956$$

$$\lambda_S = -\frac{\ln 0.9956}{10\,000} = 4.41 \times 10^{-7}.$$

The scheme 4.21(b) is more reliable than the scheme 4.21(a) at time $t = 10\,000\,\text{hr}$ if:

$$\lambda_S \leq 4.41 \times 10^{-7}\,\text{fhr}^{-1}.$$

4.3   _k_-out-of-_n_ Majority Voting Systems

Control systems and safety-critical systems must be designed such that the specified function is delivered with a very high level of reliability or availability, even in the presence of some malfunction of their parts. Suppose that a control system continuously checks the temperature of a plant, and if the temperature exceeds a critical threshold T^c the control system must switch on an alarm (or switch off the system). When considering a failure in the system, two undesirable events may occur that are usually classified as:

   Type I error or false positive: The temperature of the plant does not actually exceed the critical value, but due to a fault in the system, the control system actuates an erroneous alarm signal. This event is classified as a _false alarm_ and the failure sometimes termed _fail-safe_. The false alarm may cause erroneous actions (e.g., outage of the system) that can be detrimental for the plant and, in any case, can have a negative economic impact, but does not entail safety problems.

   Type II error or false negative: The temperature of the plant actually exceeds the critical value, but due to a fault in the system, the control system does not actuate the alarm signal. This event is _safety-critical_ and the failure is sometimes termed _fail-danger_.

   A well-designed control system must be protected from both types of faulty operation. Parallel duplication of the control line increases the probability that at least one line is correctly operating, but is not a sufficient cure: if the two lines agree on the action to be sent to the actuator (either "switch on" or "not switch on" the alarm) then the control system operates correctly. But if an error of any type occurs in either of the two lines, the controller receives conflicting commands: one line sends a "switch on" command and the other line a "no switch on" command. The controller is in an ambiguous situation and may not be in a condition to take the correct decision.

A possible solution is to use triplication of the lines so that the correct decision is based on majority voting of the outputs. This two-out-of-three configuration is usually termed triple modular redundancy (TMR), and the conventional RBD configuration for this system is shown in Figure 4.22 [14].

The above idea can be generalized to a system with n parallel components where the correct operation of the system requires that $k (\leq n)$ or more components out of n operate correctly. Such systems are usually called k-out-of-n systems. If $k = n$ the configuration reduces to a series system (n-out-of-n), whereas if $k = 1$, the configuration becomes a parallel system (1-out-of-n). The usual application of k-out-of-n logic requires that $k \geq (n + 1)/2$ so that the majority of the n components must work correctly for the system to be operating correctly. We refer to these systems as "majority voting systems."

k-out-of-n majority voting systems are also extensively considered in the IEC Standard 61508 [15], which sets out a generic approach for all safety lifecycle activities for systems comprised of programmable electronic elements that are used to perform safety functions. Since in a control system the final decision about the action to be taken must be unique, a voting element, called the *voter* (see Figure 4.22), must collect the signals from the n lines and propagate one single signal corresponding to the majority of the input signals.

To compute the reliability of the k-out-of-n system, assume for now that the voter is perfect and acts instantaneously, and assume further that all the n components are statistically identical and function independently of each other. Let R denote the reliability of an individual component. The probability of finding exactly i components working out of n can be expressed as the binomial probability mass function:

$$P\{i/n\} = \binom{n}{i} R^i (1 - R)^{n-i}, \tag{4.46}$$

where the binomial coefficient $\binom{n}{i}$ gives the number of ways to choose i elements from a set of n elements. The k-out-of-n system functions correctly in all the configurations in

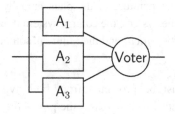

Figure 4.22   Two-out-of-three majority voting system.

which at least $i \geq k$ components function properly. Hence, the reliability of the system is:

$$R_{k/n} = \sum_{i=k}^{n} P\{i/n\}$$

$$= \sum_{i=k}^{n} \binom{n}{i} R^i (1-R)^{n-i}. \tag{4.47}$$

From Eq. (4.47) we can obtain the series system reliability when $k = n$:

$$R_{n/n} = \sum_{i=n}^{n} \binom{n}{i} R^i (1-R)^{n-i}$$

$$= \binom{n}{n} R^n (1-R)^0$$

$$= R^n, \tag{4.48}$$

and the parallel system reliability when $k = 1$:

$$R_{1/n} = \sum_{i=1}^{n} \binom{n}{i} R^i (1-R)^{n-i}$$

$$= \sum_{i=0}^{n} \binom{n}{i} R^i (1-R)^{n-i} - \binom{n}{0} R^0 (1-R)^n$$

$$= 1 - (1-R)^n. \tag{4.49}$$

For a TMR system, Eq. (4.47) becomes:

$$R_{2/3} = R_{TMR} = \sum_{i=2}^{3} \binom{3}{i} R^i (1-R)^{3-i}$$

$$= \binom{3}{2} R^2 (1-R) + \binom{3}{3} R^3 (1-R)^0$$

$$= 3R^2 (1-R) + R^3, \tag{4.50}$$

and thus

$$R_{2/3} = R_{TMR} = 3R^2 - 2R^3. \tag{4.51}$$

Note that

$$R_{TMR} \begin{cases} > R & \text{if} \quad R > \frac{1}{2}, \\ = R & \text{if} \quad R = \frac{1}{2}, \\ < R & \text{if} \quad R < \frac{1}{2}. \end{cases} \tag{4.52}$$

Thus, TMR increases the reliability over the simplex system only if the simplex reliability is greater than 0.5, which is the common situation in practice.

In deriving (4.51), the voter was assumed to be perfect and instantaneous. If the voter has a reliability R_V, then the reliability of the TMR is further degraded and becomes:

$$R_{2/3} = R_{TMR} = R_V(3R^2 - 2R^3). \tag{4.53}$$

4.3.1 The Exponential Case

If the voter is perfect and the n components have independent and identical exponentially distributed lifetimes with rate λ, we have, from Eq. (4.47),

$$R_{k/n}(t) = \sum_{i=k}^{n} \binom{n}{i} e^{-i\lambda t}(1 - e^{-\lambda t})^{n-i}, \tag{4.54}$$

and for the TMR system:

$$R_{2/3}(t) = 3R^2(t) - 2R^3(t) = 3e^{-2\lambda t} - 2e^{-3\lambda t}. \tag{4.55}$$

We compare the reliability of the TMR with that of a single component with the same rate λ as a function of time:

$$3e^{-2\lambda t} - 2e^{-3\lambda t} > e^{-\lambda t}$$

$$2e^{-3\lambda t} - 3e^{-2\lambda t} + e^{-\lambda t} < 0. \tag{4.56}$$

Substituting $e^{-\lambda t}$ with x, we have:

$$2x^2 - 3x + 1 < 0, \tag{4.57}$$

whose solution is $1/2 \le x \le 1$. Thus, Eq. (4.56) is satisfied if:

$$e^{-\lambda t} > 1/2, \quad \text{hence,} \quad t < \frac{1}{\lambda}\ln 2. \tag{4.58}$$

A comparison of the reliability of a simplex unit with the reliability of the TMR is shown in Figure 4.23 for the value $\lambda = 0.1$. The two curves cross at a value $R(t) = 0.5$,

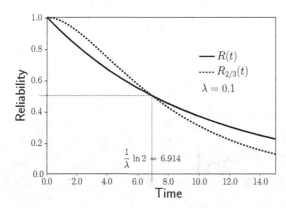

Figure 4.23 Comparison of the reliability of a simplex unit with a two-out-of-three majority voting system, with $\lambda = 0.1$.

and the reliability of the TMR is higher than the reliability of a single component only for mission times less than $t = \frac{1}{\lambda} \ln 2$.

The expected life of the TMR system is given by:

$$E[X] = \int_0^\infty 3e^{-2\lambda t}dt - \int_0^\infty 2e^{-3\lambda t}dt.$$

Thus, the TMR MTTF is

$$E[X] = \frac{3}{2\lambda} - \frac{2}{3\lambda} = \frac{5}{6\lambda}. \qquad (4.59)$$

We observe that the MTTF of the TMR is smaller than the MTTF of a single component $(= 1/\lambda)$ for any value of λ.

The last result tells us that the MTTF can be a misleading indicator in characterizing the behavior of a system. In this case, the MTTF does not account for the fact that the reliability of the TMR exceeds that of a single component for a mission time less than $\frac{\ln 2}{\lambda} \approx \frac{0.7}{\lambda} \approx 0.7 \, \text{MTTF}$.

If the lifetime of the voter is exponentially distributed with rate λ_V, then the reliability and the MTTF of the TMR become:

$$R_{2/3} = e^{-\lambda_V t}(3e^{-2\lambda t} - 2e^{-3\lambda t}),$$

$$E[X] = \int_0^\infty e^{-\lambda_V t}(3e^{-2\lambda t} - 2e^{-3\lambda t})dt = \frac{\lambda_V + 5\lambda}{(2\lambda + \lambda_V)(3\lambda + \lambda_V)}. \qquad (4.60)$$

Example 4.24 *High-Availability Platform from Sun Microsystems*
A high availability requirement in telecommunication systems is usually more stringent than most other sectors of industry. The carrier-grade platform from Sun Microsystems requires a "five-nines and better" availability (see Table 3.5).

From the availability point of view, the top-level architecture of a typical carrier-grade platform was modeled in [10] as a reliability block diagram consisting of series, parallel, and *k*-out-of-*n* subsystems, as shown in Figure 4.24. The SCSI subsystem is the dual disk subsystem block depicted in Figure 4.12. The rightmost block is a four-out-of-seven satellite block. We will refer to this RBD model in a later chapter as $M_{\text{SUN-RBD}}$.

The steady-state availability for each block will be analyzed in detail in Example 18.2. The results for the block availability, unavailability and downtime (in min yr^{-1}) and for the system as a whole are given in Table 4.7 [10].

4.3.2   Application of *k*-out-of-*n* Redundancy

The most usual form of *k*-out-of-*n* redundancy is the TMR (or two-out-of-three) system TMR is a fault-tolerant structure [16] since the occurrence of a single failure is tolerate Therefore, the TMR structure can be considered as the basic building block for

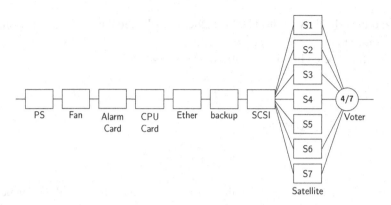

Figure 4.24   Reliability block diagram, $M_{\text{SUN-RBD}}$, of a carrier-grade platform (from [10]).

Table 4.7 Steady-state availability, unavailability and downtime (in min yr^{-1}) for the system of Figure 4.24.

Block	Availability	Unavailability	Downtime (min yr^{-1})
PS subsystem	0.999 999 840 0000	1.60×10^{-7}	8.41×10^{-2}
Fan subsystem	0.999 999 999 9955	4.48×10^{-12}	2.36×10^{-6}
Alarm card subsystem	0.999 999 970 0000	3.00×10^{-8}	1.58×10^{-2}
CPU Card subsystem	0.999 906 213 0000	9.38×10^{-5}	49.3
Ethernet subsystem	0.999 999 922 0000	7.80×10^{-8}	4.10×10^{-2}
Backplane block	0.999 995 053 0000	4.95×10^{-6}	2.6
SCSI subsystem	0.999 999 967 0000	3.30×10^{-8}	1.73×10^{-2}
Satellite subsystem (4 out of 7)	1	0	0
System as a whole	0.999 900 966 0000	9.90×10^{-5}	52.1

design of fault tolerant systems [14, 17], mainly in the area of computer and control systems devoted to the control of safety-critical applications.

Moreover, the intrinsic logic of the TMR structure offers an easy way to implement *auto-diagnose* systems, i.e., systems in which the presence of a fault is detected, recognized, and signaled by the system itself. In fact, the task of the voter is to compare the inputs and to detect possible deviations of one input from the other two. Then, it is easy to provide the voter with some additional mechanisms (acoustic or lighting alarm) to alert the control operator or the maintenance crew of the presence of a fault and to provide the exact localization of the fault among the three input lines.

Example 4.25   *Performance-Driven Two-Out-of-Three System*

A pumping station for a fire-fighting brigade is designed to provide a capacity of $V \mathrm{l\,s}^{-1}$. The system requires a redundant configuration to guarantee operation with high reliability. Two configurations are under study:

Table 4.8 Comparison of the reliability vs. time of two pumping installations.

Time hr	Configuration 1 $\lambda_a = 1.00 \times 10^{-5}\,\mathrm{hr}^{-1}$	Configuration 2 $\lambda_b = 0.50 \times 10^{-5}\,\mathrm{hr}^{-1}$
2000	9.99608×10^{-1}	9.99705×10^{-1}
4000	9.98463×10^{-1}	9.98839×10^{-1}
6000	9.96609×10^{-1}	9.97431×10^{-1}
8000	9.94089×10^{-1}	9.95508×10^{-1}
10000	9.90944×10^{-1}	9.93096×10^{-1}

1. Two parallel pumps of model type P_a with cost C_a, constant failure rate λ_a, and capacity V each.
2. Three pumps of model type P_b with cost C_b, constant failure rate, λ_b, and capacity $V/2$ each, in a two-out-of-three configuration.

In order to compare configuration 1 against configuration 2, both the reliability and the cost of the two model types need to be taken into consideration. If the failure rates and the costs of the two pump models P_a and P_b are the same, then the parallel solution is superior. However, it is reasonable to assume that a pump model with half the capacity has a lower cost and a lower failure rate. With this assumption, configuration 2 could be the proper choice.

To be more concrete, assume that the cost and failure rate of pump P_b are reduced by a factor $\beta \leq 1$ with respect to those of pump P_a. With $\beta = 1/2$, the cost of the two-out-of-three configuration 2 is $C_{\mathrm{Conf2}} = 3/2\,C_a$, while the cost of configuration 1 is $C_{\mathrm{Conf1}} = 2\,C_a$; hence, $C_{\mathrm{Conf2}} < C_{\mathrm{Conf1}}$.

Table 4.8 reports the reliability of the two configurations from $t = 0$ up to the mission time $t = 10000\,\mathrm{hr}$ with the failure rate of pump P_a equal to $\lambda_a = 1.00 \times 10^{-5}\,\mathrm{hr}^{-1}$ and the failure rate of pump P_b equal to $\lambda_b = \beta\lambda_a = 0.50 \times 10^{-5}\,\mathrm{hr}^{-1}$. With the chosen reduction factor, the two-out-of-three configuration is superior to the parallel configuration in both cost and reliability.

The previous analysis has neglected the presence of the voter, which can be very critical in safety and control systems. However, since the voter is in series in both configurations its reliability does not modify the relative figures of merit of the two structures.

Problems

4.13 Write down the reliability expression for a twin-engine aircraft that will function properly if at least one engine is functioning properly. Next, do the same for a four-engine aircraft that will work properly as long as at least two engines work properly. Assume R is the reliability of a single engine, and that engine failures are mutually independent. First write down expressions of the system reliability for the two cases. Next, determine the conditions under which a twin-engine aircraft has higher reliability than a four-engine aircraft. Next, assume a time-dependent reliability function $R(t)$ and a time-independent failure rate λ for each engine; determine the mission time t at which the two system reliabilities intersect. Finally, write down expressions for the system MTTF for the two cases.

4.14 For the base repeater of Figure 4.11 assume now that the system is up as long as both control channels are up and two-out-of-three voice channels are up. Find the reliability of the system and the MTTF, with R_c the reliability of a single control channel and R_v the reliability of a single voice channel.

4.15 In Example 4.25, with the given numerical value for the failure rate $\lambda_a = 1.00 \times 10^{-5}\,\mathrm{hr}^{-1}$, find the value of β such that the reliability of the two configurations is equal at $t = 1000\,\mathrm{hr}$ and at $t = 10\,000\,\mathrm{hr}$.

4.16 In very critical applications, a three-out-of-five control system is adopted. Find the reliability as a function of time when the five lines are i.i.d. with TTF $\sim \mathrm{ER}_2(\lambda)$.

4.3.3   The Consecutive *k*-out-of-*n* System

A consecutive k-out-of-n system is a system with n ordered components that fails whenever k consecutive components fail [18–20].

The following are examples of possible situations in which the failure condition of a system is reached when a number of consecutive objects are non-operating.

- Consider a linear wireless sensor network (WSN) where the sensors send messages to a base station. To save battery energy the sensors have a limited transmission power and each sensor can only reach the two consecutive neighboring sensors in the direction of the base station. This arrangement guarantees that the signal is transmitted from any sensor unless there are two consecutive sensors that are down.
- Civil aviation is regulated by the ICAO (International Civil Aviation Organization) standards, which require that for a landing strip to be considered operational, no more than two consecutive lamps may be out of service.
- Consider n parking spaces sequentially positioned on a street, with each space being suitable for one car. If a bus arrives, it needs to find four consecutive free spaces to park.

Two typical arrangements may be considered: a linear arrangement and a circular arrangement. When all the components have the same probability p of working and $q = 1 - p$ of failing, the system reliability is denoted as $R_L(p,k,n)$ for the linear arrangement and $R_C(p,k,n)$ for the circular arrangement. The reliability expressions have been provided in recursive form for any n and k in [18] for the linear case and in [19] for both linear and circular cases. For any n, assume now that $k = 2$, i.e., the system fails if there are more than two consecutive failures in a linear row of n components. Following [18], if the number of failed components j is greater than $j > \lfloor (n+1)/2 \rfloor$, then there are at least two consecutive failed components, and thus the cases where the number of failures is $j > [(n+1)/2]$ do not contribute to the system reliability. From [18, 19] we have:

$$R_L(p,2,n) = \sum_{j=0}^{\lfloor (n+1)/2 \rfloor} \binom{n-j+1}{j} (1-p)^j p^{n-j}. \tag{4.61}$$

For the circular case we have, from [19],

$$\begin{aligned}
R_C(p,2,n) &= p^2 \sum_{i=0}^{1} (i+1)(1-p)^i R_L(p,2,n-i-2) \\
&= p^2 R_L(p,2,n-2) + p^2 2(1-p) R_L(p,2,n-3) \tag{4.62} \\
&= \sum_{j=0}^{\lfloor (n-1)/2 \rfloor} \binom{n-1-j}{j} (1-p)^j p^{n-j} \\
&\quad + 2 \sum_{j=0}^{\lfloor (n-2)/2 \rfloor} \binom{n-2-j}{j} (1-p)^{j+1} p^{n-1-j}.
\end{aligned}$$

Example 4.26 A WSN is composed of a linear array of seven equispaced sensors. Each sensor has a transmission range that reaches two neighboring sensors only. Hence if two consecutive failures occur, the transmission of the message is blocked, and the system is considered failed. Assuming that all the sensors are statistically identical and have the same probability of being in a working state with $p = 0.9$, the reliability of the system can be derived as a two-out-of-n consecutive problem by resorting to Eq. (4.61) with $n = 7$ and $k = 2$:

$$R_L(0.9,2,7) = 3p^6 - 9p^5 + 6p^4 + p^3 = 0.945513.$$

If the sensors are deployed in a circular configuration, the reliability is calculated from Eq. (4.62):

$$R_C(0.9,2,7) = p^7 + 7p^6(1-p) + 14p^5(1-p)^2 + 7p^4(1-p)^3 = 0.937567.$$

The reliability of the circular arrangement is less than that of the linear one, since the combination of a faulty sensor in the first position and in the last position produces a failure only in the circular arrangement.

4.4 Factoring for Non-Series-Parallel Systems

Not all RBDs can be simplified to combinations of series-parallel blocks. General techniques for configurations of independent components will be developed in Chapter 5. However, by a method known as "factoring" or "conditioning," the analysis of a non-series-parallel RBD can be decomposed into the analysis of simpler blocks for which the solution may be known. Thus, a complex problem is reduced to the solution of various simpler subproblems. Let S be a system composed of n components labeled $i = 1, 2, \ldots, n$. The components and the system have binary states; as usual, we indicate by $S(X_i)$ the system (component) working and by $\overline{S}(\overline{X}_i)$ the system (component) failed. Conditioning on the status of component k, by the theorem of total probability we have:

$$P\{S\} = P\{S|X_k\}P\{X_k\} + P\{S|\overline{X}_k\}P\{\overline{X}_k\}, \tag{4.63}$$

where the term $P\{S|X_k\}$ is the conditional probability of the system S being up given that component X_k is up, and $P\{S|\overline{X}_k\}$ is the conditional probability of the system S being up given that component $(\overline{X}_k)$ is down. The above conditional probabilities may be simpler to evaluate and may be solved through series-parallel reduction.

Example 4.27   Consider the non-series-parallel system of Figure 4.25 [11]. We condition on component 2 and apply Eq. (4.63). The term $P\{S|X_2\}$ is the probability of the system functioning given that component 2 is functioning. Observe that under the assumption that component 2 is functioning the system is equivalent to the parallel composition of components 4 and 5. Therefore, using Eq. (4.24) we get

$$P\{S|X_2\} = 1 - (1 - R_4(t))(1 - R_5(t)) = R_4(t) + R_5(t) - R_4(t) + R_5(t). \tag{4.64}$$

To compute $P\{S|\overline{X}_2\}$, observe that given that component 2 is down the system is equivalent to the series of components 1 and 4 in parallel with the series of components 3 and 5. Thence:

$$\begin{aligned} P\{S|\overline{X}_2\} &= 1 - (1 - R_1(t)R_4(t))(1 - R_3(t)R_5(t)) \\ &= R_1(t)R_4(t) + R_3(t)R_5(t) - R_1(t)R_3(t)R_4(t) + R_5(t). \end{aligned} \tag{4.65}$$

Combining Eqs. (4.64) and (4.65), we have

$$\begin{aligned} R_S(t) &= [R_4(t) + R_5(t) - R_4(t) + R_5(t)]R_2(t) + [R_1(t)R_4(t) + R_3(t)R_5(t) \\ &\quad - R_1(t)R_3(t)R_4(t) + R_5(t)](1 - R_2(t)). \end{aligned}$$

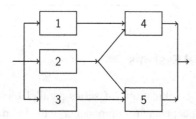

Figure 4.25   A non-series-parallel system of five components.

Problems

4.17 A multiprocessor system is composed of two parallel processors P_1 and P_2, each with its respective private memory M_1 and M_2, and a global memory M_3 shared by the two processors. The processors and memories are i.i.d. with an exponentially distributed TTF of rate λ for the processors and γ for the memories. The system is working when at least one processor is working with either its private or the shared memory working properly. Resorting to the factoring approach, find an expression for the system reliability. Next, write down an expression for the system MTTF. Also derive an expression for the sensitivity of the system reliability and system MTTF with respect to λ and γ.

4.18 For the non-series-parallel system of Example 4.25, derive expressions for the reliability importance of all the five components. Next, assume that the time to failure of component i is Weibull distributed with scale parameter η_i and common shape parameter β. Write down an expression for the system MTTF.

4.5   Non-Identical *k*-out-of-*n* System

The factoring technique of the previous section can be adopted to solve the problem of a k-out-of-n system with non-identical components. In fact, quite often n non-identical components are arranged in a k-out-of-n configuration. Let $R_1, \ldots, R_n$ be the respective reliabilities of these n components. We will still assume that failures of these components are independent. A recursive formula for the reliability of a non-identical k-out-of-n system can be derived by conditioning on the state of component n [21]. If component n has failed (with probability $1 - R_n$), the conditional reliability is given by $R_{k/n-1}$, since we need k or more functioning components out of the remaining $n-1$ components. In the case that component n is functioning properly (with probability R_n), the conditional reliability is given by $R_{k-1/n-1}$, since we need only $k-1$ or more functioning components out of the remaining $n-1$ components:

$$R_{k/n} = (1 - R_n) \cdot R_{k/n-1} + R_n \cdot R_{k-1/n-1},$$
$$R_{0/n} = 1, \tag{4.66}$$
$$R_{j/i} = 0 \text{ when } j > i,$$

where R_i is the reliability of component i. In [21] it is shown that this algorithm has a complexity of $O(n^2)$. As a special case of Eq. (4.67), let $n = 3$ and $k = 2$, that is, a non-identical two-out-of-three system. Applying Eq. (4.67), we get:

$$R_{2/3} = (1 - R_3) \cdot R_{2/2} + R_3 \cdot R_{1/2},$$
$$R_{2/2} = (1 - R_2) \cdot R_{2/1} + R_2 \cdot R_{1/1},$$
$$R_{2/1} = 0,$$
$$R_{1/1} = (1 - R_1) \cdot R_{1/0} + R_1 \cdot R_{0/0} = (1 - R_1) \cdot 0 + R_1 = R_1,$$

$$R_{1/2} = (1 - R_2) \cdot R_{1/1} + R_2 \cdot R_{0/1} = (1 - R_2) \cdot R_1 + R_2,$$

$$R_{2/2} = (1 - R_2) \cdot 0 + R_2 \cdot R_1 = R_2 \cdot R_1,$$

hence, $$R_{2/3} = (1 - R_3) \cdot R_1 \cdot R_2 + R_3 \cdot [(1 - R_2) \cdot R_1 + R_2]$$

$$= R_1 \cdot R_2 + R_1 \cdot R_3 + R_2 \cdot R_3 - 2 \cdot R_1 \cdot R_2 \cdot R_3. \tag{4.67}$$

Example 4.28   As an example of application of the formula in (4.67), consider software fault tolerance based on design diversity where different teams (say, for example, three teams) are given the same specifications and are asked to develop software to these specifications. The teams are encouraged (required?) to apply design diversity [22], which may consist in using different algorithms, different languages, different software development systems, different testing techniques and so on. As a result, the diverse versions of the software system that are produced are not likely to fail on the same set of inputs. These diverse software versions can be used in different configurations, like n-version programming and recovery blocks. We consider n-version programming here [23]. The three different versions in our example will most likely have different reliabilities (say, R_1, R_2, and R_3). Furthermore, in spite of our best efforts there will likely be some common source of bugs, so that on some inputs all the versions will fail with the associated probability R_{common}. Then, using the above expression with $k = 2$ and $n = 3$, the reliability of a two-out-of-three n-version programming system is given by:

$$R_{\text{common}}(R_1 R_2 + R_2 R_3 + R_3 R_1 - 2 R_1 R_2 R_3).$$

4.5.1 k-out-of-n System with Two Groups of Components

Having covered two extreme cases, i.e., one in which all n components are statistically identical and one in which all the components are assumed to have different reliabilities, we now consider the case of a k-out-of-n system where the components are divided into two different groups. The first group contains n_1 components, all i.i.d. with reliability R_1. The second group contains n_2 components, all i.i.d. with reliability R_2, with $n_1 + n_2 = n$.

We are allowed to choose any combination of i components out of the first group and j components out of the second group, with $i + j \geq k$. We thus have the following constraints:

$$0 \leq i \leq n_1, \quad 0 \leq j \leq n_2, \quad i + j \geq k, \quad n_1 + n_2 = n.$$

Figure 4.26 shows how to compute the reliability of the system. First, we suppose, without loss of generality, that $n_1 \geq n_2$; then, the line $i + j = k$ can be in three different

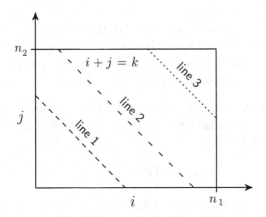

Figure 4.26 Computing the reliability of a *k*-out-of-*n* system with two groups of components.

positions according to the value of k, as depicted in Figure 4.26:

$$\text{line 1} \quad k < n_2,$$
$$\text{line 2} \quad n_2 \leq k < n_1,$$
$$\text{line 3} \quad k \geq n_1.$$

We denote by $R_{\text{Case 1}}, R_{\text{Case 2}}, R_{\text{Case 3}}$ the system reliability obtained in the three cases by enumerating all the combinations of i and j:

Case 1, $k < n_2$:

$$1 - R_{\text{Case 1}} = 1 - \sum_{i=0}^{k-1} \sum_{j=0}^{k-1-i} \binom{n_1}{i} R_1^i (1 - R_1)^{n_1-i} \binom{n_2}{j} R_2^j (1 - R_2)^{n_2-j}.$$

Case 2, $n_2 \leq k < n_1$:

$$R_{\text{Case 2}} = \left[\sum_{i=k-n_2}^{k} \sum_{j=k-i}^{n_2} \binom{n_1}{i} R_1^i (1 - R_1)^{n_1-i} \binom{n_2}{j} R_2^j (1 - R_2)^{n_2-j} \right]$$
$$+ \left[\sum_{i=k+1}^{n_1} \sum_{j=0}^{n_2} \binom{n_1}{i} R_1^i (1 - R_1)^{n_1-i} \binom{n_2}{j} R_2^j (1 - R_2)^{n_2-j} \right].$$

Case 3, $k \geq n_1$:

$$R_{\text{Case 3}} = \sum_{i=k-n_2}^{n_1} \sum_{j=k-i}^{n_2} \binom{n_1}{i} R_1^i (1 - R_1)^{n_1-i} \binom{n_2}{j} R_2^j (1 - R_2)^{n_2-j}. \qquad (4.68)$$

We will use this expression in Chapter 6 for the IBM BladeCenter availability model.

Example 4.29 *Multi-Voltage Propulsion System of a High-Speed Train (continued)*
In Example 4.14, taken from [9], the propulsion system of a high-speed train is modeled
by three identical parallel subsystems whose reliability is given in Eq. (4.36). We
elaborate this example in three different ways:

1. Only two-out-of-three identical subsystems are needed to provide the power to the
   system. Denoting by R the reliability of one subsystem (4.36) and by $R_{2/3\text{-iid}}$ the
   reliability of the system, we have:

$$R_{2/3\text{-iid}} = 3R^2 - 2R^3.$$

2. The three subsystems have different reliabilities, say R_1, R_2, R_3. The reliability of
   the system $R_{2/3\text{-dif}}$ becomes:

$$R_{2/3\text{-dif}} = R_1 R_2 + R_2 R_3 + R_3 R_1 - 2R_1 R_2 R_3.$$

3. Subsystems 1 and 2 are identical with reliability R_1, and subsystem 3 has reliability
   R_2. We apply the results of Section 4.5.1. Since $n_1 = 2, n_2 = 1$ and $k = 2$, we are in
   Case 3 above. i can range from $k - n_2$ to n_1, hence $i = 1, 2$. j can range from $k - i$ to
   n_2, hence $j = 0, 1$. We apply Eq. (4.68), exploring all the cases:

   a. $i = 1, j = 1$:

$$\binom{2}{1} R_1^1 (1 - R_1)^1 \binom{1}{1} R_2^1 (1 - R_2)^0 = 2R_1 R_2 - 2R_1^2 R_2.$$

   b. $i = 2, j = 0$:

$$\binom{2}{2} R_1^2 (1 - R_1)^0 \binom{1}{0} R_2^0 (1 - R_2)^1 = 2R_1^2 - R_1^2 R_2.$$

   c. $i = 2, j = 1$:

$$\binom{2}{2} R_1^2 (1 - R_1)^0 \binom{1}{1} R_2^1 (1 - R_2)^0 = R_1^2 R_2.$$

   Summing the three terms, we get:

$$R_{2/3\text{-group}} = R_1^2 + 2R_1 R_2 - 2R_1^2 R_2.$$

   The same result could also have been obtained from Eq. (4.67).

Problems

4.19 *Quorum in an Assembly*
In an assembly or in a board of trustees, the procedural rules require that for the validity
of a decision, a quorum of 2/3 of the legal members must be personally present in

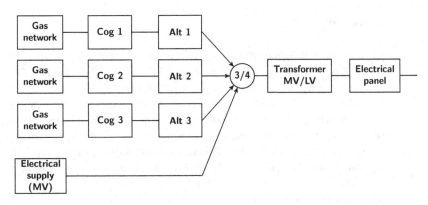

Figure 4.27   Reliability block diagram for Problem 4.22.

the assembly. Let n be the number of legal members and let the integer $k = \lceil 2/3 \cdot n \rceil$. According to historical records, each member of the assembly has a probability p of being present. Compute the probability that the quorum will be reached.

4.20   *Quorum in an Assembly (continued)*
In Problem 4.19, assume $n = 25$ and $p = 0.9$. Numerically compute the probability that the simple majority of 51% is reached, or the majority of $2/3$ is reached.

4.21   *Quorum in an Assembly (continued)*
Suppose now that according to historical records the members of the assembly can be partitioned into two groups. The first group of n_1 members attends meetings regularly with probability p_1, while the second group of $n_2 = n - n_1$ members attends meetings with probability p_2. Assuming that $n_2 < n_1$, and that the quorum is again $2/3$, compute the probability that the meeting will be valid.

4.22   A proposed configuration of an electrical supply system has four sources of supply consisting of three gas co-generators and a medium voltage (MV) electrical supply [24]. Each gas supply needs its gas network, co-generator, and alternator to all be working properly. The system is considered operational if three or more out of the four supplies are operational besides the transformer and the electrical panel (see Figure 4.27).

We assume independent failures and failure rates denoted by: gas network (λ_g), co-generator (λ_c), alternator (λ_a), electrical supply (λ_e), transformer (λ_t), and electrical panel (λ_p). Write down an expression for the system reliability. Next, assuming independent repair and similarly labeled repair rates, write down an expression for the instantaneous and steady-state system availability.

4.6   Further Reading

The IEC 1078 standard [1] formally defines reliability block diagrams. The MIL-HDBK-217 [5] also contains detailed discussions on RBDs. Another source is Shooman [11]. Similarly, Sahner *et al.* [25] contains several RBD examples.

References

[1] IEC 61078, *Reliability Block Diagram Method*. IEC Standard No. 61078, 1991.

[2] K. Trivedi, *Probability and Statistics with Reliability, Queueing and Computer Science Applications*, 2nd edn. John Wiley & Sons, 2001.

[3] Z. Birnbaum, "On the importance of different components in a multicomponent systems," in *Multivariate Analysis*, vol. II, ed. E. P. R. Krishnaiah. Academic Press, 1969, pp. 581–592.

[4] A. Bobbio and A. Premoli, "Fast algorithm for unavailability and sensitivity analysis of series-parallel systems," *IEEE Transactions on Reliability*, vol. R-31, pp. 359–361, 1982.

[5] MIL-HDBK-217F, *Reliability Prediction of Electronic Equipment*. Department of Defense, USA, 1991.

[6] K. Trivedi, "Availability analysis of Cisco GSR 12000 and Juniper M20/M40," Cisco internal technical report, 2000.

[7] D. Wang and K. S. Trivedi, "Modeling user-perceived reliability based on user behavior graphs," *International Journal of Reliability, Quality and Safety Engineering*, vol. 16, no. 04, pp. 303–329, 2009.

[8] G. Levitin, A. Lisnianski, H. Ben-Haim, and D. Elmakis, "Redundancy optimization for series-parallel multi-state systems," *IEEE Transactions on Reliability*, vol. R-47, pp. 165–172, 1998.

[9] G. Cosulich, P. Firpo, and S. Savio, "Power electronics reliability impact on service dependability for railway systems: A real case study," in *Proc. Int. Symp. Industrial Electronics, ISIE '96.*, vol. 2, Jun 1996, pp. 996–1001.

[10] K. Trivedi, R. Vasireddy, D. Trindade, S. Nathan, and R. Castro, "Modeling high availability systems," in *Proc. IEEE Pacific Rim Int. Symp. on Dependable Computing (PRDC)*, 2006.

[11] M. Shooman, *Probabilistic Reliability: An Engineering Approach*. McGraw Hill, 1968.

[12] J. Blake and K. Trivedi, "Multistage interconnection network reliability," *IEEE Transactions on Computers*, vol. C-38, pp. 1600–1604, 1989.

[13] J. Blake, "Comparative analysis of multistage interconnection networks," Ph.D. Thesis, Department of Computer Science, Duke University, 1987.

[14] W. Bouricius, W. Carter, D. Jessep, P. Schneider, and A. Wadia, "Reliability modeling for fault-tolerant computers," *IEEE Transactions on Computers*, vol. C-20, pp. 1306–1311, 1971.

[15] IEC 61508, *Functional Safety of Electrical/Electronic/Programmable Electronic Safety-Related Systems*. IEC Standard No. 61508, 2011.

[16] A. Avizienis, "Fault-tolerance, the survival attribute of digital systems," *Proceedings IEEE*, vol. 66, pp. 1109–1125, 1978.

[17] A. Bobbio, "Dependability analysis of fault-tolerant systems: A literature survey," *Microprocessing and Microprogramming*, vol. 29, pp. 1–13, 1990.

[18] D. Chiang and S. Niu, "Reliability of consecutive k-out-of-n," *IEEE Transactions on Reliability*, vol. R-30, no. 1, pp. 87–89, 1981.

[19] C. Derman, G. Lieberman, and S. Ross, "On the consecutive-k-of-n:f system," *IEEE Transactions on Reliability*, vol. R-31, no. 1, pp. 57–63, 1982.

[20] M. Chao, J. Fu, and M. Koutras, "Survey of reliability studies of consecutive-k-out-of-n:f and related systems," *IEEE Transactions on Reliability*, vol. 44, no. 1, pp. 120–127, Mar. 1995.

[21] R. Sahner, K. Trivedi, and A. Puliafito, *Performance and Reliability Analysis of Computer Systems: An Example-based Approach Using the SHARPE Software Package*. Kluwer Academic Publishers, 1996.

[22] A. Avizienis and J.-C. Laprie, "Dependable computing: From concepts to design diversity," *Proceedings of the IEEE*, vol. 74, no. 5, pp. 629–638, 1986.

[23] B. Randell and J. Xu, "The evolution of the recovery block concept," in *Software Fault Tolerance*, ed. M. R. Lyu. John Wiley & Sons, 1994, ch. 1, pp. 1–22.

[24] F. Salata, A. de Lieto Vollaro, R. de Lieto Vollaro, and L. Mancieri, "Method for energy optimization with reliability analysis of a trigeneration and teleheating system on urban scale: A case study," *Energy and Buildings*, vol. 86, pp. 118–136, 2015.

[25] R. A. Sahner and K. Trivedi, "Reliability modeling using SHARPE," *IEEE Transactions on Reliability*, vol. R-36, no. 2, pp. 186–193, Jun. 1987.

5   Network Reliability

The network abstraction has proved to be very valuable in representing many social, economic, and technological phenomena that involve complex interconnections and relationships among entities. The vertices in the network abstraction represent the entities, and the edges represent the physical or relational links between them. A natural property of this complex interconnection is the existence of multiple redundant paths between pairs of entities, so that even in the case of the failure of some elements (nodes or arcs) of the network, a connection can still be established. Dependable networks are intrinsically reliable. For this reason, the reliability of networks has always been a major concern since the beginnings of reliability engineering [1–4]. In the successive decades the literature on network reliability has grown constantly, interwoven with graph theory. In this chapter, we cannot give a complete account of the studies in this field, and we are more directly concerned with the use of decision diagrams and their variants.

In the graph abstraction of a probabilistic network the elements of the graph (both vertices and edges) are usually represented as binary entities with an up and a down state to which a probability of being up or down is assigned. However, in Section 5.4, multi-valued networks, in which the network elements can have multiple states, will also be considered.

The network reliability is defined as the probability that all nodes, or a distinguished subset of the nodes, or a designated pair of nodes of the graph, are connected through at least one path of working edges. The computational techniques proposed in the literature to tackle the network reliability problem always assume that the network elements are statistically independent, and can be classified into two main categories:

- approaches in which the desired network reliability measure is directly calculated (series-parallel reduction [5], pivotal decomposition using keystone components [6, 7]);
- approaches in which all possible ways through which two specified nodes can communicate (or not communicate) with each other are first enumerated (path/cut set search [8]) and then the reliability expression is evaluated, resorting to different techniques [3], like the inclusion–exclusion method, sum of disjoint products [9, 10] or binary decision diagrams (BDDs) [11].

Since the appearance of the Bryant's seminal paper [11], BDDs have provided an extraordinarily efficient method of representing and manipulating Boolean functions [12], and have been exploited in binary dependability problems [13] as the representation of the connectivity of Boolean networks. The present chapter discusses the main

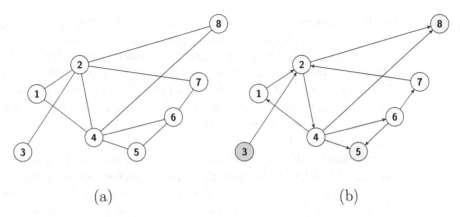

Figure 5.1   (a) An undirected network; (b) A directed network.

algorithms that have been developed to compute the reliability of a network. Starting from the representation of binary network connectedness, extensions of BDD will be discussed in relation to weighted networks in Section 5.3 and multi-valued networks in Section 5.4.

5.1   Networks and Graphs

A network can be represented by means of a graph $G = (V, E)$, where V is the set of vertices (or nodes) of cardinality $|V| = N_V$ and E is the set of edges (or arcs) of cardinality $|E| = N_E$. The edges may be undirected and can be traversed in either direction, Figure 5.1(a), or they may be directed and can be traversed only in the direction indicated by the arrow, Figure 5.1(b).

For a connected network with undirected edges, the presence of at least one edge per node guarantees that all the nodes are reachable from any node. Usually, real networks are much more connected than this minimal threshold, thus allowing multiple paths among any pair of vertices. For directed graphs the connectivity property is more cumbersome, since a connecting path must follow the directions of the arcs. In directed networks the nodes can be classified into three distinct categories: (i) nodes that form a subset of connected nodes that can be reached from any node in the subset by following the directions of the arcs – unshaded nodes in Figure 5.1(b); (ii) source nodes that have only output arcs and cannot be reached from any other node – node 3 in Figure 5.1(b); and (iii) sink nodes that have only input arcs: once reached, they can never be deserted – nodes 5 and 8 in Figure 5.1(b).

5.2   Binary Probabilistic Networks

A binary probabilistic network is a network whose elements (nodes and edges) can be in two possible mutually exclusive and collectively exhaustive states: the up state to which the binary value true $(t, 1)$ is assigned, or the down state to which the binary value false $(f, 0)$ is assigned. A binary probabilistic network can be represented in the form of a graph $G = (V,E,P)$, where P is the probability function that assigns to each element of the network (both nodes and arcs) a probability of being in the up or the down state. P can be a Bernoulli distribution (Section 3.2.8) with p the probability of the up state and $(1 - p)$ the probability of the down state. Alternatively, P can be a distribution of time to failure, or the instantaneous or steady-state availability for each arc, as indicated in the introduction to Part II.

Two nodes of the network are connected if there is a path along which all the nodes and the arcs are functioning properly. In a binary probabilistic network, the connectivity function $C(V,E)$ is a Boolean function whose value is $C(V,E) = 1$ when the network is connected and $C(V,E) = 0$ when the network is disconnected. Following tradition, we assign a failure probability only to the arcs. A similar approach can be used for cases where only the nodes fail, or cases where both the arcs and the nodes fail.

In general, we define as K-terminal reliability the probability that a given subset K $(\leq N_V)$ of vertices communicate with each other by a path of working edges [14]. With $K = 2$ we have the usual definition of two-terminal reliability, and with $K = N_V$ the all-terminal reliability. In directed networks, we usually assume a source node s and a termination node t, and the connectivity must be verified in the direction from s to t. This problem is usually referred to in the literature as (s,t)-reliability or the (s,t) connectedness problem. In the following, we present three classical techniques to evaluate the reliability of a network:

- minpath or mincut analysis,
- Shannon decomposition,
- graph-visiting algorithm.

The reliability evaluation associated with the three techniques mentioned above can be carried out by any classical method like the *inclusion–exclusion* formula or the *sum of disjoint products*. However, we emphasize the use of BDDs [11, 15], which have proven to be a very efficient technique for dealing with Boolean functions. In the next section we provide a brief review of the structure and main properties of BDDs.

5.2.1   Basic Properties of a Binary Decision Diagram

A BDD can encode Boolean expressions in a very efficient way [12], and for this reason BDDs have gained wide popularity whenever the model can be reduced to a Boolean expression. A recent survey on the use of BDDs in dependability is presented in [13], and a list of free libraries that are available for the utilization and manipulation of BDDs is given in [16].

A BDD is a compact representation of a Boolean function as a directed acyclic graph, obtained by recursively applying the Shannon decomposition formula. Let $x = (x_1, x_2, \ldots, x_n)$ be a set of n Boolean variables and $F(x) = F(x_1, x_2, \ldots, x_n)$ a Boolean function over x. Assuming x_i as the pivot variable, the Shannon decomposition formula can be written as:

$$F(x_1, x_2, \ldots, x_n) = x_i \wedge F_{x_i=1} \vee \bar{x}_i \wedge F_{x_i=0}, \tag{5.1}$$

where $F_{x_i=1}$ is derived from $F(x)$ assuming x_i is true, and $F_{x_i=0}$ is derived from $F(x)$ assuming x_i is false:

$$F_{x_i=1} = F(x_1, x_2, \ldots, x_i = 1, \ldots, x_n), \tag{5.2}$$

$$F_{x_i=0} = F(x_1, x_2, \ldots, x_i = 0, \ldots, x_n).$$

The functions $F_{x_i=1}$ and $F_{x_i=0}$ depend on the remaining $n-1$ variables; applying the Shannon decomposition (5.1) by pivoting with respect to one of the remaining $n-1$ variables, each one of the two functions $F_{x_i=1}$ and $F_{x_i=0}$ can be decomposed into two functions that depend on $n-2$ variables. Thus, iteratively applying the Shannon decomposition formula pivoting with respect to a complete sequence of all the variables, a complete decomposition can be obtained. The sequence of decompositions can be represented in graphical form using a binary tree. Each node of the tree represents the pivot variable with respect to which the decomposition is done, and from each node spawn two branches. Consider a generic node x_i: the left branch has the value $x_i = 1$ and its descendant is the term $F_{x_i=1}$, defined in (5.2), of the decomposed function; the right branch has the value $x_i = 0$ and its descendant is the term $F_{x_i=0}$, defined in (5.2), of the decomposed function. We use the convention of drawing the left branch $x_i = 1$ in a solid line and the right branch $x_i = 0$ in a dashed line.

An ordered BDD is one in which the decomposition follows an ordered sequence of variables, such that each variable is encountered no more than once in any path and always in the same order along each path. An ordered BDD can be reduced by the successive application of two rules:

- The *deletion* rule eliminates a node with both of its outgoing edges leading to the same successor.
- The *merging* rule eliminates one of two nodes with the same label as well as the same pair of successors (share the same subtree).

A BDD that is both ordered and reduced is called a reduced ordered BDD (ROBDD) [12]. In this book, all references to BDDs imply ROBDDs. The ordering of the variables in an ROBDD strongly affects the complexity of the decomposition. There is no known way to find the optimal ordering and thus we need to resort to some heuristics.

Once a complete decomposition of a Boolean function F of n variables is achieved, the ROBDD is a binary-tree-like structure with n levels, one level for each variable, with the root node of the BDD the first variable in the ordered sequence. The terminal nodes can be merged into two single nodes, denoted as 1 and 0, representing the value of the function $F(x) = 1$ and $F(x) = 0$.

Any path from the root node to the terminal node 1 gives a combination of variables for which the final value of the function $F(x)$ is equal to one, while any path from the root node to the terminal node 0 gives a combination of variables for which the final value of the function $F(x)$ is equal to zero.

All the paths in a BDD connecting the root node with the terminal node 1 are disjoint Boolean functions, since in any two paths at least one variable is true in one of the two and false in the other one. The value of the function $F(x) = 1$ can also be represented as the union of its mutually disjoint paths leading to the terminal node 1 (Section 5.2.6).

Similarly for the paths from the root node to the node 0. The value of the function $F(x) = 0$ can also be represented as the union of its mutually disjoint paths leading to the terminal node 0 (Section 5.2.6).

If each variable x_i is assigned a probability p_i of being true (thus $(1 - p_i)$ of being false), we can compute the probability $P\{F(x)\}$ of the function $F(x)$ directly from the BDD. A possible efficient algorithm is based on the recursive application of Eq. (5.3), which is the probabilistic version of the Shannon decomposition rule:

$$P\{F(x)\} = p_i P\{F_{x_i=1}\} + (1 - p_i)P\{F_{x_i=0}\}$$
$$= P\{F_{x_i=0}\} + p_i(P\{F_{x_i=1}\} - P\{F_{x_i=0}\}). \qquad (5.3)$$

5.2.2   Network Reliability Based on Minpath Analysis

Given a binary probabilistic network $G = (V, E, P)$ and two nodes (s,t), the following definitions hold:

Definition 5.1 *An* (s,t) *path is a subset of arcs that guarantees the source* s *and sink* t *are connected when all the arcs in the subset are working. A path is a* minpath *if there is no proper subset of the path that is also a path.*

Definition 5.2 *The number of edges that form a minpath is called the* order *of the minpath.*

The connectivity of a path H_i is a Boolean function expressed as the conjunction of its arcs in their up states. If $H_1, H_2, \ldots, H_h$ are the h minpaths of a network between s and t, the connectivity function of the network $C_{s,t}$ is defined as the disjunction of its minpaths:

$$C_{s,t} = H_1 \vee H_2 \vee \cdots \vee H_h = \bigvee_{i=1}^{h} H_i. \qquad (5.4)$$

The connectivity function (5.4) is in the form of a logical sum of its minpaths, where the minpaths are a logical product of Boolean variables representing the arc conditions. This sum-of-product form of a logical function is called *disjunctive normal form* (DNF). It can be shown that any Boolean function can be expressed in DNF form [17].

The disconnectivity function is then defined as:

$$\overline{C_{s,t}} = \neg C_{s,t}. \qquad (5.5)$$

The two-terminal reliability can then be calculated as the probability of the union of the minpaths:

$$R_{s,t} = P\{C_{s,t}\} = P\{H_1 \vee H_2 \vee \cdots \vee H_h\}. \qquad (5.6)$$

The probability of the union of non-disjoint events, as in Eq. (5.6), can be computed by different techniques that will be illustrated below:

- the inclusion–exclusion formula [18, 19],
- the sum of disjoint products (SDP) [9, 20–22],
- a BDD [11, 23].

The algorithm for the exhaustive search of the minpaths relies on a recursive call of a function based on the classical Dijkstra's algorithm [24]. We assume that all the edges have a cost equal to one so that Dijkstra's algorithm finds the path with the minimum number of hops (path length). For an alternative algorithm, see Section 5.2.6.

Example 5.1 *Bridge Network*
A bridge network with directed arcs is drawn in Figure 5.2 with v_1 being the source node s and v_4 the terminal node t. We wish to compute the (s,t)-reliability of this bridge network.

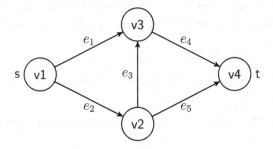

Figure 5.2   A directed bridge network.

The network of Figure 5.2 has three minpaths:

$$H_1 = e_1 \wedge e_4, \quad H_2 = e_2 \wedge e_5, \quad H_3 = e_2 \wedge e_3 \wedge e_4. \tag{5.7}$$

Substituting Eq. (5.7) into Eq. (5.4), the network connectivity is expressed as:

$$C_{s,t} = H_1 \vee H_2 \vee H_3 = (e_1 \wedge e_4) \vee (e_2 \wedge e_5) \vee (e_2 \wedge e_3 \wedge e_4). \tag{5.8}$$

The network reliability can, hence, be calculated as:

$$R_{s,t} = P\{C_{s,t}\} = P\{H_1 \vee H_2 \vee H_3\} = P\{(e_1 \wedge e_4) \vee (e_2 \wedge e_5) \vee (e_2 \wedge e_3 \wedge e_4)\}. \tag{5.9}$$

The evaluation of $R_{s,t}$ in Eq. (5.9) can be performed by the three different methods mentioned above.

Inclusion–exclusion method
We can rewrite Eq. (5.9) as:

$$
\begin{aligned}
R_{s,t} &= P\{H_1\} + P\{H_2\} + P\{H_3\} - P\{H_1 \wedge H_2\} - P\{H_1 \wedge H_3\} - P\{H_2 \wedge H_3\} \\
&\quad + P\{H_1 \wedge H_2 \wedge H_3\} \\
&= P\{e_1 \wedge e_4\} + P\{e_2 \wedge e_5\} + P\{e_2 \wedge e_3 \wedge e_4\} \\
&\quad - P\{e_1 \wedge e_4 \wedge e_2 \wedge e_5\} - P\{e_1 \wedge e_4 \wedge e_2 \wedge e_3\} - P\{e_2 \wedge e_5 \wedge e_3 \wedge e_4\} \\
&\quad + P\{e_1 \wedge e_2 \wedge e_3 \wedge e_4 \wedge e_5\} \\
&= p_1 p_4 + p_2 p_5 + p_2 p_3 p_4 - p_1 p_2 p_4 p_5 - p_1 p_2 p_3 p_4 - p_2 p_3 p_4 p_5 + p_1 p_2 p_3 p_4 p_5 .
\end{aligned}
\tag{5.10}
$$

SDP
The SDP method originated from a paper by Abraham [9]. It was then elaborated and improved in many subsequent papers [20–22, 25]. A performance comparison of the different algorithms appeared in [20, 26]. The SDP method is based on the following identity for Boolean functions, in which the union of two (or more) functions can be expressed as the union of disjoint terms (see Section 6.12). Given that a, b, and c are Boolean variables whose probabilities of being true are p_a, p_b, and p_c, respectively, we have:

$$
a \vee b = a \vee (\bar{a} \wedge b), \qquad a \vee b \vee c = a \vee (\bar{a} \wedge b) \vee (\bar{a} \wedge \bar{b} \wedge c). \tag{5.11}
$$

Since the expressions on the right-hand sides of (5.11) are disjoint, we can write:

$$
\begin{aligned}
P\{a \vee b\} &= P\{a\} + P\{\bar{a} \wedge b\} = p_a + (1 - p_a) \cdot p_b, \\
P\{a \vee b \vee c\} &= P\{a\} + P\{\bar{a} \wedge b\} + P\{\bar{a} \wedge \bar{b} \wedge c\}) = \\
&\quad P_a + (1 - p_a) \cdot p_b + (1 - p_a)(1 - p_b) p_c.
\end{aligned}
$$

Generalization to n Boolean variables is straightforward. If a, b, and c are Boolean functions (instead of variables), the same identity (5.11) can be applied, but in order to arrive at mutually disjoint terms the identity can possibly be iterated.
   Applying the identity (5.11) to Eq. (5.8), we get:

$$
\begin{aligned}
C_{s,t} &= H_1 \vee (\overline{H}_1 \wedge H_2) \vee (\overline{H}_1 \wedge \overline{H}_2 \wedge H_3) \\
&= (e_1 \wedge e_4) \vee (\overline{e_1 \wedge e_4} \wedge e_2 \wedge e_5) \vee (\overline{e_1 \wedge e_4} \wedge \overline{e_2 \wedge e_5} \wedge e_2 \wedge e_3 \wedge e_4), \\
&\qquad \text{(applying De Morgan's law and simplifying)} \\
&= (e_1 \wedge e_4) \vee ((\overline{e}_1 \vee \overline{e}_4) \wedge e_2 \wedge e_5) \vee ((\overline{e}_1 \vee \overline{e}_4) \wedge (\overline{e}_2 \vee \overline{e}_5) \wedge e_2 \wedge e_3 \wedge e_4) \\
&= (e_1 \wedge e_4) \vee (\overline{e}_1 \wedge e_2 \wedge e_5) \vee (\overline{e}_4 \wedge e_2 \wedge e_5) \vee (\overline{e}_1 \wedge \overline{e}_5 \wedge e_2 \wedge e_3 \wedge e_4).
\end{aligned}
\tag{5.12}
$$

It is easy to see that the two products $(\bar{e}_1 \wedge e_2 \wedge e_5)$ and $(\bar{e}_4 \wedge e_2 \wedge e_5)$ are not yet disjoint, and, thus, a further step of the SDP algorithm should be applied to them to yield:

$$= (\bar{e}_1 \wedge e_2 \wedge e_5) \vee (\bar{e}_4 \wedge e_2 \wedge e_5)$$

$$= (\bar{e}_1 \wedge e_2 \wedge e_5) \vee \overline{(\bar{e}_1 \wedge e_2 \wedge e_5)} \wedge \bar{e}_4 \wedge e_2 \wedge e_5$$

(applying De Morgan's law and simplifying) $\qquad(5.13)$

$$= (\bar{e}_1 \wedge e_2 \wedge e_5) \vee ((e_1 \vee \bar{e}_2 \vee \bar{e}_5) \wedge \bar{e}_4 \wedge e_2 \wedge e_5)$$

$$= (\bar{e}_1 \wedge e_2 \wedge e_5) \vee (e_1 \wedge \bar{e}_4 \wedge e_2 \wedge e_5).$$

Finally, combining Eqs. (5.12) with Eq. (5.13) we obtain a disjoint form of the connectivity $C_{s,t}$ from which we can easily compute the reliability $R_{s,t}$:

$$C_{s,t} = (e_1 \wedge e_4) \vee (\bar{e}_1 \wedge e_2 \wedge e_5) \vee (e_1 \wedge \bar{e}_4 \wedge e_2 \wedge e_5) \vee (\bar{e}_1 \wedge \bar{e}_5 \wedge e_2 \wedge e_3 \wedge e_4),$$

$$R_{s,t} = p_1 p_4 + (1 - p_1) p_2 p_5 + p_1 (1 - p_4) p_2 p_5 + (1 - p_1)(1 - p_5) p_2 p_3 p_4$$

$$= p_1 p_4 + p_2 p_5 - p_1 p_2 p_4 p_5 + p_2 p_3 p_4 - p_1 p_2 p_3 p_4 - p_2 p_3 p_4 p_5 + p_1 p_2 p_3 p_4 p_5.$$

BDD

For the construction of the BDD, assume the arbitrary ordering $e_2 \prec e_5 \prec e_3 \prec e_1 \prec e_4$. The Shannon decomposition of the Boolean connectivity expression (5.8) is built step by step in Figure 5.3(a). The binary tree of Figure 5.3(a) can be reduced to an ROBDD by merging the terminal leaves 1 and 0, and observing that there are two identical subtrees, like those generated by the duplicated node $(e_1 e_4)$. Merging the two identical subtrees we arrive at the reduced BDD of Figure 5.3(b).

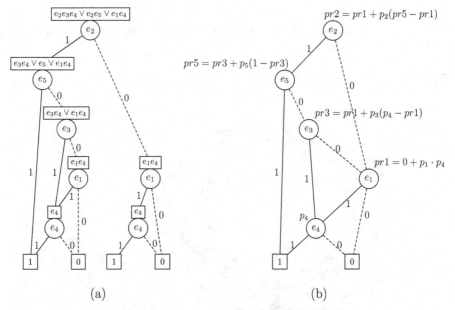

(a)   (b)

Figure 5.3 Connectivity of the bridge network of Figure 5.2. (a) BDD representation of Eq. (5.8); (b) ROBDD representation and probability computation.

The computation of the probability from the ROBDD of Figure 5.3(b) proceeds, starting from the terminal leaves 1 and 0, by recursive application of Eq. (5.3) as explicitly derived in Eq. (5.14) and pictorially shown in Figure 5.3(b). The network reliability is, from Eq. (5.14), $R_{s,t} = pr_2$.

$$pr_4 = p_4,$$
$$pr_1 = 0 + p_1 p_4,$$
$$pr_3 = pr_1 + p_3(p_4 - pr_1) = p_1 p_4 + p_3 p_4 - p_1 p_3 p_4,$$
$$pr_5 = pr_3 + p_5(1 - pr_3) = \qquad (5.14)$$
$$p_1 p_4 + p_3 p_4 - p_1 p_3 p_4 + p_5 - p_1 p_4 p_5 - p_3 p_4 p_5 + p_1 p_3 p_4 p_5,$$
$$pr_2 = pr_1 + p_2(pr_5 - pr_1) =$$
$$p_1 p_4 + p_2 p_3 p_4 - p_1 p_2 p_3 p_4 + p_2 p_5 - p_1 p_2 p_4 p_5 - p_2 p_3 p_4 p_5 + p_1 p_2 p_3 p_4 p_5.$$

If all the arcs have a probability of being in the up state equal to $p_i = 0.9$ $(i = 1, 2, \ldots, 5)$, the reliability calculated from Eq. (5.14) is $R_{s,t} = 0.971\,19$, and the probability that the network is disconnected is $1 - R_{s,t} = 0.028\,81$.

Example 5.2 *Regular Network*
Figure 5.4 shows an example of a regular directed network (all the nodes have the same number of arcs) with $n = 5$ nodes and $k = 2$ arcs out of each node. Visual inspection of Figure 5.4, assuming s=v_1 and t=v_5, shows that the graph possesses five minpaths (listed in order of their rank):

$$H_1 = 6 \wedge 8; \ H_2 = 6 \wedge 3 \wedge 4; \ H_3 = 1 \wedge 7 \wedge 4; \ H_4 = 1 \wedge 2 \wedge 8; \ H_5 = 1 \wedge 2 \wedge 3 \wedge 4.$$
$$(5.15)$$

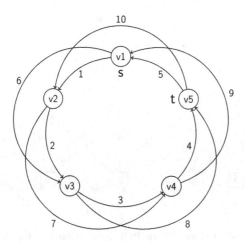

Figure 5.4   A regular network with $n = 5$ and $k = 2$.

Substituting Eq. (5.15) in Eq. (5.4), the Boolean function for the network (s,t)-connectivity is:

$$C_{s,t} = (6 \wedge 8) \vee (6 \wedge 3 \wedge 4) \vee (1 \wedge 7 \wedge 4) \vee (1 \wedge 2 \wedge 8) \vee (1 \wedge 2 \wedge 3 \wedge 4). \quad (5.16)$$

Problems

5.1 For the bridge network, derive expressions for the Birnbaum importance of (s,t)-reliability with respect to all the arc reliabilities.

5.2 For the bridge network, associate Weibull time to failure distributions (3.22) with arc i with parameters η_i and β_i. Now write down the expression for the (s,t)-reliability at time t of the network. Next, assuming $\beta_i = \beta$ for all i, write down an expression for the system MTTF.

5.3 Assuming, in the previous problem, $\beta_i = 1$ and $\eta_i = \lambda$ for all i, develop closed-form equations for the uncertainty in the (s,t)-reliability of the bridge network due to parameter λ. Assume Λ is an r-stage Erlang random variable with parameter s (as in Chapter 3).

5.4 Assuming exponentially distributed times to failure with parameter λ_i and exponentially distributed times to repair with parameter μ_i for the arc labeled i, develop closed-form equations for the steady-state availability, instantaneous availability, and expected interval availability of the bridge network. Then carry out the sensitivity analysis by writing down expressions for the derivatives of the steady-state availability with respect to each failure rate and repair rate.

5.5 Derive an expression for the (s,t)-reliability of the network of Figure 5.4, starting from the (s,t)-connectivity expression (5.16) using the SDP method and then using the BDD method.

5.2.3   Network Reliability Based on Mincut Analysis

A complementary approach to the one described in Section 5.2.2 can be followed by generating mincuts instead of minpaths, where a mincut is a minimal subset of edges whose failure disconnects the s and t nodes.

Definition 5.3 *An* (s,t) *cut is a subset of arcs such that the source* s *and sink* t *are disconnected if all the arcs of the subset have failed. A cut is a* mincut *if there is no proper subset of the arcs of the cut that is also a cut.*

Definition 5.4 *The number of edges that form a mincut is called the* order *of the mincut.*

Proposition 5.1 *Any minpath contains at least one element from each mincut, and any mincut contains at least one element from each minpath [27].*

A cut K_j is the conjunction of its arcs in their **down** state. If $K_1, K_2, \ldots, K_k$ are the k mincuts of a network, the disconnectivity function $\overline{C_{s,t}}$ is the disjunction of its mincuts:

$$\overline{C_{s,t}} = K_1 \vee K_2 \vee \cdots \vee K_k = \bigvee_{j=1}^{k} K_j. \tag{5.17}$$

We can then use one of the inclusion–exclusion, SDP or BDD methods to obtain an expression for the (s,t)-unreliability of the network. Since the disconnectivity function in (5.17) is a Boolean function, it can be represented by means of a BDD. Applying De Morgan's law to Eq. (5.17), we get for the connectivity:

$$C_{s,t} = \overline{K_1} \wedge \overline{K_2} \wedge \cdots \wedge \overline{K_k} = \bigwedge_{j=1}^{k} \overline{K_j}. \tag{5.18}$$

The list of mincuts can be generated from the list of minpaths by means of Proposition 5.1. However, a direct algorithm is given in [28], based on a property proved in [29] that states that in a two-terminal reliability problem, the removal of a mincut partitions the nodes into two mutually exclusive subsets such that:

- the source terminal belongs to one set and the sink terminal belongs to the other set,
- each set comprises a connected subnetwork,
- each arc of a mincut has one terminal in one set and the other terminal in the other set.

This principle was used in [29] to construct a systematic binary tree search that generates all minimal cutsets by sequentially adding vertices and branches until each path culminates in a minimal cutset. In [28], it is shown that the same partitioning principle can be implemented without constructing a binary search tree. Indeed, in [28] each minimal cutset is constructed in one step by adding and removing arcs in a manner that satisfies the partitioning principle of [29]. The algorithm in [28] offers opportunities for improving both computational time and computer storage requirements. An alternative algorithm is proposed in Section 5.2.6.

Example 5.3   *Bridge Network: Mincut*
The bridge network of Figure 5.2 possesses four mincuts:

$$K_1 = \overline{e_1} \wedge \overline{e_2}, \quad K_2 = \overline{e_4} \wedge \overline{e_5}, \quad K_3 = \overline{e_2} \wedge \overline{e_4}, \quad K_4 = \overline{e_1} \wedge \overline{e_3} \wedge \overline{e_5}. \tag{5.19}$$

Comparing Eq. (5.19) with Eq. (5.7), one can verify that Proposition 5.1 holds. The network disconnectivity (5.17) is expressed as:

$$\overline{C_{s,t}} = (\overline{e_1} \wedge \overline{e_2}) \vee (\overline{e_4} \wedge \overline{e_5}) \vee (\overline{e_2} \wedge \overline{e_4}) \vee (\overline{e_1} \wedge \overline{e_3} \wedge \overline{e_5}). \tag{5.20}$$

$P\{\overline{C_{s,t}}\}$ is the (s,t)-unreliability of the network that can then be computed by any available methods for the computation of the probability of Boolean functions.

Applying De Morgan's laws to (5.20), the network connectivity is expressed as:

$$C_{s,t} = (e_1 \vee e_2) \wedge (e_4 \vee e_5) \wedge (e_2 \vee e_4) \wedge (e_1 \vee e_3 \vee e_5). \qquad (5.21)$$

Manipulating Eq. (5.21) with the rules of Boolean algebra (Section 6.12), Eq. (5.8) can easily be obtained.

Problems

5.6 Consider a simple water distribution network [30] consisting of five pipes, as shown in Figure 5.5. Assume node 1 as source and node 5 as sink. If all the five pipe sections are subject to possible failure with failure probability $q = 0.05$, find the two-terminal reliability by first finding all the minpaths followed, then using SDP. Repeat by first finding all the mincuts and then using the SDP method.

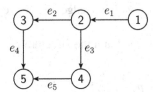

Figure 5.5 A water distribution network with five pipes [30].

5.7 In the network of Figure 5.5, assume now that the nodes can also fail (in addition to the edges) with probability $q_N = 0.1$. Find the two-terminal reliability by the minpaths followed, using the SDP method.

5.8 In [31], the water distribution network of Figure 5.6 was examined. Given that node 1 is the source and node 2 the sink, find the two-terminal network reliability using the BDD method when the reliability of pipe section i is r_i. Next, find the Birnbaum importance of each pipe section. Finally, numerically compute the network reliability when the reliability of each pipe section is $r = 0.9$.

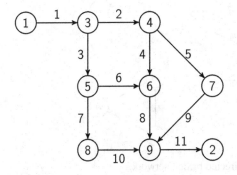

Figure 5.6 A water distribution network from [31].

5.9 In the network of Figure 5.6, assume now that the pipe sections are undirected. Find the two-terminal network reliability (using the method of your choice) when the reliability of each pipe section is $r = 0.9$.

5.2.4   Network Reliability Based on Factoring

Factoring algorithms exploit the factoring/conditioning principle described in Section 4.4, and expressed by Eq. (4.63), that can be interpreted as the probability version of the pivotal Shannon decomposition. The same factoring/conditioning principle can be applied to networks [6, 32].

Example 5.4 *Bridge Network: Factoring*

We return to the bridge network with directed arcs of Figure 5.2. Taking into account the arc direction, a possible decomposition step that gives rise to two series/parallel substructures is obtained assuming as pivot variable edge e_1, as represented in Figure 5.7.

If e_1 is up, the two nodes v_1 and v_3 are merged and the structure reduces to e_4 in parallel with the series of e_2 and e_5, hence:

$$P\{F_{e_1=1}\} = p_2 p_5 + p_4 - p_2 p_4 p_5 \,.$$

If e_1 is down, the connection is guaranteed only through edge e_2, which connects e_5 in parallel with the series of e_3 and e_4. Hence:

$$P\{F_{e_1=0}\} = p_2 (p_3 p_4 + p_5 - p_3 p_4 p_5) \,.$$

Using Eq. (5.3), we have:

$$R_{s,t} = p_1 (p_2 p_5 + p_4 - p_2 p_4 p_5) + (1 - p_1)(p_2 p_3 p_4 + p_2 p_5 - p_2 p_3 p_4 p_5)$$
$$= p_1 p_4 + p_2 p_5 + p_2 p_3 p_4 - p_1 p_2 p_4 p_5 - p_1 p_2 p_3 p_4 - p_2 p_3 p_4 p_5 + p_1 p_2 p_3 p_4 p_5 \,.$$

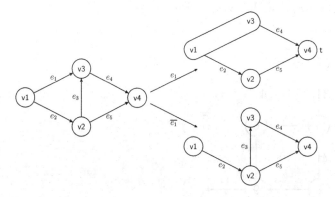

Figure 5.7   Factoring decomposition of the directed bridge network.

In the above example, one level of factoring was sufficient to obtain two series-parallel structures that can be solved directly. For more complex networks, factoring may need to be replicated to obtain the same simplification.

Note that both the factoring approach and the BDD approach rely on the pivotal decomposition. The selection of elements to factor on will determine the complexity of the algorithm [33].

The factoring algorithm is usually iterated until series-parallel structures are obtained; by contrast, the BDD method applies the pivotal decomposition repeatedly until a 0 or a 1 is reached. However, the factorization method can also be iterated on a complete sequence of variables to reach a complete decomposition of the connectivity function, allowing a direct derivation of the BDD. This method has been exploited in [34] and revisited in [23]. In the graphical representation of the factorization method, the subgraph obtained by setting the edge $e_i = 1$ is the graph of the function $F_{e_i=1}$ in Eq. (5.1). This operation is called *contraction* [34] since it corresponds to joining the source and the destination vertices of edge e_i. Similarly, the subgraph obtained by setting $e_i = 0$ is the graph of the function $F_{e_i=0}$ in Eq. (5.1). This operation is called *deletion* [34].

In terms of graph factorization, Shannon's decomposition (5.1) can be rewritten as:

$$G = e_i T \vee \overline{e}_i E,\qquad(5.22)$$

where T is the subgraph obtained from G by contracting edge e_i and E is the subgraph obtained from G by deleting edge e_i.

5.2.5 Graph-Visiting Algorithm

The BDD representation of the two-terminal connectivity of a graph can be directly derived without a search for the minpaths or mincuts, or without a pivotal decomposition, but through a direct visit of the graph. In [35], an algorithm is proposed that generates the BDD directly, via a recursive visit of the graph, without explicitly deriving the Boolean expression for the connectivity function. The algorithm is sketched in Algorithm 5.1 in pseudocode form.

Given a graph $G = (V,E)$ and two nodes (s,t), the algorithm starts from s and visits the graph according to a given but arbitrary visiting strategy until t is reached. The BDD construction starts recursively once the terminal node t is reached. The BDDs of the nodes along a path from s to t are combined with AND, while the paths starting from a node with more than one outgoing edge are combined with OR. The algorithm is followed step by step in Example 5.5.

Example 5.5 *Bridge Network: Graph-Visiting Algorithm*
The application of Algorithm 5.1 is illustrated on the directed bridge network of Figure 5.2. The graph is visited adopting a depth-first search along the progressive (but arbitrary) numbers assigned to the nodes. Starting from the source node s (node 1), the visit proceeds toward node 2 along edge e_2, toward node 3 along edge e_3 and toward

---

Algorithm 5.1 BDD-visiting algorithm for (s,t)-reliability.

    start_node = source_node,
    sink_node
1: **procedure** BDD_GEN(START_NODE)
2:    T_bdd = 0;
3:    set start_node in this_path
4:    **for** edge_i in the set of edges starting from start_node **do**
5:        next_node = the other end of edge_i;
6:        **if** next_node == sink_node **then**
7:            subpath_bdd = edge_i_bdd;
8:        **else**
9:            **if** next_node is already in this_path **then**
10:               continue;
11:            **else**
12:               subpath_bdd = bdd_gen(next_node) AND edge_i_bdd;
13:            **end if**
14:            T_bdd=T_bdd OR subpath_bdd;
15:        **end if**
16:    **end for**
17:    clear start_node in this_path;
18:    return T_bdd;
19: **end procedure**

---

the sink t (node 4) along edge e_4. Once the sink is reached, the BDD construction starts with the BDD representing arc e_4 – Figure 5.8(a).

Since e_4 is the only outgoing edge out of node 3, the visit returns to node 2. A new BDD for arc e_3 is generated, Figure 5.8(b), and the AND operator between the two generated BDDs is applied ($e_4 \wedge e_3$) – Figure 5.8(c). Coming back to node 2,

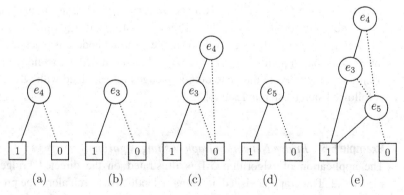

Figure 5.8   Application of the graph-visiting algorithm to the bridge network I. (a) BDD arc e_4; (b) BDD arc e_3; (c) BDD $e_4 \wedge e_3$; d) BDD arc e_5; e) BDD $((e_4 \wedge e_3) \vee e_5)$.

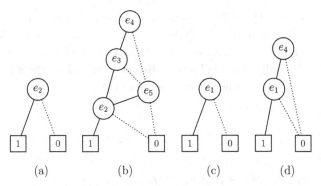

Figure 5.9   Application of the graph-visiting algorithm to the bridge network II. (a) BDD e_2; (b) BDD $(((e_4 \wedge e_3) \vee e_5) \wedge e_2)$; (c) BDD e_1; (d) BDD $e_4 \wedge e_1$.

the algorithm visits arc e_5, reaching the sink and building the corresponding BDD, Figure 5.8(d), that is combined using OR with the BDD generated so far $((e_4 \wedge e_3) \vee e_5)$ – Figure 5.8(e). Since there are no more arcs emerging from node 2, the visit returns to node 1 and the new BDD representing arc e_2, Figure 5.9(a), is combined using AND with the already constructed BDD $(((e_4 \wedge e_3) \vee e_5) \wedge e_2)$ – Figure 5.9(b).

The same procedure is used to generate the BDD representing e_1, Figure 5.9(c), and the path (e_1, e_4) between node 1 and node 4 – Figure 5.9(d). The BDD for arc e_4 exists already, and it is not generated again. Finally, the BDDs of the two paths outgoing from node 1 are combined using OR to obtain the final BDD shown in Figure 5.10, which represents the $(\mathtt{s},\mathtt{t})$-connectivity function

$$C_{\mathtt{s},\mathtt{t}} = (e_1 \wedge e_4) \vee (((e_4 \wedge e_3) \vee e_5) \wedge e_2). \tag{5.23}$$

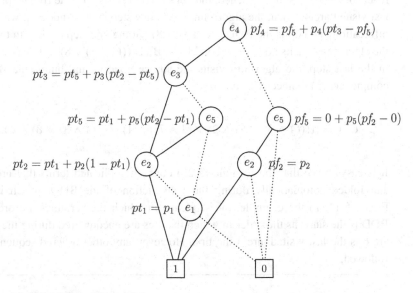

Figure 5.10   The final BDD of Figure 5.2.

The order in which the variables appear in the BDD of Figure 5.10 is different from that in Figure 5.3; hence, the two BDDs have a different structure even if they represent the same Boolean function. The network reliability can be calculated by recursively applying Eq. (5.3), as illustrated in Figure 5.10 and in Eq. (5.24), where it is shown that the result coincides with the one obtained in Eq. (5.14):

$$pt_1 = p_1,$$

$$pf_2 = p_2,$$

$$pt_2 = pt_1 + p_2(1 - pt_1) = p_1 + p_2 - p_1 p_2,$$

$$pf_5 = 0 + p_5(pf_2 - 0) = p_2 p_5, \tag{5.24}$$

$$pt_5 = pt_1 + p_5(pt_2 - pt_1) = p_1 + p_5(p_2 - p_1 p_2),$$

$$pt_3 = pt_5 + p_3(pt_2 - pt_5) = p_1 + p_2 p_5 - p_1 p_2 p_5 + p_2 p_3 - p_1 p_2 p_3 - p_2 p_3 p_5$$
$$+ p_1 p_2 p_3 p_5,$$

$$pf_4 = pf_5 + p_4(pt_3 - pf_5) = p_2 p_5 + p_1 p_4 + p_2 p_3 p_4 - p_1 p_2 p_4 p_5 - p_1 p_2 p_3 p_4$$
$$- p_2 p_3 p_4 p_5 + p_1 p_2 p_3 p_4 p_5.$$

Example 5.6 *Regular Network: Graph-Visiting Algorithm*
The construction of the BDD for the regular directed network of Figure 5.4 according to Algorithm 5.1 proceeds along the following steps. The graph is visited according to the progressive (but arbitrary) numbers assigned to the nodes. Starting from the source node $s = v_1$, the visit proceeds along nodes v_2, v_3, v_4, until the sink node $t = v_5$ is reached. Once the sink is reached, the construction starts with the BDD representing the last visited arc 4. Then, the algorithm makes one step back to node v_3, where it finds a bifurcation and builds the BDD $((4 \wedge 3) \vee 8)$. Going one step back with the recursion, the algorithm revisits node v_2 and builds the BDD $(((4 \wedge 3) \vee 8) \wedge 2) \vee (4 \wedge 7)$. Finally, in the last step, the algorithm visits the source node v_1 and builds the BDD for the complete (s,t)-connectivity function:

$$C_{s,t} = (((((4 \wedge 3) \vee 8) \wedge 2) \vee (4 \wedge 7)) \wedge 1) \vee (((4 \wedge 3) \vee 8) \wedge 6). \tag{5.25}$$

It is easy to see that the formula (5.25) contains replicated terms that are simplified and folded automatically during the construction of the BDD, which is shown in Figure 5.11. In the example, the sequence in which the variables are ordered in the BDD is the same as the order in which the arcs are encountered during the graph visit: arc 6 is the last visited, arc 4 the first. However, any other ordered sequence could be followed.

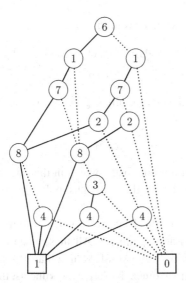

Figure 5.11   The final ROBDD of the network of Figure 5.4.

5.2.6   Deriving Minpaths and Mincuts from BDD

The algorithm based on repeated factoring and Algorithm 5.1 both construct the BDD of the Boolean connectivity function without a preliminary search of the minpaths or mincuts. Nevertheless, it is sometimes useful, or even required, to determine the list of the minpaths and/or the mincuts of the network. The minpaths provide information about the most favorable connections (in terms of number of hops) between source and sink, whereas the mincuts give useful information about the network bottlenecks, i.e., the most critical events that may disrupt the network into two non-communicating parts.

Given a BDD representing the connectivity function of a binary network, two dual properties are easily verified by searching for the sequences of nodes linking the root node of the BDD with the terminal leaf labeled 1 (for the paths) or the terminal leaf labeled 0 (for the cuts):

1. Any node sequence in a BDD linking the BDD root with the terminal leaf 1 is a path of the network, but not necessarily a minpath. To derive the minpaths, the list of paths obtained from the BDD must be simplified and minimized by expunging non-minimal paths.

   From the BDD of Figure 5.10, the sequences from the root node to the terminal leaf 1 are:

$$e_4\,e_3\,e_2, \quad e_4\,\overline{e_3}\,e_5\,e_2, \quad e_4\,\overline{e_3}\,\overline{e_5}\,e_1, \quad e_4\,e_3\,\overline{e_2}\,e_1, \quad \overline{e_4}\,e_5\,e_2. \tag{5.26}$$

All the path sequences derived from the BDD in (5.26) are disjoint products since, by construction, any two of them contain at least one node that is directed in one and negated in the other. Thus, BDD can also be seen as an SDP method and the network reliability can be calculated from the sequences in (5.26), to get:

$$R_{(s,t)} = p_4 p_3 p_2 + p_4 (1 - p_3) p_5 p_2 + p_4 (1 - p_3)(1 - p_5) p_1$$
$$+ p_4 p_3 (1 - p_2) p_1 + (1 - p_4) p_5 p_2.$$

To find the minpath list from (5.26), the first simplifying step is to remove the negated elements to get:

$$e_4 e_3 e_2, \quad e_4 e_5 e_2, \quad e_4 e_1, \quad e_4 e_3 e_1, \quad e_5 e_2.$$

The second step is to delete the non-minimal paths which in this case are $\{e_4 e_5 e_2\}$ and $\{e_4 e_3 e_1\}$, to obtain the final list of the minpaths only: $\{e_4 e_3 e_2\}$, $\{e_4 e_1\}$, $\{e_5 e_2\}$.

2. Any node sequence in a BDD linking the BDD root with the terminal leaf 0 is a cut of the network, but not necessarily a mincut. All these node sequences are disjoint products since, by construction, any two of them contain at least one node that is directed in one and negated in the other. To derive the mincuts, the list of cuts obtained from the BDD must be simplified (by deleting the non-negated terms) and minimized by expunging non-minimal cuts.

From the BDD of Figure 5.10, the sequences from the root node to the leaf 0 are:

$$\overline{e_4}\,\overline{e_5}, \quad \overline{e_4} e_5 \overline{e_2}, \quad e_4 e_3 \overline{e_2}\,\overline{e_1}, \quad e_4 \overline{e_3} e_5 \overline{e_1}, \quad e_4 \overline{e_3} e_5 \overline{e_2}\,\overline{e_1}.$$

After simplification and expunging, the list of mincuts is obtained:

$$\overline{e_1}\,\overline{e_2}, \quad \overline{e_4}\,\overline{e_5}, \quad \overline{e_2}\,\overline{e_4}, \quad \overline{e_1}\,\overline{e_3}\,\overline{e_5}. \tag{5.27}$$

An alternative approach to the search for the minpaths and mincuts was proposed by Rauzy in [36]. He introduced a BDD operator "$/$" called *without* whose recursive application reduces the BDD of a Boolean function to a new BDD that encodes the

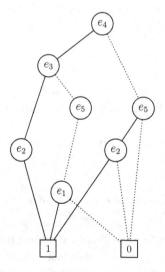

Figure 5.12   The reduced BDD with the minpaths of Figure 5.2.

minimal representation in terms of minpaths or mincuts of the original BDD. The detailed proof of this method and the related algorithms are in [36]. Application of the reduction method to the BDD of Figure 5.10 produces the BDD of Figure 5.12, where the only possible connections between the BDD root and the terminal node labeled 1 are the three minpaths in (5.7).

Example 5.7   *GARR: Italian Research and Education Network*
The GARR network operates the Italian national high-speed telecommunication network for universities and scientific institutions [37]. Its shareholders are four major Italian research and academic organizations.

The layout of the major nodes and connections of the GARR network in 2008 is shown in Figure 5.13. The network is composed of 42 nodes and 52 links. The (s,t) network reliability and the list of minpaths and mincuts are evaluated in two cases: in the first case, assuming node TO as source and node CT as sink; in the second case, node TS1 as source and node NA as sink.

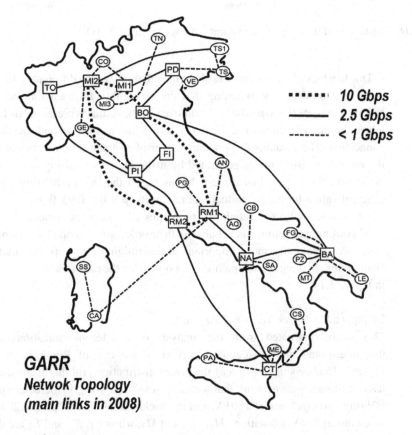

Figure 5.13   Map of the GARR network in 2008.

Table 5.1 Reliability computations on the graph of Figure 5.13.

Source node	Terminal node	Minpath	Mincut	No. of BDD nodes	Reliability single edge		
					0.9	0.95	0.99
TO	CT	196	481	1003	0.977428	0.994713	0.999797
TS1	NA	168	385	604	0.975771	0.994486	0.999795

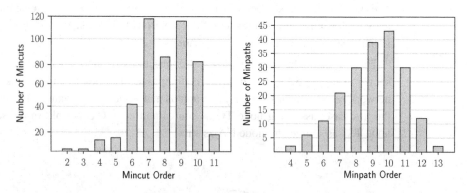

Figure 5.14 Histogram of the minpath and mincut order for the case s = TO, t = CT.

Due to a lack of consistent data about the reliability of the different links, a parametric study is carried out by assuming that the probability for each link being up varies ranging from $p = 0.9$ to $p = 0.99$. A summary of the results is presented in Table 5.1 [38], where the first row shows the TO to CT connection and the second row the TS1 to NA connection. The columns display the number of minpaths, the number of mincuts, and the number of BDD nodes generated by using the BDD-visiting algorithm described in Section 5.2.5. The final three columns report the (s,t)-reliability assuming three different values for the individual link reliability: $p = 0.9, 0.95, 0.99$.

The connectivity and reliability properties of a network depend not only on the number of minpaths/mincuts but also on their order, since minpaths or mincuts of lower order are structurally more important in determining the properties of the network. Histograms showing the minpath/mincut orders for the case s = TO, t = CT are shown in Figure 5.14.

Example 5.8 *Power Network Reliability*

This study originated from the analysis of a telecommunications node outage that interrupted the service to a portion of the city of Rome in 2004 [39, 40]. Figure 5.15 shows the portion of the power distribution grid that feeds the node of the telecommunications network that went out of service. The grid includes a portion of the HV (high voltage) grid at 150 kV, and the backbone of the MV grid at 20 kV. The node is fed through MV substations M_5, M_6, and M_7, while E_1, E_2, and E_3 are the interfaces connecting the portion of the represented grid with the power transmission grid.

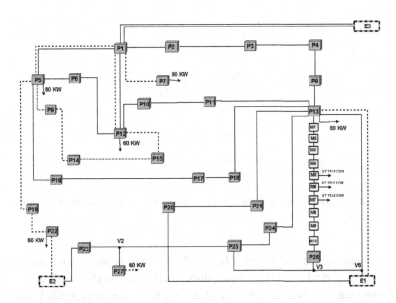

Figure 5.15   Layout of a portion of a power network.

The physical link between any two nodes is an electrical trunk. The failure rates of the various trunks have been estimated from the data available from the electrical company and range between a maximum of $\lambda_{max} = 0.148\,\mathrm{yr}^{-1}$ and a minimum of $\lambda_{min} = 0.005\,29\,\mathrm{yr}^{-1}$. To supply power to the node, substations M_j ($j = 5,6,7$) should be connected with the entry points E_i ($i = 1,2,3$). We denote by $C(E_i, M_j)$ the connectivity functions between node E_i ($i = 1,2,3$) and node M_j ($j = 5,6,7$). The reliability (unreliability) of the connection is given by:

$$R_{E_i,M_j}(t) = P\{C(E_i, M_j)\}, \qquad U_{E_i,M_j}(t) = 1 - R_{E_i,M_j}(t). \tag{5.28}$$

To compute the probability that the node is correctly fed by the power grid the following cases are analyzed:

1. The point-to-point reliability between a single node E_i and a single node M_j.
2. The probability that all the nodes M_j are fed by a single source E_i. To this end, the individual point-to-point connectivity functions are combined with AND, as in

$$R_{And} = P\{C(E_i, M_5) \wedge C(E_i, M_6) \wedge C(E_i, M_7)\}. \tag{5.29}$$

3. The probability that at least one M_j node is fed by a source E_i. To this end, the individual point-to-point connectivity functions are combined with OR:

$$R_{Or} = P\{C(E_i, M_5) \vee C(E_i, M_6) \vee C(E_i, M_7)\}. \tag{5.30}$$

4. The probability that at least two out of three M_j nodes are fed by a source E_i:

$$R_{2/3} = P_{2/3}\{C(E_i, M_5), C(E_i, M_6), C(E_i, M_7)\}. \tag{5.31}$$

Similar equations hold for the unreliability, which is plotted in Figure 5.16 assuming node E_1 as a source node.

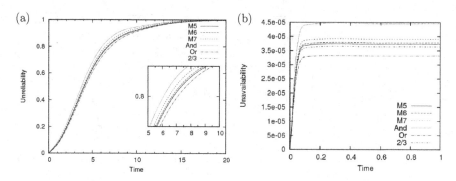

Figure 5.16 Unreliability (a) and unavailability (b) of various configurations between E_1 and M_5, M_6, M_7 (from [39]).

The curves denoted by M_5, M_6, and M_7 plot the point-to-point unreliability (5.28), the curve marked "And" the unreliability computed from (5.29), the curve marked "Or" the unreliability from Eq. (5.30) and, finally, the curve marked "2/3" the unreliability computed from Eq. (5.31). As expected, the "And" configuration is the most unreliable, while the "Or" configuration is the most reliable. These results may guide the designers to find a balance between network complexity (and hence cost) and reliability.

The unavailabilities for the same connection configurations are reported in Figure 5.16(b), assuming a repair rate equal to $\mu = 50\,\text{yr}^{-1}$ corresponding to a pessimistic upper bound for the out of service time of each trunk equal to about one week. To gain an insight on the weakest points of the network, we examine the list of the mincuts, whose total number is displayed in Table 5.2.

Table 5.2 Numbers of mincuts for the configurations examined in Figure 5.16.

Source node	M_5	And	Or	2/3
E_1	134	168	106	142

The two most critical mincuts are given by the combinations of links $(P_{26}-M_{10}, P_{13}-M_1)$ and $(P_{26}-M_{10}, M_3-M_2)$, and correspond to the interruption of the trunks that feed the series of MV stations.

This observation suggests a possible design improvement in the configuration of the grid by modifying the connection of the MV stations M_1-M_{10} to nodes P_{26} and P_{13} in parallel instead of in series, as in Figure 5.15. The analysis of this new parallel connection shows a remarkable improvement in the computed unreliability and unavailability functions, as shown in Figure 5.17. The curves of the parallel configuration are marked with "Par," and the previous curves from Figure 5.16 are also shown for comparison. Modifying the structure may improve the unavailability by an order of magnitude.

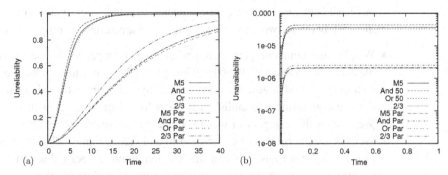

Figure 5.17   Unreliability (a) and unavailability (b) of various parallel configurations between E_1 and M_5, M_6, M_7 compared with Figure 5.16 (from [39]).

Problems

5.10   For the bridge network of Figure 5.2, determine the reliability importance with respect to all the arcs.

5.11   Apply the graph-visiting algorithm to the water distribution network of Figure 5.5 and derive the corresponding BDD. Then, find the minpath and mincut list from the BDD.

5.12   Apply the graph-visiting algorithm to the water distribution network of Figure 5.6 and derive the corresponding BDD. Then, find the minpath and mincut list from the BDD.

5.13   In the network of Figure 5.5, assume that the only source (node 1) should feed three destinations, nodes 3, 4, and 5, under three policies:

>   *Case 1:* The network is considered working if all the destinations are connected to the source.
>   *Case 2:* The network is considered working if at least one destination is connected to the source.
>   *Case 3:* The network is considered working if two out of three destinations are connected to the source.

Find the network reliability in the three cases above.

5.3   Binary Probabilistic Weighted Networks

The quality of service that a networked system can deliver is determined by its reliability as well as other attributes of its elements like capacity, delay, cost, length, resistance, etc. Some examples are the capacity or bandwidth characterizing the connections in communication or transport systems, the distance between nodes in a highway network or the message delivery delay in an information or control system. To study the quality of service (QoS) of such systems a richer representation is required [41].

Motivated by these observations, the notion of binary probabilistic networks can be extended by defining binary probabilistic weighted networks (BPWN), whose edges have been assigned weights (indicated as a label associated with the edge). We distinguish between two types of weights that characterize the properties of a network:

- Weights represent a property that is additive over the paths, like the cost, the length or the delay associated with the edge. Going from the source to the sink, the weights sum over the traversed edges. The aim of the analysis is usually to evaluate the probability that the connection is established at (or below) a given cost (or distance, or delay). We identify this category of weights that are additive over the paths as *cost*.
- Weights are interpreted as capacities or bandwidths, representing the maximum flow that an edge can transmit when up. The maximum capacity that can be transmitted through a cut is the sum of the capacities of the edges forming the cut. The aim of the analysis is to evaluate the ability of the network to transport a given amount of flow with a given probability. We identify this category of weights that are additive over cuts as *capacities*.

Weighted networks are usually interpreted in the literature as flow networks [42, 43], with weights interpreted as capacities. Analysis techniques for weighted networks are derived from those used for binary networks. In particular, BPWN can be represented and analyzed by resorting to an extension of a BDD called a multi-terminal binary decision diagram (MTBDD) [44–46] or algebraic decision diagram (ADD) [47].

MTBDDs encode a multi-valued function of Boolean variables into a binary tree [44–46], and have the advantage of inheriting properties and algorithms from BDDs. While BDDs have only two terminal leaves 0 and 1, MTBDD can have more than two terminal leaves that identify all the possible values taken by the multi-valued function along the paths from the root to the terminal leaves. Like BDDs, MTBDDs provide a compact representation of a multi-valued function by means of repeated application of Shannon's decomposition. Each node of the MTBDD represents a Boolean variable and has two successors: the left branch (usually shown with a solid line) represents the value of the variable being equal to 1, and the right branch (dashed line) represents the value of the variable being equal to 0. Similarly to a BDD, an MTBDD is constructed by imposing an ordering on the variables, and the size of the tree is extremely sensitive to the chosen ordering. An ordered MTBDD can be reduced by successive applications of the same deletion and merging rules as in the standard BDD (Section 5.2.1). The deletion rule eliminates a node that has both of its outgoing edges leading to the same successor. The merging rule eliminates one of two nodes with the same label as well as the same pair of successors. In the area of weighted networks, Hachtel and Somenzi [47] have presented a BDD-based algorithm for finding the maximum flow in a 0–1 network, and Gu and Xu [48] have introduced the use of ADDs (conceptually identical to MTBDDs) for computing the maximum flow in non-probabilistic networks. In [49], the use of MTBDDs is illustrated for various examples of weighted networks.

5.3.1   Definition of BPWN

A BPWN $G = (V, E, P, W)$ is a binary probabilistic network as defined in Section 5.2 with the addition of a weight function W that assigns to each edge e a non-negative real weight $w(e)$. $w(e)$ has the meaning of a reward assigned to the arc that qualifies the service carried by the arc, or some attribute of the arc like cost, delay, distance, capacity, etc. Given a source node s and a terminal node t, the reward accumulated passing from s to t when the network is connected is a discrete random variable $\Psi_{s,t}$ that depends on the structure of the graph, on the probability function P and on the weight function W. The computation of the pmf and the Cdf of the accumulated reward $\Psi_{s,t}$ is performed by representing the discrete values of the function as the terminal leaves of an MTBDD constructed over the n Boolean variables representing the arcs. To compute the reward function $\Psi_{s,t}$ by means of an MTBDD, new operators must be defined (in addition to the standard Boolean operators NOT, AND and OR) that depend on the type of weights: costs or capacities [49].

5.3.2   Weight as Cost

The weight $w(e)$ of arc e is a cost attached to the traversal from its input node to its output node. An intuitive and simple physical idea is the length of the arc, its electrical resistance or the time to traverse the arc.

Definition 5.5 *The* cost *of a minpath H_i ($i = 1, 2, \ldots, h$) is the sum of the costs of the edges forming the minpath. A path that is not a minpath has a cost no less than the cost of any included minpath.*

Definition 5.6 *Given a BPWN, the* reward *random variable $\Psi_{s,t}$ is defined as the minimal cost of traversing a path from s to t, for any configuration of* up *and* down *states of its edges. By Definition 5.5, $\Psi_{s,t}$ assumes the value of the minpath with minimal cost connecting s to t in the given configuration, or a value of ∞ if the network is not connected in the given configuration.*

To compute the probability mass function and the Cdf of the discrete random variable $\Psi_{s,t}$ on an MTBDD, additional operations are needed. When the weights are interpreted as costs, these new operations are introduced using the simple series and parallel networks of Figure 5.18. Nodes s and t are the source and the destination, and weights are represented in square brackets. Arc 1 has weight a and arc 2 has weight b.

The network of Figure 5.18(a) has a single minpath $H_1 = 1 \wedge 2$, the connectivity function is $C_{s,t} = H_1$ and when the network is connected ($C_{s,t} = 1$) the cost function is $\Psi_{s,t} = a + b$, otherwise it is $\Psi_{s,t} = \infty$. The MTBDD operation corresponding to the Boolean $\wedge$ is a sum of weights, and hence this operation is called "AndSum."

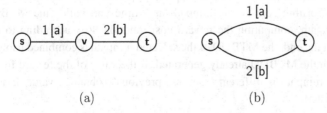

(a)   (b)

Figure 5.18   Network with two arcs: (a) in series; (b) in parallel.

Table 5.3   Truth table for AndSum and OrMin.

Arc 1	Arc 2	$C_{s,t}$ $1 \wedge 2$	AndSum	$C_{s,t}$ $1 \vee 2$	OrMin
0	0	0	0	0	0
0	1	0	0	1	b
1	0	0	0	1	a
1	1	1	$a+b$	1	$\min\{a,b\}$

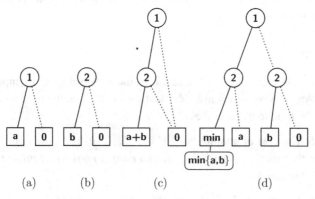

(a)   (b)   (c)   (d)

Figure 5.19   Basic MTBDD operations: (a) MTBDD for arc 1; (b) MTBDD for arc 2; (c) MTBDD for (1 AndMin 2); (d) MTBDD for (1 OrSum 2).

The network of Figure 5.18(b) has two minpaths, $H_1 = 1$ and $H_2 = 2$, and the connectivity function is $C_{s,t} = H_1 \vee H_2$. When the network is connected ($C_{s,t} = 1$) and both arcs are up, the cost is $\Psi_{s,t} = \min\{a,b\}$; when the network is connected through arc 1 (arc 2 being down), the cost is $\Psi_{s,t} = a$; when it is connected through arc 2 (arc 1 being down), the cost is $\Psi_{s,t} = b$. The MTBDD operation corresponding to the Boolean $\vee$ is called "OrMin."

The generation of the cost reward function $\Psi_{s,t}$ of Definition 5.6 requires the implementation of the AndSum and OrMin operators, whose definitions are summarized in the truth table of Table 5.3 and in the corresponding MTBDD of Figure 5.19 [49].

The construction of the MTBDD requires a preliminary search for the minpaths. Then, the algorithm selects one minpath at a time and builds the MTBDD for the selected minpath, computing the reward cost function $\Psi_{s,t}$ according to the rules of Table 5.3. At the end, the MTBDD of the selected minpath is combined using the OrMin operator with the MTBDD already generated. If the value of the reward function along the selected minpath is different from any previously obtained value, a new terminal leaf is generated.

Example 5.9 *Bridge Network with Cost Weights*
The bridge network of Figure 5.2 is augmented with weights assigned to the arcs, shown in square brackets in the labeled network of Figure 5.20. The weights are assumed to be costs. The network has $h = 3$ minpaths, as in (5.7), with the total cost given in Table 5.4.

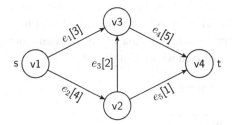

Figure 5.20 Weighted version of bridge network of Figure 5.2.

Table 5.4 Costs of the minpaths of Figure 5.20.

Minpath	Arcs	Cost
H_1	$e_1 \wedge e_4$	8
H_2	$e_2 \wedge e_5$	5
H_3	$e_2 \wedge e_3 \wedge e_4$	11

The MTBDD constructed on the weighted bridge network of Figure 5.20 is given in Figure 5.21. The algorithm generates four terminal leaves; three leaves have the accumulated reward of the minpaths given in Table 5.4, while the fourth leaf, marked ∞,

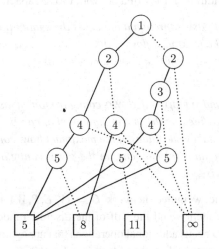

Figure 5.21 MTBDD for the cost function of the bridge network of Figure 5.20.

represents combinations of up and down edges for which the network is not connected. The values on the terminal leaves represent the minimum cost incurred along the path starting from the root node.

Assigning to all the arcs a probability of being up equal to $p = 0.9$ and recursively applying Eq. (5.3), the results are reported in Table 5.5. Column 1 reports all the values of the reward function $\Psi_{s,t}$ represented by the terminal leaves. Column 2 is the probability mass function $P\{\Psi_{s,t} = c\}$; the last row labeled ∞ is the probability that the network is not connected (which is coincident with the value of the non-weighted network of Figure 5.2). Column 3 is the cumulative distribution function, i.e., the probability that the network is connected with a cost $\Psi_{s,t} \leq c$.

Table 5.5 Probability mass function and Cdf of the accumulated reward $\Psi_{s,t}$.

Cost c	Probability mass $P\{\Psi_{s,t} = c\}$	Cdf $P\{\Psi_{s,t} \leq c\}$
5	0.8100	0.8100
8	0.1539	0.9639
11	0.0073	0.9712
∞	0.0288	—

Note that the cumulative distribution is defective, with a defect equal to the probability that the network is not connected. If a maximum allowable cost $c = 10$ is specified, the probability that s and t are connected with a cost less than this value is 0.9639.

5.3.3   Weight as Capacity

The weight $w(e)$ of the arc e represents the maximum flow that an arc is able to carry. Concrete instances are the bandwidth of a trunk of a network or the capacity of a pipe.

Definition 5.7 *The total flow that passes through a mincut is the sum of the capacities of the arcs forming the mincut. The maximum flow of the network is determined by the mincut with the minimum capacity. Any cut that is not a mincut has a flow greater than any included mincut.*

Definition 5.8 *Given a BPWN and an up and down configuration of its edges, the reward random variable $\Psi_{s,t}$ is defined as the maximum flow that can pass from s to t in the given configuration. $\Psi_{s,t}$ takes the values of the maximum flow, corresponding to the flow of the mincut with the minimal capacity, in the given configuration if the network is connected, or 0 otherwise.*

The analysis of a probabilistic weighted network $G = (V, E, P, W)$ is aimed at computing the probability mass function and the Cdf of the discrete random variable $\Psi_{s,t}$. Construction of the MTBDD needs additional operations that are again introduced with reference to Figure 5.18, where the labels a and b on arcs 1 and 2, respectively, are

Table 5.6 Truth table of AndMin and OrSum.

Arc 1	Arc 2	$C_{s,t}$ $1 \wedge 2$	AndMin	$C_{s,t}$ $1 \vee 2$	OrSum
0	0	0	0	0	0
0	1	0	0	1	b
1	0	0	0	1	a
1	1	1	min$\{a,b\}$	1	$a+b$

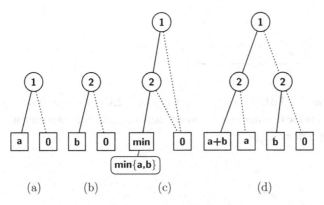

Figure 5.22 Basic MTBDD operations: (a) MTBDD for arc 1; (b) MTBDD for arc 2; (c) MTBDD for (1 AndMin 2); (d) MTBDD for (1 OrSum 2).

now interpreted as capacities. The network of Figure 5.18(a) has two mincuts, $K_1 = \overline{1}$ and $K_2 = \overline{2}$; the disconnectivity function is $\overline{C}_{s,t} = K_1 \vee K_2$ and the connectivity is $C_{s,t} = \overline{K_1} \wedge \overline{K_2}$. When the network is connected ($C_{s,t} = 1$) the flow function is $\Psi_{s,t} = \min\{a,b\}$, otherwise $\Psi_{s,t} = 0$. In the computation of the flow function the Boolean $\wedge$ operator corresponds to a min in the flows. This operation is called "AndMin."

The network of Figure 5.18(b) has a single mincut, $K_1 = \overline{1} \wedge \overline{2}$, and the connectivity function is $C_{s,t} = \overline{K_1} = 1 \vee 2$. When both arcs are up, the flow is $\Psi_{s,t} = a + b$; when the network is connected ($C_{s,t} = 1$) through arc 1 only (arc 2 being down), the flow is $\Psi_{s,t} = a$; when it is connected through arc 2 only (arc 1 being down), the flow is $\Psi_{s,t} = b$. The MTBDD operation corresponding to the Boolean $\vee$ is called "OrSum." The generation of the flow reward function $\Psi_{s,t}$ defined in Definition 5.8 requires the implementation of the AndMin and OrSum operators. The definitions of these operators are described in the truth table of Table 5.6 and in the corresponding MTBDD of Figure 5.22 [49].

When the weights are interpreted as capacities, the MTBDD is built starting from a preliminary search for the mincuts. The MTBDD generation algorithm takes one mincut at a time from the list of mincuts and builds the MTBDD for the given mincut by means of the OrSum operator over all the arcs of the mincut. The obtained MTBDD is combined using the AndMin operator with the MTBDD already generated.

Example 5.10 *Bridge Network with Capacity Weights*

Assume that the weights on the edges of the bridge network of Figure 5.20 represent capacities. The network has $k = 4$ mincuts as in (5.27), with maximum capacities reported in Table 5.7.

Table 5.7 Capacities of the mincuts.

Minpath	Arcs	Capacity
K_1	$e_1 \vee e_2$	7
K_2	$e_2 \vee e_4$	9
K_3	$e_4 \vee e_5$	6
K_4	$e_1 \vee e_3 \vee e_5$	6

The generated MTBDD is displayed in Figure 5.23, where the values on the terminal leaves represent the maximum flow that the network can carry along a path starting from the root node. For example:

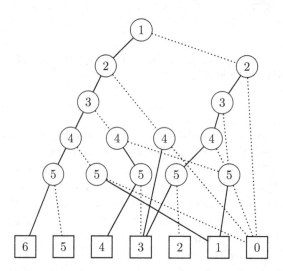

Figure 5.23   MTBDD for the flow function of the bridge network of Figure 5.20.

- The terminal leaf 6 is reached along the path $1,2,3,4,5$, in which all the arcs are up. In this case the maximum flow is carried by the weakest cut $(4,5)$.
- The terminal leaf 3 is reached along the paths $(1,2,\overline{3},4,\overline{5})$, $(1,\overline{2},3,4)$ or $(1,\overline{2},\overline{3},4)$, where the weakest cut is given by the arc 1 only, which carries a flow of 3. Another possible path is $(\overline{1},2,3,4,5)$, where the weakest cut is $(3,5)$, which also carries a maximum flow equal to the sum of the capacities of arcs 3 and 5, i.e., $2+1 = 3$.
- The terminal leaf 1 can be reached along many paths, for instance $(\overline{1},2,3,\overline{4},5)$, that are all characterized by sharing as weakest cut arc 5, which carries a capacity of 1.

As in the previous example, assigning to all the arcs a failure probability equal to $p = 0.9$, the probability mass for the different terminal values of the MTBDD of Figure 5.23 are given in Table 5.8. If there is a minimum threshold for the flow $\psi_{min} = 4$, the probability that the network is connected carrying a flow greater than the threshold is given by $P\{\Psi_{s,t} > 4\} = 1 - P\{\Psi_{s,t} \leq 4\} = 1 - 0.34390 = 0.65610$.

Table 5.8 Probability mass function and Cdf of the flow reward $\Psi_{s,t}$.

Flow f	Prob. mass $P\{\Psi_{s,t} = f\}$	$1 - Cdf$ $P\{\Psi_{s,t} > f\}$
0	0.02881	0.9712
1	0.08829	0.8829
2	0.00729	0.8756
3	0.15390	0.7217
4	0.06561	0.6561
5	0.06561	0.5905
6	0.59049	0.0000

Example 5.11 *Benchmark Network from [43]*

Figure 5.24 shows a benchmark example taken from [43], where the weights are reported, as usual, in square brackets.

Reference [43] assigns the same probability value of $p = 0.9$ to all the arcs and assumes weights as capacities, while in the present framework weights can alternatively be interpreted as capacities or costs. Table 5.9 shows the results, obtained from the generation of the MTBDD, that match those reported in [43].

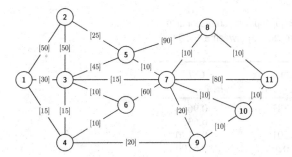

Figure 5.24 Benchmark network example.

The first column in Table 5.9 reports all the possible values of flow that can be transmitted between s and t and that give rise to a different terminal leaf in the MTBDD. The second column reports the corresponding probability mass function and the third column the probability of transmitting a flow greater than the corresponding flow value [49].

Table 5.9 Results for the network of Figure 5.24.

Flow f	Prob. mass $P\{\Psi_{s,t}=f\}$	$1-Cdf$ $P\{\Psi_{s,t}>f\}$	Flow f	Prob. mass $P\{\Psi_{s,t}=f\}$	$1-Cdf$ $P\{\Psi_{s,t}>f\}$
0	0.002 3361	0.997 6639	50	0.035 9151	0.701 4512
10	0.019 8815	0.977 7824	55	0.071 9313	0.629 5199
15	0.009 6750	0.968 1074	60	0.048 1033	0.581 4166
20	0.079 3581	0.888 7493	65	0.180 3750	0.401 0416
25	0.003 0942	0.885 6552	70	0.102 8710	0.298 1706
30	0.012 5073	0.873 1479	75	0.114 7200	0.183 4506
35	0.010 6024	0.862 5455	80	0.016 6772	0.166 7734
40	0.018 6152	0.843 9303	85	0.166 7720	0.0
45	0.106 5640	0.737 3663			

Example 5.12 *Weighted GARR: Italian Research and Education Network [38]*
With reference to Example 5.7 and Figure 5.13, we have added two further attributes to the GARR network:

- the bandwidth capacity as derived from the legend of Figure 5.13;
- the geographical distances between any two nodes.

The network is undirected and the links can be traversed in both directions. With respect to Example 5.7 we consider one single pair of source and sink nodes: TO–CT. Furthermore, we assume that all the links have a probability $p = 0.9$ of being up.

Bandwidth Probability Distribution
Given the edge capacities, the probability of assuring a given level of service between any two nodes can be evaluated. Fixing, for the connection TO–CT, a minimum tolerable capacity level of $f = 2500$ Mbps, Table 5.10 shows the probability mass

Table 5.10 Flow $f > 2500$ Mbps and corresponding mass and cumulative probability between $s = $ TO and $t = $ CT.

Flow f (Mbps)	Prob. mass $P\{\Psi_{s,t}=f\}$	$1-Cdf$ $P\{\Psi_{s,t}>f\}$
$f < 2500$	0.0387	0.9613
2500	0.3292	0.6321
2510	0.0201	0.6120
2520	0.0020	0.6100
2530	0.0027	0.6073
2655	0.0096	0.5977
2665	0.0036	0.5941
2675	0.0055	0.5886
2685	0.0069	0.5817
2810	0.0002	0.5815
2820	0.0006	0.5809
2830	0.0014	0.5795
5000	0.5795	0.0000

function $P\{\Psi_{s,t} = f\}$ and the probability $P\{\Psi_{s,t} > f\}$ of getting a capacity greater than f. The table is ordered from the minimal capacity exceeding $f = 2500$ Mbps to the maximum achievable bandwidth ($f = 5000$ Mbps).

From Table 5.10, we can read, for instance, that the probability of having the two nodes connected with a capacity $f > 2685$ Mbps is 0.5817.

Distance Probability Distribution

The edges have a weight corresponding to the length (in km) of the arc, and we calculate the probability to reach the sink node with a distance less than a preassigned threshold. We assume TO as source node and CT as sink node, and we find that the minimal distance of the connection TO–CT is 1522 km. In Table 5.11 we have quoted only the connections with a distance lower than a threshold $d = 1700$ km.

Table 5.11 Distance $d < 1700$ km and corresponding mass and cumulative probability between $s = $ TO and $t = $ CT.

Distance d (km)	Prob. mass $P\{\Psi_{s,t} = d\}$	Cdf $P\{\Psi_{s,t} \le d\}$
1522	6.56×10^{-1}	0.6561
1525	1.25×10^{-1}	0.7808
1536	5.90×10^{-2}	0.8398
1539	1.12×10^{-2}	0.8510
1561	6.91×10^{-2}	0.9202
1564	1.60×10^{-2}	0.9362
1567	9.35×10^{-3}	0.9455
1575	6.22×10^{-3}	0.9517
1578	1.44×10^{-3}	0.9532
1581	8.41×10^{-4}	0.9540
1635	1.09×10^{-3}	0.9551
1649	9.82×10^{-5}	0.9552
1659	3.14×10^{-4}	0.9555
1673	2.82×10^{-5}	0.9556
1677	9.23×10^{-5}	0.9556
1691	8.30×10^{-6}	0.9557
$d > 1700$	4.43×10^{-2}	—

The row with value ∞ gives the probability that the connection is not established, or is established with a total length greater that ψ_{max}.

5.4 Multi-State Networks

Many real-world systems are composed of multi-state elements, with different performance levels and several failure modes, and with various effects on the system's performance and dependability. Thus, the simplifying hypothesis used so far that nodes and edges of a network are binary entities must be relaxed in favor of a multi-valued representation.

Such systems are called multi-state systems (MSS), where the state of a component may be represented by a non-negative integer taking values in a discrete set. Under the multi-state assumption, the relationship between component states and system states is much more intricate and the analysis becomes more complex [50–52]. Even the term "reliability" becomes inappropriate, since the combination of different states of the components may provide different levels of performance for the whole system. In this case, a more appropriate and meaningful measure is the quality of service provided by the system as a combination of the states of its components. This measure is also referred to as *performability* [53]. The definition and formalization of MSS can be traced far back in the reliability literature, but only recently have various extensions of decision diagrams found an application in this particular field [54–60].

Weighted multi-state probabilistic networks are networks whose elements, both nodes and arcs, can have different mutually exclusive states with an associated probability of being in that state. Moreover, a performance attribute, called "weight," can be attached to the states of each element to characterize a property of the element in that state. In previous studies [54, 55], the performance attribute was assumed to be the capacity or bandwidth of the arcs, whose value decreases with the degradation of the arc.

Here, the weights can also have different physical meanings, as in Section 5.3 for binary networks. Thus, in addition to the interpretation of weights as flows, weights can represent a performance attribute, called "cost," that increases with the arc degradation and may represent a property like the resistance or the time to traverse the arc.

Examples of MSS can be found in different fields and in different problems:

- *Degradable systems:* Components or subsystems show different degradation levels with reduced performance capabilities. Possible areas of applications are power networks, communication and wireless sensor networks, pipeline networks, aqueducts [61].
- *Multi-mode failures:* Components or subsystems may manifest multi-mode failures, typically "stuck open" and "stuck closed," like valves or switching devices. In safety studies the failure mode influences the fail-safe or fail-unsafe mode of operation of the system [55, 62].
- *Optimal system design:* For each component, the technology or the market may offer different alternatives with different cost, performance, and reliability features [63]. The choice of the different alternatives for an optimal design can be translated into a multi-state weighted problem.

When the weights are interpreted as costs (or traversal times), the QoS is the total cost to traverse the network, and it is computed by a function that is additive over the paths. When the weights are assumed to be the arc capacities, the QoS function represents the total flow that the network is able to carry and is computed by a function that is additive over the cuts of the networks [64]. In both interpretations the analysis is performed by resorting to multi-valued decision diagrams (MDDs) [65, 66] and by defining suitable arithmetic/logic operations on the MDD.

5.4.1 Weighted Multi-State Networks

A multi-state network element is an element that can be found in different mutually exclusive and collectively exhaustive states, to which a probability mass function and a weight value is assigned. In the present context we assign probabilities and weights only to the arcs.

Assume a network has n arcs numbered from 1 to n. Arc i has $(m_i + 1)$ states numbered from 0 to m_i. The states are mutually exclusive and collectively exhaustive. A probability mass p_{ij} is assigned to each state, where i ($i = 1, 2, \ldots, n$) indicates the arc and j ($j = 0, 1, 2 \ldots, m_i$) the state of the arc. The probability masses satisfy the normalization condition

$$\sum_{j=0}^{m_i} p_{ij} = 1, \quad i = 1, 2 \ldots, n.$$

Similarly, a weight w_{ij} is assigned to state j of arc i. Denoting by x_i the state of component i ($x_i = 0, \ldots, m_i$), the state of the network is a vector $x = (x_1, x_2, \ldots x_n)$. Since each x_i can take $(m_i + 1)$ values, the total number of states of the system is

$$m = \prod_{i=1}^{n} (m_i + 1). \tag{5.32}$$

5.4.2 Multi-Valued Decision Diagrams

Multi-valued decision diagrams [65–67] are directed, acyclic graphs used to represent a function f of n discrete variables $(x_1, x_2, \ldots, x_n)$, where x_i can take $(m_i + 1)$ values. The function f takes the form

$$f : \{0, \ldots, m_1\} \times \cdots \times \{0, \ldots, m_n\} \to \{0, \ldots, S - 1\}.$$

Nodes in the MDD are either terminal or non-terminal. The terminal nodes correspond to the return values of the function and are labeled with a value $0, \ldots, S - 1$. Non-terminal nodes are labeled with a variable x_i, and contain $m_i + 1$ pointers to the subsequent nodes. These pointers correspond to the cofactors of f, where the cofactor for variable x_i is defined as

$$f_{x_i = c} \equiv f(x_n, \ldots, x_{i+1}, c, x_{i-1}, \ldots, x_1),$$

where the constant c is one of the possible values of x_i. A non-terminal node representing function f is then written as the $(m_i + 1)$-tuple $(x_i, f_{x_i=0}, \ldots, f_{x_i=m_i})$. As in the BDD, the paths in an ordered MDD (OMDD) visit non-terminal nodes according to some total ordering on the variables $x_n \prec \cdots \prec x_1$.

A reduced OMDD (ROMDD) has the following additional properties:

- There are no duplicate terminal nodes. That is, at most one terminal node is labeled with a given value.
- There are no duplicate non-terminal nodes. That is, given two non-terminal nodes $(x_i, f_{x_i=0}, \ldots, f_{x_i=m_i})$ and $(x_j, g_{x_j=0}, \ldots, g_{x_j=m_i})$, we must have either $x_i \neq x_j$ or $f_{x_i=k} \neq g_{x_i=k}$ for some $k \in \{0, \ldots, m_i\}$.
- All non-terminal nodes depend on the value of their variables. That is, given a non-terminal node $(x_i, f_{x_i=0}, \ldots, f_{x_i=m_i})$, we must have $f_{x_i=k} \neq f_{x_i=\ell}$ for some $k, \ell \in 0, \ldots, m_i$.

It has been shown that ROMDDs are a canonical structure: given any integer function and a variable ordering, there is exactly one ROMDD representation for that function. Binary decision diagrams are a special case of MDDs applied to binary logic functions. The size of an ROMDD depends heavily, as in the BDD case, on the variable ordering used to build the ROMDD.

In the present case, the n variables are the arcs of the network and the function $f = \Psi_{s,t}$ is the total weight from the source s to the sink t evaluated utilizing a proper adaptation of Definitions 5.6 and 5.8 according to the interpretation of the weights [64]:

- If the weights are cost functions, the terminal leaves of the MDD provide all the possible values of the function $\Psi_{s,t}$ computed along the minpaths that connect s to t (Definition 5.6).
- If the weights are capacity functions, the terminal leaves of the MDD provide all the possible values of the flows $\Psi_{s,t}$ that can be transmitted from s to t when s and t are connected. The flow values are computed from the mincuts following Definition 5.8.

5.4.3   Basic Operations for MDD Manipulation

In order to compute the function $\Psi_{s,t}$, new operators for manipulating and constructing the MDDs are needed. The new operators for manipulating weighted MDDs are introduced on the two elementary series-parallel networks of Figure 5.25. The two arcs e_1 and e_2 are assumed to have three states numbered $\{0, 1, 2\}$ with respective weights $(0, a_1, a_2)$ and $(0, b_1, b_2)$, as indicated in square brackets.

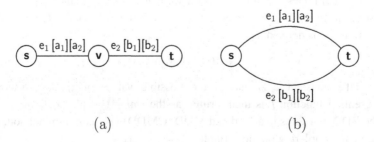

Figure 5.25   Network with two multi-valued arcs: (a) in series; (b) in parallel.

Weights Interpreted as Costs

The network of Figure 5.25(a) has a single minpath $H_1 = e_1 \wedge e_2$, and the connectivity function is $C_{s,t} = H_1$. When the network is connected ($C_{s,t} = 1$), the cost function is $\Psi_{s,t} = a_i + b_j$, where i and j are the states of e_1 and e_2, respectively; otherwise, $\Psi_{s,t} = 0$. In the computation of the cost function, the $\wedge$ operator corresponds to a sum in the costs. This operation is called "AndSum."

The network of Figure 5.25(b) has two minpaths, $H_1 = e_1$ and $H_2 = e_2$; the connectivity function is $C_{s,t} = H_1 \vee H_2 = e_1 \vee e_2$. When both arcs are up with e_1 in state i and e_2 in state j, the cost is $\Psi_{s,t} = \min\{a_i, b_j\}$; when the network is connected through arc e_1 in state i the cost is $\Psi_{s,t} = a_i$, when it is connected through arc e_2 in state j the cost is $\Psi_{s,t} = b_j$. The MDD operation corresponding to the Boolean $\vee$ is called "OrMin."

Generation of the cost function $\Psi_{s,t}$ requires the implementation of the AndSum and OrMin operators. The definitions of these operations (corresponding to the multi-valued extensions of the binary operators of Table 5.3) are given in Table 5.12. Figure 5.26 gives the basic MDDs for the two components e_1 and e_2; Figure 5.27 provides the MDDs for the two networks of Figure 5.25.

Weights Interpreted as Flows

The network of Figure 5.25(a) has two mincuts, $K_1 = \overline{e_1}$ and $K_2 = \overline{e_2}$; the connectivity function is $C_{s,t} = \overline{K_1} \wedge \overline{K_2} = e_1 \wedge e_2$. When the network is connected ($C_{s,t} = 1$), the flow function is $\Psi_{s,t} = \min\{a_i, b_j\}$, where i and j are the states of e_1 and e_2, respectively. In the computation of the flow function the $\wedge$ operator corresponds to a min in the flows. This operation is called "AndMin."

Table 5.12 Truth table of AndSum and OrMin.

State of e_1	State of e_2	$C_{s,t} = e_1 \wedge e_2$	AndSum	$C_{s,t} = e_1 \vee e_2$	OrMin
0	0	0	0	0	0
0	1	0	0	1	b_1
0	2	0	0	1	b_2
1	0	0	0	1	a_1
1	1	1	$a_1 + b_1$	1	$\min\{a_1, b_1\}$
1	2	1	$a_1 + b_2$	1	$\min\{a_1, b_2\}$
2	0	0	0	1	a_2
2	1	1	$a_2 + b_1$	1	$\min\{a_2, b_1\}$
2	2	1	$a_2 + b_2$	1	$\min\{a_2, b_2\}$

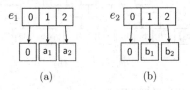

(a)   (b)

Figure 5.26   Basic MDD operations: (a) MDD for arc e_1; (b) MDD for arc e_2.

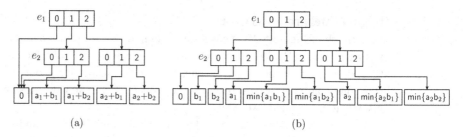

(a)   (b)

Figure 5.27   Basic MDD operations: (a) e_1 AndSum e_2; (b) e_1 OrMin e_2.

The network of Figure 5.25(b) has a single mincut, $K_1 = \overline{e_1} \wedge \overline{e_2}$, and the connectivity function is $C_{s,t} = \overline{K_1}$. When both arcs are up with e_1 in state i and e_2 in state j, the flow is $\Psi_{s,t} = a_i + b_j$. When the network is connected through arc e_1 in state i the flow is $\Psi_{s,t} = a_i$; when it is connected through arc e_2 in state j the flow is $\Psi_{s,t} = b_j$. The MDD operation corresponding to the Boolean $\vee$ is called "OrSum."

Generation of the flow function $\Psi_{s,t}$ requires the implementation of the AndMin and OrSum operators. The definitions of these operations are given in Table 5.13, and the corresponding MDD construction is shown in Figure 5.28 for the same basic components e_1 and e_2 of Figure 5.26.

Table 5.13   Truth table of AndMin and OrSum.

State of e_1	State of e_2	$C_{s,t}e_1 \wedge e_2$	AndMin	$C_{s,t}e_1 \vee e_2$	OrSum
0	0	0	0	0	0
0	1	0	0	1	b_1
0	2	0	0	1	b_2
1	0	0	0	1	a_i
1	1	1	$\min\{a_1, b_1\}$	1	$a_1 + b_1$
1	2	1	$\min\{a_1, b_2\}$	1	$a_1 + b_2$
2	0	0	0	1	a_2
2	1	1	$\min\{a_2, b_1\}$	1	$a_2 + b_1$
2	2	1	$\min\{a_2, b_2\}$	1	$a_2 + b_2$

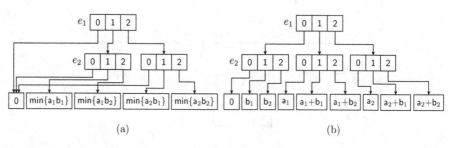

(a)   (b)

Figure 5.28   Basic MDD operations: (a) e_1 AndMin e_2; (b) e_1 OrSum e_2.

5.4.4 Algorithmic Implementation and Probability Evaluation

The implementation of the algorithm for the evaluation of the function $\Psi_{s,t}$ is based on construction of the MDD using the Iowa State University MEDDLY library [68, 69], and the goal of the analysis is to find the probability mass associated with all the possible outcomes of the function $\Psi_{s,t}$. When $\Psi_{s,t}$ is a cost function, the MDD is built from the list of the minpaths $H_1, H_2, \ldots, H_h$ and all the possible outcomes are determined by applying the operators AndSum and OrMin defined in Table 5.12 and Figure 5.27.

When $\Psi_{s,t}$ is a capacity function, the list of all the mincuts $K_1, K_2, \ldots, K_k$ is first generated and all the possible outcomes are determined by applying the operators AndMin and OrSum defined in Table 5.13 and Figure 5.28 [64].

The probability mass of the MDD terminal nodes is computed in a top-down fashion using a breadth-first search on the MDD. The algorithm starts from the root and computes the probabilities of the MDD nodes at successive levels until the terminal nodes are reached. Given a node d in the MDD, the probability of d depends only on the parent nodes and on the connecting edges, according to the following basic rule:

$$P\{\text{node}_d\} = \sum_{k \in \text{set of parent nodes of } d} P\{\text{node}_k\} \cdot p(\text{arc}_{k,d}), \tag{5.33}$$

where $p(\text{arc}_{k,d})$ is the probability associated with the edge connecting node k with node d in the MDD. For example, in the MDD of Figure 5.28(a), the probability associated with the terminal node 0 is computed as $P_0 = p_{1,0} + p_{1,1} \cdot p_{2,0} + p_{1,2} \cdot p_{2,0}$, where $p_{i,j}$ is the probability for the arc i to be in state j.

Example 5.13 *Bridge Network: Multi-Valued Version*
We consider a multi-valued version of the weighted bridge of Figure 5.20. In the present example, all the arcs have three mutually exclusive states denoted as state 1 (working), state 2 (degraded), and state 0 (failed). Each arc is assigned a probability and a weight when in state 0, 1 or 2. Weights can be interpreted either as costs or flows. All the input data are summarized in Table 5.14.

Table 5.14 State probability and weight of arcs for bridge network.

State	Probability	Weight as cost					Weight as flow				
		e_1	e_2	e_3	e_4	e_5	e_1	e_2	e_3	e_4	e_5
working (state 1)	0.65	3	4	2	5	1	3	4	2	5	1
degraded (state 2)	0.25	6	8	4	10	2	1.5	2	1	2.5	0.5
failed (state 0)	0.1	∞	∞	∞	∞	∞	0	0	0	0	0

Weights as Costs
The weights increase with the degradation level according to the values in Table 5.14:

state 1 (working): the cost of arc i is the nominal cost w_i with probability $p_{i,1} = 0.65$.

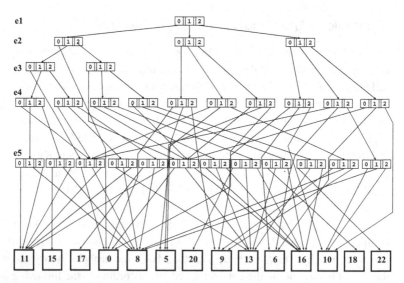

Figure 5.29   MDD of bridge network (cost).

state 2 (degraded): the cost of arc i is $2w_i$ with probability $p_{i,2} = 0.25$.
state 0 (failed): the cost of arc i is ∞ with probability $p_{i,0} = 0.1$.

The minpaths were already derived in Eqs. (5.7). The cost of each minpath is obtained by the AndSum operator, while the MDDs representing the minpaths are combined using the OrMin operator. Assuming as an arbitrary variable sequence $e_1 \prec e_2 \prec e_3 \prec e_4 \prec e_5$, the final MDD is displayed in Figure 5.29. The terminal leaves of the MDD provide all the possible outcomes of $\Psi_{s,t}$ and are shown, together with the associated probability masses, in Table 5.15(a). The probability mass for the cost value ∞ in the table is the probability that s and t are not connected (notice that this value is coincident with the value of the non-weighted network of Figure 5.2).

Weights as Flows
The weights decrease with the degradation level according to the following specification:

state 1 (working): the flow of arc i is the nominal cost w_i with probability $p_{i,1} = 0.65$.
state 2 (degraded): the flow of arc i is $0.5\,w_i$ with probability $p_{i,2} = 0.25$.
state 0 (failed): the flow of arc i is 0 with probability $p_{i,0} = 0.1$.

The list of the mincuts was given in Eq. (5.19). The OrSum operator is applied in order to compute the flow of each mincut. The MDDs representing these mincuts are combined using the AndMin operator. Assuming the same variable ordering and building the MDD, the $\Psi_{s,t}$ outcomes and their probabilities are shown in Table 5.15(b). The row with flow equal to 0 corresponds to the probability that s and t are not connected.

Table 5.15 Probabilities of the MDD terminal values.

Cost c	Prob. mass $P\{\Psi_{s,t}=c\}$	Flow f	Prob. mass $P\{\Psi_{s,t}=f\}$
5	0.42250	0.0	0.02881
6	0.16250	0.5	0.02452
8	0.17533	1.0	0.06579
9	0.09384	1.5	0.02958
10	0.03609	2.0	0.03665
11	0.03362	2.5	0.06226
13	0.03193	3.0	0.12912
15	0.00105	3.5	0.19451
16	0.01293	4.0	0.05138
17	0.00040	4.5	0.06839
18	0.00040	5.0	0.14829
20	0.00040	5.5	0.04462
22	0.00015	6.0	0.11602
∞	0.02881		

(a) Weight as cost.   (b) Weight as flow

Example 5.14 *Bridge Network: Multi-Valued Version*
A multi-state bridge network, similar to the one of Example 5.13 but with edge e_3 bidirectional, was considered in [61]. The number of states per arc and the respective weights, interpreted as flows, are taken from [61] and are reported in Table 5.16(a). The problem is to find the probability $P\{\Psi_{s,t} \geq 3\}$ that the flow between s and t is greater than 3. Reference [61] considers four cases with different probabilities; Table 5.16(a) gives the number of states for each arc and the corresponding probabilities for case 1 only. Table 5.16(b) reports the results for the four cases as in [61, 64].

Table 5.16 Multi-state bridge flow network after [61].

Arc	Arc weight				Arc probability (case 1)				Case	$P\{\Psi_{s,t} \geq 3\}$
e_1	0	1	2	3	0.050	0.025	0.025	0.9	Case 1	0.8310
e_2	0	1	—	—	0.020	0.98	—	—	Case 2	0.6776
e_3	0	1	—	—	0.050	0.95	—	—	Case 3	0.5537
e_4	0	1	2	—	0.025	0.025	0.95	—	Case 4	0.4951
e_5	0	1	2	—	0.075	0.025	0.9	—		

(a) Input data.   (b) Output results.

Example 5.15 In Example 5.11, we have considered the binary flow network of Figure 5.24 taken from [43]. The example is enriched here by considering multi-state arcs with three or four states [64]. The input data and the results are shown in Table 5.17 compared with the binary case of Example 5.11.

Table 5.17 Benchmark network: the number of MDD nodes with multi-state arcs.

Arc multiplicity	Arc probability				MDD	MDD	Terminal
(m)	state 1	state 2	state 3	state 0	nodes	peak	leaves
2 Prob. Weight	0.9 w_i	—	—	0.10 0	9730	9732	17
3 Prob. Weight	0.65 w_i	0.25 $0.5w_i$	—	0.10 0	143498	143498	34
4 Prob. Weight	0.50 w_i	0.15 $0.5w_i$	0.25 $0.25w_i$	0.10 0	621813	2904516	68

Column 1 indicates the arc multiplicity m (assumed to be the same for all the arcs). The columns from 2 to 5 give, for each multiplicity $m = 2, 3, 4$, the arc probability and weight. The probability is assumed to be the same for all the arcs, while the weight w_i is the nominal weight specified in brackets for each arc in Figure 5.24. State 0 is the down state and has a probability of 0.1 for all the multiplicities. In the binary case, with $m = 2$, state 1 is the single up state to which a probability 0.9 is assigned. With $m = 3$, states 1 and 2 are (possibly degraded) up states with probabilities 0.65 and 0.25, respectively, and in state 2 the weight is reduced. With $m = 4$, the up states are 1, 2, and 3 with probabilities 0.50, 0.15, and 0.25, respectively, and the weight is progressively reduced in state 2 and state 3.

Note that since the network has 21 arcs, the number of states of the 3 considered multiplicities is m^{21} ($m = 2, 3, 4$). The efficiency of the MDD construction is shown in columns 6 and 7, where the final number of MDD nodes is reported together with the pick number representing the maximum number of nodes reached during the MDD construction. The last column in the table indicates the number of terminal leaves of the MDD, which increases with the arc multiplicity. For each terminal leaf, the MDD algorithm computes the corresponding mass probability.

5.5   Limitation of the Exact Algorithm

The exact computation of the network reliability is known to be NP-complete [2]. Several papers have tested the ability of different exact algorithms to cope with large network sizes. For BDD-based algorithms, various benchmark networks have been tested in [70] and scalable networks in [7, 71–73]. The complexity of the analysis algorithm depends not only on the number of nodes and arcs of the network but also on the structure and the degree of connectivity. Hence, it is problematic to establish general algorithmic limitations. Further, the analysis algorithms for directed networks are far more efficient than for undirected networks, for which the dimension of the networks that can be analytically solved reduces drastically.

As an example, the largest networks reported in [73] have the following dimensions:

- Regular square grids with source and destination nodes located at diagonally opposite vertices: For the directed version the largest network has $16 \times 16 = 256$ nodes and 480 arcs; for the undirected version, $7 \times 7 = 49$ nodes and 84 arcs.
- Randomly generated networks (with low connectivity degree): For the directed version 500 nodes and 1494 arcs, for the undirected version 60 nodes and 160 arcs.

Various benchmarks and different shapes of network have also been documented in [7, 72], where it is reported that highly asymmetric directed grid networks with 7×1000 nodes can be solved exactly.

As the dimension or the degree of connectivity of the graph increases, the time and the storage required to find all minpaths or mincuts can be prohibitively large. Thus, for such large networks, techniques providing approximations or bounds appear to be the only possible solution.

5.6 Computing Upper and Lower Bounds of Reliability

Natural and simple bounding methods are based on computing the reliability (or the unreliability) with a subset of all minpaths and a subset of all mincuts [74–76]. In fact, computing the reliability from a subset of all minpaths yields a reliability lower bound, while computing the reliability from a subset of all mincuts yields a reliability upper bound. The difference between the upper and the lower bounds is the magnitude of the bounding error.

Given a binary probabilistic network $G = (V, E, P)$, and a source node s and destination node t, we have seen in Eq. (5.4) that the connectivity function $C_{s,t}$ can be expressed as the disjunction of its minpaths. Suppose now that the network has h minpaths and we take a subset containing $h' < h$ minpaths. Then, if we define the Boolean function

$$C'_{s,t} = \bigvee_{i=1}^{h'} H_i,$$ (5.34)

we have that

$$P\{C'_{s,t}\} \leq P\{C_{s,t}\} \leq R_{s,t}.$$ (5.35)

Hence, $P\{C'_{s,t}\}$ provides a lower bound for the (s,t)-reliability $R_{s,t}$.

Similarly, if we take a subset $k' < k$ of the k mincuts and we define the Boolean function

$$\overline{C''_{s,t}} = \bigvee_{j=1}^{k'} K_j,$$ (5.36)

we have that

$$P\{\overline{C''_{s,t}}\} \leq P\{\overline{C_{s,t}}\} \leq U_{s,t}.$$ (5.37)

Hence, $P\{\overline{C''_{s,t}}\}$ provides a lower bound for the (s,t)-unreliability $U_{s,t}$, and $1 - P\{\overline{C''_{s,t}}\}$ an upper bound for the (s,t)-reliability $R_{s,t}$.

Taking any subset of minpaths and any subset of mincuts and applying Eqs. (5.35) and (5.37), we obtain a lower and an upper bound for $R_{s,t}$, respectively.

The algorithms to compute upper and lower bounds can be divided into two phases: the first phase consists of finding a suitable subset of minpaths (for the lower bound) and a suitable subset of mincuts (for the upper bound). Of course, minpaths or mincuts of low order are more effective in determining the bounds. Once the subsets are determined, the actual bound computation (the second phase) can be carried out by any network reliability technique discussed in this chapter. For this second phase, an SDP algorithm was used in [76] while BDD was used in [73, 77].

The algorithm proposed in [77, 78] computes a bound on the unreliability. To narrow the gap between the upper and lower bounds as much as possible with limited time and storage space, the search for the proper subset of minpaths and mincuts is divided into two steps:

1. *Minpath/mincut search:* Find new minpath/mincut candidates for inclusion in the unreliability bounds computation.
2. *Minpath/mincut selection:* Select from the minpath/mincut candidates identified in step 1 those that contribute the most in narrowing the gap between the current upper and lower bounds.

In the second step, heuristics are first used to find the most important minpath/mincut candidates. When no such candidates can be found, exhaustive search algorithms are utilized to enumerate all the minpath/mincut candidates, and a suitable selection criterion is used to decide whether to include the newly generated minpath into the already constructed BDD or discard it. The acceptance criterion for a newly generated minpath/mincut is based on whether its contribution to the overall bound exceeds a predetermined threshold. The algorithm allows user control of the gap between the two bounds by means of the specified overall execution time. Hence, a tradeoff between bounding error and computational resources can be easily made. The algorithm is implemented in the SHARPE software package [79].

Example 5.16 *Aircraft Network*
The graph shown in Figure 5.30 is an example of a typical aircraft current return network with 82 nodes and 171 edges [77, 78]. The graph contains six sections, labeled A, B, C, D, E, and F, besides the source and the target vertices.

Each edge may fail with a constant failure rate, and the system represented by the reliability graph is up if there is at least one path (with all edges up) from the source to the target. In the numerical experiments the failure rate of each edge is $\lambda = 10^{-7} \, \text{hr}^{-1}$ and the default system exposure time is $t = 10^5$ hr (so the reliability of each edge is $r = e^{-\lambda t} = e^{-10^{-7} \cdot 10^5} \approx 0.99$).

Due to structural and symmetry properties of the graph, the exact number of minpaths from the source to the target can be calculated to be $\approx 4.2 \times 10^{12}$. It is impossible to

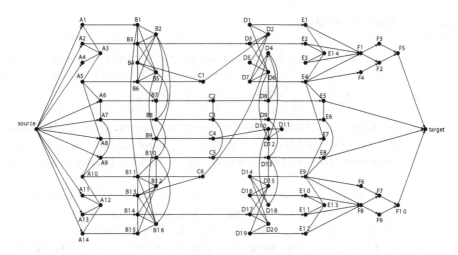

Figure 5.30   A typical aircraft electrical network from [77].

enumerate all of these 4.2 trillion minpaths, let alone construct the BDD to calculate the exact system reliability.

Table 5.18 Unreliability bounds and numbers of selected minpaths/mincuts with different execution times.

Runtime	20 s	120 s	900 s
Upper bound (U)	$1.1460365721 \times 10^{-8}$	$1.0814324701 \times 10^{-8}$	$1.0255197263 \times 10^{-8}$
Lower bound (L)	$1.0199959877 \times 10^{-8}$	$1.0199959877 \times 10^{-8}$	$1.0199959877 \times 10^{-8}$
$U - L$	$0.1260405844 \times 10^{-8}$	$0.0614364824 \times 10^{-8}$	$0.0055237386 \times 10^{-8}$
$\frac{U-L}{L}$ (%)	10.997955	5.681028	0.538628
No. of selected minpaths	28	29	33
No. of selected mincuts	113	113	113

The results of applying the bounding algorithm to system unreliability are reported in Table 5.18 for three different execution times. The first two rows show the upper and lower bounds, while the third and fourth rows show the difference between the bound and the percentage error. The last two rows show the number of minpaths/mincuts selected and used to compute the bounds for the different execution times. Compared with the case with 33 minpaths selected for 900 s execution time, there are 28 minpaths selected for 20 s execution time and 29 minpaths selected for the 120 s case. The number of mincuts selected for these three cases are all equal to 113. The number of mincuts selected not changing with the runtime is not because it is not possible to find more mincuts but because the mincuts found did not qualify for inclusion according to the inclusion criterion established in [77].

5.7   Further Reading

Those interested in binary decision diagrams should read Bryant's seminal paper [11]. Similarly, Ball's seminal paper [2] is yet another treatise on computing network reliability. Useful references are also [3] and [80]. Readers may also consult Colbourn [4] for an extensive discussion and several algorithmic approaches for combinatorics of network reliability. A more recent comprehensive coverage of network reliability can be found in [81]. Another valuable source on network reliability is the book by Rubino [82]. Fishman presents Monte Carlo techniques for estimating network reliability in [83].

Real-world networks today (the Internet, the World Wide Web, public telecommunication networks, etc.) can reach millions or even billions of vertices and overcome the ability of the algorithms (both exact and approximate) discussed in this chapter. The last decades have witnessed a shift in network research with the focus moving to large-scale statistical properties of graphs with the aim of predicting what the behavior of complex networked systems will be on the basis of structural properties and the local rules governing individual vertices. The new approaches forced by this change of scale are outside the scope of this book; the interested reader may refer to [84–88].

References

[1] A. Rosenthal, "Computing the reliability of complex networks," *SIAM Journal on Applied Mathematics*, vol. 32, pp. 384–393, 1977.

[2] M. Ball, "Computing network reliability," *Operations Research*, vol. 27, pp. 823–838, 1979.

[3] A. Agrawal and R. E. Barlow, "A survey of network reliability and domination theory," *Operations Research*, vol. 32, pp. 478–492, 1984.

[4] C. J. Colbourn, *The Combinatorics of Network Reliability*. Oxford University Press, 1987.

[5] A. Bobbio and A. Premoli, "Fast algorithm for unavailability and sensitivity analysis of series-parallel systems," *IEEE Transactions on Reliability*, vol. R-31, pp. 359–361, 1982.

[6] L. Page and J. Perry, "A practical implementation of the factoring theorem for network reliability," *IEEE Transactions on Reliability*, vol. 37, pp. 259–267, 1988.

[7] G. Hardy, C. Lucet, and N. Limnios, "Computing all-terminal reliability of stochastic networks by BDDs," in *Proc. Applied Stochastic Modeling and Data Analysis, ASMDA2005*, 2005.

[8] A. Balan and L. Traldi, "Preprocessing minpaths for sum of disjoint products," *IEEE Transaction on Reliability*, vol. 52, no. 3, pp. 289–295, Sep. 2003.

[9] J. Abraham, "An improved algorithm for network reliability," *IEEE Transactions on Reliability*, vol. 28, pp. 58–61, 1979.

[10] M. Veeraraghavan and K. Trivedi, "An improved algorithm for the symbolic reliability analysis of networks," *IEEE Transactions on Reliability*, vol. 40, pp. 347–358, 1991.

[11] R. Bryant, "Graph-based algorithms for Boolean function manipulation," *IEEE Transactions on Computers*, vol. C-35, pp. 677–691, 1986.

[12] R. Bryant, "Symbolic Boolean manipulation with ordered binary decision diagrams," *ACM Computing Surveys*, vol. 24, pp. 293–318, 1992.

[13] L. Xing and S. Amari, *Binary Decision Diagrams and Extensions for System Reliability Analysis*. Wiley-Scrivener, 2015.

[14] A. Satyanarayana and R. K. Wood, "A linear-time algorithm for computing k-terminal reliability in series-parallel networks," *SIAM Journal on Computing*, vol. 14, no. 4, pp. 818–832, 1985.

[15] K. Brace, R. Rudell, and R. Bryant, "Efficient implementation of a BDD package," in *Proc. 27th ACM/IEEE Design Automation Conf.*, 1990, pp. 40–45.

[16] List of BDD software libraries. [Online]. Available: https://github.com/johnyf/tool_lists/blob/master/bdd.md (last accessed April 12, 2017).

[17] W. Schneeweiss, *Boolean Functions with Engineering Applications and Computer Programs*. Springer Verlag, 1989.

[18] K. Dohmen, "Inclusion–exclusion and network reliability," *The Electronic Journal of Combinatorics*, vol. 5, no. R36, pp. 1–8, 1998.

[19] K. Trivedi, *Probability and Statistics with Reliability, Queueing and Computer Science Applications*, 2nd ed. John Wiley & Sons, 2001.

[20] T. Luo and K. Trivedi, "An improved algorithm for coherent-system reliability," *IEEE Transaction on Reliability*, vol. 47, pp. 73–78, 1998.

[21] K. Heidtmann, "Statistical comparison of two sum-of-disjoint product algorithms for reliability and safety evaluation," in *Proc. 21st Int. Conf. SAFECOMP 2002*. LNCS-2434, 2002, pp. 70–81.

[22] J. Xing, C. Feng, X. Qian, and P. Dai, "A simple algorithm for sum of disjoint products," in *Proc. Ann. IEEE Reliability and Maintainability Symp.*, 2012.

[23] G. Hardy, C. Lucet, and N. Limnios, "k-terminal network reliability measures with binary decision diagrams," *IEEE Transactions on Reliability*, vol. 56, pp. 506–515, 2007.

[24] T. Cormen, C. E. Leiserson, and R. L. Rivest, *Introduction to Algorithms*. MIT Press and McGraw-Hill, 1990.

[25] S. Rai, M. Veeraraghavan, and K. S. Trivedi, "A survey of efficient reliability computation using disjoint products approach," *Networks*, vol. 25, no. 3, pp. 147–163, 1995.

[26] A. Rauzy, E. Châtelet, Y. Dutuit, and C. Bérenguer, "A practical comparison of methods to assess sum-of-products," *Reliability Engineering and System Safety*, vol. 79, no. 1, pp. 33–42, 2003.

[27] A. Kaufmann, D. Grouchko, and R. Cruon, *Mathematical Models for the Study of the Reliability of Systems*. Academic Press, 1977.

[28] L. Yan, H. A.Taha, and T. L. Landers, "A recursive approach for enumerating minimal cutset in a network," *IEEE Transaction on Reliability*, vol. 43, no. 3, pp. 383–387, Sep. 1994.

[29] P. Jensen and M. Bellmore, "An algorithm to determine the reliability of a complex system," *IEEE Transactions on Reliability*, vol. R-18, no. 4, pp. 169–174, Nov. 1969.

[30] Y. Tung, L. Mays, and M. Cullinane, "Reliability analysis of systems," in *Reliability Analysis of Water Distribution Systems*, ed. L. W. Mays. American Society of Civil Engineers, 1989, ch. 9, pp. 259–298.

[31] R. Gupta and P. Bhave, "Reliability analysis of water-distribution systems," *Journal of Environmental Engineering*, vol. 120, no. 2, pp. 447–461, 1994.

[32] L. Page and J. Perry, "Reliability of directed networks using the factoring theorem," *IEEE Transactions on Reliability*, vol. 38, pp. 556–562, 1989.

[33] R. K. Wood, "Factoring algorithms for computing k-terminal network reliability," *IEEE Transactions on Reliability*, vol. R-35, pp. 269–278, 1986.

[34] K. Sekine and H. Imai, "A unified approach via BDD to the network reliability and path number," Dept. Information Science, University of Tokyo, Tech. Rep. TR-95-09, 1995.

[35] X. Zang, H. Sun, and K. Trivedi, "A BDD-based algorithm for reliability graph analysis," Dept. of Electrical Engineering, Duke University, Tech. Rep., 2000.

[36] A. Rauzy, "New algorithms for fault tree analysis," *Reliability Engineering & System Safety*, vol. 40, pp. 203–211, 1993.

[37] GARR, Italian Research, and Education Network. [Online]. Available: www.garr.it

[38] A. Bobbio and R. Terruggia, "Reliability and QoS analysis of the Italian GARR network," Tech. Rep. TR-INF-2008-06-04-UNIPMN, Dip. Informatica, Università Piemonte Orientale. [Online]. Available: www.di.unipmn.it/TechnicalReports/TR-INF-2008-06-04-UNIPMN.pdf

[39] A. Bobbio, G. Bonanni, E. Ciancamerla, R. Clemente, A. Iacomini, M. Minichino, A. Scarlatti, R. Terruggia, and E. Zendri, "Unavailability of critical SCADA communication links interconnecting a power grid and a telco network," *Reliability Engineering and System Safety*, vol. 95, pp. 1345–1357, 2010.

[40] Integrated Risk Reduction of Information-based Infrastructure Systems. [Online]. Available: www.irriis.org/

[41] M. Newman, "Analysis of weighted networks," *Phys. Rev. E*, vol. 70, p. 056131, Nov. 2004.

[42] C.-C. Jane and J. Yuan, "A sum of disjoint products algorithm for reliability evaluation flow of flow networks," *European Journal of Operational Research*, vol. 127, no. 3, pp. 664–675, Jun. 2001.

[43] S. Soh and S. Rai, "An efficient cutset approach for evaluating communication-network reliability with heterogeneous link-capacities," *IEEE Transactions on Reliability*, vol. 54, no. 1, pp. 133–144, 2005.

[44] E. M. Clarke, M. Fujita, P. C. McGeer, K. McMillan, and J. Yang, "Multi-terminal binary decision diagrams: An efficient data structure for matrix representation," in *IWLS'93: Int. Workshop on Logic Synthesis*, Carnegie Mellon University, Report 5-1993, 1993, pp. 6a:1–15.

[45] E. Clarke, M. Fujita, and X. Zhao, "Applications of multi-terminal binary decision diagrams," Technical Report CMU-CS-95-160, 1995. [Online]. Available: www.dtic.mil/dtic/tr/fulltext/u2/a296385.pdf

[46] M. Fujita, P. McGeer, and J. Y. Yang, "Multi-terminal binary decision diagrams: An efficient data structure for matrix representation," *Formal Methods in System Design*, vol. 10, no. 2–3, pp. 149–169, 1997.

[47] G. D. Hachtel and F. Somenzi, "A symbolic algorithms for maximum flow in 0-1 networks," *Form. Methods Syst. Des.*, vol. 10, no. 2–3, pp. 207–219, 1997.

[48] T. Gu and Z. Xu, "The symbolic algorithms for maximum flow in networks," *Computers and Operations Research*, vol. 34, pp. 799–816, 2007.

[49] A. Bobbio and R. Terruggia, "Reliability and quality of service in weighted probabilistic networks using algebraic decision diagrams," in *Proc. IEEE Ann. Reliability and Maintainability Symp.*, Fort Worth, TX, 2009, pp. 19–24.

[50] K. Kolowrocki, "On limit reliability functions of large multi-state systems with ageing components," *Appl. Math. Comput.*, vol. 121, no. 2–3, pp. 313–361, 2001.

[51] A. Lisnianski and G. Levitin, *Multi-State System Reliability: Assessment, Optimization and Applications*, Series on Quality, Reliability and Engineering Statistics: Volume 6. World Scientific, 2003.

[52] E. Zaitseva and V. Levashenko, "Investigation multi-state system reliability by structure function," in *DEPCOS-RELCOMEX '07: Proc. 2nd Int. Conf. on Dependability of Computer Systems*. IEEE Computer Society, 2007, pp. 81–90.

[53] J. Meyer, "On evaluating the performability of degradable systems," *IEEE Transactions on Computers*, vol. C-29, pp. 720–731, 1980.

[54] X. Zang, D. Wang, H. Sun, and K. Trivedi, "A BDD-based algorithm for analysis of multistate systems with multistate components," *IEEE Transactions on Computers*, vol. 52, no. 12, pp. 1608–1618, 2003.

[55] L. Xing and Y. Dai, "A new decision diagram based method for efficient analysis on multi-state systems," *IEEE Transactions on Dependable and Secure Computing*, vol. 6, no. 3, pp. 161–174, 2009.

[56] S. Amari, L. Xing, A. Shrestha, J. Akers, and K. Trivedi, "Performability analysis of multistate computing systems using multivalued decision diagrams," *IEEE Transactions on Computers*, vol. 59, no. 10, pp. 1419–1433, 2010.

[57] E. Zaitseva and V. Levashenko, "Multi-state system analysis based on multiple-valued decision diagram," *Journal of Reliability and Statistical Studies*, vol. 5, pp. 107–118, 2012.

[58] K. Gopal, K. Aggarwal, and J. Gupta, "Reliability analysis of multistate device networks," *IEEE Transactions on Reliability*, vol. 27, no. 3, pp. 233–236, Aug. 1978.

[59] S. Ross, "Multivalued state component systems," *The Annals of Probability*, vol. 7, no. 2, pp. 379–383, 1979.

[60] J. Hudson and K. Kapur, "Reliability analysis of multistate systems with multistate components," *IIE Transactions*, vol. 15, pp. 127–135, 1983.

[61] A. Shrestha, L. Xing, and Y. Dai, "Decision diagram based methods and complexity analysis for multi-state systems," *IEEE Transactions on Reliability*, vol. 59, no. 1, pp. 145–161, 2010.

[62] G. Levitin, "Reliability of multi-state systems with two failure-modes," *IEEE Transactions on Reliability*, vol. R-52, no. 3, pp. 340–348, 2003.

[63] G. Levitin and A. Lisnianski, "A new approach to solving problems of multi-state system reliability optimization," *Quality and Reliability Engineering International*, vol. 17, pp. 93–104, 2001.

[64] R. Terruggia and A. Bobbio, "QoS analysis of weighted multi-state probabilistic networks via decision diagrams," in *Computer Safety, Reliability, and Security*, ed. E. Schoitsch. Springer Verlag, LNCS, Vol 6351, 2010, pp. 41–54.

[65] T. Kam, T. Villa, R. Brayton, and A. Sangiovanni-Vincentelli, "Multi-valued decision diagrams: Theory and applications," *Multiple-Valued Logic*, vol. 4, no. 1, pp. 9–62, 1998.

[66] G. Ciardo, G. Lüttgen, and A. Miner, "Exploiting interleaving semantics in symbolic state-space generation," *Formal Methods in System Design*, vol. 31, no. 1, pp. 63–100, 2007.

[67] D. Miller and R. Drechsler, "On the construction of multiple-valued decision diagrams," in *Proc. IEEE 32nd Int. Symp. on Multiple-Valued Logic*, 2002, pp. 264–269.

[68] I. S. U. Research Foundation, Meddly decision diagram library. [Online]. Available: http://meddly.sourceforge.net/

[69] J. Babar and P. Miner, "Meddly: Multi-terminal and Edge-valued Decision Diagram Library," in *Proc. 7th Int. Conf. on Quantitative Evaluation of Systems (QEST'10)*, 2010, pp. 195–196.

[70] F. Yeh, S. Lu, and S. Kuo, "OBDD-based evaluation of k-terminal network reliability," *IEEE Transactions on Reliability*, vol. 51, no. 4, pp. 443–451, 2002.

[71] A. Bobbio, R. Terruggia, E. Ciancamerla, and M. Minichino, "Evaluating network reliability versus topology by means of BDD algorithms," in *Int. Probabilistic Safety Assessment and Management Conf. (PSAM-9)*, 2008.

[72] J. Herrmann, "Improving reliability calculation with augmented binary decision diagrams," in *24th IEEE Int. Conf. on Advanced Information Networking and Applications*, 2010, pp. 328–333.

[73] M. Beccuti, A. Bobbio, G. Franceschinis, and R. Terruggia, "A new symbolic approach for network reliability analysis," in *IEEE Int. Conf. on Dependable Systems and Networks, DSN2012*, 2012, pp. 1–12.

[74] M. O. Ball and J. S. Provan, "Calculating bounds on reachability and connectedness in stochastic networks," *Networks*, vol. 13, no. 2, pp. 253–278, 1983.

[75] F. Beichelt and L. Spross, "Bounds on the reliability of binary coherent systems," *IEEE Transactions on Reliability*, vol. 38, pp. 425–427, 1989.

[76] C.-C. Jane, W.-H. Shen, and Y. W. Laih, "Practical sequential bounds for approximating two-terminal reliability," *European Journal of Operational Research*, vol. 195, no. 2, pp. 427–441, Jun. 2009.

[77] S. Sebastio, K. Trivedi, D. Wang, and X. Yin, "Fast computation of bounds for two-terminal network reliability," *European Journal of Operational Research*, vol. 238, no. 3, pp. 810–823, 2014.

[78] K. Trivedi, D. Wang, T. Sharma, A. V. Ramesh, D. Twigg, L. Nguyen, and Y. Liu, "Reliability estimation for large networked systems," U.S. Patent 20 090 323 539, 11, 2011.

[79] R. Sahner, K. Trivedi, and A. Puliafito, *Performance and Reliability Analysis of Computer Systems: An Example-based Approach Using the SHARPE Software Package*. Kluwer Academic Publishers, 1996.

[80] I. Gertsbakh and Y. Shpungin, *Network Reliability Calculations Based on Structural Invariants*. John Wiley & Sons, Ltd, 2013, pp. 135–146.

[81] M. O. Ball, C. J. Colbourn, and J. S. Provan, "Network reliability," *Handbooks in Operations Research and Management Science*, vol. 7, pp. 673–762, 1995.

[82] G. Rubino, "Network reliability evaluation," in *State-of-the-Art in Performance Modeling and Simulation*, eds. K. Bagchi and J. Walrand. Gordon and Breach, 1998, ch. 11, pp. 275–302.

[83] G. S. Fishman, "A Monte Carlo sampling plan for estimating network reliability," *Operations Research*, vol. 34, no. 4, pp. 581–594, 1986.

[84] R. Albert and A. Barabasi, "Statistical mechanics of complex networks," *Review Modern Physics*, vol. 74, pp. 47–97, 2002.

[85] S. Dorogovtsev and J. Mendes, "Evolution of networks," *Advances in Physics*, vol. 51, pp. 1079–1187, 2002.

[86] M. Newman, "The structure and function of complex networks," *SIAM Review*, vol. 45, pp. 167–256, 2003.

[87] S. Boccaletti, V. Latora, Y. Moreno, M. Chavez, and D. U. Hwang, "Complex networks: Structure and dynamics," *Physics Reports*, vol. 424, no. 4–5, pp. 175–308, 2006.

[88] G. Caldarelli and M. Catanzaro, *Networks: A Very Short Introduction*. Oxford University Press, 2012.

6 Fault Tree Analysis

A fault tree (FT) is a graphical paradigm for the representation and analysis of the critical conditions whose combined occurrence causes a specific event, called the *top event* (TE), to occur. When the TE is one particular undesired event, like a catastrophic failure or a failure to accomplish a given mission, or the occurrence of a condition that may entail human, economic or environmental losses, then the analysis of the combination of elementary events that leads to the occurrence of the TE assumes the name *fault tree analysis* (FTA).

The TE is the root of the tree and the construction of the tree is based upon deductive reasoning, in a "top-down" approach, from general to specific. FTA is particularly suited to the analysis of complex systems comprising several subsystems or components that are connected in various configurations, with high levels of redundancy. FTA is commonly used by reliability and safety engineers dealing with aircraft, space, chemical, electrical, ICT, power, and nuclear systems. FTA is addressed in the IEC 1025 standard [1]. Further details of the theoretical and practical basis of the methodology can be found in [2–8]. A very recent survey of the FT methodology with an extended bibliography is given in [9].

While series-parallel RBDs and FTs without repeated events have the same modeling power, fault trees with repeated events (FTRE) have greater modeling power than either series-parallel RBDs or reliability graphs, as shown in Figure 2.7 [10]. Furthermore, FTs can also have a NOT gate, allowing them to model non-coherent structures. FTA thus has two main advantages over other non-state-space techniques like RBD (Chapter 4) and network reliability (Chapter 5); it can handle repeated events and non-coherent structures. FTA is also more flexible and open to the modeler's ability and imagination, and has witnessed a variety of applications in many different fields.

6.1 Motivation for and Application of FTA

FTA is aimed at identifying the causes or combination of causes that lead to the TE, and provides the designer with a simple technique to include in the evaluation of items such as common cause failures, human error, external factors like environmental factors, and so on. FT methods can be applied from the early stages of the system design, and the fault tree can be progressively refined and updated as the design evolves. The evolution of the FT should result in an increased understanding of the modes of failure as the design proceeds. The following steps summarize the application of the procedure in system design.

1. *Definition of the scope of the analysis:* The typical objective for FTA is the investigation of the chance of occurrence of undesirable events that can lead to economic losses or can have a potential impact on the safety of the operating and maintenance personnel or of the general public, or an impact on the environment.

2. *Familiarization with the system design and function:* This step is the most important in the preliminary phase of FT construction since systematic and detailed knowledge of the system under study is required. This phase requires the cooperation of a team comprising the reliability analysts, the designers, and the operating and maintenance engineers. It should lead to a better understanding of the physical and functional structure of the system, the boundaries between blocks or subsystems and their interactions, the possible deviations of each block from its correct mode of operation, the man–machine interaction, and so forth.

3. *Definition of the top event:* The TE is the starting point of the analysis. TE candidates are events whose occurrence may lead to unsafe operating conditions, catastrophic failure or malfunction, failure to accomplish the assigned mission, and so on. If more than one TE needs to be investigated, a different tree for each of the TEs might be generated and analyzed. Alternatively we can use a multi-state fault tree, in which case multiple TEs can be accommodated in a single FT.

4. *Construction of the fault tree:* This phase consists of the identification of the possible failure modes that induce critical conditions in the higher levels of the system. The effects of different failure modes on the higher levels are graphically represented by means of logic gates using the approach described in subsequent sections of this chapter.

5. *Qualitative analysis:* The qualitative analysis is intended to provide the logical expression of the TE as a function of the basic events. The qualitative analysis enumerates all the failure modes whose combination causes the TE to occur and locates the potential weak points of the structure.

6. *Quantitative analysis:* If probability values are assigned to the basic events of the FT, the quantitative analysis evaluates the probability of occurrence of the TE, and of any intermediate levels in the FT. Importance analysis may rank each event according to its influence on the overall system measures, and leads to the evaluation of indices that quantify the criticality of each single basic event in determining the occurrence of the TE. In any practical case, numerical evaluations must be carried out with the help of a computerized tool [9].

7. *Results of the analysis:* The final results are reported in suitable graphical and tabular forms. Most of the available FTA tools provide facilities at this level.

The enumeration of all the relevant failure modes and their effects on the TE is by far the most difficult portion of the task, and this phase is largely dependent on the skill of the analyst or the group of experts involved in the exercise. Automatic derivation of FTs from other higher-level languages like Petri nets [11], UML (unified modeling language) [12, 13] or HAZOP [14] has been attempted. On the other hand, the graphical (or textual) construction of the FT and the subsequent qualitative and quantitative analysis must rely on a suitable computer program [9, 15].

6.2   Construction of the Fault Tree

Once the TE has been defined, the construction of the FT proceeds by identifying the *causes* or events at the lower level for the occurrence of the TE, and their logical relationships. Each immediate cause is then recursively treated as another TE, the analysis proceeds to determine their immediate causes, and so on. In this way, the construction evolves recursively from events to their causes, aiming to identify events at a finer resolution, until a desired level of detail is reached.

Interactions between causes at each level of the iterative construction are represented by means of logic gates, while the output of the logic gates represents the occurrence of the higher-level event of the tree. The events at which the construction of the tree is ended are called *terminal events* or *basic events*. The basic events reflect the state of the elementary causes whose deviation from correct operation can affect the occurrence of the TE.

Fault trees can be categorized as *coherent* or *non-coherent*:

- A coherent FT, as defined in [3], must enjoy two properties: (i) all the basic events are relevant (the state of the system must depend on the state of each of its components); (ii) any deviation from the correct operation of any basic event cannot decrease the chances of the TE being reached. It has been shown that a coherent FT can only have OR and AND gates as logical connectors.
- A non-coherent FT admits the use of the NOT gate (and derived gates like XOR). The consequence is that in a non-coherent FT the system may be failed when a component is working, and may be restored to a non-failed condition when the component fails.

Basic events can be categorized as *binary* or *multi-state*:

- A binary basic event can be found in one of two mutually exclusive and exhaustive states that we denote, as usual, with E = "the functioning state" and $\overline{E}$ = "the faulty state." A binary basic event can be represented by a Boolean variable.
- A multi-state basic event is one that can be represented by more than two states, so that the state of a component may be represented by a non-negative integer taking values in a finite set (as in Section 5.4).

Unless otherwise specified, we deal with coherent, binary FTs. A brief excursion into multi-state FTs is taken in Section 6.9. Non-coherent FTs are not explicitly considered in this chapter, and appropriate references are deferred to Section 6.11.

The graphical construction of an FT is based on the representation of the events and of their relations by means of suitable symbols. The list of the most commonly used symbols (taken from Annex A of [1]) is shown in Figure 6.1.

6.2.1   The OR Logic Gate

The OR logic gate symbolically represents a logical OR operator (also denoted as logic sum and indicated with the symbol $\vee$ or $+$). The output event of an OR gate is verified if one or more of the input events are verified.

Symbol	Meaning of symbols
 $\oplus$	OR Gate
 $\odot$	AND Gate
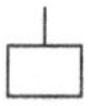 $\square$	Event
 $\bigcirc$	Basic Event
 $\diamondsuit$	Undeveloped Event

Figure 6.1   The main standard graphical symbols used in FTA construction.

Example 6.1   Consider a series system S whose RBD representation is given in Figure 6.2(a). The same logical structure can be represented by the FT of Figure 6.2(b), where the top event $TE = \bar{S}$ is the system failure event and is the output of an OR gate whose inputs are $\bar{A}$ (failure of component A), $\bar{B}$ (failure of component B), $\bar{C}$ (failure of component C) and $\bar{D}$ (failure of component D). The top event $TE = \bar{S}$ occurs if one or more of the input events occur. Symbolically,

$$\bar{S} = \bar{A} \vee \bar{B} \vee \bar{C} \vee \bar{D}.$$

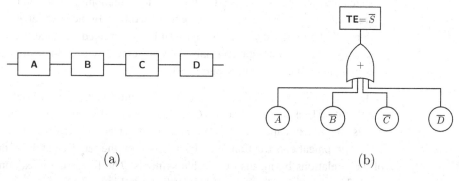

(a)   (b)

Figure 6.2   (a) RBD of a series system; (b) Corresponding representation as an FT.

Example 6.2   The FT for the system described in Example 4.1 is given in Figure 6.3, and the TE is:

$$TE = \bar{S}_{sys} = \bar{S} \vee \bar{L} \vee \bar{P} \vee \bar{V}.$$

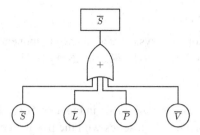

Figure 6.3 FT of the system of Figure 4.2.

The FT representation of Figure 6.3 is more flexible than the corresponding RBD of Figure 4.2. Indeed, any basic event of the FT of Figure 6.3 can be thought of as the top event of a subtree representing the conditions that lead to the occurrence of that basic event. Other basic events like human error, common cause, etc., can easily be accommodated in the picture.

6.2.2 The AND Logic Gate

The AND logic gate symbolically represents a logical AND operator (also denoted as logic multiplication and indicated with the symbol $\wedge$ or $\cdot$). The output event of an AND gate is verified if and only if all the input events are jointly verified.

Example 6.3 Consider a parallel system S whose RBD representation is given in Figure 6.4(a). The same logical structure can be represented by the FT of Figure 6.4(b), where the top event $\text{TE} = \overline{S}$ is the system failure and is the output of an AND gate whose input events are the failures of components $\overline{A}$, $\overline{B}$, $\overline{C}$, and $\overline{D}$. The TE occurs when all the components are failed. Symbolically,

$$\overline{S} = \overline{A} \wedge \overline{B} \wedge \overline{C} \wedge \overline{D}.$$

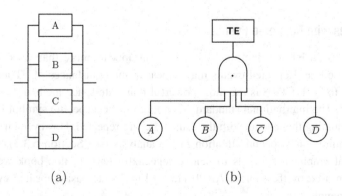

(a)   (b)

Figure 6.4 (a) RBD of a parallel system; (b) Corresponding representation as an FT.

6.2.3   Fault Trees with OR and AND Gates

An FT representing the failure of a complex system with many components connected in various redundant configurations will have both OR and AND gates in the same tree.

Example 6.4   Consider a system S whose RBD is given in Figure 6.5(a). Blocks A and B are in parallel redundancy while block C is in series with the parallel configuration. If TE is the system failure event (TE $= \bar{S}$), the FT representation for the same system is given in Figure 6.5(b).

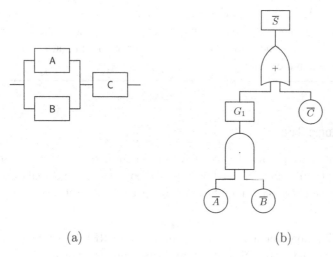

(a)   (b)

Figure 6.5   (a) RBD for a series-parallel system S; (b) Corresponding FT with TE $= \bar{S}$.

Example 6.5   Adopting the same notation, Figure 6.6(a) represents the RBD of a different system S, and Figure 6.6(b) its FT representation with TE $= \bar{S}$.

6.2.4   Fault Trees with Repeated Events

An FTRE is an FT in which at least one event appears more than once in different nodes of the tree. Repeated events may appear in many instances of FT analysis and, according to [10], FTRE is the most powerful non-state-space method as indicated in Figure 2.7. The international standards do not prescribe a specific symbol for repeated events, assuming that events with the same label are repeated. However, since repeated events require some special attention in the analysis (see Section 6.3.3), many tools use special graphical symbols to denote repeated events. In this book we adopt the convention used by the tool SHARPE [15, 16] to denote basic repeated events by an inverted triangle.

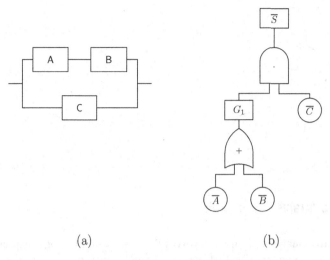

(a)   (b)

Figure 6.6   (a) RBD for another series-parallel system S; (b) Corresponding FT with TE $= \bar{S}$.

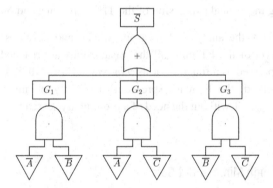

Figure 6.7   FT representation of a two-out-of-three system with AND and OR gates.

6.2.5   The *k*-out-of-*n* Node

A k-out-of-n system can be modeled by means of an FT with repeated events using only OR and AND gates. The system fails if $n - k + 1$ or more components fail. Hence, the system has $\binom{n}{n-k+1}$ combinations of k basic events that lead to system failure. A fault tree for the simple case of a two-out-of-three system is shown in Figure 6.7.

However, since a k-out-of-n construct is rather common in systems that are usually modeled by an FT, like high-availability and safety-critical systems, many software packages provide an implicit k-out-of-n gate that has n component failure events as input and one single output that occurs with the rules of Section 4.3. The implicit k-out-of-n gate is represented in Figure 6.8 and does not need to be expanded by the user as in Figure 6.7; its correct elaboration is carried out by the software package being used.

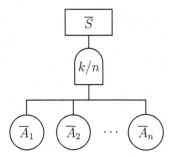

Figure 6.8   An implicit k-out-of-n gate.

6.3   Qualitative Analysis of a Fault Tree

The qualitative analysis of an FT involves the derivation of a logic expression for the TE, as a function of the terminal events, in such a way that all the combinations of events whose joint occurrence precipitates the TE are clearly evidenced. Two methods are possible for deriving the logical expression of the TE – *top-down* and *bottom-up*:

- In the *top-down* method, the analysis starts from the TE and develops the logical expression at each level of the FT until all the basic events are reached. With this method, the TE is expressed as a function of its *minimal cut sets* (MCSs).
- In the *bottom-up* method, the logical expression is directly obtained through a recursive search along the FT without the need of a preliminary search for the minimal cut sets.

6.3.1   Logical Expression through Minimal Cut Sets

Rephrasing Definition 5.3 in terms of FTs, a combination of basic events whose joint occurrence leads to the TE is called a *cut* or, following the standard terminology in FT literature, a *cut set* (CS) for the system. A CS that does not contain any subset that is also a CS is minimal and is called a *mincut*, or a *minimal cut set* (MCS).

Suppose that an FT has m MCSs denoted by $K_1, K_2, \ldots, K_m$. According to the above definition, the occurrence of any K_i ($i = 1, 2, \ldots, m$) implies the occurrence of the TE, hence [compare with the disconnectivity Eq. (5.17)]:

$$\mathrm{TE} = K_1 \vee K_2 \vee \cdots \vee K_m = \bigvee_{i=1}^{m} K_i. \tag{6.1}$$

A generic MCS K_i is the product of ℓ_i terminal events:

$$K_i = E_{i1} \wedge E_{i2} \wedge \cdots \wedge E_{i\ell_i}.$$

Note that, in general, any terminal event can appear in several MCSs even if the event is not a repeated event. This sum-of-product form of the TE in (6.1) is also known as a disjunctive normal form (DNF) [17].

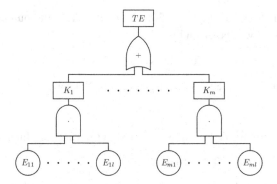

Figure 6.9 Two-level representation of an FT from its DNF.

The logical expression of a TE in DNF (6.1) can be represented as a two-level FTRE, as in Figure 6.9. The TE at the top level is the output of an OR gate whose inputs are the m MCSs, and the second level is an AND gate whose inputs are the (possibly repeated) basic events corresponding to each MCS.

The search for the MCS depends only on the logical structure of the FT and not on the probability values assigned to the occurrence of the terminal events. The list of all the MCSs provides very valuable information to the analyst since it contains all the minimal sequences of failure events that result in the TE occurrence, and allows the analyst to identify the potential weak points of the system.

Determination of the Cut Sets

The determination of the CSs proceeds iteratively in a top-down manner, starting from the TE and applying the rules of Boolean algebra (see Section 6.12), until all the terminal nodes are reached. If the FT does not contain repeated events, this search directly provides the MCSs, otherwise if the FT contains repeated events the list of MCSs must be further extracted from the obtained list of CSs.

Example 6.6 In the OR gate of Figure 6.2(b), the occurrence of any single basic event $\overline{A}, \overline{B}, \overline{C}$ or $\overline{D}$ leads to the TE, so each basic event is an MCS of the system:

$$K_1 = \overline{A}, \quad K_2 = \overline{B}, \quad K_3 = \overline{C}, \quad K_4 = \overline{D},$$
$$TE = K_1 \vee K_2 \vee K_3 \vee K_4.$$

Example 6.7 In the AND gate of Figure 6.4(b), only the joint occurrence of all the basic events leads to the TE; hence, all the basic events belong to the same MCS:

$$K_1 = \overline{A} \wedge \overline{B} \wedge \overline{C} \wedge \overline{D} \quad \text{and} \quad TE = K_1.$$

Example 6.8   Consider the FT of Figure 6.5(b). Since the TE is the output of an OR gate, its logical expression will be given by:

$$TE = G_1 \vee \overline{C}. \tag{6.2}$$

Since G_1 is a non-terminal event and is the output of an AND gate,

$$G_1 = \overline{A} \wedge \overline{B}. \tag{6.3}$$

By replacing the expression for G_1 given by Eq. (6.3) in Eq. (6.2), we finally get:

$$TE = (\overline{A} \wedge \overline{B}) \vee \overline{C}. \tag{6.4}$$

Since all the terms in Eq. (6.4) are now terminal events, we can derive the list of the MCSs:

$$K_1 = \overline{A} \wedge \overline{B}, \qquad K_2 = \overline{C}.$$

Example 6.9   With reference to the FT of Figure 6.6, the search for the MCS proceeds along the following steps:

$$TE = G_1 \wedge \overline{C},$$
$$G_1 = \overline{A} \vee \overline{B},$$
$$TE = G_1 \wedge \overline{C} = (\overline{A} \vee \overline{B}) \wedge \overline{C} = (\overline{A} \wedge \overline{C}) \vee (\overline{B} \wedge \overline{C}). \tag{6.5}$$

From (6.5) we can derive the list of the MCSs:

$$K_1 = \overline{A} \wedge \overline{C}, \qquad K_2 = \overline{B} \wedge \overline{C}.$$

We suggest the reader draw an FT based on the DNF (refer to the general form given in Figure 6.9) for this example. Such a tree will result in an FTRE as the event $\overline{C}$ will then be repeated even though the original tree (Figure 6.6) in this case has no repeated events.

Minimal Cut Set Order

An MCS is a subset of the set of all basic events. The number of basic events in an MCS is called the *order* of the MCS. The order is an important parameter in ascertaining which failure combinations are potentially more critical for the system. In fact, an MCS of order 1 means that the failure of a single basic component is sufficient to trigger the TE, indicating no level of redundancy with respect to that particular component failure. In an MCS of order 2, two failures of basic components are needed for system failure to occur. In general, if ℓ_i is the order of MCS K_i, ℓ_i failure events are necessary to verify the TE. In order to make the information contained in the MCS order easily available, many FTA tools list the MCSs in ascending order, starting from the ones of lowest order which are potentially the most critical.

In Example 6.8, MCS K_2 is of order 1 and is potentially more critical than MCS K_1, which is of order 2. In Example 6.9, both MCSs are of order 2 and their potential criticality cannot be ascertained merely from ranking them by their order.

Example 6.10   *System Redundancy vs. Component Redundancy*

We return to the problem of system redundancy vs. component redundancy considered in Section 4.2.8. Figure 6.10(a) shows the FT for the system redundancy configuration whose RBD is shown in Figure 4.13(b), while Figure 6.10(a) shows the FT for the component redundancy configuration whose RBD is shown in Figure 4.13(c).

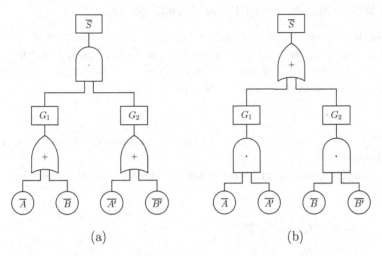

(a)   (b)

Figure 6.10   (a) FT for system redundancy; (b) FT for component redundancy.

System Redundancy

Assuming $\mathrm{TE} = \overline{S}_{\mathrm{sr}}$ (complete system failure), the DNF becomes:

$$\mathrm{TE} = \overline{S}_{\mathrm{sr}} = G_1 \wedge G_2 = (\overline{A} \vee \overline{B}) \wedge (\overline{A}' \vee \overline{B}') = (\overline{A} \wedge \overline{A}') \vee (\overline{A} \wedge \overline{B}')$$

$$\vee (\overline{B} \wedge \overline{A}') \vee (\overline{B} \wedge \overline{B}'),$$

from which the following list of four MCSs, all of order 2, is obtained:

$$K_1 = \overline{A} \wedge \overline{A}', \quad K_2 = \overline{A} \wedge \overline{B}', \quad K_3 = \overline{B} \wedge \overline{A}', \quad K_4 = \overline{B} \wedge \overline{B}'.$$

Component Redundancy

The DNF of $\mathrm{TE} = \overline{S}_{\mathrm{cr}}$ becomes:

$$\mathrm{TE} = \overline{S}_{\mathrm{cr}} = G_1 \vee G_2 = (\overline{A} \wedge \overline{A}') \vee (\overline{B} \wedge \overline{B}'),$$

from which the following two MCSs, each of order 2, are obtained:

$$K_1 = \overline{A} \wedge \overline{A}', \quad K_2 = \overline{B} \wedge \overline{B}'.$$

Comparing the list of MCSs in the two configurations, we observe that the system redundancy has failure combinations (in particular, $\overline{A} \wedge \overline{B}'$ and $\overline{B} \wedge \overline{A}'$) that do not cause failure in the component redundancy configuration. Hence, from a simple comparison of MCSs in the two cases, it is possible to assert that the component redundancy configuration is more dependable than the system redundancy configuration since all the failure modes for the former are also failure modes for the latter but not vice versa. This is the same conclusion that was drawn in Section 4.2.8 on the basis of a quantitative argument.

6.3.2   Logical Expression through Graph-Visiting Algorithm

The bottom-up analysis method is based on a recursive depth-first search (DFS) algorithm along the FT; each time an AND gate is encountered its output is expressed as the logical product of its inputs, and each time an OR gate is encountered its output is expressed as the logical sum of its inputs (see also Section 5.2.5).

Example 6.11   Figure 6.11 provides the FT representation of the RBD of Figure 4.6 in Example 4.11.

Analysis through the DFS algorithm provides the following expression:

$$G_1 = \overline{A} \wedge \overline{B},$$

$$G_2 = (\overline{A} \wedge \overline{B}) \vee \overline{C},$$

$$G_3 = \overline{D} \vee \overline{E},$$

$$\text{TE} = ((\overline{A} \wedge \overline{B}) \vee \overline{C}) \wedge (\overline{D} \vee \overline{E}) \tag{6.6}$$

$$= (\overline{A} \wedge \overline{B} \wedge \overline{D}) \vee (\overline{A} \wedge \overline{B} \wedge \overline{E}) \vee (\overline{C} \wedge \overline{D}) \vee (\overline{C} \wedge \overline{E}).$$

The DNF expression of the TE also provides the list of the four MCSs.

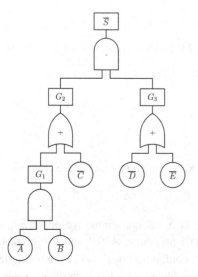

Figure 6.11   FT for the series-parallel system of Figure 4.6.

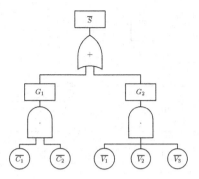

Figure 6.12   FT for the base repeater of Example 4.15.

Example 6.12   *Base Repeater in a Cellular Communication System*
We return to the base repeater in a cellular communication system with two control channels and three voice channels of Example 4.15, whose RBD was depicted in Figure 4.11. The system is considered **down** when either no control channels or no voice channels are functioning. The FT representing system failure is given in Figure 6.12. The logical expression for the TE in DNF is:

$$\overline{S} = (\overline{C_1} \wedge \overline{C_2}) \vee (\overline{V_1} \wedge \overline{V_2} \wedge \overline{V_3}).$$

Example 6.13   *Agropark Appraisal [18]*
Although FTA is commonly applied to model the propagation of failures in technical systems, the fault tree technique can also be of use in more general contexts for helping decision makers to choose among different alternatives. FT has been applied for economic or social risk assessments [19, 20] and for studying the possible failures in the implementation of a complex procedure or service system [21]. As an example, the possible causes of failure in the realization of an agropark project were investigated using an FT in [18]. From a system design perspective, the FTA provides a logical framework for understanding the way in which a project can fail that is essential for the project appraisal.

The FT of Figure 6.13 (derived from [18]) explores the main reasons that can lead to the failure of the project. Even if quantification of the FT may not be easy or may be subject to further uncertainties, the technique nevertheless helps the project developers to understand the main causes of an unsuccessful implementation and to suggest remedies. We leave it to the reader to find all the mincuts for this example.

6.3.3   Qualitative Analysis of FTRE

As mentioned in Section 6.2.4, FTs with repeated events require specific analysis algorithms. The logical expression for the TE as a function of the terminal events is derived with the same iterative procedure, but now the final terms are CSs and not

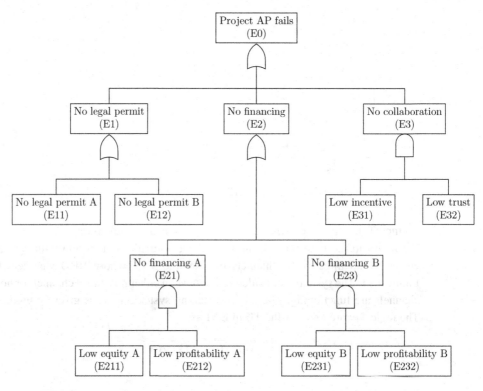

Figure 6.13   FT for an agropark appraisal process [18].

necessarily MCSs. The search for the list of MCSs from the list of CSs requires an additional step to expunge the non-minimal cuts.

Example 6.14   Figure 6.14 shows an FTRE in which $\overline{A}$ and $\overline{I}$ appear more than once and are represented as inverted triangles. The logical expression for the TE is derived in a top-down fashion, through the following steps:

$$
\begin{aligned}
\text{TE} &= G_1 \vee G_2 \qquad\qquad\qquad\qquad\qquad\qquad\qquad\qquad\qquad\qquad (6.7)\\
&= (G_4 \wedge \overline{A}) \vee (G_5 \wedge \overline{B})\\
&= ((G_8 \vee \overline{I} \vee \overline{L}) \wedge \overline{A}) \vee ((\overline{S} \vee \overline{A}) \wedge \overline{B}))\\
&= (((G_{11} \wedge G_{12}) \vee \overline{I} \vee \overline{L}) \wedge \overline{A}) \vee (\overline{S} \wedge \overline{B}) \vee (\overline{A} \wedge \overline{B})\\
&= ((((\overline{M} \vee \overline{I}) \wedge (\overline{C} \vee \overline{I})) \vee \overline{I} \vee \overline{L}) \wedge \overline{A}) \vee (\overline{S} \wedge \overline{B}) \vee (\overline{A} \wedge \overline{B})\\
&= (((\overline{M} \wedge \overline{C}) \vee (\overline{I} \wedge \overline{C}) \vee (\overline{M} \wedge \overline{I}) \vee (\overline{I} \wedge \overline{I}) \vee \overline{I} \vee \overline{L}) \wedge \overline{A}) \vee (\overline{S} \wedge \overline{B}) \vee (\overline{A} \wedge \overline{B})\\
&= (\overline{M} \wedge \overline{C} \wedge \overline{A}) \vee (\overline{I} \wedge \overline{C} \wedge \overline{A}) \vee (\overline{M} \wedge \overline{I} \wedge \overline{A}) \vee (\overline{I} \wedge \overline{I} \wedge \overline{A})\\
&\quad \vee (\overline{I} \wedge \overline{A}) \vee (\overline{L} \wedge \overline{A}) \vee (\overline{S} \wedge \overline{B}) \vee (\overline{A} \wedge \overline{B}).
\end{aligned}
$$

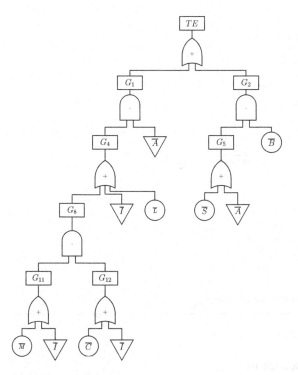

Figure 6.14 FT with repeated events.

From (6.7) the list of CSs is given by:

$$C_1 = (\overline{M} \wedge \overline{C} \wedge \overline{A}), \quad C_2 = (\overline{I} \wedge \overline{C} \wedge \overline{A}), \quad C_3 = (\overline{M} \wedge \overline{I} \wedge \overline{A}),$$
$$C_4 = (\overline{I} \wedge \overline{I} \wedge \overline{A}), \quad C_5 = \overline{I} \wedge \overline{A}, \quad C_6 = \overline{L} \wedge \overline{A},$$
$$C_7 = \overline{S} \wedge \overline{B}, \quad C_8 = \overline{A} \wedge \overline{B}.$$

Examining this list, it is easy to see that C_2, C_3, and C_4 are not minimal. In fact, C_4 reduces to C_5 by the idempotence law (Section 6.12), C_2 contains C_5 and by the absorption law (Section 6.12) is not minimal, and C_3 contains C_5. After appropriate reduction, the list of MCSs becomes: $K_1 = C_1$, $K_2 = C_5$, $K_3 = C_6$, $K_4 = C_7$, and $K_5 = C_8$. The TE can finally be written as:

$$TE = K_1 \vee K_2 \vee K_3 \vee K_4 \vee K_5$$
$$= (\overline{M} \wedge \overline{C} \wedge \overline{A}) \vee (\overline{I} \wedge \overline{A}) \vee (\overline{L} \wedge \overline{A}) \vee (\overline{S} \wedge \overline{B}) \vee (\overline{A} \wedge \overline{B}).$$

Example 6.15   We return to Example 6.12, but we assume now that one of the control channels may function as a voice channel and that the system is down when either no control channels or no voice channels are functioning. The FT representing system failure now has repeated events and is shown in Figure 6.15.

Analysis of the logical expression for the TE yields:

$$\overline{S} = G_1 \vee G_2$$
$$= (\overline{C_1} \wedge \overline{C_2}) \vee ((\overline{C_1} \vee \overline{C_2}) \wedge \overline{V_1} \wedge \overline{V_2} \wedge \overline{V_3})$$
$$= (\overline{C_1} \wedge \overline{C_2}) \vee (\overline{C_1} \wedge \overline{V_1} \wedge \overline{V_2} \wedge \overline{V_3}) \vee (\overline{C_2} \wedge \overline{V_1} \wedge \overline{V_2} \wedge \overline{V_3}). \quad (6.8)$$

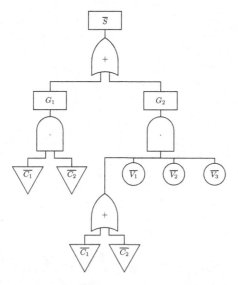

Figure 6.15   FT for the revised base repeater example.

6.3.4   Fault Tree and Success Tree

We have considered logic tree structures in which the root node TE is an undesired (or failure) event for the system as a whole and the leaves of the tree represent component failure events. This structure has been referred to as a *fault tree*. However, in a completely analogous way, it is possible to represent the success of a system rather than its failure, by assuming as TE the working condition of the system – a *success tree*. In this case, the leaves of the tree represent the working condition of the components. The general use of FTs instead of success trees is mainly due to two reasons:

- Analysts are usually more interested in the combination of component failures that can bring the system down, particularly in safety-critical systems.
- Efficient numerical solvers are more easily available for FTs than for success trees.

Example 6.16   Consider again the RBD of Figure 6.5. If we consider the success of the system S as the TE to be analyzed, we see that S occurs if C works and either A or B work. This structure can be expressed in a logical way as:

$$S = (A \lor B) \land C = (A \land C) \lor (B \land C). \tag{6.9}$$

The terms $(A \land C)$ and $(B \land C)$ represent the set of *minimal paths* (or *minpaths*), as defined in Definition 5.1. Of course, the logical expression (6.9) may be easily represented by a success tree with OR and AND gates.

Application of De Morgan's laws (see Section 6.12) allows us to convert the logical relations that express success into the equivalent relations that express failure.

Example 6.17 We negate both terms in Eq. (6.9), and we apply De Morgan's laws to get:

$$\overline{S} = \overline{(A \wedge C) \vee (B \wedge C)} = \overline{(A \wedge C)} \wedge \overline{(B \wedge C)} = (\overline{A} \vee \overline{C}) \wedge (\overline{B} \vee \overline{C})$$

$$= (\overline{A} \wedge \overline{B}) \vee (\overline{A} \wedge \overline{C}) \vee (\overline{B} \wedge \overline{C}) \vee (\overline{C} \wedge \overline{C})$$

$$= (\overline{A} \wedge \overline{B}) \vee \overline{C}, \tag{6.10}$$

where we have made use of the identity $(\overline{A} \wedge \overline{C}) \vee (\overline{B} \wedge \overline{C}) \vee (\overline{C} \wedge \overline{C}) = \overline{C}$ (see Section 6.12). The expression (6.10) coincides with the one derived from the FT of Example 6.7.

6.3.5   Structural Importance

We have seen that different basic events do not have the same criticality in determining the occurrence of the TE. Thus, it is useful to rank them according to some structural importance index that is based on the knowledge of the Boolean function associated with the TE [3, 22].

Given an FT with n basic events (or components), the TE can be expressed as a Boolean function $F(x_1, x_2, \ldots x_n)$ of n variables, whose value is 1 for the combinations of the n variables for which the TE is true and 0 otherwise. Define the structural sensitivity of component i as

$$S_i^S = F_{x_i=1} - F_{x_i=0}, \tag{6.11}$$

where $F_{x_i=1}$ and $F_{x_i=0}$ are the values of functions obtained from the Shannon decomposition of the function $F(x_1, x_2, \ldots x_n)$ with $x_i = 1$ and $x_i = 0$, respectively (see Section 5.2.1). Since we are dealing with a coherent FT (see Section 6.2), the sensitivity S_i^S is equal to 1 only in the states in which $F_{x_i=1} = 1$ and $F_{x_i=0} = 0$, i.e., in the states in which component x_i is critical and its failure changes the state of the TE from up to down.

The structural importance index I_i^S depends on the number of states in which component i is critical, and it is defined as the sum of the structural sensitivities normalized by the number of states:

$$I_i^S = \frac{\sum_\Omega S_i^S}{2^n},$$

where Ω is the ensemble of all the possible combinations of up and down states of the n components, whose cardinality is 2^n.

Example 6.18   Consider the three-component system of Example 6.8, whose Boolean representation, from Eq. (6.4), is $\text{TE} = (\overline{A} \wedge \overline{B}) \vee \overline{C}$. Table 6.1 shows the truth table of the TE, and in the last three columns, labeled S_A^S, S_B^S, and S_C^S, the structural sensitivities with respect to the variables $\overline{A}$, $\overline{B}$, and $\overline{C}$, respectively.

The structural importance indices I_i^S are then:

$$I_A^S = \frac{1}{4}, \quad I_B^S = \frac{1}{4}, \quad I_C^S = \frac{3}{4},$$

showing that component C is structurally three times more important than components A and B.

Table 6.1 Structural sensitivity of the three-component FT of Figure 6.5.

$\overline{A}$	$\overline{B}$	$\overline{C}$	$\text{TE} = \overline{S}$	S_A^S	S_B^S	S_C^S
0	0	0	0	0	0	1
0	0	1	1	0	0	1
0	1	0	0	1	0	1
0	1	1	1	0	0	1
1	0	0	0	0	1	1
1	0	1	1	0	0	1
1	1	0	1	1	1	0
1	1	1	1	0	0	0

Problems

6.1   Enumerate all the MCSs of a three-out-of-five system and draw the related expanded FT.

6.2   Find the list of the MCSs for the FT of Figure 6.13.

6.3   Return to Problem 4.17 of two parallel processors P_1 and P_2 with private memory M_1 and M_2 and shared memory M_3. Find all the MCSs and the structural importance indices of each of the five components.

6.4   An FT with repeated events for a ventilation system is represented in Figure 6.16. Find the list of MCSs and the logical expression for the TE.

6.5   In the fluid level control system of Figure 4.9, two pumps are replicated in parallel. Suppose now that the system has three pumps that work on a two-out-of-three basis. Find the FT of the system and derive the logical expression for the TE and the list of MCSs.

6.6   Convert the RBD of the propulsion system of Figure 4.10 into an FT and find the list of MCSs and the logical expression of the TE.

6.7   Suppose now that the three modules of the propulsion system of Figure 4.10 work in a two-out-of-three basis. Find the FT of the system and derive the logical expression for the TE and the list of MCSs.

6.8   Find structural importance indices for all the components in the system of Example 6.11.

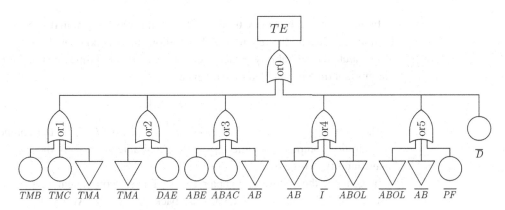

Figure 6.16 FT for a ventilation system with repeated events.

6.4 Quantitative Analysis

The quantitative analysis of an FT has the objective of evaluating the probability of occurrence (and related measures) of the TE, of the MCSs and of any other intermediate event of the FT. To accomplish this task, the probabilities of occurrence of the basic events must be supplied, and the occurrences of the basic events must be assumed to be statistically independent. With this assumption, the properties of the FT are completely specified if a probability is assigned to each single basic event according to the dependability metrics defined on a single unit as elaborated in Chapter 3. An extension of the standard (or *static*) FT known as a *dynamic fault tree* (DFT) in which new (non-Boolean) gates are defined to include dependencies is considered in Section 16.5.

6.4.1 Probability of Basic Events

Let the probability of occurrence of the basic event $\overline{E}_i$ be $q_i = P\{\overline{E}_i\}$, where q_i can be any measure associated with $\overline{E}_i$, as indicated in the introduction to Part II. In principle, any quantification that maintains the independence assumption can be assigned. The assignment of q_i follows the rules of Chapter 3 and depends on the type of information that is available about the failure and repair process of the basic components and about the maintenance and management policy of the system. It is customary that the analysis tools incorporate standard ways in which q_i can be assigned/interpreted:

Fixed probability: Basic event E_i is a binary event to which a Bernoulli pmf is assigned, so that $q_i = P\{\overline{E}_i\}$ is a constant value. As an example, q_i could be the probability of an incorrect operation by a human operator, or a failure on demand of a protective device, or the probability of failure during a test.

Self-revealing failure: A component has a self-revealing failure when it is connected in such a way that its failure is promptly recognized. Two cases arise: the component is non-repairable or it is repairable (with statistically independent repair across all components).

In the first case, the failure time distribution must be known; in the second case, both the failure and the repair time distributions must be known (see Chapter 3). To compute the probability values in either case a mission time must be fixed. Let the mission time be denoted by t_M; hence

$$q_i = P\{\overline{E}_i, t_M\}. \tag{6.12}$$

In the case of a non-repairable component, q_i in Eq. (6.12) is the component unreliability computed at time t_M; in the repairable case, q_i is the component unavailability at time t_M.

In general, the failure and repair times of basic components can have any distribution [15, 16, 23]. Some software packages like SHARPE [15, 16] allow the assignment of TTF and TTR distributions taken from a large selection of distribution functions. By contrast, many commercial tools accept as input parameters for each basic component only a constant failure rate and a constant repair rate, thus assuming that the failure and repair times of each component are exponentially distributed and restricting the analysis to just this case. Denoting the constant failure and repair rates of component i by λ_i and μ_i, respectively, (6.12) becomes [see Eq. (3.116)]:

$$q_i = \frac{\lambda_i}{\lambda_i + \mu_i} (1 - e^{-(\lambda_i + \mu_i)t_M}). \tag{6.13}$$

By letting t_M approach infinity, we assign to component i its steady-state unavailability. Setting $\mu_i = 0$, the usual unreliability expression (3.13) for a component with constant failure rate is assigned.

Non-Self-Revealing Failure: When the failure of a component does not have a prompt and visible effect on the system and remains hidden, we speak about non-self-revealing failures. Typical situations that fall under this category are dormant components in a standby configuration [24].

When a component is in a dormant condition and fails, its failure does not have any immediate effect on the functioning of the system and hence is considered to be non-self-revealing. The failure reveals itself only when the dormant component is switched into operation following failure of the active component. Most protection and safety systems are not operative during most of their life and are put into operation upon the failure of the active device.

Accurate modeling of standby configurations requires us to resort to more powerful formalisms; these are first considered in Chapter 8 and are extensively discussed in Part III of the book. However, many FT tools include the possibility of quantifying non-self-revealing failures based on the observation that the only way to ascertain whether a dormant device is in good operating condition is to perform periodic diagnostic tests. This policy allows one to check whether at the test instant the device is operating, but of course does not guarantee that device failure will not occur in the interval before the next test.

Assume θ_i is the interval between diagnostic tests of dormant component i, and λ_i its constant failure rate. If a failure is revealed at the end of a testing interval θ

Table 6.2 Parameter window to quantify a
basic event probability value in an FT.

Parameter	Value
λ	
μ	
Probability	
Inter-testing interval θ	

it can be expected to have occurred at a time equal to $\theta/2$. Then the probability of finding the component down can be approximated as [8, 24, 25]:

$$q_i = \lambda_i \theta/2 \quad \text{(first-order approximation to } 1 - e^{-\lambda_i \theta/2}\text{)}.$$

Many FT tools assume, by default, that the basic events have constant failure and repair rates or, if they can be in a dormant status, that the inter-test interval is known. In this case, in order to facilitate the entering of the appropriate parameters, a window similar to the one shown in Table 6.2 can be opened for each basic event.

6.4.2 Quantitative Analysis of an FT without Repeated Events

The probability of the TE in an FT without repeated events can be computed directly from knowledge of the failure probabilities of the basic events either in a bottom-up fashion or starting from the Boolean expression.

Bottom-Up Algorithm

A simple linear time algorithm can be used to compute the probability of TE. Let $q_i = F_i(t)$ be the unreliability or unavailability of component i; the computation of the TE probability proceeds following a bottom-up search through the FT (Section 6.3.2) by observing that due to independence the following rules can be applied to compute the probability of the output of AND, OR, and k-out-of-n gates [see Eq. (4.47)]:

$$\begin{cases} \displaystyle\prod_{i=1}^{n} F_i(t) & \text{AND gate with } n \text{ inputs,} \\[2ex] \displaystyle 1 - \prod_{i=1}^{n}(1 - F_i(t)) & \text{OR gate with } n \text{ inputs,} \\[2ex] \displaystyle\sum_{i=k}^{n} \binom{n}{i} F_i(t)^i (1 - F_i(t))^{n-i} & k\text{-out-ot-}n \text{ gate with } n \text{ identical inputs.} \end{cases}$$

$$(6.14)$$

Any probability distribution can be used in (6.14) [15, 23].

Example 6.19   Applying Eq. (6.14), the TE probability of the FT of Figure 6.11 proceeds as follows:

$$P\{G_1\} = F_A(t)\,F_B(t),$$

$$P\{G_2\} = 1 - (1 - P\{G_1\})\,(1 - F_C(t)) = F_A(t)\,F_B(t) + F_C(t) - F_A(t)\,F_B(t)\,F_C(t),$$

$$P\{G_3\} = 1 - (1 - F_D(t))\,(1 - F_E(t)) = F_D(t) + F_E(t) - F_D(t)F_E(t),$$

$$P\{\text{TE}\} = P\{G_2\}P\{G_3\}.$$

Example 6.20   The TE probability of the base repeater FT given in Figure 6.12 is computed as:

$$P\{G_1\} = F_C(t)^2,$$

$$P\{G_2\} = F_V(t)^3,$$

$$P\{\text{TE}\} = 1 - (1 - F_C(t)^2)\,(1 - F_V(t)^3).$$

Example 6.21   *A Fluid Level Controller*
Representing the system of Example 4.1 as an FT and assuming as TE the system failure event, we have:

$$P\{\text{TE}\} = F_s(t) = 1 - (1 - F_S(t))(1 - F_L(t))(1 - F_P(t))(1 - F_V(t)).$$

By denoting as $R_i(t) = 1 - F_i(t)$ ($i = $ S, L, P, V) the probability of success of a basic event, the success probability of the FT becomes (see Example 4.1):

$$P\{\overline{\text{TE}}\} = R_s(t) = R_S(t)\,R_L(t)\,R_P(t)\,R_V(t).$$

Algorithm Based on the Boolean Expression
Once the Boolean expression of the TE is determined, its probability can be computed by any technique already exploited in Chapter 5, such as inclusion–exclusion formula, SDP or BDD. The use of BDDs in FTA was proposed in [26], where their computational efficiency was demonstrated. Since then, the BDD solution technique has been incorporated in many commercial as well as academic tools [15, 27]. Surveys of the BDD method in reliability analysis can be found in [28, 29].

The evaluation of the probability of the TE does not necessarily require the preliminary search for the MCSs. If the BDD is built first, the list of the CSs can be determined by searching all the paths connecting the root node of the BDD to its terminal node 1, and then minimizing the list of the CSs to find the MCSs. An alternative approach, proposed in [26], consists in transforming the original BDD into a new BDD embedding all the MCSs only. Details of the transformation algorithm can be found in [26].

Example 6.22   The logical expression for the TE of the FT of Figure 6.5 is:

$$\text{TE} = (\overline{A} \wedge \overline{B}) \vee \overline{C}.$$

Applying the inclusion–exclusion formula,

$$
\begin{aligned}
P\{\text{TE}\} &= P\{(\overline{A} \wedge \overline{B}) \vee \overline{C}\} \\
&= P\{\overline{A} \wedge \overline{B}\} + P\{\overline{C}\} - P\{(\overline{A} \wedge \overline{B}) \wedge \overline{C}\} \\
&= P\{\overline{A}\} P\{\overline{B}\} + P\{\overline{C}\} - P\{\overline{A}\} P\{\overline{B}\} P\{\overline{C}\}.
\end{aligned}
\tag{6.15}
$$

If the basic events have the probabilities

$$P\{\overline{A}\} = q_A, \quad P\{\overline{B}\} = q_B, \quad P\{\overline{C}\} = q_C,$$

Eq. (6.15) becomes:

$$q_{\text{TE}} = q_A q_B + q_C - q_A q_B q_C.$$

Alternatively, applying the SDP approach, we have (commuting terms first):

$$
\begin{aligned}
\text{TE} &= \overline{C} \vee (\overline{A} \wedge \overline{B}) \\
&= \overline{C} \vee (C \wedge \overline{A} \wedge \overline{B}).
\end{aligned}
$$

Observing that the two terms in the last expression are disjoint, we can write:

$$
\begin{aligned}
P\{\text{TE}\} &= P\{\overline{C}\} + P\{C \wedge \overline{A} \wedge \overline{B}\} \\
&= q_C + (1 - q_C) q_A q_B = q_C + q_A q_B - q_A q_B q_C.
\end{aligned}
$$

The third approach is to use BDDs. Two possible BDDs for the TE expression are shown in Figure 6.17, adopting the sequence $C \prec A \prec B$ in Figure 6.17(a) and the sequence $A \prec C \prec B$ in Figure 6.17(b).

To compute the probability of the TE, we recursively apply Eq. (5.3) at each node of the BDDs of Figure 6.17, following the node number marked on the figure. For the sequence $C \prec A \prec B$, we have:

$$
\begin{aligned}
p_1 &= q_B, \\
p_2 &= q_A q_B, \\
P\{\text{TE}\} = p_3 &= q_A q_B + q_C (1 - q_A q_B) = q_A q_B + q_C - q_A q_B q_C.
\end{aligned}
\tag{6.16}
$$

For the sequence $A \prec C \prec B$, we have:

$$
\begin{aligned}
p_1' &= q_B, \\
p_2' &= q_C, \\
p_3' &= q_B + q_C (1 - q_B), \\
P\{\text{TE}\} = p_4' = p_2' + q_A (p_3' - p_2') &= q_C + q_A (q_B - q_B q_C) = q_A q_B + q_C - q_A q_B q_C.
\end{aligned}
\tag{6.17}
$$

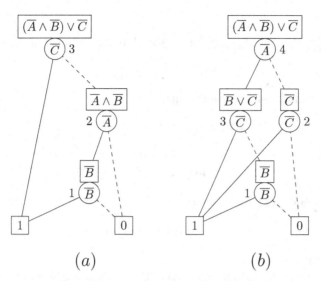

Figure 6.17   Two possible BDD representations of Figure 6.5. (a) Sequence $C \prec A \prec B$; (b) Sequence $A \prec C \prec B$.

Example 6.23   The TE of the FT of Figure 6.6(b) has the following Boolean expression (see Example 6.9):

$$\text{TE} = (\overline{A} \wedge \overline{C}) \vee (\overline{B} \wedge \overline{C}).$$

Applying the inclusion–exclusion formula, we get:

$$P\{\text{TE}\} = P\{(\overline{A} \wedge \overline{C})\} + P\{(\overline{B} \wedge \overline{C})\} - P\{(\overline{A} \wedge \overline{B} \wedge \overline{C})\}$$

$$= q_A q_C + q_B q_C - q_A q_B q_C.$$

With the SDP approach, we get:

$$\text{TE} = (\overline{A} \wedge \overline{C}) \vee (\overline{\overline{A} \wedge \overline{C}} \wedge \overline{B} \wedge \overline{C})$$

$$= (\overline{A} \wedge \overline{C}) \vee ((A \vee C) \wedge \overline{B} \wedge \overline{C})$$

$$= (\overline{A} \wedge \overline{C}) \vee (A \wedge \overline{B} \wedge \overline{C}) \vee (C \wedge \overline{B} \wedge \overline{C})$$

$$= (\overline{A} \wedge \overline{C}) \vee (A \wedge \overline{B} \wedge \overline{C}),$$

and then $q_{\text{TE}} = q_A q_C + (1 - q_A) q_B q_C = q_A q_C + q_B q_C - q_A q_B q_C.$

The BDD representation adopting the sequence $\overline{A} \prec \overline{C} \prec \overline{B}$ is shown in Figure 6.18. The probability of the TE can be computed recursively as:

$$p_1 = q_B,$$

$$p_2 = q_C,$$

$$p_3 = q_B q_C, \tag{6.18}$$

$$P\{\text{TE}\} = p_4 = p_3 + q_A (p_2 - p_3) = q_B q_C + q_A (q_C - q_B q_C)$$

$$= q_A q_C + q_B q_C - q_A q_B q_C.$$

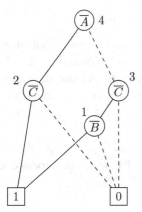

Figure 6.18   BDD structure of the FT of Figure 6.6.

It can be verified that the probability computation of the TE in Eq. (6.18) can also be obtained by evaluating the probability of the union of all the paths in the BDD connecting the root node with the terminal node 1 that by construction are mutually disjoint (see Section 5.2.1).

The paths connecting the root node $\overline{A}$ with the terminal node 1 in Figure 6.18 are:

$$Path_1 = \overline{A}\,\overline{C}, \quad Path_2 = A\,\overline{C}\,\overline{B}, \quad \text{with } Path_1 \wedge Path_2 = \phi.$$

Hence: $P\{TE\} = P\{Path_1\} + P\{Path_2\} = q_A q_C + (1 - q_A) q_C q_B.$

Problems

6.9   Compare the probability of the TE in the system redundancy configuration vs. the component redundancy configuration of Example 6.10 assuming that all components have a TTF with a Weibull distribution (3.27):

$$F_A(t) = F_{A'}(t) = 1 - e^{-\alpha_A t^{\beta_A}}, \quad F_B(t) = F_{B'}(t) = 1 - e^{-\alpha_B t^{\beta_B}},$$

with:

$$\alpha_A = 1500\,\text{hr}^{-1}, \quad \beta_A = 1.2, \quad \alpha_B = 2000\,\text{hr}^{-1}, \quad \beta_B = 1.5.$$

6.10   In Problem 6.5, a two-out-of-three scheme for the pump subsystem was assumed. Assuming as in Example 4.1 that all the components have an exponentially distributed TTF with the failure rates

$$\lambda_S = 2 \times 10^{-6}\,\text{fhr}^{-1}, \ \lambda_L = 5 \times 10^{-6}\,\text{fhr}^{-1}, \ \lambda_P = 2 \times 10^{-5}\,\text{fhr}^{-1}, \ \lambda_V = 1 \times 10^{-5}\,\text{fhr}^{-1},$$

compute the probability of the TE via a bottom-up approach as well as via the logical expression of the TE. Find the probability of each of the MCSs.

6.4.3   Quantitative Analysis of an FT with Repeated Events

The quantitative analysis of an FTRE can be based on the qualitative analysis, by computing the probability associated with the logical expression of the TE. The analysis may be based on a factoring algorithm similar to the one described for RBD in Section 4.4; however, SDP or BDD are the most commonly used methods. Notice that the SDP method is always preferred over the method of inclusion–exclusion.

Example 6.24   Figure 6.19(a) [27] shows an FT with one repeated event $\overline{C}$. The Boolean function expressing the occurrence of the TE is:

$$TE = G1 \vee G2$$
$$= (\overline{A} \wedge \overline{B} \wedge \overline{C}) \vee (\overline{C} \wedge \overline{D}).$$

The SDP method proceeds as follows:

$$TE = (\overline{C} \wedge \overline{D}) \vee (\overline{(\overline{C} \wedge \overline{D})} \wedge (\overline{A} \wedge \overline{B} \wedge \overline{C}))$$
$$= (\overline{C} \wedge \overline{D}) \vee ((C \vee D) \wedge (\overline{A} \wedge \overline{B} \wedge \overline{C})) = (\overline{C} \wedge \overline{D}) \vee (\overline{A} \wedge \overline{B} \wedge \overline{C} \wedge D).$$

The last two terms are disjoint, so that the probability becomes:

$$P\{TE\} = P\{\overline{C} \wedge \overline{D}\} + P\{\overline{A} \wedge \overline{B} \wedge \overline{C} \wedge D\}$$
$$= q_C q_D + q_A q_B q_C (1 - q_D)$$
$$= q_C q_D + q_A q_B q_C - q_A q_B q_C q_D.$$

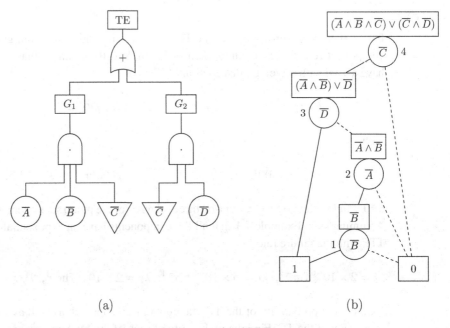

(a)   (b)

Figure 6.19   (a) FT with one repeated event; (b) The associated BDD structure.

The same FT of Figure 6.19(a) [27] can be analyzed by building the BDD shown in Figure 6.19(b). The probability computation proceeds in the usual way to give:

$$p_1 = q_B,$$
$$p_2 = q_A q_B,$$
$$p_3 = q_A q_B + q_D(1 - q_A q_B),$$
$$P\{TE\} = p_4 = q_A q_B q_C + q_C q_D(1 - q_A q_B) = q_A q_B q_C + q_C q_D - q_A q_B q_C q_D.$$
(6.19)

Example 6.25 We return to the modified base repeater Example 6.15 where the control channels may replace one of the voice channels. The FT was drawn in Figure 6.15. The application of the SDP method to the TE expression derived in (6.8) proceeds as follows:

$$TE = (\overline{C}_1 \wedge \overline{C}_2) \vee ((\overline{(\overline{C}_1 \wedge \overline{C}_2)}) \wedge (\overline{C}_1 \wedge \overline{V}_1 \wedge \overline{V}_2 \wedge \overline{V}_3))$$
$$\vee ((\overline{(\overline{C}_1 \wedge \overline{C}_2)}) \wedge \overline{(\overline{C}_1 \wedge \overline{V}_1 \wedge \overline{V}_2 \wedge \overline{V}_3)} \wedge (\overline{C}_2 \wedge \overline{V}_1 \wedge \overline{V}_2 \wedge \overline{V}_3))$$
$$= (\overline{C}_1 \wedge \overline{C}_2) \vee (C_1 \wedge \overline{C}_2 \wedge \overline{V}_1 \wedge \overline{V}_2 \wedge \overline{V}_3) \vee (C_2 \wedge \overline{C}_1 \wedge \overline{V}_1 \wedge \overline{V}_2 \wedge \overline{V}_3).$$

The last three terms are disjoint, so that the probability becomes:

$$P\{TE\} = q_{C_1} q_{C_2} + (1 - q_{C_1}) q_{C_2} q_{V_1} q_{V_2} q_{V_3} + (1 - q_{C_2}) q_{C_1} q_{V_1} q_{V_2} q_{V_3}$$
$$= q_{C_1} q_{C_2} + q_{C_1} q_{V_1} q_{V_2} q_{V_3} + q_{C_2} q_{V_1} q_{V_2} q_{V_3} - 2 q_{C_1} q_{C_2} q_{V_1} q_{V_2} q_{V_3}.$$

The BDD representing the Boolean function of the TE of Eq. (6.8) is shown in Figure 6.20. The number on the left of each node simply indicates its order.

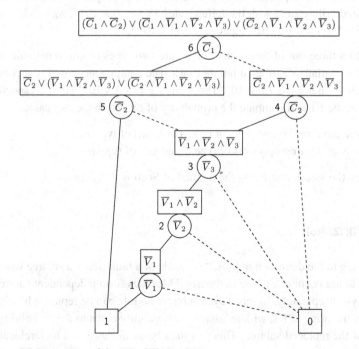

Figure 6.20   BDD structure of the FT of Figure 6.15.

The quantitative analysis proceeds in a bottom-up fashion applying Eq. (5.3).

$$p_1 = q_{V_1},$$
$$p_2 = q_{V_1} q_{V_2},$$
$$p_3 = q_{V_1} q_{V_2} q_{V_3},$$
$$p_4 = q_{C_2} q_{V_1} q_{V_2} q_{V_3},$$
$$p_5 = q_{V_1} q_{V_2} q_{V_3} + q_{C_2} (1 - q_{V_1} q_{V_2} q_{V_3}),$$
$$p_6 = p_4 + q_{C_1} (p_5 - p_4)$$
$$= q_{C_1} q_{C_2} + q_{C_1} q_{V_1} q_{V_2} q_{V_3} + q_{C_2} q_{V_1} q_{V_2} q_{V_3} - 2 q_{C_1} q_{C_2} q_{V_1} q_{V_2} q_{V_3}.$$

Problems

6.11 For the multiprocessor system of Problem 6.3, assume that the processors and memories have i.i.d. exponentially distributed TTF with rates λ and γ, respectively. The system is working when at least one processor is working with either its private memory or the shared memory. Find the Cdf of the system TTF by means of fault tree analysis, factoring the solution as in Section 4.4, assuming the common memory as a pivot component.

6.12 For the multiprocessor system of Problem 6.11, find the probability of the TE adopting the SDP and then the BDD approaches.

6.13 With reference to Problem 6.1, compute the TE of a three-out-of-five system where the basic components A, B, C, D, and E are non-repairable with an exponentially distributed failure time with the failure rates $\lambda_A = 1.0 \times 10^{-4}$, $\lambda_B = 3.0 \times 10^{-4}$, $\lambda_C = 5.0 \times 10^{-4}$, $\lambda_D = 7.0 \times 10^{-4}$, and $\lambda_E = 9.0 \times 10^{-4}$.

6.14 In a three-out-of-five system, there are two types of non-repairable components with exponentially distributed failure times. The components are of type A or of type B, with failure rates $\lambda_A = 1.0 \times 10^{-4} \, \text{fhr}^{-1}$ and $\lambda_B = 3.0 \times 10^{-4} \, \text{fhr}^{-1}$, respectively.
   Derive the FT and compute the probability of the TE in the two cases:

(a) There are three components of type A and two of type B.
(b) There are four components of type A and one of type B.

Check the solution utilizing the method of Section 4.5.1.

6.5   Modularization

According to the definition in [30], "a module of a fault tree is a subtree whose terminal events do not occur elsewhere in the tree." Modules form independent subtrees that can be analyzed separately. In this way, a subtree module can be replaced by a single basic event in the original FT, with an assigned probability equal to the probability of the top event of the replaced subtree. This will then be an instance of a hierarchical model we will consider in a more general setting in Chapter 16 in which all submodels are of fault tree type.

In an FT with repeated events, a module must contain all the occurrences of a repeated event, and different modules cannot share repeated events. A linear-time algorithm to detect modules of an FT has been described in [30].

With this modularization procedure, a fault tree with repeated events can be reduced to a top-level standard FT by replacing each module with a single basic event whose probability is equal to the probability of the TE of the module [31]. By decomposing an FT into simpler modules, the modularization procedure reduces the computational cost of solving the overall fault tree by employing a hierarchical model solution. Many more examples of such multi-level models will be considered in a more general setting in Part V of this book. What is unique here is that the submodels are automatically determined from the overall fault tree model, while in Part V of the book we will assume that submodels and their interactions are already developed by the modeler.

Example 6.26 We return to Example 6.14 and to the FT of Figure 6.14. According to the definition of Section 6.5, the subtree emerging from gate G_4 is a module, since its terminal events ($\overline{M}, \overline{I}, \overline{C},$ and $\overline{L}$) do not occur elsewhere in the tree. Hence, the subtree emerging from gate G_4 can be analyzed separately, and gate G_4 replaced in the original FT by a terminal event whose probability equals the probability of the subtree emerging from gate G_4. After this replacement, the resulting structure is an FT with repeated events that cannot be modularized any further.

Problems

6.15   Find two modules in the fault tree of the ventilation system of Figure 6.16.

6.6   Importance Measures

Importance measures are quantitative measures that can be assigned to basic events (or to MCSs) to rank their contribution to the occurrence of the TE. Importance measures help to identify critical parts in the system, and to guide in the choice of the corrective actions that are most effective in improving its dependability. Moreover, the relative importance measures may indicate the components that need additional development or design improvements, and may suggest efficient criteria for adding redundancy, for planning preventive maintenance strategies or for generating repair checklists. We have earlier considered structural importance measures that do not need the specification of probabilities for the basic events. We now consider more refined importance measures that assume the specification of probabilities for the basic events.

Since importance measures are in general time dependent, their relative ranking may depend upon the mission time. More detailed surveys of importance indices (and related measures) and their computation are given in [32] and, more recently, in [33].

6.6.1   Birnbaum Importance Index

Denoting by $q_{TE}(t)$ the time-dependent probability of occurrence of the TE, and by $q_i(t)$ the time-dependent probability of occurrence of the basic component i, the sensitivity of the system to component i is defined as (see also Sections 4.1.2 and 4.2.4):

$$I_i^B(t) = \frac{\partial q_{TE}(t)}{\partial q_i(t)}. \tag{6.20}$$

The importance index (6.20) is known in the FT literature as the Birnbaum importance index $I_i^B(t)$, after Birnbaum's original paper [34]. A practical way to compute $I_i^B(t)$ can be obtained by decomposing $q_{TE}(t)$ according to Eq. (5.3), pivoting with respect to basic event i:

$$q_{TE}(t) = q_i(t)\, q_{TE\,|x_i=1}(t) + (1 - q_i(t))\, q_{TE\,|x_i=0}(t), \tag{6.21}$$

where $q_{TE\,|x_i=1}(t)$ and $q_{TE\,|x_i=0}(t)$ represent the probability of the TE when the ith basic event is stuck at 1 or 0, respectively. Due to the decomposition in Eq. (6.21), $q_{TE\,|x_i=1}(t)$ and $q_{TE\,|x_i=0}(t)$ do not depend on x_i; hence, applying (6.20), we obtain:

$$I_i^B(t) = \frac{\partial q_{TE}(t)}{\partial q_i(t)} = q_{TE\,|x_i=1}(t) - q_{TE\,|x_i=0}(t). \tag{6.22}$$

The physical meaning of the importance index $I_i^B(t)$ is that it measures the rate at which an increase in the component unavailability (or unreliability) affects the increase in the system unavailability (or unreliability). The component with the highest $I_i^B(t)$ is the one whose increase in unavailability has the maximum effect on the system unavailability. But the importance index $I_i^B(t)$ may also be interpreted as the probability that component i is *relevant*, i.e., its failure, from a system up state, causes the TE to occur.

An algorithm for the computation of $I_i^B(t)$ based on the SDP method is given in [35], while one based on the BDD method is described in [27, 32]. If the $q_i(t)$ are fixed failure probabilities, steady-state unavailabilities, time-independent values for the Birnbaum indices are obtained. More generally, from (6.22) we see that the Birnbaum importance indices may be time dependent. To arrive at a numerical result, a value of the time t must be specified.

Example 6.27   Consider the FT in Figure 6.5, with TE $= (\overline{A} \wedge \overline{B}) \vee \overline{C}$. Applying Eq. (6.22) with respect to $\overline{A}, \overline{B}$, and $\overline{C}$, respectively, we obtain:

$$\overline{A} \quad \Longrightarrow \quad \begin{cases} q_{TE,\,|\overline{A}=1}(t) &= P\{\overline{B} \vee \overline{C}\} = q_B(t) + q_C(t) - q_B(t)q_C(t), \\ q_{TE,\,|\overline{A}=0}(t) &= P\{\overline{C}\} = q_C(t), \\ I_{\overline{A}}^B(t) &= q_B(t)\,(1 - q_C(t)); \end{cases}$$

$$
\overline{B} \quad \Longrightarrow \quad \begin{cases} q_{\text{TE},|\overline{B}=1}(t) &= P\{\overline{A} \vee \overline{C}\} = q_A(t) + q_C(t) - q_A(t)q_C(t), \\ q_{\text{TE},|\overline{B}=0}(t) &= P\{\overline{C}\} = q_C(t), \\ I_B^B(t) &= q_A(t)(1 - q_C(t)); \end{cases}
$$

$$
\overline{C} \quad \Longrightarrow \quad \begin{cases} q_{\text{TE},|\overline{C}=1}(t) &= 1, \\ q_{\text{TE},|\overline{C}=0}(t) &= P\{\overline{A} \wedge \overline{B}\} = q_A(t)q_B(t), \\ I_C^B(t) &= 1 - q_A(t)q_B(t). \end{cases}
$$

$$(6.23)$$

Since the failure probabilities $q_A(t)$, $q_B(t)$, and $q_C(t)$ are (should be!) very small numbers, it is clear that $I_C^B(t)$ is close to (but less than) 1, while $I_A^B(t)$ and $I_B^B(t)$ are of the same orders of magnitude as $q_B(t)$ and $q_A(t)$, respectively. Component C has the highest index of importance, and between A and B the most important is the most reliable (the one with lower failure probability) – see also Section 4.2.4.

A numerical example follows in Example 6.30.

6.6.2 Criticality Importance Index

The Birnbaum index does not depend on the unavailability of the component itself. In order to incorporate this information into an importance coefficient, a new measure, called the *criticality index*, is defined as:

$$
I_i^C(t) = \frac{\partial q_{\text{TE}}(t)}{\partial q_i(t)} \frac{q_i(t)}{q_{\text{TE}}(t)} = I_i^B \frac{q_i(t)}{q_{\text{TE}}(t)}
$$

$$
= \frac{(q_{\text{TE}|x_i=1}(t) - q_{\text{TE}|x_i=0}(t))\, q_i(t)}{q_{\text{TE}}(t)}. \tag{6.24}
$$

The criticality index $I_i^C(t)$ can be interpreted as the probability of the joint event that component i is relevant (its failure in a working state causes the TE to occur) and component i failed before time t conditioned on the system failure by time t.

Example 6.28 Consider again the FT of Figure 6.5. Applying Eq. (6.24), we obtain, respectively,

$$
I_A^C(t) = \frac{[q_B(t)(1 - q_C(t))]q_A(t)}{q_A(t)q_B(t) + q_C(t) - q_A(t)q_B(t)q_C(t)},
$$

$$
I_B^C(t) = \frac{[q_A(t)(1 - q_C(t))]q_B(t)}{q_A(t)q_B(t) + q_C(t) - q_A(t)q_B(t)q_C(t)},
$$

$$
I_C^C(t) = \frac{[1 - q_A(t)q_B(t)]q_C(t)}{q_A(t)q_B(t) + q_C(t) - q_A(t)q_B(t)q_C(t)}.
$$

We note that, in this case, $I_A^C(t) = I_B^C(t)$. Even if components A and B have different unavailabilities, they have the same criticality index.

If $q_A(t)$, $q_B(t)$, and $q_C(t)$ are of the same order of magnitude, it is easy to verify that, also in this case, component C has a criticality index greater than that of A and B.

A numerical example follows in Example 6.30.

6.6.3   Vesely–Fussell Importance Index

The criticality index accounts only for states in which the component is relevant. The Vesely–Fussell importance index (I_i^V) measures the probability that component i contributes to system failure given that a system failure has already occurred by time t [36]. Hence, I_i^V is calculated from the probability of the union of the MCSs containing i divided by the total probability of occurrence of the TE. Let $K_1, K_2, \ldots, K_m$ be the m MCSs of the system and let $c_j^{(i)}$ be an indicator variable that is equal to 1 if component i belongs to MCS K_j. Then:

$$I_i^V(t) = \frac{P\{\bigvee_{j=1}^m (c_j^{(i)} K_j)\}}{q_{TE}(t)}. \tag{6.25}$$

If component i belongs to a single MCS K_j, the formula in (6.25) reduces to

$$I_i^V(t) = \frac{P\{K_j\}}{q_{TE}(t)}. \tag{6.26}$$

Example 6.29   Consider, again, the FT of Figure 6.5. Since each component appears in one single MCS only, the formula in (6.26) applies:

$$I_A^V(t) = \frac{q_A(t)\,q_B(t)}{q_A(t)q_B(t) + q_C(t) - q_A(t)q_B(t)q_C(t)},$$

$$I_B^V(t) = \frac{q_A(t)\,q_B(t)}{q_A(t)q_B(t) + q_C(t) - q_A(t)q_B(t)q_C(t)},$$

$$I_C^V(t) = \frac{q_C(t)}{q_A(t)q_B(t) + q_C(t) - q_A(t)q_B(t)q_C(t)}.$$

We note again that $I_A^V(t) = I_B^V(t)$. Even if components A and B have different unavailabilities, they have the same Vesely–Fussell importance index.

A numerical comparison follows in Example 6.30.

Example 6.30   We assume that components A, B, and C in Examples 6.27, 6.28, and 6.29 are non-repairable, with an exponentially distributed TTF with respective rates $\lambda_A = 1.0 \times 10^{-6}\,\mathrm{fhr}^{-1}$, $\lambda_B = 2.0 \times 10^{-6}\,\mathrm{fhr}^{-1}$, and $\lambda_C = 1.0 \times 10^{-5}\,\mathrm{fhr}^{-1}$.

For three different mission times, we compute, in Table 6.3, the unreliabilities of the individual components and of the TE. The numerical values of the importance indices defined in Examples 6.27, 6.28, and 6.29 are compared in Table 6.4.

Table 6.3 Unreliabilities of components A, B, and C in Examples 6.27, 6.28, and 6.29, and of the TE.

Time (hr)	$q_A(t)$	$q_B(t)$	$q_C(t)$	$q_{TE}(t)$
1 000	0.001 000	0.001 998	0.009 950	0.009 952
5 000	0.004 987	0.009 950	0.048 770	0.048 817
10 000	0.009 950	0.019 801	0.095 162	0.095 340

Table 6.4 Importance indices of components A, B, and C in Examples 6.27, 6.28, and 6.29.

Time (hr)	I_A^B	I_B^B	I_C^B	$I_A^C = I_B^C$	I_C^C	$I_A^V = I_B^V$	I_C^V
1 000	0.001 978	0.000 989	0.999 998	0.000 198	0.999 799	0.000 201	0.999 801
5 000	0.009 465	0.004 744	0.999 950	0.000 967	0.998 983	0.001 016	0.999 033
10 000	0.017 917	0.009 003	0.999 803	0.001 870	0.997 933	0.002 066	0.998 130

Example 6.31 Consider the FT in Figure 6.6, with TE $= (\overline{A} \wedge \overline{C}) \vee (\overline{B} \wedge \overline{C})$ and $q_{TE}(t) = [q_A(t) + q_B(t) - q_A(t)q_B(t)]q_C(t)$. The three importance indices turn out to be:

$$I_A^B(t) = (1 - q_B(t))q_C(t),$$

$$I_B^B(t) = (1 - q_A(t))q_C(t),$$

$$I_C^B(t) = q_A(t) + q_B(t) - q_A(t)q_B(t);$$

$$I_A^C(t) = \frac{(1 - q_B(t))q_C(t)q_A(t)}{q_{TE}(t)},$$

$$I_B^C(t) = \frac{(1 - q_A(t))q_C(t)q_B(t)}{q_{TE}(t)},$$

$$I_C^C(t) = \frac{(q_A(t) + q_B(t) - q_A(t)q_B(t))q_C(t)}{q_{TE}(t)} = 1;$$

$$I_A^V(t) = \frac{q_A(t)q_C(t)}{q_{TE}(t)},$$

$$I_B^V(t) = \frac{q_B(t)q_C(t)}{q_{TE}(t)},$$

$$I_C^V(t) = \frac{P\{(\overline{A} \wedge \overline{C}) \vee (\overline{B} \wedge \overline{C})\}}{q_{TE}(t)} = 1.$$

It is noteworthy to observe that for this particular system configuration both the criticality and the Vesely–Fussell importance indices for component C are equal to 1, since component C must fail in order for the TE to occur.

Assuming for components A, B, and C the same TTF distribution as in Example 6.30, the $q_{TE}(t)$ and the importance indices are given in Table 6.5 (the values of $I_C^C(t)$ and $I_C^V(t)$ are omitted, being equal to 1).

Looking at the table, we can appreciate the increasing weight of component B with respect to component A, passing from the Birnbaum to the criticality and Vesely–Fussell

Table 6.5 Importance indices of components A, B, and C in Example 6.31.

Time (hr)	$q_{TE}(t)$	I_A^B	I_B^B	I_C^B	I_A^C	I_B^C	I_A^V	I_B^V
1 000	2.98×10^{-5}	0.009 93	0.009 94	0.002 99	0.333 00	0.666 33	0.000 99	0.001 99
5 000	0.000 72	0.048 28	0.048 53	0.014 89	0.331 67	0.665 00	0.004 98	0.009 94
10 000	0.002 81	0.093 28	0.094 21	0.029 55	0.330 00	0.663 33	0.009 93	0.019 76

importance indices, due to the higher failure rate of component B.

Problems

6.16 For the system of Problem 6.3, derive the Birnbaum, criticality, and Vesely–Fussell importance indices for one processor, one private memory, and the shared memory.

6.17 A software system consists of two proxy servers and three application servers. For the system to function properly, at least one proxy server and at least two out of three application servers should be functioning properly. Assume α and β are the failure rates of the proxy and application servers, respectively. Derive an expression for the software system reliability, first using an RBD and then using an FT with BDD representation; then derive an expression for the software system mean time to failure. Make any independence assumptions that are required.

6.18 In the system of Problem 6.17, derive the Birnbaum, criticality, and Vesely–Fussell importance indices with respect to the proxy server and the application server.

6.7   Case Studies

This section presents three large case studies to reveal the extensive application of FTA in real-life examples.

Example 6.32 *A Programmable Logic Controller (PLC) system*
We consider a PLC system with a two-out-of-three majority voting configuration whose layout is shown in Figure 6.21. The PLC system is intended to process a digital signal by means of suitable processing units, and to provide an output signal. To achieve a high level of fault tolerance, three different and independent channels (identified as channels Ch_A, Ch_B, and Ch_C, respectively) are used to process the signals and a two-out-of-three majority voting hardware device (voter) is used to vote on individual channels to produce the output. For each channel, a digital input unit (DI), a processing unit (CPU) and a digital output unit (DO) are employed. Redundancy is also present at the CPU level; indeed, each CPU receives the signal elaborated by the other CPUs

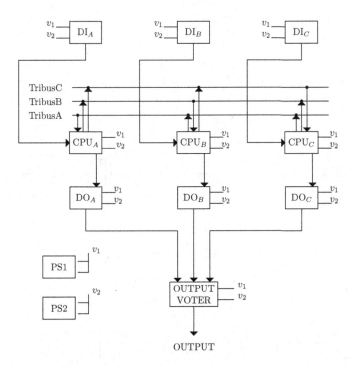

Figure 6.21   The PLC case study.

through special buses called Tribus$_A$, Tribus$_B$, and Tribus$_C$, respectively. Each CPU then performs a two-out-of-three majority vote to determine the correct input signal to be processed. A very similar fault tolerant design for a digital electronic control system intended to control an aircraft gas turbine engine is described in [37].

Finally, the system is completed by a redundant power supply system with two parallel independent power supply units (PS$_1$ and PS$_2$) connected to all the components. The FT for this PLC system is shown in Figure 6.22.

To model the signal input to each CPU, a two-out-of-three gate is used. In Figure 6.22, only the subtree related to Ch$_A$ is explicitly represented. The subtrees related to Ch$_B$ and Ch$_C$ are obtained in the same way by a change of symbols.

Since this system is meant to be applied in safety-critical applications, the analysis specifically addresses the system reliability, i.e., the probability that during a defined mission time the system operates continuously without failures. We assume non-repairable components with exponentially distributed TTF, with the failure rates given in Table 6.6. This table also shows the unreliability of each component at a mission time $t_M = 4 \times 10^5$ hr of system operation. Looking at the table, the CPUs are the most unreliable components.

The FT of Figure 6.22 has 65 MCSs, one of order 1 (indicating the voter being faulty) and the remaining of order 2. For instance, an MCS of order 2 that can immediately be derived from the FT of Figure 6.22 is {PS$_1$, PS$_2$} corresponding to the joint breakdown of both power supplies. The 65 MCSs are ranked in order of their unreliability in

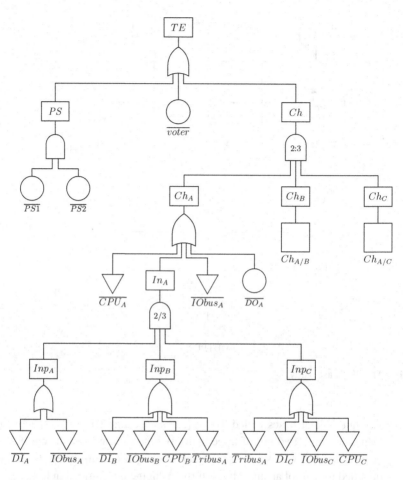

Figure 6.22   Fault tree for the PLC system.

Table 6.6 Failure rates and probabilities for the PLC components.

Component	Failure rate (f hr^{-1})	Unreliability $t_M = 4 \times 10^5$ hr
IObus	$\lambda_{IO} = 2.0 \times 10^{-9}$	0.00080
Tribus	$\lambda_{Tri} = 2.0 \times 10^{-9}$	0.00080
Voter	$\lambda_V = 6.6 \times 10^{-8}$	0.02605
DO	$\lambda_{DO} = 2.45 \times 10^{-7}$	0.09335
DI	$\lambda_{DI} = 2.8 \times 10^{-7}$	0.10595
PS	$\lambda_{PS} = 3.37 \times 10^{-7}$	0.12611
CPU	$\lambda_{CPU} = 4.82 \times 10^{-7}$	0.17535

Table 6.7. The second column in Table 6.7 gives the number of MCSs that can be obtained by rotating the symbols $(X, Y = A, B, C)$.

Table 6.7 The 65 MCSs for the PLC system ($X, Y = A, B, C$).

Order	Number	MCS	Unreliability
2	3	CPU_X, CPU_Y	0.03075
1	1	Voter	0.02605
2	6	CPU_X, DI_Y	0.01858
2	6	CPU_X, DO_Y	0.01637
2	1	PS_1, PS_2	0.01590
2	3	DI_X, DI_Y	0.01123
2	3	DO_X, DO_Y	0.00871
2	6	$CPU_X, IObus_Y$	1.40×10^{-4}
2	6	$CPU_X, Tribus_Y$	1.40×10^{-4}
2	6	$DI_X, IObus_Y$	8.47×10^{-5}
2	6	$DI_X, IObus_Y$	7.46×10^{-5}
2	6	$DO_X, Tribus_Y$	7.46×10^{-5}
2	12	$IObus_X, Tribus_Y$	6.39×10^{-7}

The probability of system failure at time $t_M = 1 \times 10^5$ hr is $P\{TE\} = 0.03031$. If we compute the importance indices, we see that for the structural importance index the basic events IObus and CPU have the same importance, while for the Birnbaum and criticality indices the CPU are the most important components.

We can analyze intermediate nodes of the FT, for instance node Inp_A. The failure probability of node Inp_A at the same mission time of $t_M = 1 \times 10^5$ hr is $P\{In_A = faulty\} = 0.9398 \times 10^{-2}$. Even if components CPU_B and CPU_C are the most unreliable basic events, if we compute the importance indices of the subtree emerging from node Inp_A, the ranking changes. For the structural as well as the Birnbaum indices the most important component is $Tribus_A$. For the criticality importance index the ranking varies as a function of the mission time. For short mission times (e.g., $t_M = 1000$ hr) the most critical component is $Tribus_A$, while for longer mission times (e.g., $t_M = 1 \times 10^5$ hr) the most critical components are the two CPUs, since their unreliability prevails.

Example 6.33 *Availability Analysis of a Blade Server System [38]*
The IBM BladeCenter blade server system is a commercial, high-availability system. A typical configuration consists of up to $n = 14$ separate blade servers mounted on a chassis with a cooling system with internal redundancy, two power domains each containing two redundant power supply modules and redundant network switches that are plugged into a mid-plane module [39]. The multiple network links are configured as two pairs of identical network switches each consisting of a pair of Ethernet switches and a pair of fiber channel switches.

Four major subsystems are the midplane (Md), n blade servers, cooling subsystem (C) and a power domain (P_1). The system fails if any one of these subsystems fail, so if the midplane, the cooling subsystem, or power domain 1 or k or more of the blade servers fail, the system fails. This is indicated by the top OR gate in the top level FT for the system shown in Figure 6.23. The inputs to the k-out-of-n gate are the 14 blade servers' failure events, divided into two groups: blades Bl_1 to Bl_6 are powered by power

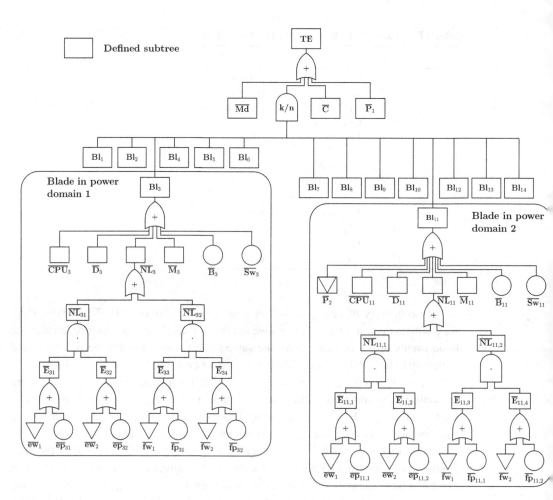

Figure 6.23   The top-level fault tree, $M_{\text{Top-BladeCenter}}$, for the IBM BladeCenter.

domain 1 and blades from Bl_7 to Bl_{14} are powered by power domain 2. Hence, the k-out-of-n redundancy of the BladeCenter is composed of two groups of distinct types of blocks, as considered in Section 4.5.1. Further, the cooling subsystem and the midplane are supplied by power domain 1, and hence the failure of P_1 results in the failure of the system as a whole. A blade server labeled Bl_i, $i = 1, 2, \ldots, 6$, goes down if its network link NL_i, its disk subsystem D_i, its memory subsystem M_i, its CPU subsystem CPU_i, its hardware base B_i or its operating system software Sw_i goes down. This is indicated by the OR gate in Figure 6.23 inside the box labeled "Blade in power domain 1." The remaining blade servers are powered by power domain 2, and hence the OR gate inside the box labeled "Blade in power domain 2" in Figure 6.23 has one more input labeled $\overline{P_2}$. In the figure, only blade failure events $\overline{Bl_3}$ and $\overline{Bl_{11}}$ are developed in full detail.

The square block representing the network link failure event $\overline{NL_3}$ for blade server failure event $\overline{Bl_3}$ in the FT of Figure 6.23 is developed further as a subtree. The two ethernet ports (e.g., ep_{31} and ep_{32}) and the two fiber channel ports (e.g., fp_{31} and fp_{32}) are specific for each blade, while the two ethernet switches (ew_1 and ew_2) and the two fiber channel switches (fw_1 and fw_2) are shared by all n blades (and are therefore repeated events). Each switch failure event is ORed with a port failure event on the blade. Thus, if either port on the blade or its switch goes down, communication on that link is down. If the blade cannot communicate through at least one port to each network type, then the blade is considered down. The resulting fault tree, $M_{\text{Top-BladeCenter}}$, shown in Figure 6.23 is the point of departure for several different elaborations of BladeCenter availability models.

Among the leaf nodes in the fault tree $M_{\text{Top-BladeCenter}}$, $\overline{B}_i$ is the failure event of the hardware base for blade i, and $\overline{Sw}_i$ is the software failure event for blade i. These units do not have any redundancy. The cooling subsystem, midplane, power supply subsystem, CPU subsystem, memory subsystem and hard disk subsystem are all internally redundant. For now we will assume that their redundancy can be modeled by a simple fault tree as shown in Figure 6.24, where Y denotes a generic block that stands successively for C, Md, P_1, P_2, CPU, M or D.

The models presented in this chapter are based on the simplified assumption of statistical independence across all components. The square blocks in Figure 6.23 for events $\overline{C}$, $\overline{Md}$, $\overline{P}_1$, $\overline{P}_2$, $\overline{CPU}$, $\overline{M}$, and $\overline{D}$ are solved separately for their steady-state unavailabilities. These are supplied to the leaf nodes in the fault tree $M_{\text{Top-BladeCenter}}$.

The independence assumption within the above subsystems ($\overline{C}$, $\overline{Md}$, $\overline{P}_1$, $\overline{P}_2$, $\overline{CPU}$, $\overline{M}$, and $\overline{D}$) will be dropped in Chapter 9, where more refined state-space models for the different blocks are presented. The solution of the complete system will be obtained hierarchically in Example 18.3, where the Markov models of these subsystems from Section 9 will be solved and their solutions supplied to the top-level fault tree $M_{\text{Top-BladeCenter}}$. Such hierarchical model specification *and solution* are supported by the software package SHARPE [15].

The overall fault tree (with the independence assumption for now) is solved by varying the parameters k and n in the k-out-of-n gate. In particular, assuming $k = 1$ and $n = 1$, the availability of a single blade is studied in [38]. Note that since there are two types of input to the k-out-of-n gate, at present the way to solve it is to assume that all the n inputs are non-identical. To efficiently solve for such fault trees, our algorithm, described in Section 4.5.1, that considers two different groups of inputs to such a gate is implemented in the SHARPE software package.

The estimated MTTFs for each basic event (component) have been found to range between a low and a high value [38], as reported in Table 6.8.

For all the components we assume that the downtime is composed of two elements: a logistic time, called mean service response time (MTTRSP), and a component

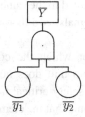

Figure 6.24   Subtree for the redundant parallel blocks $\overline{Md}$, $\overline{C}$, $\overline{P_1}$, $\overline{P_2}$, $\overline{CPU}$, $\overline{M}$, and $\overline{D}$.

Table 6.8 Low and High MTTFs for the elementary blocks of the FT of Figure 6.23.

Component		Low MTTF$_l$ (hr)	High MTTF$_h$ (hr)
Md	Midplane	310 000	420 000
C	Cooling	3 100 000	4 200 000
P	Power supply	670 000	910 000
B	Base blade	220 000	300 000
CPU	Microprocessor unit	2 500 000	3 400 000
M	Memory unit	480 000	660 000
D	RAID disk unit	200 000	350 000
ew	Ethernet switch	120 000	160 000
ep	Ethernet port	6 200 000	8 400 000
fw	Fiber channel switch	320 000	440 000
fp	Fiber channel port	1 300 000	1 800 000

replacement time (MTTRC). The MTTRSP accounts for the time needed for the repair service to arrive on site after notification, and depends on the maintenance contract signed by the organization. Here we have assumed that MTTRSP = 2.5 hr. The replacement times, instead, depend on the specific component type, but for now we assume MTTRC = 0.5 hr, leading to MTTR = 3 hr for all components. For each component we can compute a low and a high value for the steady-state availability as a function of the low and high MTTF values of Table 6.8:

$$A_{i(l/h)} = \frac{\text{MTTF}_{i(l/h)}}{\text{MTTF}_{i(l/h)} + \text{MTTR}_i}.$$

We leave the actual computation of the overall system availability for various values of k and n as an exercise for the reader.

Example 6.34   *Safety Analysis for a Turbine Control System [40]*
This case study is intended to assess the safety characteristics of a digital controller for a gas turbine. The digital control system is composed of two subsystems, the *main controller* (MC) and the *backup* (BU), whose hardware structures are depicted in Figure 6.25.

The main controller provides control and shutdown functions, and carries out two main tasks: *control*, to ensure dependability, and *protection*, to ensure safety. The backup unit provides redundant protection with respect to two critical events only: over-speed and over-temperature. The backup unit has a CPU independent of the main controller and uses a separate power supply circuit (connected to the same supply inlet). The MC and BU units share the following transducer signals: two thermocouples and one speed probe.

A safety-critical situation is assumed to be one for which the controller does not provide the correct control function, and the backup unit does not provide the protection function. We adopt as TE the occurrence of a safety-critical failure. The FT representation of the TE is shown in Figure 6.26. We assume that the elementary blocks have exponentially distributed TTFs with the respective failure rates shown in Table 6.9.

Table 6.9 Failure rates for the elementary blocks of the digital controller of Figure 6.25.

Component		Failure Rate (f hr^{-1})
I/O	I/O bus	$\lambda_{IO} = 2.0 \times 10^{-9}$
DI	Digital input	$\lambda_{DI} = 3.0 \times 10^{-7}$
AI	Analog input	$\lambda_{AI} = 3.0 \times 10^{-7}$
MEM	Memory	$\lambda_{M} = 5.0 \times 10^{-8}$
CPU	32-bit microprocessor	$\lambda_{CPU} = 5.0 \times 10^{-7}$
DO	Digital output	$\lambda_{DO} = 2.5 \times 10^{-7}$
AO	Analog output	$\lambda_{AO} = 2.5 \times 10^{-7}$
RO	Relay output	$\lambda_{RO} = 2.5 \times 10^{-7}$
WD	Watchdog relay	$\lambda_{WD} = 2.5 \times 10^{-7}$
therm	Thermocouple	$\lambda_{th} = 2.0 \times 10^{-9}$
speed	Speed probe	$\lambda_{sp} = 2.0 \times 10^{-9}$
PS	Power supply inlet	$\lambda_{PS} = 3.0 \times 10^{-7}$
S_{MC}	Supply circuit (MC)	$\lambda_{Sm} = 3.0 \times 10^{-7}$
S_{BU}	Supply circuit (BU)	$\lambda_{Sb} = 3.0 \times 10^{-7}$

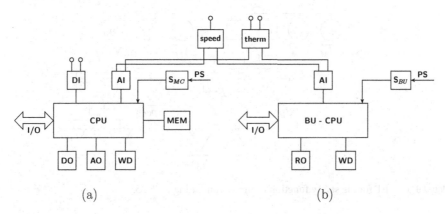

(a)   (b)

Figure 6.25   Hardware structure of (a) the main controller, and (b) the backup unit.

The analysis found $M = 43$ MCSs, 2 of them of order 3 and 41 of order 4. The 2 MCSs of order 3 contribute 20.2% to the TE unreliability, while the 41 of order 4 contribute the remaining 79.8%. The seven most critical MCSs sorted by their unreliability are reported in the inset of Figure 6.26.

The TE unreliability is reported in Figure 6.27. However, since the failure rate values are uncertain, we have computed a sensitivity band for the TE unreliability by incrementing the failure rates by a quantity equal to (+10%) [curve labeled λ_{max} (+10%)] and by decrementing the failure rates by a quantity equal to (−10%) [curve labeled λ_{min} (−10%)]. The results obtained are comparatively plotted in Figure 6.27. The area between the curves λ_{max} and λ_{min} represents the uncertainty band in the TE unreliability due to the uncertainty in the input parameters.

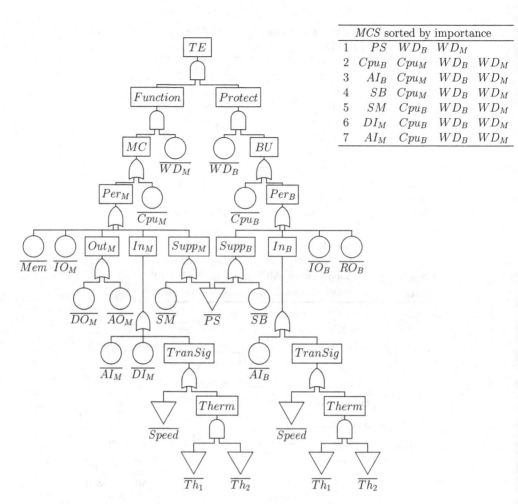

MCS sorted by importance			
1	PS	WD_B	WD_M
2	Cpu_B Cpu_M	WD_B	WD_M
3	AI_B Cpu_M	WD_B	WD_M
4	SB Cpu_M	WD_B	WD_M
5	SM Cpu_B	WD_B	WD_M
6	DI_M Cpu_B	WD_B	WD_M
7	AI_M Cpu_B	WD_B	WD_M

Figure 6.26   FT for the safety function of the system of Figure 6.25.

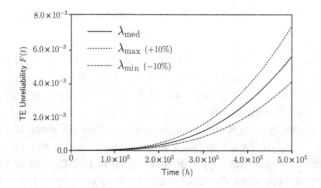

Figure 6.27   Unreliability and uncertainty band for the system of Figure 6.25.

Problems

6.19 Reproduce the results for the PLC (Example 6.32) using the SHARPE software package.

6.20 Solve the fault tree of the BladeCenter (Example 6.33) using the SHARPE software package. Obtain the system availability for all values of k.

6.21 Reproduce the results for the turbine (Example 6.34) using the SHARPE software package.

6.8 Attack and Defense Tree

Dependability and security have been considered in the past as separate disciplines, each one with its specific language, attributes, modeling, and analysis techniques [41]. However, as the two disciplines evolved [42], the security community has progressively moved toward more formal and quantitative modeling techniques inspired by or derived from the dependability community [43, 44], augmenting the interchange and the cross-fertilization between the two areas [45].

Attack trees (AT) and their variants and extensions utilize the logic and analysis capabilities of FTs; the different name is simply a historical and terminological question. Attack trees were introduced in [46], as a visual and systematic methodology for security assessment. Since then, they have been widely used both in research [47, 48] and in industrial and control applications [49].

The root node of the AT (the TE in an FT) is the final goal of the attack, and the AT structure is a binary AND–OR tree that represents different ways of achieving that goal. Each basic event (or leaf) of the AT is an atomic exploit. When each leaf is labeled with the probability that the corresponding exploit is successful, then the model can be used to evaluate the probability that the attacker attains the final goal. Attack trees can also be extended by adding weights or rewards in terms of costs or impacts of an attack. The model can be further enriched by means of a sequence of detection and mitigation events that taken together constitute *countermeasures*. Such structures are called attack and countermeasure trees (ACTs) [48]. An attack can propagate toward the top of the ACT only if the countermeasure designed to block it fails. Implementing the countermeasures has an investment cost that can be compared with the monetary benefit in terms of probability of reduced damage. An optimal strategy to select the countermeasures on an ACT is discussed in [50].

6.8.1 Countermeasures

To respond to an attack, the attack should first be detected and then actions should be taken to prevent or reduce its effects. The defender action can be divided into a detection phase and a mitigation phase, so that the success of an individual attack A can be represented as the ACT of Figure 6.28(a), where D is the detection event and M is the mitigation event. Denoting by p_A the probability of an atomic exploit, by p_d the

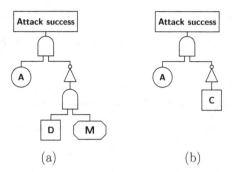

Figure 6.28 Attack and countermeasure tree. (a) Detection and mitigation; (b) countermeasure.

probability of a successful detection and by p_m the probability of a successful mitigation action, the probability of the TE (successful attack) becomes

$$P_{TE} = p_A (1 - p_d p_m).$$

The combined action of detection and mitigation is what we call countermeasure [48]. With this definition, the ACT of Figure 6.28(a) can be redrawn as in Figure 6.28(b), where the terminal leaf C is the countermeasure.

The qualitative and quantitative analysis of an ACT follows the same pattern described in Sections 6.3 and 6.4 for an FT. The ordered list of the MCSs provides the minimal combinations of basic exploits that lead to a successful attack. Further, if the probability of individual basic events is known, the probability of each MCS and of a globally successful attack in an ACT may be computed with or without countermeasures. Furthermore, criticality indices with or without countermeasures can be evaluated. The real benefit of implementing a countermeasure can be evaluated by computing the gain in the probability of reaching the TE without and with the countermeasure. Given a countermeasure C_i, the gain $\Delta P_{TE_{C_i}}$ is calculated as the difference between the probability of reaching the TE without the countermeasure C_i and with the countermeasure C_i.

6.8.2 Weighted Attack Tree (WAT)

A more realistic and effective analysis of an attack sequence can be obtained by including in the scenario the cost of implementing the attack and the impact (in terms of monetary damage) that the attack may cause on the attacked system and/or on the services it delivers. Attack cost, or simply cost, is the cost of implementing a specific attack, and impact cost, (or simply impact), is the monetary damage caused by the attack. Cost and impacts can be assigned at any level of an ACT, and as the attack proceeds they combine to determine the final cost and the final impact of a successful attack.

In [48], the following propagation rules for the cost and impact in an ACT are defined and implemented:

- The cost (impact) attached to the output of an AND gate is the sum of the costs (impacts) of its input elements. The rationale behind this propagation rule is that all the inputs must be true for an AND gate to be true, and hence their costs (impacts) are additive.

Table 6.10 Rules for propagating cost/impact in a WAT.

Gate type	Attack cost	Attack impact
AND	$\sum_{i=1}^{n} r_i$	$\sum_{i=1}^{n} r_i$
OR	$\min_{i=1}^{n} r_i$	$\max_{i=1}^{n} r_i$

- The cost of the output of an OR gate is the minimum of the costs among its inputs, while the impact of the output of an OR gate is, on the contrary, the maximum of the impacts among its inputs. The rationale behind this propagation rule is that in front of a choice represented by an OR gate, the most convenient strategy for the attacker and the worst scenario for the defender is the combination of minimum cost with maximum impact.

Given a gate (AND or OR) with n inputs whose cost/impact is r_i ($i = 1, 2, \ldots, n$), the above rules can be formalized as in Table 6.10. The analysis of a WAT may follow the same techniques used for weighted networks in Section 5.3, for instance by using MTBDD [51].

Combining cost and impact with a probabilistic analysis provides an enriched view of the attack–defense strategy. Implementing countermeasures on an ACT has its own cost, which we refer to as security investment. WATs allow us to make an economic balance of the investment, balancing the damage that a successful attack will cause to the system and its services with the probability that the attack is successful with or without countermeasures.

An index, called the *return on investment* (ROI), is defined in [48] with respect to a single countermeasure C_i as follows:

$$\mathrm{ROI}_{C_i} = \frac{\text{profit from } C_i - \text{cost implementing } C_i}{\text{cost implementing } C_i}$$
$$= \frac{I_{\mathrm{TE}} \times \Delta P_{\mathrm{TE}_{C_i}} - \mathrm{cost}_{C_i}}{\mathrm{cost}_{C_i}}. \tag{6.27}$$

In Eq. (6.27), I_{TE} is the maximum impact produced by an attack, which can be evaluated by propagating along the ACT the individual impacts according to the rules of Table 6.10. $\Delta P_{\mathrm{TE}_{C_i}}$ is the probability gain obtained by implementing countermeasure C_i, and cost_{C_i} is the investment cost of implementing countermeasure C_i.

Example 6.35 *ACT for a SCADA System*

Analysis of the security of a SCADA system in the form of an AT was first proposed in [52], and the subsequent analysis of the ACT obtained by the addition of countermeasures was discussed in [48]. The countermeasures are of two types. The *detection* event prevents the attack exploit being successful with an assigned probability; the *mitigation* event reduces the probability of an undetected attack exploit being successful. In the ACT of Figure 6.29 [48], only the mitigation events are explicitly

Table 6.11 Countermeasure probabilities for the ACT of Figure 6.29.

ACT countermeasure	Probability	ACT countermeasure	Probability
M_{sw}	0.25	M_{G_1}	0.4
M_{G_2}	0.5	M_{G_3}	0.6

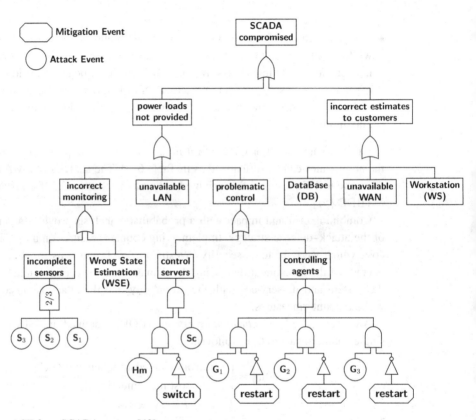

Figure 6.29   ACT for a SCADA system [48].

represented, but a detection probability of $p_{det} = 0.6$ is assumed for all the atomic attack exploits.

A parametric sensitivity analysis was also performed, assuming that all the atomic attack exploits have the same success probability ranging from 0 to 1. The probabilities assigned to the mitigation countermeasures are shown in Table 6.11.

The probability of a successful attack for the ACT of Figure 6.29 is given in Figure 6.30. The solid curve gives the probability of a successful attack without countermeasures, while the dashed curve gives the probability of a successful attack when the countermeasures (detection plus mitigation) are activated.

An economic balance between security investment and reduced damage is provided in [48], where it is shown how the ROI index increases when the cost of a countermeasure decreases or its probability of being active increases.

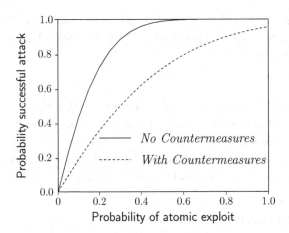

Figure 6.30   Probability of successful attack for the ACT of Figure 6.29 [48].

6.9   Multi-State Fault Tree

We have already observed in Section 5.4 that there are many situations in which individual components are better modeled with more than two states. Similarly to what was done in Section 5.4.1 for networks, we can define a multi-state FT (MFT) as an FT in which the basic events have multiple exhaustive and mutually exclusive states to which a probability value is attached. The system as a whole (the TE) can similarly have more than two states.

Definition and analysis of MFTs can be found back in [53–56]. More recently, analysis of MFTs has been carried out by resorting to BDDs in [57] and to MDDs in [58, 59].

By adopting the same notation as in Section 5.4.1, we assume n basic events numbered from 1 to n, and we indicate with x_i the ith basic component that has $(m_i + 1)$ states numbered from 0 to m_i, with p_{ij} the probability of basic event i ($i = 1, 2, \ldots, n$) in state j ($j = 0, 1, 2, \ldots, m_i$). The MFT states can be represented by means of the component state vector $x = (x_1, x_2, \ldots, x_n)$ that has n entries (one for each variable) where each entry x_i indicates the state of the ith variable. Hence, the component state vector has m values with m given by:

$$m = \prod_{i=1}^{n} (m_i + 1).$$

The whole system can be represented by s discrete conditions or states, numbered from 1 to s. To completely specify the model, a multi-valued function needs to be defined that maps the m values of the vector x into the s states of the system. The MFT provides a method to define this mapping, adopting a rather intuitive extension of the way in which binary FTs are constructed and analyzed.

We will discuss MFTs through a benchmark example that was first introduced in [57] and solved by adopting a BDD-based algorithm. The same benchmark example has been solved in subsequent papers [58, 59] by resorting to MDD (defined in Section 5.4.2). In [60], the same example is solved using MFT, CTMC, and SRN. We only discuss the MDD representation here.

Example 6.36   Figure 6.31 shows the system, taken from [57], with two boards, B_1 and B_2, where each board has a processor and a memory. The memories, M_1 and M_2, are shared by both the processors, P_1 and P_2.

If the processor and memory on the same board can fail separately, but s-dependently, each board B_i ($i = 1, 2$) can be considered as a single component with $m_i + 1 = 4$ states:

- State B_{i0}: both P and M are down.
- State B_{i1}: P is functional, but M is down.
- State B_{i2}: M is functional, but P is down.
- State B_{i3}: both P and M are functional.

We then assume that the system can be in $s = 3$ possible mutually exclusive conditions or states that we define as follows:

- State s_1: at least one processor and both of the memories are functional.
- State s_2: at least one processor and exactly one memory are functional.
- State s_3: no processor or no memory is functional.

Figure 6.32 shows three MFTs, one for each system state s_i as the top event. The basic events B_{ij} represent the board B_i being in state j. The structure of the MFT indicates the combination of events whose occurrence entails the TE.

Each MFT of Figure 6.32 can be converted into a corresponding MDD with two terminal leaves: terminal leaf 1 indicates the logical combination of basic events for which the TE is true; terminal leaf 0 indicates the logical combination of basic events for which the TE is false.

The MDD representation for state s_1 of Figure 6.32(a) is shown in Figure 6.33. We adopt the same style as in Section 5.4.1, where each node of the MDD encodes a multi-state variable B_i. The order of the variables in the construction of the MDD affects the complexity of the MDD. In the case of Figure 6.33, the chosen sequence is $B_1 \prec B_2$.

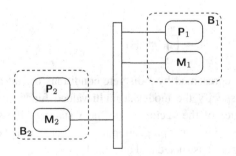

Figure 6.31   The system structure of Example 6.36.

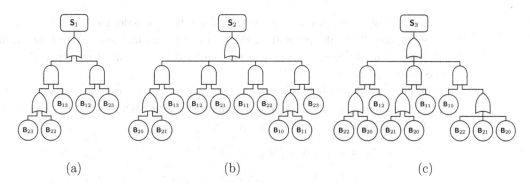

(a)   (b)   (c)

Figure 6.32   MFTs for the three states of the system of Figure 6.31.

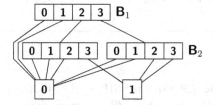

Figure 6.33   MDD representing the MFT of state s_1 in Figure 6.32(a).

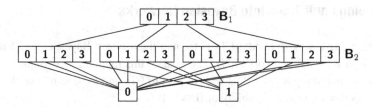

Figure 6.34   MDD representing the MFT of state s_2 in Figure 6.32(b).

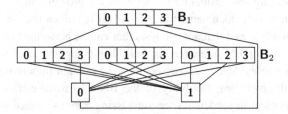

Figure 6.35   MDD representing the MFT of state s_3 in Figure 6.32(c).

Similarly, from the MFT of Figure 6.32(b) we obtain the MDD of Figure 6.34, and from the MFT of Figure 6.32(c) we obtain the MDD of Figure 6.35.

The probability of the event TE = true in the MFT is represented by the probability of terminal leaf 1 in the corresponding MDD. The probabilities of the nodes of an MDD can be computed following the procedure illustrated in Section 5.4.4, proceeding recursively in a top-down fashion, starting from the root node, until the terminal node 1 is reached. Since the paths from the root to the terminal node 1 are disjoint, the probability of the terminal node can be computed as the sum of the probabilities of the disjoint paths.

In this way, the probability of the system states P_{s_1}, P_{s_2}, and P_{s_3} can be computed as the probability of terminal node 1 in the respective MDDs of Figures 6.33, 6.34, and 6.35. The state probabilities are given by:

$$P_{s_1} = p_{12}p_{23} + p_{13}(p_{22} + p_{23}),$$

$$P_{s_2} = p_{10}p_{23} + p_{11}(p_{22} + p_{23}) + p_{12}p_{21} + p_{13}(p_{20} + p_{21}),$$

$$P_{s_3} = p_{10}(p_{20} + p_{21} + p_{22}) + p_{11}(p_{20} + p_{21}) + p_{12}(p_{20} + p_{22}),$$

where the indices (i, j) in p_{ij} are ordered as the variables of the MDD and indicate the probability of board B_i in state j.

Problems

6.22   Solve the MFTs in Example 6.36 using the SHARPE software package. Either use some reasonable parameter values or refer to [60] for parameter values.

6.10   Mapping Fault Trees into Bayesian Networks

An alternative way to handle FTs is to map them into a Bayesian belief network, or, simply, a Bayesian network (BN). The goal of this mapping is twofold: to exploit the greater modeling power of BNs with respect to FTs and to benefit from the augmented analysis flexibility and capability of BNs [61–64].

A classical introduction to BNs is given in [65], and an extensive overview of possible applications in various technical fields is provided in [66]. The use of BNs in dependability modeling and evaluation has witnessed a growing interest in the past few years and is extensively documented in [67–70]. It is known that any FT can be directly mapped into a BN, and that any analysis that can be performed on the FT can be performed by means of BN inference, as well. The computation of the posterior probabilities given some evidence can be specialized to obtain importance measures [61]. Furthermore, the modeling flexibility of the BN formalism can accommodate various kinds of statistical dependencies, encompassing various extensions that have been proposed for FTs like the DFT formalism [71] that will be presented in Section 16.5.

6.10.1   Bayesian Network Definition

A BN is a pair $N = \langle\langle V, E\rangle, P\rangle$, where $\langle V, E\rangle$ are the nodes and the edges of a directed acyclic Graph (DAG), respectively [61]. Discrete random variables $[X_1, X_2, \ldots, X_n]$ are assigned to the n nodes, while the edges E represent the causal probabilistic relationships among the nodes. With these assignments, $P = p_{X_1,X_2,...,X_n}\{x_1,x_2,...,x_n\}$ is the joint pmf of the n random variables $[X_1,X_2,...,X_n]$ over V.

In a BN, we can then identify a qualitative part (the structure represented by the DAG) and a quantitative part (the set of conditional probabilities). The qualitative part represents a set of conditional independence assumptions that can be captured through a graph-theoretic notion called *d-separation* [65]. This notion has been shown to model the usual set of independence assumptions that a modeler assumes when considering each edge from variable X to variable Y as a direct dependence (or as a cause–effect relationship) between the events represented by the variables.

The quantitative analysis is based on the conditional independence assumption. Given three random variables X, Y and Z, X is said to be conditionally independent from Y given Z if the conditional joint pmf can be written as

$$p_{X,Y|Z}\{x,y|z\} = p_{X|Z}\{x|z\} \times p_{Y|Z}\{y|z\}.$$

Because of these assumptions, the quantitative part is completely specified by considering the probability of each value of a variable conditioned by every possible instantiation of its parents (i.e., by considering only local conditioning). These local conditional probabilities are specified by defining, for each node, a *conditional probability table* (CPT). The CPT contains, for each possible value of the variables X_i associated with node i, all the conditional probabilities with respect to all the combinations of values of the variables Parent(X_i) associated with the parent nodes. Variables having no parents are called *root variables*, and marginal prior probabilities are associated with them. According to these assumptions (d-separation and conditional independence), the joint pmf of the n random variables of the BN can be factorized as in Eq. (6.28):

$$p_{X_1,X_2,...,X_n}\{x_1,x_2,...,x_n\} = \prod_{i=1}^{n} p_{X_i|\text{Parent}(X_i)}\{x_i|\text{Parent}(x_i)\}. \qquad (6.28)$$

The joint pmf P in (6.28) provides the complete probabilistic description of the BN. Then, the basic inference task of a BN consists in computing the posterior probability of any given subset of *query* variables Q given the observation of another subset of variables E, called the *evidence*, represented as instantiation of some of the variables to one of their admissible values. If E is the empty set, the prior or unconditional probability of Q is computed.

The general problem of computing posterior probabilities on an arbitrary BN is known to be NP-hard [72]; however, the problem reduces to polynomial complexity if the structure of the net is such that the underlying undirected graph contains no cycles. Even if this restriction is not often satisfied in practice, research on efficient probabilistic computation has produced considerable results showing that acceptable computation can be performed even in networks with general structure and hundreds of nodes [73]. Several commercial and freeware software tools are available for the construction and analysis of BNs [70].

6.10.2   Mapping an Algorithm from FT to BN

We recall that the basic assumptions of the standard FT formalism are that events (usually binary) are statistically independent, and events and their causes are connected by logical AND and OR gates.

We first show how the Boolean gates can be converted into an equivalent BN, and then how some restrictive assumptions behind the FT can be relaxed, and what kind of quantitative analysis can be performed.

Figure 6.36(a) shows the conversion of an OR gate into equivalent nodes in a BN, and Figure 6.36(b) shows the similar conversion of an AND gate. The root nodes $\overline{A}$ and $\overline{B}$ are assigned prior probabilities (coincident with the probability values assigned to the corresponding basic nodes in the FT), and child node $\overline{S}$ is assigned its CPT. Since the OR and AND gates represent deterministic causal relationships, all the entries of the

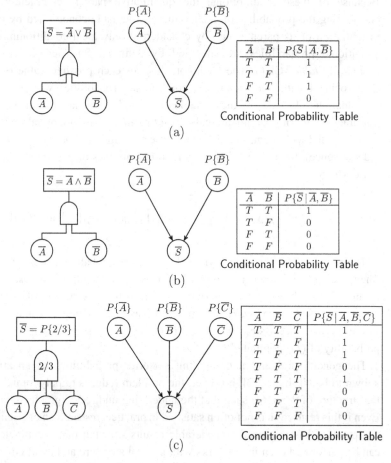

(a)

(b)

(c)

Figure 6.36   Conversion of Boolean gates into the corresponding BN. (a) OR gate; (b) AND gate; (c) two-out-of-three gate.

CPT corresponding to $P\{\overline{S}|\overline{A},\overline{B}\}$ are either 0's or 1's. Figure 6.36(c) shows the similar conversion of a k-out-of-n gate with $n = 3$ and $k = 2$.

The conversion algorithm from an FT to a BN proceeds along the following steps:

- For each *leaf node* of the FT, create a *root node* in the BN; in the case of repeated events create just one root node in the BN.
- Assign to root nodes in the BN the prior probability of the corresponding basic events in the FT.
- For each *gate* of the FT, create a corresponding *node* in the BN.
- Connect nodes in the BN as corresponding gates are connected in the FT.
- For each gate (OR, AND or k-out-of-n) in the FT, assign the equivalent CPT to the corresponding node in the BN as in Figure 6.36.

Due to the very special nature of the gates appearing in an FT, non-root nodes of the BN are actually deterministic nodes and not random variables; the corresponding CPT can be assigned automatically following the examples of Figure 6.36.

Example 6.37   *A Multiprocessor System*

The above mapping algorithm is illustrated in this example concerning a multiprocessor system with redundancy as shown in Figure 6.37 and taken from [11, 61]. The system is composed of a bus N connecting two processors P_1 and P_2 each having access to a local memory bank (respectively M_1 and M_2) and, through the bus, to a shared memory bank M_3, so that if the local memory bank fails, the processor can use the shared one. Each processor is connected to a mirrored disk unit. If one of the disks fails, the processor switches on the mirror. The whole system is functional if and only if the bus N is functional and one of the processing subsystems is functional. Figure 6.37 also shows the partitioning into logical subsystems, i.e., the processing subsystems S_i ($i = 1, 2$), the mirrored disk units D_i ($i = 1, 2$) and the memory subsystems M_{i3} ($i = 1, 2$).

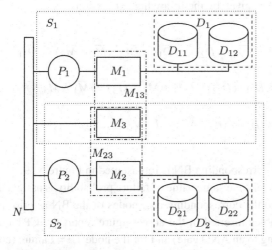

Figure 6.37   A redundant multiprocessor system [11, 61].

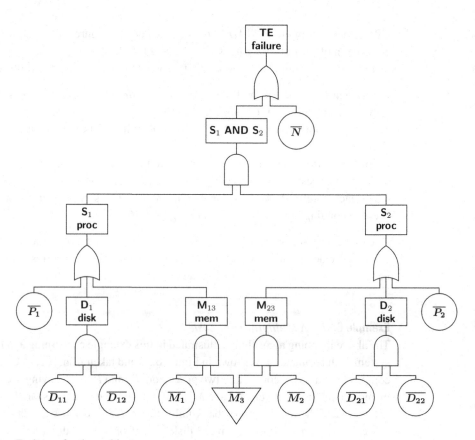

Figure 6.38   Fault tree for the multiprocessor system.

The FT for this system is shown in Figure 6.38. The logical expression of the TE as a function of the MCSs is given by the following expression:

$$\text{TE} = \overline{N} \lor (\overline{D}_{11} \land \overline{D}_{12} \land \overline{D}_{21} \land \overline{D}_{22}) \lor (\overline{D}_{11} \land \overline{D}_{12} \land \overline{M}_2 \land \overline{M}_3) \lor (\overline{D}_{11} \land \overline{D}_{12} \land \overline{P}_2)$$

$$\lor (\overline{M}_1 \land \overline{M}_3 \land \overline{D}_{21} \land \overline{D}_{22}) \lor (\overline{M}_1 \land \overline{M}_2 \land \overline{M}_3) \lor (\overline{M}_1 \land \overline{M}_3 \land \overline{P}_2)$$

$$\lor (\overline{P}_1 \land \overline{D}_{21} \land \overline{D}_{22}) \lor (\overline{P}_1 \land \overline{M}_2 \land \overline{M}_3) \lor (\overline{P}_1 \land \overline{P}_2).$$

The structure of the corresponding BN is represented in Figure 6.39.

To quantify both models, the same value of failure probability is assigned to the basic events of the FT and to the corresponding root nodes of the BN as prior probabilities. As an example, the tables inserted in the above figure report the CPT entries for the node S_{12} (corresponding to an AND gate) and for the node TE = Failure (corresponding to an OR gate).

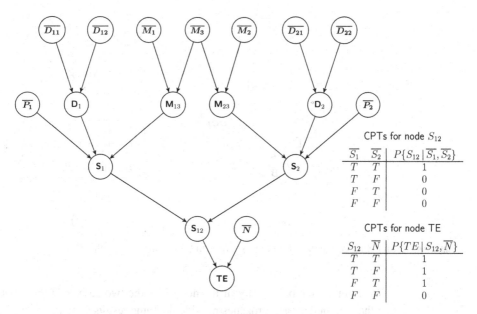

CPTs for node S_{12}

$\overline{S_1}$	$\overline{S_2}$	$P\{S_{12} \mid \overline{S_1}, \overline{S_2}\}$
T	T	1
T	F	0
F	T	0
F	F	0

CPTs for node TE

S_{12}	$\overline{N}$	$P\{TE \mid S_{12}, \overline{N}\}$
T	T	1
T	F	1
F	T	1
F	F	0

Figure 6.39   Bayesian network for the multiprocessor system.

The mapping procedure we have described shows that any FT can be naturally converted into a BN. However, BNs are a more general formalism than FTs; for this reason, there are several modeling aspects underlying BNs that can make them very appealing for dependability analysis [61, 70]. In the following, we examine a number of extensions to the standard FT methodology, and we show how they can be captured in the BN framework.

6.10.3 Probabilistic Gates: Common Cause Failures

Differently from FTs, the dependence relations among variables in a BN are not restricted to being deterministic, so that BNs are able to model uncertainty in the behavior of the gates by suitably specifying the conditional probabilities in the CPT entries. Probabilistic gates may reflect imperfect knowledge of the system behavior, or may avoid the construction of a more detailed and refined model. A typical example is the incorporation of CCF analysis [74].

CCF refers to dependent failures where simultaneous (or near-simultaneous) multiple failures result from a single shared cause. Common cause failures are usually modeled in FTs by adding an OR gate, directly connected to the TE, in which one input is the system failure, and the other input the CCF leaf to which the probability of failure due to common causes is assigned. In the BN formalism such additional constructs are not necessary, since the probabilistic dependence can be included in the CPT. Figure 6.40 shows an AND gate with CCF and the corresponding BN. The value q_{CCF} is the probability of failure of the system due to common causes, when one or both components are up.

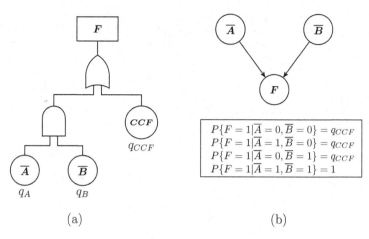

Figure 6.40   Modeling common cause failures: (a) in an FT; (b) in a BN (with CPT).

We compute the probability of the node F in the two cases of Figure 6.40, to show that the two modeling approaches provide the same results:

$$P\{F_{FT} = 1\} = P\{(\overline{A} \wedge \overline{B}) \vee \overline{CCF}\} = q_{CCF} + q_A q_B - q_{CCF} q_A q_B,$$

$$P\{F_{BN} = 1\} = q_A q_B + q_{CCF}(q_A(1 - q_B) + (1 - q_A)q_B + (1 - q_A)(1 - q_B))$$

$$= q_{CCF} + q_A q_B - q_{CCF} q_A q_B.$$

6.10.4   Noisy Gates

Of particular interest for reliability aspects is one peculiar modeling feature often used in building BN models: *noisy gates*. As mentioned in Section 6.10.1, when specifying CPT entries one has to condition the value of a variable on every possible instantiation of its parent variables, making the number of required entries exponential with respect to the number of parents. By assuming that the node variable is influenced by any single parent independently of the others (causal independence), noisy gates reduce this effort by requiring a number of parameters linear in the number of parents.

Consider for example the subsystem S_1 in Figure 6.37: it fails if the disk unit D_1 or the processor P_1 or the memory subsystem M_{13} fail. Since node S_1 in the BN of Figure 6.39 has three parent nodes, this implies that the modeler has to provide eight CPT entries in order to completely specify this local model.

Of course, if this local model is a strictly deterministic Boolean gate (as the OR gate in the example), the construction of the CPT can be done automatically following the rules of Figure 6.36. Consider now the case where the logical OR interaction is *noisy* or probabilistic: even if one of the components of S_1 fails, there is a (possibly small) positive probability that the subsystem works. This corresponds to the fact that the system may maintain some functionality, or may be able to reconfigure (with some probability of reconfiguration success) in the presence of particular faults. Suppose, for example, that S_1 can work with probability 0.01 even if the disk unit D_1 has failed.

Moreover, suppose that the system can also recover if other components of S_1 fail, but with a smaller probability (e.g., 0.005).

By adopting a noisy-OR model we can then set the following inhibitor probabilities:

$$q_{D_1} = P\{S_1|\overline{D_1},P_1,M_{13}\} = 0.01,$$

$$q_{P_1} = P\{S_1|D_1,\overline{P_1},M_{13}\} = 0.005,$$

$$q_{M_{13}} = P\{S_1|D_1,P_1,\overline{M_{13}}\} = 0.005.$$

Due to the causal independence hypothesis, we can compute the probability that the subsystem S_1 works (or fails) when both D_1 and P_1 have failed and M_{13} is still working as:

$$P\{S_1|\overline{D_1},\overline{P_1},M_{13}\} = 0.01 \times 0.005 = 0.00005,$$

$$P\{\overline{S_1}|\overline{D_1},\overline{P_1},M_{13}\} = 1 - (0.01 \times 0.005) = 0.99995.$$

As one might expect, the probability that the subsystem S_1 fails when both components D_1 and P_1 are down is larger (but less than 1) than the probability that S_1 fails given that only D_1 (or only P_1) has failed.

Noisy-AND constructs can similarly be used to introduce uncertainty in AND gates. Consider, for instance, the mirrored disk subsystem D_1 that fails if both disks D_{11} and D_{12} fail. However, in a more refined view of the model, we can suppose that the mirrored connection is not perfect, and there is a small probability (e.g., 0.001) that the disk subsystem D_1 fails when a single disk is up (e.g., $p_{\overline{D_{11}}} = P\{\overline{D_1}|D_{11},\overline{D_{12}}\} = p_{\overline{D_{12}}} = P\{\overline{D_1}|\overline{D_{11}},D_{12}\} = 1 \times 10^{-3}$). By exploiting noisy-AND interaction, we can then compute the probability of D_1 failing when both disks are functional as $P\{\overline{D_1}|D_{11},D_{12}\} = 10^{-3} \times 10^{-3} = 1 \times 10^{-6}$. Noisy gates can be easily combined with common cause failures by properly adjusting the entries of the CPT.

6.10.5 Multi-State Variables

Multi-state components, as considered in Section 6.9, can be easily accommodated in the framework of BN. A component with $m_i + 1$ states can be represented by a discrete BN variable with $m_i + 1$ values and with a marginal probability assigned to each value equal to the probability of the corresponding state. The CPT of a non-terminal node has a number of entries equal to the cross product of the values of the parent nodes. In the case of n-ary variables, it is also possible to generalize the noisy gate constructs by avoiding, in some cases, the specification of the whole CPT.

Example 6.38 *A Safety-Critical Switching Device*
Consider the component with two competing failure modes introduced in Example 3.1. The working condition is identified with the symbol w, the failure mode *fail-open* with (o) and the failure mode *fail-short* with (s). The BN of Figure 6.41(b) can include n-ary variables by assigning the proper marginal probability to each variable value and

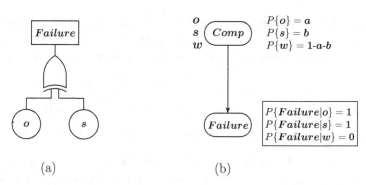

Figure 6.41   A component with two exclusive failure modes. (a) Fault tree; (b) Bayesian network (with marginal probabilities and CPT).

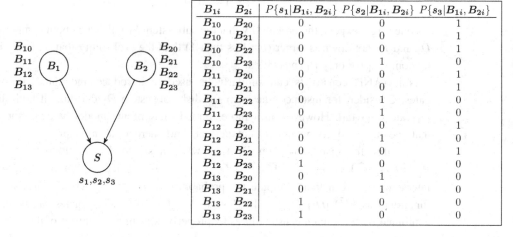

B_{1i}	B_{2i}	$P\{s_1|B_{1i}, B_{2i}\}$	$P\{s_2|B_{1i}, B_{2i}\}$	$P\{s_3|B_{1i}, B_{2i}\}$
B_{10}	B_{20}	0	0	1
B_{10}	B_{21}	0	0	1
B_{10}	B_{22}	0	0	1
B_{10}	B_{23}	0	1	0
B_{11}	B_{20}	0	0	1
B_{11}	B_{21}	0	0	1
B_{11}	B_{22}	0	1	0
B_{11}	B_{23}	0	1	0
B_{12}	B_{20}	0	0	1
B_{12}	B_{21}	0	1	0
B_{12}	B_{22}	0	0	1
B_{12}	B_{23}	1	0	0
B_{13}	B_{20}	0	1	0
B_{13}	B_{21}	0	1	0
B_{13}	B_{22}	1	0	0
B_{13}	B_{23}	1	0	0

Figure 6.42   Bayesian network for the multi-state system of Example 6.36, and the corresponding CPT.

adjusting the entries of the CPT. The marginal probabilities to assign to the events are those computed in Example 3.1.

In contrast, the same behavior is difficult to represent with an FT. A possible solution is shown in Figure 6.41(a), where the component failure modes are modeled as two independent binary events o and s in input to an XOR gate since they are mutually exclusive. The resulting structure is a non-coherent FT and is not covered in this chapter. Further, as also observed in [8], from a system reliability perspective the XOR gate would imply that the TE cannot be reached when no failure modes are active (correct behavior), but also when both failure modes are active (wrong behavior). Thus, the two models do not represent the same situation, and the BN of Figure 6.41(b) provides the correct behavior while the FT of Figure 6.41(a) does not.

Example 6.39   We return to the MFT of Example 6.36 where the two boards B_1 and B_2 have four states numbered from 0 to 3. In Example 6.36 three possible system states were identified, indicated as s_1, s_2, and s_3, and for each state the corresponding MFT was constructed in Figure 6.32.

We can represent the three system states s_1, s_2, and s_3 by the single BN of Figure 6.42. The two root nodes model the boards B_1, and B_2 and have associated a variable with four values; the child node S has associated a variable with three values. The CPT (with 16 entries) shown to the right in Figure 6.42 realizes the same logical dependencies described by the MFT.

6.10.6   Sequentially Dependent Failures

The FT techniques discussed in this chapter are based on the assumption that component failures are independent. BNs can overcome the independence assumption in many ways [75]. Noisy gates are an example of dependence that cannot be represented in FTs. A further possibility is to add new arcs in a BN that accommodate new conditional dependencies among components. This possibility is illustrated in the following example.

Example 6.40   *A Multiprocessor System with Power Supply*
Suppose the modeler wants to refine the description of the multiprocessor system of Figure 6.38 by adding the power supply PS. When PS fails it causes a system failure, but it may also induce the processors to break down. In the FT representation of Figure 6.43 a new basic event PS is added, which is connected to the OR leading to TE to represent a new cause of system failure. However, modeling the dependence between the failure of PS and the failure of processor P_i ($i = 1, 2$) is not possible in the FT formalism.

BNs are better suited to model this case. In addition, the model may be made even more accurate by resorting to a multi-state model for the power supply. Indeed, a more realistic situation could be the following: PS is modeled with three possible modes, working (w), defective (d), and failed (f), where the first mode w corresponds to a correct behavior, the mode d to a reduced voltage condition and mode f to a complete loss of power. Of course, the f mode causes the whole system to be down, but we want to model the fact that in the d mode, the processors increase their conditional attitude to break down. This can be modeled by associating with the variable PS three values corresponding to the above modes and by adding two arcs connecting the node PS to the nodes P_1 and P_2, respectively. Figure 6.44 shows the modified BN structure for this model. In this case, P_1 and P_2 are no longer root nodes with marginal probabilities as in Figure 6.39, but they conditionally depend on PS and the corresponding CPT needs to be assigned. If we assume that the two processors P_1 and P_2 are EXP(λ_{p_i}) when PS is in the working mode of operation, and that the degraded mode of operation d increases the failure rate λ_p by a factor $\alpha > 1$, we can construct the CPT of P_i ($i = 1, 2$) as in Table 6.12.

Note that the other CPT that needs to be modified is the one related to node TE = Failure, which is the OR of $S_{12} = $ T, $\overline{N} = $ T, and $PS = $ f.

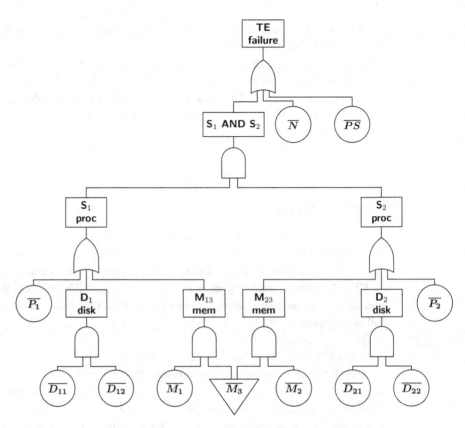

Figure 6.43   Fault tree for the redundant multiprocessor system with power supply (*PS*).

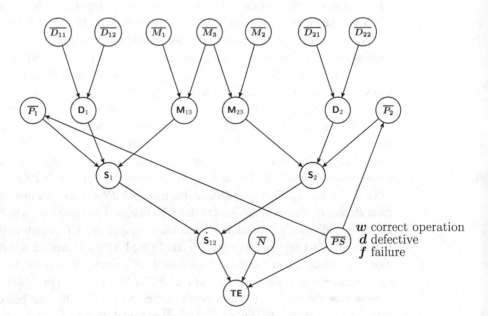

Figure 6.44   Bayesian network for the redundant multiprocessor system with power supply (*PS*).

Table 6.12 Conditional probability table of processor nodes conditioned by power supply *PS*.

	$\overline{PS}$	w	d	f
$P\{P_i|\overline{PS}\}$	F	$e^{-\lambda_{P_i}t}$	$e^{-\alpha\lambda_{P_i}t}$	0
	T	$1-e^{-\lambda_{P_i}t}$	$1-e^{-\alpha\lambda_{P_i}t}$	1

6.10.7 Dependability Analysis through BN Inference

Any quantitative analysis performed on an FT can then be performed on the corresponding BN; moreover, other interesting measures can be obtained from the BN that cannot be directly evaluated in an FT. Let us first consider the basic analyses of an FT and how they are performed in the corresponding BN:

- *Unavailability of the TE:* this measure corresponds to computing the prior probability of the variable TE = Failure, that is, $P\{Q|E\}$ with Q = Failure and $E = \emptyset$ (the empty set).
- *Unavailability of a given subsystem:* this measure corresponds to computing the prior probability of a node of the BN, that is, $P\{Q|E\}$ with Q = any node and $E = \emptyset$.

Further, BNs have the possibility of characterizing and evaluating some standard importance measures like the Birnbaum importance index I^B (Section 6.6.1) and the Vesely–Fussell importance index I^V (Section 6.6.3).

The Birnbaum importance index of component x_i at time t, defined in Eq. (6.22), can be rewritten in terms of a BN as:

$$I_i^B(t) = P\{\text{TE}(t) = \text{Failure}|x_i(t)\} - P\{\text{TE}(t) = Failure|\overline{x}_i(t)\},$$

showing that computation of the $I_i^B(t)$ index entails posterior probability computation, in particular the variation in the probability of the TE given that the component $x_i(t)$ is down or up at time t, respectively.

The Vesely–Fussell importance index $I_i^V(t)$, as defined in Eq. (6.25), gives the probability that at least one minimal cut set containing component x_i is failed at time t, given that the system is failed at time t. In other words, $I_i^V(t)$ is exactly the probability that component x_i is down at time t given that the system is down at time t. In terms of a BN we can write

$$I_i^V(t) = P\{\overline{x}_i(t)|\text{TE}(t) = \text{Failure}\},$$

so that $I_i^V(t)$ can be computed as the marginal posterior probability of the failure of $x_i(t)$, given the occurrence of system failure.

The above remarks introduce another aspect that is peculiar to the use of BNs with respect to FTs: the possibility of performing diagnostic problem-solving on the modeled system. In fact, in many cases the system analyst may be interested in determining the possible explanations of an exhibited failure in the system. Cut set determination is a step in this direction, but it may not be sufficient in certain situations. Classical diagnostic inference on a BN involves:

- computation of the posterior marginal probability distribution on each component;
- computation of the posterior joint probability distribution on subsets of components;
- computation of the posterior joint probability distribution on the set of all nodes, but the evidence ones.

The first kind of computation is perhaps the most popular one when using BNs for diagnosis, and is based on well-established algorithms, like the *belief propagation* algorithm [65, 70], that can compute the marginal posterior probability of each node.

Example 6.41   *Backward Inference on the Multiprocessor System*

Let us consider again the multiprocessor system whose FT is shown in Figure 6.38 and BN in Figure 6.39. For the purpose of a quantitative analysis, the failure distributions of all the components are assumed to be exponential with the failure rates given in the second column of Table 6.13. The failure probability of each component, evaluated at $t = 5000$ hr, is reported in the third column of Table 6.13. Forward propagation on the BN allows us to compute the unreliability of the TE as the *a priori* probability of system failure, i.e., $P\{TE = \text{Failure}\} = 0.012313$.

If we observe that the system is failed at time t, the marginal posterior fault probabilities of each component can be computed, and are reported in the fourth column of Table 6.13. It is worth remembering that such probabilities are exactly the Vesely–Fussell index of the components. Observe that the severity ranking based on posterior probabilities is different from the one based on prior probabilities.

From Table 6.13, we can notice that the most critical component is in this case a single disk. However, this information is not completely significant from the diagnostic point of view; indeed, by considering the disk units, the only way of having the failure is to assume that all the disks $D_{11}, D_{12}, D_{21}, D_{22}$ have failed at the same time (indeed, it is the only MCS involving only disks). This information (the fact that all disks have to be jointly faulty to get the failure) is not directly derived by marginal posteriors on components. In fact, for diagnostic purposes a more suitable analysis should consider the posterior joint probability of all the components given the system failure as evidence.

Table 6.13 Failure rates and prior and posterior probabilities for the multiprocessor system at $t = 5000$ hr.

Component C	Failure rate λ_C	Prior $P\{\overline{C}\}$	Posterior $P\{\overline{C}|\text{Failure}\}$
Disk D_{ij}	$\lambda_D = 8.0 \times 10^{-5}$	0.329680	0.98438
Proc P_i	$\lambda_P = 5.0 \times 10^{-7}$	0.002497	0.02250
Mem M_j	$\lambda_M = 3.0 \times 10^{-8}$	0.000150	0.00015
Bus N	$\lambda_N = 2.0 \times 10^{-9}$	0.000010	0.00081

This analysis corresponds to searching for the most probable state, given the failure, over the state space represented by all the possible instantiations of the root variables (i.e., components).

In this case, we can determine by means of BN inference that the most probable diagnosis (i.e., the most probable state given the system failure) is exactly the one corresponding to the faulty value of all the disks and the working value of all the other components; in particular, we obtain:

$$P\{N,M_1,M_2,M_3,P_1,P_2,\overline{D_{11}},\overline{D_{12}},\overline{D_{21}},\overline{D_{22}},|\text{Failure}\} = 0.954\,22.$$

Notice that the above diagnosis does not correspond to the cut set $\{\overline{D_{11}},\overline{D_{12}},\overline{D_{21}},\overline{D_{22}}\}$, since the latter does not imply that the unmentioned components are working. For instance, we can compute

$$P\{\overline{D_{11}},\overline{D_{12}},\overline{D_{21}},\overline{D_{22}},|\text{Failure}\} = 0.959\,51,$$

where we do not specify the condition of the unmentioned components.

An interesting possibility offered by BNs is that there exist algorithms able to produce diagnoses (either viewed as only-root assignments or all-variable assignments) in order of their probability of occurrence, without exploring the whole state space; they are usually called *any-time* algorithms, since the user can stop the algorithm at any time, getting an approximate answer that is improved if more time is allocated to the algorithm. For example, an algorithm based on the model described in [76] is able to provide the most probable diagnoses, given the observation of the failure, in the multiprocessor system at any desired level of precision. By specifying a maximum admissible error ϵ in the posterior probability of the diagnoses, the algorithm is able to produce, every diagnosis d, in decreasing order of their occurrence probability, with an estimate $P'(d|\text{Failure})$ such that its actual posterior probability is $P\{d|\text{Failure}\} = P'\{d|\text{Failure}\} \pm \epsilon$.

In the example, by requiring diagnoses to be root assignments and $\epsilon = 1 \times 10^{-6}$ the first three most probable diagnoses are, in order:

$$d_1 : (N,M_1,M_2,M_3,P_1,P_2,\overline{D_{11}},\overline{D_{12}},\overline{D_{21}},\overline{D_{22}}),$$

$$d_2 : (N,M_1,M_2,M_3,P_1,\overline{P_2},\overline{D_{11}},\overline{D_{12}},D_{21},D_{22}),$$

$$d_3 : (N,M_1,M_2,M_3,\overline{P_1},P_2,D_{11},D_{12},\overline{D_{21}},\overline{D_{22}}).$$

The first represents the already mentioned most probable diagnoses with all disks faulty, while the second and the third are two symmetrical diagnoses corresponding to a disk failure in one subsystem and processor failure in the other subsystem.

Posterior probabilities are then computed within the given error level as $P\{d1|\text{Failure}\} = 0.954\,22$, $P\{d2|\text{Failure}\} = P\{d3|\text{Failure}\} = 0.009\,887$. The algorithm guarantees that any further diagnosis has a posterior probability smaller than or equal to $P\{d2|\text{Failure}\}$. This result is in general obtained without exploring the whole state space, which in this case is equal to $2^{10} = 1024$ states, 10 being the number of components. Similar results may be obtained for complete variable assignments.

Problems

6.23   Convert the FT of Figure 6.11 into the corresponding BN.

6.24   In the previous problem, assume now that all the components are three-state components with states numbered from 0 to 2. For components A, B, and C, States 0 and 1 are down states and State 2 is the only up state. For components D and E, State 0 is the only down state and States 1 and 2 are the up states. Derive the corresponding multi-state FT and the MDD. Then convert the multi-state FT into the corresponding BN.

6.11   Further Reading

The literature on the theory and applications of FTs is immense, since the technique has been used in many technical fields for dependability analysis, safety analysis and risk assessment [4, 9]. In this section we mention a few additional topics not covered in the chapter.

Other Exact and Approximate Measures

For very large FTs the exact computation of the probability of the TE may become very time consuming or even impossible. In those cases, cutoff techniques may be applied by either eliminating MCSs with a probability lower than a (user-defined) cutoff bound, or with an order greater than a (user-defined) cutoff bound.

In systems with repairable components, when all the basic events are quantified with their unavailability, the standard FT analysis provides the TE unavailability. In this case, the exact value of the TE unreliability cannot be determined by means of standard analysis [77]. However, (good) approximated conservative results can be obtained through determination of the expected number of failures (ENF). This bound is based on the unconditional failure frequency, which is the time derivative of the ENF. Derivation of such measures can be found in [33, 78].

Inhibit Gate

The inhibit (INH) gate is sometimes used in risk assessment and safety studies, where a protective device is inserted into a system to block the propagation of a potentially catastrophic failure [2, 8].

The INH gate has only two input variables: the initiating event A and the inhibiting event H. The output event takes place if at the time instant when event A occurs the condition defined by event H has also occurred. The INH gate is not a Boolean gate since the two events are both necessary, but must occur in sequence, first H and then

A [79]. The INH and AND gates are often confused, giving rise to modeling and analysis errors [79].

Extensions of standard FTs to include various kinds of dependencies (like the temporal sequence implied by the INH gate) are called dynamic fault trees and are considered in Section 16.5.

Non-Coherent FT

A non-coherent FT contains the NOT Boolean operator in the tree structure and may have a rather counterintuitive behavior [80, 81], since the failure of a basic event may reduce the probability of reaching the TE. Non-coherent structures may be encountered in safety and security studies, as extensively detailed in [82], where a large number of examples are reviewed, with pointers to the corresponding literature. General analysis techniques are investigated in [83, 84], and more specialized studies in [85–88].

6.12 Useful Properties of Boolean Algebra

Let x, y, and z be Boolean variables. The following relations hold.

Idempotence property: $\quad x \vee x = x$

$\qquad\qquad\qquad\qquad\quad x \wedge x = x$

Forcing function: $\quad\qquad x \vee 1 = 1$

$\qquad\qquad\qquad\qquad\quad x \wedge 0 = 0$

Commutative property: $\quad x \vee y = y \vee x$

$\qquad\qquad\qquad\qquad\quad x \wedge y = y \wedge x$

Associative property: $\quad\;\; x \vee (y \vee z) = (x \vee y) \vee z = x \vee y \vee z$

$\qquad\qquad\qquad\qquad\quad x \wedge (y \wedge z) = (x \wedge y) \wedge z = x \wedge y \wedge z$

Distributive property: $\quad\;\; (x \wedge y) \vee (x \wedge z) = x \wedge (y \vee z)$

$\qquad\qquad\qquad\qquad\quad (x \vee y) \wedge (x \vee z) = x \vee (y \wedge z)$

Absorption property: $\quad\;\;\; x \vee (x \wedge y) = x$

$\qquad\qquad\qquad\qquad\quad x \wedge (x \vee y) = x$

$\qquad\qquad\qquad\qquad\quad x \vee y = x \vee (\bar{x} \wedge y)$

$\qquad\qquad\qquad\qquad\quad x \wedge y = x \wedge (\bar{x} \vee y)$

De Morgan's law: $\qquad\;\; \overline{x \wedge y} = \bar{x} \vee \bar{y}$

$\qquad\qquad\qquad\qquad\quad \overline{x \vee y} = \bar{x} \wedge \bar{y}$

Some common identities:

$$x \wedge 1 = x \qquad x \vee 0 = x$$
$$x \wedge \bar{x} = 0 \qquad x \vee \bar{x} = 1$$
$$x \vee 1 = 1 \qquad \bar{x} \vee 1 = 1$$
$$x \wedge 0 = 0 \qquad \bar{x} \wedge 0 = 0$$
$$x \wedge (y \vee 1) = x \qquad \bar{x} \wedge (y \vee 1) = \bar{x}$$
$$x \wedge (y \vee \bar{y}) = x \qquad \bar{x} \wedge (y \vee \bar{y}) = \bar{x}$$

References

[1] IEC 61025, *Fault Tree Analysis*. IEC Standard No. 61025, 2nd edn., 2006.

[2] A. Hixenbaugh, *Fault Tree for Safety*. The Boeing Company, 1968.

[3] R. Barlow and F. Proschan, *Statistical Theory of Reliability and Life Testing*. Holt, Rinehart, and Winston, 1975.

[4] E. Henley and H. Kumamoto, *Reliability Engineering and Risk Assessment*. Prentice Hall, 1981.

[5] W. Lee, D. Grosh, F. Tillman, and C. Lie, "Fault tree analysis, methods and applications: A review," *IEEE Transactions on Reliability*, vol. R-34, pp. 194–203, 1985.

[6] S. Contini and A. Poucet, "Advances on fault tree and event tree techniques," in *System Reliability Assessment*, eds. A. Colombo and A. S. de Bustamante. Kluwer Academic P.G., 1990, pp. 77–102.

[7] W. Schneeweiss, *The Fault Tree Method*. LiLoLe Verlag, 1999.

[8] M. Stamatelatos and W. Vesely, *Fault Tree Handbook with Aerospace Applications*. NASA Office of Safety and Mission Assurance, 2002, vol. 1.1.

[9] E. Ruijters and M. Stoelinga, "Fault tree analysis: A survey of the state of the art in modeling, analysis and tools," *Computer Science Review*, vol. 15–16, pp. 29–62, 2015.

[10] M. Malhotra and K. Trivedi, "Power-hierarchy among dependability model types," *IEEE Transactions on Reliability*, vol. R-43, pp. 493–502, 1994.

[11] M. Malhotra and K. Trivedi, "Dependability modeling using Petri nets," *IEEE Transactions on Reliability*, vol. R-44, pp. 428–440, 1995.

[12] J. Jürjens, *Developing Safety-Critical Systems with UML*. Springer, 2003, pp. 360–372.

[13] S. Bernardi, J. Merseguer, and D. Petriu, "Dependability modeling and analysis of software systems specified with UML," *ACM Computing Surveys*, vol. 45, no. 1, pp. 2:1–2:48, Dec. 2012.

[14] N. Piccinini and I. Ciarambino, "Operability analysis devoted to the development of logic trees," *Reliability Engineering and System Safety*, vol. 55, pp. 227–241, 1997.

[15] R. Sahner, K. Trivedi, and A. Puliafito, *Performance and Reliability Analysis of Computer Systems: An Example-Based Approach Using the SHARPE Software Package*. Kluwer Academic Publishers, 1996.

[16] K. S. Trivedi and R. Sahner, "SHARPE at the age of twenty-two," *SIGMETRICS Perform. Eval. Rev.*, vol. 36, no. 4, pp. 52–57, Mar. 2009.

[17] W. Schneeweiss, *Boolean Functions with Engineering Applications and Computer Programs*. Springer Verlag, 1989.

[18] L. Ge, M. van Asseldonk, and M. van Galen, "Stochastic fault tree analysis for agropark project appraisal," in *Proc. 21st IFAMA – Internationl Food and Agribusiness Management Association*, 2011.

[19] B. Ayyub, *Risk Analysis in Engineering and Economics*. Chapman and Hall/CRC, 2003.

[20] P. Lacey, "An application of fault tree analysis to the identification and management of risks in government funded human service delivery," in *Proc. 2nd Int. Conf. on Public Policy and Social Sciences*, eds. K. Singh and B. Singh, 2011.

[21] G. Youngjung, S. Hyeonju, L. Sungjoo, and P. Yongtae, "Application of fault tree analysis to the service process: Service tree analysis approach," *Journal of Service Management*, vol. 20, no. 4, p. 433–454, 2009.

[22] R. Fricks and K. Trivedi, "Importance analysis with Markov chains," in *Proc. IEEE Ann. Reliability and Maintainability Symp.*, 2003.

[23] G. Bucci, L. Carnevali, and E. Vicario, "A tool supporting evaluation of non-Markovian fault trees," *Proc. Int. Conf. on Quantitative Evaluation of Systems*, pp. 115–116, 2008.

[24] J. Vaurio, "Treatment of general dependencies in system fault-tree and risk analysis," *IEEE Transactions on Reliability*, vol. 51, pp. 278–287, 2002.

[25] A. V. Ramesh, D. W. Twigg, U. R. Sandadi, T. C. Sharma, K. S. Trivedi, and A. K. Somani, "An integrated reliability modeling environment," *Reliability Engineering and System Safety*, vol. 65, no. 1, pp. 65–75, 1999.

[26] A. Rauzy, "New algorithms for fault tree analysis," *Reliability Engineering and System Safety*, vol. 40, pp. 203–211, 1993.

[27] R. Sinnamon and J. Andrews, "Improved accuracy in quantitative fault tree analysis," *Quality and Reliability Engineering International*, vol. 13, pp. 285–292, 1997.

[28] A. Rauzy, "A brief introduction to binary decision diagrams," *Journal Européen des Systèmes Automatisés (RAIRO-APII-JESA)*, vol. 30, no. 8, pp. 1033–1051, 1996.

[29] L. Xing and S. Amari, *Binary Decision Diagrams and Extensions for System Reliability Analysis*. Wiley-Scrivener, 2015.

[30] Y. Dutuit and A. Rauzy, "A linear-time algorithm to find modules of fault tree," *IEEE Transactions on Reliability*, vol. 45, pp. 422–425, 1996.

[31] R. Gulati and J. Dugan, "A modular approach for analyzing static and dynamic fault-trees," in *Proc. IEEE Ann. Reliability and Maintainability Symp.*, 1997, pp. 57–63.

[32] Y. Dutuit and A. Rauzy, "Efficient algorithms to assess components and gates importance in fault tree analysis," *Reliability Engineering and System Safety*, vol. 72, pp. 213–222, 2000.

[33] S. Contini and V. Matuzas, "New methods to determine the importance measures of initiating and enabling events in fault tree analysis," *Reliability Engineering and System Safety*, vol. 96, no. 7, pp. 775–784, 2011.

[34] Z. Birnbaum, "On the importance of different components in a multicomponent systems," in *Multivariate Analysis - II*, ed. E. P. R. Krishnaiah. Academic Press, 1969, pp. 581–592.

[35] M. Veeraraghavan and K. Trivedi, "An improved algorithm for the symbolic reliability analysis of networks," *IEEE Transactions on Reliability*, vol. 40, pp. 347–358, 1991.

[36] F. C. Meng, "Relationships of Fussell–Vesely and Birnbaum importance to structural importance in coherent systems," *Reliability Engineering and System Safety*, vol. 67, no. 1, pp. 55–60, 2000.

[37] K. Hjelmgren, S. Svensson, and O. Hannius, "Reliability analysis of a single-engine aircraft FADEC," in *Proc. Ann. Reliability and Maintainability Symp.*, 1998, pp. 401–407.

[38] W. E. Smith, K. S. Trivedi, L. Tomek, and J. Ackaret, "Availability analysis of blade server systems," *IBM Systems Journal*, vol. 47, no. 4, pp. 621–640, 2008.

[39] R. Credle, D. Brown, L. Davis, D. Robertson, T. Ternau, and D. Green, "The cutting edge: IBM E-Server BladeCenter," *IBM Redpaper REDP-3581-01*, 2003.

[40] A. Bobbio, S. Bologna, E. Ciancamerla, P. Incalcaterra, C. Kropp, M. Minichino, and E. Tronci, "Advanced techniques for safety analysis applied to the gas turbine control system of ICARO co-generative plant," in *X TESEC (Genova)*, 2001, pp. 339–350.

[41] D. Nicol, W. Sanders, and K. Trivedi, "Model-based evaluation: From dependability to security," *IEEE Transactions on Dependable and Secure Computing*, vol. 1, no. 1, pp. 48–65, 2004.

[42] A. Avizienis, J. Laprie, B. Randell, and C. Landwehr, "Basic concepts and taxonomy of dependable and secure computing," *IEEE Transactions on Dependable and Secure Computing*, vol. 1, no. 1, pp. 11–33, 2004.

[43] R. Ortalo, Y. Deswarte, and M. Kaaniche, "Experimenting with quantitative evaluation tools for monitoring operational security," *IEEE Transactions on Software Engineering*, vol. 25, no. 5, pp. 633–650, Sep./Oct. 1999.

[44] K. S. Trivedi, D. S. Kim, A. Roy, and D. Medhi, "Dependability and security models," in *Proc. 7th Int. Workshop on Design of Reliable Communication Networks*, 2009, pp. 11–20.

[45] L. Pietre-Cambacedes and M.Bouissou, "Cross-fertilization between safety and security engineering," *Reliability Engineering and System Safety*, vol. 110, pp. 110–126, 2013.

[46] B. Schneier, "Attack trees," *Dr. Dobb's Journal of Software Tools*, vol. 24, no. 12, pp. 21–29, 1999.

[47] I. Fovino, M. Masera, and A. D. Cian, "Integrating cyber attacks within fault trees," *Reliability Engineering and System Safety*, vol. 94, pp. 1394–1402, 2009.

[48] A. Roy, D. S. Kim, and S. Trivedi, "Act: Towards unifying the constructs of attack and defense trees," *Security and Communication Networks*, vol. 3, pp. 1–15, 2011.

[49] J. Byres, M. Franz, and D. Miller, "The use of attack trees in assessing vulnerabilities in SCADA systems," in *Int. Infrastructure Survivability Workshop (IISW'04)*, Lisbon, 2004.

[50] A. Roy, D. S. Kim, and S. Trivedi, "Scalable optimal countermeasure selection using implicit enumeration on attack countermeasure trees," in *Proc. Int. Conf. on Dependable Systems and Networks (DSN 2012)*. IEEE Computer Society, 2012, pp. 1–12.

[51] A. Bobbio, L. Egidi, and R. Terruggia, "A methodology for qualitative/quantitative analysis of weighted attack trees," in *4th IFAC Workshop on Dependable Control of Discrete Systems (DCDS13)*, York (UK), 4–6 Sep. 2013, p. 6.

[52] S. Zonouz, H. Khurana, W. Sanders, and T. Yardley, "RRE: A game-theoretic intrusion response and recovery engine," in *IEEE/IFIP Int. Conf. on Dependable Systems Networks*, 2009, pp. 439–448.

[53] L. Caldarola, "Fault tree analysis with multistate components," Kernforschungszentrum Karlsruhe, Tech. Rep. KfK 2761 – EUR 5756e, 1979.

[54] A. Wood, "Multistate block diagrams and fault trees," *IEEE Transactions on Reliability*, vol. R-34, pp. 236–240, 1985.

[55] Y. Kai, "Multistate fault-tree analysis," *Reliability Engineering and System Safety*, vol. 28, pp. 1–7, 1990.

[56] M. Veeraraghavan and K. S. Trivedi, "A combinatorial algorithm for performance and reliability analysis using multistate models," *IEEE Transactions on Computers*, vol. 43, no. 2, pp. 229–234, 1994.

[57] X. Zang, D. Wang, H. Sun, and K. Trivedi, "A BDD-based algorithm for analysis of multistate systems with multistate components," *IEEE Transactions on Computers*, vol. 52, no. 12, pp. 1608–1618, 2003.

[58] L. Xing and Y. Dai, "A new decision diagram based method for efficient analysis on multi-state systems," *IEEE Transactions on Dependable and Secure Computing*, vol. 6, no. 3, pp. 161–174, 2009.

[59] S. Amari, L. Xing, A. Shrestha, J. Akers, and K. Trivedi, "Performability analysis of multistate computing systems using multivalued decision diagrams," *IEEE Transactions on Computers*, vol. 59, no. 10, pp. 1419–1433, 2010.

[60] K. Trivedi, X. Yin, and D. S. Kim, "Recent advances in system reliability," in *Multi-State Availability Modeling in Practice*, eds. A. Lisnianski and I. Frenkel. Springer-Verlag, 2011.

[61] A. Bobbio, L. Portinale, M. Minichino, and E. Ciancamerla, "Improving the analysis of dependable systems by mapping fault trees into Bayesian networks," *Reliability Engineering and System Safety*, vol. 71, pp. 249–260, 2001.

[62] H. Boudali and J. B. Dugan, "A discrete-time Bayesian network reliability modeling and analysis framework," *Reliability Engineering and System Safety*, vol. 87, pp. 337–349, 2005.

[63] M. Neil and D. Marquez, "Availability modelling of repairable systems using Bayesian networks," *Engineering Applications of Artificial Intelligence*, vol. 25, no. 4, pp. 698–704, Jun. 2012.

[64] A. Bobbio, D. Codetta-Raiteri, S. Montani, and L. Portinale, "Reliability analysis of systems with dynamic dependencies," in *Bayesian Networks: A Practical Guide to Applications*. John Wiley & Sons, 2008, pp. 225–238.

[65] J. Pearl, *Probabilistic Reasoning in Intelligent Systems: Networks of Plausible Inference*. Morgan Kaufmann, 1988.

[66] O. Pourret, P. Naïm, and B. Marcot, *Bayesian Networks: A Practical Guide to Applications*. John Wiley & Sons, 2008.

[67] J. Torres-Toledano and L. Sucar, "Bayesian networks for reliability analysis of complex systems," in *Lecture Notes in Artificial Intelligence*, vol. 1484. Springer Verlag, 1998, pp. 195–206.

[68] H. Langseth and L. Portinale, "Bayesian networks in reliability," *Reliability Engineering and System Safety*, vol. 92, pp. 92–108, 2007.

[69] P. Weber, G. Medina-Oliva, C. Simon, and B. Iung, "Overview on Bayesian network applications for dependability, risk analysis and maintenance areas," *Engineering Applications of Artificial Intelligence*, vol. 25, no. 4, pp. 671–682, Jun. 2012.

[70] L. Portinale and D. Codetta-Raiteri, *Modeling and Analysis of Dependable Systems: A Probabilistic Graphical Model Perspective*. World Scientific, 2015.

[71] J. B. Dugan, S. Bavuso, and M. Boyd, "Fault-trees and Markov models for reliability analysis of fault-tolerant digital systems," *Reliability Engineering and System Safety*, vol. 39, pp. 291–307, 1993.

[72] G. Cooper, "The computation complexity of probabilistic inference using Bayesian belief networks," *Artificial Intelligence*, vol. 33, pp. 393–405, 1990.

[73] N. Zhang and D. Poole, "Exploiting causal independence in Bayesian network inference," *Journal of Artifical Intelligence Research*, vol. 5, pp. 301–328, 1996.

[74] A. Moslehl, D. M. Rasmuson, and F. M. Marshall, *Guidelines on Modeling Common-Cause Failures in Probabilistic Risk Assessment*, NUREG/CR-5485. U.S. Nuclear Regulatory Commission, 1998.

[75] D. Codetta-Raiteri, A. Bobbio, S. Montani, and L. Portinale, "A dynamic Bayesian network based framework to evaluate cascading effects in power grids," *Engineering Applications of Artificial Intelligence*, vol. 25, pp. 683–697, 2012.

[76] L. Portinale and P. Torasso, "A comparative analysis of Horn models and Bayesian networks for diagnosis," in *Lecture Notes in Artificial Intelligence*, vol. 1321. Springer, 1997, pp. 254–265.

[77] C. Clarotti, "Limitations of minimal cut-set approach in evaluating reliability of systems with repairable components," *IEEE Transactions on Reliability*, vol. R-30, no. 4, pp. 335–338, 1981.

[78] H. Kumamoto and E. Henley, *Probabilistic Risk Assessment and Management for Engineers and Scientists*. IEEE Press, 1996.

[79] M. Demichela, N. Piccinini, I. Ciarambino, and S. Contini, "On the numerical solution of fault trees," *Reliability Engineering and System Safety*, vol. 82, no. 2, pp. 141–147, 2003.

[80] J. D. Andrews, "The use of not logic in fault tree analysis," *Quality and Reliability Engineering International*, vol. 17, no. 3, pp. 143–150, 2001.

[81] S. Oliva, "Non-coherent fault trees can be misleading," *Journal of System Safety*, vol. 42, no. 3, pp. 1–5, 2006.

[82] S. Contini, G. Cojazzi, and G. Renda, "On the use of non-coherent fault trees in safety and security studies," *Reliability Engineering and System Safety*, vol. 93, no. 12, pp. 1886–1895, 2008.

[83] T. Chu and G. Apostolakis, "Methods for probabilistic analysis of noncoherent fault trees," *IEEE Transactions on Reliability*, vol. R-29, no. 5, pp. 354–360, Dec. 1980.

[84] S. Beeson, "Non coherent fault tree analysis," Ph.D. Thesis, Loughborough University, 2002.

[85] A. Rauzy and Y. Dutuit, "Exact and truncated computations of prime implicants of coherent and non-coherent fault trees within aralia," *Reliability Engineering and System Safety*, vol. 58, no. 2, pp. 127–144, 1997.

[86] S. Beeson and J. Andrews, "Birnbaum measure of component importance for non-coherent systems," *IEEE Transactions on Reliability*, vol. 52, pp. 213–219, 2003.

[87] S. Beeson and J. Andrews, "Importance measures for non-coherent-system analysis," *IEEE Transactions on Reliability*, vol. 52, pp. 301–310, 2003.

[88] D. Wang and K. Trivedi, "Computing steady-state mean time to failure for non-coherent repairable systems," *IEEE Transactions on Reliability*, vol. 54, pp. 506–516, 2005.

7 State Enumeration

The *state enumeration* method is based on the concept of a *state* of the system, on the enumeration of all the possible states of the system (the *state space*) [1] and on the definition of a system *structure function* that identifies the states that correspond to a working condition for the system and those that correspond to a failed condition. This method is, in principle, applicable to all the model types discussed so far and to many others.

If the component failures (and repairs) are statistically independent, and the probability of each component being up or down is known, the probability of occurrence of each state may be computed by simple multiplication. However, if dependencies among components are present, more complex quantitative analysis techniques need to be considered and these are deferred to the next part of the book dealing with state-space methods. This chapter can be considered to be a bridge between the current part of the book and the next.

In Section 7.1, we discuss how to define the state of a system and the concept of state space. In Section 7.2, a probabilistic model of the evolution of the system in the state space is built up, and the related dependability measures are defined. In Section 7.3, the dependability measures for various systems, including non-series-parallel systems, are computed under the assumption of statistical independence of the component failures (and repairs).

7.1 The State Space

Consider a system with ℓ components each of which is in one of two mutually exclusive conditions referred to as working (or up) and failed (or down). We assign a binary state indicator variable X_i ($i = 1, 2, \ldots, \ell$) to each of the ℓ components. The state indicator variable X_i identifies the state of the ith component and assumes the following values:

$$X_i = \begin{cases} 1 & \text{component } i \text{ up,} \\ 0 & \text{component } i \text{ down.} \end{cases}$$

The state of the system can then be represented as a vector $X = (X_1, X_2, \ldots, X_\ell)$ [2], where X_i indicates the state of component i. The state space of the system, denoted by Ω, is the set of all the possible values of X, i.e., the set of all the possible combinations of the ℓ components in working or failed condition. Let $n = |\Omega|$ be the cardinality of Ω; since each X_i is a binary variable, $n = 2^\ell$. Hence, the states of the system can be numbered $\{s_1, s_2, \ldots, s_n\}$, where each state s_j corresponds to a unique instantiation of vector X. If simultaneous events are not considered, transitions can happen only between states that differ in the value of a single binary variable in X.

The state space can be conveniently represented by a labeled directed graph whose nodes are the states of the system and whose edges represent the direct transitions between states. The label on a node indicates the state number, and the label on an edge indicates the component number i whose change of value triggers the state transition.

The concept of state space can be generalized to systems made of *multi-state* components, as we have already seen in Section 5.4 for network reliability and in Section 6.9 for FTs. In multi-state systems, each component i $(i = 1, 2, ..., \ell)$ can be characterized by $(m_i + 1)$ possible states so that the component indicator variable X_i is an integer that can assume any value from 0 to m_i. The cardinality $n = |\Omega|$ of the state space is then given by:

$$n = \prod_{i=1}^{\ell} (m_i + 1).$$

In the subsequent sections of this chapter we maintain the assumption that the state indicator variable X_i is a binary variable. The extension to multi-state components is straightforward and we refer the reader to Sections 5.4 and 6.9 for examples. We further assume that components are non-repairable. However, under independence conditions, repairable components can also be considered, as illustrated in Section 7.4, even if a more extensive analysis of dependent repair is deferred to the subsequent parts of the book.

Example 7.1 *A Non-Repairable Two-Component System*

Consider a system with $\ell = 2$ components X_1 and X_2. Independent of the way in which the two components are connected in the system, there are $n = 2^2 = 4$ possible states for the system and they can be identified as s_1, s_2, s_3, and s_4. State s_1 corresponds to vector $X = (1, 1)$ and represents the condition in which both components 1 and 2 are up; state s_2 corresponds to $X = (0, 1)$ and represents the condition in which component 1 is down and component 2 is up; state s_3 corresponds to $X = (1, 0)$ and represents the condition in which component 1 is up and component 2 is down; finally, state s_4 corresponds to $X = (0, 0)$ and represents the condition in which both components 1 and 2 are down. The state space of the system is listed in Table 7.1, where the states are enumerated and ordered according to the number of failed components. A graphical representation of the state space is shown in Figure 7.1, where the label on the transitions indicates which component has undergone a change of state.

Table 7.1 State enumeration of a two-component system.

	State vector X	System state #
0 failures	1 1	1
1 failure	0 1	2
	1 0	3
2 failures	0 0	4

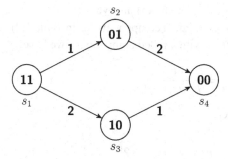

Figure 7.1   State space of a non-repairable two-component system.

Example 7.2   *A Non-Repairable Three-Component System*
In a system with $\ell = 3$ components, the state space has cardinality $n = 2^3 = 8$. The enumeration of all the possible states, ordered according to the number of failed components, is shown in Table 7.2, and a graphical representation of the state space with the corresponding labeled transitions is shown in Figure 7.2.

Table 7.2 State enumeration of a three-component system.

	State vector X	System state #
0 failures	1 1 1	1
	0 1 1	2
1 failure	1 0 1	3
	1 1 0	4
	0 0 1	5
2 failures	0 1 0	6
	1 0 0	7
3 failures	0 0 0	8

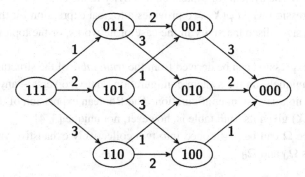

Figure 7.2   State space of a non-repairable three-component system.

The states $\{s_1, s_2, \ldots, s_n\} = \Omega$ form an exhaustive and mutually exclusive set. Grouping the states by the number of failed components (as in Table 7.2), and denoting by $n^{(k)}$ the number of system states with k failed components, we have:

$$n^{(k)} = \binom{\ell}{k} = \frac{\ell!}{k!\,(\ell - k)!}, \quad k = 0, 1, \ldots, \ell. \tag{7.1}$$

The total number of states is then:

$$n = \sum_{k=0}^{\ell} n^{(k)} = 2^{\ell}.$$

If we consider only failure events and discard the possibility of simultaneous failures, each node belonging to the class $n^{(k)}$ (grouping the $n^{(k)}$ states with k failed components) can have input arcs only from nodes of the class $n^{(k-1)}$ and output arcs only toward nodes of the class $n^{(k+1)}$.

7.1.1   Characterization of System States: Truth Table and Structure Function

We assume that the system as a whole also has a binary behavior: working or failed. Hence, we introduce a binary indicator variable Y for the system [2, 3]:

$$Y = \begin{cases} 1 & \text{system is in a working state}, \\ 0 & \text{system is in a failed state}. \end{cases} \tag{7.2}$$

For each state $s \in \Omega$, corresponding to a single value of vector $X = (X_1, X_2, \ldots, X_\ell)$, the indicator variable Y assumes either the value 0 or 1. We can write:

$$Y = \varphi(X) \quad \text{or} \quad Y = \varphi(X_1, X_2, \ldots, X_\ell). \tag{7.3}$$

The function $Y = \varphi(X)$ in Eq. (7.3) is called the *structure function* of the system and, with binary components, the structure function is a Boolean function. The Boolean expression of the structure function when $\varphi(X) = 1$ provides a logical expression for the correct operation of the system (or system connectivity in the case of networks), while the Boolean expression when $\varphi(X) = 0$ provides a logical expression for the failure of the system (or system disconnection in the case of networks, or the top event in case of a fault tree).

The Boolean expression can be derived from the *truth table* of the structure function, which depends uniquely on the system configuration, and vice versa – any truth table corresponds to a unique system configuration. The Boolean expression of the structure function $Y = \varphi(X)$ given its truth table is, however, not unique [3, 4].

The state space Ω can be partitioned into two collectively exhaustive and mutually exclusive subsets Ω_u and Ω_d.

$$\Omega_u = \{s \in \Omega : \varphi(X) = 1\}, \qquad \Omega_d = \{s \in \Omega : \varphi(X) = 0\}.$$

Ω_u groups the system up states for which $Y = \varphi(X) = 1$, while Ω_d groups the system down states for which $Y = \varphi(X) = 0$. Define by $n_u = |\Omega_u|$ the cardinality of Ω_u, and by $n_d = |\Omega_d|$ the cardinality of Ω_d. The following relations hold:

$$\Omega = \Omega_u \cup \Omega_d, \quad \Omega_u \cap \Omega_d = 0, \quad n = n_u + n_d. \tag{7.4}$$

While the state enumeration depends only on the number of components ℓ, the truth table of the structure function $Y = \varphi(X)$ reflects the system configuration. The truth table can be reported conveniently in tabular form by adding one column to the state enumeration table indicating the value (either 0 or 1) of the structure function.

7.1.2   Structural Importance and Frontier States

The structural importance and the structural importance index were defined in Section 6.3.5, in the framework of FT. The structural importance for component X_i was computed looking at the system states for which the structural sensitivity S_i^S defined in Eq. (6.11) is equal to one. The subset of states for which $S_i^S = 1$ is the subset of *frontier states* for component i, i.e., the states that belong to subset Ω_u when $X_i = 1$ and to Ω_d when $X_i = 0$. In terms of the structure function $\varphi(X)$, the frontier states for component i are the states for which:

$$\varphi(X)_{X_i=1} = 1, \qquad \varphi(X)_{X_i=0} = 0.$$

The structural importance index for component i is then computed as the number of frontier states for component i divided by the number of states of the whole state space, and thus requires the exploration of the truth table of the system (see Example 6.18).

7.1.3   Boolean Expression of the Structure Function

The Boolean expression of the structure function is not uniquely defined by its truth table. However, given the truth table, it is possible to determine a special form of the structure function, called the canonical disjunctive normal form (CDNF) [4], which, instead, is unique. In each row of the truth table, we construct a term of the function (called the *minterm*) obtained as the conjunction of ℓ Boolean variables, one for each component. In a minterm the ith Boolean variable appears in the direct form X_i if the corresponding ith entry of vector X is equal to 1, and in the negated form $\overline{X}_i$ if the corresponding ith entry of vector X is equal to 0.

The minterms are mutually disjoint: the conjunction of any two minterms is equal to 0. In fact, taking any two minterms, there is at least one Boolean variable that appears in the direct form X_i in one of them and in the negated form $\overline{X}_i$ in the other one.

The CDNF of a structure function is not minimal, and can be simplified into a DNF (see Section 6.3.1) using the rules of Boolean algebra reported in Section 6.12. The terms in the simplified DNF form, for which an example is given in (6.1), usually do not enjoy the property of being disjoint.

Example 7.3 *Series-Parallel System with Two Components*
Table 7.1 represents the state space for any system with $\ell = 2$ components.

The structure function depends on the way the components are connected in the system. For a series connection [Figure 7.3(a)], the only system up state is state s_1 corresponding to $X = (1, 1)$. Hence,

$$n_u = 1, \quad \Omega_u = \{1\}; \qquad n_d = 3, \quad \Omega_d = \{2, 3, 4\}. \tag{7.5}$$

The truth table for the series system is reported in the fourth column of Table 7.3. It is easy to verify that a possible expression for the structure function is:

$$\varphi(X) = X_1 \wedge X_2. \tag{7.6}$$

For the parallel connection [Figure 7.3(b)], the only system down state is state s_4 corresponding to $X = (0, 0)$. Hence,

$$n_u = 3, \quad \Omega_u = \{1, 2, 3\}; \qquad n_d = 1, \quad \Omega_d = \{4\}. \tag{7.7}$$

The corresponding truth table is reported in the fifth column of Table 7.3. It is easy to verify that a possible expression for the structure function is:

$$\varphi(X) = X_1 \vee X_2. \tag{7.8}$$

Example 7.4 *System with Three Components*
Table 7.2 represents the state space for a system with $\ell = 3$ components. The system configuration is reflected in the truth table. Consider the series-parallel system of

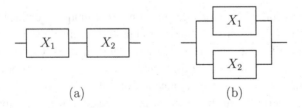

(a)   (b)

Figure 7.3   Series (a) and parallel (b) system with two components.

Table 7.3   Truth table for a series-parallel system with two components.

	State vector X	System state #	Series system $Y = \varphi(X)$	Parallel system $Y = \varphi(X)$
0 failure	1 1	1	1	1
1 failure	0 1	2	0	1
	1 0	3	0	1
2 failures	0 0	4	0	0

Figure 7.4, already considered in Figures 6.5 and 6.6. Observing Figure 7.4(a), the working states for the system correspond to states for which there exists a path of working components connecting the input to the output. Therefore, in the system up states component X_3 must be working and at least one among X_1 and X_2 must be working. Hence, the system up states are $(1, 1, 1)$, $(0, 1, 1)$, and $(1, 0, 1)$, while the remaining ones are down states. In this case:

$$n_u = 3, \quad \Omega_u = \{1, 2, 3\};$$

$$n_d = 5, \quad \Omega_d = \{4, 5, 6, 7, 8\}.$$

The truth table for Figure 7.4(a) is reported in the fourth column of Table 7.4. The minterms related to the truth table are reported in the last column. The structure function of configuration (a) can be written in CDNF as the sum of the minterms for each of which $\varphi(X = 1)$:

$$\varphi(X) = (X_1 \wedge X_2 \wedge X_3) \vee (\overline{X}_1 \wedge X_2 \wedge X_3) \vee (X_1 \wedge \overline{X}_2 \wedge X_3). \tag{7.9}$$

The Boolean function in (7.9) can be simplified using the rules of Boolean algebra to get the DNF:

$$\varphi(X) = (X_1 \wedge X_3) \vee (X_2 \wedge X_3). \tag{7.10}$$

Note that in the reduced form the two terms $(X_1 \wedge X_3)$ and $(X_2 \wedge X_3)$ are the minpaths of the structure and are no longer disjoint.

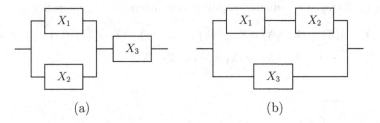

(a)   (b)

Figure 7.4   Series-parallel system configurations with three components.

Table 7.4   Truth table for the series-parallel systems of Figure 7.4.

	State vector X	System state #	Config. (a) $Y = \varphi(X)$	Config. (b) $Y = \varphi(X)$	Config. 2-out-of-3 $Y = \varphi(X)$	Minterms
0 failures	1 1 1	1	1	1	1	$X_1 X_2 X_3$
	0 1 1	2	1	1	1	$\overline{X}_1 X_2 X_3$
1 failure	1 0 1	3	1	1	1	$X_1 \overline{X}_2 X_3$
	1 1 0	4	0	1	1	$X_1 X_2 \overline{X}_3$
	0 0 1	5	0	1	0	$\overline{X}_1 \overline{X}_2 X_3$
2 failures	0 1 0	6	0	0	0	$\overline{X}_1 X_2 \overline{X}_3$
	1 0 0	7	0	0	0	$X_1 \overline{X}_2 \overline{X}_3$
3 failures	0 0 0	8	0	0	0	$\overline{X}_1 \overline{X}_2 \overline{X}_3$

Consider now the configuration depicted in Figure 7.4(b). The state space of the system remains the one reported in Table 7.2, but the partition of the system up and down states is, in this case:

$$n_u = 5, \quad \Omega_u = \{1, 2, 3, 4, 5\};$$

$$n_d = 3, \quad \Omega_d = \{6, 7, 8\}.$$

The structure function is reported in the fifth column of Table 7.4, and its CDNF can be written as:

$$\varphi(X) = (X_1 \wedge X_2 \wedge X_3) \vee (\overline{X}_1 \wedge X_2 \wedge X_3) \vee (X_1 \wedge \overline{X}_2 \wedge X_3)$$

$$\vee (X_1 \wedge X_2 \wedge \overline{X}_3) \vee (\overline{X}_1 \wedge \overline{X}_2 \wedge X_3). \tag{7.11}$$

The structure function in (7.11) can be simplified to get:

$$\varphi(X) = (X_1 \wedge X_2) \vee X_3, \tag{7.12}$$

where the two terms in the reduced DNF (7.12) are the minpaths of the structure.

Example 7.5 *Two-out-of-Three Majority Voting System*

The k-out-of-n majority voting system was introduced in Section 4.3 and, in the formalism of FTs, in Section 6.2.5.

The two-out-of-three system of Figure 7.5, neglecting the voter, is a three-component system whose truth table is reported in the sixth column of Table 7.4. The CDNF for the structure function can be obtained by summing the minterms, and is reported in Eq. (7.13) together with the simplified form in terms of the minpaths:

$$Y = \varphi(X) = (X_1 \wedge X_2 \wedge X_3) \vee (\overline{X}_1 \wedge X_2 \wedge X_3) \vee (X_1 \wedge \overline{X}_2 \wedge X_3) \vee (X_1 \wedge X_2 \wedge \overline{X}_3)$$

$$= (X_1 \wedge X_2) \vee (X_1 \wedge X_3) \vee (X_2 \wedge X_3). \tag{7.13}$$

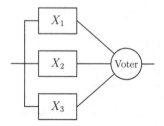

Figure 7.5 Two-out-of-three majority voting system.

7.2 The Failure Process Defined on the State Space

The evolution of the system in time can be represented by means of the succession of states traversed by the system due to failure or repair events of its components.

Consider component i. The state variable of component i can be considered as a function of time $X_i(t)$, defined as:

$$X_i(t) = \begin{cases} 1 & \text{component } i \text{ working at time } t, \\ 0 & \text{component } i \text{ failed at time } t. \end{cases}$$

The transition of component i from state $X_i(t) = 1$ to state $X_i(t) = 0$ (or vice versa) at time t is reflected in a transition of the system from a state s_j corresponding to $X = (X_1, \ldots, X_i = 1, \ldots, X_\ell)$ to the state $s_{j'}$ corresponding to $X = (X_1, \ldots, X_i = 0, \ldots, X_\ell)$ at time t – notice the analogy with the Shannon decomposition (5.1). Define

$$\begin{cases} P\{X_i(t) = 1\} &= R_i(t), \\ P\{X_i(t) = 0\} &= 1 - R_i(t). \end{cases} \tag{7.14}$$

In (7.14), $X_i(t)$ is the indicator random variable for component i and $R_i(t)$ is the probability that component i is working at time t. $R_i(t)$ coincides with the reliability of component i in the case of non-repairable components or with the instantaneous availability of component i in the case of repairable components. Note that $R_i(t)$ can be the survival function of any TTF distribution (see Section 3.2).

Denote by $Z(t)$ the state occupied by the system at time t. The expression $Z(t) = s_j$, means that the system is in state $s_j \in \Omega$ at time t. For any value of t, $Z(t)$ is a random variable that assumes non-negative values in the states of Ω. The probability that the system is in state $s_j \in \Omega$ at time t is denoted by $\pi_j(t)$ and is given by:

$$\pi_j(t) = P\{Z(t) = s_j\} \tag{7.15}$$

under the normalization condition

$$\sum_{j=1} \pi_j(t) = 1 \quad \text{for any} \quad t \geq 0.$$

$\{Z(t) : t \geq 0\}$ is a stochastic process defined over the discrete state space Ω and with continuous time parameter t. The quantitative evaluation of the state probabilities expressed by Eq. (7.15) completely determines the stochastic process $Z(t)$ and, hence, the behavior of the system. If the components are statistically independent, evaluation of (7.15) can be performed by resorting to combinatorial formulas and will be considered in Section 7.3; the case of statistically dependent components will be considered in the third part of this book.

7.2.1 Dependability Measures Defined on the State Space

Reliability: Non-Repairable Components Case

$$R_S(t) = \sum_{s_i \in \Omega_u} \pi_i(t), \tag{7.16}$$

$$F_S(t) = \sum_{s_i \in \Omega_d} \pi_i(t) = 1 - R_S(t).$$

System MTTF: Non-Repairable Components Case

$$\text{MTTF} = \int_0^\infty R_S(t)\,dt = \sum_{s_i \in \Omega_u} \int_0^\infty \pi_i(t)\,dt.$$

Instantaneous Availability: Repairable Components Case

$$A(t) = \sum_{s_i \in \Omega_u} \pi_i(t), \tag{7.17}$$

$$U(t) = \sum_{s_i \in \Omega_d} \pi_i(t) = 1 - A(t).$$

From the above, the steady-state availability and the expected interval availability of the system can also be determined.

7.3   System Reliability with Independent Components

If the structure function $\varphi(X)$ is expressed in canonical form, the system reliability can be expressed as the sum of the probabilities of the states $s_j \in \Omega_u$. Further, if the components are statistically independent, the probability $\pi_j(t)$ of being in a generic state s_j with characteristic vector $X = (X_1, X_2, \ldots, X_\ell)$ at time t can be expressed as the product of the probabilities of the individual variables:

$$\pi_j(t) = P\{X_1(t)\} \cdot P\{X_2(t)\} \cdots P\{X_\ell(t)\}, \tag{7.18}$$

where each term $P\{X_i(t)\}$ takes the value given by Eq. (7.14) according to whether the actual value of the indicator random variable $X_i(t)$ is 1 or 0, respectively. Hence, since the component failures are independent, the contribution of component i to the state probability is $R_i(t)$ in all the states in which $X_i(t) = 1$, while it is $1 - R_i(t)$ in all the states in which $X_i(t) = 0$.

In the usual case in which the time to failure distribution of each individual component is considered exponentially distributed with failure rate λ_i, Eq. (7.14) takes the form:

$$P\{X_i(t)\} = \begin{cases} R_i(t) = e^{-\lambda_i t} & \text{if } X_i(t) = 1, \\ 1 - R_i(t) = 1 - e^{-\lambda_i t} & \text{if } X_i(t) = 0. \end{cases} \tag{7.19}$$

According to Eq. (7.14), the state enumeration tables can be updated by indicating the probability of each state in a new column.

Example 7.6   *Series-Parallel System with Two Components (continued)*
Table 7.1 can be updated as in Table 7.5 by adding a column with the probability of the corresponding system state.

Table 7.5 State probabilities of a two-component system.

	State vector X	System state #	State probability $\pi_i(t)$
0 failure	1 1	1	$R_1(t)R_2(t)$
1 failure	0 1	2	$[1 - R_1(t)]R_2(t)$
	1 0	3	$R_1(t)[1 - R_2(t)]$
2 failures	0 0	4	$[1 - R_1(t)][1 - R_2(t)]$

Table 7.6 State probabilities of a three-component system.

	State vector X	System state #	State probability $\pi_i(t)$
0 failure	1 1 1	1	$R_1(t)R_2(t)R_3(t)$
	0 1 1	2	$[1 - R_1(t)]R_2(t)R_3(t)$
1 failure	1 0 1	3	$R_1(t)[1 - R_2(t)]R_3(t)$
	1 1 0	4	$R_1(t)R_2(t)[1 - R_3(t)]$
	0 0 1	5	$[1 - R_1(t)][1 - R_2(t)]R_3(t)$
2 failures	0 1 0	6	$[1 - R_1(t)]R_2(t)[1 - R_3(t)]$
	1 0 0	7	$R_1(t)[1 - R_2(t)][1 - R_3(t)]$
3 failures	0 0 0	8	$[1 - R_1(t)][1 - R_2(t)][1 - R_3(t)]$

The reliability of the series system $R_S(t)$ – see (4.2) – can be expressed as:

$$R_S(t) = \sum_{s_i \in \Omega_u} \pi_i(t) = \pi_1(t) = R_1(t)R_2(t). \tag{7.20}$$

For the reliability of the parallel system (4.23) we get:

$$R_S(t) = \sum_{s_i \in \Omega_u} \pi_i(t) = \pi_1(t) + \pi_2(t) + \pi_3(t)$$

$$= R_1(t)R_2(t) + [1 - R_1(t)]R_2(t) + R_1(t)[1 - R_2(t)]$$

$$= R_1(t) + R_2(t) - R_1(t)R_2(t).$$

Example 7.7 *Three-Component System (continued)*
The updated table with the state probability expressions for a three-component system is given in Table 7.6. Given the configuration, we can compute the system reliability.
   In the case of the system configuration of Figure 7.4(a), we have:

$$R_S(t) = \sum_{s_i \in \Omega_u} \pi_i(t) = \pi_1(t) + \pi_2(t) + \pi_3(t)$$

$$= R_3(t)[R_1(t) + R_2(t) - R_1(t) \cdot R_2(t)].$$

For the system configuration in Figure 7.4(b), we have:

$$R_S(t) = \sum_{s_i \in \Omega_u} \pi_i(t) = \pi_1(t) + \pi_2(t) + \pi_3(t) + \pi_4(t) + \pi_5(t)$$

$$= R_1(t) \cdot R_2(t) + R_3(t) - R_1(t) \cdot R_2(t) \cdot R_3(t).$$

For the two-out-of-three majority voting system with non-identical components we can derive the system reliability using the state enumeration approach, as follows (see Section 4.5):

$$R_S(t) = \sum_{s_i \in \Omega_u} \pi_i(t) = \pi_1(t) + \pi_2(t) + \pi_3(t) + \pi_4(t) \tag{7.21}$$

$$= R_1(t) \cdot R_2(t) + R_1(t) \cdot R_3(t) + R_2(t) \cdot R_3(t) - 2 R_1(t) \cdot R_2(t) \cdot R_3(t).$$

If components are i.i.d., $R_1(t) = R_2(t) = R_3(t) = R(t)$, Eq. (7.21) becomes:

$$R_S(t) = \pi_1(t) + \pi_2(t) + \pi_3(t) + \pi_4(t) \tag{7.22}$$

$$= R^3(t) + 3R^2(t)[1 - R(t)]$$

$$= 3R^2(t) - 2R^3(t).$$

Example 7.8   *Non-Series-Parallel System*
We return to Example 4.27, and we show how the reliability of the non-series-parallel system of Figure 4.25 can be computed by means of the state enumeration technique. Since $\ell = 5$, the total number of system states is $n = 2^5 = 32$. The structure function and the probability of each system state, under the assumption of component independence, are reported in the truth table in Table 7.7.

If the components are i.i.d., we derive from the data of Table 7.7 that the system reliability is given by:

$$R_S(t) = R^5(t) + 5R^4(t)(1 - R(t)) + 9R^3(t)(1 - R(t))^2 + 4R^2(t)(1 - R(t))^3$$

$$= R^5(t) - R^4(t) - 3R^3(t) + 4R^2(t).$$

Example 7.9   *Series-Parallel System with Five Components*
Consider again the system examined in Example 4.11 (Figure 4.6). Looking at Table 7.7, the subset of up states for the system is

$$\Omega_u = \{1, 2, 3, 4, 5, 6, 7, 8, 9, 10, 11, 12, 13, 16, 17, 22, 25\}.$$

Hence, if components are i.i.d., the system reliability becomes:

$$R_S(t) = R^5(t) + 5R(t)^4(1 - R(t)) + 8R^3(t)(1 - R(t))^2 + 3R^2(t)(1 - R(t))^3$$

$$= R^5(t) - 2R^4(t) - R^3(t) + 3R^2(t),$$

which coincides with the result obtained in Eq. (4.33).

Table 7.7 State enumeration for the system of Figure 4.25.

Failures	State vector X	System state #	State probability	Structure function $\varphi(X)$
0	1 1 1 1 1	1	$R_1 R_2 R_3 R_4 R_5$	1
	0 1 1 1 1	2	$[1 - R_1] R_2 R_3 R_4 R_5$	1
	1 0 1 1 1	3	$R_1 [1 - R_2] R_3 R_4 R_5$	1
1	1 1 0 1 1	4	$R_1 R_2 [1 - R_3] R_4 R_5$	1
	1 1 1 0 1	5	$R_1 R_2 R_3 [1 - R_4] R_5$	1
	1 1 1 1 0	6	$R_1 R_2 R_3 R_4 [1 - R_5]$	1
	0 0 1 1 1	7	$[1 - R_1] [1 - R_2] R_3 R_4 R_5$	1
	0 1 0 1 1	8	$[1 - R_1] R_2 [1 - R_3] R_4 R_5$	1
	0 1 1 0 1	9	$[1 - R_1] R_2 R_3 [1 - R_4] R_5$	1
	0 1 1 1 0	10	$[1 - R_1] R_2 R_3 R_4 [1 - R_5]$	1
2	1 0 0 1 1	11	$R_1 [1 - R_2] [1 - R_3] R_4 R_5$	1
	1 0 1 0 1	12	$R_1 [1 - R_2] R_3 [1 - R_4] R_5$	1
	1 0 1 1 0	13	$R_1 [1 - R_2] R_3 R_4 [1 - R_5]$	1
	1 1 0 0 1	14	$R_1 R_2 [1 - R_3] [1 - R_4] R_5$	1
	1 1 0 1 0	15	$R_1 R_2 [1 - R_3] R_4 [1 - R_5]$	1
	1 1 1 0 0	16	$R_1 R_2 R_3 [1 - R_4] [1 - R_5]$	0
	0 0 0 1 1	17	$[1 - R_1] [1 - R_2] [1 - R_3] R_4 R_5$	0
	0 0 1 0 1	18	$[1 - R_1] [1 - R_2] R_3 [1 - R_4] R_5$	1
	0 0 1 1 0	19	$[1 - R_1] [1 - R_2] R_3 R_4 [1 - R_5]$	0
	0 1 0 0 1	20	$[1 - R_1] R_2 [1 - R_3] [1 - R_4] R_5$	1
3	0 1 0 1 0	21	$[1 - R_1] R_2 [1 - R_3] R_4 [1 - R_5]$	1
	0 1 1 0 0	22	$[1 - R_1] R_2 R_3 [1 - R_4] [1 - R_5]$	0
	1 0 0 0 1	23	$R_1 [1 - R_2] [1 - R_3] [1 - R_4] R_5$	0
	1 0 0 1 0	24	$R_1 [1 - R_2] [1 - R_3] R_4 [1 - R_5]$	1
	1 0 1 0 0	25	$R_1 [1 - R_2] R_3 [1 - R_4] [1 - R_5]$	0
	1 1 0 0 0	26	$R_1 R_2 [1 - R_3] [1 - R_4] [1 - R_5]$	0
	0 0 0 0 1	27	$[1 - R_1] [1 - R_2] [1 - R_3] [1 - R_4] R_5$	0
	0 0 0 1 0	28	$[1 - R_1] [1 - R_2] [1 - R_3] R_4 [1 - R_5]$	0
4	0 0 1 0 0	29	$[1 - R_1] [1 - R_2] R_3 [1 - R_4] [1 - R_5]$	0
	0 1 0 0 0	30	$[1 - R_1] R_2 [1 - R_3] [1 - R_4] [1 - R_5]$	0
	1 0 0 0 0	31	$R_1 [1 - R_2] [1 - R_3] [1 - R_4] [1 - R_5]$	0
5	0 0 0 0 0	32	$[1 - R_1] [1 - R_2] [1 - R_3] [1 - R_4] [1 - R_5]$	0

7.4   Repairable Systems with Independent Components

A repairable component can be described by a two-state model, as in Figure 3.24, in which the system availability $A(t)$ is the probability of being in the up state at time t and the unavailability $U(t) = 1 - A(t)$ is the probability of being in the down state at time t. If a system is composed of independently repairable components, and each component i is characterized by a two-state model like the one of Figure 3.24, Eq. (7.14) becomes:

$$\begin{cases} P\{X_i(t) = 1\} & = & A_i(t), \\ P\{X_i(t) = 0\} & = & 1 - A_i(t). \end{cases} \qquad (7.23)$$

The state transition diagram of a system of ℓ components has $n = 2^\ell$ states, but from each state failure as well as repair arcs emanate. The system availability is then computed from Eq. (7.17). Note that the assumption of independent repair implies that if in a state there are $k > 1$ failed components, all the failed components are repaired concurrently by k independent repair persons.

Example 7.10 *A Repairable Two-Component System*
The state space of a two-component repairable system with independent repair is given in Figure 7.6. The arcs labeled *if* $(i = 1,2)$ denote failure transitions while the arcs labeled *ir* $(i = 1,2)$ denote repair transitions. The truth table is the same as any two-component system, and the system availabilities for the series connection $A_S(t)$ and the parallel connection $A_P(t)$ are obtained as in Example 7.6, and are given by:

$$A_S(t) = \sum_{s_i \in \Omega_u} \pi_i(t) = \pi_1(t) = A_1(t) A_2(t),$$

$$A_P(t) = \sum_{s_i \in \Omega_u} \pi_i(t) = \pi_1(t) + \pi_2(t) + \pi_3(t)$$

$$= A_1(t) A_2(t) + [1 - A_1(t)] A_2(t) + A_1(t)[1 - A_2(t)]$$

$$= A_1(t) + A_2(t) - A_1(t) A_2(t).$$

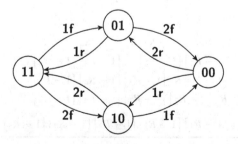

Figure 7.6   State space of a repairable two-component system.

Problems

7.1 Consider the system redundancy and the component redundancy configurations of Figure 4.13. Derive the truth table and the CDNF in the two cases.

7.2 Using the rules of Boolean algebra in Section 6.12, simplify the CDNF in Problem 7.1 to obtain a minimal form as a disjunction of its minpaths. Derive the reliability expression from the minimal form and verify that the results are equal to those obtained in Section 4.2.8.

7.3 For the same system redundancy and component redundancy configurations of Figure 4.13, derive the CDNF when the structure function is $\varphi(X) = 0$. Show that in this case the structure function expresses the system failure as a function of its mincuts.

7.4 With the state enumeration technique, compute the reliability of a three-out-of-five system using the data of Problem 6.13.

7.5 With the state enumeration technique, compute the reliability of a three-out-of-five system with two groups of components using the data of Problem 6.14.

7.6 Consider the multiprocessor system of Problem 4.17 and compute the system reliability using the state enumeration method.

7.7 The structural importance index defined in Section 7.1.2 is based on the exploration of the state space. Find the structural importance index for the components of the system of Example 7.9 (Figure 4.6).

References

[1] M. Shooman, *Probabilistic Reliability: An Engineering Approach*. McGraw Hill, 1968.

[2] R. Barlow and F. Proschan, *Statistical Theory of Reliability and Life Testing*. Holt, Rinehart and Winston, 1975.

[3] A. Kaufmann, D. Grouchko, and R. Cruon, *Mathematical Models for the Study of the Reliability of Systems*. Academic Press, 1977.

[4] W. Schneeweiss, *Boolean Functions with Engineering Applications and Computer Programs*. Springer Verlag, 1989.

8   Dynamic Redundancy

The type of redundancy that we have studied so far does not require the system to take any action upon the failure of a component. This approach is often referred to as *static* redundancy. Alternatively, we can expect the *fault management* system to take overt action dynamically to deal with component failure. We refer to this as *dynamic* redundancy. A simple example will illustrate the point. Cars usually carry a spare tire in their trunk to deal with the occurrence of a flat tire. Upon the failure of one of the four active tires, a series of steps are needed to deal with the failure: (1) detection of the failure, (2) location of the failure, (3) raising the car using a car jack to facilitate the replacement of the flat tire, (4) removal of the failed tire, and finally (5) mounting the spare tire which will now become an active tire. The failed tire will have to be repaired and take the place of the spare tire in the trunk. Such standby sparing systems are ubiquitous. The reliability analysis of such systems is the objective of this chapter.

Modeling the dynamic redundancy behavior requires using a random variable to represent each step of dealing with the failure. The overall system behavior will then be expressed as the sum of the random variables. The techniques that we examine in this chapter are primarily the sums of exponentially distributed random variables and mixtures of random variables. Note that the Markov chain modeling approach, considered at length in Part III, can be used for such reliability studies, but the simpler methods outlined in this chapter can be used for non-repairable standby sparing systems with exponential distributed component lifetimes. This chapter may be seen as a bridge between the non-state-space methods presented in the second part of the book and the state-space methods to be examined in Part III.

Standby redundancy is commonly of three types: cold standby, warm standby, and hot standby. In the first case, a component with a spare status is assumed not to fail, while in the other two cases the spare can fail even when not being actively used. In some situations it may be reasonable to assume that if the spare is not actually in use, its failure rate is lower than that of the active unit. In this case, the spare is said to be a warm standby. Otherwise the spare fails at the same rate as the active unit and hence is considered a hot spare, and this is considered a hot standby redundancy.

8.1   Cold Standby Case

Consider a system with one active unit and a cold standby spare. We denote this structure as (CS-2). As a concrete instance, when I go out of town I take a spare battery for my hearing aid, which has one active battery. I wish to study the probability that my hearing aid functions throughout my two-week trip. Through experience I have found that the average lifetime of a battery can be assumed to be one week.

Assume that the time to failure of an active unit is exponentially distributed with rate λ. It is clear that the overall time to failure of the two-component standby system is the sum of the two random variables, since after the failure of the original active component, the spare replaces the failed active component and hence becomes active. If we assume that the time to failure of the newly active component is independent and identically distributed as the original active component, the overall lifetime is a two-stage Erlang random variable with parameter λ (see Section 3.2.5). If we let the TTF random variable for the originally active unit be denoted by X_1 and that of the spare after it becomes active by X_2, the overall lifetime is $Z = X_1 + X_2$, as shown in Figure 8.1.

We know from probability theory that the density function of a sum of random variables is given by the convolution integral of the densities of the random variables forming the sum. We use this property here to determine the density function of Z. Mindful of the fact that X_1, X_2 are both non-negative i.i.d. EXP(λ) random variables, the convolution integral becomes:

$$f_Z(t) = \int_0^t f_{X_1}(x) f_{X_2}(t-x) dx$$

$$= \int_0^t \lambda e^{-\lambda x} \lambda e^{-\lambda(t-x)} dx$$

$$= \lambda^2 t e^{-\lambda t}.$$

Hence, the reliability $R_{CS\text{-}2}(t)$ is given by

$$R_{CS\text{-}2}(t) = \int_t^\infty f_Z(x) dx = (1 + \lambda t) e^{-\lambda t}. \tag{8.1}$$

Also,

$$\text{MTTF}_{CS\text{-}2} = \frac{2}{\lambda}.$$

Note that in this derivation the detection of the failed component and the switching to the spare takes no time and is assumed to be perfect.

Since in this example we have $\frac{1}{\lambda} = 7$ days, the probability that I will have a working hearing aid over a trip of two weeks is given by

$$R_{CS\text{-}2}(14) = (1 + \frac{14}{7}) e^{-\frac{14}{7}} = 3e^{-2} = 0.406.$$

I can certainly increase the reliability by carrying multiple spare batteries.

Figure 8.1   A cold standby system with one equal spare.

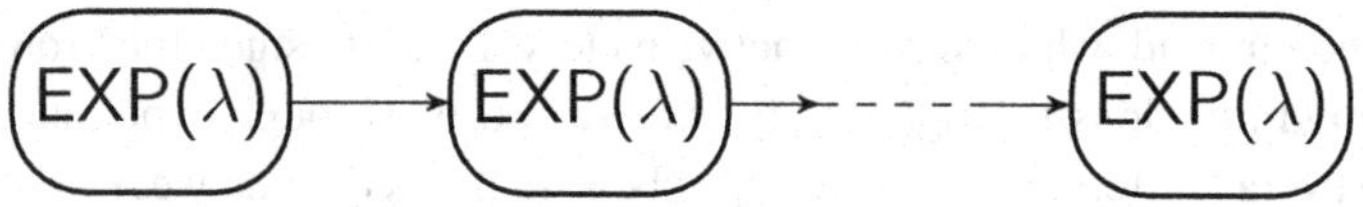

Figure 8.2   A cold standby system with $n-1$ equal spares.

The above formula is easily generalized to the case where there are $(n-1)$ cold spares, as represented in Figure 8.2.

For this n-component cold standby system, the reliability is n-stage Erlang and can be written [see (3.45)] as:

$$R_{\text{CS-}n}(t) = \sum_{k=0}^{n-1} \frac{(\lambda t)^k}{k!} e^{-\lambda t},$$

$$\text{MTTF}_{\text{CS-}n} = \frac{n}{\lambda}.$$

It is easy to generalize to the case where the spare will have a different failure rate with respect to the initially active component. Thus, assume that the initially active component has a failure rate λ_1, and when the spare is switched in its failure rate is λ_2. Now, in this case the overall lifetime of the system is still the sum of two independent exponentially distributed random variables. However, unlike the previous case, the two random variables have different rate parameters. Hence the convolution will result in a two-stage hypoexponential (see Section 3.2.5) rather than an Erlang overall lifetime:

$$f_Z(t) = \int_0^t f_{X_1}(x) f_{X_2}(t-x) dx$$

$$= \int_0^t \lambda_1 e^{-\lambda_1 x} \lambda_2 e^{-\lambda_2(t-x)} dx$$

$$= \frac{\lambda_1 \lambda_2}{\lambda_1 - \lambda_2} (e^{-\lambda_2 t} - e^{-\lambda_1 t}), \tag{8.2}$$

and

$$R_{\text{CS-}2}(t) = \int_t^\infty f_Z(u) du = \frac{\lambda_1}{\lambda_1 - \lambda_2} e^{-\lambda_2 t} - \frac{\lambda_2}{\lambda_1 - \lambda_2} e^{-\lambda_1 t},$$

$$\text{MTTF}_{\text{CS-}2} = \frac{1}{\lambda_1} + \frac{1}{\lambda_2}. \tag{8.3}$$

This can clearly be generalized to the case of $(n-1)$ cold spares and one active component. The distribution of system time to failure is an n-stage hypoexponential (see Section 3.2.5). Hence,

$$R_{\text{CS-}n}(t) = \int_t^\infty f_Z(u) du,$$

where $f_Z(u) = \sum_{i=1}^{n} a_i \lambda_i e^{-\lambda_i u}$ and $a_i = \prod_{\substack{j=1 \\ i \neq j}}^{k} \frac{\lambda_j}{\lambda_j - \lambda_i}$, $1 \leq i \leq n$, and

$$\text{MTTF}_{\text{CS-}n} = \sum_{i=1}^{n} \frac{1}{\lambda_i}.$$

Further generalization is possible where we relax the restriction of an exponential lifetime distribution of each component. So long as independence still holds, we can apply the convolution formula.

Problems

8.1   For the hearing aid example, with $1/\lambda = 7\,\text{d}$, determine the minimum number of spare batteries I have to carry in order to have the probability of a working hearing aid for a two-week trip greater than 0.9.

8.2   Consider a two-component cold standby system in which the component i TTF has a mass at origin, so that

$$F_{X_i}(t) = q_i + (1 - q_i)(1 - e^{-\lambda_i t}), \quad t \geq 0.$$

Thus, q_i is the probability that the component is in a failed state at the time it is put in active mode. Now determine an expression for the system reliability and system MTTF.

8.3   Consider again a two-component cold standby system such that both components have identically distributed TTFs (when active) given by the two-stage hypoexponential HYPO(λ_1, λ_2) (see Section 3.2.5). Now determine an expression for the system reliability and system MTTF.

8.4   A component A has two spares B and C in cold standby that are activated in sequence. When A fails, B is switched on; when B fails, C is switched on. The three components have exponentially distributed TTFs with failure rates α, β, and γ, respectively. Find the system reliability and the MTTF in the following cases (using Laplace transforms):

(a) $\alpha \neq \beta \neq \gamma$
(b) $\alpha = \beta \neq \gamma$
(c) $\alpha = \beta = \gamma$

8.2   Warm Standby

Quite often, it is unreasonable to assume that a component in standby status will never fail. For example, when a car tire develops a flat, the spare tire in the trunk may on occasion also be found to be flat. To capture such scenarios, we will assume that a component in standby status has a failure rate that is likely much smaller than the failure rate λ of the same component while in an active state. We introduce a *dormancy factor* α (with $0 \leq \alpha \leq 1$), such that $\alpha \lambda$ is the failure rate of the component when dormant. With $\alpha = 0$ we return to the cold standby case of the previous section.

As an example, consider the tire subsystem of an automobile where the failure rate of an individual tire is λ while it is in an active state and it is $\alpha\lambda$ when in spare status. Assume for now that the detection and switching actions are instantaneous and perfect. To evaluate the overall time to failure of the tire subsystem we consider two cases: (1) one of the active tires fails first, is replaced by the spare and then a second tire fails; (2) the spare tire fails first, and when an active tire fails, there are no more available spares and the tire subsystem fails.

In both case (1) and case (2) the time to failure of the tire subsystem can be divided into two sequential phases. In the first phase either one of the four active tires or the spare fails. Consider case (1). This first phase duration, say Y_1, can be seen as $Y_1 = min\{X_1,X_2,X_3,X_4,W\}$, where $X_1,\ldots,X_4$ are the time to failure random variables for the four initially active tires while W is the TTF of the spare tire. By our assumptions, X_i are i.i.d. EXP(λ) and W is EXP($\alpha\lambda$). Furthermore, W is independent of each X_i. Then, from a result of Section 3.3.1, we know that $Y_1 \sim$ EXP($4\lambda + \alpha\lambda$) = EXP($(4+\alpha)\lambda$).

If one of the four active tires has failed, the spare will be switched in, its failure rate will thence be λ. Now the tire subsystem will fail when one of the four active tires fail, as there is no longer a working spare. The remaining time to failure of each of the three non-failed tires is EXP(λ), due to the memoryless property of the exponential distribution, and thus the duration of the second phase, say Y_2, of the overall TTF is the minimum of four independent EXP(λ) random variables. Hence, we know from Section 3.3.1 that $Y_2 \sim$ EXP(4λ).

In case (2), if the spare is the first one to fail, the remaining TTF of each of the active tires is EXP(λ) and, once again, we can conclude that $Y_2 \sim$ EXP(4λ). Thus the overall TTF for the tire subsystem is represented in Figure 8.3 and is given by

$$Y = Y_1 + Y_2.$$

Since Y_1 and Y_2 are independent, Y is a two-stage hypoexponentially distributed random variable HYPO($(4+\alpha)\lambda, 4\lambda$). Plugging into Eqs. (8.2) and (8.3), we get

$$f_{\text{Tire}}(t) = \frac{4(4+\alpha)\lambda}{\alpha}[e^{-4\lambda t} - e^{-(4+\alpha)\lambda t}]$$

and

$$R_{\text{Tire}}(t) = \frac{(4+\alpha)}{\alpha}e^{-4\lambda t} - \frac{4}{\alpha}e^{-(4+\alpha)\lambda t}.$$

Also,

$$\text{MTTF}_{\text{Tire}} = \frac{1}{(4+\alpha)\lambda} + \frac{1}{4\lambda}.$$

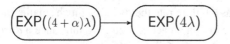

Figure 8.3   A warm standby representation of the tire system with one spare.

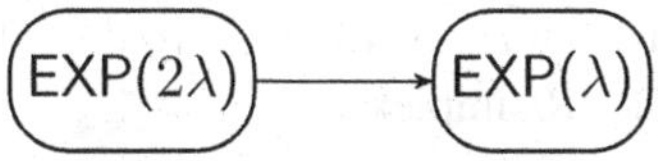

Figure 8.4   Two-unit hot standby TTF (two-component parallel system).

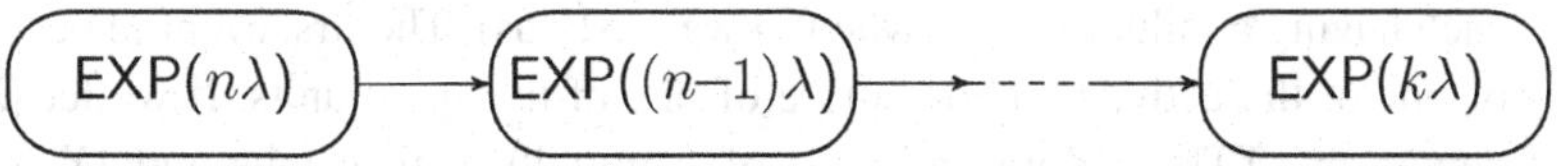

Figure 8.5   *k*-out-of-*n* system TTF as a hypoexponential distribution.

8.3   Hot Standby and *k*-out-of-*n*

This case has been covered in Chapter 4, on reliability block diagrams, and in Chapter 6 on fault trees, where we considered parallel redundant and *k*-out-of-*n* systems. However, it is instructive to reconsider these cases as a special case of hypoexponentially distributed time to failure.

First, consider a two-component parallel redundant system, which can also be viewed as a two-component hot standby (HS) system. Since both the active and the standby units are assumed to have an $EXP(\lambda)$ distribution for TTF, the first event to occur is the failure of one of these two units, whose distribution is the minimum of two independent $EXP(\lambda)$ random variables. After the first failure, the system continues to function and the remaining time to failure is $EXP(\lambda)$, due to the memoryless property of the exponential distribution. Hence, the overall time to failure is seen as a $HYPO(2\lambda, \lambda)$ random variable (Figure 8.4).

From Eqs. (8.2) and (8.3), and from inspection of Figure 8.4, we get (as expected):

$$R_{HS-2}(t) = 2e^{-\lambda t} - e^{-2\lambda t}, \tag{8.4}$$

$$MTTF_{HS-2} = \frac{1}{2\lambda} + \frac{1}{\lambda}.$$

The extension to an *n*-unit hot standby system is straightforward, and is $HYPO(n\lambda, (n-1)\lambda, \ldots, \lambda)$; its MTTF is

$$MTTF_{HS-n} = \sum_{i=1}^{n} \frac{1}{i\lambda}.$$

Generalization to a *k*-out-of-*n* system is also quite easy. The time to failure distribution of such a system is seen to be $HYPO(n\lambda, (n-1)\lambda, \ldots, k\lambda)$, as in Figure 8.5.

Problems

8.5   Consider a TMR system where each component has $EXP(\lambda)$ TTF. Derive the reliability and MTTF of the TMR system using a two-stage hypoexponential $HYPO(3\lambda, 2\lambda)$ distribution.

8.6   Consider a modification of TMR where, upon the first failure, the failed and one of the non-failed components are removed from service. Such a system, known as

TMR/simplex, has a HYPO($3\lambda, \lambda$) distribution as its TTF. Using this idea, derive the reliability and MTTF expressions for the TMR/simplex.

Extension to the case of k-out-of-n with warm standby can also be carried out. Suppose we have n initially active units, each with a failure rate λ, and m warm spares each having a failure rate γ (when in warm status). The first event to occur is the failure of one of the active units or failure of one of the spare units. If we let $\{X_1, X_2, \ldots, X_n\}$ denote the TTF random variables of n initially active units and $\{Y_1, Y_2, \ldots, Y_m\}$ the TTF random variables of m initially spare units, then the time to the first event is given by $min\{X_1, X_2, \ldots, X_n, Y_1, Y_2, \ldots, Y_m\}$, which is EXP($n\lambda + m\gamma$). Once the first failure occurs, we will have n active units and $(m-1)$ spares. Due to the memoryless property of the exponential distribution, the time to the next event will be given by an EXP($n\lambda + (m-1)\gamma$) random variable. Extending this argument, the overall TTF random variable is seen to be HYPO($n\lambda + m\gamma, n\lambda + (m-1)\gamma, \ldots, n\lambda + \gamma, n\lambda, \ldots, k\lambda$), as shown in Figure 8.6.

8.4   Imperfect Fault Coverage

A seminal paper by Bouricius *et al.* [1] first defined the notion of fault coverage, also called the coverage factor (CF), as a conditional probability to account for the efficacy of fault/error-handling mechanisms (FEHMs). More formally, CF = $P\{$system recovery | fault occurs$\}$. This concept has been rapidly and widely recognized as a major concern in dependability evaluation studies. Since then, a vast amount of work has been devoted to refining the notion of coverage [2–4], and to the identification or estimation of relevant parameters [5] and associated component-level and system-level reliability models. As a result, several modeling techniques and software packages have been developed that incorporate imperfect coverage [6–10]. FEHM models will be further elaborated in later chapters (for example, see Example 10.28). Here we assume that the recovery process subsequent to the occurrence of a component failure takes place in phases and that the time spent in each phase is negligible (or not considered for now). A typical three-phase FEHM with detection, location, and recovery can be modeled as the simple series RBD of Figure 8.7. Each of the three FEHM phases is associated with a probability of success. Hence, the overall probability of successful system recovery c is given by the product of the success probabilities of the individual phases $c = c_d c_l c_r$.

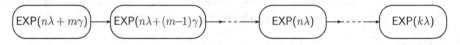

Figure 8.6   Hybrid k-out-of-$n + m$ system with n initially active units and m spares.

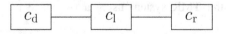

Figure 8.7   Reliability block diagram for a three-phase FEHM.

In other applications, the phase names and the number of phases may be different but the idea is still the same.

Example 8.1 Returning to our car tire example, once one of the working tires is flat, detection will be followed by stopping the car, locating and using the jack, locating and using the wrench to remove the flat tire, and mounting the spare tire. Each of these phases may not work correctly with some probability. Let the respective probabilities of correct working of these phases be given by c_d, c_j, c_w, and c_s, so that the overall coverage is $c = c_d c_j c_w c_s$.

Recall that the overall lifetime of the tire subsystem was denoted by Y. Define a random variable X with three possible values as follows:

- $X = 1$ if one of the active tires fails first and each of the set of coverage steps above works correctly so that the tire subsystem continues to function with the original spare being made active.
- $X = 0$ if an active tire fails first and one of the steps above does not work correctly so that the tire subsystem fails (due to a coverage failure).
- $X = 2$ if the spare fails first. In this case the tire subsystem continues to function until one of the active tires fails next.

It is clear that the conditional distribution of Y given $X = 1$ or $X = 2$ is HYPO$((4 + \alpha)\lambda, 4\lambda)$, as pictured in Figure 8.3, while the conditional distribution of Y given $X = 0$ is EXP$((4 + \alpha)\lambda)$. Thus the conditional LST of the time to failure Y is:

$$\mathscr{L}_{Y|X}(s \,|\, X = 1 \text{ or } 2) = \frac{(4+\alpha)\lambda}{s + (4+\alpha)\lambda} \frac{4\lambda}{s + 4\lambda},$$

$$\mathscr{L}_{Y|X}(s \,|\, X = 0) = \frac{1}{s + (4+\alpha)\lambda}. \tag{8.5}$$

The conditional reliability is:

$$R(t|X = 1 \text{ or } 2) = \frac{4+\alpha}{\alpha} e^{-4\lambda t} - \frac{4}{\alpha} e^{-(4+\alpha)\lambda t},$$

$$R(t|X = 0) = e^{-(4+\alpha)\lambda t},$$

and the conditional MTTF is:

$$E[Y|X = 1 \text{ or } 2] = \frac{1}{(4+\alpha)\lambda} + \frac{1}{4\lambda},$$

$$E[Y|X = 0] = \frac{1}{(4+\alpha)\lambda}.$$

Now, the pmf of the discrete random variable X is easily derived:

$$p_X(1) = \frac{4\lambda c}{(4+\alpha)\lambda} = \frac{4c}{(4+\alpha)},$$

$$p_X(2) = \frac{\alpha\lambda}{(4+\alpha)\lambda} = \frac{\alpha}{(4+\alpha)},$$

$$p_X(0) = \frac{4\lambda(1-c)}{(4+\alpha)\lambda} = \frac{4(1-c)}{(4+\alpha)}.$$

Hence, the unconditional LST, reliability, and MTTF are obtained using the theorem of total probability:

$$\mathcal{L}_Y(s) = \frac{\lambda(4c+\alpha)}{(4+\alpha)\lambda}\frac{(4+\alpha)\lambda}{s+(4+\alpha)\lambda}\frac{4\lambda}{s+4\lambda} + \frac{4\lambda(1-c)}{(4+\alpha)\lambda}\frac{1}{s+(4+\alpha)\lambda},$$

$$R_Y(t) = \frac{4c+\alpha}{4+\alpha}\left[\frac{4+\alpha}{\alpha}e^{-4\lambda t} - \frac{4}{\alpha}e^{-(4+\alpha)\lambda t}\right] + \frac{4(1-c)}{(4+\alpha)}e^{-(4+\alpha)\lambda t}$$

$$= \frac{4c+\alpha}{\alpha}e^{-4\lambda t} - \frac{4c}{\alpha}e^{-(4+\alpha)\lambda t}, \tag{8.6}$$

$$E[Y] = \frac{4c+\alpha}{4+\alpha}\left[\frac{1}{(4+\alpha)\lambda} + \frac{1}{4\lambda}\right] + \frac{4(1-c)}{4+\alpha}\frac{1}{(4+\alpha)\lambda}$$

$$= \frac{1}{(4+\alpha)\lambda} + \frac{4c+\alpha}{4+\alpha}\frac{1}{4\lambda}.$$

Example 8.2 Consider the safety of a skydiver where the lifetime distributions of the active and spare parachutes are considered to be Bernoulli [11]. Let the reliability of the primary parachute be R_1 and that of the secondary (spare) parachute be R_2. In the case that the primary fails to operate, the skydiver will try to use the secondary parachute. The skydiver will be successful in cutting off the primary parachute and opening the secondary with probability c, while with probability $1-c$ this operation may not be successful. Then the overall probability of success is:

$$R_p = R_1 + (1-R_1)cR_2.$$

The sensitivities of the two components are:

$$I_1 = \frac{\partial R_p}{\partial R_1} = 1 - cR_2,$$

$$I_2 = \frac{\partial R_p}{\partial R_2} = c - cR_1.$$

If we assume that the per unit cost of improving the primary and the secondary chutes is the same, it is optimal to have: $\frac{dR_p}{dR_1} = \frac{dR_p}{dR_2}$ or $1 - cR_2 = c - cR_1$ or $R_2 = R_1 + \frac{1-c}{c}$. Since reliability is a probability, we have the expression:

$$R_2 = \min\left\{R_1 + \frac{1-c}{c}, 1\right\}. \tag{8.7}$$

So far we have assumed that each of the steps in an FEHM takes zero time to execute. In practice, each step takes some time. Consider a single-stage FEHM for the sake of convenience. Let D be the amount of time taken by this stage. Assume also that if allowed to complete, the stage will always succeed. Often, however, there may be an upper limit on the time allowed for this stage as the system is down during that period. This could be the case if the equipment under consideration is for the healthcare of an individual. Thus if τ is the upper bound on the time allowed to complete the activity with FEHM then the probability of successful completion, or the coverage, is now given by:

$$c = F_D(\tau) = P\{D \le \tau\}.$$

With a complementary probability the system experiences a coverage failure. Yet another possibility considered in the literature is that of a near-coincident (or nearly concurrent) fault. Suppose that the system is in the process of handling a component failure and before it completes its actions, another component in the system fails. In many systems, such a near-coincident fault will bring the system down [2, 9]. Assume that the time to occurrence of a near-coincident fault is a random variable X with distribution $EXP(\gamma)$. Then a coverage failure due to a near-coincident fault occurs if the time to the next fault occurrence satisfies $X \le D$. Hence, from (3.58),

$$c = P\{X > D\} = \int e^{-\gamma t} f_D(t) dt. \tag{8.8}$$

Thus if D is deterministic with a value d, $c = e^{-\gamma d}$, while if D is $EXP(\delta)$ then $c = \frac{\delta}{\delta + \gamma}$. These different views of coverage can clearly be combined [2].

Problems

8.7 For the tire subsystem in Example 8.1, derive expressions for the sensitivity of the MTTF to the three parameters λ, α, and c.

8.8 Some cars have a spare tire that is different from the active tires. Derive an expression for the tire subsystem reliability and MTTF assuming that the original active tire failure rate is λ while the spare tire failure rate is $\alpha\gamma$ when in spare status and γ when the spare tire is made active.

8.9 Plot the function R_2 of Eq. (8.7) as a function of c for values of $R_1 = 0.5, 0.8, 0.9$, 0.95, and 0.99.

8.10 Derive an expression for c using Eq. (8.8) in the following cases:

(a) D has a modified exponential distribution with an upper bound d_u and parameter δ;
(b) D is uniformly distributed between d_l and d_u.

8.5 Epistemic Uncertainty Propagation

The reliability and related results for the different examples above were obtained assuming that the failure rates of the components and other parameters such as the coverage probability have a fixed and known value. In real life, all the parameter values are derived from measurements followed by statistical inference. Due to the finite sample size, the parameters have (epistemic) uncertainty associated with them and hence the results previously derived can be considered as obtained by conditioning each parameter to a specific single value. In this section we derive the unconditioned expected values for the MTTF and for the reliability in the different cases, according to the uncertainty propagation approach developed in Section 3.4 [12, 13].

8.5.1   Cold Standby with Identical Components

Given a two-component system in cold standby configuration, and assuming $f_\Lambda(\lambda)$ to be the density function of the failure rate parameter Λ (now considered to be a random variable), we may obtain the unconditional expected reliability of the system at time t resorting to the theorem of total expectation:

$$E[R_{CS\text{-}2}(t)] = \int_0^\infty R_{CS\text{-}2}(t|\lambda)f_\Lambda(\lambda)d\lambda, \qquad (8.9)$$

where $R_{CS\text{-}2}(t|\lambda)$ is the conditional reliability of a two-component system for a fixed value of the rate parameter equal to λ, as obtained in Eq. (8.1).

Since the TTF of the component is exponentially distributed, we know from Section 3.4 that the failure rate will be r-stage Erlang distributed with rate parameter s, where r is the number of observations or sample size in the lifetime experiments for estimating λ, and s is the observed total life during the test. Therefore,

$$f_\Lambda(\lambda) = \underbrace{\frac{\lambda^{r-1}s^r e^{-\lambda s}}{(r-1)!}}_{\text{Erlang pdf}} .$$

Hence, the expected reliability of the two-component system at time t is given by:

$$
\begin{aligned}
E[R_{CS\text{-}2}(t)] &= \int_0^\infty R_{CS\text{-}2}(t|\lambda)f_\Lambda(\lambda)d\lambda \\
&= \int_0^\infty (1+\lambda t)e^{-\lambda t}\frac{\lambda^{r-1}s^r e^{-\lambda s}}{(r-1)!}d\lambda \\
&= \int_0^\infty e^{-\lambda t}\frac{\lambda^{r-1}s^r e^{-\lambda s}}{(r-1)!}d\lambda + \frac{tr}{s}\int_0^\infty e^{-\lambda t}\frac{\lambda^r s^{r+1}e^{-\lambda s}}{r!}d\lambda \\
&= \left(\frac{s}{s+t}\right)^r + \frac{tr}{s}\left(\frac{s}{s+t}\right)^{r+1} \\
&= \left(\frac{1}{1+\hat\lambda t/r}\right)^r + \hat\lambda t\left(\frac{1}{1+\hat\lambda t/r}\right)^{r+1},
\end{aligned}
$$

where $\hat\lambda = r/s$ is the maximum likelihood estimator for the rate parameter λ with r observations over a total time s. Note that in the limit as the sample size r approaches infinity, we get

$$\lim_{r\to\infty} E[R_{CS\text{-}2}(t)] = (1+\hat\lambda t)e^{-\hat\lambda t}.$$

To find the variance of the two-component system, we first find $E[R_{CS-2}(t)^2]$. Now,

$$[R_{CS-2}(t|\lambda)]^2 = ((1+\lambda t)e^{-\lambda t})^2 = (1+2\lambda t + (\lambda t)^2)e^{-2\lambda t}.$$

Therefore, $E[R_{CS-2}(t)^2]$ is given by:

$$E[R_{CS-2}(t)^2] = \int_0^\infty (1+2\lambda t + (\lambda t)^2)e^{-2\lambda t} \frac{\lambda^{r-1}s^r e^{-\lambda s}}{(r-1)!} d\lambda$$

$$= \int_0^\infty e^{-2\lambda t} \frac{\lambda^{r-1}s^r e^{-\lambda s}}{(r-1)!} d\lambda + \frac{2tr}{s}\int_0^\infty e^{-2\lambda t} \frac{\lambda^r s^{r+1} e^{-\lambda s}}{r!} d\lambda$$

$$+ \frac{t^2 r(r+1)}{s^2} \int_0^\infty e^{-2\lambda t} \frac{\lambda^{r+1}s^{r+2}e^{-\lambda s}}{(r+1)!} d\lambda$$

$$= \left(\frac{s}{s+2t}\right)^r + \frac{2tr}{s}\left(\frac{s}{s+2t}\right)^{r+1} + \frac{t^2 r(r+1)}{s^2}\left(\frac{s}{s+2t}\right)^{r+2}$$

$$= \left(\frac{1}{1+2\hat{\lambda}t/r}\right)^r + 2\hat{\lambda}t\left(\frac{1}{1+2\hat{\lambda}t/r}\right)^{r+1} + \hat{\lambda}t^2\left(\hat{\lambda}+\frac{\hat{\lambda}}{r}\right)\left(\frac{1}{1+2\hat{\lambda}t/r}\right)^{r+2}.$$

Finally,

$$\text{Var}[R_{CS-2}(t)] = E[R_{CS-2}(t)^2] - E[R_{CS-2}(t)]^2.$$

To find the unconditional expected reliability of an n-component system, we follow the same approach. We first consider the unconditional expectation for the probability $P_k(t) = \frac{(\lambda t)^k}{k!}e^{-\lambda t}$, adapting the relation (8.9):

$$E[P_k(t)] = \int_0^\infty \left(\frac{(\lambda t)^k}{k!}e^{-\lambda t}\right)\frac{\lambda^{r-1}s^r e^{-\lambda s}}{(r-1)!} d\lambda$$

$$= \frac{t^k (r+k-1)!}{s^k(r-1)!(k)!}\int_0^\infty e^{-\lambda t}\frac{\lambda^{r+k-1}s^{r+k}e^{-\lambda s}}{(r+k-1)!} d\lambda$$

$$= \binom{r+k-1}{k}\left(\frac{t}{s}\right)^k \left(\frac{s}{s+t}\right)^{r+k}$$

$$= \binom{r+k-1}{k}\left(\frac{\hat{\lambda}t}{r}\right)^k \left(\frac{1}{1+\hat{\lambda}t/r}\right)^{r+k}.$$

Since $R_{CS-n}(t|\lambda) = \sum_{k=0}^{n-1}P_k(t|\lambda)$, the unconditional expected reliability of $R_{CS-n}(t)$ can be derived as follows:

$$E[R_{CS-n}(t)] = \sum_{k=0}^{n-1}\binom{r+k-1}{k}\left(\frac{t}{s}\right)^k \left(\frac{s}{s+t}\right)^{r+k}$$

$$= \sum_{k=0}^{n-1}\binom{r+k-1}{k}\left(\frac{\hat{\lambda}t}{r}\right)^k \left(\frac{1}{1+\hat{\lambda}t/r}\right)^{r+k}.$$

8.5.2   Warm Standby

To find the unconditional expected reliability of the warm standby case, we refer to the car tire example of Section 8.2, where λ is the value of the failure rate of an active unit while the spare unit failure rate is $\gamma = \alpha\lambda$.

To compute the unconditional expected reliability we again use the equation

$$E[R_{\text{Tire}}(t)] = \int_0^\infty R_{\text{Tire}}(t|\lambda,\gamma)f(\lambda,\gamma).$$

Here, $f(\lambda,\gamma)$ is the joint density function of the two failure rate parameters Λ and Γ. Assume that these two random variables are independent (likely because the lifetime experiments to determine these parameters were carried out independently); we get:

$$f(\lambda,\gamma) = \left(\frac{\lambda^{r-1}s^r e^{-\lambda s}}{(r-1)!}\right)\left(\frac{\gamma^{b-1}g^b e^{-\gamma g}}{(b-1)!}\right),$$

where b is the number of observed failures and g is the total lifetime observed in the test for estimating γ. Therefore, the expected reliability is given by:

$$E[R_{\text{Tire}}(t)] = \int_0^\infty \int_0^\infty R_{\text{Tire}}(t|\lambda,\gamma)f(\lambda,\gamma)d\lambda d\gamma$$

$$= \int_0^\infty \int_0^\infty \left(\frac{4\lambda+\gamma}{\gamma}e^{-4\lambda t} - \frac{4\lambda}{\gamma}e^{-(4\lambda+\gamma)t}\right)\left(\frac{\lambda^{r-1}s^r e^{-\lambda s}}{(r-1)!}\right)\left(\frac{\gamma^{b-1}g^b e^{-\gamma g}}{(b-1)!}\right)d\lambda d\gamma.$$

The above integral can be decomposed into three separate integrals as follows:

$$E[R_{\text{Tire}}(t)] = I_1 + I_2 + I_3,$$

$$I_1 = \int_0^\infty \int_0^\infty e^{-4\lambda t}\left(\frac{\lambda^{r-1}s^r e^{-\lambda s}}{(r-1)!}\right)\left(\frac{\gamma^{b-1}g^b e^{-\gamma g}}{(b-1)!}\right)d\lambda d\gamma$$

$$= \left(\frac{s}{s+4t}\right)^r,$$

$$I_2 = \int_0^\infty \int_0^\infty \frac{4\lambda}{\gamma}e^{-4\lambda t}\left(\frac{\lambda^{r-1}s^r e^{-\lambda s}}{(r-1)!}\right)\left(\frac{\gamma^{b-1}g^b e^{-\gamma g}}{(b-1)!}\right)d\lambda d\gamma$$

$$= \frac{4gr}{(b-1)s}\left(\frac{s}{s+4t}\right)^{r+1},$$

$$I_3 = \int_0^\infty \int_0^\infty \frac{-4\lambda}{\gamma}e^{-(4\lambda+\gamma)t}\left(\frac{\lambda^{r-1}s^r e^{-\lambda s}}{(r-1)!}\right)\left(\frac{\gamma^{b-1}g^b e^{-\gamma g}}{(b-1)!}\right)d\lambda d\lambda$$

$$= \frac{-4gr}{(b-1)s}\left(\frac{s}{s+4t}\right)^{r+1}\left(\frac{g}{g+t}\right)^{b-1}.$$

Hence, the sum of all three integrals gives us the expected reliability:

$$E[R_{\text{Tire}}(t)] = I_1 + I_2 + I_3$$

$$= \left(\frac{s}{s+4t}\right)^r + \frac{4gr}{(b-1)s}\left(\frac{s}{s+4t}\right)^{r+1} - \frac{4gr}{(b-1)s}\left(\frac{s}{s+4t}\right)^{r+1}\left(\frac{g}{g+t}\right)^{b-1}$$

$$= \left(\frac{1}{1+4\hat{\lambda}t/r}\right)^r + \frac{4\hat{\lambda}}{(\hat{\gamma}-\hat{\gamma}/b)}\left(\frac{1}{1+4\hat{\lambda}t/r}\right)^{r+1}$$

$$- \frac{4\hat{\lambda}}{(\hat{\gamma}-\hat{\gamma}/b)}\left(\frac{1}{1+4\hat{\lambda}t/r}\right)^{r+1}\left(\frac{1}{1+\hat{\gamma}t/b}\right)^{b-1}.$$

8.5.3 Hot Standby and k-out-of-n

For the two-component parallel redundant system, the conditional reliability $R_{\text{HS-2}}(t|\lambda)$ is derived from (8.4) as

$$R_{\text{HS-2}}(t|\lambda) = 2e^{-\lambda t} - e^{-2\lambda t}.$$

The unconditional expected reliability can be found using the linearity property of expectation as follows:

$$E[R_{\text{HS-2}}(t)] = E[2e^{-\Lambda t} - e^{-2\Lambda t}]$$

$$= 2E[e^{-\Lambda t}] - E[e^{-2\Lambda t}]$$

$$= 2\left(\frac{s}{s+t}\right)^r - \left(\frac{s}{s+2t}\right)^r$$

$$= 2\left(\frac{1}{1+\hat{\lambda}t/r}\right)^r - \left(\frac{1}{1+2\hat{\lambda}t/r}\right)^r.$$

Similarly, the variance can be found as follows:

$$\text{Var}[R_{\text{HS-2}}(t)] = \text{Var}[2e^{-\Lambda t} - e^{-2\Lambda t}]$$

$$= 4\text{Var}[e^{-\Lambda t}] + \text{Var}[e^{-2\Lambda t}]$$

$$= 4\left(\left(\frac{s}{s+2t}\right)^r - \left(\frac{s}{s+t}\right)^{2r}\right) + \left(\left(\frac{s}{s+4t}\right)^r - \left(\frac{s}{s+2t}\right)^{2r}\right)$$

$$= 4\left(\left(\frac{1}{s+2\hat{\lambda}t/r}\right)^r - \left(\frac{1}{1+\hat{\lambda}t/r}\right)^{2r}\right)$$

$$+ \left(\left(\frac{1}{s+4\hat{\lambda}t/r}\right)^r - \left(\frac{1}{1+2\hat{\lambda}t/r}\right)^{2r}\right).$$

For a k-out-of-n system, we know that $R_{k\text{-of-}n}(t|\lambda) = \sum_{i=k}^{n}\binom{n}{i}\binom{i-1}{k-1}(-1)^{i-k}e^{-i\lambda t}$ (see Eq. (3.78) in [14]). To find the expected reliability, we again use the linearity property of expectation. Therefore, the expected reliability of the system is:

$$E[R_{k\text{-of-}n}(t)] = \sum_{i=k}^{n} \binom{n}{i}\binom{i-1}{k-1}(-1)^{i-k}\left(\frac{s}{s+it}\right)^r$$

$$= \sum_{i=k}^{n} \binom{n}{i}\binom{i-1}{k-1}(-1)^{i-k}\left(\frac{1}{1+i\hat{\lambda}t/r}\right)^r.$$

Similarly, the variance can be calculated as follows:

$$\text{Var}[R_{k\text{-of-}n}(t)] = \sum_{i=k}^{n}\left(\binom{n}{i}\binom{i-1}{k-1}(-1)^{i-k}\right)^2 \text{Var}[e^{-i\Lambda t}]$$

$$= \sum_{i=k}^{n}\left(\binom{n}{i}\binom{i-1}{k-1}\right)^2$$

$$\left(\left(\frac{1}{1+2i\hat{\lambda}t/r}\right)^r - \left(\frac{1}{1+i\hat{\lambda}t/r}\right)^{2r}\right).$$

References

[1] W. Bouricius, W. Carter, D. Jessep, P. Schneider, and A. Wadia, "Reliability modeling for fault-tolerant computers," *IEEE Transactions on Computers*, vol. C-20, pp. 1306–1311, 1971.

[2] J. Bechta Dugan, and K. Trivedi, "Coverage modeling for dependability analysis of fault-tolerant systems," *IEEE Transactions on Computers*, vol. 38, no. 6, pp. 775–787, Jun. 1989.

[3] H. Amer and E. McCluskey, "Calculation of coverage parameters," *IEEE Transactions on Reliability*, vol. R-36, pp. 194–198, 1987.

[4] S. Amari, J. Dugan, and R. Misra, "A separable method for incorporating imperfect fault-coverage into combinatorial models," *IEEE Transactions on Reliability*, vol. 48, pp. 267–274, 1999.

[5] M. Hsueh, T. Tsai, and R. Iyer, "Fault injection techniques and tools," *IEEE Computer*, vol. 30, no. 4, pp. 75–82, Apr. 1997.

[6] K. Trivedi and R. Geist, *A Tutorial on the CARE III Approach to Reliability Modeling*, NASA Contractor Report 3488. National Aeronautics and Space Administration, Scientific and Technical Information Branch, 1981.

[7] S. Bavuso, "A user's view of CARE III," in *Proc. Ann. Symp. on Reliability and Maintainability*, 1984, pp. 382–389.

[8] S. Makam and A. Avizienis, "ARIES 81: A reliability and life-cycle evaluation tool for fault-tolerant systems," in *Proc. 12th Int. Symp. on Fault-Tolerant Computing*, 1982, pp. 267–274.

[9] S. J. Bavuso, J. Bechta Dugan, K. Trivedi, E. M. Rothmann, and W. E. Smith, "Analysis of typical fault-tolerant architectures using HARP," *IEEE Transactions on Reliability*, vol. R-36, no. 2, pp. 176–185, Jun. 1987.

[10] R. Sahner, K. Trivedi, and A. Puliafito, *Performance and Reliability Analysis of Computer Systems: An Example-based Approach Using the SHARPE Software Package*. Kluwer Academic Publishers, 1996.

[11] G. Hoffmann, Private communication.

[12] K. Mishra and K. S. Trivedi, "Uncertainty propagation through software dependability models," in *IEEE 22nd Int. Symp. on Software Reliability Engineering*, 2011, pp. 80–89.

[13] K. Mishra and K. S. Trivedi, "Closed-form approach for epistemic uncertainty propagation in analytic models," in *Stochastic Reliability and Maintenance Modeling*, vol. 9. Springer, 2013, pp. 315–332.

[14] K. Trivedi, *Probability and Statistics with Reliability, Queueing and Computer Science Applications*, 2nd edn. John Wiley & Sons, 2001.

Part III

State-Space Models with Exponential Distributions

Consider the scenario of a typical data center which contains many server racks, each of which holds multiple server blades. With the passage of time, the server blades experience failures and need to be repaired or replaced. Typically the data center may employ a service person to go round the data center replacing the failed blades. The failures of the servers can be considered as independent of each other. However, the replacement of the servers by the (shared) repair person introduces statistical dependence in the failure/repair behavior of the servers.

Many of the non-state-space techniques presented in Part II of this book can compute the reliability or availability of a system under the assumption of statistical independence among the components. On the other hand, if the components behave in a statistically dependent manner, e.g., if the failure or repair of any one of them is dependent on the state of other components, like in the example scenario above, more sophisticated paradigms (or model types) are necessary; these model types should be able to incorporate the possible conditional dependence of each component on the state of the other components.

The continuous-time Markov chain (CTMC) is the most widely used model type that is capable of handling this dependence. This versatile modeling formalism can easily represent dependencies among components, while lending itself to a solution approach based on constructing mathematical equations that describe its dynamic behavior as a function of time or in steady state. Furthermore, various methods are available for computing a closed-form or numerical solution of the model.

State-space-based models like CTMC present significant flexibility in modeling system behavior where statistical independence among the components cannot be assumed. However, state-space-based models suffer from two major drawbacks, namely, model *largeness* and model *stiffness*.

Model largeness arises due to the need to explicitly represent the complex dependencies that arise in modeling real systems. The resulting state-space explosion often leads to very large and complex models. Manual generation of the large state space is often an error-prone process. Thus a higher-level formalism that enables concise specification of such complex system behavior in terms of component-level behavior and their interdependencies followed by automatic generation of the state space becomes essential. One such *largeness tolerance* technique based on the Petri net formalism, extended with the ability to represent the system evolution as a function of time, will be delineated in this part of the book.

Part III will cover the following areas:

- continuous-time Markov chain: availability models (in Chapter 9);
- continuous-time Markov chain: reliability models (in Chapter 10);
- continuous-time Markov chain: queuing systems (in Chapter 11);
- stochastic Petri nets (in Chapter 12).

All the above techniques are characterized by common fundamental assumptions:

Assumption 1: The system behavior can be represented as a set of discrete states.

Assumption 2: Transitions among the states can take an arbitrary amount of time. The distribution of the transition time is always exponential.

9 Continuous-Time Markov Chain: Availability Models

This chapter introduces the use of the CTMC approach for the computation of availability and related measures for a system. Subsequent chapters will deal with the use of CTMC models for computing the reliability and performance of systems. For a deeper understanding of CTMC theory and applications, the reader may refer to [1–6].

The chapter is organized as follows. Section 9.1 introduces the basic concepts and the mathematical formulation for the CTMC. Then, the classification of the states of a CTMC and its limiting distribution are explained in Section 9.2. Various dependability measures defined on a CTMC are delineated in Section 9.3. The Markov reward modeling (MRM) framework is introduced in Section 9.4, and the availability measures are redefined in terms of Markov reward models in Section 9.5. In Section 9.6, the availability of the IBM BladeCenter submodels (whose FT version was discussed in Example 6.33) is examined in some detail. Subsequently, parametric sensitivity analysis is presented in Section 9.7. Finally, steady-state solution methods for CTMC are given in Section 9.8.

9.1 Introduction

A formal definition of a CTMC and the associated mathematical equations are briefly reviewed here. Let $Z(t)$ be a stochastic process defined over a discrete state space Ω of cardinality $n = |\Omega|$ (where Ω may possibly be countably infinite). $Z(t)$ is a continuous-time Markov chain if, given any ordered sequence of time instants $(0 < t_1 < t_2 < \cdots < t_m)$, the probability of being in state s_{j_m} at time t_m depends only on the state $s_{j_{m-1}}$ occupied by the CTMC at the previous instant of time t_{m-1} and not on the complete sequence of state occupancies of the system at times t_i, $(0 < i < m-1)$ [3, 4]. This can be rephrased as saying that the future evolution of the process depends only on the present state and not on the past sequence of traversed states prior to the current state. This property is usually referred to as the *Markov property* and is sometimes erroneously confused with the memoryless property (see Section 3.2.1). The distinction, and related implications, between the memoryless and Markov properties will be further discussed in Part IV.

Formally, the Markov property may be written as:

$$P\{Z(t_m) = s_{j_m} \,|\, Z(t_{m-1}) = s_{j_{m-1}}, \ldots, Z(t_1) = s_{j_1}\}$$

$$= P\{Z(t_m) = s_{j_m} \,|\, Z(t_{m-1}) = s_{j_{m-1}}\}. \tag{9.1}$$

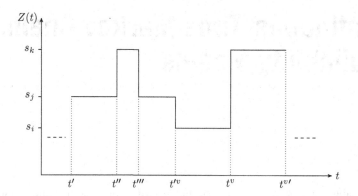

Figure 9.1   A possible trajectory of a Markov chain.

A possible trajectory of the Markov chain $Z(t)$ is represented in Figure 9.1, where s_i, s_j, and s_k are possible states of the CTMC and t', t'', t''', t^{iv}, t^v, and t^{vi} are the random instants of time at which the Markov chain $Z(t)$ jumps from one state to another.

9.1.1   Chapman–Kolmogorov Equations

Let us introduce the following notation:

$$p_{ij}(u,t) = P\{Z(t) = j \,|\, Z(u) = i\} \quad (u \le t), \tag{9.2}$$

where $p_{ij}(u,t)$ represents the conditional probability that the Markov chain $Z(t)$ is in state j at time t, given that it was in state i at time u; it is called the *transition probability* from state i to j. For the transition probabilities $p_{ij}(u,t)$, the following initial conditions hold:

$$p_{ii}(t,t) = 1, \qquad p_{ij}(t,t) = 0 \ (i \neq j). \tag{9.3}$$

Further, let $\pi_i(t)$ be the (unconditional) probability that the process $Z(t)$ is in state i at time t. $\pi_i(t)$ is the state occupancy probability, or simply the *state probability*, and is defined as:

$$\pi_i(t) = P\{Z(t) = i\}. \tag{9.4}$$

From the previous definitions, the following normalization conditions must hold:

$$\sum_{j=1}^{n} p_{ij}(u,t) = 1, \qquad \sum_{i=1}^{n} \pi_i(t) = 1. \tag{9.5}$$

Let u, t, and v be three instants of time such that $u \le t \le v$. The Markov property (9.1), combined with the theorem of total probability, leads to the following fundamental

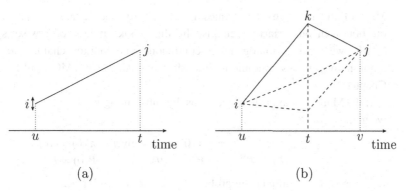

Figure 9.2 Pictorial representation of the Chapman–Kolmogorov equations. (a) State probability; (b) transition probability.

equations, called the *Chapman–Kolmogorov* (CK) equations.

$$\pi_j(t) = \sum_i \pi_i(u) \cdot p_{ij}(u,t), \tag{9.6}$$

$$p_{ij}(u,v) = \sum_k p_{ik}(u,t) \cdot p_{kj}(t,v) \quad \text{for } u \le t \le v. \tag{9.7}$$

The CK equations are represented pictorially in Figure 9.2.

It is convenient to introduce a matrix notation. Let $P(u,t) = [p_{ij}(u,t)]$ denote the square matrix of dimension $(n \times n)$ whose entries are the transition probabilities $p_{ij}(u,t)$ defined in Eq. (9.2). $P(u,t)$ is called the *transition probability matrix* of the CTMC. Let $\pi(t) = [\pi_i(t)]$ denote the row vector of dimension $(1 \times n)$ whose entries are the state probabilities $\pi_i(t)$ defined in Eq. (9.4). $\pi(t)$ is called the *(transient) state probability vector* of the CTMC. In matrix notation, the initial condition (9.3) assumes the form

$$P(t,t) = I,$$

where I is the identity matrix of appropriate dimensions, and the normalization condition (9.5) takes the form

$$P(t,t)e^T = e^T, \qquad \pi(0)e^T = 1,$$

where e is a row vector of appropriate dimension with all entries equal to one, and ^{T} means transposition.

Furthermore, in matrix notation, the CK equations (9.6) and (9.7) can be written as:

$$\begin{aligned} \pi(t) &= \pi(u) \cdot P(u,t), \\ P(u,v) &= P(u,t) \cdot P(t,v). \end{aligned} \tag{9.8}$$

A Markov chain is said to be homogeneous when the transition probabilities in matrix $P(u,v)$ depend only on the length of the time interval $(v-u)$ and not on the values of the time instants u and v. Formally, given time instants t_1 and t_2, the time homogeneity property is written as:

$$P(t_1,t_1+x) = P(t_2,t_2+x) = P(0,x). \tag{9.9}$$

Most of the modeling techniques in availability and reliability analysis are based on homogeneous Markov chains. In this book, unless otherwise specified, by CTMC we mean a homogeneous continuous-time Markov chain. The definition of a non-homogeneous continuous-time Markov chain (NHCTMC) will be dealt with in Chapter 13.

If the Markov chain is homogeneous, by substituting $u = 0$ and $\theta = v - t$ in Eqs. (9.8), we get:

$$\begin{aligned} \pi(t) &= \pi(0) \cdot P(t) \quad \text{given} \quad \pi(0) = \pi_0, \\ P(t + \theta) &= P(t) \cdot P(\theta) \qquad\qquad P(0) = I, \end{aligned} \tag{9.10}$$

where $\pi(0)$ is the initial state probability vector of the CTMC.

9.1.2   The Infinitesimal Generator Matrix

Define, for $i \neq j$ and for $\Delta t \geq 0$,

$$q_{ij} = \frac{dp_{ij}(t)}{dt}\bigg|_{t=0} = \lim_{\Delta t \to 0} \frac{p_{ij}(\Delta t) - p_{ij}(0)}{\Delta t} = \lim_{\Delta t \to 0} \frac{p_{ij}(\Delta t)}{\Delta t}. \tag{9.11}$$

From Eq. (9.11) it is easy to see that $q_{ij} \geq 0$. Similarly, for $i = j$ and for $\Delta t \geq 0$, define

$$q_{ii} = \frac{dp_{ii}(t)}{dt}\bigg|_{t=0} = \lim_{\Delta t \to 0} \frac{p_{ii}(\Delta t) - p_{ii}(0)}{\Delta t} = -\lim_{\Delta t \to 0} \frac{1 - p_{ii}(\Delta t)}{\Delta t}. \tag{9.12}$$

From Eq. (9.12) it can be seen that $q_{ii} \leq 0$. The physical interpretation of the quantities q_{ij} can be inferred by rewriting Eqs. (9.11) and (9.12) in the following form:

$$p_{ij}(\Delta t) = P\{Z(t + \Delta t) = j \,|\, Z(t) = i\} = q_{ij}\,\Delta t + o(\Delta t),$$

$$p_{ii}(\Delta t) = P\{Z(t + \Delta t) = i \,|\, Z(t) = i\} = 1 + q_{ii}\,\Delta t + o(\Delta t), \tag{9.13}$$

where $o(\Delta t)$ indicates a quantity having an order of magnitude smaller than Δt, i.e., $\lim_{\Delta t \to 0} o(\Delta t)/\Delta t = 0$. From (9.13), $q_{ij}\,\Delta t$ is the conditional probability of jumping to state j in a small interval Δt, given that the CTMC was in state i at the beginning of the interval. The quantities q_{ij} are called the *transition rates* of the CTMC.

Since the transition from state i to some state $j \in \Omega$ in the interval $(t, t + \Delta t]$ is a certain event:

$$1 = \sum_j p_{ij}(\Delta t) = 1 + q_{ii}\,\Delta t + \sum_{j:j\neq i} q_{ij}\,\Delta t.$$

Hence,

$$q_{ii} = -\sum_{j:j\neq i} q_{ij}.$$

The values q_{ij} can be grouped into the matrix $Q = [q_{ij}]$, called the *infinitesimal generator matrix* of the CTMC. The matrix Q is defined as:

$$Q = [q_{ij}] \quad \text{where} \quad \begin{aligned} q_{ij} &\geq 0 & i \neq j \\ q_{ii} &\leq 0 & q_{ii} = -\sum_{j:j\neq i} q_{ij}. \end{aligned} \tag{9.14}$$

Equation (9.14) shows that the off-diagonal entries are non-negative, the diagonal entries are non-positive and the row sum is equal to 0. Further, the off-diagonal entries of row i represent the transitions out of state i, while the off-diagonal entries of column i represent the transitions into state i. The diagonal entry of a generic state i is minus the sum of the rates outgoing from state i. Any matrix sharing the properties of Eq. (9.14) can be viewed as the infinitesimal generator of a CTMC.

9.1.3 Kolmogorov Differential Equation

In a small interval $(t, t + \Delta t]$, the CK equations can be written as:

$$
\begin{aligned}
p_{ij}(t + \Delta t) &= \sum_k p_{ik}(t) p_{kj}(\Delta t) \\
&= p_{ij}(t) p_{jj}(\Delta t) + \sum_{k:k \neq j} p_{ik}(t) p_{kj}(\Delta t) \\
&= p_{ij}(t)(1 + q_{jj}\,\Delta t) + \sum_{k:k \neq j} p_{ik}(t) q_{kj}\,\Delta t + o(\Delta t).
\end{aligned}
\tag{9.15}
$$

Rearranging Eq. (9.15), we get:

$$
\begin{aligned}
\frac{p_{ij}(t + \Delta t) - p_{ij}(t)}{\Delta t} &= p_{ij}(t) q_{jj} + \sum_{k:k \neq j} p_{ik}(t) q_{kj} + \frac{o(\Delta t)}{\Delta t} \\
&= \sum_k p_{ik}(t) q_{kj} + \frac{o(\Delta t)}{\Delta t}.
\end{aligned}
$$

Taking the limit as $\Delta t \to 0$,

$$
\frac{dp_{ij}(t)}{dt} = \sum_k p_{ik}(t) q_{kj} \quad \text{with initial condition} \quad p_{ij}(0) = \begin{cases} 1 & i = j \\ 0 & i \neq j \end{cases}.
\tag{9.16}
$$

In matrix notation, Eq. (9.16) becomes:

$$
\frac{d\boldsymbol{P}(t)}{dt} = \boldsymbol{P}(t) \cdot \boldsymbol{Q}, \qquad \boldsymbol{P}(0) = \boldsymbol{I}.
\tag{9.17}
$$

Let $\boldsymbol{\pi}(t)$ be the (transient) state probability vector at time t. Differentiating both sides of the first equation of (9.10), we have:

$$
\frac{d\boldsymbol{\pi}(t)}{dt} = \boldsymbol{\pi}(0) \cdot \frac{d\boldsymbol{P}(t)}{dt} = \boldsymbol{\pi}(0) \cdot \boldsymbol{P}(t) \cdot \boldsymbol{Q},
\tag{9.18}
$$

from which we derive the state probability equation:

$$
\frac{d\boldsymbol{\pi}(t)}{dt} = \boldsymbol{\pi}(t) \cdot \boldsymbol{Q} \quad \text{with initial condition} \quad \boldsymbol{\pi}(0) = \boldsymbol{\pi}_0.
\tag{9.19}
$$

Equation (9.19) can be expanded in one equation per state in the following form:

$$\begin{cases} \dfrac{d\pi_1(t)}{dt} = \pi_1(t)\,q_{11} + \pi_2(t)\,q_{21} + \cdots + \pi_n(t)\,q_{n1} \\ \dfrac{d\pi_2(t)}{dt} = \pi_1(t)\,q_{12} + \pi_2(t)\,q_{22} + \cdots + \pi_n(t)\,q_{n2} \\ \quad\vdots \end{cases} \tag{9.20}$$

Equations (9.17) and (9.19) are the fundamental equations for CTMC, known as Kolmogorov differential equations. The state probability equation (9.19) consists of a set of n first-order differential equations with constant coefficients, whose solutions provide the state occupancy probabilities at time t. Equation (9.19) is in the form of an *ordinary differential equation initial value problem* (ODE IVP). The ODE IVP in Eq. (9.19) has the formal solution:

$$\boldsymbol{\pi}(t) = \boldsymbol{\pi}(0) \cdot e^{\boldsymbol{Q}t}, \tag{9.21}$$

where $e^{\boldsymbol{Q}t} = \boldsymbol{P}(t)$ is defined by the following series expansion (see Section 10.4):

$$e^{\boldsymbol{Q}t} = \boldsymbol{I} + \boldsymbol{Q}t + \frac{1}{2}(\boldsymbol{Q}t)^2 + \frac{1}{3!}(\boldsymbol{Q}t)^3 + \cdots = \sum_{i=0}^{\infty} \frac{1}{i!}(\boldsymbol{Q}t)^i. \tag{9.22}$$

The state probabilities are usually the quantities of interest in a CTMC model. However, if the problem requires knowledge of each conditional transition probability, i.e., the probability of being in state j at time t given that the CTMC started in state i time $t = 0$, Eq. (9.17) must be solved. Equation (9.17) consists of a set of $n \times n$ first-order differential equations with constant coefficients. These are also ODE IVPs.

Various closed-form and numerical solution techniques are available for solving the fundamental CTMC equations, and will be discussed in Sections 9.8 and 10.4.

Example 9.1   *Physical Interpretation of Transition Rates*

Consider the system of binary components whose state space was derived in Section 7.1. Suppose a state i is directly connected to a state j, and the two states differ by the value of a single binary variable X_k. Assume that in the source state i, $X_k = 1$ (kth component up), and in the destination state j, $X_k = 0$ (kth component down). Then the transition $i \rightarrow j$ represents the failure of component k. From Eq. (9.13), $q_{ij}\,\Delta t$ is the probability of a transition to state j in the interval $(t, t + \Delta t]$ given that the CTMC was in state i at time t, but because of the physical meaning of transition $i \rightarrow j$, in our case q_{ij} coincides with the definition of the failure rate of component k in state i. Since in a homogeneous CTMC the transition rates are time independent, we can model components with a time-independent failure (or repair) rate. From the model we can arrange the values of the failure rates to be the proper entries of a matrix that we call the *transition rate matrix* $\boldsymbol{Q}_R$. The infinitesimal generator $\boldsymbol{Q}$ is obtained by adjusting the diagonal entries so that in each row the row sum is equal to zero – see Eq. (9.14). Similarly, if the transition q_{ji}, from a down state to an up state, exists and is different from 0, its physical meaning is the repair rate of component k in state j.

Example 9.2   *One Repairable Component: Availability*

Consider the behavior of a single blade server in a data center. The blade server can fail with TTF EXP(λ). Further assume that the blade is repairable, and that the TTR is an exponentially distributed random variable EXP(μ). The two-state transition diagram of this system, introduced in Section 3.5.6, is shown in Figure 9.3 as a CTMC. State 1 is the up state and state 0 is the down state. The transition from the up state 1 to the down state 0 represents the failure of the blade with rate λ, and the transition from the down state 0 to the up state 1 represents the repair of the failed blade server at the rate μ. The infinitesimal generator matrix of the CTMC of Figure 9.3 is:

$$Q = \begin{array}{c} \\ (1) \\ (0) \end{array} \begin{array}{cc} (1) & (0) \\ \begin{bmatrix} -\lambda & \lambda \\ \mu & -\mu \end{bmatrix} \end{array}. \tag{9.23}$$

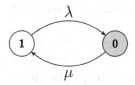

Figure 9.3   Continuous-time Markov chain state diagram of a repairable blade server.

Applying Eq. (9.19), we get:

$$\begin{bmatrix} \dfrac{d\pi_1(t)}{dt} & \dfrac{d\pi_0(t)}{dt} \end{bmatrix} = \begin{bmatrix} \pi_1(t) & \pi_0(t) \end{bmatrix} \cdot \begin{bmatrix} -\lambda & \lambda \\ \mu & -\mu \end{bmatrix}. \tag{9.24}$$

From the vector–matrix equation (9.24), we obtain the scalar equations:

$$\begin{cases} \dfrac{d\pi_1(t)}{dt} & = & -\lambda\pi_1(t) + \mu\pi_0(t), \\ \dfrac{d\pi_0(t)}{dt} & = & \lambda\pi_1(t) - \mu\pi_0(t). \end{cases} \tag{9.25}$$

We can obtain a fully symbolic solution of Eq. (9.25) by resorting to the Laplace transform method (see Section 10.1.2). Assuming that the initial state probability vector is $\pi(0) = [1, 0]$, and taking the Laplace transform on both sides of Eq. (9.25), we can write:

$$\begin{cases} s\pi_1^*(s) - 1 & = & -\lambda\pi_1^*(s) + \mu\pi_0^*(s), \\ s\pi_0^*(s) & = & \lambda\pi_1^*(s) - \mu\pi_0^*(s). \end{cases} \tag{9.26}$$

Solving the algebraic set of equations (9.26) in the Laplace domain, we obtain:

$$\begin{cases} \pi_1^*(s) & = & \dfrac{s + \mu}{s(s + \lambda + \mu)}, \\ \pi_0^*(s) & = & \dfrac{\lambda}{s(s + \lambda + \mu)}. \end{cases} \tag{9.27}$$

Taking the inverse Laplace transform of Eqs. (9.27) via partial fraction expansion [3], we finally obtain the solution in the time domain, in the form:

$$\begin{cases} \pi_1(t) & = & \dfrac{\mu}{\lambda + \mu} + \dfrac{\lambda}{\lambda + \mu}\, e^{-(\lambda + \mu)t}, \\[2mm] \pi_0(t) & = & \dfrac{\lambda}{\lambda + \mu} - \dfrac{\lambda}{\lambda + \mu}\, e^{-(\lambda + \mu)t}. \end{cases} \qquad (9.28)$$

Recalling from Section 3.5.6 that the instantaneous system availability $A(t)$ is defined as the probability that the system is up at time t, we have $A(t) = \pi_1(t)$. Correspondingly, the instantaneous system unavailability $U(t)$ is defined as $U(t) = 1 - A(t) = \pi_0(t)$. From Eq. (9.28) it is easily seen that an asymptotic solution exists, and is given by:

$$\begin{cases} \lim\limits_{t \to \infty} \pi_1(t) & = & \pi_1 & = & \dfrac{\mu}{\lambda + \mu}, \\[2mm] \lim\limits_{t \to \infty} \pi_0(t) & = & \pi_0 & = & \dfrac{\lambda}{\lambda + \mu}. \end{cases} \qquad (9.29)$$

Clearly, π_1 and π_0 are the asymptotic (or limiting) system availability and unavailability, respectively.

While the state probabilities are of immense interest in understanding the behavior of a CTMC, other features of the process can be just as useful to the modeler. Such measures of interest include the MTTF and the MTTR, the duration of a sojourn in a state, the mean number of transitions between a given pair of states in a given interval of time, and the mean passage time from a subset of states to another subset of states. We define various interesting measures of a CTMC in the following sections, and we show how to compute these measures starting from the specification of the process.

9.1.4   Distribution of the Sojourn Time in a Given State

The sojourn time (also known as the holding time) of the CTMC in a state is a random variable defined as the time from entry into the state until exit from the state. Suppose the (homogeneous) CTMC enters a generic state i at time $t = 0$ and we wish to study the distribution of the sojourn time of the CTMC in state i. For this purpose, we isolate state i by deleting all the transitions entering state i (while keeping all the outgoing transitions), as in Figure 9.4.

Then the probability of state i satisfies the following differential equation:

$$\frac{d\pi_i(t)}{dt} = -\pi_i(t)\, q_i, \qquad \pi_i(0) = 1,$$

where $q_i = -q_{ii} = \sum_{j:j \neq i} q_{ij}$ is a constant equal to the sum of the rates out of state i. The solution of the above equation is:

$$\pi_i(t) = e^{-q_i t}.$$

Thus, the sojourn time in a generic state i is exponentially distributed with a rate equal to the sum of the exit rates. From the results of Section 3.3.1, the distribution of the

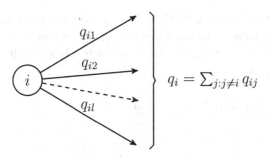

Figure 9.4   Sojourn time in CTMC state i.

sojourn time corresponds to the minimum of the m exponentially distributed random variables representing the transition times out of state i. The transition from a state to the next one occurs according to a *race model*. The statistically shortest transition time is the one that wins the race.

The probability that the sojourn time in state i terminates by time t due to a transition to state j is given by:

$$p_{ij}(t) = \frac{q_{ij}}{q_i}\left(1 - e^{-q_i t}\right),\tag{9.30}$$

where $\frac{q_{ij}}{q_i}$ is the conditional probability of the next jump being to state j given that a transition out of state i has occurred.

9.2   Classification of States and Stationary Distribution

The states of a CTMC can be partitioned into two classes: *recurrent* states and *transient* states [1, 4]. A state is recurrent if the CTMC will eventually return to that state with probability 1; otherwise the state is transient. A recurrent state can be *positive recurrent* (or *non-null recurrent*) if the expected time to reenter the state is finite, or *null recurrent* if the expected time to reenter the state is infinite.

The state space of a CTMC can always be partitioned into a set (possibly empty) of transient states and one or more closed sets of recurrent states. A state j is *reachable* from i for some $t > 0$ if $p_{ij}(t) > 0$. A closed set of recurrent states is a set in which all pairs of states are mutually reachable. Once a CTMC has reached a state in a closed set of recurrent states, it never leaves the set again.

If a closed set of recurrent states contains a single state, this state cannot have outgoing transitions and is called an *absorbing* state. If state i is absorbing, $q_{ij} = 0$ for any $j \neq i$, and the corresponding row of the infinitesimal generator matrix has only 0 entries. Continuous-time Markov chains with absorbing states will be considered in Chapter 10.

9.2.1   Irreducible Markov Chain

A CTMC is said to be *irreducible* [1, 4] if its state space is formed by a single set of recurrent states, so that every state is reachable from every other state. We state, without proof [1, 4], that for an irreducible CTMC the state probabilities reach an asymptotic value as the time goes to infinity, and this asymptotic value is independent of the initial condition. We call the asymptotic solution the *steady-state* solution. If the steady-state solution exists, then, for any i:

$$\lim_{t \to \infty} \pi_i(t) = \pi_i, \qquad \lim_{t \to \infty} \frac{d \pi_i(t)}{dt} = 0, \qquad (9.31)$$

and, in vector form,

$$\lim_{t \to \infty} \pi(t) = \pi, \qquad \lim_{t \to \infty} \frac{d \pi(t)}{dt} = 0, \qquad (9.32)$$

where π is the steady-state value of the state probability vector and 0 is the vector of appropriate dimension with all entries equal to 0. Taking into account (9.32), Eq. (9.19) becomes:

$$\pi \cdot Q = 0 \quad \text{with} \quad \pi e^{T} = 1. \qquad (9.33)$$

Equation (9.33) is a linear homogeneous set of n equations with constant coefficients. It is easy to verify that $\pi_i = 0$ for all i is a solution. If a non-zero solution exists, multiplication of the solution by any constant will still be a solution to the equation. But since π is a probability vector, it should also satisfy the normalization condition represented in matrix form in Eq. (9.33). The normalized steady-state solution is then unique.

For finite-state irreducible CTMCs, positive and unique limiting probabilities can be determined in this way. For an infinite-state irreducible CTMC, the additional condition that its states are all positive recurrent is required. To force the set of equations to have a single positive solution, we can incorporate the normalization condition into the set of equations by replacing any one of the n equations with the normalization condition. Without loss of generality we choose to replace the nth equation with the normalization condition by substituting the last column of the Q matrix with all entries equal to 1 and the last entry of the right-hand side vector with 1. Thus, we obtain:

$$\begin{bmatrix} \pi_1 \pi_2 \cdots \pi_{n-1} \pi_n \end{bmatrix} \begin{bmatrix} q_{11} & q_{12} & \cdots & q_{1,n-1} & 1 \\ q_{21} & q_{22} & \cdots & q_{2,n-1} & 1 \\ \vdots & \vdots & \ddots & \vdots & \vdots \\ q_{n-1,1} & q_{n-1,2} & \cdots & q_{n-1,n-1} & 1 \\ q_{n,1} & q_{n,2} & \cdots & q_{n,n-1} & 1 \end{bmatrix} = [0\,0\cdots 0\,1]. \qquad (9.34)$$

It is evident that the last equation in (9.34) is simply the normalization condition. Several different solution techniques for the homogeneous linear system (9.33) [7] will be reviewed in Section 9.8.

The steady-state probability vector π has the following properties:

- For all initial conditions, the occupancy state probability $\pi_i(t)$ tends to the constant value π_i as $t \to \infty$, and the π_i form a probability mass function over the set of states.
- If the initial probability is $\pi_i \, \forall i$, then $\pi_i(t) = \pi_i$ for all t.

- The fraction of time spent in state i during the interval $(0,t]$ tends to π_i as $t \to \infty$. In steady state, the fraction of time spent by the CTMC in state i is equal to π_i.

The steady-state equation can be interpreted as a probability *balance equation*. The balance equation means that for every state the probability flow in equals the probability flow out, and this equation can be written directly from observation of the CTMC transition graph, without the need to derive the infinitesimal generator. In fact, solving Eq. (9.33) for state i, we can write

$$-\pi_i q_{ii} = \pi_1 q_{1i} + \pi_2 q_{2i} + \cdots + \pi_n q_{ni}, \qquad (9.35)$$

where the left-hand side is the probability flow out of state i and the right-hand side is the probability flow into state i.

Example 9.3 Returning to Example 9.2, we note that Eq. (9.29) gives the steady-state solution for the state probabilities of the CTMC. The same result can be obtained directly, by replacing the second equation with the normalization condition in (9.34), to get:

$$[\pi_1 \ \pi_0] \begin{bmatrix} -\lambda & 1 \\ \mu & 1 \end{bmatrix} = [0 \ 1] . \qquad (9.36)$$

Expanding (9.36), we get:

$$\begin{cases} -\lambda \pi_1 + \mu \pi_0 &= 0, \\ \pi_1 + \pi_0 &= 1. \end{cases} \qquad (9.37)$$

Note that the first equation in (9.37) is the probability balance equation of state 1, which could have been written directly without resorting to Eq. (9.36). From Eq. (9.37) we obtain the steady-state availability and unavailability, respectively, as:

$$\begin{cases} A_{ss} &= \pi_1 = \dfrac{\mu}{\lambda + \mu} = \dfrac{\text{MTTF}}{\text{MTTR} + \text{MTTF}}, \\ U_{ss} &= \pi_0 = \dfrac{\lambda}{\lambda + \mu} = \dfrac{\text{MTTR}}{\text{MTTR} + \text{MTTF}}, \end{cases} \qquad (9.38)$$

since the component $\text{MTTF} = 1/\lambda$ and $\text{MTTR} = 1/\mu$.

9.2.2 Expected State Occupancy

The total time spent by the CTMC $Z(t)$ in a generic state i during the interval $(0,t]$ is a random variable whose expected value is denoted by $b_i(t)$. Note that such a sojourn can be collected over multiple visits to the state. To compute the expected value $b_i(t)$, we introduce an indicator variable $I_i(t)$ defined as follows:

$$\begin{cases} I_i(t) = 1 & \text{if} \quad Z(t) = i, \\ I_i(t) = 0 & \text{if} \quad Z(t) \neq i. \end{cases}$$

Then, by construction, the total time spent by the CTMC in state i during the interval $(0, t]$ can be reformulated as:

$$\text{total time spent in state } i = \int_0^t I_i(u)\, du,$$

with initial condition $b_i(0) = 0$. Hence, the expected value $b_i(t)$ becomes:

$$
\begin{aligned}
b_i(t) = E\left[\int_0^t I_i(u)\, du \right] &= \int_0^t E[I_i(u)]\, du \\
&= \int_0^t [0 \cdot P\{I_i(u) = 0\} + 1 \cdot P\{I_i(u) = 1\}]\, du \\
&= \int_0^t \pi_i(u)\, du.
\end{aligned}
\tag{9.39}
$$

Equation (9.39) shows that the expected state occupancy during the interval $(0, t]$ can be obtained by integrating the corresponding state probability over the same interval. Define the vector $b(t) = [b_i(t)]$ whose entries are the expected state occupancies defined in Eq. (9.39):

$$b(t) = \int_0^t \pi(u)\, du. \tag{9.40}$$

Direct integration of Eq. (9.19) yields:

$$\frac{db(t)}{dt} = b(t)Q + \pi(0) \tag{9.41}$$

under the initial condition $b(0) = 0$. We are often interested in the time-averaged expected state occupancy denoted by the vector $g(t) = (1/t)b(t)$. Using Eq. (9.41) we get:

$$\frac{dg(t)}{dt} = g(t)\left(Q - \frac{I}{t}\right) + \frac{\pi(0)}{t}, \tag{9.42}$$

where we define $g(0) = 0$. From the normalization condition on $\pi(t)$, and from Eq. (9.40), we obtain the following normalization condition for $b(t)$ and $g(t)$, respectively:

$$b(t)e^{\mathrm{T}} = t, \qquad g(t)e^{\mathrm{T}} = 1. \tag{9.43}$$

Equations (9.43) simply state that the sum of the expected times spent over all states equals the length of the interval of observation. Equation (9.42) has a singularity at $t = 0$; however, as t increases, solving Eq. (9.42) is numerically more convenient than Eq. (9.41) because of the following asymptotic property [8]:

$$\lim_{t \to \infty} g_i(t) = \lim_{t \to \infty} \frac{1}{t} b_i(t) = \lim_{t \to \infty} \pi_i(t) = \pi_i, \tag{9.44}$$

where π is the steady-state probability vector defined in Eq. (9.33). The systems of linear differential equations (9.41) and (9.42) can be solved for $b(t)$ and $g(t)$ by the same methods used for solving Eq. (9.19) for $\pi(t)$ [9]. In particular, integrating (9.22), we obtain for the vector $b(t)$ the following series expansion:

$$b(t) = b(0) + \pi(0)\left(It + \frac{t^2}{2}Q + \cdots + \frac{t^{i+1}}{(i+1)!}Q^i + \cdots\right).$$

The limiting behavior of $b_i(t)$ as $t \to \infty$ will be considered in Section 10.2.2.

Problems

9.1 For the two-state availability model of Figure 9.3, derive the transient state probability expressions (and hence instantaneous availability) assuming that at time $t = 0$ the system is in the down state with probability q and in the up state with probability $1 - q$.

9.2 For the two-state availability model of Figure 9.3, derive expressions for $b_0(t)$ and $b_1(t)$ by solving the differential equations (9.41). Compare the results with those obtained from direct integration as in Eqs. (9.40) using Eqs. (9.28).

9.3 In the two-state availability model of Figure 9.3, assume now that the failure transition is $ER_2(\lambda)$ and the repair transition $HYPER(\mu_1, \mu_2, q)$. Find the steady-state availability and the expected uptime and downtime in the interval $(0, t]$.

9.3 Dependability Models Defined on a CTMC

Now we wish to specialize the use of CTMCs to compute dependability measures. From Section 7.1.1, the state space Ω of a dependability model can be partitioned into a subset Ω_u of cardinality n_u, which groups the up states in which the structure function of the system is equal to 1, and a subset Ω_d of cardinality n_d, which groups the down states in which the structure function of the system is equal to 0. We have $\Omega_u \cup \Omega_d = \Omega$, $\Omega_u \cap \Omega_d = \emptyset$, and $n_u + n_d = n$.

We adopt the following notation. A vector with a subscript u means the partition of the vector in the subset Ω_u, while a subscript d means the partition of the vector in the subset Ω_d. If a matrix has a double subscript (among u and d), the first subscript is the row dimension and the second subscript the column dimension. From the above, the infinitesimal generator matrix of the CTMC can be partitioned in the following way (see Figure 9.5):

$$Q = \begin{bmatrix} Q_{uu} & Q_{ud} \\ Q_{du} & Q_{dd} \end{bmatrix}, \tag{9.45}$$

where matrix Q_{uu} is an $n_u \times n_u$ matrix that groups the transition rates inside the states in Ω_u, Q_{ud} is an $n_u \times n_d$ matrix that groups the transition rates from states in Ω_u to states in Ω_d, Q_{du} is an $n_d \times n_u$ matrix that groups the transition rates from states in Ω_d to states in Ω_u, and, finally, Q_{dd} is an $n_d \times n_d$ matrix that groups the transition rates inside the states in Ω_d.

The non-zero transition rates in matrix Q_{ud} represent failure transitions for the system, i.e., transitions that lead the system directly from an up state to a down state,

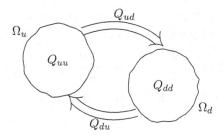

Figure 9.5   Partition of the state space into up and down states.

and, conversely, the non-zero transition rates in matrix Q_{du} represent repair transitions for the system, i.e., transitions that lead the system directly from a down state to an up state.

In an availability model both Q_{ud} and Q_{du} must have non-zero entries, while in a reliability model the states in Ω_d are absorbing so that Q_{du} and Q_{dd} are zero matrices (matrices with all entries equal to 0).

The instantaneous system availability at time t is defined as the sum of the state probabilities at time t over the up states, and the instantaneous system unavailability as the sum of the state probabilities at time t over the down states:

$$A(t) = \sum_{i \in \Omega_u} \pi_i(t), \qquad U(t) = \sum_{j \in \Omega_d} \pi_j(t). \tag{9.46}$$

The steady-state availability and unavailability are defined in a similar way utilizing the steady-state probability vector computed from Eq. (9.33), or from the balance equations (9.35).

$$A = \sum_{i \in \Omega_u} \pi_i, \qquad U = \sum_{j \in \Omega_d} \pi_j. \tag{9.47}$$

The expected interval availability $A_I(t)$ is defined as the average proportion of time that the CTMC spends in up states Ω_u during an interval $(0,t]$, and is given by:

$$A_I(t) = \frac{1}{t} \sum_{i \in \Omega_u} b_i(t) = \sum_{i \in \Omega_u} g_i(t), \tag{9.48}$$

where $b_i(t)$ and $g_i(t)$ are computed according to Eqs. (9.41) and (9.42), respectively. These and other measures will be defined in a uniform way by resorting to Markov reward models in Section 9.4.

9.3.1   Expected Uptime and Expected Downtime

The expected uptime $U_I(t)$ is defined as the expected total amount of time that the CTMC spends in the up states Ω_u during an interval $(0,t]$, and is given by:

$$U_I(t) = \sum_{i \in \Omega_u} b_i(t).$$

In steady state, the expected uptime over an interval T_I will then turn out to be (see also Example 3.9)

$$U_I = T_I \sum_{i \in \Omega_u} \pi_i = T_I A.$$

Similarly, the expected downtime $D_I(t)$ is defined as the total amount of time that the CTMC spends in the down states Ω_d during an interval $(0, t]$, and is given by:

$$D_I(t) = \sum_{i \in \Omega_d} b_i(t);$$

in steady state, the expected downtime over any interval T_I turns out to be:

$$D_I = T_I \sum_{i \in \Omega_d} \pi_i = T_I U.$$

These results have been already used in Examples 3.9–3.11.

Example 9.4 *Two Identical Repairable Components*
As an example of application of the probability balance equation, we consider a two-server parallel redundant system that consists of two blade servers with identical failure rate λ and repair rate μ. When both the blades have failed, the system is considered to have failed. We wish to examine the effects of non-shared (one repairer for each blade server) vs. shared repair (only one repairer) on the steady-state availability. The non-shared case can be modeled and solved using a model for independent components, but the shared case needs the use of Markov chains.

Non-Shared (Independent) Repair
The CTMC transition graph for this system is shown in Figure 9.6. In state 2 both components are working, in state 1 only one component is working and state 0 is the system down state.

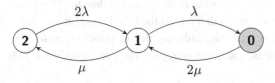

Figure 9.6   Continuous-time Markov chain of a two-component system with non-shared repair facilities.

Non-shared repair means that each component undergoes repair as soon as it fails, independent of the state of the other components. This repair policy implies that there are as many repair persons as failed components, so that in state 0 both components are under repair at the same time. This means that two independent EXP(μ) random variables will be competing. Hence, the time to first repair completion will be EXP(2μ), implying a transition with rate 2μ from state 0 to state 1. The steady-state balance equations can be written as:

$$
\begin{cases}
2\lambda\pi_2 & = & \mu\pi_1, \\
(\lambda+\mu)\pi_1 & = & 2\lambda\pi_2+2\mu\pi_0, \\
2\mu\pi_0 & = & \lambda\pi_1, \\
\pi_2+\pi_1+\pi_0 & = & 1,
\end{cases}
$$

$$
\begin{cases}
\pi_2 & = & (\mu/2\lambda)\pi_1, \\
\pi_1 & = & (2\mu/\lambda)\pi_0, \\
\dfrac{\mu}{2\lambda}\dfrac{2\mu}{\lambda}\pi_0+\dfrac{2\mu}{\lambda}\pi_0+\pi_0 & = & 1,
\end{cases}
$$

from which the steady-state system unavailability $U^{(\mathrm{ns})}$ can be derived as:

$$
U^{(\mathrm{ns})} = \pi_0 = \cfrac{1}{1+\dfrac{2\mu}{\lambda}+\dfrac{2\mu^2}{2\lambda^2}} = \cfrac{1}{\left(1+\dfrac{\mu}{\lambda}\right)^2} = \left(\dfrac{\lambda}{\lambda+\mu}\right)^2,
$$

and the steady-state availability $A^{(\mathrm{ns})}$ becomes

$$
A^{(\mathrm{ns})} = 1 - U^{(\mathrm{ns})} = 1 - \left(\dfrac{\lambda}{\lambda+\mu}\right)^2 = \dfrac{\mu(2\lambda+\mu)}{(\lambda+\mu)^2}.
$$

Note that the same result could have been obtained from an RBD by observing that the steady-state availability of one single component is $A = \mu/(\lambda+\mu)$, see Eq. (3.116), and, for a parallel system with identical components, $A^{(\mathrm{ns})} = 2A - A^2$.

Shared (Dependent) Repair

Shared repair means that the number of repair persons is less than the possible maximum number of failed components, so that not all the failed components can be under repair at the same time. In this example, only one repair person is available, so that from state 0 only one component can undergo repair. The corresponding CTMC is shown in Figure 9.7.

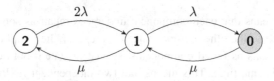

Figure 9.7   Continuous-time Markov chain of a two-component system with shared repair facilities.

The steady-state balance equations can be written as:

$$\begin{cases} 2\lambda\pi_2 & = & \mu\pi_1, \\ (\lambda+\mu)\pi_1 & = & 2\lambda\pi_2+\mu\pi_0, \\ \mu\pi_0 & = & \lambda\pi_1, \\ \pi_2+\pi_1+\pi_0 & = & 1, \end{cases}$$

$$\begin{cases} \pi_2 & = & (\mu/2\lambda)\pi_1, \\ \pi_1 & = & (\mu/\lambda)\pi_0, \\ \dfrac{\mu}{2\lambda}\dfrac{\mu}{\lambda}\pi_0+\dfrac{\mu}{\lambda}\pi_0+\pi_0 & = & 1, \end{cases}$$

from which the steady-state system unavailability $U^{(sh)}$ is obtained as:

$$U^{(sh)} = \pi_0 = \cfrac{1}{1+\dfrac{\mu}{\lambda}+\dfrac{\mu^2}{2\lambda^2}} = \frac{2\lambda^2}{2\lambda^2+2\lambda\mu+\mu^2},$$

and the steady-state availability $A^{(sh)}$ is given as

$$A^{(sh)} = 1 - U^{(sh)} = \frac{\mu(2\lambda+\mu)}{2\lambda^2+2\lambda\mu+\mu^2}.$$

Comparing the availabilities of the non-shared and shared repair, we obtain:

$$\frac{A^{(ns)}}{A^{(sh)}} = \frac{2\lambda^2+2\lambda\mu+\mu^2}{\lambda^2+2\lambda\mu+\mu^2} = 1+(\frac{\lambda}{\lambda+\mu})^2 > 1$$

$$A^{(ns)} > A^{(sh)}.$$

The non-shared (independent) case provides higher availability.

Example 9.5 *Two Identical Repairable Components with Imperfect Coverage*
We continue with the two-server parallel redundant system with shared repair that consists of two blade servers with identical failure rate λ and repair rate μ. When a blade server failure occurs, the system must detect which blade has failed and reconfigure to continue operation with the single remaining blade. However, the detection and recovery process may complete successfully with a coverage probability c, and with probability $(1-c)$ the recovery process does not complete successfully, and the system incurs a complete failure and moves to the down state 0. When both the blades have failed, the system is considered to have failed. The corresponding CTMC availability model is shown in Figure 9.8.

The steady-state balance equations can be written as:

$$\begin{cases} 2\lambda\pi_2 & = & \mu\pi_1, \\ (\lambda+\mu)\pi_1 & = & 2\lambda c\pi_2+\mu\pi_0, \\ \mu\pi_0 & = & 2\lambda(1-c)\pi_2+\lambda\pi_1, \\ \pi_2+\pi_1+\pi_0 & = & 1, \end{cases}$$

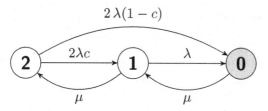

Figure 9.8   Continuous-time Markov chain availability model of a two-component system with imperfect coverage.

from which the steady-state system unavailability $U^{(sh)}$ is obtained as:

$$U^{(sh)} = \pi_0 = \frac{2\lambda^2 + 2\lambda\mu(1-c)}{2\lambda^2 + 2\lambda\mu(2-c) + \mu^2},$$

and the steady-state availability $A^{(sh)}$ is given as

$$A^{(sh)} = 1 - U^{(sh)} = \frac{\mu(2\lambda + \mu)}{2\lambda^2 + 2\lambda\mu(2-c) + \mu^2}.$$

Example 9.6   *Two Identical Repairable Components with Common Cause Failure*
Common cause failures were introduced in Section 2.7 and defined as the result of one or more events causing the concurrent failure of one or more components. An example of modeling the effect of a CCF on a parallel system of two non-repairable components was presented in Section 6.10.3, in the framework of an FT converted into a BN.

The presence of CCF is a typical manifestation of statistical dependence among components, and needs to be modeled resorting to state-space techniques. Reconsider Example 9.5, a system with two identical components but now with perfect coverage and shared repair, but we add an exponentially distributed CCF with rate λ_{CCF}. The state-space representation of the system is given in Figure 9.9.

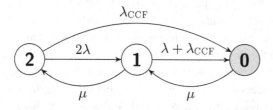

Figure 9.9   Availability model of a two-component system with common cause failure.

The steady-state balance equations can be written as:

$$\begin{cases} (2\lambda + \lambda_{CCF})\pi_2 & = & \mu\pi_1, \\ (\lambda + \lambda_{CCF} + \mu)\pi_1 & = & 2\lambda\pi_2 + \mu\pi_0, \\ \mu\pi_0 & = & \lambda_{CCF}\pi_2 + (\lambda + \lambda_{CCF})\pi_1, \\ \pi_2 + \pi_1 + \pi_0 & = & 1, \end{cases}$$

from which an expression for the steady-state system unavailability $U^{(\text{CCF})} = \pi_0$ can be easily derived.

Example 9.7 *Availability of Combined Hardware and Software System*
We return to Example 4.6, but now we introduce the dependencies between the hardware and software components shown in Figure 9.10, where state hs is the only up state where both hardware and software are working, while states $\overline{\text{h}}$ and $\overline{\text{s}}$ are down states, since components h or s, respectively, are not working (the indices in parentheses are a state identification number to be used in the equations).

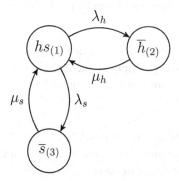

Figure 9.10 System with a hardware and a software component and dependent failures.

When the system is down caused by the failure of either of the two components, the only possible action is the repair of the failed component. In other words, software cannot fail when the hardware is down and vice versa. The balance equations in the steady state are:

$$(\lambda_s + \lambda_h)\pi_1 = \mu_h \pi_2 + \mu_s \pi_3,$$

$$\mu_h \pi_2 = \lambda_h \pi_1,$$

$$\mu_s \pi_3 = \lambda_s \pi_1.$$

Taking into account the normalization condition, the steady-state availability is:

$$A_{\text{CTMC}} = \pi_1 = \frac{1}{1 + \dfrac{\lambda_h}{\mu_h} + \dfrac{\lambda_s}{\mu_s}} = \frac{\mu_h \mu_s}{\mu_h \mu_s + \lambda_h \mu_s + \mu_h \lambda_s}. \qquad (9.49)$$

Comparing this result with the result obtained from the RBD model of Example 4.6, we see that

$$A_{\text{CTMC}} = \frac{\mu_h \mu_s}{\mu_h \mu_s + \lambda_h \mu_s + \mu_h \lambda_s} > A_{\text{RBD}} = \frac{\mu_h \mu_s}{\mu_h \mu_s + \lambda_h \mu_s + \mu_h \lambda_s + \lambda_h \lambda_s}. \qquad (9.50)$$

To shed further insight, we develop the CTMC for the independent case in Figure 9.11, where the only up state is still state hs. Solving the CTMC of Figure 9.11 would provide the same A_{RBD} expression. Note that the independent case has an

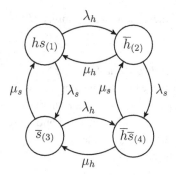

Figure 9.11   System with a hardware and a software component and different dependent failures.

additional failed state $\overline{hs}$ that increases the system unavailability as compared with the dependent case. This additional state is probably an incorrect interpretation of the real system since it means that the software component can fail when the hardware component is down, and vice versa.

Example 9.8   *Telecommunication Switching System Model*
We take from [10] the example of a telecommunications switching system with n trunks. We carry out traditional availability analysis of the system. In practice, all trunks are subject to failure for various reasons such as hardware/software faults, human error in system maintenance and repair, and impairment/damage from adverse environments. For ease of illustration, we assume the failure and repair times of each trunk are exponentially distributed with rates γ and τ, respectively. Assume that a single repair facility is shared by all trunks in the system. Then, the pure availability model of the system is a homogeneous CTMC as shown in Figure 9.12, where state i indicates that there are i non-failed trunks in the system [3].

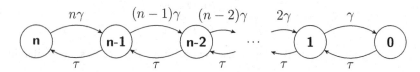

Figure 9.12   Continuous-time Markov chain availability model of a telecommunication switching system.

The steady-state probability of the above CTMC being in state k can be computed by solving the following balance equations:

$$n\gamma\pi_n = \tau\pi_{n-1},$$

$$(k\gamma + \tau)\pi_k = (k+1)\gamma\pi_{k+1} + \tau\pi_{k-1}, \ k = 1, 2, \ldots, n-1,$$

$$\gamma\pi_1 = \tau\pi_0.$$

Solving this linear system of equations, the steady-state probability of the Markov chain of Figure 9.12 being in state i is given by [3]:

$$\pi_i = \frac{(\frac{\tau}{\gamma})^i/i!}{\sum_{k=0}^{n}(\frac{\tau}{\gamma})^k/k!}. \tag{9.51}$$

If we assume that the system is up as long as at least l trunks are functioning properly, then the system steady-state availability is given by [3]:

$$A(l) = \sum_{i=l}^{n} \pi_i. \tag{9.52}$$

Example 9.9 *A Repairable Fluid Level Controller*
We assume now that the components in Example 4.1 are repairable, but only a single shared crew is available. If more than one component is failed, the repairer must decide which component to repair first. The set of rules and actions that the repair person follows to accomplish repair is known as the *repair policy*. A repair policy that gives priority to one component over another is a *priority* policy, while one that repairs in the order of arrival is a *first-come-first-served* (FCFS) policy. A priority repair policy is *preemptive* if an ongoing repair action on a lower-priority component is interrupted in order to repair a higher-priority component that has failed. A priority repair policy is *non-preemptive* if an ongoing repair (even on a lower-priority component) is always completed once initiated. Different preemptive and non-preemptive repair policies can be envisaged, and they can be easily represented as a CTMC on the whole state space. While adopting a preemptive priority repair policy, a priority list is defined and the repair crew repairs the components according to the order defined in the priority list. When a component fails, if a lower-priority component is under repair then that repair is preempted and the higher-priority component undergoes repair [11].

Table 9.1 Failure and repair rates for the system of Example 4.1.

Component	MTTF $\frac{1}{\lambda}$ (hr)	MTTF^{-1} λ (f hr^{-1})	MTTR $\frac{1}{\mu}$ (hr)	MTTR^{-1} μ (f hr^{-1})
P	5000	2×10^{-4}	16	0.0625
V	10000	1×10^{-4}	8	0.125
S	2500	4×10^{-4}	4	0.25
L	2000	5×10^{-4}	2	0.5

Given the failure and repair rates of Table 9.1, we assume that the priority list is ordered according to the decreasing values of the MTTR, i.e., the component with highest MTTR is repaired first; hence, the list becomes $P \succ V \succ S \succ L$.

The CTMC with a shared preemptive priority repair policy is shown in Figure 9.13. The labels inside the states indicate a vector of ordered components (S, L, P, V) where, as usual, 1 indicates the up state and 0 the down state. From each state one single

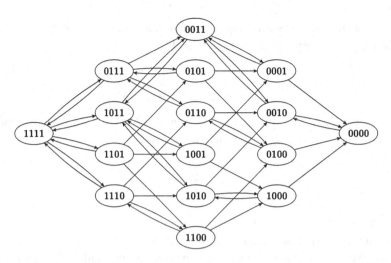

Figure 9.13 Continuous-time Markov chain model for shared repair with decreasing MTTR priority.

component only can undergo repair according to the priority list. The steady-state availability computed from the CTMC of Figure 9.13 is $A_{Sys} = 0.993414494$.

Example 9.10 *A Redundant Repairable Fluid Level Controller*
Continuing Example 9.9 [11], we now consider redundant repairable subsystems that are modeled individually by the CTMCs shown in Figure 9.14.

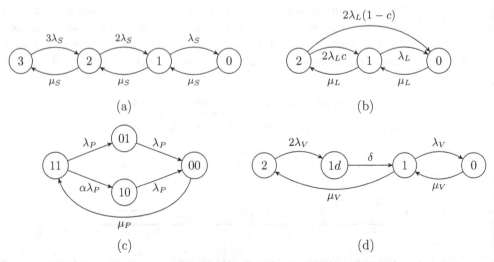

Figure 9.14 Continuous-time Markov chains for redundant units: (a) sensor; (b) control logic; (c) pump; (d) valve.

Subsystem S is made up of three sensors that work in a two-out-of-three logic with a single repair person. The corresponding CTMC availability model is shown in Figure 9.14(a), and the steady-state availability is given by $A_S = \pi_3 + \pi_2$.

Subsystem L contains two units duplicated in a parallel redundant configuration, and the recovery mechanism has a coverage probability c. The CTMC is shown in Figure 9.14(b), and its steady-state availability is given by $A_L = \pi_2 + \pi_1$.

Subsystem P contains two pump systems duplicated in a warm standby configuration [12], with dormancy factor α for the standby unit (Section 8.2). The subsystem is repaired only upon the failure of both pumps. The CTMC is shown in Figure 9.14(c), and its steady-state availability is given by $A_P = \pi_{11} + \pi_{01} + \pi_{10}$.

Subsystem V is made up of two duplicated valve units in a parallel redundant configuration. The reconfiguration upon failure of one component takes an exponentially distributed reconfiguration time with rate δ. During the reconfiguration delay the valve subsystem is not available. The CTMC is shown in Figure 9.14(d), and its steady-state availability is given by $A_V = \pi_2 + \pi_1$.

The monolithic CTMC for the whole system has $n = 4 \times 3 \times 4 \times 4 = 192$ states. The direct description of the states, state transitions and the infinitesimal generator matrix is time-consuming and error prone. Further, if the whole system has a single shared repair person then the correct generation of the state space requires that in each failure state of the system only one repair transition can emerge for the unit with the highest priority. In Chapter 12, we show how to automate the generation of such a large state space by using a higher-level paradigm known as a stochastic reward net, which is a flavor of stochastic Petri net.

Example 9.11 *A Redundant Repairable Fluid Level Controller with Travel Time*
In a more realistic setting, if the repair person is not on site then he/she must be summoned and his/her *travel time* must be accounted for [11, 13]. The travel time (or arrival time) accounts for the delay from the instant at which the request for service is notified to the instant at which the repair person is on site and ready to start the repair operation. Note that the travel time is often much larger than the actual repair time. We assume that the travel time is exponentially distributed with rate μ_t.

To include the travel time, we must duplicate all the states in which the repair service is required, to distinguish whether the repairer is already on site or not. The individual CTMCs for each subsystem are drawn in Figure 9.15, where we have denoted by a subscript u the states in which a repair is requested but the repair person is not on site and by a subscript t the states in which a repair is requested and the repair person is already on site. From the states labeled t, when the repair person is on site the repair is accomplished with the given repair rate, while from the states labeled u, the repairer should be notified and a travel transition with rate μ_t needs to be considered.

If a single repair person is shared among all the subsystems with the preemptive priority order from earlier, the monolithic CTMC model has $5 \times 6 \times 5 \times 7 = 1050$ states. Automated generation of the state space of the CTMC starting from a stochastic reward net model specification is examined in Example 12.18.

Problems

9.4 Derive the state space of a two-component system with identical components, shared repair, coverage and CCF. Find the steady-state availability in this case.

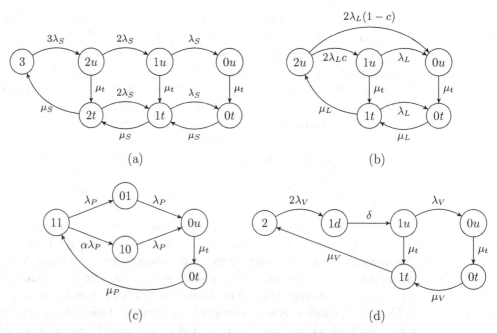

Figure 9.15   Continuous-time Markov chains for redundant units with travel time: (a) sensor; (b) control logic; (c) pump; (d) valve.

9.5   A parallel redundant system is composed of two different components A and B, with exponentially distributed failure times with rates λ_A and λ_B, respectively. Derive expressions for its steady-state availability under different repair policies:

(a) Independent repair with constant repair rates μ_A and μ_B.
(b) A single repairer with preemptive priority with $A \succ B$ and constant repair rates μ_A and μ_B.
(c) A deferred repair policy where repair occurs only after a system failure with priority list $A \succ B$ and constant repair rates μ_A and μ_B.
(d) Repair only at system failure with an FCFS repair policy (i.e., at system failure start repairing the component that failed first) and constant repair rates μ_A and μ_B.
(e) Repair only after system failure, but both components are repaired at once with constant repair rate $(\mu_A + \mu_B)/2$.

Compare and discuss the effects of the different alternatives.

9.6   In Problem 9.5, consider the inclusion of CCF with the second repair policy (single repairer with preemptive priority with $A \succ B$ and constant repair rates μ_A and μ_B). Compute the steady-state availability.

9.7   In a repair shop, failed components arrive according to a Poisson process of rate λ. The shop has the resources to repair up to n components at the same time. When n components are under repair, incoming failed components are rejected. The rate of repair of a component is μ.

(a) Depict the transition graph of the repair shop.

(b) Derive the balance equations for the steady-state probabilities.

(c) With $n = 3$, derive closed-form expressions for the state probabilities in the steady state. Using the state probabilities, derive expressions for the rejection probability, the rate of rejection and the average number of components in the repair shop.

9.8 A computational process consists of three modules, A, B, and C, with their execution times being exponentially distributed with respective rate parameters a, b, and c. Modules A and B execute in sequence, while module C executes in parallel with them. As soon as either module C or the sequence AB finishes, the process completes. A CTMC model of this computation is shown in Figure 9.16.

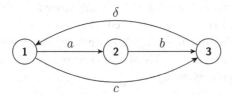

Figure 9.16   An iterative process with three modules.

If the overall process is iterated indefinitely after an exponentially distributed delay of rate δ (Figure 9.16), derive expressions for the steady-state probabilities of all the states. Write down an expression for the throughput of the computational process.

9.9 Jobs arrive at a resource according to a Poisson process of rate λ, and are executed with an exponentially distributed service time of rate μ. The resource may accommodate up to n jobs. Further, the resource alternates between up and down states. The sojourn time in the up state is $EXP(\alpha)$, and in the down state it is $EXP(\beta)$. The process between the up and down states is independent of the arrival and service process. While the resource is up, jobs may arrive and be served; when the resource is down, incoming jobs are rejected and cannot be served.

(a) Depict the transition graph of the complete model.

(b) Derive the balance equations for the state probabilities in the steady state.

(c) With $n = 1$ (buffer with a single position), find the asymptotic probability that a single job is in execution.

9.4   Markov Reward Models

A Markov reward model (MRM) extends the modeling capabilities of a CTMC by adding to the states or the transitions or both an attribute called *reward* [14, 15]. The reward often represents a performance level or a cost associated with the state, or some property of the state or of the transition. Further, the use of a reward attached to CTMC models provides a unified framework to define and compute measures that characterize the system behavior of interest to the modeler. Two ways can be envisaged to assign rewards: *reward rates* are non-negative real-valued constants associated with each state of a Markov chain [14, 15]; *impulse rewards* are non-negative real-valued constants associated with states or transitions [16, 17].

The reward rate of a state indicates the reward gained per unit time during a sojourn of the CTMC in that state. An impulse reward defines the reward gained by entering (or exiting) a specific state or traversing a transition. The CTMC will be referred to as the structure state process, while the reward variables are called the *reward structure*. Changing the reward structure on the same structure state process provides different views of the model and enables the computation of different (reward-based) measures for the CTMC.

9.4.1   MRM with Reward Rates

Let r_i be a non-negative real-valued reward rate attached to state i, and define $r = [r_i]$ to be the reward vector of dimension n defined over the state space Ω of a CTMC. This definition implies that a reward $r_i \Delta t$ is accumulated when the process $Z(t)$ stays in state i for a time duration Δt. The structure state Markov process $Z(t)$, together with the reward rates attached to each state, form the MRM. Instantaneous as well as cumulative reward measures are defined in the following.

Expected Instantaneous Reward Rate

Let $X(t)$ denote the instantaneous reward rate at time t. By definition:

$$X(t) = r_i \quad \text{if} \quad Z(t) = i. \tag{9.53}$$

Consider a CTMC $\{Z(t) \mid t \geq 0\}$ with the sample path shown in Figure 9.1. The corresponding sample path of the process $\{X(t) \mid t \geq 0\}$ is shown in Figure 9.17(a), where r_i, r_j, and r_k are the respective reward rate values associated with states i, j, and k of Figure 9.1. We have assumed that $r_i > r_j$ and $r_k = 0$. The expected instantaneous reward rate at time t is computed as:

$$E[X(t)] = \sum_{i=1}^{n} r_i P\{Z(t) = i\} = \sum_{i=1}^{n} r_i \pi_i(t), \tag{9.54}$$

and the expected reward rate in steady state is:

$$E[X] = \sum_{i=1}^{n} r_i \pi_i. \tag{9.55}$$

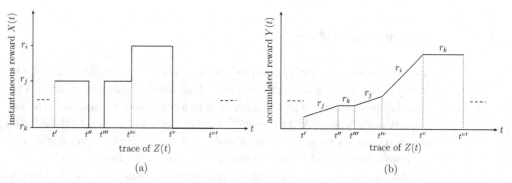

Figure 9.17   Continuous-time Markov chain model with reward rate. (a) Instantaneous reward $X(t)$; (b) Total accumulated reward $Y(t)$.

In matrix notation, Eqs. (9.54) and (9.55) are written as:

$$E[X(t)] = \pi(t)r^T, \quad E[X] = \pi r^T. \tag{9.56}$$

The closed-form or numerical techniques, as well as the complexity, for solving Eqs. (9.54) and (9.55) are the same as those for solving the standard CTMC equations for the state probability vector, in transient (9.19) or in steady state (9.33), respectively. Different reward structures combined with Eq. (9.56) provide different reliability or availability measures [3, 18–21].

Expected Accumulated Reward

Cumulative measures are related to the accumulation of reward during a finite time interval. The total accumulated reward up to time t is the random variable $Y(t)$ defined as:

$$Y(t) = \int_0^t X(u)\,du. \tag{9.57}$$

Corresponding to the sample path of the CTMC $\{Z(t) \mid t \geq 0\}$ of Figure 9.1, the sample path of the accumulated reward process $\{Y(t) \mid t \geq 0\}$ is shown in Figure 9.17(b). During the sojourn in state i, the accumulated $Y(t)$ grows linearly with slope r_i. During the sojourn in state k, $Y(t)$ remains constant since $r_k = 0$.

The distribution of the accumulated reward $Y(t)$ can be computed, but we leave that for an advanced study of the literature [21–26]. On the other hand, from Eq. (9.57), the expected accumulated reward is simply given by:

$$E[Y(t)] = E[\int_0^t X(u)\,du] = \sum_i r_i \int_0^t \pi_i(u)du = \sum_i r_i b_i(t), \tag{9.58}$$

so that $E[Y(t)]$ of Eq. (9.58) can be computed by solving Eq. (9.41). A sometimes useful related measure is the time-averaged expected accumulated reward $E[W(t)] = E[Y(t)]/t$, which can also be seen as the average rate at which the reward is accumulated in the interval $(0,t]$. The expected values of the accumulated reward $Y(t)$ and the averaged accumulated reward $W(t)$ can be written in matrix notation as:

$$E[Y(t)] = b(t)r^T, \quad E[W(t)] = g(t)r^T, \tag{9.59}$$

and can be computed by solving Eqs. (9.41) and (9.42). The notion of accumulated reward in the presence of absorbing states in the CTMC will be considered in Chapter 10.

Example 9.12 *Two Repairable Components: Expected Cost of Repair*
In Example 9.4 we considered a two-component parallel system with two possible repair policies. We have shown that the non-shared repair policy, in which there are as many repairers as failed components, provides higher availability than a shared repair policy with a single repairer. Now we compare the expected cost of the repair for the same two policies by means of an MRM. We assign to each state a reward rate equal to the cost of

repairing the components that are failed in that state. Further, we assume that the cost of repair is based on the amount of time spent in actual repair. The per unit time cost is assumed to be c.

For the non-shared (independent) repair (Figure 9.6), the following reward rates are assigned:

$$r_2 = 0, \quad r_1 = c, \quad r_0 = 2c.$$

The expected repair cost in steady state per unit time becomes, from Eq. (9.55),

$$E[C^{(ns)}] = c\pi_1 + 2c\pi_0 = \frac{2\lambda c(\lambda + \mu)}{(\lambda + \mu)^2}. \tag{9.60}$$

Thus, for example, the expected cost of repair over an operational period of duration τ will be $\tau E[C^{(ns)}]$.

In the shared repair case (Figure 9.7), the following reward rates are assigned:

$$r_2 = 0, \quad r_1 = c, \quad r_0 = c.$$

The steady-state expected repair cost per unit time becomes, from Eq. (9.55),

$$E[C^{(sh)}] = c\pi_1 + c\pi_0 = \frac{2\lambda c(\lambda + \mu)}{2\lambda^2 + 2\lambda\mu + \mu^2}. \tag{9.61}$$

Comparing Eq. (9.60) with Eq. (9.61), it is easy to see that $C^{(ns)} > C^{(sh)}$. The non-shared repair policy provides a higher availability but at a higher expected repair cost.

9.4.2   MRM with Impulse Reward

An impulse reward defines the reward accrued on executing a specific activity such as visiting a state or traversing a transition [16, 17, 27].

Impulse Reward Associated with Visiting a State

Given that $R_i \geq 0$ is a non-negative real constant representing the impulse reward associated with state i, a reward equal to R_i is gained immediately each time the CTMC visits state i (we assume that the impulse is gained at the entrance to the state), and the accumulated impulse reward is the sum of the impulses gained at each visit.

Recalling the sample path of the CTMC $\{Z(t) \mid t \geq 0\}$ of Figure 9.1, the impulse rewards are represented in Figure 9.18(a), where R_i, R_j, and R_k are the impulse rewards associated with the states i, j, and k, and where we have assumed that $R_i > R_j$ and $R_k = 0$.

The accumulated reward in this case is a staircase function that increases by the value of the impulse each time the CTMC enters a new state. Corresponding to the sample path of Figure 9.1, the sample path of the accumulated reward $\{Y(t) \mid t \geq 0\}$ is shown in Figure 9.18(b).

Example 9.13   In a CTMC, if we assume $R_i = 1$ and $R_j = 0$ $(j \neq i)$, the accumulated impulse reward up to time t counts the number of visits in state i up to time t.

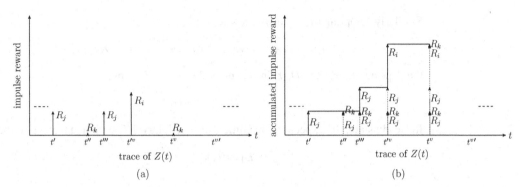

Figure 9.18 Continuous-time Markov chain with impulse reward. (a) Impulse reward; (b) Accumulated impulse reward.

9.5 Availability Measures Defined on an MRM

The availability measures defined in Section 9.3 can be restated in a more uniform and consistent way by resorting to MRMs.

9.5.1 Instantaneous, Steady-State, and Interval Availability

For a dependability model, we define the following reward rates:

$$\begin{cases} r_i = 1 & \text{if } i \in \Omega_\mathsf{u}, \\ r_i = 0 & \text{if } i \in \Omega_\mathsf{d}. \end{cases} \tag{9.62}$$

In vector form, Eq. (9.62) becomes:

$$r = [r_\mathsf{u}\, r_\mathsf{d}] \quad \text{with} \quad r_\mathsf{u} = e \quad \text{and} \quad r_\mathsf{d} = 0. \tag{9.63}$$

The transient availability $A(t)$ defined in Eq. (9.46) and the steady-state availability A defined in Eq. (9.47) can be rewritten as:

$$A(t) = \pi(t)\, r^\mathsf{T}, \qquad A = \pi\, r^\mathsf{T}. \tag{9.64}$$

Using the reward rates as in Eq. (9.62), the expected interval availability in the interval $(0, t]$ (9.48) becomes:

$$A_\mathrm{I}(t) = g(t)\, r^\mathsf{T}.$$

9.5.2 Expected Uptime and Expected Downtime

The expected uptime, defined in Section 9.3.1, adopting the reward rates in Eq. (9.62), becomes:

$$U_\mathrm{I}(t) = b(t)\, r^\mathsf{T}.$$

In steady state, the expected uptime over any interval T_I will then turn out to be

$$U_\mathrm{I} = T_\mathrm{I}\, \pi\, r^\mathsf{T} = T_\mathrm{I} A.$$

Similarly, adopting the following reward rates:

$$r = [r_u \; r_d] \quad \text{with} \quad r_u = 0 \quad \text{and} \quad r_d = e,$$

the expected downtime $D_I(t)$ in the interval $(0,t]$ is given by:

$$D_I(t) = b(t)r^T.$$

In steady state, the expected downtime over any interval T_I will then turn out to be

$$D_I = T_I \pi r^T = T_I U.$$

9.5.3 Expected Number of Transitions

Let q_{ij} $(i \neq j)$ be a generic non-zero, non-diagonal entry of the infinitesimal generator Q. State i is directly connected to state j by means of the transition rate q_{ij}. The expected number of transitions $E[N_{ij}(t)]$ from state i to state j in the interval $(0,t]$ is given by:

$$E[N_{ij}(t)] = q_{ij} \int_0^t \pi_i(u) \, du = q_{ij} b_i(t). \tag{9.65}$$

In steady state, the expected number of transitions $i \to j$ per unit time is:

$$\eta_{ij} = \lim_{t\to\infty} E[N_{ij}(t)]/t = \lim_{t\to\infty} q_{ij} b_i(t)/t = q_{ij} \pi_i. \tag{9.66}$$

Defining the reward structure as:

$$\begin{cases} r_k = q_{ij} & \text{if} \quad k = i, \\ r_k = 0 & \text{if} \quad k \neq i, \end{cases} \tag{9.67}$$

the measures in (9.65) and (9.66) are examples of the expected accumulated reward in the interval $(0,t]$ and the expected reward rate in steady state.

9.5.4 Expected Number of Visits

When computing the expected number of visits into a state we need to distinguish whether the visits occur entering or exiting the state. The expected number of input visits to state j in the interval $(0,t]$ (denoted as $E[N_j(t)]$) can be computed as the sum of the expected number of transitions entering state j. The transition rates of the transitions entering state j are located in the jth column of the infinitesimal generator Q (excluding the diagonal element), hence:

$$E[N_j(t)] = \sum_{i=1,i\neq j}^n q_{ij} \int_0^t \pi_i(u) \, du = \sum_{i=1,i\neq j}^n q_{ij} b_i(t),$$

and in steady state the expected number of visits to state j per unit time is given by:

$$\eta_j = \sum_{i=1,i\neq j}^n q_{ij} \pi_i.$$

Alternatively, the number of visits out of state j can be evaluated as:

$$E[N_j(t)] = \sum_{i=1,i\neq j}^{n} q_{ji} \int_0^t \pi_j(u)\,du = \sum_{i=1,i\neq j}^{n} q_{ji}\,b_j(t) = q_j b_j(t),$$

and in steady state the expected number of visits to state j per unit time is given by:

$$\eta_j = \sum_{i=1,i\neq j}^{n} q_{ji}\pi_j = q_j\pi_j.$$

9.5.5 Expected Number of System Failures/Repairs

The expected number of system failures in the interval $(0,t]$, $E[N_F(t)]$, is given by the total expected number of transitions from any state in Ω_u to any state in Ω_d. The transitions that contribute to the expected number of system failures are those for which an arc exists that directly connects an up state with a down state. These transitions correspond to the non-zero entries in the partition Q_{ud} of the infinitesimal generator matrix (9.45). The row sum of matrix Q_{ud} is an n_u-dimensional (column) vector whose entries give the total rate out of each state in Ω_u toward a state in Ω_d. Finally, applying Eq. (9.65), we obtain for the expected number of system failures in the interval $(0,t]$ $E[N_F(t)]$ (in matrix form):

$$E[N_F(t)] = b_u(t)\,Q_{ud}\,e_d^T. \tag{9.68}$$

Defining the following reward structure:

$$\begin{cases} r_u = Q_{ud}\,e_d^T, \\ r_d = 0, \end{cases} \tag{9.69}$$

the expected number of system failures in the interval $(0,t]$ is cast as an instance of the expected accumulated reward.

To compute the expected number of system repairs $E[N_R(t)]$ we proceed in a similar way. If the matrix Q_{du} has non-zero entries, then

$$E[N_R(t)] = b_d(t)\,Q_{du}\,e_u^T. \tag{9.70}$$

Taking the limit of Eqs. (9.68) and (9.70), after dividing by t, as $t \to \infty$ we obtain the expected number of system failures η_F and system repairs η_R, respectively, per unit time when the system is in steady state. The quantities η_F and η_R can also be seen as the unconditional failure intensity (failure frequency) and the unconditional repair intensity (repair frequency) – see also (3.5.6):

$$\eta_F = \pi_u\,Q_{ud}\,e_d^T, \qquad \eta_R = \pi_d\,Q_{du}\,e_u^T. \tag{9.71}$$

Example 9.14 *One Repairable Component*

Consider a system with one repairable component as in Example 9.2. The only up state is state 1, and application of Eq. (9.68) provides:

$$E[N_F(t)] = b_1(t)\lambda = \lambda \int_0^t \left[\frac{\mu}{\lambda + \mu} + \frac{\lambda}{\lambda + \mu} e^{-(\lambda + \mu)u} \right] du$$

$$= \frac{\lambda \mu}{\lambda + \mu} t + \frac{\lambda^2}{(\lambda + \mu)^2} (1 - e^{-(\lambda + \mu)t}).$$

In this case, for $t \to \infty$ the expected number of system failures grows linearly with slope $\lambda \mu / (\lambda + \mu)$. However, in the limit we can apply Eq. (9.71) to get the expected number of failures per unit time in steady state as $\eta_F = \lambda \mu / (\lambda + \mu)$.

Example 9.15 *Telecommunication Switching System Model with Fault Detection/Reconfiguration Delay [28]*
The availability model shown in Figure 9.12 was developed under the assumption that a fault in the switching system is detected instantaneously and the repair process starts immediately after the failure. This assumption is relaxed in the availability model shown in Figure 9.19. We assume that there are n channels and the MTTF of a channel is $1/\gamma$. After a channel failure, the system goes to a detection/reconfiguration state, denoted by d_k, where k is the number of non-failed channels before the system goes to the detection/reconfiguration state. The time to fault detection/reconfiguration is assumed to be exponentially distributed with mean $1/\delta$. After detection/reconfiguration, channel failure is either recovered with probability c (coverage factor) or the fault stays within the system with probability $(1 - c)$. A failed channel that cannot be recovered by detection/reconfiguration is finally repaired with MTTR $= 1/\tau$.

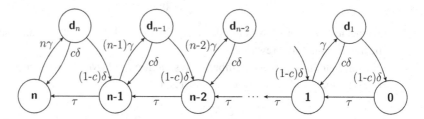

Figure 9.19   Availability model of switching system with fault detection/reconfiguration delay.

The steady-state probability vector $\boldsymbol{\pi}$ for the CTMC in Figure 9.19 is denoted by:

$$\boldsymbol{\pi} = [\pi_n, \pi_{d_n}, \dots, \pi_0]. \tag{9.72}$$

The generator matrix for the CTMC in Figure 9.19 is denoted by $\boldsymbol{Q}$. The steady-state probabilities can be computed by solving the system of linear equations:

$$\boldsymbol{\pi}\boldsymbol{Q} = 0 \qquad \text{with} \qquad \sum_{k=0}^n \pi_k + \sum_{k=1}^n \pi_{d_k} = 1. \tag{9.73}$$

We use the Markov reward approach described in Section 9.4.1 to compute the steady-state unavailability. We assume that detection/reconfiguration states are considered to be down if the sojourn times in those states are longer than a predefined threshold t_h. In a detection/reconfiguration state, the probability that the sojourn time is longer than t_h is given by $e^{-\delta t_h}$. So, we assign reward rate $e^{-\delta t_h}$ to the detection/reconfiguration states d_k ($k = 1, \ldots, n$), reward rate 1 to state 0 and reward rate 0 to all other states. Thus, the steady-state unavailability is given by:

$$UA_d = e^{-\delta t_h} \sum_{k=1}^{n} \pi_{d_k} + \pi_0. \tag{9.74}$$

The transient solution is obtained by solving the ODE system

$$\frac{d\pi(t)}{dt} = \pi(t) Q, \tag{9.75}$$

and the unavailability vs. time is given by:

$$UA_d(t) = e^{-\delta t_h} \sum_{k=1}^{n} \pi_{d_k}(t) + \pi_0(t). \tag{9.76}$$

9.5.6 Equivalent Failure and Repair Rate

In steady-state availability modeling it is sometimes useful to have a high-level view of the system, and to reduce the system to a two-state model in which one single up state replaces the aggregate of all the up states of the original model, and one single down state replaces the aggregate of all the down states of the original model. The single up state and the single down state communicate via an *equivalent failure rate* and an *equivalent repair rate*. For this aggregation, it is necessary to define values for the equivalent failure rate λ_{eq} and the equivalent repair rate μ_{eq} such that the steady-state availability of the original model can be written as:

$$A = \frac{\mu_{eq}}{\lambda_{eq} + \mu_{eq}} = \frac{\text{MTTF}_{eq}}{\text{MTTF}_{eq} + \text{MTTR}_{eq}},$$

where $\text{MTTF}_{eq} = 1/\lambda_{eq}$ and $\text{MTTR}_{eq} = 1/\mu_{eq}$.

Figure 9.20 depicts the concept of aggregating a general CTMC model into a two-state model such that the two-state model is equivalent to the original one with respect to the steady-state availability. λ_{eq} is defined as:

$$\lambda_{eq} = P\{\text{transition u} \rightarrow \text{d takes place} \,|\, \text{system is up}\} = \frac{\eta_F}{A}, \tag{9.77}$$

where η_F is the unconditional failure intensity (or the failure frequency) defined in Eq. (9.71) and A is the steady-state availability (9.47). Similarly, the equivalent repair rate is:

$$\mu_{eq} = P\{\text{transition d} \rightarrow \text{u takes place} \,|\, \text{system is down}\} = \frac{\eta_R}{1 - A},$$

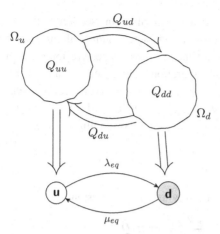

Figure 9.20   Construction of an equivalent two-state model.

where η_R is the unconditional repair intensity (or the repair frequency) defined by Eq. (9.71).

Example 9.16   *High-Availability Platform from Sun Microsystems*
We revisit Example 4.24, where we considered a higher-level RBD model of the top-level architecture of a typical carrier grade platform that was examined in [29]. For that model, the steady-state availabilities of the various subsystems were displayed in Table 4.7. In the present example, we consider one of the subsystems, the fan subsystem, and we develop the detailed lower-level CTMC model in Figure 9.21. For future reference we refer to this CTMC model as M_{Fan}. In this model, λ_F is the failure rate of a single fan, and μ_{F1} and μ_{F2} are the respective repair rates when one or both of the fan units have failed.

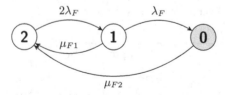

Figure 9.21   Continuous-time Markov chain model, M_{Fan}, of the fan subsystem of the carrier-grade platform [29].

States 2 and 1 are **up** states, while state 0 is a **down** state. For this model, writing out the balance equations we get:

$$
\begin{cases}
2\lambda_F \pi_2 & = & \mu_{F1}\pi_1 + \mu_{F2}\pi_0, \\
(\lambda_F + \mu_{F1})\pi_1 & = & 2\lambda_F \pi_2, \\
\mu_{F2}\pi_0 & = & \lambda_F \pi_1, \\
\pi_0 + \pi_1 + \pi_2 & = & 1.
\end{cases}
$$

Solving the balance equations, we get

$$
\begin{cases}
\pi_2 &= \dfrac{(\lambda_F \mu_{F2} + \mu_{F1} \mu_{F2})}{(2\lambda_F^2 + 3\lambda_F \mu_{F2} + \mu_{F1} \mu_{F2})}, \\[2mm]
\pi_1 &= \dfrac{2\lambda_F \mu_{F2}}{(2\lambda_F^2 + 3\lambda_F \mu_{F2} + \mu_{F1} \mu_{F2})}, \\[2mm]
\pi_0 &= \dfrac{2\lambda_F^2}{(2\lambda_F^2 + 3\lambda_F \mu_{F2} + \mu_{F1} \mu_{F2})}.
\end{cases}
$$

Thus, the steady-state system unavailability U_{ss} is given by:

$$
U_{ss} = \pi_0 = \frac{2\lambda_F^2}{(2\lambda_F^2 + 3\lambda_F \mu_{F2} + \mu_{F1} \mu_{F2})},
$$

and the steady-state availability A_{ss} is given by:

$$
A_{ss} = 1 - U_{ss} = \frac{(3\lambda_F \mu_{F2} + \mu_{F1} \mu_{F2})}{(2\lambda_F^2 + 3\lambda_F \mu_{F2} + \mu_{F1} \mu_{F2})}.
$$

The expected downtime D_I of the system in minutes per year is given by:

$$
D_I = 60 \times 24 \times 365 \times (1 - A_{ss}).
$$

Substituting $\lambda_F = 3.4905147 \times 10^{-7}\,\mathrm{f\,hr^{-1}}$, $\mu_{F1} = 0.24\,\mathrm{rep\,hr^{-1}}$, and $\mu_{F2} = 0.226415\,\mathrm{rep\,hr^{-1}}$ as given in [29], we get the steady-state availability of the fan subsystem $A_{ss} = 0.999\,999\,999\,9955$, the steady-state unavailability $U_{ss} = 4.484\,256\,28 \times 10^{-12}$ and the expected steady-state downtime $D_I = 60 \times 24 \times 365 \times U_{ss} = 2.356\,925\,10 \times 10^{-6}\,\mathrm{min\,yr^{-1}}$ (see Example 3.9), which matches with the value given in Table 4.7. A problem at the end of this section is about computing the failure and repair frequencies as well as the equivalent failure and repair rates of this model.

Example 9.17 *Availability of OS in IBM WebSphere Model*
This example considers the availability modeling of the Linux operating system (OS) used in the IBM SIP WebSphere model that was presented in [30].

From the UP state, the model enters the down state DN with failure rate λ_{OS}. After failure detection with a mean time of $1/\delta_{OS}$, the system enters the failure-detected state DT. The OS is then rebooted with the mean time to reboot given by $1/\beta_{OS}$. With probability b_{OS} the reboot is successful and the system returns to the UP state. With probability $1 - b_{OS}$ the reboot is unsuccessful, and the system enters the DW state where a repair person is summoned. The travel time of the repair person is exponentially distributed with rate α_{sp}. The system then moves to state RP. The repair (which in this case could well be a patch of the OS) takes a mean time of $1/\mu_{OS}$, and after its completion the system returns to the UP state.

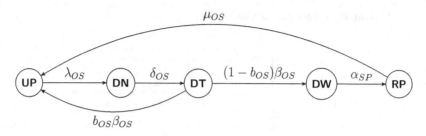

Figure 9.22 Continuous-time Markov chain model, M_{OS}, of the node OS in an IBM WebSphere system.

Setting up the balance equations for this model, we get:

$$
\begin{cases}
\lambda_{OS}\,\pi_{UP} &= \mu_{OS}\pi_{RP} + b_{OS}\,\beta_{OS}\,\pi_{DT}, \\
\delta_{OS}\,\pi_{DN} &= \lambda_{OS}\,\pi_{UP}, \\
\beta_{OS}\,\pi_{DT} &= \delta_{OS}\,\pi_{DN}, \\
\alpha_{sp}\,\pi_{DW} &= (1-b_{OS})\,\beta_{OS}\,\pi_{DT}, \\
\mu_{OS}\,\pi_{RP} &= \alpha_{sp}\,\pi_{DW}, \\
\pi_{UP} + \pi_{DN} + \pi_{DT} + \pi_{DW} + \pi_{RP} &= 1.
\end{cases}
$$

After rearranging and solving the equations, the steady-state availability of the OS is:

$$
A_{ss} = \pi_{UP} = \frac{1}{\lambda_{OS}}\left[\frac{1}{\lambda_{OS}} + \frac{1}{\delta_{OS}} + \frac{1}{\beta_{OS}} + (1-b_{OS})\left(\frac{1}{\alpha_{sp}} + \frac{1}{\mu_{OS}}\right)\right]^{-1}.
$$

Substituting the values for the parameters from [30], $1/\lambda_{OS} = 4000\,\text{hr}$, $1/\delta_{OS} = 1\,\text{hr}$, $1/\beta_{OS} = 10\,\text{min}$, $b_{OS} = 0.9$, $1/\mu_{OS} = 1\,\text{hr}$, and $1/\alpha_{sp} = 2\,\text{hr}$, we get $A_{ss} = 0.9999633$ and the expected downtime D_I in minutes per year is $D_I = 60 \times 24 \times 365 \times (1 - A_{ss}) = 19.29\,\text{min yr}^{-1}$. A problem at the end of this section is about computing the failure and repair frequencies as well as the equivalent failure and repair rates of this model.

Problems

9.10 For the CTMC model of Figure 9.21, find the expressions for the failure and repair frequencies in steady state and the equivalent failure and repair rates.

9.11 For the CTMC model of Figure 9.22, find the expressions for the failure and repair frequencies in steady state and the equivalent failure and repair rates.

9.5.7   Defects per Million

Defects per million (DPM) is a commonly used service (un)reliability measure for telecommunication systems [31, 32]. In simple terms, the metric counts the number of unsuccessful attempts per million attempts at obtaining a service. A simple approach to calculate the DPM is to multiply the unavailability by a constant of proportionality [33]. This simple approach makes intuitive sense, because if the system is unavailable at the moment the service is requested, then the request can be considered lost or unsuccessful. Given the system unreliability U_{ss} in steady state, the DPM metric can be approximated by

$$DPM = U_{ss} \times 10^6.$$

A more accurate approach will be considered in the next chapter.

Example 9.18 *Defects Per Million for an Application Server*
This example, adapted from [34], presents the modeling of an application server that services SIP request messages. One method of providing fault tolerance is to use application server replication in cold standby mode that acts as the backup for the primary server. When the primary server fails, the standby (cold) server starts to take over from the (failed) primary server.

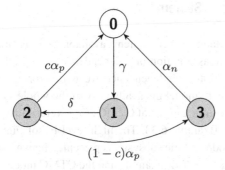

Figure 9.23 Continuous-time Markov chain model for server with cold standby replication.

Figure 9.23 shows the CTMC model for an application server with cold standby replication. In this model, the server is available for handling incoming requests in state 0. The server may fail at the rate γ, upon which the system enters state 1, in which the server is considered unavailable to handle incoming requests. The unavailability in state 1 is not observable until the failure is detected. This takes an exponentially distributed time with mean $1/\delta$. The system subsequently enters state 2 in which recovery is initiated. If the threshold of the number of failures within a certain interval has not been reached, the process is restarted on the same node (i.e., the primary server). This takes an exponentially distributed time with mean $1/\alpha_p$. If the threshold is exceeded, the process is failed over to a different node (i.e., the cold standby server). The time to perform this failover is assumed to be exponentially distributed with mean $1/\alpha_n$. The threshold-crossing behavior is captured by the use of a coverage parameter, denoted by c.

Writing the balance equations for this model, we get:

$$\begin{cases} \gamma \pi_0 & = & c\alpha_p \pi_2 + \alpha_n \pi_3, \\ \delta \pi_1 & = & \gamma \pi_0, \\ \alpha_p \pi_2 & = & \delta \pi_1, \\ \alpha_n \pi_3 & = & (1-c)\alpha_p \pi_2, \\ \pi_0 + \pi_1 + \pi_2 + \pi_3 & = & 1. \end{cases}$$

Solving the balance equations, we get for the steady-state availability:

$$A_{ss} = \pi_0 = \frac{1}{\gamma}\left[\frac{1}{\gamma} + \frac{1}{\delta} + \frac{1}{\alpha_p} + \frac{(1-c)}{\alpha_n}\right]^{-1}.$$

For this system, the DPM can be approximated as

$$DPM = (1 - A_{ss}) \times 10^6. \tag{9.78}$$

Substituting $1/\gamma = 10\,\mathrm{d}$, $1/\delta = 1\,\mathrm{s}$, $1/\alpha_p = 30\,\mathrm{s}$, $1/\alpha_n = 15\,\mathrm{s}$, and $c = 0.95$, we get $A_{ss} = 0.999\,963\,25$, and DPM $= 36.75$. We will discuss a more accurate method of computing the DPM in the next chapter.

9.6   Case Study: IBM Blade Server System

An FT model of the IBM blade server system availability was introduced in Example 6.33, where, clearly, the assumption of independence among all components was made. To relax the assumption of independence we will resort to the use of a CTMC. However, using a CTMC to model the availability of the whole system will be too complex. Instead, we will illustrate the use of CTMCs for several subsystems within the IBM blade server system of Example 6.33. The high-level FT of Figure 6.23 will be combined with the CTMC models of the subsystems presented here as a hierarchical modeling case study in Example 18.3. This example and the CTMC models are derived from [35], where the modeling of the entire system is discussed in detail.

Example 9.19   *Midplane Availability Model of the Blade Server System*
In this example we consider the failure/repair behavior of the midplane of a blade server [35]. The corresponding Markov model is shown in Figure 9.24, and is named M_{Md} for later reference.

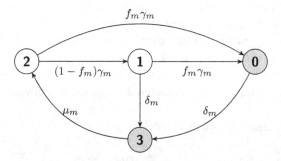

Figure 9.24   Continuous-time Markov chain model, M_{Md}, of the midplane of the blade server system.

In the model, state 2 is the fault-free state of the midplane. The midplane fails with a mean time to failure of $1/\gamma_m$. Upon occurrence of a failure, the midplane may enter state 1, which represents the condition where it is still operational after a fault has been detected due to redundant paths. At this point, failover to an alternate communication path, if necessary, was successful and a midplane replacement was requested. A failure not covered (due to inability to failover to the alternate path) results in a transition to state 0. State 0 is a **down** state and the transition rate to state 0 is determined by the common mode factor, f_m.

In both states 0 and 1, the system is awaiting the arrival of the service person to perform the repair. Upon the arrival of the service person with a mean travel time equal to $1/\delta_m$, the system moves to state 3 from either state 0 or state 1. The midplane is repaired with a mean time to repair equal to $1/\mu_m$. State 3 is a **down** state because the chassis must be taken out of service to replace the midplane. The steady-state availability of the midplane submodel is the probability that the system is in one of the operational states 1 or 2.

Writing out the balance equations for the CTMC model, we get:

$$\begin{cases} \gamma_m \pi_2 & = & \mu_m \pi_3, \\ (f_m \gamma_m + \delta_m) \pi_1 & = & (1 - f_m) \gamma_m \pi_2, \\ \delta_m \pi_0 & = & f_m \gamma_m (\pi_2 + \pi_1), \\ \mu_m \pi_3 & = & \delta_m (\pi_0 + \pi_1), \\ \pi_0 + \pi_1 + \pi_2 + \pi_3 & = & 1. \end{cases}$$

Solving the above balance equations, the steady-state availability is given as $A_{Md} = \pi_2 + \pi_1$. A closed-form expression for the midplane steady-state availability is:

$$A_{Md} = \frac{\delta_m \mu_m (\gamma_m + \delta_m)}{(\delta_m + f_m \gamma_m)(\gamma_m \mu_m + \delta_m \mu_m + \gamma_m \delta_m)}.$$

Example 9.20 *Cooling and Power Domain Availability Model of the Blade Server System [35]*
Continuing the availability modeling of the IBM BladeCenter, we now consider the failure/repair behavior of the chassis, cooling and power domain subsystems [35]. The corresponding Markov model is shown in Figure 9.25 and referred to as M_P and M_C

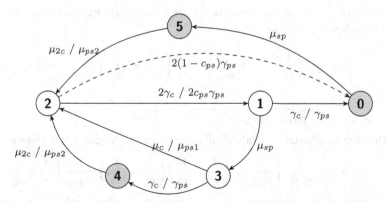

Figure 9.25 Continuous-time Markov chain model of the cooling (M_C) and power domains (M_P) of the blade server system.

for later reference. The models for the chassis cooling and both the power domains are similar in structure, and hence a single CTMC reflecting both the availability models is presented in Figure 9.25, where two rates are specified: the first is for the cooling submodel, while the second is for the power domain submodel. The dashed transition applies only to the power domain submodel.

In the model, states 1, 2, and 3 are up states. An individual blower fails with an MTTF of $1/\gamma_c$. Blower failures are detected by a management module that monitors tachometers on each blower. When one of the blowers or its monitoring hardware fails, the remaining blower enters a full-speed mode. Because of this fail-safe mode, no coverage factor is included in the cooling subsystem model and the dashed transition from state 2 to state 0 is not used in this submodel. State 2 represents the condition when both blowers are operating normally. If one of the two blowers or its monitoring hardware fails, the cooling subsystem transitions from state 2 to state 1. A service person is summoned with mean service response time of $1/\mu_{sp}$. When the service person arrives, the subsystem enters state 3. Since the blower is hot-pluggable and redundant, it can be replaced and the cooling subsystem returns to state 2 with a mean time of $1/\mu_c$ without any downtime. While in state 1, the second blower could fail before the arrival of the service person, resulting in a shutdown of the cooling subsystem and hence of the entire blade server. In this case, a transition to state 0 occurs. The cooling subsystem then transitions to state 5 on the arrival of the service person with a mean service response time of $1/\mu_{sp}$. The repair process completes and the subsystem is restored to state 2 with a mean repair time for both blowers of $1/\mu_{2c}$. Note that, for simplicity, state 5 could have been omitted and a transition from state 0 to state 2 with a transition rate of $1/(1/\mu_{sp} + 1/\mu_{2c})$ utilized. Modeled in this way, the transition time from state 0 to state 2 is not exponentially distributed but rather two-stage hypoexponentially distributed. Strictly speaking, the model will then be a semi-Markov process (SMP) (see Chapter 14), that, for the steady-state solution, can be treated as a Markov chain. This is because the steady-state probabilities of an SMP depend only on the mean sojourn times in states. The second blower could also fail while the subsystem is under repair in state 3. Here, the transition occurs to state 4 and, since the repair person is already on-site, only the repair time for both blowers is required to return the subsystem to state 2. Let d_C represent the denominator for the closed-form expression for the steady-state availability of this submodel:

$$d_C = \frac{1}{\gamma_c^2}\left(\frac{1}{\gamma_c} + \frac{1}{\mu_c} + \frac{3}{\mu_{sp}}\right) + \frac{1}{\gamma_c}\left(\frac{2}{\mu_{sp}^2} + \frac{2}{\mu_{2c}}\left(\frac{1}{\mu_c} + \frac{1}{\mu_{sp}}\right) + \frac{3}{\mu_c\mu_{sp}}\right)$$
$$+ \frac{2}{\mu_c\mu_{sp}}\left(\frac{1}{\mu_{2c}} + \frac{1}{\mu_{sp}}\right).$$

Then the steady-state availability of the cooling subsystem, A_C, is given as

$$A_C = \frac{1}{\gamma_c\, d_C}\left(\frac{2}{\mu_c\gamma_c} + \left(\frac{1}{\gamma_c} + \frac{1}{\mu_c}\right)\left(\frac{1}{\gamma_c} + \frac{1}{\mu_{sp}}\right) + \frac{2}{\mu_{sp}}\left(\frac{1}{\gamma_c} + \frac{1}{\mu_c}\right)\right).$$

The reader is urged to verify the above expression by writing the balance equations and solving them.

The BladeCenter contains two identical power domain subsystems. Each power domain contains redundant, hot-pluggable power supply modules and is also represented by the Markov submodel shown in Figure 9.25. States 1, 2, and 3 are up states. Each power supply module fails with an MTTF of $1/\gamma_{ps}$, and they have mean times to repair of $1/\mu_{ps1}$ and $1/\mu_{ps2}$ for one or two power supplies, respectively. This model differs from the cooling subsystem model in that when one of the power supply modules fails in state 2, a not covered fault can bring both of the power supplies down with probability $(1 - c_{ps})$, as depicted by the dashed transition from state 2 to state 0. Let d_P represent the expression for the denominator:

$$d_P = \frac{1}{\gamma_{ps}^2}\left(\frac{1}{\gamma_{ps}} + \frac{2c_{ps}}{(\mu_{ps1} - \mu_{ps2})} + \frac{1}{\mu_{ps1}} + \frac{2}{\mu_{ps2}} + \frac{3}{\mu_{sp}}\right) +$$

$$\frac{1}{\gamma_{ps}}\left(\frac{2}{\mu_{sp}^2} + \frac{2}{\mu_{ps2}}\left(\frac{1}{\mu_{ps2}} + \frac{1}{\mu_{sp}}\right) + \frac{3}{\mu_{ps1}\mu_{sp}}\right) + \frac{2}{\mu_{ps1}\mu_{sp}}\left(\frac{1}{\mu_{ps2}} + \frac{1}{\mu_{sp}}\right).$$

Then, a closed-form expression for the steady-state availability of the power supply subsystem, A_P, is given by:

$$A_P = \frac{1}{\gamma_{ps}\, d_P}\left(\frac{2c_{ps}}{\mu_{ps1}\gamma_{ps}} + \left(\frac{1}{\gamma_{ps}} + \frac{1}{\mu_{ps1}}\right)\left(\frac{1}{\gamma_{ps}} + \frac{1}{\mu_{sp}}\right) + \frac{2c_{ps}}{\mu_{sp}}\left(\frac{1}{\gamma_{ps}} + \frac{1}{\mu_{ps1}}\right)\right).$$

The reader is urged to verify the above expression by writing out the balance equations and solving them.

Example 9.21 *Processor and Memory Availability Model of the Blade Server [35]* Continuing the availability modeling of the IBM BladeCenter, we now consider the failure/repair behavior of the processor and memory subsystems of the blade server [35]. The corresponding CTMC is shown in Figure 9.26, and is labeled M_{CPU} and M_M for later reference. The models for the processor and memory subsystems are both similar in structure, and hence a single CTMC reflecting them both is presented in Figure 9.26. Where two rates are specified, the first one applies to the processor submodel, while

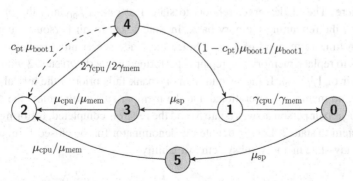

Figure 9.26 Continuous-time Markov chain model of the processor (M_{CPU}) and memory (M_M) of the blade server system.

the second applies to the memory submodel. The dashed transition applies only to the processor submodel.

Each blade server has two processors. States 2 and 1 are up states representing a blade server with two or one operational processors. While in state 2, the processor subsystem may experience a failure represented by the transition to state 4. This event may be due to a transient hardware fault with probability c_{pt}. In this case, the fault clears on reboot and the processor subsystem recovers to state 2, as depicted by the dashed transition, when the blade server reboots. Otherwise, the failure is a hard fault that is detected on blade server reboot and the processor subsystem recovers in degraded mode to state 1 with a single functioning processor. In state 1, a service person is summoned with a mean time to respond of $1/\mu_{sp}$. When this person arrives the blade server must be removed for repair, so the processor subsystem enters the down state 3 for the repair and then returns to state 2 once it is completed. If the second processor fails while the processor subsystem is still in state 1, then no functioning processors are left and the subsystem transitions to the down state 0 – a repair person is summoned and the repair process is completed, restoring the subsystem to state 2. Let d_{CPU} denote the denominator for the closed-form expression of the steady-state availability:

$$d_{CPU} = \left(\frac{2}{\mu_{boot1}} + \frac{1}{\gamma_{cpu}} + \frac{2(1 - c_{pt})}{\mu_{cpu}} + \frac{2(1 - c_{pt})}{\mu_{sp}} \right).$$

A closed-form expression for the steady-state availability of the processor subsystem, A_{CPU}, is:

$$A_{CPU} = \frac{2(1 - c_{pt})}{(\gamma_{cpu} + \mu_{sp})d_{CPU}} + \frac{1}{\gamma_{cpu}\, d_{CPU}}.$$

Each blade server has two banks of memory. Each bank comprises two memory DIMMs. The memory subsystem is also represented by the Markov submodel in Figure 9.26. States 2 and 1 are up states representing a blade server with two or one operational banks of memory, respectively. While in state 2, the memory subsystem may experience an unrecoverable multi-bit error represented by the transition to state 4. The error is processed by the blade server BIOS and the memory experiencing the multi-bit error is always reconfigured so that the dashed transition from state 4 to state 2 is not used here. The blade server reboots to state 1 at rate μ_{boot1} and the blade recovers, utilizing the remaining memory bank. In state 1, the repair person is summoned with a mean time to respond of $1/\mu_{sp}$. Since the blade server must be removed from the chassis to replace memory, the repair is performed in the down state 3 with a mean time to repair of $1/\mu_{mem}$. If the second memory bank fails prior to the arrival of the repair person, no memory is operational and the memory subsystem enters the down state 0 until the repair person arrives (state 5) and the repair is completed, restoring the memory subsystem to state 2. Let d_M denote the denominator for the closed-form expression of the steady-state memory subsystem availability:

$$d_M = \left(\frac{2}{\mu_{boot1}} + \frac{1}{\gamma_{mem}} + \frac{2}{\mu_{mem}} + \frac{2}{\mu_{sp}} \right).$$

A closed-form expression for the steady-state availability of the memory subsystem, A_M, is:

$$A_M = \frac{2}{(\gamma_{mem} + \mu_{sp})\, d_M} + \frac{1}{\gamma_{mem}\, d_M}.$$

Example 9.22 *Disk and RAID Availability Model of the Blade Server System [35]*
Continuing the availability modeling of the IBM blade server, we now consider the failure/repair behavior of the disk subsystem of the blade server [35]. The corresponding Markov model, M_D, is shown in Figure 9.27.

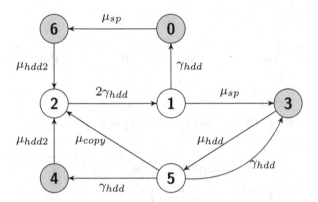

Figure 9.27   Continuous-time Markov chain model, M_D, of the disk subsystem of the blade server system.

Two configurations for the disk subsystem are considered. One configuration uses a single disk in each blade, while the other uses a mirrored RAID 1 disk per blade. In the first case, a disk is modeled as an alternating renewal process. The closed-form solution for the availability of this simple disk subsystem is

$$A_D = \frac{1}{\gamma_{hdd}} \left(\frac{1}{\gamma_{hdd}} + \frac{1}{\mu_{hdd}} + \frac{1}{\mu_{sp}} \right)^{-1}.$$

For the second case, the Markov submodel captures the behavior of dual disk drives in a mirrored configuration (i.e., RAID 1) where one disk drive can fail and all the data and programs are still accessible via the remaining disk drive. The RAID 1 CTMC availability model has six states (see Figure 9.27). Both disk drives are in operation in the up state 2. The MTTF of each disk drive is $1/\gamma_{hdd}$. States 1 and 5 are up states representing a blade server with one operational disk drive. When the RAID controller chip on the blade server detects that a drive has failed, the RAID subsystem enters state 1. The remaining drive supplies all the data to the blade server. A repair person is summoned with a mean time to respond of $1/\mu_{sp}$, and the subsystem enters the down state 3 since the blade server must be removed from service to replace the drive. If the second drive fails before the arrival of the repair person, the RAID subsystem transitions from state 1 to down state 0, with no remaining drives, where it remains until the repair person arrives and the repair is completed. In state 3, the disk drive is replaced with a mean time to repair of $1/\mu_{hdd}$ and the subsystem enters up state 5. Then, the data must be copied onto the new disk drive with a mean time to completion of $1/\mu_{copy}$. If the new drive fails during the copy process, then the subsystem returns to state 3 and that drive is replaced a second time. From state 5, it is also possible for the disk drive holding the data to fail before the copy is completed. In that case, the subsystem enters the down state 4. In both states 0 and 4, both disk drives are replaced with fresh preloaded drives with a mean time to repair of $1/\mu_{hdd2}$. Let d_{RAID} denote the denominator for a closed-form expression for the steady-state availability of the disk subsystem utilizing RAID 1:

$$d_{RAID} = \frac{1}{\gamma_{hdd}^2}\left(\frac{1}{\gamma_{hdd}} + \frac{3}{\mu_{copy}} + \frac{2}{\mu_{hdd}} + \frac{3}{\mu_{sp}}\right)$$
$$+ \frac{1}{\gamma_{hdd}}\left(\frac{2}{\mu_{sp}^2} + \frac{2}{\mu_{copy}\mu_{hdd2}} + \frac{4}{\mu_{copy}\mu_{hdd}} + \frac{3}{\mu_{copy}\mu_{sp}} + \frac{2}{\mu_{hdd2}\mu_{sp}}\right)$$
$$+ \frac{2}{\mu_{copy}\mu_{sp}}\left(\frac{1}{\mu_{sp}} + \frac{1}{\mu_{hdd2}}\right).$$

Then the steady-state RAID availability, A_{RAID}, is given by:

$$A_{RAID} = \frac{1}{\gamma_{hdd}d_{RAID}}\left(\frac{2}{\mu_{copy}\gamma_{hdd}} + \left(\frac{1}{\mu_{sp}} + \frac{1}{\gamma_{hdd}}\right)\left(\frac{1}{\gamma_{hdd}} + \frac{1}{\mu_{copy}}\right)\right.$$
$$\left. + 2\left(\frac{1}{\mu_{sp}\gamma_{hdd}} + \frac{1}{\mu_{copy}\mu_{sp}}\right)\right).$$

Example 9.23 *Software Availability Model of the Blade Server [35]*
Continuing the availability modeling of the IBM blade server, we now consider the failure/repair behavior of the software subsystem of the blade server [35]. The corresponding Markov model is shown in Figure 9.28 and is named M_{Sw} for later reference.

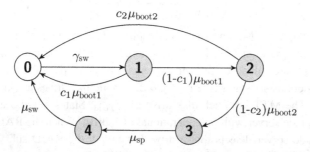

Figure 9.28   Continuous-time Markov chain model, M_{Sw}, of the software of the blade server system.

The model has five states. State 0 is the only up state. Even without a hardware failure, the blade server software can crash or hang and the blade server enters the down state 1 until the operating system can perform a fast reboot and the middleware and applications can restart. A software failure is covered by a fast reboot with probability

c_1. If the failure is not covered by a fast reboot, then the system moves to state 2, where a longer reboot and recovery action such as consistency checks and automated data recovery are attempted. This step will be successful with a coverage factor of c_2 and the software returns to state 0. Otherwise, the software enters state 3, where a repair person is required to restore corrupted data or to perform a software repair or other recovery action. Conceptually, manual recovery will be required to deal with residual Bohrbugs (i.e., a bug that manifests itself reliably under a well-defined but possibly unknown set of conditions [36]), while the faster automated recovery will succeed if the failure is caused by a Mandelbug (i.e., a bug whose causes are so complex that its behavior appears chaotic). A closed-form expression for the steady-state availability of the software subsystem is:

$$A_{Sw} = \frac{1}{\gamma_{sw}} \left(\frac{1}{\mu_{boot1}} + (1 - c_1)\frac{1}{\mu_{boot2}} + \frac{1}{\gamma_{sw}} + (1 - c_1 + c_1 c_2 - c_2) \left(\frac{1}{\mu_{sw}} + \frac{1}{\mu_{sp}} \right) \right)^{-1}.$$

Problems

9.12 Returning to the midplane availability model presented in Example 9.19, derive the expected number of system failures and repairs.

9.13 For each of the CTMC subsystem models for the IBM BladeCenter presented above, write out the balance equations.

9.14 Using the balance equations derived in the previous problem, derive the equations for computing the subsystem availability. The final results in this case are already presented in the examples above.

9.15 Derive expressions for the subsystem downtime in minutes per year for each of the CTMC models of the IBM BladeCenter system constructed above.

9.16 In the RAID submodel of Figure 9.27, we note that in state 5, the new disk that replaces a failed disk may experience failure with a greater rate than can be attributed to infant mortality. Similarly, the source disk may experience a higher failure rate since it is being actively used both for servicing requests and as the source for copying the data to the new disk. Considering these issues, suppose the failure rate of the new disk is γ'_{hdd} and the source disk is now γ''_{hdd}; modify the CTMC accordingly and derive the expression for the steady-state RAID availability.

9.17 In the processor submodel of Figure 9.26, we have ignored the possibility of the second processor failing in states 4 and 3. Add a transition at the rate γ_{cpu} from state 4 to state 0 and a transition from state 3 to state 5 at the rate γ_{cpu}. Derive the expression for the steady-state subsystem availability. Repeat for the memory subsystem.

9.7   Parametric Sensitivity Analysis

As stated in Section 2.2.7, parametric sensitivity analysis is focused on pinpointing the particular factor(s) or component(s) that are the most influential in affecting the output behavior of the system. Parametric sensitivity analysis has already been exploited in Section 3.5.7. Further, importance indices and Birnbaum indices, introduced in Section 4.2.7 in the framework of RBDs and in Section 6.6.1 in the framework of FTs, are examples of sensitivity measures.

Sensitivity analysis is performed by computing the partial derivatives of the output metric of interest with respect to each input parameter. The derivatives are referred to as sensitivity functions. The sensitivity function of a given measure Y, which depends on a parameter θ, is computed as in Eq. (9.79) for unscaled and scaled sensitivity:

$$S_\theta(Y) = \frac{\partial Y}{\partial \theta}, \tag{9.79}$$

$$SS_\theta(Y) = \frac{\partial Y}{\partial \theta}\left(\frac{\theta}{Y}\right). \tag{9.80}$$

The adoption of scaled or unscaled sensitivity analysis depends on the type of the output metric, the range of values of output as well as the input parameter and on the cost of collecting accurate information regarding an input parameter [37, 38].

The absolute values of sensitivity functions provide the ranking that is used to compare the degree of influence of each input parameter on the output metric of interest.

Example 9.24 *DPM for an Application Server: Sensitivity Analysis*
Returning to Example 9.18, the steady-state availability can be expressed as

$$A_{ss}(\gamma, \delta, \alpha_p, \alpha_n, c) = \pi_0 = \frac{1}{\gamma}\left[\frac{1}{\gamma} + \frac{1}{\delta} + \frac{1}{\alpha_p} + \frac{(1-c)}{\alpha_n}\right]^{-1}.$$

As is clear from the example, the steady-state availability is a function of the five parameters $(\gamma, \delta, \alpha_p, \alpha_n, c)$. The unscaled and scaled sensitivity of A_{ss} to all the parameters is shown in Table 9.2.

Substituting in the numerical values taken from Example 9.18 (i.e., $1/\gamma = 10\,$d, $1/\delta = 1\,$s, $1/\alpha_p = 30\,$s, $1/\alpha_n = 15\,$s, and $c = 0.95$), we get the numerical values, for

Table 9.2 Sensitivity of A_{ss}.

Parameter θ	$S_\theta(A_{ss})$	$SS_\theta(A_{ss})$
γ	$-\frac{1}{\gamma^2}\left[\frac{1}{\delta} + \frac{1}{\alpha_p} + \frac{(1-c)}{\alpha_n}\right] \times \left[\frac{1}{\gamma} + \frac{1}{\delta} + \frac{1}{\alpha_p} + \frac{(1-c)}{\alpha_n}\right]^{-2}$	$-\left[\frac{1}{\delta} + \frac{1}{\alpha_p} + \frac{(1-c)}{\alpha_n}\right] \times \left[\frac{1}{\gamma} + \frac{1}{\delta} + \frac{1}{\alpha_p} + \frac{(1-c)}{\alpha_n}\right]^{-1}$
δ	$\frac{1}{\gamma\delta^2}\left[\frac{1}{\gamma} + \frac{1}{\delta} + \frac{1}{\alpha_p} + \frac{(1-c)}{\alpha_n}\right]^{-2}$	$\frac{1}{\delta}\left[\frac{1}{\gamma} + \frac{1}{\delta} + \frac{1}{\alpha_p} + \frac{(1-c)}{\alpha_n}\right]^{-1}$
α_p	$\frac{1}{\gamma\alpha_p^2}\left[\frac{1}{\gamma} + \frac{1}{\delta} + \frac{1}{\alpha_p} + \frac{(1-c)}{\alpha_n}\right]^{-2}$	$\frac{1}{\alpha_p}\left[\frac{1}{\gamma} + \frac{1}{\delta} + \frac{1}{\alpha_p} + \frac{(1-c)}{\alpha_n}\right]^{-1}$
α_n	$\frac{1-c}{\gamma\alpha_n^2}\left[\frac{1}{\gamma} + \frac{1}{\delta} + \frac{1}{\alpha_p} + \frac{(1-c)}{\alpha_n}\right]^{-2}$	$\frac{1-c}{\alpha_n}\left[\frac{1}{\gamma} + \frac{1}{\delta} + \frac{1}{\alpha_p} + \frac{(1-c)}{\alpha_n}\right]^{-1}$
c	$\frac{1}{\gamma\alpha_n}\left[\frac{1}{\gamma} + \frac{1}{\delta} + \frac{1}{\alpha_p} + \frac{(1-c)}{\alpha_n}\right]^{-2}$	$\frac{c}{\alpha_n}\left[\frac{1}{\gamma} + \frac{1}{\delta} + \frac{1}{\alpha_p} + \frac{(1-c)}{\alpha_n}\right]^{-1}$

Table 9.3 Ranking of scaled sensitivities for steady-state availability.

Parameter θ	$SS_\theta(A_{ss})$
γ	-3.6746×10^{-5}
α_p	3.4721×10^{-5}
c	1.6492×10^{-5}
δ	1.1574×10^{-6}
α_n	8.6802×10^{-7}

the scaled sensitivity only, reported in Table 9.3. From Table 9.3 we observe that the scaled sensitivity with respect to γ is negative, while the scaled sensitivities with respect to all the other parameters are positive. An increase in the parameter γ (the failure rate) decreases the availability, while the opposite behavior is verified for all the other parameters. Further, a small variation in the parameters with the highest absolute value (γ or α_p) will have more influence on the system availability than a similar variation in the parameters with the lowest absolute value (δ or α_n).

9.7.1 Parametric Sensitivity Analysis for a CTMC

Let π and Q both depend on a parameter θ. By differentiating the steady-state balance equation (9.33) with respect to θ the following equation, for the sensitivities of the steady-state probabilities, is obtained:

$$\frac{\partial \pi}{\partial \theta} Q = -\pi \frac{\partial Q}{\partial \theta}, \qquad \sum_i \frac{\partial \pi_i}{\partial \theta} = 0. \tag{9.81}$$

The sensitivities of the state probabilities as functions of t with respect to θ are obtained by differentiating both sides of Eq. (9.19), assuming that the initial probability vector $\pi(0)$ is independent of θ:

$$\frac{\partial}{\partial \theta} \frac{d\pi(t)}{dt} = \frac{\partial \pi(t)}{\partial \theta} Q + \pi(t) \frac{\partial Q}{\partial \theta} \quad \text{with initial condition} \quad \frac{\partial \pi(0)}{\partial \theta} = \mathbf{0}. \tag{9.82}$$

In system modeling, useful measures can often be defined in terms of MRMs (Section 9.5). In order to compute the sensitivity of the dependability measures defined through an MRM, both the reward rate vector r and its derivative $\frac{\partial}{\partial \theta} r$ need to be specified. The derivatives of the instantaneous or cumulative reward measures are then computed resorting to Eqs. (9.54) and (9.58), to get:

$$\frac{\partial E[X]}{\partial \theta} = \sum_i \left(r_i \frac{\partial \pi_i}{\partial \theta} + \frac{\partial r_i}{\partial \theta} \pi_i \right)$$

for the steady-state expected reward rate,

$$\frac{\partial E[X(t)]}{\partial \theta} = \sum_i \left(r_i \frac{\partial \pi_i(t)}{\partial \theta} + \frac{\partial r_i}{\partial \theta} \pi_i(t) \right)$$

for the instantaneous expected reward rate and

$$\frac{\partial E[Y(t)]}{\partial \theta} = \sum_i \left(r_i \frac{\partial \int_0^t \pi_i(u)du}{\partial \theta} + \frac{\partial r_i}{\partial \theta} \int_0^t \pi_i(u)du \right)$$

for the expected accumulated reward in the interval $(0, t]$.

For further exploration of this topic including its applications see [37–40].

Problems

9.18   In Example 9.24 compute the numerical values for the unscaled sensitivities for the steady-state availability and compare the results with the scaled values of Table 9.3.

9.19   Returning to the midplane availability model presented in Example 9.19, conduct a parametric sensitivity analysis for the subsystem availability A_{ss} as shown in the example above. Repeat for each of the submodels of the BladeCenter.

9.20   In Example 9.12, compute the sensitivity of the expected system cost for the non-shared and shared repair with respect to the three parameters λ, μ, and c.

9.21   Assuming now in Example 9.12 that the cost c is a decreasing function of λ (the cost increases for more reliable components), compute the sensitivity of the expected system cost for the non-shared and shared repair with respect to the three parameters λ, μ, and c. Assume as an example $c = a/\lambda$ with $a > 0$.

9.8   Numerical Methods for Steady-State Analysis of Markov Models

A CTMC is completely specified once its infinitesimal generator Q and the initial probability vector $\pi(0)$ are specified. Instead, the steady-state probability vector π does not depend on $\pi(0)$ and is obtained by solving Eq. (9.33). Numerical techniques for solving Eq. (9.33) are illustrated and discussed in [7]. A brief survey is given in the following.

9.8.1   Power Method

The equation for steady-state probabilities (9.33) may be rewritten as

$$\pi = \pi(I + Q/q) = \pi Q^\star, \tag{9.83}$$

where $q \geq \max_i |q_{ii}|$ and $Q^\star = I + Q/q$. The derivation and meaning of matrix $Q^\star$ is extensively discussed in Section 10.3.1. The entries $q^\star_{ij}$ represent the ultimate probability of jumping into state j when a transition out of state i occurs and $t \to \infty$. Note that all the entries of matrix $Q^\star$ are probabilities, i.e., real numbers between 0 and 1, and that its row sums are equal to 1. $Q^\star$ is a stochastic matrix and is called the generator matrix of the discrete-time Markov chain (DTMC) embedded in the uniformized CTMC generated by matrix Q (Section 10.3.1) [3, 7]. Hence, Eq. (9.83) expresses the steady-state solution of the embedded DTMC [3].

Equation (9.83) can be written in an iterative form as:

$$\pi^{(i)} = \pi^{(i-1)} Q^\star, \tag{9.84}$$

where $\pi^{(i)}$ is the value of the iterate at the end of the ith step. We start off the iteration by initializing $\pi^{(0)}$ to an initial guess (whose value does not influence the final solution), for instance:

$$\pi^{(0)} = \pi(0).$$

It is well known that this iteration converges to a fixed point [7], and the number of iterations, k, needed for convergence is governed by the second largest eigenvalue of Q^*. This method is referred to as the power method. In order to ensure convergence of the power iteration Eq. (9.84), we require that

$$q > \max_i |q_{ii}|, \qquad (9.85)$$

since this assures that the DTMC is aperiodic [7, 18].

9.8.2 Successive Over-Relaxation

The equation for steady-state probabilities (9.34) defines a linear system of equations:

$$\pi Q = 0.$$

Thus, standard numerical techniques for the solution of a linear system of equations will be applicable in this case. Direct methods such as Gaussian elimination can be used to solve these equations, but for large Markov chains with sparse generator matrices, iterative methods such as SOR are more convenient [7].

The matrix Q is split into three summands [3, 41]:

$$Q = D - L - U,$$

where L and U are strictly upper triangular and lower triangular, respectively, and D is a diagonal matrix. Then the SOR iteration equation can be written as:

$$\pi^{(k+1)} = \pi^{(k)}[(1 - \omega) + \omega L D^{-1}] + \omega \pi^{(k+1)} U D^{-1},$$

where $\pi^{(k)}$ is the kth iterate for π, and ω is the relaxation parameter. Note that Gauss–Seidel is a special case of SOR with $\omega = 1$. Further note that there are methods available for the optimal choice of the relaxation parameter ω. Further details of this method may be found in [7, 41].

9.9 Further Reading

A basic introduction to the theory of CTMC with several examples illustrating the computation of various metrics can be found in [3]. For a deeper understanding of CTMC theory and applications, the reader may refer to [1, 4–6]. Howard [14] presents the theory behind MRMs in detail. Other resources for the use of MRMs in availability and performance analysis can be found in [3, 15, 21, 42–45]. Numerical solutions of Markov chains can be found presented in great detail in [7]. Also, the recent book by Dayar [46] is a good resource on analyzing Markov chains.

References

[1] D. Cox and H. Miller, *The Theory of Stochastic Processes*. Chapman and Hall, 1965.

[2] J. G. Kemeny and J. L. Snell, *Finite Markov Chains*. Springer, 1983.

[3] K. Trivedi, *Probability and Statistics with Reliability, Queueing and Computer Science Applications*, 2nd edn. John Wiley & Sons, 2001.

[4] V. G. Kulkarni, *Modeling and Analysis of Stochastic Systems*. Chapman and Hall, 1995.

[5] B. Sericola, *Markov Chains: Theory and Applications*. John Wiley & Sons, 2013.

[6] G. Rubino and B. Sericola, *Markov Chains and Dependability Theory*. Cambridge University Press, 2014.

[7] W. Stewart, *Introduction to the Numerical Solution of Markov Chains*. Princeton University Press, 1994.

[8] E. Cinlar, *Introduction to Stochastic Processes*. Prentice-Hall, 1975.

[9] A. Reibman and K. Trivedi, "Transient analysis of cumulative measures of Markov chain behavior," *Stochastic Models*, vol. 5, pp. 683–710, 1989.

[10] Y. Liu and K. Trivedi, "Survivability quantification: The analytical modeling approach," *International Journal of Performability Engineering*, vol. 2, no. 1, p. 29, 2006.

[11] H. Sukhwani, A. Bobbio, and K. Trivedi, "Largeness avoidance in availability modeling using hierarchical and fixed-point iterative techniques," *International Journal of Performability Engineering*, vol. 11, no. 4, pp. 305–319, 2015.

[12] R. Peng, Q. Zhai, L. Xing, and J. Yang, "Reliability of 1-out-of-$(n+1)$ warm standby systems subject to fault level coverage," *International Journal of Performability Engineering*, vol. 9, no. 1, pp. 117–120, 2013.

[13] L. Tomek and K. Trivedi, "Fixed point iteration in availability modeling," in *Proc. 5th Int. GI/ITG/GMA Conf. on Fault-Tolerant Computing Systems, Tests, Diagnosis, Fault Treatment*. Springer-Verlag, 1991, pp. 229–240.

[14] R. Howard, *Dynamic Probabilistic Systems, Volume II: Semi-Markov and Decision Processes*. John Wiley & Sons, 1971.

[15] A. Reibman, R. Smith, and K. Trivedi, "Markov and Markov reward model transient analysis: An overview of numerical approaches," *European Journal of Operational Research*, vol. 40, pp. 257–267, 1989.

[16] W. Sanders and J. Meyer, "A unified approach for specifying measures of performance, dependability and performability," in *Dependable Computing for Critical Applications*, ser. Dependable Computing and Fault-Tolerant Systems, eds. A. Avižienis and J.-C. Laprie. Springer, 1991, vol. 4, pp. 215–237.

[17] K. Lampka and M. Siegle, "A symbolic approach to the analysis of multi-formalism Markov reward models," in *Theory and Application of Multi-Formalism Modeling*, eds. M. Gribaudo and M. Iacono. IGI-Global, 2014, ch. 91, pp. 170–195.

[18] A. Goyal, S. Lavenberg, and K. Trivedi, "Probabilistic modeling of computer system availability," *Annals of Operations Research*, vol. 8, pp. 285–306, 1987.

[19] U. Sumita, J. Shanthikumar, and Y. Masuda, "Analysis of fault tolerant computer systems," *Microelectronics and Reliability*, vol. 27, pp. 65–78, 1987.

[20] D. Heimann, N. Mittal, and K. Trivedi, "Availability and reliability modeling for computer systems," in *Advances in Computers*, ed. M. Yovits, vol. 31. Academic Press, 1990, pp. 176–231.

[21] K. Trivedi, J. Muppala, S. Woolet, and B. Haverkort, "Composite performance and dependability analysis," *Performance Evaluation*, vol. 14, pp. 197–215, 1992.

[22] V. Kulkarni, V. Nicola, and K. Trivedi, "On modeling the performance and reliability of multi-mode computer systems," *The Journal of Systems and Software*, vol. 6, pp. 175–183, 1986.

[23] R. Smith, K. Trivedi, and A. Ramesh, "Performability analysis: Measures, an algorithm and a case study," *IEEE Transactions on Computers*, vol. C-37, pp. 406–417, 1988.

[24] V. Nicola, A. Bobbio, and K. Trivedi, "A unified performance reliability analysis of a system with a cumulative down time constraint," *Microelectronics and Reliability*, vol. 32, pp. 49–65, 1992.

[25] H. Nabli and B. Sericola, "Performability analysis: A new algorithm," *IEEE Transactions on Computers*, vol. 45, pp. 491–494, 1996.

[26] B. R. Haverkort, R. Marie, G. Rubino, and K. S. Trivedi (eds.), *Performability Modelling: Techniques and Tools*. John Wiley & Sons, 2001.

[27] R. German, A. van Moorsel, M. Qureshi, and W. Sanders, "Expected impulse rewards in Markov regenerative stochastic petri nets," in *Application and Theory of Petri Nets 1996*, ser. Lecture Notes in Computer Science, eds. J. Billington and W. Reisig. Springer, 1996, vol. 1091, pp. 172–191.

[28] R. Ghosh, D. Kim, and K. Trivedi, "System resiliency quantification using non-state-space and state-space analytic models," *Reliability Engineering and System Safety*, vol. 116, pp. 109–125, 2013.

[29] K. Trivedi, R. Vasireddy, D. Trindade, S. Nathan, and R. Castro, "Modeling high availability systems," in *Proc. IEEE Pacific Rim Int. Symp. on Dependable Computing (PRDC)*, 2006.

[30] K. S. Trivedi, D. Wang, J. Hunt, A. Rindos, W. E. Smith, and B. Vashaw, "Availability modeling of SIP protocol on IBM© WebSphere©," in *Proc. Pacific Rim Int. Symp. on Dependable Computing (PRDC)*, 2008, pp. 323–330.

[31] A. Oodan, K. Ward, C. Savolaine, M. Daneshmand, and P. Hoath, "Telecommunications quality of service management: From legacy to emerging services," *info*, vol. 5, no. 4, p. 45, 2003.

[32] C. Johnson, Y. Kogan, Y. Levy, F. Saheban, and P. Tarapore, "VOIP reliability: A service provider's perspective," *IEEE Communications Magazine*, vol. 42, no. 7, pp. 48–54, Jul. 2004.

[33] Cisco Systems, Inc., *Introduction to Performance Management*, Cisco Systems, 2004.

[34] S. Mondal, X. Yin, J. Muppala, J. Alonso Lopez, and K. S. Trivedi, "Defects per million computation in service-oriented environments," *IEEE Transactions on Services Computing*, vol. 8, no. 1, pp. 32–46, Jan. 2015.

[35] W. E. Smith, K. S. Trivedi, L. Tomek, and J. Ackaret, "Availability analysis of blade server systems," *IBM Systems Journal*, vol. 47, no. 4, pp. 621–640, 2008.

[36] M. Grottke and K. S. Trivedi, "Fighting bugs: Remove, retry, replicate, and rejuvenate," *IEEE Computer*, vol. 40, no. 2, pp. 107–109, 2007.

[37] J. T. Blake, A. L. Reibman, and K. S. Trivedi, "Sensitivity analysis of reliability and performability measures for multiprocessor systems," *SIGMETRICS Performance Evaluation Review*, vol. 16, no. 1, pp. 177–186, May 1988.

[38] R. de S. Matos, P. R. M. Maciel, F. Machida, D. S. Kim, and K. S. Trivedi, "Sensitivity analysis of server virtualized system availability," *IEEE Transactions on Reliability*, vol. 61, no. 4, pp. 994–1006, 2012.

[39] N. Sato and K. S. Trivedi, "Stochastic modeling of composite web services for closed-form analysis of their performance and reliability bottlenecks," in *Proc. 5th Int. Conf. Service-Oriented Computing (ICSOC)*, 2007, pp. 107–118.

[40] Z. Zheng, K. Trivedi, K. Qiu, and R. Xia, "Semi-Markov models of composite web services for their performance, reliability and bottlenecks," *IEEE Transactions on Services Computing*, vol. 6, no. 1, 2016.

[41] G. Ciardo, A. Blakemore, P. F. Chimento, J. K. Muppala, and K. Trivedi, "Automated generation and analysis of Markov reward models using stochastic reward nets," in *Linear Algebra, Markov Chains, and Queueing Models*, ser. IMA Volumes in Mathematics and its Applications, eds. C. D. Meyer and R. J. Plemmons, Springer, 1993, vol. 48, pp. 145–191.

[42] M. Beaudry, "Performance-related reliability measures for computing systems," *IEEE Transactions on Computers*, vol. C-27, pp. 540–547, 1978.

[43] G. Bolch, S. Greiner, H. de Meer, and K. S. Trivedi, *Queueing Networks and Markov Chains: Modeling and Performance Evaluation with Computer Science Applications*, 2nd ed. Wiley-Interscience, Apr. 2006.

[44] J. Meyer, "On evaluating the performability of degradable systems," *IEEE Transactions on Computers*, vol. C-29, pp. 720–731, 1980.

[45] R. Sahner, K. Trivedi, and A. Puliafito, *Performance and Reliability Analysis of Computer Systems: An Example-based Approach Using the SHARPE Software Package*. Kluwer Academic Publishers, 1996.

[46] T. Dayar, *Analyzing Markov chains using Kronecker Products: Theory and Applications*. Springer Science & Business Media, 2012.

10 Continuous-Time Markov Chain: Reliability Models

Chapter 9 dealt with continuous-time Markov chain models where the CTMC is an irreducible Markov chain, i.e., the state space is formed by a single set of recurrent states. Thus every state is reachable from every other state. In such circumstances, the steady-state solution which yields the probability vector of the CTMC being in any of the states is meaningful and possible. However, there are many situations where the CTMC may not be irreducible. In these cases, the state space can be divided into two non-intersecting sets: a set of *transient* states and a set of *recurrent* or *absorbing* states. For these CTMCs, a set of metrics that examine the CTMC behavior in terms of the time spent in the transient states until absorption into one of the absorbing states is more meaningful. This chapter thus concentrates on evaluating metrics for *non-irreducible* Markov chains.

From the dependability perspective, non-irreducible Markov chains enable the modeling of systems which upon occurrence of total failure cannot be repaired or restored back to normal operating condition. As an example, a spacecraft mission is better modeled using CTMCs with absorbing states. Here we are interested in the probability that the system does not experience catastrophic failure during its mission, so metrics like reliability and mean time to failure are appropriate. As another example, a chemical plant may reach a safety-critical state; in this case, a metric of interest may be the distribution (or the expected value) of the time to reach the critical state for the first time. Even for those systems where repair from a system failure state is enabled, these could be made absorbing for the purpose of computing system reliability and system MTTF until the first system failure.

Further, non-irreducible CTMCs may also be invoked to study the recovery process after a failure has occurred; in this case, the failure is the initiating state of the CTMC and the absorbing state is the final outcome of the recovery process.

Thus, in this chapter we examine the general behavior and solution of CTMCs with absorbing states first. Thereafter, dependability metrics such as reliability, the Cdf of the time to failure, mean time to failure, expected number of failures and repairs, and first-passage time are introduced for CTMCs with absorbing states.

10.1 Continuous-Time Markov Chain Reliability Models

A CTMC model is completely specified once the infinitesimal generator matrix Q and the initial probability vector $\pi(0)$ are defined. In fact, given Q and $\pi(0)$, the

CTMC can be solved by applying Eqs. (9.18) and (9.19). The infinitesimal generator must be derived from the physical description of the problem. Building a CTMC model means deriving the corresponding infinitesimal generator. In a reliability model, once the system reaches one of the down states it does not return to an up state. With reference to the partitioned infinitesimal generator (9.45), all the down states in Ω_d are absorbing states, and only the matrices $\boldsymbol{Q}_{uu}$ and $\boldsymbol{Q}_{ud}$ have non-zero entries. The system reliability $R(t)$ and the system unreliability $F(t)$ as functions of time are computed as:

$$R(t) = \sum_{i \in \Omega_u} \pi_i(t), \qquad F(t) = \sum_{i \in \Omega_d} \pi_i(t) = 1 - R(t). \tag{10.1}$$

Note that $F(t)$ is the distribution function of the time to system failure. By adopting the reward rate vector defined in (9.63), the reliability can be computed as an instantaneous expected reward rate, in matrix form, as:

$$R(t) = \boldsymbol{\pi}(t)\boldsymbol{r}^{\mathsf{T}}. \tag{10.2}$$

Example 10.1   *A Single-Component System (A Two-State CTMC)*
Let us return to the single blade server model that we considered in Example 6.33. Suppose that upon the failure of the blade server the system can no longer be repaired. Consequently we are interested in finding out if the system is still functioning at time t, given that it begins operation at time 0. In other words, we require the computation of the reliability of the system. The CTMC of the reliability model of this system has only two states, as shown in Figure 10.1. In state 1 the blade is up; in state 0 the blade is down. The transition rate from state 1 to state 0 is the blade failure rate λ.

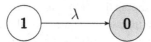

Figure 10.1   State diagram of a CTMC for a single blade server system.

From the problem specification, the transition rate matrix is:

$$\boldsymbol{Q}_{\mathrm{R}} = \begin{bmatrix} 0 & \lambda \\ 0 & 0 \end{bmatrix}. \tag{10.3}$$

By inserting the diagonal entries, the infinitesimal generator matrix is obtained:

$$\boldsymbol{Q} = \begin{bmatrix} -\lambda & \lambda \\ 0 & 0 \end{bmatrix}. \tag{10.4}$$

Applying Eq. (9.19), we get:

$$\left[\frac{d\pi_1(t)}{dt}, \frac{d\pi_0(t)}{dt} \right] = [\pi_1(t), \pi_0(t)] \begin{bmatrix} -\lambda & \lambda \\ 0 & 0 \end{bmatrix}. \tag{10.5}$$

Expanding Eq. (10.5), we obtain:

$$\begin{cases} \dfrac{d\pi_1(t)}{dt} & = & -\lambda\pi_1(t), \\[2mm] \dfrac{d\pi_0(t)}{dt} & = & \lambda\pi_1(t). \end{cases} \qquad (10.6)$$

In order to solve the system of equations in (10.6), the initial probability vector must be specified. If we assume that the CTMC is in state 1 (component up) at time $t = 0$ with probability 1, the initial probability vector is:

$$\pi(0) = [1, 0]. \qquad (10.7)$$

Solving the system of equations in (10.6) under the initial condition (10.7), we obtain:

$$\begin{cases} \pi_1(t) & = & e^{-\lambda t}, \\[2mm] \pi_0(t) & = & 1 - e^{-\lambda t}. \end{cases} \qquad (10.8)$$

Recalling that the system reliability $R(t)$ is the probability that the system is in the up state throughout the interval $(0, t]$, we have

$$R(t) = \pi_1(t). \qquad (10.9)$$

As expected, the obtained result gives the reliability of a component with constant failure rate λ. The Cdf of the time to system failure or the system unreliability, $F(t)$, at time t is given by $F(t) = \pi_0(t) = 1 - R(t)$. Furthermore, the MTTF can be computed – see (3.15) – as:

$$\text{MTTF} = \int_0^\infty R(t)dt = \int_0^\infty e^{-\lambda t}dt = 1/\lambda.$$

Example 10.2 *A Single-Component System: Mass at Origin*
It is certainly possible that the system above is initially in a failed state with some probability q. We can solve the system of two differential equations in (10.6) with the initial probability vector $\pi(0) = [1 - q, q]$ to get:

$$\begin{cases} R(t) & = & \pi_1(t) & = & (1-q)e^{-\lambda t}, \\[2mm] F(t) & = & \pi_0(t) & = & 1 - (1-q)e^{-\lambda t}. \end{cases} \qquad (10.10)$$

The reliability function in this case starts at $1 - q$ at $t = 0$ and then decays to zero as t approaches infinity. The unreliability (or the time to failure distribution) function has a mass at origin equal to q. The MTTF in this case is:

$$\text{MTTF} = \int_0^\infty R(t)dt = \int_0^\infty (1-q)e^{-\lambda t}dt = (1-q)/\lambda.$$

Example 10.3 *Defective Distribution: Two Failure Modes (Section 3.2.11)*
Recall Figure 3.16, which depicts a three-state CTMC where two failure modes are

considered: open and short. The CTMC state probability vector at time t can be obtained by solving:

$$\left[\frac{d\pi_u(t)}{dt}, \frac{d\pi_o(t)}{dt}, \frac{d\pi_c(t)}{dt}\right] = [\pi_u(t), \pi_o(t), \pi_c(t)] \begin{bmatrix} -(\lambda_o + \lambda_c) & \lambda_o & \lambda_c \\ 0 & 0 & 0 \\ 0 & 0 & 0 \end{bmatrix}. \quad (10.11)$$

Solving the system of equations in (10.11), we get:

$$\begin{cases} \pi_u(t) &= e^{-(\lambda_o + \lambda_c)t}, \\ \pi_o(t) &= \dfrac{\lambda_o}{\lambda_o + \lambda_c} - \dfrac{\lambda_o}{\lambda_o + \lambda_c}e^{-(\lambda_o + \lambda_c)t}, \\ \pi_c(t) &= \dfrac{\lambda_c}{\lambda_o + \lambda_c} - \dfrac{\lambda_c}{\lambda_o + \lambda_c}e^{-(\lambda_o + \lambda_c)t}. \end{cases}$$

Note that the distribution of time to reach state o (starting in state u at time 0 with probability 1) approaches $\frac{\lambda_o}{\lambda_o + \lambda_c}$ as t approaches infinity. Hence this distribution is defective with a defect (i.e., a mass at infinity) equal to $\frac{\lambda_c}{\lambda_o + \lambda_c}$. Similarly, the time to reach state c (starting in state u at time 0) is defective with a defect equal to $\frac{\lambda_o}{\lambda_o + \lambda_c}$.

Example 10.4   *A Two-Component System: Dependent Components*
Consider a system with two components. Its state space consists of four states and was examined in Example 7.1.

Suppose that the time to failure of component 1 is exponentially distributed, but its failure rate depends on the system state. This means that the failure rate of component 1 in state $s_1 = (1, 1)$ is different from the failure rate of component 1 in state $s_3 = (1, 0)$. Similarly, we assume that the failure rate of component 2 depends on the state of the system, and hence the failure rate of component 2 in state $s_1 = (1, 1)$ is different from the failure rate of component 2 in state $s_2 = (0, 1)$. We indicate with λ_1 the failure rate of component 1 in state s_1 and with $\hat{\lambda}_1$ ($\lambda_1 \neq \hat{\lambda}_1$) the failure rate of component 1 in state s_3. Similarly, we indicate with λ_2 the failure rate of component 2 in state s_1 and with $\hat{\lambda}_2$ ($\lambda_2 \neq \hat{\lambda}_2$) the failure rate of component 2 in state s_2. The state transition diagram of the CTMC model of the system is shown in Figure 10.2, from which the infinitesimal

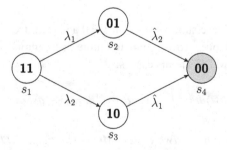

Figure 10.2   Continuous-time Markov chain representation of a two-component system.

generator can be obtained:

$$Q = \begin{array}{c} \\ (1,1) \\ (0,1) \\ (1,0) \\ (0,0) \end{array} \begin{array}{cccc} (1,1) & (0,1) & (1,0) & (0,0) \\ \begin{bmatrix} -(\lambda_1 + \lambda_2) & \lambda_1 & \lambda_2 & 0 \\ 0 & -\hat{\lambda}_2 & 0 & \hat{\lambda}_2 \\ 0 & 0 & -\hat{\lambda}_1 & \hat{\lambda}_1 \\ 0 & 0 & 0 & 0 \end{bmatrix} \end{array}. \qquad (10.12)$$

Equation (9.19) becomes:

$$\left[\frac{d\pi_1(t)}{dt}, \frac{d\pi_2(t)}{dt}, \frac{d\pi_3(t)}{dt}, \frac{d\pi_4(t)}{dt} \right] = \qquad (10.13)$$

$$[\pi_1(t), \pi_2(t), \pi_3(t), \pi_4(t)] \begin{bmatrix} -(\lambda_1 + \lambda_2) & \lambda_1 & \lambda_2 & 0 \\ 0 & -\hat{\lambda}_2 & 0 & \hat{\lambda}_2 \\ 0 & 0 & -\hat{\lambda}_1 & \hat{\lambda}_1 \\ 0 & 0 & 0 & 0 \end{bmatrix}.$$

From Eq. (10.13), the following set of coupled differential equations in scalar form is obtained:

$$\begin{cases} \dfrac{d\pi_1(t)}{dt} &= -(\lambda_1 + \lambda_2)\pi_1(t), \\[2mm] \dfrac{d\pi_2(t)}{dt} &= \lambda_1 \pi_1(t) - \hat{\lambda}_2 \pi_2(t), \\[2mm] \dfrac{d\pi_3(t)}{dt} &= \lambda_2 \pi_1(t) - \hat{\lambda}_1 \pi_3(t), \\[2mm] \dfrac{d\pi_4(t)}{dt} &= \hat{\lambda}_2 \pi_2(t) + \hat{\lambda}_1 \pi_3(t). \end{cases} \qquad (10.14)$$

For a complete specification of the problem, the initial state probability vector should be given. Assume that the system is initially in state $s_1 = (1, 1)$ in which all the components are up:

$$\pi(0) = [1, 0, 0, 0]. \qquad (10.15)$$

The solution of the above equations will be pursued later in this section.

Example 10.5 *System with Two Identical Components*
If the components are identical and independent, we can assume in Example 10.4 that $\lambda_1 = \lambda_2 = \hat{\lambda}_1 = \hat{\lambda}_2 = \lambda$. In this case, the CTMC state diagram of Figure 10.2 can be simplified as shown in Figure 10.3.

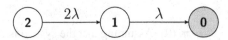

Figure 10.3 Continuous-time Markov chain state diagram of a system with two identical components.

The label inside each state indicates the number of working components. The Kolmogorov differential equations (10.13) in this case become:

$$\left[\frac{d\pi_2(t)}{dt}, \frac{d\pi_1(t)}{dt}, \frac{d\pi_0(t)}{dt}\right] = [\pi_2(t), \pi_1(t), \pi_0(t)]\begin{bmatrix} -2\lambda & 2\lambda & 0 \\ 0 & -\lambda & \lambda \\ 0 & 0 & 0 \end{bmatrix}. \qquad (10.16)$$

The state-by-state version of Eq. (10.16) is:

$$\begin{cases} \dfrac{d\pi_2(t)}{dt} &= -2\lambda\pi_2(t), \\[2mm] \dfrac{d\pi_1(t)}{dt} &= 2\lambda\pi_2(t) - \lambda\pi_1(t), \\[2mm] \dfrac{d\pi_0(t)}{dt} &= \lambda\pi_1(t). \end{cases} \qquad (10.17)$$

Assuming that state 2 is the initial state, we obtain, by direct integration:

$$\begin{cases} \pi_2(t) &= e^{-2\lambda t}, \\[1mm] \pi_1(t) &= 2(e^{-\lambda t} - e^{-2\lambda t}), \\[1mm] \pi_0(t) &= 1 - \pi_1(t) - \pi_2(t) = 1 - 2e^{-\lambda t} + e^{-2\lambda t}. \end{cases}$$

We can check that the reliability is HYPO$(2\lambda, \lambda)$,

$$R(t) = \pi_2(t) + \pi_1(t) = 2e^{-\lambda t} - e^{-2\lambda t}, \qquad (10.18)$$

and is exactly the same expression that has been obtained in the previous chapters since we have assumed independence across the two components – see (8.4). The MTTF can be computed to yield:

$$\mathrm{MTTF} = \int_0^\infty R(t)dt = \int_0^\infty [2e^{-\lambda t} - e^{-2\lambda t}]dt = \frac{1.5}{\lambda}.$$

Example 10.6   *Telecommunication Switching System Model*
Returning to the telecommunications switching system of Example 9.8, we carry out the instantaneous and expected interval availability analysis of the system at time t.

The transient probability of being in state i at time t, $\pi_i(t)$, can be computed by solving the following differential equations:

$$\begin{cases} \dfrac{d\pi_0(t)}{dt} &= -\tau\pi_0(t) + \gamma\pi_1(t), \\[2mm] \dfrac{d\pi_k(t)}{dt} &= -(\tau + k\gamma)\pi_k(t) + \tau\pi_{k-1}(t) + (k+1)\gamma\pi_{k+1}(t), \quad k = 1, 2, \ldots, n-1, \\[2mm] \dfrac{d\pi_n(t)}{dt} &= -n\gamma\pi_n(t) + \tau\pi_{n-1}(t), \end{cases}$$

$$(10.19)$$

with the initial condition $\pi_k(t=0) = \pi_k(0)$. Assuming that the system is up as long as at least l trunks are functioning, the instantaneous availability $A(l,t)$ and the expected interval availability $A_I(l,t)$ are given as:

$$A(l,t) = \sum_{k=l}^{n} \pi_k(t), \qquad A_I(l,t) = \frac{\sum_{k=l}^{n} \int_0^t \pi_k(x)\,dx}{t}. \tag{10.20}$$

Numerical computation using a software package is recommended in this case as a closed-form solution will be very complex except for very small values of n.

10.1.1 Convolution Integration Method for Transient Probabilities

The Kolmogorov differential equations (9.7) can be put in the form of a coupled system of integral equations as follows (for a proof, see [1], Vol. II, pp. 483–488):

$$p_{ij}(t) = \delta_{ij}e^{q_{ii}t} + \int_0^t \sum_k p_{ik}(x)q_{kj}e^{q_{jj}(t-x)}\,dx,$$

where $p_{ij}(t)$ are the entries of the transition probability matrix $\mathbf{P}(t)$, and δ_{ij} is the Kronecker delta function defined by $\delta_{ij} = 1$ if $i = j$ and 0 otherwise. The above equation can be specialized to obtain the unconditional probabilities of states at time t:

$$\pi_j(t) = \pi_j(0)e^{q_{jj}t} + \int_0^t \sum_k \pi_k(x)q_{kj}e^{q_{jj}(t-x)}\,dx.$$

These equations can be solved relatively easily for acyclic CTMCs; the method is known as the *convolution integration method*. Acyclic CTMCs naturally arise from reliability models of systems without repair. Such systems are sometimes known as closed fault-tolerant systems [2–4].

Example 10.7 *A Two-Component System: Solution of Eqs. (10.14)*
Applying the above convolution integration method to the two-component system model introduced in Example 10.4, we first obtain an explicit solution for state 1, which does not have any transitions leading into it, and then we solve for the successive state probabilities:

$$\pi_1(t) = \pi_1(0)e^{-(\lambda_1+\lambda_2)t} = e^{-(\lambda_1+\lambda_2)t}, \tag{10.21}$$

$$\pi_2(t) = \pi_2(0)e^{-\hat{\lambda}_2 t} + \int_0^t \pi_1(x)\lambda_1 e^{-\hat{\lambda}_2(t-x)}\,dx$$

$$= \frac{\lambda_1}{\lambda_1+\lambda_2-\hat{\lambda}_2}(e^{-\hat{\lambda}_2 t} - e^{-(\lambda_1+\lambda_2)t}),$$

$$\pi_3(t) = \frac{\lambda_2}{\lambda_1+\lambda_2-\hat{\lambda}_1}(e^{-\hat{\lambda}_1 t} - e^{-(\lambda_1+\lambda_2)t}),$$

$$\pi_4(t) = 1 - \pi_1(t) - \pi_2(t) - \pi_3(t).$$

Problems

10.1 Solve the system of differential equations in (10.11) using the convolution integration method.

10.2 Solve the system of differential equations in (10.16) using the convolution integration method.

10.3 Solve for the instantaneous and expected interval availability, Eq. (10.20), of the telecommunication switching model using a software package such as SHARPE. First assume that the system is in state n at time 0. Next, carry out a steady-state analysis of the same model and then use the steady-state probability vector as the initial state probability vector. Vary l from 1 to $n-1$.

10.1.2 Solution with Laplace Transforms

Given a generic function of time $f(t)$, and indicating its Laplace transform by $\mathscr{L}[f(t)] = f^*(s)$, the following relation holds [5]:

$$\mathscr{L}\left[\frac{f(t)}{dt}\right] = f^*(s) - f(0). \tag{10.22}$$

By taking Laplace transforms on both sides of Eq. (9.20) and using Eq. (10.22), we get:

$$\begin{cases} s\pi_1^*(s) - \pi_1(0) &= \pi_1^*(s)\,q_{11} + \pi_2^*(s)\,q_{21} + \cdots + \pi_n^*(s)\,q_{n1}, \\ s\pi_2^*(s) - \pi_2(0) &= \pi_1^*(s)\,q_{12} + \pi_2^*(s)\,q_{22} + \cdots + \pi_n^*(s)\,q_{n2}, \\ \vdots \end{cases} \tag{10.23}$$

After rearranging terms in (10.23):

$$\begin{cases} \pi_1^*(s)\,(s - q_{11}) - \pi_2^*(s)\,q_{21} - \cdots - \pi_n^*(s)\,q_{n1} &= \pi_1(0), \\ -\pi_1^*(s)\,q_{12} + \pi_2^*(s)\,(s - q_{22}) - \cdots - \pi_n^*(s)\,q_{n2} &= \pi_2(0), \\ \vdots \end{cases} \tag{10.24}$$

The set of equations in (10.24) is a linear algebraic system of equations in the variable s, and can be rewritten in matrix form as:

$$\boldsymbol{\pi}^*(s)\,(s\boldsymbol{I} - \boldsymbol{Q}) = \boldsymbol{\pi}(0),$$

whose formal solution is given by:

$$\boldsymbol{\pi}^*(s) = \boldsymbol{\pi}(0)\,(s\boldsymbol{I} - \boldsymbol{Q})^{-1}. \tag{10.25}$$

Solving (10.25) and taking the inverse Laplace transform via partial fraction expansion, we obtain a solution for the state probability vector in the time domain.

Example 10.8 *A Two-Component System (continued)*
We return to Example 10.4, a system with two components. An alternative way of obtaining a fully symbolic solution of Eq. (10.14) instead of the convolution integration method exploited in Example 10.7 is to resort to the Laplace transform method

described above. Applying the Laplace transform to both sides of Eq. (10.14), according to Eq. (10.23), we get:

$$
\begin{cases}
s\pi_1^*(s) - 1 &= -(\lambda_1 + \lambda_2)\pi_1^*(s), \\
s\pi_2^*(s) &= \lambda_1 \pi_1^*(s) - \hat{\lambda}_2 \pi_2^*(s), \\
s\pi_3^*(s) &= \lambda_2 \pi_1^*(s) - \hat{\lambda}_1 \pi_3^*(s), \\
s\pi_4^*(s) &= \hat{\lambda}_2 \pi_2^*(s) + \hat{\lambda}_1 \pi_3^*(s).
\end{cases}
\tag{10.26}
$$

The linear system of equations in (10.26) can be solved to get a symbolic solution in the transform domain:

$$
\begin{cases}
\pi_1^*(s) &= \dfrac{1}{s + \lambda_1 + \lambda_2}, \\[2mm]
\pi_2^*(s) &= \dfrac{\lambda_1}{(s + \lambda_1 + \lambda_2)(s + \hat{\lambda}_2)}, \\[2mm]
\pi_3^*(s) &= \dfrac{\lambda_2}{(s + \lambda_1 + \lambda_2)(s + \hat{\lambda}_1)}, \\[2mm]
\pi_4^*(s) &= \dfrac{\lambda_1\hat{\lambda}_2(s + \hat{\lambda}_1) + \hat{\lambda}_1\lambda_2(s + \hat{\lambda}_2)}{s(s + \lambda_1 + \lambda_2)(s + \hat{\lambda}_1)(s + \hat{\lambda}_2)}.
\end{cases}
\tag{10.27}
$$

Inverting $\pi_1^*(s)$ is immediate by Laplace table lookup. In order to invert the successive Laplace transforms in (10.27) we use partial fraction expansion. For example, we rewrite:

$$
\pi_2^*(s) = \frac{A_1}{s + \lambda_1 + \lambda_2} + \frac{A_2}{s + \hat{\lambda}_2}.
$$

Equating the two expressions for $\pi_2^*(s)$, we have:

$$
A_2 = \frac{\lambda_1}{\lambda_1 + \lambda_2 - \hat{\lambda}_2}, \qquad A_1 = -A_2.
$$

In a similar way, the partial fraction form of each of the above transforms can be written. Subsequent inversion by a table lookup yields the time domain state probabilities:

$$
\begin{cases}
\pi_1(t) &= e^{-(\lambda_1 + \lambda_2)t}, \\[2mm]
\pi_2(t) &= \dfrac{\lambda_1}{\lambda_1 + \lambda_2 - \hat{\lambda}_2}(e^{-\hat{\lambda}_2 t} - e^{-(\lambda_1 + \lambda_2)t}), \\[2mm]
\pi_3(t) &= \dfrac{\lambda_2}{\lambda_1 + \lambda_2 - \hat{\lambda}_1}(e^{-\hat{\lambda}_1 t} - e^{-(\lambda_1 + \lambda_2)t}), \\[2mm]
\pi_4(t) &= 1 - \pi_1(t) - \pi_2(t) - \pi_3(t).
\end{cases}
\tag{10.28}
$$

Note that if we assume in Eq. (10.28) that the two components are statistically independent (i.e., $\lambda_1 = \hat{\lambda}_1$ and $\lambda_2 = \hat{\lambda}_2$), the above state probabilities can be expressed as the product of the individual probabilities of the single components and can be derived directly by means of non-state-space techniques (see Chapters 4 and 7).

Example 10.9   *Series-Parallel System with Two Components*

The expressions in (10.28) represent the state probabilities for any system with $n = 2$ dependent components. If we now assume a specific system configuration (possibly through the structure function $\varphi(x)$ discussed in Chapter 7), we can partition the state space into up or down states, as in Section 9.3. If the two components are connected in series (Figure 10.4(a), reproduced from Figure 7.3), the only system up state is state 1, so that:

$$R_{\text{ser}}(t) = \pi_1(t) = e^{-(\lambda_1+\lambda_2)t}. \tag{10.29}$$

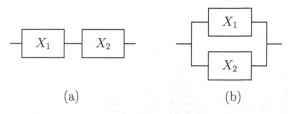

(a)	(b)

Figure 10.4   Series (a) and parallel (b) system with two components.

If the components are connected in parallel, Figure 10.4(b), the only system down state is state 4, so that:

$$R_{\text{par}}(t) = \pi_1(t) + \pi_2(t) + \pi_3(t)$$

$$= e^{-(\lambda_1+\lambda_2)t}$$

$$+ \frac{\lambda_1}{\lambda_1+\lambda_2-\hat{\lambda}_2}(e^{-\hat{\lambda}_2 t} - e^{-(\lambda_1+\lambda_2)t}) \tag{10.30}$$

$$+ \frac{\lambda_2}{\lambda_1+\lambda_2-\hat{\lambda}_1}(e^{-\hat{\lambda}_1 t} - e^{-(\lambda_1+\lambda_2)t}).$$

Note again that in the case of statistical independence between the two components (i.e., $\lambda_i = \hat{\lambda}_i$, $i = 1, 2$), we obtain the same equation derived using RBD (Chapter 4) or state enumeration (Chapter 7). Equations (10.29) and (10.30) can be obtained in terms of expected reward rate at time t from Eq. (10.2) by assigning the following reward rate vectors:

$$r_{\text{ser}} = [1, 0, 0, 0], \qquad r_{\text{par}} = [1, 1, 1, 0].$$

From Eq. (3.7) we know that once the system reliability function is obtained, the system mean time to failure can be computed by integration. However, it will be shown in Section 10.2.1 that a simpler method exists that does not require the solution of the Kolmogorov differential equations. For the time being, applying Eq. (3.7) we obtain:

$$\text{MTTF}_{\text{ser}} = \int_0^\infty R_{\text{ser}}(t)dt = \frac{1}{(\lambda_1+\lambda_2)}, \tag{10.31}$$

and

$$\text{MTTF}_{par} = \int_0^\infty R_{par}(t)dt = \frac{1}{(\lambda_1 + \lambda_2)}$$

$$+ \frac{\lambda_1}{\lambda_1 + \lambda_2 - \hat{\lambda}_2} \left(\frac{1}{\hat{\lambda}_2} - \frac{1}{(\lambda_1 + \lambda_2)} \right) \quad (10.32)$$

$$+ \frac{\lambda_2}{\lambda_1 + \lambda_2 - \hat{\lambda}_1} \left(\frac{1}{\hat{\lambda}_1} - \frac{1}{(\lambda_1 + \lambda_2)} \right)$$

$$= \frac{1}{(\lambda_1 + \lambda_2)} + \frac{\lambda_1}{\hat{\lambda}_2 (\lambda_1 + \lambda_2)} + \frac{\lambda_2}{\hat{\lambda}_1 (\lambda_1 + \lambda_2)}.$$

Example 10.10 *Standby System Configuration*
Standby configurations have already been introduced in Chapter 8. The standby configuration is typical for fault tolerance, safety, emergency and protection devices, where the protection is normally dormant and is put in operation only in the case of a breakdown of the principal component. A block diagram representation of a standby system of two components is shown in Figure 10.5, where A is the main component, B the dormant component and S the switch that commutes operation to the dormant component B in the case of failure of A. We assume that the TTF of A is EXP(λ_A) and the TTF of B is EXP(λ_B).

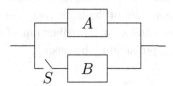

Figure 10.5   Block diagram representation of a standby system.

To compute the reliability of a standby redundant configuration we introduce the following assumptions:

- Component B has a reduced failure rate $\alpha \lambda_B$ when dormant, with dormancy factor $0 \leq \alpha \leq 1$ (Section 8.2).
- The switch S is perfect (does not fail) and instantaneous (but we will soon relax these assumptions).

Based on the value of α we distinguish the following three cases:

$\alpha = 0$, *cold standby*: The failure rate of the standby component B is equal to 0 when dormant.

$0 < \alpha < 1$, *warm standby*.

$\alpha = 1$, *hot standby*: The failure rate of the standby component B is the same when dormant and when in operation. This case is the same as the parallel configuration considered in Example 10.9. This configuration is also referred to as *active–active*.

In the case of a standby system, the state diagram of Figure 10.2 can be specialized as in Figure 10.6.

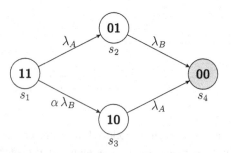

Figure 10.6 Continuous-time Markov chain state diagram of a warm standby system with two components.

Cold Standby

Since $\alpha = 0$, the failure rate of component B in state $s_1 = (1,1)$ is equal to zero, and hence state $s_3 = (1,0)$ is not reachable. The component dependencies introduced by the cold standby configuration reduce the state space of the system to three states of Figure 10.7(a) only, as reported in the state transition diagram of Figure 10.7(b). By substituting in Figure 10.2 $\lambda_1 = \lambda_A$ (failure rate of the main component A), $\lambda_2 = 0$ (failure rate of component B when dormant) and $\hat{\lambda}_2 = \lambda_B$ (failure rate of component B when operating), from Eq. (10.28) the state probabilities become:

$$
\begin{cases}
\pi_1(t) & = & e^{-\lambda_A t}, \\[2mm]
\pi_2(t) & = & \dfrac{\lambda_A}{\lambda_A - \lambda_B}(e^{-\lambda_B t} - e^{-\lambda_A t}), \\[2mm]
\pi_3(t) & = & 0 \text{ (not reachable)}, \\[2mm]
\pi_4(t) & = & 1 - \pi_1(t) - \pi_2(t) - \pi_3(t).
\end{cases}
\tag{10.33}
$$

Finally, the cold standby system reliability is:

$$
R_{CS}(t) = \pi_1(t) + \pi_2(t) = \frac{\lambda_A}{\lambda_A - \lambda_B}e^{-\lambda_B t} - \frac{\lambda_B}{\lambda_A - \lambda_B}e^{-\lambda_A t},
\tag{10.34}
$$

and the system MTTF, by integration, is:

$$
\mathrm{MTTF_{CS}} = \int_0^\infty R_{CS}(t)dt = \frac{\lambda_A}{\lambda_A - \lambda_B}\frac{1}{\lambda_B} - \frac{\lambda_B}{\lambda_A - \lambda_B}\frac{1}{\lambda_A} = \frac{1}{\lambda_A} + \frac{1}{\lambda_B}.
\tag{10.35}
$$

If the two components have equal failure rates ($\lambda_A = \lambda_B = \lambda$), Eqs. (10.33) and (10.34) are undefined. One method of resolution is to use L'Hopital's rule. We use the alternative of returning to the Laplace transform equation (10.27) and substituting the values $\lambda_1 = \hat{\lambda}_2 = \lambda$ and $\lambda_2 = 0$. With this substitution, the state probability transforms for states s_1

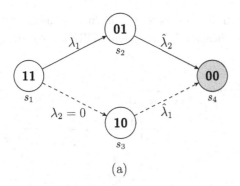

(a)

(b)

Figure 10.7 Continuous-time Markov chain state diagram of a cold standby system with two components: (a) the complete state space, (b) the reduced state space.

and s_2 become (compare with the results of Section 8.1):

$$\begin{cases} \pi_1^*(s) & = & \dfrac{1}{s+\lambda}, \\ \pi_2^*(s) & = & \dfrac{\lambda}{(s+\lambda)^2}. \end{cases} \tag{10.36}$$

Taking the inverse Laplace transform of (10.36), we obtain:

$$\begin{cases} \pi_1(t) & = & e^{-\lambda t}, \\ \pi_2(t) & = & \lambda t e^{-\lambda t}. \end{cases} \tag{10.37}$$

Hence, the system reliability is:

$$R_{\text{CS-2}}(t) = \pi_1(t) + \pi_2(t) = (1 + \lambda t)\, e^{-\lambda t},$$

and the system mean time to failure is:

$$\text{MTTF}_{\text{CS-2}} = \int_0^\infty R_{\text{CS-2}}(t)dt = \frac{2}{\lambda}.$$

Warm Standby
By substituting the proper values of the failure rates of Figure 10.6 into Eq. (10.28), we obtain:

$$\begin{aligned} R_{\text{WS}}(t) &= \pi_1(t) + \pi_2(t) + \pi_3(t) \\ &= e^{-(\lambda_A + \alpha\lambda_B)t} + \frac{\lambda_A}{\lambda_A - (1-\alpha)\lambda_B}(e^{-\lambda_B t} - e^{-(\lambda_A + \alpha\lambda_B)t}) \\ &\quad + e^{-\lambda_A t} - e^{-(\lambda_A + \alpha\lambda_B)t} \\ &= e^{-\lambda_A t} + \frac{\lambda_A}{\lambda_A - (1-\alpha)\lambda_B}(e^{-\lambda_B t} - e^{-(\lambda_A + \alpha\lambda_B)t}). \end{aligned} \tag{10.38}$$

It is easily verified that Eq. (10.38) reduces to the reliability of a parallel system with two independent components when $\alpha = 1$. The system MTTF is obtained by integration of the reliability expression:

$$
\begin{aligned}
\mathrm{MTTF_{WS}} = \int_0^\infty R_{WS}(t)dt &= \frac{1}{(\lambda_A + \alpha\lambda_B)} \\
&+ \frac{\lambda_A}{\lambda_A - (1-\alpha)\lambda_B}\left(\frac{1}{\lambda_B} - \frac{1}{(\lambda_A + \alpha\lambda_B)}\right) \\
&+ \frac{1}{\lambda_A} - \frac{1}{(\lambda_A + \alpha\lambda_B)} \\
&= \frac{\lambda_A\lambda_B + \lambda_A^2 + \alpha\lambda_B^2}{\lambda_A\lambda_B(\lambda_A + \alpha\lambda_B)}.
\end{aligned}
\tag{10.39}
$$

Example 10.11   *Unavailability On Demand (Imperfect Coverage)*
In Example 10.10 the switching mechanism from the active component to the spare component is considered perfect. But in practice it is often the case that upon the failure of the active component the switching device does not work properly and the spare component cannot be switched into operation. The probability that the spare component does not enter into operation upon demand is often called *unavailability on demand*. Unavailability on demand can be viewed as a manifestation of an imperfect coverage mechanism and can be modeled as in Figure 10.8, where $(1 - d)$ is the unavailability on demand. Note that we have used the symbol c for imperfect coverage, while we use the symbol d here to match the notation in common use in the respective community. Also note that we leave the solution of the model, to obtain expressions for system reliability and system MTTF, as an exercise.

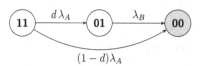

Figure 10.8   Continuous-time Markov chain state diagram of a two-component standby system with imperfect coverage.

Example 10.12   *Two-Component Reliability Model with Repair*
We return to the two-server parallel redundant system with identical components that we considered in Example 10.5. Suppose that upon the failure of both the blades, i.e., when the system reaches state 0, the system cannot be repaired. Note that even if repair from state 0 is allowed, as in Example 9.4, by making state 0 an absorbing state we can study the system's reliability. The corresponding CTMC state diagram is shown in Figure 10.9. Note that a reliability model without repair is obtained by deleting the repair arc from state 1 to state 2. It is worth emphasizing that the model in Figure 10.9 is a reliability model since the (only) system failure state (0) is an absorbing state.

In state 1, one component has failed but the system is operational. Assuming that the system starts operation in one of the up states (1 or 2), and if at time t the system is in one of the up states, we are sure that the system could not have failed at any time prior to t. Thus the sum of the state probabilities of all the up states at time t will yield the system reliability.

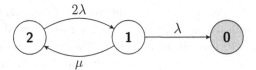

Figure 10.9   Two-component reliability model with repair.

The Kolmogorov differential equations in matrix form for this example are:

$$\left[\frac{d\pi_2(t)}{dt}, \frac{d\pi_1(t)}{dt}, \frac{d\pi_0(t)}{dt}\right] = [\pi_2(t), \pi_1(t), \pi_0(t)] \begin{bmatrix} -2\lambda & 2\lambda & 0 \\ \mu & -(\lambda+\mu) & \lambda \\ 0 & 0 & 0 \end{bmatrix}.$$

The scalar version of the system of equations is:

$$\begin{cases} \dfrac{d\pi_2(t)}{dt} &= -2\lambda\pi_2(t) + \mu\pi_1(t), \\ \dfrac{d\pi_1(t)}{dt} &= 2\lambda\pi_2(t) - (\lambda+\mu)\pi_1(t), \\ \dfrac{d\pi_0(t)}{dt} &= \lambda\pi_1(t). \end{cases}$$

Taking Laplace transforms on both sides, we get:

$$\begin{cases} s\pi_2^*(s) - 1 &= -2\lambda\pi_2^*(s) + \mu\pi_1^*(s), \\ s\pi_1^*(s) &= 2\lambda\pi_2^*(s) - (\lambda+\mu)\pi_1^*(s), \\ s\pi_0^*(s) &= \lambda\pi_1^*(s). \end{cases}$$

Solving this system of algebraic equations in the s-domain, we get:

$$\pi_0^*(s) = \frac{2\lambda^2}{s[s^2 + (3\lambda+\mu)s + 2\lambda^2]},$$

and by an inversion via partial fraction expansion, we get an expression for the system reliability:

$$R(t) = 1 - \pi_0(t) = \frac{\alpha_1}{\alpha_1 - \alpha_2}e^{-\alpha_2 t} - \frac{\alpha_2}{\alpha_1 - \alpha_2}e^{-\alpha_1 t}, \tag{10.40}$$

where α_1 and α_2 are the roots of the equation

$$s^2 + (3\lambda+\mu)s + 2\lambda^2 = (s+\alpha_1)(s+\alpha_2).$$

So,

$$\alpha_1, \alpha_2 = \frac{(3\lambda + \mu) \pm \sqrt{\lambda^2 + 6\lambda\mu + \mu^2}}{2}.$$

We are now in a position to compare the effect of redundancy combined with repair over redundancy without repair.

In Figure 10.10 we plot the reliability of a two-component parallel redundant system with repair, Eq. (10.40), with the reliability of a two-component parallel redundant system without repair, Eq. (10.18) and the reliability of a single-component system without redundancy, Eq. (10.9). We have used $\lambda = 1/8760\,\mathrm{f\,hr^{-1}}$ (one failure per year) and $\mu = 1/2\,\mathrm{r\,hr^{-1}}$.

Example 10.13 *Two-Component Reliability Model with Common Cause Failure*
If we assume that the time to occurrence of a CCF is a random variable with distribution EXP(λ_{CCF}), the presence of a CCF in a two-component system can be modeled by the CTMC of Figure 10.11(a).

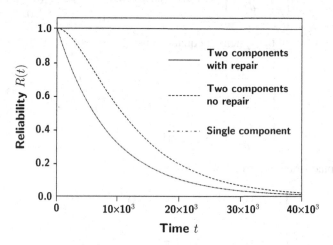

Figure 10.10 Reliability of the two-component parallel redundant system with and without repair, and a single-component system.

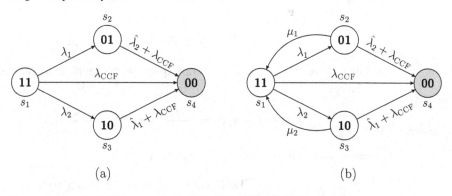

(a)   (b)

Figure 10.11 Two-component reliability model. (a) With CCF; (b) with CCF and repair.

In state 11 the occurrence of CCF causes a direct transition to the system down state 00, so that the sojourn time in state 11 is a random variable with distribution $EXP(\lambda_1 + \lambda_2 + \lambda_{CCF})$. In state 01 or state 10 the CCF is still active and increases the rates to the down state 00.

Even in the presence of CCF the two components may be repaired without interrupting the operation of the system, as represented in Figure 10.11(b).

The two Markov graphs of Figure 10.11 have been derived from IEC Standard 61508 [6]. The solution of the CTMCs of Figure 10.11 is left as a problem.

10.1.3 Other Dependability Measures Defined on a CTMC

In Section 9.5 we defined in terms of MRM various measures that can characterize different aspects of a dependable system, like the expected number of failures or repairs and the expected number of visits into a state, both as a function of t and in steady state.

Example 10.14 *Expected Number of Failures in a Two-Component Series System*
For the series system of two components examined in Example 10.9, the only up state is state 1, while states 2, 3, and 4 are down states. The infinitesimal generator Q (10.12) can be partitioned as:

$$Q = \begin{bmatrix} Q_{uu} & Q_{ud} \\ Q_{du} & Q_{dd} \end{bmatrix} = \left[\begin{array}{c|ccc} -(\lambda_1 + \lambda_2) & \lambda_1 & \lambda_2 & 0 \\ 0 & -\hat{\lambda}_2 & 0 & \hat{\lambda}_2 \\ 0 & 0 & -\hat{\lambda}_1 & \hat{\lambda}_1 \\ 0 & 0 & 0 & 0 \end{array} \right]. \tag{10.41}$$

We can compute from (9.68) the expected number of failures up to time t, $E[N_F(t)]$. We observe that the transitions that lead from a state in Ω_u to a state in Ω_d are contained in the row vector $Q_{ud} = [\lambda_1, \lambda_2]$. Applying Eq. (9.68), we get:

$$E[N_F(t)] = b_1(t)(\lambda_1 + \lambda_2) = (\lambda_1 + \lambda_2)\int_0^t e^{-(\lambda_1 + \lambda_2)u}\, du$$

$$= 1 - e^{-(\lambda_1 + \lambda_2)t}.$$

The expected number of failures coincides, in this case, with the system unreliability. Hence, for $t \to \infty$ the expected number of failures tends to 1 (failure is certain).

Problems

10.4 For Example 10.11, solve for the state probabilities at time t using first the convolution integral approach and then using the Laplace transform approach. Assume that the system is in state $(1, 1)$ at time 0. Next, derive expressions for the system reliability at time t and then the sensitivities (derivatives) of system reliability with respect to the three parameters λ_A, λ_B, and d.

10.5 For Example 10.12, derive expressions for the expected number of failures and the expected number of repairs in the interval $(0, t]$.

10.6   A critical component A is provided with two spares, S_1 and S_2, in warm standby with dormancy factor α. When A fails, spare S_1 is put into operation with a perfect and instantaneous commutation; also, when S_1 fails, spare S_2 is put into operation with a perfect and instantaneous commutation. Assume that A, S_1, and S_2 have a constant failure rate λ when in operation. Derive the state transition diagram and the infinitesimal generator matrix. Then solve for the state probabilities and hence derive an expression for the system reliability at time t.

10.7   A non-repairable pumping station is composed of two pumps A and B with independent but non-identically distributed TTFs $EXP(\lambda_A)$ and $EXP(\lambda_B)$, respectively. We wish to compare two possible configurations, a parallel redundant configuration and a cold standby configuration, from three different points of view:

(a) The reliability of the configurations as functions of time.
(b) The expected instantaneous flow rate produced by the two configurations as functions of time, assuming that pump A has a flow rate ν_A and pump B has a flow rate ν_B. Clearly, these flow rates apply when the respective pump is active; the flow rate of a spare pump is 0.
(c) The expected accumulated flow produced by the two configurations as functions of time.

10.8   We come back to Problem 9.8 describing a computational process decomposed into three modules A, B, and C with execution times distributed as $EXP(a)$, $EXP(b)$, and $EXP(c)$, respectively. The CTMC for a single execution is depicted in Figure 10.12.

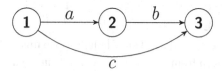

Figure 10.12   Continuous-time Markov chain for Problem 10.8.

Assuming that the initial state is state 1 and the final state of completion is state 3, derive expressions for:

1. The Cdf of the time needed to complete the computational process (probability of being in state 3).
2. The expected time needed to complete the computational process.

10.9   Derive the infinitesimal generator and the expression for the system reliability for the CTMCs of Figures 10.11(a) and 10.11(b).

10.2   Continuous-Time Markov Chains with Absorbing States

As defined in Section 9.2, an absorbing state is a state with no outgoing arcs and, consequently, the corresponding row of the infinitesimal generator matrix has only zero entries. Thus, in the partitioned infinitesimal generator matrix Q of Eq. 9.45, only the matrices Q_{uu} and Q_{ud} have non-zero entries.

Absorbing states appear in many problems, in particular when the reliability is concerned. In effect, since the reliability of a system is the probability of continuous operation in an interval $(0, t]$, the failure states for the system in the corresponding CTMC model must be absorbing states. Absorbing states are also of interest since it will be shown that some properties like the expected time to absorption (and its Cdf or higher moments) or the absorption probability in a CTMC with multiple absorbing states can be computed by solving linear equations, without the need to solve the complete transient behavior of the CTMC.

10.2.1 Single Absorbing State: Cdf of Time to Absorption

Suppose that a CTMC has a single absorbing state a, as in Figure 10.13, while all the other states are transient states. We can partition the infinitesimal generator of the CTMC as follows:

$$Q = \left[\begin{array}{c|c} Q_u & a^T \\ \hline 0 & 0 \end{array} \right], \tag{10.42}$$

where Q_u is the partition of the Q matrix over the transient states in Ω_u, a is a column vector grouping the transition rates from any transient state to the absorbing state, and the row corresponding to the absorbing state has only zero entries. Since the row sum of the infinitesimal generator is equal to zero, it follows that $a^T = -Q_u e^T$.

The Markov equation in partitioned form can be written as:

$$\left[\dfrac{d\pi_u(t)}{dt} \quad \dfrac{d\pi_a(t)}{dt} \right] = [\pi_u(t) \; \pi_a(t)] \left[\begin{array}{c|c} Q_u & a^T \\ \hline 0 & 0 \end{array} \right], \tag{10.43}$$

where $\pi_u(t)$ refers to the partition of the state probability vector over the states in Ω_u, and $\pi_a(t)$ is the probability of the absorbing state a.

Solving Eqs. (10.43) in partitioned form, we get:

$$\begin{cases} \dfrac{d\pi_u(t)}{dt} &= \pi_u(t) \, Q_u, \\ \dfrac{d\pi_a(t)}{dt} &= \pi_u(t) \, a^T; \end{cases} \tag{10.44}$$

$$\begin{cases} \pi_u(t) &= \pi_u(0) \, e^{Q_u t}, \\ \dfrac{d\pi_a(t)}{dt} &= \pi_u(0) \, e^{Q_u t} a^T, \end{cases}$$

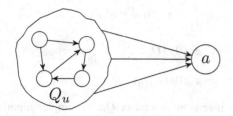

Figure 10.13 Continuous-time Markov chain with one absorbing state a.

where $\pi_u(0)$ is the partition of the initial probability vector $\pi(0)$ over the transient states. Here, it is assumed for the sake of simplicity that $\pi_u(0)\,e^T = 1$, i.e., the initial probability in the absorbing state is $\pi_a(0) = 0$.

Given that T_a is the random variable denoting the time to reach the absorbing state a starting in one of the up states at time $t = 0$ (time to absorption), we can compute its Cdf $F_a(t)$ as:

$$F_a(t) = P\{T_a \leq t\} = P\{Z(t) = a\} = \pi_a(t)$$

$$= 1 - \pi_u(0)\,e^{Q_u t}e^T. \tag{10.45}$$

T_a is called a *continuous phase-type* (PH) random variable and $F_a(t)$ is a PH distribution [7–9]. PH distributions are very relevant in dependability modeling, and Chapter 15 is completely devoted to this subject. In fact, all the time to failure distributions in Chapter 8 were PH.

Taking Laplace transforms, Eq. (10.43) becomes:

$$[s\pi_u^*(s) - \pi_u(0) \quad s\pi_a^*(s)] = [\pi_u^*(s) \quad \pi_a^*(s)]\left[\begin{array}{c|c} Q_u & a^T \\ \hline 0 & 0 \end{array}\right],$$

$$\begin{cases} s\pi_u^*(s) - \pi_u(0) &= \pi_u^*(s)\,Q_u, \\ s\pi_a^*(s) &= \pi_u^*(s)\,a^T; \end{cases}$$

$$\begin{cases} \pi_u^*(s) &= \pi_u(0)(sI - Q_u)^{-1}, \\ \pi_a^*(s) &= \dfrac{1}{s}\pi_u^*(s)\,a^T = \dfrac{1}{s}\pi_u(0)(sI - Q_u)^{-1}a^T. \end{cases}$$

Hence, the transform of the Cdf and the density of T_a are:

$$F_a^*(s) = \frac{1}{s}\pi_u(0)(sI - Q_u)^{-1}a^T,$$

$$f_a^*(s) = sF_a^*(s) = \pi_u(0)(sI - Q_u)^{-1}a^T. \tag{10.46}$$

10.2.2   Single Absorbing State: Expected Time to Absorption

Since T_a is the random variable representing the time to absorption, the expected time to absorption can be computed as:

$$E[T_a] = \int_0^\infty (1 - F_a(t))\,dt$$

$$= \int_0^\infty \pi_u(0)\,e^{Q_u t}e^T\,dt$$

$$= \pi_u(0)\left[Q_u^{-1}\,e^{Q_u t}e^T\right]_0^\infty \tag{10.47}$$

$$= \pi_u(0)\,(-Q_u)^{-1}e^T.$$

Equation (10.47) involves the inversion of matrix Q_u. A simpler algorithm for the computation of $E[T_a]$ can be obtained by considering the expected state occupancy in the transient states before absorption. Equation (9.41) can be rewritten in partitioned form as:

$$\left[\frac{d\boldsymbol{b}_u(t)}{dt} \ \frac{d\boldsymbol{b}_a(t)}{dt}\right] = [\boldsymbol{b}_u(t) \ \boldsymbol{b}_a(t)] \left[\begin{array}{c|c} \boldsymbol{Q}_u & \boldsymbol{a}^T \\ \hline \boldsymbol{0} & 0 \end{array}\right] + [\boldsymbol{\pi}_u(0) \ \pi_a(0)]. \tag{10.48}$$

By considering only the partition over the transient states,

$$\frac{d\boldsymbol{b}_u(t)}{dt} = \boldsymbol{b}_u(t)\,\boldsymbol{Q}_u + \boldsymbol{\pi}_u(0),$$

and letting $(t \to \infty)$, since states in $\boldsymbol{Q}_u$ are transient we get:

$$\lim_{t\to\infty} b_i(t) = \tau_i \quad \text{and} \quad \lim_{t\to\infty} \frac{db_i(t)}{dt} = 0, \quad i \in \boldsymbol{Q}_u,$$

where $\tau_i = E[T_i]$ is the expected total (possibly over many sojourns) time spent in the transient state i before absorption. Grouping the expected times τ_i into a vector $\boldsymbol{\tau} = [\tau_i]$, we have:

$$\boldsymbol{\tau}\,\boldsymbol{Q}_u = -\boldsymbol{\pi}_u(0). \tag{10.49}$$

The total expected time till absorption is the sum of the expected times in the transient states:

$$E[T_a] = \boldsymbol{\tau}\,\boldsymbol{e}^T. \tag{10.50}$$

The expected time to absorption can be calculated directly starting from the CTMC specification by means of Eq. (10.49), without computing the reliability $R(t)$ followed by an integration. Also, in practice no matrix inversion, as in Eq. (10.47), needs to be carried out as the linear system of equations in (10.49) is easier to solve using SOR and related methods for solving a linear system of equations.

Example 10.15 *Parallel System of Two Different Components: MTTF*
The MTTF for a parallel system with two different components was derived in Example 10.9 by integrating the reliability function. As noted earlier, the same results can be obtained in a simpler way by solving the linear equations (10.49) over the transient states to get:

$$-\tau_1(\lambda_1 + \lambda_2) = -1 \quad \tau_1 = \frac{1}{\lambda_1 + \lambda_2},$$

$$\lambda_1\tau_1 - \hat{\lambda}_2\tau_2 = 0 \quad \tau_2 = \frac{\lambda_1}{\hat{\lambda}_2(\lambda_1 + \lambda_2)},$$

$$\lambda_2\tau_1 - \hat{\lambda}_1\tau_3 = 0 \quad \tau_3 = \frac{\lambda_2}{\hat{\lambda}_1(\lambda_1 + \lambda_2)},$$

$$\text{MTTF} = \tau_1 + \tau_2 + \tau_3 = \frac{1}{\lambda_1 + \lambda_2} + \frac{\lambda_1}{\hat{\lambda}_2(\lambda_1 + \lambda_2)} + \frac{\lambda_2}{\hat{\lambda}_1(\lambda_1 + \lambda_2)}.$$

The three summands in the MTTF expression above are the expected times spent in states 1, 2, and 3, respectively, before absorption.

Example 10.16 *Warm Standby System: MTTF*

The warm standby system was described in Example 10.10, and its state transition graph is represented in Figure 10.6. States s_1, s_2, and s_3 are transient states, and state s_4 is an absorbing state. Notice that the absorbing state is also the only down state for the system. Hence, the mean time to absorption is the system MTTF. Partitioning the infinitesimal generator according to (10.42), we obtain:

$$Q = \left[\begin{array}{c|c} Q_u & a \\ \hline 0 & 0 \end{array} \right] = \left[\begin{array}{ccc|c} -(\lambda_A + \alpha\lambda_B) & \lambda_A & \alpha\lambda_B & 0 \\ 0 & -\lambda_B & 0 & \lambda_B \\ 0 & 0 & -\lambda_A & \lambda_A \\ \hline 0 & 0 & 0 & 0 \end{array} \right]. \tag{10.51}$$

The linear equations implied by the vector–matrix equation (10.49) become:

$$[\tau_1 \ \tau_2 \ \tau_3] \left[\begin{array}{ccc} -(\lambda_A + \alpha\lambda_B) & \lambda_A & \alpha\lambda_B \\ 0 & -\lambda_B & 0 \\ 0 & 0 & -\lambda_A \end{array} \right] = -[1\,0\,0], \tag{10.52}$$

whose solution yields:

$$\tau_1 = \frac{1}{\lambda_A + \alpha\lambda_B},$$

$$\tau_2 = \frac{\lambda_A}{\lambda_B(\lambda_A + \alpha\lambda_B)}, \tag{10.53}$$

$$\tau_3 = \frac{\alpha\lambda_B}{\lambda_A(\lambda_A + \alpha\lambda_B)}, \tag{10.54}$$

and finally [compare this with Eq. (10.39)],

$$\text{MTTF} = \tau_1 + \tau_2 + \tau_3 = \frac{\lambda_A\lambda_B + \lambda_A^2 + \alpha\lambda_B^2}{\lambda_A\lambda_B(\lambda_A + \alpha\lambda_B)}. \tag{10.55}$$

It is easy to check that Eq. (10.55) provides the already known results in the limiting cases $\alpha = 1$ and $\alpha = 0$:

$\alpha = 1$, *hot standby* or independent parallel case (Example 10.10):

$$\text{MTTF}_{(\alpha=1)} = \frac{1}{\lambda_A} + \frac{1}{\lambda_B} - \frac{1}{\lambda_A + \lambda_B}.$$

$\alpha = 0$, *cold standby* or simply standby (Example 10.10):

$$\text{MTTF}_{(\alpha=0)} = \frac{1}{\lambda_A} + \frac{1}{\lambda_B}.$$

Alternatively, we can apply Eq. (10.47) by first computing $(-Q_u^{-1})$,

$$-Q_u^{-1} = \frac{1}{\lambda_A\lambda_B(\lambda_A + \alpha\lambda_B)} \left[\begin{array}{ccc} \lambda_A\lambda_B & \lambda_A^2 & \alpha\lambda_B^2 \\ 0 & \lambda_A(\lambda_A + \alpha\lambda_B) & 0 \\ 0 & 0 & \lambda_A(\lambda_A + \alpha\lambda_B) \end{array} \right];$$

then, application of Eq. (10.47), gives:

$$\text{MTTF}_{\text{WS}} = \begin{bmatrix} 1 & 0 & 0 \end{bmatrix} \begin{bmatrix} -Q_u^{-1} \end{bmatrix} \begin{bmatrix} 1 \\ 1 \\ 1 \end{bmatrix}$$

$$= \frac{\lambda_A \lambda_B + \lambda_A^2 + \alpha \lambda_B^2}{\lambda_A \lambda_B (\lambda_A + \alpha \lambda_B)}. \tag{10.56}$$

Example 10.17 *Two-Component System with Repair: MTTF (continued from Example 10.12)*

Returning to the two-server parallel redundant system of Example 10.12, whose corresponding CTMC is shown in Figure 10.9, when both blades have failed the system has failed and it reaches the absorbing state 0.

The infinitesimal generator can be partitioned as follows:

$$Q = \left[\begin{array}{c|c} Q_u & a^T \\ \hline 0 & 0 \end{array} \right] = \left[\begin{array}{cc|c} -2\lambda & 2\lambda & 0 \\ \mu & -(\lambda + \mu) & \lambda \\ \hline 0 & 0 & 0 \end{array} \right].$$

Assuming as initial probability $\pi_2(0) = 1$, the linear system of equations implied by Eq. (10.49) becomes:

$$\begin{bmatrix} \tau_2 & \tau_1 \end{bmatrix} \begin{bmatrix} -2\lambda & 2\lambda \\ \mu & -(\lambda + \mu) \end{bmatrix} = -[1\,0], \tag{10.57}$$

whose solution is:

$$\tau_2 = \frac{\lambda + \mu}{2\lambda^2}, \qquad \tau_1 = \frac{1}{\lambda}. \tag{10.58}$$

Finally,

$$\text{MTTF} = \tau_2 + \tau_1 = = \frac{3\lambda + \mu}{2\lambda^2}. \tag{10.59}$$

If $\mu = 0$ (two parallel components without repair) we get:

$$E[T_a] = \text{MTTF} = \frac{3}{2\lambda}.$$

Alternatively, by applying Eq. (10.47):

$$-Q_u^{-1} = \frac{1}{2\lambda(\lambda + \mu) - 2\lambda\mu} \begin{bmatrix} (\lambda + \mu) & 2\lambda \\ \mu & 2\lambda \end{bmatrix} = \frac{1}{2\lambda^2} \begin{bmatrix} (\lambda + \mu) & 2\lambda \\ \mu & 2\lambda \end{bmatrix},$$

$$E[T_a] = |1\,0| \frac{1}{2\lambda^2} \begin{bmatrix} (\lambda + \mu) & 2\lambda \\ \mu & 2\lambda \end{bmatrix} \begin{bmatrix} 1 \\ 1 \end{bmatrix} = \frac{3\lambda + \mu}{2\lambda^2}.$$

Note that by using redundancy by itself, the MTTF increases by 50% from $1/\lambda$ to $1.5/\lambda$, but by adding repair on top of redundancy the increase in the MTTF is by several orders of magnitude: $1.5/\lambda + \mu/(2\lambda^2)$, since μ will generally be several orders of magnitude larger than λ in practice.

With the data used in Figure 10.10, (i.e., $\lambda = 1/8760\,\mathrm{f\,hr^{-1}}$ and $\mu = 1/0.5\,\mathrm{r\,hr^{-1}}$), we get:

$$\mathrm{MTTF_{single}} = 8760\,\mathrm{hr}, \quad \mathrm{MTTF_{par\text{-}no\text{-}rep}} = 13\,140\,\mathrm{hr}, \quad \mathrm{MTTF_{par\text{-}with\text{-}rep}} = 1.92 \times 10^7\,\mathrm{hr}.$$

Example 10.18   *Two-Component System with Repair and Imperfect Coverage: MTTF*
We continue the previous example by assuming, more realistically, that upon a failure the system must detect which blade has failed and reconfigure to continue operation with the single remaining blade. However, the detection and recovery process may complete successfully with a coverage probability c; with probability $(1-c)$ the recovery process does not complete successfully, and the system incurs a complete failure and moves to the down state 0. Figure 10.14 shows the state space of the system with the coverage taken into account.

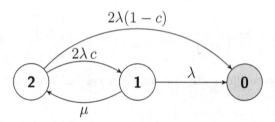

Figure 10.14   State diagram of a parallel two-component system with repair and coverage.

The partitioned infinitesimal generator is:

$$Q = \left[\begin{array}{c|c} Q_u & a^T \\ \hline 0 & 0 \end{array} \right] = \left[\begin{array}{cc|c} -2\lambda & 2\lambda c & 2\lambda(1-c) \\ \mu & -(\lambda+\mu) & \lambda \\ \hline 0 & 0 & 0 \end{array} \right].$$

Solving the linear system (10.49) with initial probability $\pi_2(0) = 1$, we obtain:

$$\tau_2 = \frac{\lambda+\mu}{2\lambda(\lambda+\mu(1-c))},$$

$$\tau_1 = \frac{2\lambda c}{2\lambda(\lambda+\mu(1-c))}. \tag{10.60}$$

Finally,

$$\mathrm{MTTF} = E[T_a] = \tau_2 + \tau_1 = = \frac{\lambda+\mu+2\lambda c}{2\lambda(\lambda+\mu(1-c))}. \tag{10.61}$$

In the case of perfect coverage, $c = 1$, we again obtain the result of Example 10.15:

$$E[T_a] = \mathrm{MTTF} = \frac{3\lambda+\mu}{2\lambda^2}.$$

10.2.3 Single Absorbing State: Moments of Time to Absorption

We can derive the higher moments of the time to absorption T_a by means of the moment theorem for Laplace transforms applied to the density in (10.46). We obtain:

$$E[T_a^i] = (-1)^i \frac{d^i f_a^*(s)}{ds^i}\bigg|_{s=0} = (-1)^i n! \, \pi_u(0) \, (Q_u)^{-i} e^T. \tag{10.62}$$

From Eq. (10.62) we can compute the second moment $E[T_a^2]$ and then the variance:

$$\text{Var}[T_a] = E[T_a^2] - E[T_a]^2 = 2\pi_u(0) \, (Q_u)^{-2} e^T - \left[\pi_u(0) \, (Q_u)^{-1} e^T\right]^2. \tag{10.63}$$

Problems

10.10 Compute the reliability and the MTTF of the cold standby system with imperfect coverage depicted in Figure 10.8. Next, determine the variance of the time to absorption using Eq. (10.63).

10.11 Compute the reliability and the MTTF of the cold standby system with imperfect coverage (as depicted in Figure 10.8), but in this case the detection and commutation mechanism from the main component A to the spare B takes an exponentially distributed delay of rate δ. Next, determine the variance of the time to absorption using Eq. (10.63).

10.12 Compute the reliability and MTTF of the same configuration as Problem 10.11, but in this case the delay and commutation mechanism has a coverage c_d, i.e., with probability $(1 - c_d)$ the detection and commutation process fails.

10.13 Compute the MTTF of a warm standby system with imperfect coverage where A with TTF $EXP(\lambda_A)$ is the main component, and B with TTF $EXP(\lambda_B)$ the warm spare.

10.14 A system is composed of two subsystems, X and Y. The system operates correctly until one of the two subsystems operates correctly. The TTF of X is $EXP(\alpha)$ and the TTF of Y is $EXP(\beta)$. Further, the two subsystems are subject to CCF failure that is exponentially distributed with rate γ and acts simultaneously on the two subsystems (Figure 10.11). Compute:

(a) The system MTTF when the components are non-repairable.
(b) The system MTTF when the components are repairable, before system failure, with rates μ_X and μ_Y, respectively.

Example 10.19 *DPM for an Application Server (continued from Example 9.18)*
In Example 9.18, we computed the DPM for a system with cold replication. While that computation of DPM was straightforward and simple, it is an approximation and it underestimates the real DPM. A detailed approach for accurately computing the DPM taking into account the failure/repair behavior of the system and retry is given in [10]. We continue with the case for cold replication. When a new call arrives at the system and finds it in the down state, then the call is not immediately considered lost. If the system

recovers within a *retry window* then the call can be processed successfully. Assume that the new call arrival process is Poisson with rate λ (Section 3.5.3).

Given that a failure has occurred, assume T_d is the random time for the system to recover and be restored to fully working condition. Let w_i represent the maximum length of the retry window for a call. If $T_d < w_i$, no new calls will be dropped due to this failure. If, on the other hand, $T_d \geq w_i$, then the mean number of new calls dropped is $\lambda(T_d - w_i)$. This follows from the property of a Poisson arrival process where the mean number of arrivals in a duration of length t is the arrival rate multiplied by t (Section 3.5.3). Figure 10.15 shows how the newly arriving calls are affected by T_d.

Figure 10.16 is a slightly modified version of Figure 9.23 that shows the state transitions after a failure has occurred. State 0 is now an absorbing state, which represents the state in which the application server has successfully recovered from failure. Suppose a failure occurs at time $t = 0$ and the CTMC of Figure 10.16 has just entered state 1. The cumulative distribution function for the time to absorption T_d can be computed from the CTMC. With probability c the time distribution is $\mathrm{HYPO}(\delta, \alpha_p)$, while with probability $1 - c$ the time distribution is $\mathrm{HYPO}(\delta, \alpha_p, \alpha_n)$. Therefore,

$$
F_{T_d}(t) = \pi_0(t) = c\left(1 - \frac{\alpha_p}{\alpha_p - \delta}e^{-\delta t} + \frac{\delta}{\alpha_p - \delta}e^{-\alpha_p t}\right)
$$
$$
+ (1 - c)\left(1 - \frac{\alpha_p}{\alpha_p - \delta}\frac{\alpha_n}{\alpha_n - \delta}e^{-\delta t} - \frac{\delta}{\delta - \alpha_p}\frac{\alpha_n}{\alpha_n - \alpha_p}\right.
$$
$$
\left. e^{-\alpha_p t} - \frac{\delta}{\delta - \alpha_n}\frac{\alpha_p}{\alpha_p - \alpha_n}e^{-\alpha_n t}\right), \quad t \geq 0.
$$

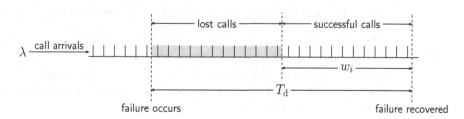

Figure 10.15 Newly arriving calls at a SIP application server.

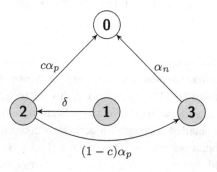

Figure 10.16 Loss model for newly arriving calls and stable calls for the cold replication case.

The mean number of new calls dropped due to a server failure is:

$$n_a = \int_{w_i}^{\infty} \lambda(t - w_i)d\pi_0(t)$$

$$= \left[\text{MTTA} - w_i + \int_0^{w_i} \pi_0(t)dt \right]\lambda,$$

where MTTA denotes the mean time to absorption at state 0, which is given by:

$$\text{MTTA} = \int_0^{\infty} t\, d\pi_0(t)$$

$$= c\left(\frac{1}{\delta} + \frac{1}{\alpha_p}\right) + (1 - c)\left(\frac{1}{\delta} + \frac{1}{\alpha_p} + \frac{1}{\alpha_n}\right)$$

$$= \frac{1}{\delta} + \frac{1}{\alpha_p} + \frac{1 - c}{\alpha_n}. \tag{10.64}$$

The result in Eq. (10.64) could also be obtained by solving the linear system in Eq. (10.65):

$$[\tau_1\ \tau_2\ \tau_3] \begin{bmatrix} -\delta & \delta & 0 \\ 0 & -\alpha_p & (1-c)\alpha_p \\ 0 & 0 & -\alpha_n \end{bmatrix} = -[1\ 0\ 0], \tag{10.65}$$

whose solution is:

$$\tau_1 = \frac{1}{\delta}, \qquad \tau_2 = \frac{1}{\alpha_p}, \qquad \tau_3 = \frac{1 - c}{\alpha_n}.$$

So,

$$n_a = \frac{1}{\delta - \alpha_p}\left[\frac{\delta}{\alpha_p}e^{-\alpha_p w_i} - \frac{\alpha_p}{\delta}e^{-\delta w_i} \right]\lambda$$

$$+ \frac{1 - c}{\delta - \alpha_p}\left[\frac{\delta}{\alpha_n - \alpha_p}e^{-\alpha_p w_i} + \frac{\alpha_p}{\delta - \alpha_n}e^{-\delta w_i} \right.$$

$$\left. - \frac{\delta - \alpha_p}{\alpha_n}\frac{\delta}{\delta - \alpha_n}\frac{\alpha_p}{\alpha_n - \alpha_p}e^{-\alpha_n w_i} \right]\lambda. \tag{10.66}$$

From the availability model for the application server with cold replication in Figure 9.23, the failure frequency for the application server is given by the expected number of transitions out of the up state 0 (see Section 9.5.3) $f = \pi_0\gamma$, where π_0 is the steady-state probability that the availability model (Figure 9.23) is in state 0 (up state). The mean number of lost calls n_a is given above. Using these values, we get the DPM caused by application server failure as:

$$\text{DPM} = \pi_0\gamma n_a \frac{10^6}{\lambda}. \tag{10.67}$$

Substituting the values of the parameters as given in Example 9.18, we get DPM = 32.29 from the above equation. Compare this more accurate value with that obtained in Example 9.18 (DPM = 36.75).

By setting the retry window interval $w_i = 0$ we can model the situation where there is no retry of calls. As we mentioned earlier, in this situation, when a new call arrives at the system and finds it in the down state, then the call is immediately considered lost. Substituting $w_i = 0$ in Eq. (10.66), the average number of lost calls is obtained as

$$n_a = \lambda \times \text{MTTA}.$$

Using this value in the expression for the DPM in Eq. (10.67), we obtain

$$\text{DPM} = \pi_0 \gamma n_a \frac{10^6}{\lambda}$$

$$= \pi_0 \gamma \text{MTTA} \lambda \frac{10^6}{\lambda}$$

$$= \frac{1}{\gamma} \left[\frac{1}{\gamma} + \frac{1}{\delta} + \frac{1}{\alpha_p} + \frac{(1-c)}{\alpha_n} \right]^{-1} \gamma \left(\frac{1}{\delta} + \frac{1}{\alpha_p} + \frac{1-c}{\alpha_n} \right) \times 10^6$$

$$= \left(1 - \frac{1}{\gamma} \left[\frac{1}{\gamma} + \frac{1}{\delta} + \frac{1}{\alpha_p} + \frac{(1-c)}{\alpha_n} \right]^{-1} \right) \times 10^6$$

$$= (1 - A_{ss}) \times 10^6.$$

This is the same as Eq. (9.78).

Example 10.20 *Renal Disease Model [11]*
Reliability modeling techniques provide new frontiers in predicting healthcare outcomes. Several pathologies are known to progress in well-defined, clinically observable states, and are therefore well adapted to Markov reward methods. As an example, a simple model for progressive renal system failure is proposed in Figure 10.17.

Here, five discrete conditions are enumerated in keeping with clinical classifications of kidney function:

- Healthy: normal renal function
- CKD: chronic kidney disease without renal failure
- ESRD: end-stage renal disease, an administrative term for patients with renal failure
- Transplant: renal failure patients who have successfully received a transplant
- Deceased: an absorbing state.

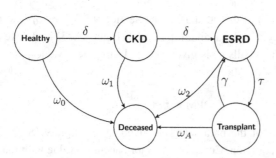

Figure 10.17 Renal disease progression using Markov reward formalisms.

As formulated, each parameter corresponds to measures tracked by public health organizations, namely:

- ω_i: prognosis rate at state i, where $i = 0, 1, 2, A$
- δ: rate of decline or condition deterioration, assuming for now a constant rate as patients progress from healthy to ESRD
- τ: transplantation rate for a given community
- γ: graft rejection rate for active transplants.

Closed-form analytic solutions for the transient probability of each condition can be obtained by transforming the corresponding Kolmogorov differential equations into the Laplace domain, solving the Laplace domain equations and finally inverting the solution back to the time domain by partial fraction expansion (Section 10.1.2). Assuming all patients begin healthy, the solution for healthy states is straightforward. In the Laplace transform domain, we get (where H indicates the healthy state and C the CKD state):

$$s\pi_H^*(s) - 1 = -(\delta + \omega_0)\pi_H^*(s) = \frac{1}{s + \delta + \omega_0},$$

$$s\pi_C^*(s) = -(\delta + \omega_1)\pi_C^*(s) + \delta\pi_H^*(s).$$

From the above equations: $\pi_C^*(s) = \frac{\delta}{s + \delta + \omega_1}\pi_H^*(s) = \frac{\delta}{(s + \delta + \omega_0)(s + \delta + \omega_1)}$

$$= \frac{\delta}{\omega_0 - \omega_1}\left[\frac{1}{s + \delta + \omega_1} - \frac{1}{s + \delta + \omega_0)}\right].$$

Inverting to the time domain, we get:

$$\pi_H(t) = e^{-(\delta + \omega_0)t},$$

$$\pi_C(t) = \frac{\delta}{\omega_0 - \omega_1}\left[e^{-(\delta + \omega_1)t} - e^{-(\delta + \omega_0)t}\right].$$

The parameter values reported in Table 10.1 are derived from [11] as estimates for a 65-year-old Medicare patient, and are based on the latest available statistics from the United States Renal Data System (USRDS) annual report [12].

We leave the expressions for the other state probabilities as an exercise.

Table 10.1 Parameter estimates for a 65-year-old Medicare patient [12].

Description	Symbol	Value (event yr^{-1})
Decline	δ	0.1887
Transplant	τ	0.1786
Graft rejection	γ	0.0050
Prognosis: Healthy	ω_0	0.0645
Prognosis: CKD	ω_1	0.1013
Prognosis: ESRD	ω_2	0.2174
Prognosis: Transplant	ω_A	0.0775

Example 10.21   *Reliability of a Multi-Voltage High-Speed Train [13]*

We return to Example 4.14, which depicted the RBD of a multi-voltage propulsion system for a high-speed train [13] for which the reliability expression was derived. Here we consider only one single module consisting of a series of four components (transformer, filter, inverter and motor) and two parallel converters. This single module can be represented by a three-state MRM model, which accounts for the power level delivered in each configuration:

- a fully operational state delivering maximum power ($r_2 = 2200\,\text{kW}$) when all components are working;
- a degraded state delivering half of the power ($r_1 = 1100\,\text{kW}$) when all the series components and one converter out of the two are working;
- a failed state delivering no power.

The MRM model, combining the structure state CTMC and the reward rates, is depicted in Figure 10.18. State 2 is the fully operational state with reward rate r_2, state 1 is the degraded state with reward rate r_1 and state 0 is the failed state with reward rate equal to 0. As in Example 4.14, we denote by λ the sum of the failure rates of the series components and by γ the failure rate of each of the converters.

The generator matrix of the CTMC of Figure 10.18 is:

$$Q = \begin{bmatrix} -(2\gamma + \lambda) & 2\gamma & \lambda \\ 0 & -(\gamma + \lambda) & \gamma + \lambda \\ 0 & 0 & 0 \end{bmatrix}.$$

Solving the transient equations by the convolution integration method of Section 10.1.1 (or using the Laplace transform method) we obtain:

$$\begin{cases} \pi_2(t) &= e^{-(2\gamma+\lambda)t}, \\ \pi_1(t) &= 2e^{-(\gamma+\lambda)t} - 2e^{-(2\gamma+\lambda)t}, \\ \pi_0(t) &= 1 - \pi_2(t) - \pi_1(t) = 1 + e^{-(2\gamma+\lambda)t} - 2e^{-(\gamma+\lambda)t}. \end{cases}$$

The result is the same as the one obtained from the RBD solution. However, in this model we can combine the system reliability with its performance in terms of power delivered. From the state probability expressions, we can write down the expected power $E[X(t)]$ available at time t as an example of expected reward rate at time t:

$$E[X(t)] = \sum_{i=0}^{2} r_i \pi_i(t) = r_2 \pi_2(t) + r_1 \pi_1(t),$$

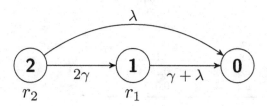

Figure 10.18   Markov reward model representing the propulsion module as a three-state model.

and the expected accumulated energy $E[Y(t)]$ delivered in the interval $(0, t]$ as an example of expected accumulated reward in the interval:

$$E[Y(t)] = \sum_{i=0}^{2} \int_0^t r_i \pi_i(x) dx = r_2 \int_0^t \pi_2(x) dx + r_1 \int_0^t \pi_1(x) dx. \tag{10.68}$$

The expected accumulated reward (energy delivered) until absorption (failure) is then easily computed:

$$E[Y(\infty)] = r_2 \int_0^\infty \pi_2(x) dx + r_1 \int_0^\infty \pi_1(x) dx$$

$$= \frac{r_2}{2\gamma + \lambda} + \frac{2r_1}{\gamma + \lambda} - \frac{2r_1}{2\gamma + \lambda}. \tag{10.69}$$

We can also obtain the Cdf of $Y(\infty)$, the reward accumulated till absorption, abbreviated here as Y, using a method proposed by Beaudry in [14]. Beaudry's method consists in dividing the transition rates of the CTMC by the corresponding reward rates. In fact, in a state with constant reward rate r, the reward accumulated in a time interval of length t is $Y = rt$; by differentiation,

$$dY = r \, dt, \qquad dt = dY/r.$$

Given a transition rate λ_i out of state i with reward rate r_i, the transition rate in reward units becomes λ_i/r_i. This technique transforms a time domain representation of the CTMC into a reward domain representation. The scaled CTMC of Figure 10.18 is shown in Figure 10.19.

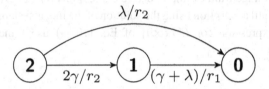

Figure 10.19 Scaled CTMC of Figure 10.18 according to Beaudry's method.

We derive expressions for the state probabilities $\pi_2^{(R)}(r)$ and $\pi_1^{(R)}(r)$ in the scaled CTMC of Figure 10.19 to obtain:

$$\begin{cases} \pi_2^{(R)}(r) = e^{-\frac{(2\gamma + \lambda)}{r_2} r}, \\[2mm] \pi_1^{(R)}(r) = \frac{2r_1\gamma}{\gamma(2r_1 - r_2) + \lambda(r_1 - r_2)} \cdot (e^{-\frac{\gamma + \lambda}{r_1} r} - e^{-\frac{2\gamma + \lambda}{r_2} r}), \\[2mm] \pi_0^{(R)}(r) = 1 - \pi_2^{(R)}(r) - \pi_1^{(R)}(r), \end{cases}$$

and, finally:

$$P(Y(\infty) \ge a) = \pi_1^{(R)}(a) + \pi_2^{(R)}(a).$$

Further, the expected energy delivered before system failure can be computed as the expected time to absorption in the scaled CTMC of Figure 10.19.

Problems

10.15   With reference to the renal disease model of Figure 10.17, compute the transient probabilities of the states E, T, and D using the method of Laplace transform.

10.16   State D is an absorbing state: compute the expected time to absorption starting from state H with probability 1.

10.17   A real-time system must retrieve data from a database in a fixed temporal window of length d. The search is accomplished in two steps. In the first step, whose duration is $EXP(\lambda)$, the specific archive is selected; in the second step, whose duration is $EXP(\mu)$, the data inside the archive is selected. Due to the unreliability of the system, the first step has a coverage c_1 (i.e., the first step fails with probability $(1 - c_1)$), and the second step has a coverage c_2. The searching operation can be modeled by a CTMC.

(a) Draw the state transition diagram and derive the infinitesimal generator matrix.
(b) Find the probability that in the temporal window $(0, d]$ the search is successful.
(c) If there are no real-time constraints, find the probability that the search is successful in any case.
(d) Assume now that if both the first and the second step fail, they can be iterated indefinitely with a duration $EXP(\tau)$. Compute the expected time to complete a successful search.

10.18   For the multi-voltage train example, carry out the integrals in Eq. (10.68) to obtain a closed-form solution for the expected accumulated reward.

10.19   For the multi-voltage train example, derive expressions for the expected times spent in states 2 and 1 until absorption using the solution of the linear system Eq. (10.49) and thus verify the expression for $E[Y(\infty)]$ of Eq. (10.69) using the alternative expression $E[Y(\infty)] = r_2 \tau_2 + r_1 \tau_1$.

10.2.4   Multiple Absorbing States

Multiple absorbing states in dependability modeling arise for two principal reasons: the presence of multiple failure causes or failure modes, and the different effects that some faults can have on the system [15]. In a state-space model each cause of failure corresponds to an absorbing state or to a subset of absorbing states. Thus, for example, one cause of failure can be exhaustion of redundancy while another may be due to imperfect coverage. It may be useful to separate these two causes of failure in order to better understand how to improve the system behavior. Another reason is not so much what caused the failure as what the consequence of failure is. Thus, for example, we may wish to separate failure states into those that correspond to a safe shutdown versus those that cause unsafe behavior. In the security context, we may distinguish security failures by those that compromise confidentiality from those that compromise integrity. To simplify the analysis, we assume in the following that the CTMC has two absorbing states only, say a and b, as in Figure 10.20. The generalization to multiple absorbing states is straightforward [15].

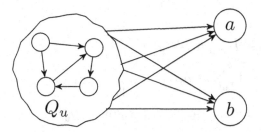

Figure 10.20 Continuous-time Markov chain with multiple absorbing states.

We partition the Kolmogorov differential equation as follows (the rows corresponding to the two absorbing states have all 0 entries):

$$\left[\frac{d\pi_u(t)}{dt} \ \frac{d\pi_a(t)}{dt} \ \frac{d\pi_b(t)}{dt} \right] = [\pi_u(t) \ \pi_a(t) \ \pi_b(t)] \left[\begin{array}{c|cc} Q_u & a^T & b^T \\ \hline 0 & 0 & 0 \\ 0 & 0 & 0 \end{array} \right], \quad (10.70)$$

where the square matrix Q_u groups the transition rates inside the transient states and the column vectors a^T and b^T consist of the rates from the transient states to the absorbing states a and b, respectively. We further assume that the initial probability is concentrated in the transient states, so that the partitioned initial probability vector is $\pi(0) = [\pi_u(0) \ 0 \ 0]$. Solving the above equations in partitioned form, we get:

$$\begin{cases} \dfrac{d\pi_u(t)}{dt} &= \pi_u(t) \, Q_u, \\ \dfrac{d\pi_a(t)}{dt} &= \pi_u(t) \, a^T, \\ \dfrac{d\pi_b(t)}{dt} &= \pi_u(t) \, b^T; \end{cases}$$

$$\begin{cases} \pi_u(t) &= \pi_u(0) \, e^{Q_u t}, \\ \dfrac{d\pi_a(t)}{dt} &= \pi_u(0) \, e^{Q_u t} a^T, \\ \dfrac{d\pi_b(t)}{dt} &= \pi_u(0) \, e^{Q_u t} b^T. \end{cases}$$

Define T_a and T_b as the times to absorption for states a and b, respectively. The corresponding Cdfs $F_a(t)$ and $F_b(t)$ are now defective distributions, and the mean times to absorption in either state a or state b are undefined (infinity):

$$F_a(t) = P\{\tau_a \le t\} = P\{Z(t) = a\} = \pi_a(t),$$

$$[6pt]F_b(t) = P\{\tau_b \le t\} = P\{Z(t) = b\} = \pi_b(t).$$

The Cdf $F_a(t)$ can be obtained by integrating the differential equation

$$\frac{d\pi_a(t)}{dt} = \pi_u(0) \, e^{Q_u t} a^T,$$

whose solution is:

$$\pi_a(t) = \int_0^t \pi_u(0) \, e^{Q_u x} a^{\mathrm{T}} \, dx$$

$$= \left[\pi_u(0) \left(e^{Q_u x} Q_u^{-1} \right) a^{\mathrm{T}} \right]_0^t \tag{10.71}$$

$$= \pi_u(0) (-Q_u^{-1}) a^{\mathrm{T}} + \pi_u(0) \, e^{Q_u t} Q_u^{-1} a^{\mathrm{T}}.$$

As $t \to \infty$, the first term gives the probability of final absorption in state a:

$$\pi_a(\infty) = \lim_{t \to \infty} \pi_a(t) = \pi_u(0) (-Q_u^{-1}) a^{\mathrm{T}}. \tag{10.72}$$

We define the conditional Cdf of the absorption time in a as $\pi_a(t|a)$ and the conditional expected time to absorption in state a as $E[T_{a|a}]$:

$$\pi_a(t|a) = \frac{\pi_a(t)}{\pi_a(\infty)},$$

$$E[T_{a|a}] = \int_0^\infty (1 - \pi_a(x|a)) \, dx$$

$$= \frac{\int_0^\infty - \left(\pi_u(0) \, e^{Q_u x} Q_u^{-1} a^{\mathrm{T}} \right) dx}{\pi_a(\infty)} \tag{10.73}$$

$$= \frac{\pi_u(0) Q_u^{-2} a^{\mathrm{T}}}{\pi_a(\infty)}.$$

By defining, from Eq. (10.49), the vector of the expected time to absorption in the transient states $\tau = -\pi_u(0) Q_u^{-1}$, Eq. (10.73) can be rewritten as:

$$E[T_{a|a}] = \frac{-\tau \, Q_u^{-1} a^{\mathrm{T}}}{\tau \, a^{\mathrm{T}}}. \tag{10.74}$$

The same result was obtained in [15] in the Laplace transform domain. The Laplace transform of Eq. (10.70) is:

$$[\pi_u^*(s) \, \pi_a^*(s) \, \pi_b^*(s)] \begin{bmatrix} s I_u - Q_u & -a^{\mathrm{T}} & -b^{\mathrm{T}} \\ 0 & s & 0 \\ 0 & 0 & s \end{bmatrix} = [\pi_u(0) \, \pi_a(0) \, \pi_b(0)]. \tag{10.75}$$

From Eq. (10.75) we get:

$$\pi_u^*(s) = \pi_u(0) \, [s I_u - Q_u]^{-1},$$

$$\pi_a^*(s) = \frac{1}{s} \pi_u(0) \, [s I_u - Q_u]^{-1} a^{\mathrm{T}}.$$

The probability of final absorption in state a can be computed from the final value theorem of Laplace transforms:

$$\pi_a(\infty) = \lim_{t \to \infty} \pi_a(t) = \lim_{s \to 0} s \pi_a^*(s) = \pi_u(0) (-Q_u)^{-1} a^{\mathrm{T}} = \tau \, a^{\mathrm{T}}.$$

The Laplace transform of the conditional Cdf in Eq. (10.73) is

$$\pi_{a|a}^*(s) = \frac{\pi_a^*(s)}{\pi_a(\infty)} = \frac{1}{s} \frac{\pi_u(0) \, [s I_u - Q_u]^{-1} \, a^T}{\pi_a(\infty)}, \tag{10.76}$$

and applying the moment theorem for Laplace transforms, we obtain:

$$E[T_{a|a}] = \lim_{s \to 0} -\frac{d(s\,\pi_{a|a}^*(s))}{ds} = \frac{\pi_u(0) Q_u^{-2} a^T}{\pi_a(\infty)}.$$

Example 10.3 already considered a single component with two failure modes and derived the defective distribution of the time to reach each one of the failure modes. The following example shows how different failure states can be defined due to different causes.

Example 10.22   *Different Failure Causes (continued from Example 10.18)*
In Example 10.18 we considered a system whose failure was caused either by exhaustion of the redundancy or due to imperfect coverage in detecting/recovering from a failure. These two causes are separated out in Figure 10.21, where the state labeled 0e indicates a failure caused by exhaustion of the redundancy, while the state labeled 0c indicates a failure caused by imperfect coverage.

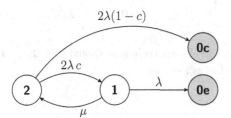

Figure 10.21   Two-component system with different causes of failure.

Applying Eq. (10.72), the eventual probability of failing due to one of the two failure causes becomes [15]:

$$\pi_{0e} = \frac{\lambda c}{\lambda + (1-c)\mu}, \qquad \pi_{0c} = \frac{(1-c)(\lambda+\mu)}{\lambda + (1-c)\mu},$$

with $\pi_{0e} + \pi_{0c} = 1$.

Example 10.23   *Warm Standby: Safety Analysis (continued from Example 10.16)*
In a safety analysis we wish to distinguish two cases: when the primary unit A fails whether or not the standby unit B is up. The safe state 0s (see Figure 10.22) is reached when the standby unit B can recover the failure of the primary unit A, while the unsafe state 0u is reached if the standby unit had failed prior to the occurrence of the primary unit failure.

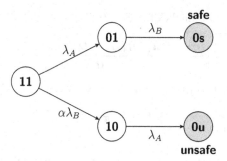

Figure 10.22 Safety model for a warm standby system.

We partition the generator matrix of this CTMC:

$$Q = \begin{bmatrix} Q_u & a^T & b^T \\ 0 & 0 & 0 \\ 0 & 0 & 0 \end{bmatrix} = \left[\begin{array}{ccc|cc} -(\lambda_A + \alpha\lambda_B) & \lambda_A & \alpha\lambda_B & 0 & 0 \\ 0 & -\lambda_B & 0 & \lambda_B & 0 \\ 0 & 0 & -\lambda_A & 0 & \lambda_A \\ \hline 0 & 0 & 0 & 0 & 0 \\ 0 & 0 & 0 & 0 & 0 \end{array} \right],$$

$$-Q_u^{-1} = \frac{1}{\lambda_A \lambda_B (\lambda_A + \alpha\lambda_B)} \begin{bmatrix} \lambda_A \lambda_B & \lambda_A^2 & \alpha\lambda_B^2 \\ 0 & \lambda_A(\lambda_A + \alpha\lambda_B) & 0 \\ 0 & 0 & \lambda_B(\lambda_A + \alpha\lambda_B) \end{bmatrix}.$$

We separate the Cdfs of the times to absorption in state 0s and state 0u, and we compute the eventual absorption probability as $t \to \infty$:

$$F_{0s}(\infty) = \begin{bmatrix} 1 & 0 & 0 \end{bmatrix} \begin{bmatrix} -Q_u^{-1} \end{bmatrix} \begin{bmatrix} 0 \\ \lambda_B \\ 0 \end{bmatrix}$$

$$= \frac{\lambda_A^2 \lambda_B}{\lambda_A \lambda_B (\lambda_A + \alpha\lambda_B)} = \frac{\lambda_A}{\lambda_A + \alpha\lambda_B}, \tag{10.77}$$

$$F_{0u}(\infty) = \begin{bmatrix} 1 & 0 & 0 \end{bmatrix} \begin{bmatrix} -Q_u^{-1} \end{bmatrix} \begin{bmatrix} 0 \\ 0 \\ \lambda_A \end{bmatrix}$$

$$= \frac{\alpha\lambda_B^2 \lambda_A}{\lambda_A \lambda_B (\lambda_A + \alpha\lambda_B)} = \frac{\alpha\lambda_B}{\lambda_A + \alpha\lambda_B}. \tag{10.78}$$

Next we derive the conditional expected absorption times into states 0s and 0u according to Eq. (10.74), assuming the initial state probability vector $\pi_u(0) = [1\,0\,0]$.

The vector τ is the solution of the linear system $\tau Q_u = -\pi_u(0)$ whose explicit equations and related solutions are:

$$
\begin{cases}
\tau_{11}(\lambda_A + \alpha\lambda_B) = -1 \\
\lambda_A \tau_{11} - \lambda_B \tau_{01} = 0 \\
\alpha\lambda_B \tau_{11} - \lambda_A \tau_{10} = 0
\end{cases}
\qquad
\begin{cases}
\tau_{11} = \dfrac{1}{\lambda_A + \alpha\lambda_B} \\[2mm]
\tau_{01} = \dfrac{\lambda_A}{\lambda_B(\lambda_A + \alpha\lambda_B)} \\[2mm]
\tau_{10} = \dfrac{\alpha\lambda_B}{\lambda_A(\lambda_A + \alpha\lambda_B)}.
\end{cases}
\tag{10.79}
$$

For the conditional expected absorption time into state 0s, $E[T_{0s|0s}]$, the numerator of Eq. (10.74) becomes:

$$
-\tau Q_u^{-1} a^T = \frac{1}{\lambda_A \lambda_B (\lambda_A + \alpha\lambda_B)} \frac{\lambda_A^2 \lambda_B + \lambda_A^2(\lambda_A + \alpha\lambda_B)}{\lambda_A + \alpha\lambda_B}.
$$

The denominator of Eq. (10.74) is given in Eq. (10.77). Combining numerator and denominator we get:

$$
E[T_{0s|0s}] = \frac{\lambda_A \lambda_B + \lambda_A(\lambda_A + \alpha\lambda_B)}{\lambda_A \lambda_B (\lambda_A + \alpha\lambda_B)} = \frac{1}{(\lambda_A + \alpha\lambda_B)} + \frac{1}{\lambda_B}.
\tag{10.80}
$$

For the conditional expected absorption time into state 0u, $E[T_{0u|0u}]$, the numerator of Eq. (10.74) becomes:

$$
-\tau Q_u^{-1} b^T = \frac{1}{\lambda_A \lambda_B (\lambda_A + \alpha\lambda_B)} \frac{\lambda_A \alpha\lambda_B^2 + \alpha\lambda_B^2(\lambda_A + \alpha\lambda_B)}{\lambda_A + \alpha\lambda_B}.
$$

The denominator of Eq. (10.74) is given in Eq. (10.78). Combining numerator and denominator we get:

$$
E[T_{0u|0u}] = \frac{\lambda_A \lambda_B + \lambda_B(\lambda_A + \alpha\lambda_B)}{\lambda_A \lambda_B (\lambda_A + \alpha\lambda_B)} = \frac{1}{(\lambda_A + \alpha\lambda_B)} + \frac{1}{\lambda_A}.
\tag{10.81}
$$

The expected absorption time (to any of the absorbing states) can be obtained by unconditioning the above results.

$$
E[T_a] = E[T_{0s|0s}] F_{0s}(\infty) + E[T_{0u|0u}] F_{0u}(\infty)
$$

$$
= \left(\frac{1}{\lambda_A + \alpha\lambda_B} + \frac{1}{\lambda_B}\right) \frac{\lambda_A}{\lambda_A + \alpha\lambda_B} + \left(\frac{1}{\lambda_A + \alpha\lambda_B} + \frac{1}{\lambda_A}\right) \frac{\alpha\lambda_B}{\lambda_A + \alpha\lambda_B}
$$

$$
= \frac{\lambda_A \lambda_B + \lambda_A^2 + \alpha\lambda_B^2}{\lambda_A \lambda_B(\lambda_A + \alpha\lambda_B)}.
\tag{10.82}
$$

The result in Eq. (10.82) is the same as that in Eq. (10.55), where the absorbing states were lumped into a single state.

Example 10.24 *Warm Standby with Repair: Safety Analysis*
Coming back to Example 10.23, we assume now that the system can recover from the *safe* failure condition and return to the fully operational State 11. We can model

the system operation as the CTMC of Figure 10.23, whose partitioned infinitesimal generator is:

$$Q = \left[\begin{array}{cccc|c} -(\lambda_A + \alpha\lambda_B) & \lambda_A & \alpha\lambda_B & 0 & 0 \\ 0 & -\lambda_B & 0 & \lambda_B & 0 \\ 0 & 0 & -\lambda_A & 0 & \lambda_A \\ \mu & 0 & 0 & -\mu & 0 \\ \hline 0 & 0 & 0 & 0 & 0 \end{array}\right]. \qquad (10.83)$$

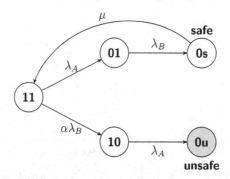

Figure 10.23 Safety model for a warm standby system with repair.

The CTMC of Figure 10.23 has a single absorbing state, and the expected time to the first unsafe failure can be computed by means of the linear system in (10.49):

$$\begin{cases} -(\lambda_A + \alpha\lambda_B)\tau_{11} + \mu\tau_{0s} & = & 1, \\ \lambda_A\tau_{11} - \lambda_B\tau_{01} & = & 0, \\ \alpha\lambda_B\tau_{11} - \lambda_A\tau_{10} & = & 0, \\ \lambda_B\tau_{01} - \mu\tau_{0s} & = & 0; \end{cases}$$

$$E[T_a] = \tau_{11} + \tau_{01} + \tau_{10} + \tau_{0s}.$$

Example 10.25 *Software Fault Tolerance Recovery Block (from [16])*
The two well-documented techniques for tolerating software design faults are *recovery blocks* (RBs) [17] and *N-version programming* [18]. Figure 10.24 shows the CTMC model for an RB architecture implemented on a dual processor system (in hot standby) for tolerating a single fault [16]. The architecture is obtained by the duplication of two variants and an acceptance test on two hardware components; accordingly, independent faults in two variants are tolerated, related faults between the variants are detected, while related faults between each variant and the acceptance test cannot be tolerated nor detected, and lead to an unsafe state. In Figure 10.24, state 2 is the up state with both hardware components and the RB working properly. State 1 is a state in which a hardware failure has been tolerated (covered) and RB is operational. State 0d is a safe state in which a failure is detected, while state 0u is an unsafe state in which a failure is undetected. The detailed meanings of the transitions and the transition rate values are

Table 10.2 Parameters for recovery block example.

Transition rate	Value	Meaning
λ_{21}	$2c\lambda_{\mathrm{h}}$	Covered hardware server failure
λ_{2d}	$2(1-c)\lambda_{\mathrm{h}}+\lambda_{\mathrm{sd}}$	Not covered hardware server failure or detected RB failure
λ_{2u}	λ_{su}	Undetected RB failure
λ_{1d}	$c\lambda_{\mathrm{h}}+\lambda_{\mathrm{sd}}$	Detected RB failure or covered hardware server failure
λ_{1u}	$(1-c)\lambda_{\mathrm{h}}+\lambda_{\mathrm{su}}$	Not covered hardware server failure or undetected RB failure

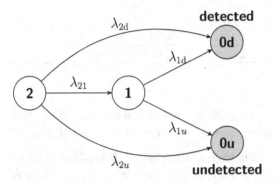

Figure 10.24 Recovery block for a hot standby system.

given in Table 10.2. λ_{h} denotes the hardware component failure rate and c the hardware coverage factor; λ_{sd} and λ_{su} denote respectively the detected and undetected failure rates of the recovery block software.

The CTMC of Figure 10.24 has two absorbing states, and its infinitesimal generator is:

$$
Q = \left[\begin{array}{c|c}
\boldsymbol{Q}_{\mathrm{u}} & \boldsymbol{a}^{\mathrm{T}} \quad \boldsymbol{b}^{\mathrm{T}} \\
\hline
\boldsymbol{0} & 0 \quad 0 \\
\boldsymbol{0} & 0 \quad 0
\end{array}\right] = \left[\begin{array}{cc|cc}
-(\sum \lambda_{2i}) & \lambda_{21} & \lambda_{2d} & \lambda_{2u} \\
0 & -(\lambda_{1d}+\lambda_{1u}) & \lambda_{1d} & \lambda_{1u} \\
\hline
0 & 0 & 0 & 0 \\
0 & 0 & 0 & 0
\end{array}\right]. \quad (10.84)
$$

To compute the absorption probabilities into states 0d and 0u we apply Eq. (10.72), assuming as initial probability values $\pi_2(0) = 1$, $\pi_k(0) = 0$ ($k = 1, 0d, 0u$). We first derive the inverse of the transient partition of the infinitesimal generator (10.84), to get:

$$
-\boldsymbol{Q}_{\mathrm{u}}^{-1} = \frac{1}{(\lambda_{21}+\lambda_{2d}+\lambda_{2u})(\lambda_{1d}+\lambda_{1u})} \left[\begin{array}{cc}
\lambda_{1d}+\lambda_{1u} & \lambda_{21} \\
0 & \lambda_{21}+\lambda_{2d}+\lambda_{2u}
\end{array}\right].
$$

Finally, writing $\lambda_s = \lambda_{sd} + \lambda_{su}$, we obtain the eventual absorption probabilities:

$$F_{0d}(\infty) = \pi_{0d}(\infty) = \begin{bmatrix} 1 & 0 \end{bmatrix} \begin{bmatrix} -Q_u^{-1} \end{bmatrix} \begin{bmatrix} \lambda_{2d} \\ \lambda_{1d} \end{bmatrix}$$

$$= \frac{(\lambda_{1d} + \lambda_{1u})\lambda_{2d} + \lambda_{21}\lambda_{1d}}{(\lambda_{21} + \lambda_{2d} + \lambda_{2u})(\lambda_{1d} + \lambda_{1u})}$$

$$= \frac{2(1-c)\lambda_h + \lambda_{sd}}{2\lambda_h + \lambda_s} + \frac{2c\lambda_h(c\lambda_h + \lambda_{sd})}{(2\lambda_h + \lambda_s)(\lambda_h + \lambda_s)},$$

$$F_{0u}(\infty) = \pi_{0u}(\infty) = \begin{bmatrix} 1 & 0 \end{bmatrix} \begin{bmatrix} -Q_u^{-1} \end{bmatrix} \begin{bmatrix} \lambda_{2u} \\ \lambda_{1u} \end{bmatrix}$$

$$= \frac{(\lambda_{1d} + \lambda_{1u})\lambda_{2u} + \lambda_{21}\lambda_{1u}}{(\lambda_{21} + \lambda_{2d} + \lambda_{2u})(\lambda_{1d} + \lambda_{1u})}$$

$$= \frac{\lambda_{su}}{2\lambda_h + \lambda_s} + \frac{2c\lambda_h((1-c)\lambda_h + \lambda_{su})}{(2\lambda_h + \lambda_s)(\lambda_h + \lambda_s)}.$$

Problems

10.20 Using the Laplace transform method, compute the reliability $R(t) = \pi_2(t) + \pi_1(t)$ of the software fault tolerance recovery block of Figure 10.24.

10.21 A hearing aid system has one active battery. When leaving for a four-week trip, the owner takes an identical spare battery. The lifetime of the battery is exponentially distributed with MTTF of 1 week. The spare battery can be considered in warm standby mode with dormancy factor α. Switching from the active to the spare battery has a coverage factor of c. Compute:

(a) The hearing aid reliability over the four-week trip.
(b) The hearing aid MTTF.

10.22 Continuing with Problem 10.21, now assume that the batteries are from two different manufacturers A and B with different failure rates, λ_A and λ_B. Assume that $\lambda_A < \lambda_B$. Determine the order in which the batteries should be used (main battery and standby battery) to obtain the greater MTTF in the following cases:

(a) with dormancy factor $\alpha = 0$ and perfect coverage $c = 1$
(b) with dormancy factor α and perfect coverage $c = 1$
(c) with dormancy factor $\alpha = 0$ and coverage c
(d) with dormancy factor α and coverage c.

10.23 Consider a duplex processor in a critical application, modeled by the CTMC in Figure 10.25.

Upon the failure of one component while in state 2, the system transits to the operational state 1 with coverage probability c_2, and with the complementary probability the failure leads to an unsafe system state UF. From state 1, a failure can lead to the safe shutdown state SF with probability c_1, or lead to the unsafe state UF with probability $(1 - c_1)$. Find the absorption probabilities in states SF and UF, and find

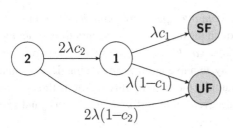

Figure 10.25   Duplex system with safe and unsafe failure states.

the ratio of c_1 to c_2 for which the redundant system has a catastrophic failure probability (state UF) greater than a single component with the same failure rate. The CTMC for a single-component system is obtained by removing state 2 and the transition out of that state.

10.24   Go back to the two-component non-repairable system with CCF of Example 10.13. We want now to distinguish whether a system failure is due to hardware component failure or to CCF. Describe the CTMC and its infinitesimal generator, then derive the Cdf of the time to ultimate system failure due to component failure or to CCF.

10.2.5   Expected First-Passage Time

In a dependable system, the state space can be partitioned into two exhaustive and mutually exclusive subsets Ω_u and Ω_d (Section 9.3). Evaluating the distribution and the moments of the first time that a CTMC jumps from Ω_u to Ω_d can be considered as a special case of a CTMC with absorbing states. This problem is usually referred to as the *first-passage time problem* [19]. The system MTTF can be computed by making all the states in subset Ω_d absorbing, and using Eq. (10.50). Conversely, to compute the system MTTR we start with the assumption that the system is in a down state in Ω_d with probability 1. We make the states in Ω_u absorbing and we apply Eq. (10.50). In a more general setting the first-passage time can be computed from any subset of initial states to any subsets of final states. Example 10.26 considers a survivability problem, where the metric to study is the first-passage time problem from a down state to the up state, while Example 10.27 considers a first-passage time problem from a set of up states to a down state for a fire protection system in warm standby.

Example 10.26   *A Detailed Survivability Study of a Power Smart Grid [20–22]*
Survivability refers to the time-varying system behavior after a failure occurs, and survivability quantification will be extensively studied in Section 16.8. In this example, we consider the survivability assessment of a smart grid distribution network.

The optimization of distribution automation power grids requires the development of scalable performability models that are able to capture the complexity of both the smart-grid power distribution network and the cyber-physical infrastructure that supports it. Traditionally, the reliability of power systems has been quantified using average metrics, such as the *system average interruption duration index* (SAIDI) [23]. Therefore, there is a need to extend the metrics for the accurate computation of customer interruption indices, based on *survivability*, where the survivability of a mission-critical application is the ability of the system to continue functioning during and after a failure or disturbance [24]. In this example, the survivability metric is the *average energy not supplied* (AENS) after a failure event until full system recovery, and is computed by means of an MRM.

The CTMC in Figure 10.26 depicts the stages that the system goes through after a failure. The initial state consists of a failure state. Then, based on manual and automated interventions, the system goes through different steps until reaching full recovery. With each state we associate its corresponding reward rate, defined as the energy not supplied at that state per time unit. Solving the MRM model, survivability-related metrics such as AENS in kW h can be computed [22].

The meanings of the states in Figure 10.26 are shown in Table 10.3.

As a failure occurs in section S of the grid (entrance in state F of Figure 10.26), power might become unavailable in a set of sections indicated by S^+. The time for section S to be isolated is indicated by ε and is negligible compared to the other time parameters of the model. In states 1 and 2 there is enough active power to supply the upstream sections, whereas in state 3 the backup active power does not suffice. In addition, in state 1 the reactive power also suffices to supply the upstream sections. Therefore, if the system transitions to state 1, it is amenable to automatic recovery, which occurs with rate $\alpha = 30\,\mathrm{hr}^{-1}$. In that case, the system transitions to state 6. Otherwise (states 2 and 3), the demand response (DR) and distributed generation (DG) programs are activated, and it takes on average $1/\beta$ units of time for them to start up. Such programs are effective with probability d_R and d_A at states 2 and 3, respectively. In the case that they are effective, a switch is closed and the system transitions to state 4, which is amenable to automatic recovery. In all other states the system is amenable to manual repair, which takes mean time $1/\gamma$ and leads the system to the full recovery state with no failures (NF).

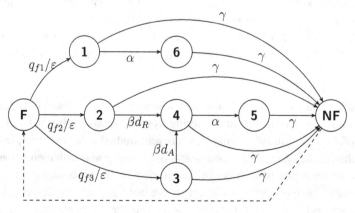

Figure 10.26 Survivability model.

Table 10.3 The physical meanings of the states, parameters and rates of Figure 10.26.

F	Failure at section S
1	Active and reactive power for S^+ OK
2	Active power for S^+ OK, reactive power for S^+ not OK
3	Active power for S^+ not OK
4	Active and reactive power for S^+ OK due to DR and DG
5	S^+ fixed due to DR and DG
6	S^+ fixed
NF	No failure: recovery completed
q_{1f}	Probability enough active and reactive power for S^+
q_{2f}	Probability enough active (but not reactive) power for S^+
q_{3f}	Probability not enough active backup power for S^+
d_R	Probability that demand response program for reactive power is effective
d_A	Probability that demand response program for active power is effective
α	Automatic repair rate
β	Demand response rate
γ	Manual repair rate
ε	Circuit isolation time ($\varepsilon \ll 1$)

Since ε is very short with respect to the other timed parameters, we can assume that the system enters at time $t = 0$ states 1, 2 or 3 with initial probabilities $\pi_1(0) = q_{f1}$, $\pi_2(0) = q_{f2}$, and $\pi_3(0) = q_{f3}$, respectively.

The infinitesimal generator, where the rows are ordered from 1 to 6, the last row being the absorbing state NF, is:

$$
\boldsymbol{Q} = \begin{bmatrix}
-(\alpha+\gamma) & 0 & 0 & 0 & 0 & \alpha & \gamma \\
0 & -(\beta d_R+\gamma) & 0 & \beta d_R & 0 & 0 & \gamma \\
0 & 0 & -(\beta d_A+\gamma) & \beta d_A & 0 & 0 & \gamma \\
0 & 0 & 0 & -(\alpha+\gamma) & \alpha & 0 & \gamma \\
0 & 0 & 0 & 0 & -\gamma & 0 & \gamma \\
0 & 0 & 0 & 0 & 0 & -\gamma & \gamma \\
0 & 0 & 0 & 0 & 0 & 0 & 0
\end{bmatrix}.
$$

The transient equation can be solved in closed form [22] to give:

$$\pi_1(t) = e^{-(\alpha+\gamma)t}\pi_1(0),$$

$$\pi_2(t) = e^{-(\beta d_R+\gamma)t}\pi_2(0),$$

$$\pi_3(t) = e^{-(\beta d_A+\gamma)t}\pi_3(0),$$

$$\pi_4(t) = \frac{\beta d_R}{\alpha - \beta d_R}(e^{-\beta d_R t} - e^{-\alpha t})e^{-\gamma t}\pi_2(0) + \frac{\beta d_A}{\alpha - \beta d_A}(e^{-\beta d_A t} - e^{-\alpha t})e^{-\gamma t}\pi_3(0),$$

$$\pi_5(t) = \frac{\alpha(1 - e^{-\beta d_R t}) - \beta d_R(1 - e^{-\alpha t})}{\alpha - \beta d_R}e^{-\gamma t}\pi_2(0)$$

$$+ \frac{\alpha(1 - e^{-\beta d_A t}) - \beta d_A(1 - e^{-\alpha t})}{\alpha - \beta d_A}e^{-\gamma t}\pi_3(0),$$

Table 10.4 The reward rates for active and reactive power for the states of Figure 10.26.

State	1–3	4	5	6	NF
Active (ENS hr^{-1})	593.58	568.98	112.45	112.45	0
Reactive (ENS hr^{-1})	361.77	308.93	26.76	26.76	0

$$\pi_6(t) = (1 - e^{-\alpha t}) e^{-\gamma t} \pi_1(0),$$

$$\pi_{NF}(t) = 1 - e^{-\gamma t}.$$

Two reward rates are assigned in Table 10.4 to measure both the active and reactive energy not supplied per hour from the occurrence of the failure (state F) to complete recovery (state NF). The reward rates may be either estimated from statistical data or computed using power flow algorithms. The reward values in Table 10.4 are taken from [22].

The expected accumulated reward up to time t is

$$E[Y(t)] = \sum_i r_i \int_0^t \pi_i(u) \, du. \tag{10.85}$$

Example 10.27　*Safety Analysis of a Fire-Fighting Pumping Station [25]*
Safety analysis and experience reveal fire to be one the most serious causes of accidents in industrial plants that store or handle a large amount of flammable material. A fire protection system requires a high level of performance coupled with high reliability. Safety analysis of a pumping station is required to demonstrate that the system provides a sufficient amount of flow (either water or foam) for a sufficient time upon demand. Furthermore, during emergency operations no repair action can take place.

A large petrochemical plant is divided into zones separated by empty areas such that the fire extinguishing system can confine the accident in the zone where it starts [26]. Each zone may request a different water flow rate and time to extinguish the fire. The following assumptions are made about the system operation:

- To apply design diversity concepts, the pumping system is formed by n_e pumps with electrical motors (EP) and n_d pumps with a diesel motor (DP) (we neglect here a common cause failure due to electrical shortage, which can be very likely and very risky in case of fire).
- The pumps are dormant in a cold standby configuration until a demand arrives prescribing a given flow rate.
- The minimum number of pumps is started upon demand to satisfy the requested flow rate, the other pumps being in (cold) standby.
- Pumps are put into operation sequentially according to a prescribed order, first the electrical pumps and then the diesel pumps.
- All the EPs have the same failure rate λ_e and capacity r_e, and all the DPs have the same failure rate λ_d and capacity $r_d \geq r_e$.
- The availability on demand is c_e for the EPs and c_d for the DPs (Example 10.11).

The operation of the pumping station is considered successful if it provides the required flow for a sufficiently long time when a fire accident is detected and the protection system is requested to operate. We assume that upon demand the requested flow rate from the plant is q_{req}. Given that the failure rates are constant, the system operation can be modeled as an MRM. In each state i the reward rate r_i is defined as the capacity delivered by the system in that state. Given that $\pi_i(t)$ is the transient state probability of finding the system in state i at time t, two measures can be defined to characterize the system operation:

- The success probability $S(t)$ of correct operation in the interval $(0, t]$ conditioned on the arrival of a demand at time $t = 0$, computed by summing the probabilities in the states in which the flow rate delivered by the system exceeds or is equal to q_{req}:

$$S(t) = \sum_{i: r_i \geq q_{req}} \pi_i(t).$$

- The total expected capacity (total accumulated reward) $Y(t)$ delivered in the interval $(0, t]$ conditioned on the arrival of a demand at time $t = 0$:

$$Y(t) = \int_0^t \sum_i^n r_i \pi_i(u) \, du.$$

According to a previous study [26], an optimal configuration, with respect to installation costs and expected losses, for the plant under consideration is to configure the system with $n_e = 2$ EPs and $n_d = 1$ DP. The initial system configuration depends on the flow demand q_{req}. We study three possible scenarios:

Case 1: $q_{req} \leq r_e$

Case 2: $r_e \leq q_{req} \leq 2r_e$

Case 3: $2r_e \leq q_{req} \leq 2r_e + r_d$

Case 1

The instantaneous requested flow rate is equal to the flow rate of a single pump $q_{req1} = r_e$. The MRM representing the system operation is given in Figure 10.27 and comprises four states:

State 1: The first EP is in operation; the other pumps are dormant ($r_1 = r_e$).

State 2: The first EP has failed; the second EP is in operation; the DP is dormant ($r_2 = r_e$).

State 3: Both EPs have failed; the DP is in operation ($r_3 = r_d$).

State 4: Both the EPs and the DP have failed ($r_4 = 0$).

Upon demand, the system starts with the following initial probability vector:

$$\pi_1(0) = c_e,$$
$$\pi_2(0) = c_e(1 - c_e),$$
$$\pi_3(0) = c_d(1 - c_e)^2,$$
$$\pi_4(0) = (1 - c_e)^2(1 - c_d).$$

The value $\pi_4(0) \neq 0$ indicates that there is a non-zero probability that the system does not start upon demand. The infinitesimal generator for the CTMC of Figure 10.27 is:

$$Q = \begin{bmatrix} -\lambda_e & c_e\lambda_e & (1-c_e)c_d\lambda_e & (1-c_e)(1-c_d)\lambda_e \\ 0 & -\lambda_e & c_d\lambda_e & (1-c_d)\lambda_e \\ 0 & 0 & -\lambda_d & \lambda_d \\ 0 & 0 & 0 & 0 \end{bmatrix}.$$

The operational states in Figure 10.27 in which the provided flow rate satisfies the request are states 1, 2, and 3, so that the probability of safe operation and the total expected capacity are:

$$S_{\text{Case 1}}(t) = \pi_1(t) + \pi_2(t) + \pi_3(t),$$

$$Y_{\text{Case 1}}(t) = \int_0^t \sum_{i=1}^3 r_i \pi_i(u)\,du.$$

Case 2
The requested instantaneous flow rate requires both EPs to be put into operation upon demand: $q_{\text{req2}} = 2 \times r_e$. The MRM representing the system operation is shown in Figure 10.28 and comprises five states.

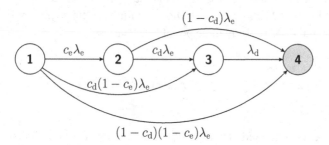

Figure 10.27 Pumping station: Case 1.

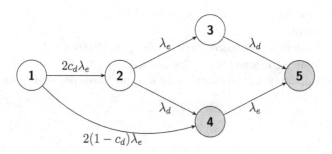

Figure 10.28 Pumping station: Case 2.

State 1:   Both EPs are in operation; the DP is dormant ($r_1 = 2r_e$).
State 2:   One EP has failed; the other EP and the DP are in operation ($r_2 = r_e + r_d$).
State 3:   Both EPs have failed; the DP is in operation ($r_3 = r_d$).
State 4:   One EP and the DP have failed; the other EP is in operation ($r_4 = r_e$).
State 5:   Both the EPs and the DP have failed ($r_5 = 0$).

The initial probability vector has the following expression:

$$\pi_1(0) = c_e^2,$$
$$\pi_2(0) = 2(1 - c_e) c_e c_d,$$
$$\pi_3(0) = c_d (1 - c_e)^2,$$
$$\pi_4(0) = 2 c_e (1 - c_e)(1 - c_d),$$
$$\pi_5(0) = (1 - c_e)^2 (1 - c_d).$$

The operational states in Figure 10.28 in which the provided flow rate satisfies the request are states 1, 2, and possibly 3 if $r_d \geq q_{req}$. Hence, the probability of safe operation and the total expected capacity are:

$$S_{\text{Case2}}(t) = \pi_1(t) + \pi_2(t) + \pi_3(t),$$

$$Y_{\text{Case2}}(t) = \int_0^t \sum_{i=1}^4 r_i \pi_i(u) \, du.$$

Case 3
The requested instantaneous flow rate requires that all three pumps are put in operation upon demand: $q_{req3} = 2 \times r_e + r_d$. The MRM representing the system operation is shown in Figure 10.29, and comprises six states.

State 1:   Both the EPs and the DP are in operation ($r_1 = 2r_e + r_d$).
State 2:   One EP has failed; the other EP and the DP are in operation ($r_2 = r_e + r_d$).
State 3:   The DP has failed; both the EPs are in operation ($r_3 = 2r_e$).
State 4:   Both the EPs have failed; the DP is in operation ($r_4 = r_d$).
State 5:   One EP and the DP have failed; the other EP is in operation ($r_5 = r_e$).
State 6:   Both the EPs and the DP have failed ($r_6 = 0$).

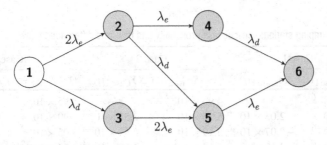

Figure 10.29   Pumping station: Case 3.

The initial probability vector has the following expression:

$$\pi_1(0) = c_e^2 c_d,$$
$$\pi_2(0) = (1 - c_e) c_e c_d,$$
$$\pi_3(0) = c_e^2 (1 - c_d),$$
$$\pi_4(0) = c_d (1 - c_e)^2,$$
$$\pi_5(0) = c_e (1 - c_e) (1 - c_d),$$
$$\pi_6(0) = (1 - c_e)^2 (1 - c_d).$$

The probability of safe operation and the total expected capacity are:

$$S_{Case\,3}(t) = \pi_1(t),$$

$$Y_{Case\,3}(t) = \int_0^t \sum_{i=1}^5 r_i \pi_i(u) \, du.$$

Calculations

For the three cases ($i = 1, 2, 3$) and as functions of time we have calculated the non-success probability ($1 - S_{Case\,i}(t)$) and the relative flow rate difference between the total accumulated flow rate provided by the system and the total requested flow rate $Y_{req\,i}(t)$:

$$\Delta Y_{Case\,i}(t) = \frac{Y_{Case\,i}(t) - Y_{req\,i}(t)}{Y_{req\,i}(t)} \quad \text{with} \quad Y_{req\,i}(t) = q_{req\,i} \times t.$$

In the computation we have assumed the following numerical values:

$$c_e = 0.99, \quad \lambda_e = 0.6 \times 10^{-5}\,\text{fhr}^{-1}, \quad c_d = 0.98, \quad \lambda_d = 0.3 \times 10^{-4}\,\text{fhr}^{-1},$$

with two possible levels of the pump flow rates [27]:

Level 1: $r_e = 500\,\text{m}^3\,\text{hr}^{-1}$, $r_d = 500\,\text{m}^3\,\text{hr}^{-1}$,
Level 2: $r_e = 500\,\text{m}^3\,\text{hr}^{-1}$, $r_d = 1000\,\text{m}^3\,\text{hr}^{-1}$.

The results have been obtained using the SHARPE tool and are shown in Table 10.5 for level 1 and Table 10.6 for level 2. In both tables the data have been computed at $t = 0$ (showing the initial unavailability on demand) and after 1/2, 1, and 3 d of continuous operation. The unavailability weakly depends on the chosen timescale and the major

Table 10.5 Pumping station: non-success probability and flow rate difference for level 1.

t	Case 1		Case 2		Case 3	
(hr)	$(1 - S_{Case\,1}(t))$	$\Delta Y_{Case\,1}(t)$	$(1 - S_{Case\,2}(t))$	$\Delta Y_{Case\,2}(t)$	$(1 - S_{Case\,3}(t))$	$\Delta Y_{Case\,3}(t)$
0	2.00×10^{-6}	—	4.96×10^{-4}	—	3.95×10^{-2}	—
12	2.05×10^{-6}	-2.03×10^{-6}	5.05×10^{-4}	-2.51×10^{-4}	3.99×10^{-2}	-1.34×10^{-2}
24	2.13×10^{-6}	-2.07×10^{-6}	5.19×10^{-4}	-2.55×10^{-4}	4.05×10^{-2}	-1.35×10^{-2}
72	2.39×10^{-6}	-2.19×10^{-6}	5.64×10^{-4}	-2.66×10^{-4}	4.24×10^{-2}	-1.38×10^{-2}

Table 10.6 Pumping station: non-success probability and flow rate difference for level 2.

t	Case 1		Case 2		Case 3	
(hr)	$(1 - S_{Case\,1}(t))$	$\Delta Y_{Case\,1}(t)$	$(1 - S_{Case\,2}(t))$	$\Delta Y_{Case\,2}(t)$	$(1 - S_{Case\,3}(t))$	$\Delta Y_{Case\,3}(t)$
0	2.00×10^{-6}	—	4.96×10^{-4}	—	3.95×10^{-2}	—
12	2.05×10^{-6}	9.65×10^{-5}	5.05×10^{-4}	9.53×10^{-3}	3.99×10^{-2}	-1.51×10^{-2}
24	2.13×10^{-6}	9.73×10^{-5}	5.19×10^{-4}	9.56×10^{-3}	4.05×10^{-2}	-1.52×10^{-2}
72	2.39×10^{-6}	9.99×10^{-5}	5.64×10^{-4}	9.68×10^{-3}	4.24×10^{-2}	-1.56×10^{-2}

effect is the unavailability on demand. The relative reduction of the delivered flow with respect to the requested flow is inside reasonable limits.

Problems

10.25 With reference to Example 10.26:

(a) Derive the closed-form expression for the expected accumulated reward up to time t using Eq. (10.85).
(b) Compute the expected accumulated reward up to absorption.
(c) Using the Beaudry technique of Example 10.21, derive the state space transition graph to evaluate the distribution of expected accumulated reward up to absorption.

10.26 With reference to the renal disease model of Figure 10.17, compute the first-passage time to state T starting from state H with probability 1.

10.27 Evaluate the MTTF for the pumping station of Figures 10.27, 10.28, and 10.29.

10.28 Assume that in case 1 of Figure 10.27 the first electric pump that fails can be repaired with constant repair rate μ_e. Derive the corresponding CTMC and its infinitesimal generator, and compute the MTTF.

10.29 Throughout Example 10.27 we have explicitly neglected a possible common cause failure due to electrical shortage. With reference to case 1 of Figure 10.27, we want to include the CCF. Electrical shortage has two possible manifestations: either the power is already down with probability d when the demand for the pumping station arrives, or the shortage occurs during the operation with a constant failure rate γ. Derive the CTMC in this case, the infinitesimal generator and the MTTF.

10.3 Continuous-Time Markov Chains with Self-Loops

Self-loops in a CTMC are transitions that exit from a state and return to the same state. Self-loops are possible and in some cases are very consistent with the nature of the system to be modeled. However, due to the structure of the infinitesimal generator, self-loops are not visible in the infinitesimal generator of a CTMC even if they are present. For this reason, they can be neglected without affecting the solution equations. To make them visible we need to distinguish between the transition rate matrix and the infinitesimal generator. The transition rate matrix groups all the transition rates including the self-loops on the diagonal; in the infinitesimal generator the diagonal is the opposite of the sum of all the rates (including the self-loop) outgoing from the state. Figure 10.30 reproduces Figure 9.4 with the addition of a self-loop in state i of rate γ_i. The transition rate matrix contains the value γ_i in the diagonal entry of row i. The ith diagonal entry q_i of the infinitesimal generator is minus the sum of all the rates outgoing from state i plus γ_i, (see Figure 10.30):

$$q_i = \gamma_i - \sum_j q_{ij} = \gamma_i - (\gamma_i + \sum_{j:j\neq i} q_{ij}) = \sum_{j:j\neq i} q_{ij}.$$

The net result is that the diagonal entry of the infinitesimal generator is minus the sum of the off-diagonal entries. It turns out that the sojourn time in state i of Figure 10.30 with the self-loop is the same as the sojourn time in state i of Figure 9.4 without the self-loop.

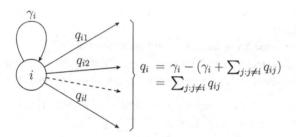

Figure 10.30   Continuous-time Markov chain with a self-loop on state i.

Example 10.28   *A Fault/Error-Handling Model for a Fault-Tolerant System [28]*
Highly reliable systems that use extensive redundancy are highly reconfigurable and have complex recovery management techniques that are handled by a *fault error handling model*. A schematic functional representation of an FEHM is given in Figure 10.31; it allows for the modeling of permanent, intermittent and transient faults,

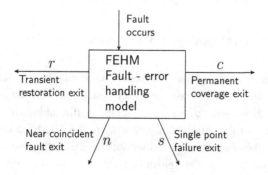

Figure 10.31   Functional representation of an FEHM illustrating the four possible exits.

and models the online recovery procedure necessary for each type. We show that the handling model naturally generates a self-loop on the state to which it is applied. The FEHM has four possible exits for which a corresponding exit probability must be evaluated.

Transient restoration occurs with probability r and represents correct recognition and recovery from a transient fault/error. The permanent coverage exit (with probability c) denotes determination of the permanent nature of the fault, and successful isolation and removal of the faulty component. The single-point failure exit (with probability s) is reached when a single fault (or a common mode failure) causes the system to crash and leads the system to a permanent faulty state denoted FSFP (failure single point failure). In highly reliable systems, such as those used for flight control, the probability of a second fault occurring while attempting recovery from a given fault cannot be ignored since it may cause immediate system failure (which occurs with probability n) and leads the system to the absorbing state denoted FNCF (failure near coincident failure). For each FEHM, it should be noted that

$$r + c + s + n = 1.$$

To evaluate the exit probabilities the internal structure of the FEHM must be made explicit [29]. A very simple CTMC model for an FEHM is given in Figure 10.32.

State A is the entry state as a fault occurs. Then a detection and isolation action starts that is assumed to take an exponentially distributed time with rate δ. The detection has two exits: with probability v the failure is correctly isolated, and the FEHM enters state C, where the system can correctly restart operation with one component fewer. With probability $(1 - v)$ the fault is transient, the fault is recovered and the FEHM enters state R from which the faulty component is resumed to operation. However, if the detection action fails, the fault propagates to the error state E with an exponentially distributed delay of rate ρ [30]. The error handling in state E takes an exponentially

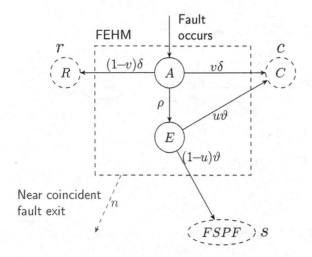

Figure 10.32 A simple CTMC model for an FEHM.

distributed time with rate ϑ and has two possible exits. With probability u the error is correctly isolated and the FEHM enters state C, and with probability $(1 - u)$ the error is not recovered and the system enters the permanent error state FSPF. The CTMC of Figure 10.32 has two transient states, A and E, and three absorbing states, R, C, and FSPF. The probability of the FEHM reaching each one of the absorbing states given that the CTMC starts with probability 1 in state A can be computed as illustrated in Section 10.2.4. The following results are obtained (with $c + r + s = 1$):

$$c = \frac{v\delta + u\rho}{\delta + \rho}, \quad r = \frac{(1 - v)\delta}{\delta + \rho}, \quad s = \frac{(1 - u)\rho}{\delta + \rho}.$$

Up to now, the FEHM was considered in isolation. If we now consider the FEHM integrated into a real system, a further effect has to be considered due to the possible manifestation of a second fault during the handling of the first one, which can cause an immediate permanent failure. To be more concrete, suppose that the system is in a state with m redundant components in operation. As a failure occurs with rate $m\lambda$, state A of the FEHM is entered as shown in Figure 10.33.

During the handling of the fault in states A and E, $(m - 1)$ redundant components are still working and they may fail at rate $(m - 1)\lambda$, leading to the permanent failure state denoted by FNCF. The CTMC of Figure 10.33 has two transient states (A and E) and four absorbing states (C, R, FSPF, FNCF). The probability of ultimate absorption in one of the four absorbing states starting from state A with probability 1 can be computed as illustrated in Section 10.2.4. Computation provides the following results (with $c + r + s + n = 1$):

$$c = \frac{((m - 1)\lambda + \vartheta)v\delta + u\rho\vartheta}{((m - 1)\lambda + \delta + \rho)((m - 1)\lambda + \vartheta)},$$

$$r = \frac{(1 - v)\delta}{(m - 1)\lambda + \delta + \rho}, \tag{10.86}$$

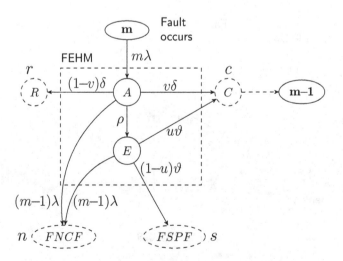

Figure 10.33   Model for a second fault during the action of an FEHM.

$$s = \frac{(1-u)\rho\vartheta}{((m-1)\lambda + \delta + \rho)((m-1)\lambda + \vartheta)},$$

$$n = \frac{(m-1)\lambda((m-1)\lambda + \vartheta) + (m-1)\lambda\rho}{((m-1)\lambda + \delta + \rho)((m-1)\lambda + \vartheta)}.$$

Since the FEHM is composed of events that occur in rapid succession, once a fault has occurred each FEHM can be replaced by an instantaneous branch point [31] obtaining the *imperfect coverage* model of Figure 10.34. The instantaneous imperfect coverage model is known to be conservative [32]. The transient restoration probability r gives rise to a self-loop on state m.

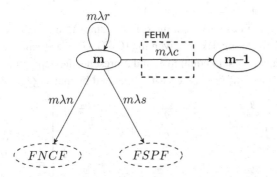

Figure 10.34   Branch point replacing an FEHM.

10.3.1   Uniformization of a CTMC

Uniformization provides the most efficient numerical technique for solving a CTMC, as will be illustrated in Section 10.4. A simple argument for explaining how the uniformization works can be introduced by resorting to the use of self-loops [33]. Consider a CTMC with infinitesimal generator Q given by:

$$Q = \begin{bmatrix} q_{11} & q_{12} & \cdots & q_{1i} & \cdots & q_{1n} \\ q_{21} & q_{22} & \cdots & q_{2i} & \cdots & q_{2n} \\ \vdots & \vdots & \ddots & \vdots & \vdots & \vdots \\ q_{i1} & q_{i2} & \cdots & q_{ii} & \cdots & q_{in} \\ \vdots & \vdots & \vdots & \vdots & \ddots & \vdots \\ q_{n1} & q_{n2} & \cdots & q_{ni} & \cdots & q_{nn} \end{bmatrix}.$$

We search for the maximum diagonal entry in absolute value and we assume that the maximum value is in row i,

$$q = \max_j |q_{jj}| = |q_{ii}|.$$

In each state of the CTMC the outgoing rate $|q_{jj}|$ is less than or equal to q. We add to each state j a self-loop with rate

$$\gamma_j = q - \sum_{k=1,n;k\neq j} q_{jk} = q - |q_{jj}| \geq 0,$$

as in Figure 10.35. The addition of the self-loops does not alter the state probability vector of the original CTMC, but with this addition the outgoing rates from each state are all identical and equal to q. The CTMC with self-loops is the *uniformized* CTMC. In the uniformized CTMC, all the departures out of any state occur with the same rate q, and hence the sequence of departures forms a Poisson process of rate q (Section 3.5.3). Given a departure from state j (at rate q), the CTMC jumps to the next state (say $k \neq j$) with probability $q_{jk}/q \geq 0$, and with probability $\gamma_j = (q - |q_{jj}|)/q = 1 - |q_{jj}|/q \geq 0$ the jump is toward the same state j – see Eq. (9.30).

Given that one transition has occurred, the CTMC jumps to the next state according to the conditional probability matrix $Q^\star$ given in (10.87). $Q^\star$ is a stochastic matrix with all entries $0 \leq q_{ij}^\star \leq 1$ and row sum equal to 1, and represents the generator matrix of the DTMC embedded into the uniformized CTMC generated by matrix Q.

$$Q^\star = \begin{bmatrix} 1 - \frac{|q_{11}|}{q} & \frac{q_{12}}{q} & \cdots & \frac{q_{1i}}{q} & \cdots & \frac{q_{1n}}{q} \\ \frac{q_{21}}{q} & 1 - \frac{|q_{22}|}{q} & \cdots & \frac{q_{2i}}{q} & \cdots & \frac{q_{2n}}{q} \\ \vdots & \vdots & \ddots & \vdots & \vdots & \vdots \\ \frac{q_{i1}}{q} & \frac{q_{i2}}{q} & \cdots & 1 - \frac{|q_{ii}|}{q} & \cdots & \frac{q_{in}}{q} \\ \vdots & \vdots & \vdots & \vdots & \ddots & \vdots \\ \frac{q_{n1}}{q} & \frac{q_{n2}}{q} & \cdots & \frac{q_{ni}}{q} & \cdots & 1 - \frac{|q_{nn}|}{q} \end{bmatrix} = I + \frac{Q}{q}. \quad (10.87)$$

The number $N(t)$ of transitions at time t in the uniformized CTMC is given by the number of events in a Poisson process of rate q, hence (3.93):

$$P\{N(t) = k\} = \frac{(qt)^k}{k!} e^{-qt}.$$

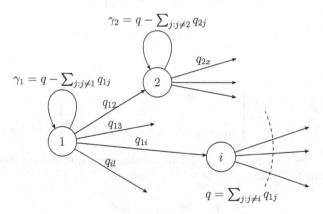

Figure 10.35   The principle of the uniformization technique.

The state probability vector after exactly $(j = 0, 1, 2, \ldots, k)$ jumps is then:

0 transitions $\qquad \pi\,(t|N(t) = 0) = \pi\,(0)\,P\{N(t) = 0\} = \pi\,(0)\,e^{-qt},$

1 transition $\qquad \pi\,(t|N(t) = 1) = \pi\,(0)\,Q^*\,P\{N(t) = 1\} = \pi\,(0)\dfrac{qt}{1!}e^{-qt}Q^*,$

2 transitions $\qquad \pi\,(t|N(t) = 2) = \pi\,(0)\,Q^{*2}P\{N(t) = 2\} = \pi\,(0)\dfrac{(qt)^2}{2!}e^{-qt}Q^{*2},$

$\qquad\qquad\quad \vdots \qquad\qquad\qquad\qquad \vdots$

k transitions $\qquad \pi\,(t|N(t) = k) = \pi\,(0)\,Q^{*k}\,P\{N(t) = k\} = \pi\,(0)\,Q^{*k}\dfrac{(qt)^k}{k!}e^{-qt}.$

Summing over k, the final equation for the computation of the state probability vector at time t is obtained:

$$\pi\,(t) = \pi\,(0)\sum_{k=0}^{\infty}Q^{*k}\frac{(qt)^k}{k!}e^{-qt}.$$

Details of the numerical method based on uniformization are given in Section 10.4.

Example 10.29 *A Redundant System with Spare [33]*
A system is composed of two active processors in parallel redundancy and one spare in cold standby. The system is part of a satellite control system for which remote recovery is possible, but not repair. The system is working when at least one processor is working: exhaustion of the components leads to a permanent failure.

Each state is labeled with indices (i,j) where i denotes the number of active processors and j the number of active spares. The failure rate of each component is λ and the recovery rate is μ. The CTMC representing the system operation is given in Figure 10.36(a), where the labels on the arcs are the rates of the transitions; the

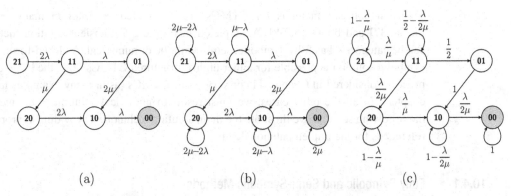

Figure 10.36 The uniformization procedure for a CTMC. (a) The original CTMC (labels are rates); (b) The uniformized CTMC (labels are rates); (c) The embedded DTMC (labels are probabilities).

infinitesimal generator is:

$$
Q = \begin{bmatrix}
-2\lambda & 2\lambda & 0 & 0 & 0 & 0 \\
0 & -(\lambda+\mu) & \lambda & \mu & 0 & 0 \\
0 & 0 & -2\mu & 2\mu & 0 & 0 \\
0 & 0 & 0 & -2\lambda & 2\lambda & 0 \\
0 & 0 & 0 & 0 & -\lambda & \lambda \\
0 & 0 & 0 & 0 & 0 & 0
\end{bmatrix}. \tag{10.88}
$$

Since the recovery rates are orders of magnitude larger than the failure rates, the largest diagonal entry in absolute value is $q = 2\mu$. The uniformized CTMC is shown in Figure 10.36(b), where each state has a self-loop added so that the total exit rate is equal to 2μ. The labels on the arcs in Figure 10.36(b) are the rates of the transitions. The CTMCs of Figures 10.36(a) and 10.36(b) are equivalent and have the same state probability vector.

Figure 10.36(c) shows, instead, the embedded DTMC from which the conditional next state transition probability matrix $Q^\star = I + \frac{Q}{q}$ can be derived. In Figure 10.36(c) the labels are the probabilities associated with the corresponding arcs.

Problems

10.30 The uniformization technique can also be applied to irreducible CTMCs. In fact, a solution method based on the embedded uniformized DTMC was proposed in Section 9.8.1. With reference to Problem 9.8, assume that the transition rate values for the CTMC of Figure 9.16 are: $a = 0.1\,\mathrm{s}^{-1}$, $b = 0.2\,\mathrm{s}^{-1}$, $c = 0.3\,\mathrm{s}^{-1}$, and $d = 1\,\mathrm{s}^{-1}$. Derive the uniformized CTMC and the embedded DTMC, as in Figure 10.36. Solve for the steady-state probability using the power method of Section 9.8.1.

10.4   Transient Solution Methods

Transient solution methods for CTMCs are considered in detail in many papers [4, 34–37] and books [5, 38]. As depicted in Figure 2.5, transient solution methods can be categorized as fully symbolic, semi-symbolic or numerical. A closed-form, fully symbolic solution is possible for either highly structured CTMCs (e.g., the birth–death process considered in Chapter 11) or very small CTMCs (as in many examples in this chapter). In all the other cases, we must resort to numerical solution techniques. In the following sections we illustrate the main solution techniques, providing appropriate references for the implementation details.

10.4.1   Fully Symbolic and Semi-Symbolic Methods

The convolution integral method in Section 10.1.1 and the Laplace transform method (10.25) offer the possibility of a fully symbolic solution that has been used in many examples in this chapter. Fully symbolic solutions are only applicable to CTMCs with a very small number of states.

The Laplace transform Eq. (10.25) is a set of algebraic equations whose solution provides the Laplace transform of each state probability $\pi_i^*(s)$ in the form

$$\pi_i^*(s) = \frac{N(s)}{D(s)}, \tag{10.89}$$

where the denominator $D(s)$ is a polynomial in s of order n and the numerator $N(s)$ is a polynomial in s of order less than n. The inverse Laplace transform of (10.89) can be obtained by a partial fraction decomposition. However, the partial fraction decomposition requires the determination of all the poles of the function (i.e., the zeros of the denominator $D(s)$), which correspond to the eigenvalues of the infinitesimal generator matrix Q of the CTMC. Numerical evaluation of the eigenvalues of a matrix requires, in general, complex algorithms [39] that tend to be unstable when two or more poles tend to be coincident.

In the particular case when a CTMC has an acyclic graph, the infinitesimal generator is an upper triangular matrix and the eigenvalues of the matrix coincide with the entries of the main diagonal. In this case, the Laplace transform method may be simpler to apply since the roots of the denominator $D(s)$ are known without computation. However, for an acyclic CTMC there also exist direct solution methods in the time domain [3, 4].

In a semi-symbolic (or semi-numerical) solution the entries in the Q matrix are numerical but the final solution $\pi(t)$ is a symbolic function of time. The semi-symbolic method is simpler than the fully symbolic method and has been implemented in the SHARPE software package [40].

10.4.2 Transient Solution via Series Expansion

The Taylor series expansion around $t = 0$ of the transition probability matrix $P(t)$ in (9.22) is:

$$P(t) = I + \frac{dP(t)}{dt}\bigg|_{t=0} t + \frac{1}{2!}\frac{d^2 P(t)}{dt^2}\bigg|_{t=0} t^2 + \cdots = \sum_{i=0}^{\infty} \frac{1}{i!} \frac{d^i P(t)}{dt^i}\bigg|_{t=0} t^i. \tag{10.90}$$

By definition:

$$\frac{dP(t)}{dt} = P(t)Q \qquad\qquad \implies \qquad \frac{dP(t)}{dt}\bigg|_{t=0} = Q,$$

$$\frac{d^2 P(t)}{dt^2} = \frac{d}{dt}(P(t)Q) = P(t)Q^2 \qquad \implies \qquad \frac{d^2 P(t)}{dt^2}\bigg|_{t=0} = Q^2,$$

$$\vdots \qquad\qquad\qquad \vdots \tag{10.91}$$

$$\frac{d^i P(t)}{dt^i} = \frac{d}{dt}(P(t)Q^{(i-1)}) = P(t)Q^i \qquad \implies \qquad \frac{d^i P(t)}{dt^i}\bigg|_{t=0} = Q^i,$$

$$\vdots \qquad\qquad\qquad \vdots$$

Using (10.91), the expansion (10.90) can finally be written as:

$$P(t) = I + Qt + \frac{1}{2}(Qt)^2 + \cdots = \sum_{i=0}^{\infty} \frac{1}{i!}(Qt)^i. \tag{10.92}$$

By analogy with scalar expansion, the series in (10.92) is written in compact form as a matrix exponential:

$$P(t) = e^{Qt}, \tag{10.93}$$

from which Eq. (9.21) is obtained. Equation (9.21) can be utilized for implementing the numerical solution of a CTMC. There are, however, practical problems in the implementation of this approach [41]:

- Q has both negative and positive entries and hence the computation has both additions and subtractions (with poor numerical behavior).
- Raising the matrix Q to its powers is both costly and fills in the zeros in the matrix (which, in general, is very large and very sparse).
- The infinite series needs to be truncated, and the approximation error is not known.

A practical implementation can, however, use the following pattern: Given that t_M is the final mission time of the transient analysis, we need to compute $\pi(t)$ in the interval $(0, t_M]$. We take a small time interval $h = t_M/n$ and we exploit the property

$$\pi(t+h) = \pi(t) \cdot P(h) = \pi(t) \cdot e^{Qh}.$$

Then, we adopt the following iterative procedure:

$$
\begin{aligned}
\pi(h) &= \pi(0) \cdot e^{Qh}, \\
\pi(2h) &= \pi(0) \cdot e^{Q2h} = \pi(h) \cdot e^{Qh}, \\
&\vdots \qquad \vdots \\
\pi(ih) &= \pi((i-1)h) \cdot e^{Qh}, \\
&\vdots \qquad \vdots
\end{aligned}
\tag{10.94}
$$

With this scheme we need to compute the matrix exponential via the series expansion in (10.92) only once at the time point h. Since h can be small, the expansion uses few terms.

To avoid matrix exponentiation in the implementation, the following iterative scheme can be used:

$$\frac{1}{i!}(Qt)^i = \frac{(Qt)}{i} \frac{1}{(i-1)!}(Qt)^{(i-1)}.$$

The series expansion can be truncated when the absolute difference between all the homologous entries in two successive terms in the series are less than a given ϵ.

Once e^{Qt} has been computed with the required accuracy, the successive time iterations in (10.94) require only matrix–vector multiplication.

In any case, we cannot avoid subtractions and the matrix e^{Qh} fills up.

10.4.3 Transient Solution via Uniformization (Jensen's method) [42]

The modeling intepretation of the uniformization of a CTMC was described in Section 10.3.1. Here we follow the analytical approach suggested in [34, 35]. We search for the maximum diagonal entry in absolute value of the infinitesimal generator,

$$q = \max_j |q_{jj}|,$$

and we multiply both sides of Eq. (9.21) to get

$$\boldsymbol{\pi}(t) e^{qt} = \boldsymbol{\pi}(0) \cdot e^{qt} e^{Qt}$$

$$= \boldsymbol{\pi}(0) \cdot e^{[Q/q+I]qt}$$

$$= \sum_{k=0}^{\infty} \boldsymbol{\pi}(0) \frac{(qt)^k}{k!} \cdot (Q^\star)^k.$$

Finally,

$$\boldsymbol{\pi}(t) = \boldsymbol{\pi}(0) e^{-qt} \sum_{k=0}^{\infty} \frac{(qt)^k}{k!} (Q^\star)^k. \tag{10.95}$$

Matrix $Q^\star$ is a DTMC matrix, and hence its entries are non-negative and ≤ 1. This avoids the numerical problems as no subtractions are involved.

To avoid the problem of raising matrix Q to its powers, we rewrite Eq. (10.95) as

$$\boldsymbol{\pi}(t) = \sum_{k=0}^{\infty} \boldsymbol{\theta}(k) e^{-qt} \frac{(qt)^k}{k!}, \tag{10.96}$$

where $\boldsymbol{\theta}(0) = \boldsymbol{\pi}(0)$ and

$$\boldsymbol{\theta}(k) = \boldsymbol{\theta}(k-1) Q^\star, \qquad k = 1, 2, \ldots \tag{10.97}$$

The term $\boldsymbol{\theta}(k)$ in (10.97) can be interpreted as the kth step state probability vector of a DTMC with transition probability matrix $Q^\star$, while the term $e^{-qt}(qt)^k/k!$ is the Poisson pmf with parameter qt. Thus, the uniformization method expresses the state probabilities of a CTMC in terms of the sum of the DTMC state probabilities of a series of steps weighted by a Poisson pmf. Observe that matrix $Q^\star$ is the same as the one used in the power method for computing the CTMC steady-state vector in Section 9.8.1.

In a numerical implementation, the infinite series must be left-/right-truncated between a minimal value k_{m} (that can be equal to 0) and a maximum value k_{M}. The uniformization method offers a natural way to control the approximation error implied by the truncation. Given a precision requirement (a truncation error tolerance ϵ), the series in (10.96) is limited to:

$$\boldsymbol{\pi}(t) \approx e^{-qt} \sum_{k=k_{\mathrm{m}}}^{k_{\mathrm{M}}} \boldsymbol{\theta}(k) \frac{(qt)^k}{k!}. \tag{10.98}$$

The values k_{m} and k_{M} can be determined from the specified truncation error tolerance ϵ by

$$\sum_{k=0}^{k_{\text{m}}-1} e^{-qt}\frac{(qt)^k}{k!} \le \frac{\epsilon}{2}, \quad 1-\sum_{k=0}^{k_{\text{M}}} e^{-qt}\frac{(qt)^k}{k!} \le \frac{\epsilon}{2}.$$

For stiff Markov chains, qt is typically very large and the term e^{-qt} almost always runs into *underflow problems*. To avoid underflow, the left and right truncation points k_{m} and k_{M} can be computed following the method proposed by Fox and Glynn [43]. The same method can also be used to compute the Poisson probabilities $e^{-qt}(qt)^k/k!$ for all $k = k_{\text{m}}, k_{\text{m}}+1,\ldots,k_{\text{M}}-1, k_{\text{M}}$.

A simpler way to estimate the truncation points k_{m} and k_{M} was suggested in [33, 35]. Let K be a discrete random variable distributed according to a Poisson distribution of parameter qt. For large qt (as is usually the case), as a consequence of the central limit theorem, K tends to be normally distributed with mean qt and standard deviation qt (Section 3.2.9). Hence, for large qt, $K - qt/\sqrt{qt}$ tends to the standardized normal random variable $Z \sim N(0,1)$. From the property of the standardized normal random variable Z, we can define k_{m} and k_{M} as:

$$k_{\text{m}} = qt - z_{\alpha/2}\sqrt{qt}, \qquad k_{\text{M}} = qt + z_{\alpha/2}\sqrt{qt}.$$

With this assignment, the area outside the interval $(k_{\text{m}}, k_{\text{M}}]$ is the truncation error tolerance ϵ of the randomization method and is given by

$$P\{Z \le -z_{\alpha/2}\} + P\{Z \ge z_{\alpha/2}\} = \alpha.$$

With $z_{\alpha/2} = 6$, the area outside the interval $(k_{\text{m}}, k_{\text{M}}]$ is of the order of 2×10^{-9}. Increasing $z_{\alpha/2}$ causes the error to decrease, and thus the randomization method provides a simple way to control the truncation error.

Note that the computational complexity of the uniformization method rises linearly with q and t. A large value of qt also implies a large number of matrix vector multiplications, which results in large roundoff errors. CTMCs with a large value of qt are hence said to be stiff.

To address the stiffness problem, variants and improvements of the uniformization methods have been proposed in the literature, such as the *uniformized power method* of Abdallah and Marie [44], the *adaptive uniformization* of van Moorsel and Sanders [45, 46] or the *regenerative randomization* of Carrasco [47].

10.4.4   ODE-Based Solution Methods

Equation (9.21) can be solved by resorting to standard techniques for ODE IVPs [48]. Different methods can be used for different kinds of problem. For example, stiff methods can be used for stiff systems (or stiff Markov chains). Methods also differ in the accuracy of the solution yielded and computational complexity.

ODE solution methods discretize the solution interval into a finite number of time intervals $\{t_1, t_2,\ldots,t_i,\ldots,t_n\}$. Given a solution at t_i, the solution at $t_i + h$ $(= t_{i+1})$ is computed. Advancement in time is made with step size h until the time at which the desired solution (the *mission time*) is reached. Commonly the step size is not constant, but may vary from step to step.

Let π_i be the approximation of the theoretical solution $\pi(t_i)$ at time t_i. In computing π_{i+1}, we may incorporate the values from previously computed approximations π_j for $j = 0, 1, \dots, i$, or even π_{i+1} itself. A method that only uses (t_i, π_i) to compute π_{i+1} is said to be an *explicit single-step method*. A *multi-step method* uses approximations at several previous steps to compute the new approximation. *An implicit method* requires a value for π_{i+1} in computing π_{i+1}.

For a solution method, the *local truncation error* (LTE) of a given step is $|\pi_i - \pi(t_i)|$. The method's *order* is the largest integer p such that the LTE is $O(h^{p+1})$.

Most of the conventional ODE methods perform acceptably for transient analysis of non-stiff Markov models [49], i.e., in cases where $\max_i\{|q_{ii}|\}t$ is not large. One of the commonly used methods is the fourth-order Runge–Kutta [50, 51], which is an explicit single-step method.

For stiff systems, the step size of an explicit method may need to be extremely small to achieve the desired accuracy [52]. However, when the step size becomes very small, the roundoff effects become significant and computational cost increases greatly (as many more time steps are needed). Implicit ODE methods, on the other hand, are inherently stable as they do not force a decrease in the step size to maintain stability. The stability of implicit methods can be characterized by the following definitions.

A method is said to be *A-stable* if all numerical approximations to the actual solution tend to zero as $n \to \infty$ when it is applied to the differential equation $\dot{y} = \lambda y$ with a fixed positive h and a (complex) constant λ with a negative real part ($\mathrm{Re}(\lambda) < 0$) [52]; n is the number of grid points that divide the solution interval. For extremely stiff problems, even A-stability does not suffice to ensure that rapidly decaying solution components decay rapidly in the numerical approximation as well, without a large decrease in the step size. A method is *L-stable* when it is A-stable and, in addition, when applied to the equation $\dot{y} = \lambda y$, where λ is a complex constant with $\mathrm{Re}(\lambda) < 0$, it yields $|y_{i+1} - y_i| \to 0$ as $\mathrm{Re}(h\lambda) \to -\infty$ [48].

TR-BDF2 Method [34]

This is a composite method that uses one step of the trapezoidal rule (TR) and one step of second-order backward difference equation (BDF2) [53]. A single step of TR-BDF2 is composed of a TR step from t_i to $t_i + \gamma h$ and a BDF2 step from $t_i + \gamma h$ to t_{i+1}, where $0 < \gamma < 1$ [34]. The trapezoidal rule applied to interval $(t_i, t_i + \gamma h_i]$ is:

$$\pi_{i+\gamma} - \pi_i = \gamma h_i \cdot \frac{\pi_{i+\gamma} Q + \pi_i Q}{2}. \tag{10.99}$$

Hence, application of the TR step is computationally the same as solving the first-order linear algebraic system

$$\pi_{i+\gamma}\left(I - \frac{\gamma h_i}{2}Q\right) = \pi_i\left(I + \frac{\gamma h_i}{2}Q\right). \tag{10.100}$$

After getting $\pi_{i+\gamma}$, the TR-BDF2 method uses BDF2 to step from $t_i + \gamma h_i$ to t_{i+1}:

$$\pi_{i+1}((2-\gamma)I - (1-\gamma)h_i Q) = \frac{1}{\gamma}\pi_{i+\gamma} - \frac{(1-\gamma)^2}{\gamma}\pi_i. \tag{10.101}$$

Most implementations of ODE methods adjust the step size at each step based on the amount of error in the solution computed at the end of the previous step. The LTE of the TR-BDF2 method is $O(h^3)$. If too much error is introduced, the step is repeated with smaller h. After a successful step, h may be increased.

The TR-BDF2 method provides reasonable accuracy for error tolerances up to 10^{-8}, and excellent stability for stiff Markov chains [34]; for tighter error tolerances, however, the computational complexity of this method rises sharply. Computation of state probabilities with a precision requirement as high as 10^{-10} is not uncommon in practice. In such cases, this method becomes computationally expensive. Thus, the need arises for stable methods with higher orders of accuracy.

Implicit Runge–Kutta Method

Implicit Runge–Kutta methods of different orders of accuracy are possible [54], as proposed in [55]. In general, these methods involve higher powers of the generator matrix Q. A third-order L-stable method is given by:

$$\pi_{i+1}\left(I - \frac{2}{3}hQ + \frac{1}{6}h^2 Q^2\right) = \pi_i\left(I + \frac{1}{3}hQ\right). \tag{10.102}$$

In principle, we could derive methods of fifth order or even higher. However, with higher orders we also need to compute higher powers of the Q matrix, which means increased computational complexity. We restrict ourselves to the third-order method, described by Eq. (10.102). This method involves only squaring the generator matrix and it is reasonable to expect that there will not be much fill-in.

Various possibilities exist for solving the system in Eq. (10.102). One possibility is to compute the matrix polynomial directly. It was found in various experiments that the fill-in of the squared generator matrix was reasonably low [56].

The other option is to factorize the matrix polynomial. We then need to solve two successive linear algebraic systems. For example, the left-hand side polynomial in Eq. (10.102) can be factorized as

$$\pi_{i+1}(I - r_1 hQ)(I - r_2 hQ) = \pi_i\left(I + \frac{1}{3}hQ\right), \tag{10.103}$$

where r_1 and r_2 are the roots of the polynomial $1 - \frac{2}{3}x + \frac{1}{6}x^2$. This system can be solved by solving two systems:

$$X(I - r_2 hQ) = \pi_i\left(I + \frac{1}{3}hQ\right), \tag{10.104}$$

$$\pi_{i+1}(I - r_1 hQ) = X. \tag{10.105}$$

Unfortunately, the roots r_1 and r_2 are complex conjugate; hence this approach will require the use of complex arithmetic.

For the third-order implicit Runge–Kutta method, the LTE vector is $O(h^4)$. Implementation details for this technique are given in [56].

Hybrid Methods

These methods combine explicit (non-stiff) and implicit (stiff) ODE methods for numerical transient analysis of Markov models. This approach [57] is based on the property that stiff Markov chains are non-stiff for an initial phase of the solution interval. A non-stiff ODE method is used to solve the model for this phase, and a stiff ODE method for the rest of the duration until the mission time. A formal criterion to determine the length of the non-stiff phase is described. A significant outcome of this approach is that the accuracy requirement automatically becomes a part of model stiffness. Two specific methods based on this approach are implemented in [57]. Both the methods use the fourth-order Runge–Kutta–Fehlberg method as the non-stiff method. One uses the TR-BDF2 method as the stiff method, whereas the other uses an implicit Runge–Kutta method. Results from solving several models show that the resulting methods are much more efficient than the corresponding stiff methods (TR-BDF2 and implicit Runge–Kutta). The implementation details are similar to those of the standard ODE implementation, with some minor modifications required to be able to switch from the non-stiff ODE method to the stiff ODE method upon detection of stiffness in the Markov chain.

10.5   Further Reading

Application of CTMC models for modeling the reliability behavior of computer systems is presented with several examples in [5]. Readers may also refer to [58–60] for detailed coverage of CTMC theory. Numerical approaches to transient analysis of CTMC can be found in [38, 56].

References

[1] W. Feller, *An Introduction to Probability Theory and its Applications*. John Wiley & Sons, 1968.

[2] Y. Ng and A. Avizienis, "A unified reliability model for fault-tolerant computers," *IEEE Transactions on Computers*, vol. C-29, pp. 1002–1011, 1980.

[3] R. Marie, A. Reibman, and K. Trivedi, "Transient solution of acyclic Markov chains," *Performance Evaluation*, vol. 7, pp. 175–194, 1987.

[4] C. Lindemann, M. Malhotra, and K. Trivedi, "Numerical methods for reliability evaluation of Markovian closed fault-tolerant systems," *IEEE Transactions on Reliability*, vol. 44, pp. 694–704, 1995.

[5] K. Trivedi, *Probability and Statistics with Reliability, Queueing and Computer Science Applications*, 2nd ed. John Wiley & Sons, 2001.

[6] IEC 61508, *Functional Safety of Electrical/Electronic/Programmable Electronic Safety-Related Systems*. IEC Standard No. 61508, 2011.

[7] M. Neuts, *Matrix Geometric Solutions in Stochastic Models*. Johns Hopkins University Press, 1981.

[8] D. Assaf and B. Levikson, "Closure of phase type distributions under operations arising in reliability theory," *The Annals of Probability*, vol. 10, pp. 265–269, 1982.

[9] D. Aldous and L. Shepp, "The least variable phase type distribution is Erlang," *Stochastic Models*, vol. 3, pp. 467–473, 1987.

[10] S. Mondal, X. Yin, J. Muppala, J. Alonso Lopez, and K. S. Trivedi, "Defects per million computation in service-oriented environments," *IEEE Transactions on Services Computing*, vol. 8, no. 1, pp. 32–46, Jan. 2015.

[11] R. Fricks, A. Bobbio, and K. Trivedi, "Reliability models of chronic kidney disease," in *Proc. IEEE Ann. Reliability and Maintainability Symp.*, 2016, pp. 1–6.

[12] United States Renal Data System, *2014 Annual Data Report: An Overview of the Epidemiology of Kidney Disease in the United States*. National Institute of Health / National Institute of Diabetes and Digestive and Kidney Diseases, 2014.

[13] G. Cosulich, P. Firpo, and S. Savio, "Power electronics reliability impact on service dependability for railway systems: A real case study," in *proc. IEEE Int. Symp. on Industrial Electronics, ISIE '96.*, vol. 2, Jun 1996, pp. 996–1001.

[14] M. Beaudry, "Performance-related reliability measures for computing systems," *IEEE Transactions on Computers*, vol. C-27, pp. 540–547, 1978.

[15] H. Choi, W. Wang, and K. Trivedi, "Analysis of conditional MTTF for fault tolerant systems," *Microelectronics and Reliability*, vol. 38, no. 3, pp. 393–401, 1998.

[16] J. C. Laprie, J. Arlat, C. Beounes, and K. Kanoun, "Architectural issues in software fault tolerance," in *Software Fault Tolerance*, ed. M. R. Lyu. John Wiley & Sons, 1994, ch. 3, pp. 47–80.

[17] B. Randell and J. Xu, "The evolution of the recovery block concept," in *Software Fault Tolerance*, ed. M. R. Lyu. John Wiley & Sons, 1994, ch. 1, pp. 1–22.

[18] A. Avizienis, "The methodology of n-version programming," in *Software Fault Tolerance*, ed. M. R. Lyu. John Wiley & Sons, 1994, ch. 2, pp. 23–46.

[19] G.-H. Hsu and X.-M. Yuan, "First passage times and their algorithms for Markov processes," *Stochastic Models*, vol. 11, no. 1, pp. 195–210, 1995.

[20] A. Koziolek, A. Avritzer, S. Suresh, D. Sadoc Menasche, K. Trivedi, and L. Happe, "Design of distribution automation networks using survivability modeling and power flow equations," in *Proc. IEEE 24th Int. Symp. on Software Reliability Engineering (ISSRE)*, Nov. 2013, pp. 41–50.

[21] A. Avritzer, S. Suresh, D. S. Menasché, R. M. M. Leão, E. de Souza e Silva, M. C. Diniz, K. S. Trivedi, L. Happe, and A. Koziolek, "Survivability models for the assessment of smart grid distribution automation network designs," in *Proc. 4th ACM/SPEC Int. Conf. on Performance Engineering*. ACM, 2013, pp. 241–252.

[22] D. S. Menasché, R. M. Meri Leäo, E. de Souza e Silva, A. Avritzer, S. Suresh, K. Trivedi, R. A. Marie, L. Happe, and A. Koziolek, "Survivability analysis of power distribution in smart grids with active and reactive power modeling," *SIGMETRICS Performance Evaluation Review*, vol. 40, no. 3, pp. 53–57, Jan. 2012.

[23] IEEE 1366, *IEEE Guide for Electric Power Distribution Reliability Indices*. IEEE Std. 1366-2003, IEEE Standards Board, 2003.

[24] Z. Ma, "Towards a unified definition for reliability, survivability and resilience (I): The conceptual framework inspired by the handicap principle and ecological stability," in *Aerospace Conference, 2010 IEEE*, Mar. 2010, pp. 1–12.

[25] A. Bobbio and A. Verna, "A performance oriented reliability model of a pumping station in a fire protection system," in *Proc. 5th EUREDATA Conf.*, ed. H. Wingender. Springer-Verlag, 1986, pp. 606–614.

[26] N. Piccinini, A. Verna, and A. Bobbio, "Optimum design of a fire extinguishing pumping installation in a chemical plant," in *Proc. World Congress III of Chemical Engineering*, 1986, Vol. II, pp. 1112–1115.

[27] S. Avogadri, G. Bello, and V. Colombari, "The ENI reliability data bank: Scope, organization and example of report," in *Proc. 4th EUREDATA Conference*, 1983, p. 7.3.

[28] S. J. Bavuso, J. Bechta Dugan, K. Trivedi, E. M. Rothmann, and W. E. Smith, "Analysis of typical fault-tolerant architectures using HARP," *IEEE Transactions on Reliability*, vol. R-36, no. 2, pp. 176–185, Jun. 1987.

[29] K. Trivedi and R. Geist, "Decomposition in reliability analysis of fault-tolerant systems," *IEEE Transactions on Reliability*, vol. R-32, no. 5, pp. 463–468, Dec. 1983.

[30] A. Avizienis, J. Laprie, B. Randell, and C. Landwehr, "Basic concepts and taxonomy of dependable and secure computing," *IEEE Transactions on Dependable and Secure Computing*, vol. 1, no. 1, pp. 11–33, 2004.

[31] A. Bobbio and K. S. Trivedi, "An aggregation technique for the transient analysis of stiff Markov chains," *IEEE Transactions on Computers*, vol. C-35, pp. 803–814, 1986.

[32] J. McGough, M. Smotherman, and K. Trivedi, "The conservativeness of reliability estimates based on instantaneous coverage," *IEEE Transactions on Computers*, vol. C-34, pp. 602–609, 1985.

[33] R. Marie, "Transient numerical solutions of stiff Markov chains," in *Proc. 20th ISATA Symp.*, 1989, pp. 255–270.

[34] A. Reibman and K. Trivedi, "Numerical transient analysis of Markov models," *Computers and Operations Research*, vol. 15, pp. 19–36, 1988.

[35] W. Grassman, "Finding transient solutions in Markovian event systems through randomization," in *Numerical Solution of Markov Chains*. Marcel Dekke, 1991.

[36] J. Muppala, M. Malhotra, and K. Trivedi, "Markov dependability models of complex systems: Analysis techniques," in *Reliability and Maintenance of Complex Systems*, NATO ASI Series, ed. S. Özekici. Springer, 1996, vol. 154, pp. 442–486.

[37] E. de Souza e Silva and H. Gail, "Transient solutions for Markov chains," in *Computational Probability*, ed. W. Grassmann. Springer, 2000, ch. 3, pp. 49–85.

[38] W. Stewart, *Introduction to the Numerical Solution of Markov Chains*. Princeton University Press, 1994.

[39] W. Grassmann, "The use of eigenvalues for finding equilibrium probabilities of certain Markovian two-dimensional queueing problems," *INFORMS Journal on Computing*, vol. 15, no. 4, pp. 412–421, 2003.

[40] R. Sahner, K. Trivedi, and A. Puliafito, *Performance and Reliability Analysis of Computer Systems: An Example-Based Approach Using the SHARPE Software Package*. Kluwer Academic Publishers, 1996.

[41] C. Moler and C. Van Loan, "Nineteen dubious ways to compute the exponential of a matrix," *SIAM Review*, vol. 20, pp. 801–835, 1978.

[42] A. Jensen, "Markoff chains as an aid in the study of Markoff processes," *Scandinavian Actuarial Journal*, vol. 1953, Supplement 1, pp. 87–91, 1951.

[43] B. L. Fox and P. W. Glynn, "Computing Poisson probabilities," *Communications of the ACM*, vol. 31, no. 4, pp. 440–445, Apr. 1988.

[44] H. Abdallah and R. Marie, "The uniformized power method for transient solutions of Markov processes," *Computers and Operations Research*, vol. 20, no. 5, pp. 515–526, 1993.

[45] W. S. A. van Moorsel, "Adaptive uniformization," *Communications in Statistics: Stochastic Models*, vol. 10, no. 3, pp. 619–648, 1994.

[46] A. van Moorsel and W. Sanders, "Transient solution of Markov models by combining adaptive and standard uniformization," *IEEE Transactions on Reliability*, vol. 46, no. 3, pp. 430–440, Sep. 1997.

[47] J. Carrasco, "Transient analysis of rewarded continuous time Markov models by regenerative randomization with Laplace transform inversion," *The Computer Journal*, vol. 46, no. 1, pp. 84–99, 2003.

[48] J. D. Lambert, *Computational Methods in Ordinary Differential Equations*. John Wiley & Sons, 1973.

[49] W. Grassmann, "Transient solution in Markovian queueing systems," *Computers and Operations Research*, vol. 4, pp. 47–56, 1977.

[50] L. F. Shampine, "Stiffness and nonstiff differential equation solvers, II: Detecting stiffness with Runge–Kutta methods," *ACM Transactions on Mathematical Software*, vol. 3, no. 1, pp. 44–53, 1977.

[51] E. Hairer, S. P. Nørsett, and G. Wanner, *Solving Ordinary Differential Equations I: Nonstiff Problems*, 2nd edn., Springer Series in Computational Mathematics. Springer, 1993, vol. 8.

[52] C. Gear, *Numerical Initial Value Problems in Ordinary Differential Equations*. Prentice-Hall, 1971.

[53] R. Bank, W. M.. Coughran, W. Fichtner, E. Grosse, D. Rose, and R. Smith, "Transient simulation of silicon devices and circuits," *IEEE Transactions on Computer-Aided Design of Integrated Circuits and Systems*, vol. 4, no. 4, pp. 436–451, Oct. 1985.

[54] O. Axelsson, "A class of A-stable methods," *BIT*, vol. 9, no. 3, pp. 185–199, 1969.

[55] M. Malhotra, J. K. Muppala, and K. S. Trivedi, "Stiffness-tolerant methods for transient analysis of stiff Markov chains," *Journal of Microelectronics and Reliability*, vol. 34, pp. 1825–1841, 1994.

[56] M. Malhotra, J. K. Muppala, and K. S. Trivedi, "Stiffness-tolerant methods for transient analysis of stiff Markov chains," *Microelectronics and Reliability*, vol. 34, pp. 1825–1841, 1994.

[57] M. Malhotra, "A computationally efficient technique for transient analysis of repairable Markovian systems," *Performance Evaluation*, vol. 24, no. 4, pp. 311–331, 1996.

[58] A. Papoulis, *Probability, Random Variables and Stochastic Processes*. McGraw Hill, 1965.

[59] D. Cox and H. Miller, *The Theory of Stochastic Processes*. Chapman and Hall, 1965.

[60] V. G. Kulkarni, *Modeling and Analysis of Stochastic Systems*. Chapman and Hall, 1995.

11 Continuous-Time Markov Chain: Queuing Systems

Continuous-time Markov chains are by nature agnostic to the actual system behavior being modeled. The previous two chapters leveraged this formalism to model the failure/repair behavior of a system, where the different states of the CTMC reflected the different operational or failed states of the system. A CTMC can definitely be used to model the performance of a system, where each state of the CTMC may represent the different performance states of the system. As an example, if we model the behavior of a system that services incoming requests, a CTMC state may reflect the number of requests currently in service and/or waiting for service. The transitions of the CTMC may reflect the arrival of requests and completion of service. This will enable us to compute interesting performance measures such as the average number of requests completed per unit time, the average number of customers waiting for service and the average response time for requests. In cases where there is an upper limit on the number of customers (for example due to limited buffer space), measures like the request rejection probability can be computed. Thus, in this chapter we illustrate several system performance-oriented scenarios that can be represented using a CTMC. The computation of several performance measures of interest will be explained. Classical performance analysis deals with the interaction between workload and system resources in the absence of resource failures. Classical reliability/availability analysis considers the failure and recovery of resources while generally ignoring the different levels of performance of each resource and the system as a whole – each is considered to be either failed or not-failed. Real systems, however, have degradable levels of performance giving rise to the need for a combined analysis of performance and failure (recovery) – that is, combined performance and reliability (availability) analysis. This topic has evolved under three separate yet connected threads. One thread has arisen from operations research and applied probability under the umbrella of queues with breakdown. The second thread originated in the fault-tolerant computing community under the aegis of performability modeling. The third thread is the completion time of a task; this started in operations research but has been studied in the computer performance and dependability community. We limit the present chapter to elementary notions in queuing theory that will be used in the rest of the book; interested readers should refer to the classical literature on the subject [1–3].

11.1   Continuous-Time Markov Chain Performance Models

A typical performance model consists of several states that reflect the operating conditions of the system. Several performance-related events will trigger the transitions among the states. Such events could include the arrival of requests to a system, the completion of a request or the movement of a request among the various subsystems of a system as it requires the corresponding service. When all the timed activities are exponentially distributed the performance model can be conceptualized into a CTMC.

The CTMC solution techniques that we have discussed in the previous chapter yield the state probabilities, both as a function of time or in steady state, and various performance metrics can then be computed from these quantities. Fortunately, the Markov reward model framework that was introduced in Section 9.5 provides the most appropriate framework to express several of these performance measures in terms of defining appropriate reward rates with the states, and instantaneous rewards with the transitions, of the CTMC.

11.2   The Birth–Death Process

The birth–death (BD) process is a CTMC with state space defined on the integers $\{0, 1, 2, \ldots\}$; it models the arrival and departure of jobs at a server (or resource). State i identifies the condition of the system in which there are i jobs in the resource. The rate of transition from state i to state $(i+1)$ is λ_i (the birth rate) and gives the rate at which new jobs arrive at the resource while in state i; the rate from state i to state $(i-1)$ is μ_i (the death rate) and gives the rate at which jobs leave the resource while in state i. The state transition diagram is given in Figure 11.1.

Let $N(t)$ be the number of elements (or jobs, requests, messages) in the system at time t. Figure 11.2 shows the way in which the event $N(t + \Delta t) = i$ can occur. By conditioning on the state of the system at time t, we can write, for $i > 0$:

$$P\{N(t+\Delta t) = i \mid N(t) = i-1\} = \lambda_{i-1}\,\Delta t + o(\Delta t),$$

$$P\{N(t+\Delta t) = i \mid N(t) = i+1\} = \mu_{i+1}\,\Delta t + o(\Delta t), \qquad (11.1)$$

$$P\{N(t+\Delta t) = i \mid N(t) = i\} = 1 - \lambda_i\,\Delta t - \mu_i\,\Delta t + o(\Delta t),$$

and for $i = 0$, we can write:

$$P\{N(t+\Delta t) = 0 \mid N(t) = 1\} = \mu_1\,\Delta t + o(\Delta t),$$

$$P\{N(t+\Delta t) = 0 \mid N(t) = 0\} = 1 - \lambda_0\,\Delta t + o(\Delta t), \qquad (11.2)$$

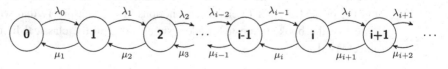

Figure 11.1   State diagram of a birth–death process.

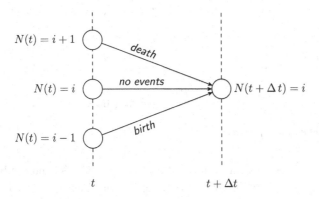

Figure 11.2 Events generating a BD process.

where $\lim_{\Delta t \to 0} \frac{o(\Delta t)}{\Delta t} = 0$.

Recalling our notation $\pi_i(t) = P\{N(t) = i\}$, we apply the theorem of total probability to the conditional probabilities in (11.1) and (11.2) to get:

$$\begin{cases} \pi_0(t + \Delta t) = \mu_1 \Delta t \pi_1(t) + (1 - \lambda_0 \Delta t) \pi_0(t) + o(\Delta t) \quad (i = 0), \\ \pi_i(t + \Delta t) = \lambda_{i-1} \Delta t \pi_{i-1}(t) + \mu_{i+1} \Delta t \pi_{i+1}(t) + (1 - \lambda_i \Delta t - \mu_i \Delta t) \pi_i(t) \\ \qquad\qquad + o(\Delta t) \quad (i > 0); \end{cases}$$

$$\begin{cases} \dfrac{\pi_0(t + \Delta t) - \pi_0(t)}{\Delta t} = -\lambda_0 \pi_0(t) + \mu_1 \pi_1(t) + \frac{o(\Delta t)}{\Delta t} \quad (i = 0), \\ \dfrac{\pi_i(t + \Delta t) - \pi_i(t)}{\Delta t} = -(\lambda_i + \mu_i) \pi_i(t) + \lambda_{i-1} \pi_{i-1}(t) + \mu_{i+1} \pi_{i+1}(t) + \frac{o(\Delta t)}{\Delta t} \quad (i > 0). \end{cases}$$

Rearranging, and taking the limit $\Delta t \to 0$, the following set of linear differential equations is derived:

$$\begin{cases} \dfrac{d\pi_0(t)}{dt} = -\lambda_0 \pi_0(t) + \mu_1 \pi_1(t) \quad (i = 0), \\ \dfrac{d\pi_i(t)}{dt} = -(\lambda_i + \mu_i) \pi_i(t) + \lambda_{i-1} \pi_{i-1}(t) + \mu_{i+1} \pi_{i+1}(t) \quad (i > 0), \end{cases}$$

(11.3)

with initial conditions:

$$\begin{cases} \pi_0(0) = 1 \quad (i = 0), \\ \pi_i(0) = 0 \quad (i > 0). \end{cases}$$

(11.4)

Equation (11.3) can also be interpreted as the transient probability balance equation in state i, as represented in Figure 11.3. The flow variation in state i equals the difference between the incoming flow and the outgoing flow:

$$\text{variation of flow} = \frac{d\pi_i(t)}{dt},$$

$$\text{incoming flow} = \lambda_{i-1} \pi_{i-1}(t) + \mu_{i+1} \pi_{i+1}(t) \quad (i > 0),$$

$$\text{outgoing flow} = (\lambda_i + \mu_i) \pi_i(t).$$

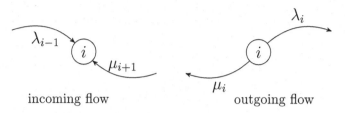

incoming flow   outgoing flow

Figure 11.3   Balance equation for a BD process.

A BD process with constant birth and death rates is a CTMC and can be solved by means of the Kolmogorov differential equations as in Chapters 9 and 10, where the infinitesimal generator Q is given by:

$$
Q = \begin{array}{c}
\\ 0 \\ 1 \\ 2 \\ \vdots \\ i-1 \\ i \\ i+1 \\ \vdots
\end{array}
\begin{array}{ccccccccc}
0 & 1 & 2 & 3 & \cdots & i-1 & i & i+1 & \cdots \\
\hline
-\lambda_0 & \lambda_0 & & & & & & & \\
\mu_1 & -(\lambda_1+\mu_1) & \lambda_1 & & & & & & \\
0 & \mu_2 & -(\lambda_2+\mu_2) & \lambda_2 & & & & & \\
\vdots & \vdots & \vdots & \vdots & \vdots & \vdots & \vdots & \vdots & \vdots \\
& & & & & & & & \\
& & & & & \mu_i & -(\lambda_i+\mu_i) & \lambda_i & \\
& & & & & & & & \\
\end{array}
\quad . \tag{11.5}
$$

The CTMC of a BD process is irreducible assuming that $\lambda_i > 0$ and $\mu_i > 0$ for each i. If a steady-state solution exists, it is characterized by:

$$
\lim_{t\to\infty} \frac{d\pi_i(t)}{dt} = 0 \qquad (i=0,1,2,\ldots).
$$

Let us write $\pi_i = \lim_{t\to\infty} \pi_i(t)$. The steady-state equations become:

$$
\begin{cases}
0 &= -\lambda_0\pi_0 + \mu_1\pi_1 \quad (i=0), \\
0 &= -(\lambda_i+\mu_i)\pi_i + \lambda_{i-1}\pi_{i-1} + \mu_{i+1}\pi_{i+1} \quad (i>0),
\end{cases}
$$

which can be rewritten as balance equations (incoming flow equals outgoing flow) as:

$$
\begin{cases}
\lambda_0\pi_0 &= \mu_1\pi_1 \quad (i=0), \\
(\lambda_i+\mu_i)\pi_i &= \lambda_{i-1}\pi_{i-1} + \mu_{i+1}\pi_{i+1} \quad (i>0).
\end{cases}
$$

The steady-state equation can be rearranged as:

$$
\begin{cases}
& \lambda_0\pi_0 - \mu_1\pi_1 & = 0 \\
\lambda_1\pi_1 - \mu_2\pi_2 &= \lambda_0\pi_0 - \mu_1\pi_1 & = 0 \\
\quad \vdots & \quad\vdots \\
\lambda_i\pi_i - \mu_{i+1}\pi_{i+1} &= \lambda_{i-1}\pi_{i-1} - \mu_i\pi_i & = 0 \\
\quad \vdots
\end{cases}
$$

From the above, the ith term becomes:

$$\lambda_{i-1}\pi_{i-1} = \mu_i \pi_i \quad \Longrightarrow \quad \pi_i = \frac{\lambda_{i-1}}{\mu_i}\pi_{i-1} \quad (i \geq 1),$$

and, finally,

$$\pi_i = \frac{\lambda_{i-1}}{\mu_i}\frac{\lambda_{i-2}}{\mu_{i-1}}\pi_{i-2} = \frac{\lambda_0 \lambda_1 \cdots \lambda_{i-1}}{\mu_1 \mu_2 \cdots \mu_i}\pi_0 = \pi_0 \prod_{j=0}^{i-1}\frac{\lambda_j}{\mu_{j+1}}. \qquad (11.6)$$

For the π_i to form a probability vector, the following normalization condition must hold:

$$\sum_{i \geq 0}\pi_i = 1;$$

hence, from Eq. (11.6):

$$\pi_0 = \frac{1}{1 + \displaystyle\sum_{i \geq 1}\prod_{j=0}^{i-1}\frac{\lambda_j}{\mu_{j+1}}}.$$

The steady-state probability vector exists, with $\pi_i > 0$, if the series $\displaystyle\sum_{i \geq 1}\prod_{j=0}^{i-1}\frac{\lambda_j}{\mu_{j+1}}$ converges. This is the case when all the states of the CTMC are recurrent non-null.

11.3 The Single Queue

A single queue is characterized by incoming jobs (messages, requests) arriving at a service center that can be composed of one or multiple servers. When a server is available the incoming job is served, otherwise it is stored in a buffer that can be of limited or of infinite capacity. The standard notation to identify the main elements that define the structure of a single queue, also called Kendall's notation [4], is $A/B/c/K/N/d$, where:

  A indicates the nature of the distribution of the inter-arrival times;
  B indicates the nature of the distribution of the service times;
  c is the number of servers;
  K is the storage capacity of the system (number of servers plus the storage capacity of the buffer) – if omitted, it is assumed to be infinite;
  N is the population that can submit jobs – if omitted, it is assumed to be infinite;
  d is the scheduling discipline such as FCFS.

The usual possibilities for the inter-arrival and service times A and B are:

  M memoryless (or exponentially distributed);
  PH phase type;
  GI generally distributed and independent;
  G generally distributed.

This chapter is limited to memoryless (or exponential) inter-arrival and service time distributions. Phase-type distributions are the subject of Chapter 15.

11.3.1   *M/M/*1 Queue

The $M/M/1$ queue represents a system in which a single server is available to service incoming requests. A buffer associated with the server holds incoming requests until it is their turn to access the server. The request arrival process in an $M/M/1$ queue is a Poisson process, i.e., the inter-arrival times are all i.i.d. with an exponential distribution of rate λ. For now, the only constraint on the scheduling discipline is that the server is not left idle if there are requests in the system and that no knowledge of the service times of individual requests is used in scheduling. The server takes a random amount of time that is exponentially distributed with rate μ to service a request. The typical graphical representation of an $M/M/1$ queue is shown in Figure 11.4.

Let $N(t)$ represent the number of requests in the system. This includes one request that is currently in service and the remaining $N(t) - 1$ requests waiting in the queue for the server. We can represent the behavior of the system using the homogeneous CTMC shown in Figure 11.5. In this CTMC, state i denotes the condition with i requests in the system.

The CTMC is a special case of the general BD process of Section 11.2 [5] for which in each state i, $\lambda_i = \lambda$, and $\mu_i = \mu$. This CTMC is irreducible if $\lambda > 0$ and $\mu > 0$.

Let π_i be the probability of being in state i in the steady state. Then, writing out the balance equations for the CTMC, we get

$$
\begin{cases}
\begin{aligned}
\lambda \pi_0 &= \mu \pi_1 \\
(\lambda + \mu) \pi_1 &= \lambda \pi_0 + \mu \pi_2 \\
(\lambda + \mu) \pi_2 &= \lambda \pi_1 + \mu \pi_3 \\
&\vdots \\
(\lambda + \mu) \pi_i &= \lambda \pi_{i-1} + \mu \pi_{i+1} \\
&\vdots
\end{aligned}
\end{cases}
$$

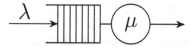

Figure 11.4   Schematic representation of an $M/M/1$ queue.

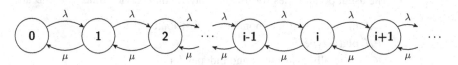

Figure 11.5   Continuous-time Markov chain state diagram of the $M/M/1$ queue.

Solving the balance equations, or directly from Eq. (11.6), we get:

$$\pi_i = \left(\frac{\lambda}{\mu}\right)\pi_0 = \rho^i \pi_0,$$

where $\rho = \lambda/\mu$ is called the *traffic intensity* of the system. Imposing the normalization condition $\sum_{i \geq 0} \pi_i = \pi_0 \sum_{i \geq 0} \rho^i = 1$, we get

$$\pi_0 = \frac{1}{\sum_{i \geq 0} \rho^i} = 1 - \rho. \tag{11.7}$$

The geometrical series in the denominator of Eq. (11.7) converges if $\rho < 1$; hence, for the $M/M/1$ queue to be stable, the traffic intensity must be less than unity or the arrival rate should be less than the service rate. In this case, all the states of the CTMC are recurrent non-null. If $\lambda \geq \mu$ (i.e., $\rho \geq 1$), the denominator in the expression for π_0 diverges. If $\lambda = \mu$, then all the states of the CTMC are recurrent null, and if $\lambda > \mu$, then all the states of the CTMC are transient. Thus, if $\lambda \geq \mu$ the system is unstable, and the number of requests in the queue increases without bound.

Server Utilization

Note that the server is busy as long as the system is not in state 0. Thus, the *utilization* of the server, meaning the fraction of the time that the server is busy in steady state, is given by:

$$U_0 = \sum_{j=1}^{\infty} \pi_j = 1 - \pi_0 = \rho.$$

This is the first metric of interest for the $M/M/1$ queue.

Expected Number of Customers $E[N]$

We denote by N the total number of customers in the system (including the customer under service) in steady state. In state i we have $N = i$. The steady-state expected number of customers in the system, $E[N]$, is derived as:

$$E[N] = \sum_{i=0}^{\infty} i \cdot \pi_i = \pi_0 \sum_{i=0}^{\infty} i \cdot \rho^i = (1 - \rho) \sum_{i=0}^{\infty} i \cdot \rho^i = \frac{\rho}{1 - \rho}. \tag{11.8}$$

To derive Eq. (11.8) we have made use of a property of the sum of the geometric series:

$$\sum_{i=0}^{\infty} i \cdot \rho^i = \rho \frac{\partial}{\partial \rho} \sum_{i=0}^{\infty} \rho^i = \rho \frac{\partial}{\partial \rho} \frac{1}{1 - \rho} = \frac{\rho}{(1 - \rho)^2}.$$

The variance of the number of customers in the queue, $\text{Var}[N]$, is

$$\text{Var}[N] = \sum_{i=0}^{\infty} i^2 \pi_i - (E[N])^2 = \frac{\rho}{(1 - \rho)^2}.$$

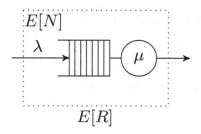

Figure 11.6   Graphical representation of Little's law.

Expected Response Time E[R]
Let the random variable R represent the *response time*, defined as the time elapsed from the instant of the request arrival until the instant of its completion and departure from the system, in steady state. *Little's law* [6] defines the relationship between the average response time, the average number of customers in the system and the request arrival rate (see Figure 11.6). Little's law states that the expected number of customers in the system $E[N]$ is equal to the arrival rate (λ in this case) times the expected response time (the expected time the customer spends in the system) $E[R]$:

$$E[N] = \lambda E[R].$$

Knowing the expression for $E[N]$, we can now get an expression for $E[R]$:

$$E[R] = E[N]/\lambda = \frac{1}{\lambda}\frac{\rho}{1-\rho} = \frac{1/\mu}{1-\rho} = \frac{\text{average service time}}{\text{prob. that the server is idle}}.$$

Expected Waiting Time E[W]
Let us define the waiting time $W = R - S$ as the time a customer waits in the queue before service, where R is the response time and S the service time. The expected waiting time $E[W]$ is given by:

$$E[W] = E[R] - E[S] = \frac{1}{\mu(1-\rho)} - \frac{1}{\mu} = \frac{\rho}{\mu(1-\rho)}.$$

Expected Number of Customers in the Line E[N_Q]
The expected number of customers in the line (awaiting for service) is obtained by applying Little's law to the queue only, as in Figure 11.7:

$$E[N_Q] = \lambda \cdot E[W] = \frac{\rho^2}{1-\rho}.$$

Expected Number of Customers in Service E[N_S]
The expected number of customers in service is:

$$E[N_S] = E[N] - E[N_Q] = \rho.$$

Table 11.1 $M/M/1$ metrics as expected steady-state reward rates $E[X] = \sum_{i=0}^{\infty} r_i \pi_i$.

Mean no. in system $E[N]$	$r_i = i$	$\dfrac{\rho}{1-\rho}$
Mean response time $E[R]$	$r_i = i/\lambda$	$\dfrac{1/\mu}{1-\rho}$
Throughput $E[T]$	$\begin{cases} r_0 = 0 \\ r_i = \mu \;\; (i > 0) \end{cases}$	λ

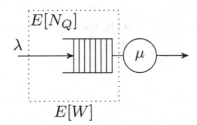

Figure 11.7 Little's law applied to the queue only.

From Little's law applied to the server only:

$$E[N_S] = \lambda \cdot E[S] = \frac{\lambda}{\mu} = \rho.$$

Expected Throughput $E[T]$
In all the states $i > 0$ the throughput is μ, but in state 0 the throughput is 0. Hence the expected throughput, $E[T]$, is [see (9.66)]:

$$E[T] = \sum_{i=1}^{\infty} \mu \pi_i = \mu \sum_{i=1}^{\infty} \pi_i = \mu (1 - \pi_0) = \mu \rho = \lambda.$$

All the above metrics can be expressed in the framework of MRM (Section 9.4), by associating a suitable reward rate r_i with state i of the CTMC. Table 11.1 gives in column 2 the reward rates to obtain the expected measures reported in column 1.

We have carried out only steady-state analysis of the $M/M/1$ queue. We note, however, that a fully symbolic transient solution for the state probabilities of this queuing system is known [7]. From this knowledge, the expected throughput at time t, expected number of customers in the system at time t and probability that the server is busy at time t can all be expressed as expected reward rate at time t using the same reward rate assignment as in Table 11.1. Cumulative reward measures can similarly also be obtained.

$M/M/1$ **FCFS: Response Time Distribution**
The distribution of the response time random variable R is derived in steady-state conditions by means of the so-called *tagged job approach*. All the derivations for $M/M/1$ thus far did not assume a specific scheduling discipline, but to derive the

distribution of response time we need to assume a specific discipline. We assume an FCFS scheduling discipline and we follow the dynamics of a particular tagged job. We assume that when the tagged job arrives, there are $n \geq 0$ jobs in the system, so the tagged job is the $(n+1)$th job and will experience a response time R given by:

$$R = S^R + S_2 + \cdots + S_n + S^T,$$

where S^R is the residual service time for the job currently (if any) undergoing service, S_i, $(i = 2, \ldots, n)$ are the service times of the jobs already in the queue and S^T is the service time for the $(n+1)$th (i.e., the tagged) job. Due to the memoryless property, S^R is $\text{EXP}(\mu)$, while all the others are also $\text{EXP}(\mu)$ by the assumption on service time distributions. Furthermore, they are all independent.

Hence, for $N = n$, the conditional LST of R is

$$\mathscr{L}_{R|N}(s|n) = \left(\frac{\mu}{s+\mu}\right)^{n+1}. \tag{11.9}$$

Note that for a Poisson arrival process, it is known that the state of the system as seen by an arriving customer is statistically the same as the state of the system at a random point in time. This property is known as "Poisson arrivals see time averages" (PASTA) [8]. Hence the probability that the tagged job finds n customers in the system is equal to the probability that there are n jobs in steady state, which is equal to:

$$\pi_n = (1-\rho)\rho^n. \tag{11.10}$$

Applying the theorem of total probability to the Laplace transforms:

$$\mathscr{L}_R(s) = \sum_{n=0}^{\infty} \mathscr{L}_{R|N}(s|n)\,\pi_n. \tag{11.11}$$

Combining Eqs. (11.11) and (11.10) with (11.9), we get:

$$\mathscr{L}_R(s) = \sum_{n=0}^{\infty} \left(\frac{\mu}{s+\mu}\right)^{n+1}(1-\rho)\rho^n = \frac{\mu(1-\rho)}{s+\mu}\sum_{n=0}^{\infty}\left(\frac{\mu\rho}{s+\mu}\right)^n$$

$$= \frac{\mu(1-\rho)}{s+\mu}\,\frac{1}{1 - \dfrac{\mu\rho}{s+\mu}} = \frac{\mu(1-\rho)}{s+\mu(1-\rho)}. \tag{11.12}$$

Inverting (11.12), the response time distribution $F_R(t)$ is seen to be exponential with rate $\mu(1-\rho) = \mu - \lambda$:

$$F_R(t) = 1 - e^{-\mu(1-\rho)t} = 1 - e^{-(\mu-\lambda)t}.$$

Problems

11.1 For the $M/M/1$ queue, write down fully symbolic expressions for the expected number of jobs in the system, the throughput and server utilization at time t, the expected number of jobs completed and the expected fraction of time the server is busy during the interval $(0, t]$.

11.2 Compute and plot the quantities in the above example as functions of time, assuming $\lambda = 0.8$ and $\mu = 1.2$.

11.3 Assume that we sample the response times of n jobs submitted to an $M/M/1$ FCFS queue. Let the sample be denoted by $R_1, R_2, \ldots, R_n$. Assume that the samples are well separated in time so that independence can be assumed. Derive the distribution function of $\sum_{i=1}^{n} R_i$ and thence that of the sample mean $\overline{R} = \sum_{i=1}^{n} R_i/n$ [9].

11.4 M/M/m: Single Queue with m Servers

This queuing system has a Poisson arrival process of rate λ, a single shared queue and m identical servers, each with service rate μ. The structure of the queuing system is shown in Figure 11.8.

The state diagram of the system is shown in Figure 11.9; the service rate grows linearly from μ in state 1 up to $m\mu$ in state m, where all the m servers are busy, and from then on retains the value $m\mu$ for states $i > m$.

The general BD process of Section 11.2 can be particularized as follows:

$$\lambda_i = \lambda \ (i \geq 0); \qquad \mu_i = \begin{cases} i\mu & (0 < i < m), \\ m\mu & (i \geq m). \end{cases}$$

The state probabilities thus satisfy:

$$\begin{cases} \pi_i = \pi_0 \prod_{j=0}^{i-1} \frac{\lambda}{(j+1)\mu} = \pi_0 \cdot \left(\frac{\lambda}{\mu}\right)^i \frac{1}{i!} & (i < m), \\ \pi_i = \pi_0 \prod_{j=0}^{m-1} \frac{\lambda}{(j+1)\mu} \cdot \prod_{k=m}^{i-1} \frac{\lambda}{m\mu} = \pi_0 \left(\frac{\lambda}{\mu}\right)^i \frac{1}{m! \, m^{i-m}} & (i \geq m). \end{cases}$$

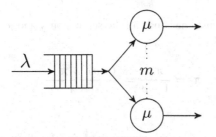

Figure 11.8 Single queue with m identical servers.

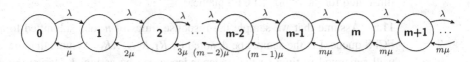

Figure 11.9 State diagram of an $M/M/m$ queue.

Define $\rho = \dfrac{\lambda}{m\mu}$; the stability condition requires $\rho < 1$. Rewriting the state probabilities in terms of ρ, we obtain:

$$
\pi_i = \begin{cases}
\pi_0 \dfrac{(m\rho)^i}{i!} & (i < m), \\[2ex]
\pi_0 \dfrac{\rho^i m^m}{m!} & (i \geq m).
\end{cases}
$$

Applying the normalization condition, we obtain:

$$
\pi_0 = \left\{ \sum_{i=0}^{m-1} \frac{(m\rho)^i}{i!} + \sum_{i=m}^{\infty} \frac{\rho^i m^m}{m!} \right\}^{-1}. \tag{11.13}
$$

The second summand in Eq. (11.13) can be rewritten as:

$$
\sum_{i=m}^{\infty} \frac{\rho^i m^m}{m!} = \frac{\rho^m m^m}{m!} \sum_{k=0}^{\infty} \rho^k = \frac{(m\rho)^m}{m!} \frac{1}{1-\rho},
$$

so that Eq. (11.13) becomes:

$$
\pi_0 = \left\{ \sum_{i=0}^{m-1} \frac{(m\rho)^i}{i!} + \frac{(m\rho)^m}{m!} \frac{1}{1-\rho} \right\}^{-1}.
$$

The expected number of customers in the system $E[N]$ is:

$$
E[N] = \sum_{i=0}^{\infty} i\pi_i = m\rho + \rho \frac{(m\rho)^m}{m!} \frac{\pi_0}{(1-\rho)^2}.
$$

The expected number of busy servers $E[M]$ is:

$$
E[M] = \sum_{i=0}^{m-1} i\pi_i + m \sum_{i=m}^{\infty} \pi_i = m\rho = \frac{\lambda}{\mu}.
$$

The probability that an arriving customer finds all the servers busy and joins the queue is given by:

$$
\pi_{[\text{queue}]} = \sum_{i=m}^{\infty} \pi_i = \frac{\pi_m}{1-\rho} = \frac{(m\rho)^m}{m!} \frac{\pi_0}{1-\rho}.
$$

Problems

11.4 For the $M/M/m$ FCFS queue, derive expressions for the distributions of the waiting time W and the response time R in the steady state using the tagged job approach illustrated for the $M/M/1$ FCFS queue.

11.5 Assume that we sample the response times of n jobs submitted to an $M/M/m$ FCFS queue. Let the sample be denoted by $R_1, R_2, \ldots, R_n$. Assume that the samples are well separated in time so that independence can be assumed. Derive the distribution function of $\sum_{i=1}^{n} R_i$ and thence that of the sample mean $\overline{R} = \sum_{i=1}^{n} R_i/n$.

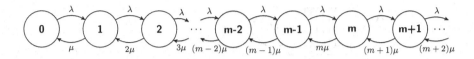

Figure 11.10 State diagram of an $M/M/\infty$ queue.

11.4.1 $M/M/\infty$: An Infinite Number of Servers

A special case of the $M/M/m$ queuing system of Section 11.4 is the $M/M/\infty$ queuing system where there are an infinite number of servers and hence each arriving customer goes immediately into service without waiting in the queue. The state diagram of the queuing system is shown in Figure 11.10.

The general BD process of Section 11.2 can be particularized as follows:

$$\begin{cases} \lambda_i &= \lambda \quad (i \geq 0), \\ \mu_i &= i\mu \quad (i \geq 0). \end{cases}$$

The state probabilities become:

$$\pi_i = \pi_0 \prod_{j=0}^{i-1} \frac{\lambda}{(j+1)\mu} = \pi_0 \frac{1}{i!} \left(\frac{\lambda}{\mu}\right)^i.$$

The normalization condition provides:

$$\pi_0 = \frac{1}{1 + \sum_{i=1}^{\infty} \frac{1}{i!} \left(\frac{\lambda}{\mu}\right)^i} = e^{-\lambda/\mu}.$$

Hence, the state probabilities assume the following form and are clearly seen to follow a Poisson pmf with parameter λ/μ:

$$\pi_i = e^{-\lambda/\mu} \frac{(\lambda/\mu)^i}{i!},$$

$$E[N] = \lambda/\mu, \quad E[R] = \frac{E[N]}{\lambda} = \frac{1}{\mu}.$$

Since each arriving customer finds an available server, no waiting time is involved and the response time distribution is the same as the service time distribution:

$$F_R(t) = 1 - e^{-\mu t}.$$

11.5 $M/M/1/K$: Finite Storage

The storage capacity of the system is K (one customer in service and $K-1$ customers in the waiting line), and later customers are refused. The structure of the queue is shown in Figure 11.11, and its state diagram is given in Figure 11.12.

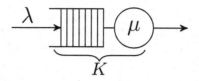

Figure 11.11 The $M/M/1/K$ queue.

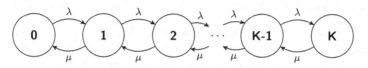

Figure 11.12 State diagram of an $M/M/1/K$ queue.

The general BD process can be particularized as follows:

$$\lambda_i = \begin{cases} \lambda & (i < K), \\ 0 & (i \geq K); \end{cases} \qquad \mu_i = \mu.$$

The state probabilities satisfy

$$\begin{cases} \pi_i & = & \pi_0 \displaystyle\prod_{j=0}^{i-1} \frac{\lambda}{\mu} = \pi_0 \cdot \rho^i \quad (i \leq K), \\ \pi_i & = & 0 \qquad\qquad\qquad (i > K). \end{cases}$$

From the normalization condition:

$$\pi_0 = \frac{1}{1 + \displaystyle\sum_{j=1}^{K} \rho^j} = \frac{1}{1 + \dfrac{\rho(1-\rho^K)}{1-\rho}} = \frac{1-\rho}{1-\rho^{K+1}}.$$

Since the $M/M/1/K$ queue is a finite-state CTMC, it is stable for any value of the traffic intensity ρ. The state probabilities are:

$$\begin{cases} \pi_i & = & \dfrac{(1-\rho)\rho^i}{1-\rho^{K+1}} \quad (i \leq K), \\ \pi_i & = & 0 \qquad\qquad (i > K). \end{cases} \tag{11.14}$$

For $\rho \to 1$ the above formula is undefined. We find the limit resorting to L'Hôpital's rule:

$$\lim_{\rho \to 1} \pi_i = \lim_{\rho \to 1} \frac{(1-\rho)\rho^i}{1-\rho^{K+1}} = \lim_{\rho \to 1} \frac{-\rho^i + i(1-\rho)\rho^{i-1}}{-(K+1)\rho^K} = \frac{1}{K+1}.$$

Let us define the *rejection probability* as the probability that an arriving customer will find the queue full and be rejected (or blocked). Since the queue is full when in state K, the rejection or blocking probability is:

$$\pi_K = \frac{(1-\rho)\rho^K}{1-\rho^{K+1}}.$$

The rate at which jobs are rejected (blocked) is given by $\lambda \pi_K$, and the rate at which jobs are accepted is given by $\lambda_{\text{acc}} = \lambda(1 - \pi_K)$.

The expected number of customers $E[N]$ is:

$$E[N] = \sum_{i=0}^{K} i \cdot \pi_i = \sum_{i=0}^{K} i \cdot \frac{(1-\rho)\rho^i}{1-\rho^{K+1}} = \frac{1-\rho}{1-\rho^{K+1}} \sum_{i=0}^{K} i \cdot \rho^i$$

$$= \frac{\rho}{1-\rho^{K+1}} \frac{1 - (K+1)\rho^K + K\rho^{K+1}}{(1-\rho)}. \tag{11.15}$$

The formula in (11.15) is based on the following finite series:

$$\sum_{i=0}^{K} i \cdot \rho^i = \rho \frac{\partial}{\partial \rho} \sum_{i=1}^{K} \rho^i = \rho \frac{\partial}{\partial \rho} \frac{\rho(1-\rho^K)}{1-\rho} = \rho \frac{1 - (K+1)\rho^K + K\rho^{K+1}}{(1-\rho)^2}.$$

From Eq. (11.15), it follows that:

$$\lim_{\rho \to 0} E[N] = 0, \quad \lim_{\rho \to \infty} E[N] = K, \quad \lim_{\rho \to 1} E[N] = \frac{K}{2},$$

where the last limit ($\rho \to 1$) is obtained by applying L'Hôpital's rule twice.

From the formula for the expected number of jobs in the system, we can get a formula for the mean response time (for accepted jobs) in the steady state using Little's law as $E[R_{\text{acc}}] = E[N]/\lambda_{\text{acc}}$. We note that the denominator in the last expression is not λ but λ_{acc}, that is, the arrival rate seen by the queue, which is also the steady-state throughput as can be seen by attaching a reward rate μ to all states in $\{1, 2, \ldots, K\}$ and reward rate zero to state 0, then computing the expected reward rate in the steady state:

$$E[T] = \sum_{i=1}^{K} \mu \pi_i = \mu(1 - \pi_0) = \lambda(1 - \pi_K) = \lambda_{\text{acc}}.$$

The distribution of response time can also be derived using the tagged job approach. We refer the reader to [5] for details, but note that the response time distribution will be defective in this case with the defect at infinity being equal to the loss probability π_K.

We could also carry out transient analysis (numerically) and compute:

- the expected number of jobs in the system at time t (an example of the expected reward rate at time t):

$$E[N(t)] = \sum_{i=0}^{K} i \cdot \pi_i(t);$$

- the expected number of jobs completed in the interval $(0, t]$:

$$E[C(t)] = \sum_{i=0}^{K} \mu \cdot \int_0^t \pi_i(x)dx;$$

- the expected number of jobs blocked (rejected) in the same interval:

$$E[B(t)] = \lambda \cdot \int_0^t \pi_K(x)dx.$$

The latter two are both examples of expected accumulated reward in the interval $(0, t]$. It is certainly possible to compute the distributions of these quantities, but we leave these to Problem 11.6.

In deriving the formulas in this section, and in particular the rejection probability, we have assumed that the customers keep on arriving even when the queue is full, and are rejected. Hence, from a modeling point of view, a more suitable representation would entail the presence in Figure 11.12 of a self-loop of rate λ originating from state K. However, since self-loops do not affect the CTMC state probability vector (Section 10.3), the self-loop is usually omitted.

11.5.1 $M/M/m/m$ Queue

The $M/M/m/m$ queue does not have a waiting line; an arriving customer enters service if and only if one of the m servers is idle. The state diagram of the system is shown in Figure 11.13.

The general BD can be expressed mathematically as follows:

$$\lambda_i = \begin{cases} \lambda & (i < m), \\ 0 & (i \ge m); \end{cases} \qquad \mu_i = \begin{cases} i\mu & (i \le m), \\ 0 & (i > m); \end{cases}$$

$$\begin{cases} \pi_i = \pi_0 \prod_{j=0}^{i-1} \dfrac{\lambda}{\mu_{j+1}} = \pi_0 \cdot \dfrac{\lambda^i}{i!\,\mu^i} & (i \le m), \\ \pi_i = 0 & (i > m). \end{cases}$$

From the normalization condition:

$$\pi_0 = \frac{1}{1 + \displaystyle\sum_{j=1}^{m} \frac{\lambda^i}{i!\,\mu^i}} = \frac{1}{1 + \displaystyle\sum_{j=1}^{m} \frac{\rho^i}{i!}},$$

where the traffic intensity ρ is defined as $\rho = \lambda/\mu$. The state probabilities are:

$$\begin{cases} \pi_i = \dfrac{\rho^i}{i!} \left(\displaystyle\sum_{j=0}^{m} \frac{\rho^j}{j!} \right)^{-1} & (i \le m), \\ \pi_i = 0 & (i > m), \end{cases} \qquad (11.16)$$

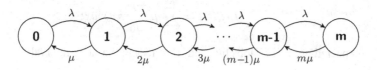

Figure 11.13   State diagram of an $M/M/m/m$ queue.

and the rejection probability is obtained by substituting $i = m$ in (11.16),

$$\pi_m = \frac{\rho^m}{m!} \left(\sum_{j=0}^{m} \frac{\rho^j}{j!} \right)^{-1}. \tag{11.17}$$

The expected number of customers in the system is:

$$E[N] = \sum_{j=0}^{m} j\pi_j = \pi_0 \sum_{j=0}^{m} j\frac{\rho^i}{i!}$$

$$= \rho - \rho\frac{\rho^m}{m!}\pi_0 = \rho(1 - \pi_m).$$

With $m = 1$, the system reduces to an $M/M/1/1$ queue that corresponds to a two-state CTMC with the state diagram shown in Figure 10.1. The steady-state probabilities are then given by:

$$\begin{cases} \pi_0 = \dfrac{1}{1+\rho} = \dfrac{\mu}{\lambda+\mu}, \\[2mm] \pi_1 = \dfrac{\rho}{1+\rho} = \dfrac{\lambda}{\lambda+\mu}. \end{cases}$$

Example 11.1 *Telecommunication Switching System Model*
We again return to the example of the telecommunications switching system [10] whose steady-state and transient availability models were presented in Examples 9.8 and 10.6, respectively.

We now consider the performance model of the system. Assume that the system consists of n trunks (or channels) with an infinite caller population. A call will be lost (referred to as blocking) when it finds all n trunks are busy upon its arrival. The call arrival process is assumed to be Poisson with rate λ. We assume call holding times are exponentially distributed with rate μ. Without considering link failure, the pure performance model is an $M/M/n/n$ queue, whose analysis using a CTMC was presented at the beginning of this section.

The steady-state probability of the CTMC being in state j is given by Eq. (11.16). The steady-state probability of state $m = n$ is the probability that an incoming call will find all trunks busy and hence will be blocked. This blocking probability P_{bk} is given by Eq. (11.17), with $m = n$, and is known as the Erlang B formula [5]. Extensions to the above model are known that relax many of the assumptions made here (see, for instance, [2, 11]).

The transient probability of being in state j can be computed by solving the following set of differential equations:

$$\frac{d\pi_0(t)}{dt} = -\lambda\pi_0(t) + \mu\pi_1(t),$$

$$\frac{d\pi_k(t)}{dt} = -(\lambda + k\mu)\pi_k(t) + \lambda\pi_{k-1}(t) + (k+1)\mu\pi_{k+1}(t), \quad k = 1, 2, \ldots, n-1,$$

$$\frac{d\pi_n(t)}{dt} = -n\mu\pi_n(t) + \lambda\pi_{n-1}(t), \tag{11.18}$$

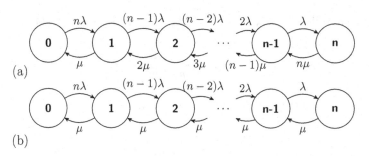

(a)

(b)

Figure 11.14   State diagram of machine repairer model: (a) independent repair, (b) shared repair.

with the initial condition $\pi_k(t=0) = \pi_k(0)$.

Similarly, the transient blocking probability is $P_{bk}(t) = \pi_n(t)$, and the expected number of lost calls in the interval $(0, t]$ is given as $N_{loss}(t) = \int_0^t \lambda P_{bk}(x)\,dx$.

Example 11.2   *n Identical Components Failure/Repair Model*

In the queuing examples of the previous paragraphs, the arrival rate was constant and independent of the number of arrivals. In a general case, both the arrival and the service rates may depend on the state. An example is the n identical components failure/repair model.

Consider a system with n identical components, each with a failure rate λ and repair rate μ. This is known as the machine repairer model. The components are repaired with two possible repair policies: (i) independent repair; (ii) shared repair with a single repair person (see Example 9.4). The failure/repair process of the system can be modeled as a BD process with $n+1$ states. We assume as a state index i the number of failed components.

Figure 11.14(a) shows the case of independent repair (as many repairers as failed components), while Figure 11.14(b) shows the case of shared repair with a single repair person. Derivation of the steady-state probability expressions is left as an exercise.

Problems

11.6   For the $M/M/1/K$ queue, numerically compute the expected number of jobs in the system, the throughput and server utilization at time t, the expected number of jobs completed, the expected number of jobs blocked and the expected fraction of the server busy during the interval $(0, t]$. Plot the quantities in the above example as functions of time, assuming $\lambda = 0.8$ and $\mu = 1.2$. Vary K over a range of values $\{1, 10, 100, 1000\}$.

11.7   Derive a formula for the response time distribution in the steady state for the $M/M/1/K$ FCFS queue and hence show that this distribution is defective.

11.8   A node of a network has a buffer with a capacity of three packets. The packets arrive as a Poisson process of rate λ and are retransmitted with an exponentially distributed delay of rate μ.

(a) Derive the steady-state probability that there are i packets in the node (with $i = 0, 1, 2, 3$).

(b) Compute the expected time to fill up the buffer for the first time starting from an empty buffer at $t = 0$ with probability 1.

11.9   For the failure/repair model of Example 11.2, with n identical components, derive the steady-state probability expressions for the independent repair policy and for the shared repair policy with a single repair person. Compute the throughput in the two cases and discuss the results.

11.10   In the failure/repair model of Example 11.2, with n identical components, we adopt a shared repair policy with $i < n$ repair persons. Derive the state space for this system and compute the steady-state probability expressions.

11.6   Closed *M/M/*1 Queue

A closed $M/M/1$ queuing system is a system where the total number of jobs in the system is constant. In this system, represented in Figure 11.15, K is the total number of jobs in the system; inactive jobs are queued in the upper queue, and are sent to the server with rate λ. The active jobs, which are queued for service in the lower queue, are served at the rate μ.

We use the number of jobs in the server subsystem, $j = K - n$, as the index of a state in the state space, where n is the number of jobs in the waiting queue. With this assignment, the state diagram of the closed $M/M/1$ queue of Figure 11.15 is similar to the state diagram of the $M/M/1/K$ queue of Figure 11.12, and a similar solution approach can be adopted. However, while in the $M/M/1/K$ the appropriate model should display a self-loop on state K since jobs keep on arriving (though rejected in this state) even when the system is full, in the closed $M/M/1$ queue case no jobs can arrive in state K since the total number of jobs is fixed.

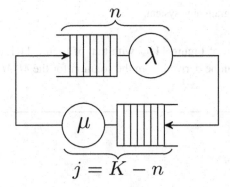

Figure 11.15   Closed $M/M/1$ queue with K jobs.

Example 11.3   *Cyclic Queuing System*

Consider the cyclic queuing model for a multiprogramming system depicted in Figure 11.16. Successive CPU bursts are exponentially distributed with rate μ, and successive I/O bursts are exponentially distributed with rate λ.

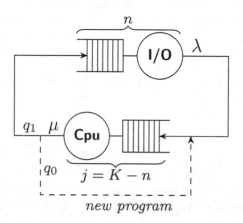

Figure 11.16   A cyclic multiprogramming system.

At the end of a CPU burst the job requires an I/O operation with probability q_1 and leaves the system with probability q_0 ($q_1 + q_0 = 1$). At the end of a job a statistically identical job enters the system, leaving the number of jobs constant (level of multiprogramming). The state diagram is shown in Figure 11.17.

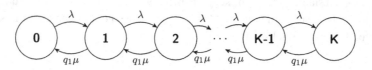

Figure 11.17   State diagram of a cyclic multiprogramming system.

Comparing the state diagram of Figure 11.17 with that of Figure 11.12, we can see that the solution of this case can be derived from the solution for the $M/M/1/K$ queue by setting $\rho = \dfrac{\lambda}{\mu q_1}$.

11.7   Queues with Breakdown

A number of situations occur in real systems where the single service station is incapacitated from time to time and unable to render service to the incoming customers.

Even if the service station may be out of service for different reasons, service interruption in a queue is generally referred to in the literature as server breakdown, and the related problems as queuing systems with breakdown [12, 13].

When a server failure occurs, the job (if any) in service will certainly be interrupted. Depending upon the system, the job may be dropped or may simply be preempted. In the former case, the job may be retried later. In the case of preemption, the job will be executed once the server is repaired. As was discussed in Section 3.7, there are several possibilities to consider in this case, such as, for instance, the preemptive resume (prs) policy or preemptive repeat (prt) policy. Further, in the prt case, there are possible subcategories depending on whether the interrupted job is restarted with a different time request (but distributed with the same Cdf) – *preemptive repeat different* (prd) – or with an identical time request – *preemptive repeat identical* (pri) [12].

Example 11.4   *M/M/1/K Queue with Breakdown*
We return to the $M/M/1/K$ queue from Section 11.5, but we now allow the server to fail at the rate γ and be repaired at the rate τ [12]. Figure 11.18 is the state diagram of the underlying homogeneous CTMC for this queuing system, where the state label (i, b) indicates that there are i jobs in the system and b indicates the state of the server (with the usual assignment: $b = 1$, server up; $b = 0$, server down).

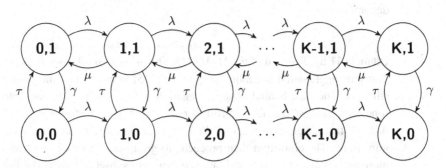

Figure 11.18   State diagram of an $M/M/1/K$ queue with server breakdown.

The steady-state balance equations can be set up similarly to the $M/M/1/K$ queue. However, obtaining a closed-form solution for the steady-state probabilities as in Section 11.5 is difficult. The solution for the CTMC can be obtained numerically given the values of the various parameters. Then we can obtain several measures of interest by an appropriate assignment of reward rates to the states of the CTMC. For a customer, some interesting measures to compute could be the blocking probability, the rate of blocking and the mean response time (for completed jobs). For the whole system, an important measure is the expected number of jobs in the system. We could also carry out transient analysis (numerically) and compute the expected number of jobs in the system at time t (an example of the expected reward rate at time t):

$$E[N(t)] = \sum_{i=0}^{K} i \cdot (\pi_{i,1}(t) + \pi_{i,0}(t)).$$

A cumulative transient analysis will yield the expected number of jobs completed in the interval $(0,t]$:

$$E[C(t)] = \sum_{i=0}^{K} \mu \cdot \int_0^t \pi_{i,1}(x)dx,$$

and the expected number of jobs blocked (rejected) in the same interval:

$$E[B(t)] = \lambda \cdot \int_0^t (\pi_{K,1}(x) + \pi_{K,0}(x))dx,$$

which are both examples of expected accumulated reward in the interval $(0,t]$.

The $M/M/1/K$ queue with breakdown is an excellent example of a stiff Markov chain due to the presence of transitions with rates of vastly differing orders of magnitude. As can be observed, the arrival and departure transitions are fast transitions, while the failure and repair transitions are slow transitions since their rates are much smaller $(\lambda, \mu \gg \gamma, \tau)$. One way to deal with such a stiff CTMC is to use stiffly stable numerical methods [14] as illustrated in Section 10.4. The following examples illustrate two other approaches to deal with the solution of stiff Markov chains, through aggregation and decomposition.

Example 11.5 *Solution of the $M/M/1/K$ Queue with Breakdown by Aggregation*
The state aggregation technique proposed in [15] grapples with the problem of CTMC stiffness. The idea behind state aggregation is to classify each state into a fast state (one with at least one fast transition rate out of it) or a slow state, and to consider that the fast states have reached their asymptotic value in the timescale of the slow transitions. The algorithm then proceeds to group the set of all fast states into one or more *fast recurrent* subsets and one *fast transient* subset of states. The fast subsets can then be analyzed in isolation to find their steady-state condition. The algorithm then replaces each fast recurrent subset with a slow state, and the fast transient subset with a probabilistic switch to construct a smaller non-stiff Markov chain that can be analyzed using conventional techniques. Here we illustrate the application of the technique to solve the $M/M/1/K$ queue with breakdown.

Looking at the CTMC in Figure 11.18, we notice that state $(K,0)$ is the only slow state in the model. The states $\{(0,1),\ldots,(K,1)\}$ form a fast recurrent subset of states, while the states $\{(0,0),\ldots,(K-1,0)\}$ form a fast transient subset. Deleting the slow transitions, the fast recurrent subset turns out to be equivalent to the $M/M/1/K$ queue, whose steady-state probability vector, now denoted as $\boldsymbol{x_1} = [x_{0,1}, x_{1,1}, \ldots, x_{K,1}]$, can be obtained, using (11.14), as:

$$x_{i,1} = \rho^i \frac{1-\rho}{1-\rho^{K+1}},$$

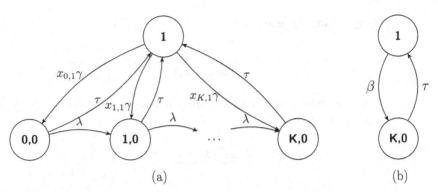

Figure 11.19 State diagram of $M/M/1/K$ queue with breakdown: (a) after aggregation of the fast recurrent subset; (b) after full aggregation.

where $\rho = \lambda/\mu$ is the traffic intensity. We can then replace the fast recurrent subset with an aggregate state, yielding the CTMC shown in Figure 11.19(a), where the rate out of the aggregate state toward state $(i,0)$ is given by γ times the probability of being in state $(i,1)$ (in steady state) in the original CTMC.

Further replacing the fast transient subset with a branch point yields the two-state CTMC shown in Figure 11.19(b). In the figure, the rate β from state 1 to the only slow state $(K,0)$ is given by

$$\beta = \gamma \sum_{i=0}^{K} x_{i,1} \delta^{K-i},$$

where $\delta = \lambda/(\lambda+\tau)$ is the probability that a new customer arrives before the repair is completed. Each term in β represents the probability that the server fails when i customers are in the system, and that $(K-i)$ customers arrive before the server is repaired.

Solving the CTMC in Figure 11.19(b) in steady state yields:

$$\tilde{\pi}_1 = \frac{\tau}{\beta+\tau}, \qquad \tilde{\pi}_{K,0} = \frac{\beta}{\beta+\tau}.$$

The disaggregation step starts by computing the unnormalized state probabilities for the fast transient states; we get:

$$\tilde{\pi}_{i,0} = \frac{\gamma}{\lambda+\tau} \tilde{\pi}_1 \sum_{j=0}^{i} x_{j,1} \delta^{i-j} = \frac{\gamma}{\lambda+\tau} \frac{\tau}{\beta+\tau} \sum_{j=0}^{i} x_{j,1} \delta^{i-j}, \quad i=0,1,\ldots,K-1.$$

The state probabilities need to be normalized using as normalization factor $s_{\text{nor}} = \tilde{\pi}_1 + \sum_{i=0}^{K} \tilde{\pi}_{i,0}$ [15]. The resulting approximate probability vector $\pi^* = \tilde{\pi}/s_{\text{nor}}$ allows us to compute interesting metrics. The probability π_1^* can then be used to obtain the approximate (normalized) probability values for the fast recurrent states as:

$$\pi_{i,1}^* = x_{i,1} \pi_1^*, \quad i=0,1,\ldots,K.$$

Using the above equations, we arrive at the following expression for π_1^* in steady state:

$$\pi_i^* = \frac{1}{1 + \frac{\gamma}{\lambda+\tau} + \sum_{i=0}^{K-1} \sum_{j=0}^{i} x_{j,1}\delta^{i-j} + \frac{\beta}{\tau}}.$$

From the above expressions, the expected number of jobs in the system in steady state $E[N]$ can be derived as:

$$E[N] = \frac{\sum_{i=1}^{K-1} \frac{\gamma}{\lambda+\tau} \sum_{j=0}^{i} i x_{j,1}\delta^{i-j} + \sum_{i=0}^{K} i x_{i,1} + \frac{K\beta}{\tau}}{1 + \frac{\gamma}{\lambda+\tau} + \sum_{i=0}^{K-1} \sum_{j=0}^{i} x_{j,1}\delta^{i-j} + \frac{\beta}{\tau}},$$

and the steady-state probability of a job being rejected is given by

$$\pi_{\text{rej}} = \left(x_{K,1} + \frac{\beta}{\tau} \right) \pi_1^*.$$

Example 11.6 *Solution of the M/M/1/K Queue with Breakdown by Decomposition*
The decomposition approach has been adopted in a pioneering work by Meyer [16] to model and solve the system *performability*. The performability model separates the overall system representation into an upper-level dependability model that typically encompasses all the slow events, and a lower-level performance model that captures all the fast events in the system. The performance metrics obtained from the performance models corresponding to each dependability state are then incorporated into the higher-level dependability model in the form of reward rates. Applying this approach to the $M/M/1/K$ queue with breakdown of Figure 11.18, we get the two-state availability model shown in Figure 11.20.

Solving the CTMC in Figure 11.20 in steady state yields

$$\hat{\pi}_1 = \frac{\tau}{\gamma+\tau}, \qquad \hat{\pi}_0 = \frac{\gamma}{\gamma+\tau}.$$

To compute the steady-state rejection probability, we assign a reward rate $r_1 = \rho^K(1 - \rho)/(1 - \rho^{K+1})$ to state 1, the probability that the buffer is full while the server is up, and $r_0 = 1$ to the **down** state. Thus, the overall rejection probability can be computed as

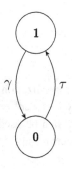

Figure 11.20 Two-state availability model for the $M/M/1/K$ queue with breakdown.

(an instance of expected steady-state reward rate):

$$\pi_{\text{rej}} = r_1 \hat{\pi}_1 + r_0 \hat{\pi}_0 = \rho^K \frac{(1-\rho)\tau}{(1-\rho^{K+1})(\gamma+\tau)} + \frac{\gamma}{\gamma+\tau}.$$

Similarly, assigning the reward rate of $r_1 = E[N]$ of an $M/M/1/K$ queue, and $r_0 = K$, yields the average number of customers $E[N_b]$ in the system:

$$E[N_b] = \frac{\rho}{1-\rho^{K+1}} \frac{1-(K+1)\rho^K + K\rho^{K+1}}{(1-\rho)} \frac{\tau}{\gamma+\tau} + \frac{K\gamma}{\gamma+\tau}.$$

Example 11.7   *Telecommunications Switching System Model*
We return to the example of the telecommunications switching system with n trunks whose availability model was presented in Examples 9.8 and 10.6; its performance model, based on the $M/M/n/n$ queue, was presented in Example 11.1. In this example we consider the combined performance and availability model of the system. The composite performance–availability model is shown in Figure 11.21, where state (i,j) indicates that there are i non-failed trunks in the system and j of them are carrying ongoing calls [5]. λ and μ are the time-independent arrival and service rates, while γ and τ are the time-independent failure and repair rates. We assume a single repair person. The transitions due to the arrival of new calls and completion of existing calls are self-explanatory. From each state (i,j) two failure transitions emanate, since we

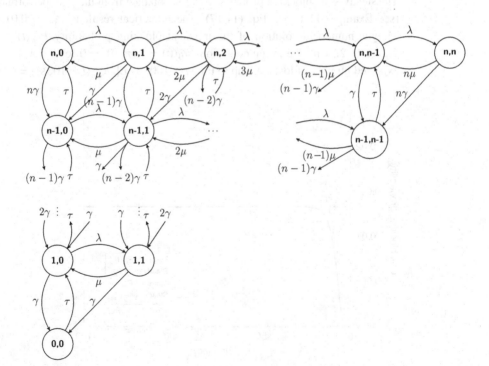

Figure 11.21   Composite CTMC model for the telecommunications switching system.

distinguish whether the failure involves a busy trunk (with rate $j\gamma$) with loss of the current call, or an idle trunk (with rate $(i-j)\gamma$) with no loss of calls.

The steady-state probability of being in state (i,j) is denoted as $\pi_{i,j}$. The steady-state and transient blocking probabilities P'_{bk} and $P'_{bk}(t)$ can be computed as

$$P'_{bk} = \sum_{k=0}^{n} \pi_{k,k}, \qquad P'_{bk}(t) = \sum_{k=0}^{n} \pi_{k,k}(t), \tag{11.19}$$

where state (k,k) in this model represents the situation where all the functioning trunks are busy or all trunks are down. The expected number of lost calls up to time t is given as

$$E[N'_{loss}(t)] = \int_{0}^{t} \lambda P'_{bk}(x)\,dx.$$

To obtain numerical results, the parameters are set at $n=25$, $\lambda=5\,\mathrm{s}^{-1}$, $\mu=0.3\,\mathrm{s}^{-1}$, $\gamma=0.002\,\mathrm{s}^{-1}$ and $\tau=0.1\,\mathrm{s}^{-1}$. The steady-state and transient probabilities of the availability, the performance and the composite models can be either solved in closed form, as in Eqs. (9.51) and (11.18), or through the use of SHARPE.

Performance

The steady-state blocking probability P_{bk} is obtained from the pure performance model (see Example 11.1) via Eq. (11.17). The numerical result is $P_{bk} = 0.01376$. The analytic-numerical solution of the transient blocking probability $P_{bk}(t)$ is shown in Figure 11.22 with the initial conditions $\pi_0(0)=1$, $\pi_k(0)=0$ for $k=1,2,3,\ldots,n$. The expected number of lost calls up to time $t=100\,\mathrm{s}$ is $E[N_{loss}(t=100\,\mathrm{s})]=6.063$.

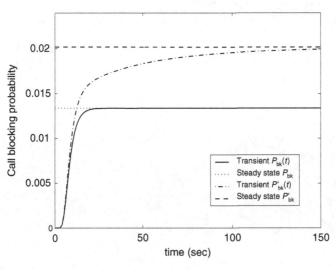

Figure 11.22 Transient blocking probabilities $P_{bk}(t)$ and $P'_{bk}(t)$.

Availability

Suppose the system is up as long as there are at least l trunks functioning properly. We assume here $l = 1$. The steady-state availability $A(1)$ is computed from the pure availability model (see Example 9.8). According to Eqs. (9.51) and (9.52), $A(1) = \sum_{k=1}^{25} \pi_k$ and the numerical result is $A(1) = 1 - 2.6935 \times 10^{-18}$. Since the change of $A(1,t)$ over time is not significant for illustration, we assume $l = 20$, i.e., the condition that at least $l = 20$ trunks be functioning for system to be functioning. In Figure 11.23, we plot the transient unavailability $U(20,t) = 1 - A(20,t)$, as well as steady-state unavailability $U(20) = 1 - A(20) = 1 - 0.9933 = 0.0067$ for reference.

Performability

First, consider the blocking probability P'_{bk} in the combined performance and availability model (see Figure 11.21) as a performability measure. The numerical result is $P'_{bk} = 0.020178$. The transient blocking probability $P'_{bk}(t)$ is shown in Figure 11.22. The expected number of lost calls up to time $t = 100$s is $E[N_{loss}(t = 100\text{s})] = 8.211$. As expected, the blocking probabilities with trunk failure included, P'_{bk} and $P'_{bk}(t)$, and $E[N'_{loss}(t)]$ are larger than the pure performance values P_{bk}, $P_{bk}(t)$, and $E[N_{loss}(t)]$.

There exist other performability measures besides P'_{bk}. Suppose resource N represents the number of available trunks in the system. Initially, $N = n$ before any failure occurs. Performability measures such as the expected fraction of available trunks $\frac{E[N]}{n}$ and the probability of having the r-percentile of all trunks functioning P_{N_r} can be

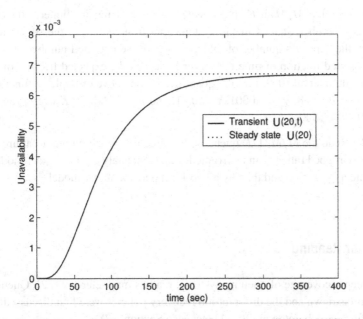

Figure 11.23 Transient unavailability $U(20,t)$ and steady-state unavailability $U(20)$.

defined as:

$$\frac{E[N]}{n} = \sum_{i=1}^{n} \frac{i}{n} \pi_i, \qquad P_{N_r} = \sum_{i \le n \cdot r\%} \pi_i, \qquad (11.20)$$

which are called "expected survivability" and "r-percentile survivability" respectively in [17].

The numerical results are as follows:

$$\frac{E[N]}{n} = 0.964\,968,$$

$$P_{N_0} = 2.693\,515 \times 10^{-18},$$

$$P_{N_{20}} = 7.775\,417 \times 10^{-12},$$

$$P_{N_{40}} = 9.007\,691 \times 10^{-8},$$

$$P_{N_{60}} = 8.878\,974 \times 10^{-5},$$

$$P_{N_{80}} = 1.724\,949 \times 10^{-2}.$$

Note that $E[N]/n$ has been called "capacity-oriented availability" by Heimann *et al.* in their performability study of multiprocessor systems [5, 18], which is essentially the same as the performability definition given by Huslende [19].

Problems

11.11 For the $M/M/1/K$ queue with failure and repair, numerically compute the expected number of jobs in the system, the throughput and the server utilization at time t, the expected number of jobs completed, the expected number of jobs blocked, the expected fraction of time the server is up and the expected fraction of server busy during the interval $(0, t]$. Plot the quantities in the above example as functions of time, assuming $\lambda = 0.8$, $\gamma = 0.0001$, $\tau = 0.01$, and $\mu = 1.2$. Vary K over a range of values $\{1, 10, 100, 1000\}$.

11.12 Redo the $M/M/1/K$ queue with failure and repair but upon failure the job will now be dropped rather than interrupted. Carry out analysis of the exact model, followed by state aggregation and then by a two-level performability model.

11.8   Further Reading

For deeper coverage of queuing systems, readers may refer to [1, 2]. Queuing systems with breakdown and the different job recovery policies were first discussed in [12], and then further elaborated in [20, 21] (see also Section 3.7).

References

[1] L. Kleinrock, *Queuing Systems, Volume 1: Theory*. Wiley Interscience, 1975.

[2] G. Bolch, S. Greiner, H. de Meer, and K. S. Trivedi, *Queueing Networks and Markov Chains: Modeling and Performance Evaluation with Computer Science Applications*, 2nd edn. Wiley Interscience, 2006.

[3] L. Lakatos, L. Szeid, and M. Telek, *Introduction to Queueing Systems with Telecommunication Applications*. Springer, 2013.

[4] D. Kendall, "Stochastic processes occurring in the theory of queues and their analysis by the method of the imbedded Markov chain," *The Annals of Mathematical Statistics*, vol. 24, no. 3, pp. 338–354, 1953.

[5] K. Trivedi, *Probability and Statistics with Reliability, Queueing and Computer Science Applications*, 2nd edn. John Wiley & Sons, 2001.

[6] J. Little, "A proof of the queueing formula $L = \lambda W$," *Operations Research*, vol. 9, pp. 383–387, 1961.

[7] P. Leguesdron, J. Pellaumail, G. Rubino, and B. Sericola, "Transient analysis of the M/M/1 queue," *Advances in Applied Probability*, vol. 25, no. 3, pp. 702–713, 1993.

[8] R. Wolff, "Poisson arrivals see time averages," *Operations Research*, vol. 30, no. 2, pp. 223–231, 1982.

[9] A. Avritzer, A. Bondi, M. Grottke, K. Trivedi, and E. Weyuker, "Performance assurance via software rejuvenation: Monitoring, statistics and algorithms," in *Proc. Int. Conf. on Dependable Systems and Networks, DSN 2006*, Jun. 2006, pp. 435–444.

[10] Y. Liu and K. Trivedi, "Survivability quantification: The analytical modeling approach," *International Journal of Performability Engineering*, vol. 2, no. 1, p. 29, 2006.

[11] S. Dharmaraja, K. S. Trivedi, and D. Logothetis, "Performance modelling of wireless networks with generally distributed hand-off interarrival times," *Computer Communications*, vol. 26, pp. 1747–1755, 2003.

[12] D. Gaver, "A waiting line with interrupted service, including priorities," *Journal of the Royal Statistical Society*, vol. B24, pp. 73–90, 1962.

[13] B. Avi-Itzhak and P. Naor, "Some queuing problems with the service station subject to breakdown," *Operations Research*, vol. 11, no. 3, pp. 303–320, 1963.

[14] M. Malhotra, J. K. Muppala, and K. S. Trivedi, "Stiffness-tolerant methods for transient analysis of stiff Markov chains," *Journal of Microelectronics and Reliability*, vol. 34, pp. 1825–1841, 1994.

[15] A. Bobbio and K. S. Trivedi, "An aggregation technique for the transient analysis of stiff Markov chains," *IEEE Transactions on Computers*, vol. C-35, pp. 803–814, 1986.

[16] J. Meyer, "On evaluating the performability of degradable systems," *IEEE Transactions on Computers*, vol. C-29, pp. 720–731, 1980.

[17] S. C. Liew and K. W. Lu, "A framework for characterizing disaster-based network survivability," *IEEE Journal on Selected Areas in Communications*, vol. 12, no. 1, pp. 52–58, Jan. 1994.

[18] D. Heimann, N. Mittal, and K. Trivedi, "Availability and reliability modeling for computer systems," in *Advances in Computers*, ed. M. Yovits, vol. 31. Academic Press, 1990, pp. 176–231.

[19] R. Huslende, "A combined evaluation of performance and reliability for degradable systems," in *Proc. ACM/SIGMETRICS Conf.*, 1981, pp. 157–164.

[20] V. Nicola, V. Kulkarni, and K. Trivedi, "Queueing analysis of fault-tolerant computer systems," *IEEE Transactions on Software Engineering*, vol. SE-13, pp. 363–375, 1987.

[21] P. Chimento and K. Trivedi, "The performance of block structured programs on processors subject to failure and repair," in *High Performance Computer Systems*, ed. E. Gelenbe. North Holland, 1988, pp. 269–280.

12 Petri Nets

As stated in the introduction to Part III, the state explosion problem implies that the infinitesimal generator matrix of a CTMC can be of very large dimensions. The manual specification/generation of a large matrix is a time-consuming and error-prone process that can be alleviated by utilizing a higher-level formalism that enables concise specification of complex system behavior. Petri nets (PNs), and their variants, are the most popular high-level formalisms used for this purpose since they allow the automated generation (and solution) of the underlying CTMC.

We start with a quick introduction to the Petri net formalism through a simple example in Section 12.1. Thereafter a formal description of Petri nets and structural extensions to PNs are presented in Section 12.2. Timing (stochastic) extensions to PNs resulting in stochastic Petri nets and their subsequent generalization to generalized stochastic Petri nets are then delineated. Several examples of the application of these formalisms to system modeling are also given. Then, stochastic reward nets, which introduce both structural extensions and reward-based specification and computation of metrics, are presented in Section 12.3.

12.1 Introduction

We first consider some motivating examples to get an overview of the modeling power of the Petri net formalism. Using these examples, we give a brief informal introduction to Petri net concepts and also illustrate how we can model a system using the PN constructs. These examples will be revisited and modified in the subsequent sections to illustrate additional concepts.

Example 12.1 *Queue with Breakdown and Repair*
Returning to the $M/M/1/K$ queue with breakdown that was introduced in Section 11.4, we note that the state space of the CTMC expands with an increase in the value of K. The model, however, mixes together behavior (performance and dependability) occurring at two timescales that are orders of magnitude apart. Clear separation of the performance and dependability behavior is unclear from the structure of the CTMC. The same behavior expressed as a (stochastic) Petri net model is shown in Figure 12.1. In this model, we have clearly marked out the parts of the model that express the performance

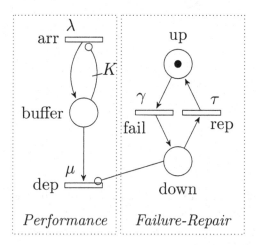

Figure 12.1   Petri net model of the $M/M/1/K$ queue with breakdown and repair.

behavior and the failure/repair behavior, respectively. Furthermore, the buffer size K is specified as a parameter in the model without requiring any modification to the structure of the model (we will see shortly how the specification of the buffer size is done in practice).

We will use the Petri net shown in Figure 12.1 to informally introduce some of the terminology of the formalism. These will be formally introduced later. In particular, we notice that the three *places* (drawn as circles: "buffer," "up," and "down") represent the number of customers in the queue, and whether the server is up or down, respectively. Similarly, the four *transitions* (drawn as rectangles: "arr," "dep," "fail," and "rep") represent the events corresponding to customer arrival or departure and the server failing and getting repaired, respectively. The *input arcs* connecting the places to transitions represent the pre-conditions for the occurrence of the events, and the output arcs from the transitions to places represent the consequences (post-conditions) of the occurrence of the events in the system. The *marking* of the PN, represented by the number of *tokens* in each place (usually depicted as black dots in the places), enables us to represent the current state of the system. The evolution of the system can thus be captured by the sequence of markings of the PN that results from the *firing* of the transitions. Firing results in removing tokens from the *input places* of a transition and depositing them into its *output places*, (potentially) resulting in a new marking of the PN.

Example 12.2   *Queue with Breakdown, Repair and Imperfect Coverage*
We now introduce a minor variation to the $M/M/1/K$ queue with breakdown example. Assume that with probability $1 - c$ the failure of the server will result in the loss of the job currently being served. This can easily be represented by modifying the PN in Figure 12.1 by adding two more transitions, "cov" and "uncov," resulting in the PN in Figure 12.2. These new *immediate* transitions (drawn as black bars) are useful in representing probabilistic branches and events that take zero time.

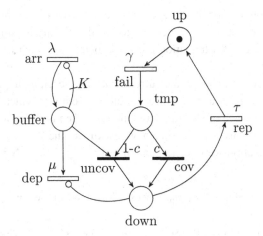

up

arr λ

γ fail

K

tmp

buffer

τ rep

$1-c$ c

dep μ uncov cov

down

Figure 12.2 Petri net of $M/M/1/K$ queue with breakdown and repair, and with covered and not-covered server failures.

In the present case, if the "uncov" transition fires, the current customer is removed from service.

Having examined informally some of the salient features of Petri nets, we will formally discuss the theoretical framework and analysis techniques in the subsequent sections.

12.2 From Petri Nets to Stochastic Reward Nets

In this section, we first review the basics of Petri nets. Thereafter we discuss how the PN is extended with timing to yield stochastic Petri nets, subsequently generalized stochastic Petri nets, and finally stochastic reward nets.

12.2.1 Petri Nets

The Petri net formalism and its theory were first conceived in the doctoral thesis of C. A. Petri in 1962 [1], hence the name. Since then, the formal language of PNs has been developed and used in many theoretical and practical application areas. A forum related to all aspects of PNs with an updated bibliographic database can be found at www.informatik.uni-hamburg.de/TGI/PetriNets/.

A PN is a bipartite directed graph whose nodes are divided into two disjoint sets called *places* and *transitions*. Directed arcs in the graph connect places to transitions (*input* arcs) and transitions to places (*output* arcs). A *marked* Petri net is obtained by associating *tokens* with places. A *marking* of a PN is a vector listing the number of tokens in the places of the PN. In a graphical representation of a PN, places are represented by circles, transitions are represented by bars or rectangles and tokens are represented by black dots or positive integers inside the places. The input places of a transition are the set of places that are connected to the transition using input arcs. Similarly, the output places of a transition are those places to which output arcs are drawn from the transition.

A transition is considered *enabled* in the current marking if each input place contains at least one token. The *firing* of a transition is an atomic action in which one token is removed from each input place of the transition and one token is added to each output place of the transition, possibly resulting in a new marking of the PN.

For definitions and notation we refer in general to [2]. A marked PN is a quintuple (P,T,I,O,M), where:

- $P = \{p_1, p_2, \ldots, p_{n_p}\}$ is the set of n_p places (drawn as circles in the graphical representation);
- $T = \{t_1, t_2, \ldots, t_{n_t}\}$ is the set of n_t transitions (drawn as bars or rectangles);
- I is the transition input relation and is represented by means of arcs directed from places to transitions;
- O is the transition output relation and is represented by means of arcs directed from transitions to places;
- $M = (m_1, m_2, \ldots, m_{n_p})$ is the initial marking. The generic entry m_i is the number of tokens (typically drawn as black dots) in place p_i in the marking M ($M \in \mathbb{Z}_{\geq 0}^{|P|}$).

Example 12.3   *Two Repairable Components Availability Model*

Returning to the two-server parallel redundant system considered in Example 9.4, we now construct the corresponding PN availability model as shown in Figure 12.3(a). In this figure, the places "up" and "down" represent the number of servers in the up and down states, respectively.

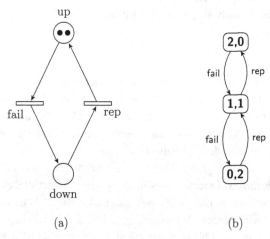

(a)   (b)

Figure 12.3   System with two repairable components. (a) PN availability model; (b) the corresponding reachability graph.

The servers are represented by the tokens in the places. The initial marking of the PN corresponds to both the servers being in the up state. The transitions "fail" and "rep," respectively represent the failure and repair events in the system.

The Reachability Set and Reachability Graph

Each distinct marking of a PN constitutes a separate state of the PN. A marking is reachable from another marking if there is a sequence of transition firings starting from the original marking that results in the new marking. In any marking of a PN, a number of transitions may be simultaneously enabled. Given an initial marking M_1, the *reachability set (graph)* $\mathscr{R}(M_1)$ of a PN is the set (graph) of markings that are reachable from marking M_1. The initial marking M_1 is the root of the reachability set. Starting from the root we search for all the enabled transitions; the firing of an enabled transition may produce a new marking which is represented as a new element in the reachability set, from which the procedure is iterated. More formally, we can say that t_k is enabled in marking M if:

$$\text{for any } p_i \in I(t_k), \quad m_i \geq 1.$$

The marking M' obtained from M by firing t_k is said to be *immediately reachable* from M, and the firing operation is denoted by the symbol $(M - t_k \rightarrow M')$. The token count in M' is given by the following relationship:

$$M'(p_i) = \begin{cases} M(p_i) + 1 & \text{if} \quad p_i \in O(t_k), \quad p_i \notin I(t_k), \\ M(p_i) - 1 & \text{if} \quad p_i \notin O(t_k), \quad p_i \in I(t_k), \\ M(p_i) & \text{otherwise.} \end{cases}$$

Note that it is certainly possible for $M' = M$. The reachability set can be represented as a labeled directed graph whose vertices are the elements of $\mathscr{R}(M_1)$ and such that for each possible transition firing $M_i - t_k \rightarrow M_j$ there exists an arc (i,j) labeled t_k. The reachability graph associated with a reachability set $\mathscr{R}(M_1)$ will be denoted by $\mathscr{G}_R(M_1)$. Methods for generating the reachability graph can be found in [2, 3], and have been implemented in many software packages [4, 5].

An *execution sequence* $\mathscr{E}$ in a marked PN is a sequence of legal markings obtained by firing a sequence of enabled transitions:

$$\mathscr{E} = \{(M_{(1)}, t_{(1)}); (M_{(2)}, t_{(2)}); \ldots; (M_{(j)}, t_{(j)}); \ldots\}.$$

An execution sequence $\mathscr{E}$ can be viewed as a connected path in the reachability graph $\mathscr{G}_R(M_1)$ of the PN.

Example 12.4 *Two Repairable Components Availability Model: Reachability Graph*
The reachability graph corresponding to the PN of Figure 12.3(a) is illustrated in Figure 12.3(b). The labels inside the states denote the corresponding marking, where

the first digit is the number of tokens in place "up" and the second digit the number of tokens in place "down." The arcs in the reachability graph are labeled with the PN transition whose firing causes the change of marking (state).

Problems

12.1 Redo the two-component PN availability model of Figure 12.3 for an n-component availability model. Also, generate the reachability graph from the PN that you constructed.

12.2.2   Structural Extensions to Petri Nets

In order to increase its modeling power and enable complex interactions to be compactly specified, extensions to the PN formalism have been considered in the literature. Some of the extensions are discussed below [6, 7].

Arc Multiplicity

A *multiplicity* may be associated with each arc, and is graphically indicated on the PN by crossing the arc with a dash labeled with an integer indicating the arc multiplicity. Consider a multiplicity k_i specified for an input arc. The transition on which this input arc with multiplicity k_i is incident will be considered enabled in the current marking if the number of tokens in the corresponding input place is at least equal to the multiplicity of the input arc k_i from that place. Furthermore, upon firing of the transition, k_i tokens are removed from the input place. Similarly, a multiplicity k_o associated with an output arc implies that upon firing the transition from which the output arc originates, the number of tokens deposited in the output place upon which the output arc is incident is equal to the multiplicity of the output arc k_o.

Example 12.5 *Two Repairable Components: Common Cause Failure*
Next, suppose that the two-component system of Example 12.3 is also affected by a common cause failure that acts simultaneously on the two components.

This new situation is represented in the PN of Figure 12.4(a) by an additional transition "ccf" with input place "up" and output place "down." The arcs connecting place "up" to transition "ccf" and from transition "ccf" to place "down" have a multiplicity equal to 2, so that the firing of transition "ccf" removes two tokens from the "up" place and deposits two tokens in the "down" place. As compared with the reachability graph of Figure 12.3(b), the reachability graph in Figure 12.4(b) has an additional arc from marking $(2,0)$ to marking $(0,2)$ labeled "ccf."

Inhibitor Arc

An *inhibitor* arc drawn from a place to a transition indicates that the transition is not enabled if the place contains at least as many tokens as the multiplicity of the inhibitor

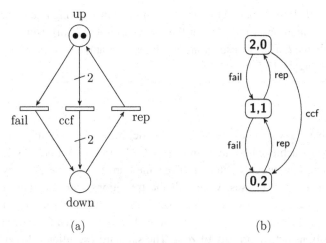

up

down

(a)   (b)

Figure 12.4   System with two repairable components and a common cause failure. Petri net model (a) and corresponding reachability graph (b) of the two components reliability model.

arc. Inhibitor arcs are graphically represented as arcs ending with a small circular head (rather than an arrow), and their multiplicity is specified with the same notation as for the normal arcs.

Example 12.6   *Queues with Breakdown (continued)*
In Figure 12.1 we can observe that there are two inhibitor arcs. The arc from place "down" to transition "dep" prevents the service from occurring when the server is down. The arc from place "buffer" to transition "arr," with multiplicity K, blocks new job arrivals when the buffer has reached its maximum capacity.

Example 12.7   *Two Repairable Components (continued)*
Consider now a modification to the PN of Example 12.3 as represented in Figure 12.5. In this case, the assumption is that if both the components have failed (and hence the

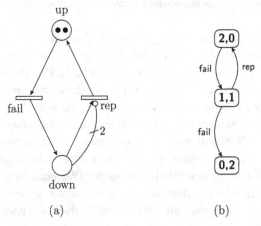

up

down

(a)   (b)

Figure 12.5   Petri net model (a) and corresponding reachability graph (b) of the two components reliability model.

system has failed), no repair action will be initiated thus giving rise to a PN reliability model. This condition is modeled in the PN of Figure 12.5(a) using an inhibitor arc with multiplicity 2. The corresponding reachability graph has an absorbing marking, as illustrated in Figure 12.5(b).

Priority Levels

An alternative, but equivalent, way [6] to model the same behavior as represented by inhibitor arcs is achieved by attaching to each PN transition a priority level, typically specified using non-negative integers. Thus, the transitions in the PN may be classified into different priority classes, where all the transitions with the same priority value belong to the same priority class. Whenever a transition with a priority k is enabled, all transitions with priorities less than k are automatically inhibited from firing. By default, all the transitions have a priority of zero. The standard execution rules are modified in the sense that, among all the transitions enabled in a given marking, only those at the highest priority level are allowed to fire.

Problems

12.2   Redo the two-component PN reliability model of Figure 12.5 for an n-component reliability model. Also, generate the reachability graph from the PN that you construct.

12.3   Generate the reachability graph for the $M/M/1/K$ queue with breakdown and repair from the PN shown in Figure 12.1.

12.2.3   Stochastic Petri Nets

Petri nets *per se* do not consider any notion of time, as they were meant for logical analysis. In order to use the PN formalism for quantitative analysis of the performance and reliability of systems, extensions to PNs have been considered by associating *firing times* with transitions. Early work in timed PNs with deterministic timing can be found in [8–11]. Applications of deterministic PN models are available in different areas, including communication protocols, performance evaluation and manufacturing.

However, stochastic modeling is more appropriate in dependability applications, and therefore we will consider from now on only timed PNs in which the timing mechanism is stochastic; we will refer to this class of models as stochastic PNs (SPNs). They were initially proposed in two doctoral theses [12, 13]. In these models, time was naturally associated with activities that induce state changes, hence with the delay incurred before firing transitions. Although other possibilities have been explored, the choice of associating time with PN transitions is the most common in the literature.

The delay between the time instant that a transition is enabled and the time it fires (in isolation) is the *firing time* of the transition. (We need to add the phrase "in isolation" since when several other concurrent transitions are enabled, firing of any of the concurrent transitions may disable the specific transition under discussion.)

When the firing time random variables associated with PN transitions are exponentially distributed, the dynamic behavior of the PN can be mapped into a homogeneous continuous-time Markov chain with state space isomorphic to the reachability graph of the PN. This case will be considered in detail in the following sections.

A timed execution sequence $\mathscr{T}_E$ of a marked PN starting from an initial marking $M_{(1)}$ is an execution sequence $\mathscr{E}$ augmented by a non-decreasing sequence of real values representing the epochs of firing of each transition, such that consecutive transitions $(t_{(j)}; t_{(j+1)})$ in $\mathscr{E}$ correspond to ordered epochs $\tau_j \leq \tau_{j+1}$ in $\mathscr{T}_E$. Thus, formally, [14, 15]:

$$\mathscr{T}_E = \{(M_{(1)}, t_{(1)}, \tau_1); (M_{(2)}, t_{(2)}, \tau_2); \ldots; (M_{(j)}, t_{(j)}, \tau_j); \ldots\}.$$

The time interval $\tau_{j+1} - \tau_j$ between consecutive epochs represents the period that the PN sojourns in marking $M_{(j)}$. In the following we always assume the initial epoch to be $\tau_1 = 0$.

Assume that the activity modeled by a PN transition takes an exponentially distributed random amount of time to complete once initiated. This means that an exponentially distributed random variable T_j with parameter $\lambda_j(M)$ is associated with each PN transition t_j. The firing time of an enabled transition t_j in marking M is a random variable with a time-independent (but possibly marking-dependent) firing rate $\lambda_j(M)$. Therefore, knowing the transitions enabled in a given marking and the associated firing rates, we can uniquely generate the stochastically timed sequence $\mathscr{T}_E$ from each execution sequence $\mathscr{E}$. In other words, the reachability graph $\mathscr{G}_R(M_1)$ of a marked PN can be mapped into a discrete-state continuous-time homogeneous Markov chain by letting each marking of $\mathscr{G}_R(M_1)$ correspond to a state in the Markov chain, and by substituting the label of the PN transition in each edge of $\mathscr{G}_R(M_1)$ with the firing rate of the corresponding transition. With this definition, we can speak without distinction between *marking M_i* and *state i*.

We can now formally define the SPN as a sextuple:

$$\text{SPN} = (P, T, I, O, M, L),$$

where P, T, I, O, M have the meanings introduced in Section 12.2.1, and $L = \{\lambda_1(M), \lambda_2(M), \ldots, \lambda_{n_t}(M)\}$ is a set of n_t non-negative real numbers representing the (possibly marking-dependent) firing rates of the exponentially distributed random variables associated with each PN transition. Note that in the formal definition of the SPN model the firing rates associated with each transition are considered *marking dependent*. This possibility increases the flexibility of the model and is often used to make the models more compact in the case of the presence of multiple identical resources.

From the reachability graph it is thus possible to automatically generate the infinitesimal generator matrix Q of the associated homogeneous CTMC. Here, Q is an $n \times n$ matrix, where n is the cardinality of the reachability set $\mathscr{R}(M_1)$. An updated list of tools that are available to help the user in editing, handling, and solving PNs is available at www.informatik.uni-hamburg.de/TGI/PetriNets/tools/db.html, where the main features of each tool are briefly described. In this chapter, we refer to the SPNP

[4, 16] and SHARPE [17] software packages for the purpose of automated generation of the CTMC starting with an SPN description.

Example 12.8 *Stochastic Petri Net Model of M/M/1/K Queue with Breakdown*
Returning to the PN model of the $M/M/1/K$ queue with breakdown of Figure 12.1, we now associate the rates λ, μ, γ, and τ with the transitions "arr," "dep," "fail," and "rep," respectively. With this, the PN now becomes an SPN. Given a value of K, we can then generate the corresponding reachability graph of the SPN, which will be isomorphic with the CTMC of Figure 11.18.

Example 12.9 *Two Repairable Components (continued)*
With reference to Figure 12.3, we assign to transition "fail" a marking-dependent firing rate given by #(up) λ, where #(up) indicates the number of tokens in place "up." Furthermore, by appropriate assignment of the rate associated with transition "rep," we can model the two repair policies of Example 9.4:

> *Non-shared (independent) repair:* The firing rate associated with transition "rep" is defined as #(down) μ. The corresponding SPN is shown in Figure 12.6(a). The CTMC derived from the reachability graph of Figure 12.3(a) is then isomorphic to the one in Figure 9.6.
>
> *Shared repair:* The firing rate associated with transition "rep" is independent of the marking and equal to μ. The corresponding SPN is shown in Figure 12.6(b). The CTMC derived from the reachability graph of Figure 12.3(b) is then isomorphic to the one in Figure 9.7.

Example 12.10 *Stochastic Petri Net Model of M/M/m/K Queue*
The model of the $M/M/m/K$ queue illustrates the modeling power of SPN using the marking-dependent firing rate. Interestingly, the structure of the SPN model of the

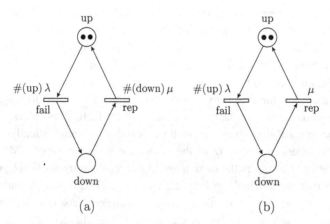

Figure 12.6 Two repairable component system SPN availability model: (a) independent repair, (b) shared repair.

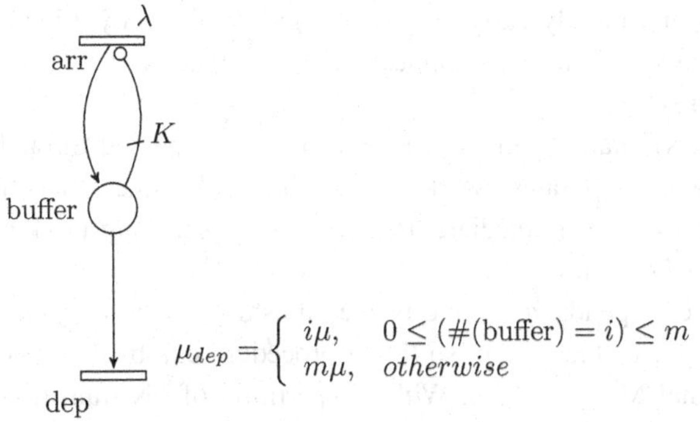

Figure 12.7   Stochastic Petri net model of the $M/M/m/K$ queue.

$M/M/m/K$ queue, shown in Figure 12.7, is identical to the performance part of the SPN model of the $M/M/1/K$ queue with breakdown reported in Figure 12.1. To complete the specification, define the firing rate μ_{dep} of transition "dep" as follows:

$$\mu_{\text{dep}} = \begin{cases} i\mu & 0 \le (\#(\text{buffer}) = i) \le m, \\ m\mu & \text{otherwise,} \end{cases}$$

where #(buffer) represents the number of tokens in the place "buffer."

This specification indicates that the firing rate of the transition "dep" is proportional to the number of tokens in the buffer (corresponding to the number of customers in the queue) below m, and is constant at the rate $m\mu$ thereafter.

Problems

12.4   Develop an SPN model of an $M/M/m/K$ queue with breakdown and repair, essentially combining the ideas from Examples 12.8 and 12.10. Generate the reachability graph by hand and thence the underlying CTMC.

12.5   Develop SPN availability models of n repairable components, thus generalizing the two models of Example 12.9. Generate the reachability graphs by hand and thence the underlying CTMCs.

12.6   Develop an SPN reliability model of n repairable components, thus generalizing the model of Example 12.7. Generate the reachability graph by hand and thence the underlying CTMC with an absorbing state.

12.2.4   Generalized Stochastic Petri Nets

The generalization of SPNs introduced by Ajmone Marsan *et al.* [3] allows transitions to have either zero firing times (*immediate* transitions) or exponentially distributed firing times (*timed* transitions), giving rise to a model called generalized stochastic Petri nets (GSPNs). The starting assumption in the GSPN model [3] is that it is desirable to associate a zero time with those transitions that do not have a significant computation attached to them and merely carry out some logic. Graphically, timed transitions are typically represented by empty rectangles while immediate transitions are represented by thin black bars.

Markings (states) enabling immediate transitions are passed through in zero time and are called *vanishing* states. Markings enabling only timed transitions are called *tangible*. Conflicts among immediate transitions in a vanishing marking are resolved using a *random switch* [3].

Since the process spends zero time in the vanishing states, they do not contribute to the time behavior of the system so that a procedure can be envisaged to eliminate them from the final Markov chain. With the partition of PN transitions into a timed and an immediate class, we introduce greater flexibility in the modeling power without increasing the dimensions of the final state space of the underlying CTMC. Analysis of a GSPN involves construction of the underlying reachability graph, referred to as the extended reachability graph (ERG), which contains both tangible and vanishing states. An arc of the ERG that is associated with the firing of an immediate transition will have a probability of firing, while an arc that is associated with a timed transition will have a firing rate. Thus the ERG is neither a DTMC nor a CTMC, but it is possible to treat the ERG as a semi-Markov process (see Chapter 14). Solution of an ERG is possible in the SPNP software package using the *preservation* option, which treats the ERG as an SMP [4, 16]. The alternative is to systematically eliminate the vanishing markings, thus producing a CTMC. This is the most common procedure for solving a GSPN, as originally proposed in [3]; in SPNP you can exercise this option by choosing *elimination* [4, 16].

Given a marking $M \in \mathscr{G}_R(M_1)$ of a GSPN, three different situations may arise:

Situation 1: Only timed transitions are enabled – Figure 12.8(a) – so that only tangible markings are generated. The reachability graph in Figure 12.8(b) is already a CTMC.

Situation 2: One or more timed transitions are enabled simultaneously with one immediate transition t_2 – Figure 12.9(a). Only the immediate transition is allowed to fire, generating the ERG of Figure 12.9(b). However, marking M_2 is vanishing and can be eliminated, producing the reduced reachability graph that gives rise to the CTMC of Figure 12.9(c), in which all the markings are tangible.

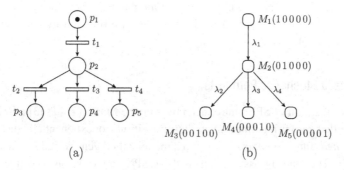

(a)   (b)

Figure 12.8   (a) Stochastic Petri net with timed transitions only; (b) The corresponding CTMC.

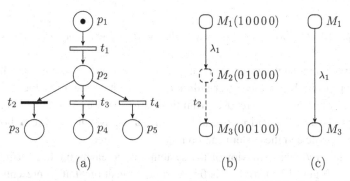

Figure 12.9 (a) Stochastic Petri net with one immediate transition; (b) The ERG; (c) The reduced RG that is a CTMC defined over tangible markings.

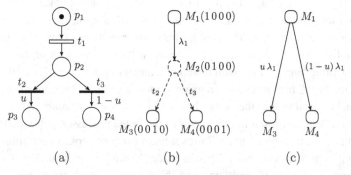

Figure 12.10 (a) Stochastic Petri net with multiple immediate transitions; (b) The ERG; (c) The reduced RG that is a CTMC defined over tangible markings.

Situation 3: Several immediate transitions are simultaneously enabled in a marking. In this case, in order to determine which immediate transition fires first, a probability mass function needs to be specified: in the language of GSPN this construct is called a *random switch*. In Figure 12.10(a) the immediate transition t_2 fires with probability u, and the immediate transition t_3 with the complementary probability $1 - u$. The ERG is shown in Figure 12.10(b). Marking M_2 is vanishing and can be eliminated, incorporating the switching probabilities into the rates leading to marking M_2. Elimination of the vanishing marking leads to the CTMC of Figure 12.10(c), which contains only tangible markings as its states.

Algorithms discussed in [3, 5, 18] eliminate vanishing markings to produce a reduced reachability graph. This reduced reachability graph will thus be a homogeneous CTMC, defined over tangible states only. The reduction procedure is automatically activated by the software packages and becomes completely transparent to the modeler.

Example 12.11   *Two-Component Reliability Model with Imperfect Coverage (continued)*

The two-component system in Example 10.18 introduces a coverage factor into the failure process. When the server failure occurs, with probability c the failure is covered and the system is able to recover from this failure. However, with probability $1 - c$, the server failure causes an irrecoverable failure of the system. Failure of both the servers is catastrophic and the system can no longer be repaired.

The failure/repair behavior of the system can be easily modeled using a GSPN, as shown in Figure 12.11(a). In this figure, timed transition "fail" represents the failure of a server. The firing rate of this transition, indicated by #(up)λ, is a marking-dependent firing rate. Here, #(up) represents the number of tokens in place "up," that is, the number of servers still functioning. When transition "fail" fires, it removes a token from its input place "up" and deposits a token in its output place "tmp." At this point the two immediate transitions "cov" and "uncov" are enabled. We need to define a random switch to resolve the conflict between transitions "cov" and "uncov." This is accomplished by declaring that the transition "cov" fires with probability c and that the transition "uncov" fires with the complementary probability $1 - c$. If the immediate transition "cov" fires it removes a token from place "tmp" and deposits a token in place "down," indicating that the server has suffered a covered failure. At this stage, timed transition "rep," representing the repair of the failed processor, gets enabled. If, instead, the immediate transition "uncov" fires it takes away one token each from places "tmp" and "up" and deposits two tokens in place "down," as indicated by the single input arcs from places "tmp" and "up" and the multiple output arc to place "down." This represents a not-covered failure in the system. We notice that when both the servers have failed, the repair transition is disabled by the multiple inhibitor arc from place

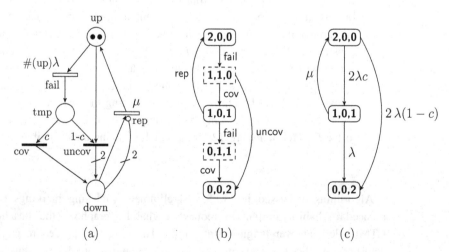

(a)   (b)   (c)

Figure 12.11   Two-component reliability model with repair and imperfect coverage. (a) The GSPN model; (b) the corresponding ERG; (c) the reduced reachability graph that is a CTMC.

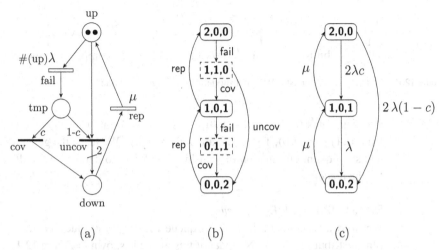

Figure 12.12 Two-component availability model with full repair and coverage. (a) The GSPN model; (b) the corresponding ERG; (c) the reduced CTMC.

"down" to transition "rep." This represents the fact that the failure of both the processors is catastrophic. The corresponding ERG is also illustrated, in Figure 12.11(b). Tangible markings are represented in the reachability graph as rounded rectangles, and vanishing markings are represented by dashed rectangles. The state label (i,j,k) represents the marking and gives the number of tokens in places "up," "tmp," and "down," respectively. This example can also be viewed as a modification of Example 12.7. By applying the reduction rules for eliminating the vanishing markings, the CTMC of Figure 12.11(c) is obtained, which is the same as that depicted in Figure 10.14.

Example 12.12 *Two-Component Availability Model with Imperfect Coverage*
Continuing the two-component system in Example 12.11, we now assume that the system can be repaired even when both the servers have failed. The CTMC availability model for this example was considered in Example 9.5, where the steady-state availability was computed. Thus, we remove the multiple inhibitor arc from place "down" to transition "rep" in Figure 12.11(a), yielding the GSPN shown in Figure 12.12(a). The corresponding ERG is also illustrated, in Figure 12.12(b), and the CTMC resulting after the elimination of vanishing markings is shown in Figure 12.12(c) – compare this CTMC with Figure 9.8. This is an irreducible CTMC that was solved to obtain the system availability in Example 9.5.

Example 12.13 *Queue with Server Failure, Repair and Imperfect Coverage*
Returning to the $M/M/1/K$ queue with breakdown and considering not-covered server failures, we see that the corresponding PN model shown in Figure 12.2 is indeed a GSPN model of the system behavior that uses the immediate transitions "cov" and "uncov" to represent the covered and not-covered failure of the server. The modification does not introduce any additional states in the CTMC, only additional transitions

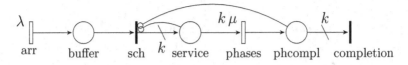

Figure 12.13   Generalized stochastic Petri net model of the $M/E_k/1$ queue.

between states $(i, 1) \longrightarrow (i - 1, 0)$, $1 \leq i \leq K$, with rate $(1 - c)\gamma$, while the transitions from $(i, 1) \longrightarrow (i, 0)$, $1 \leq i \leq K$, are at the rate $c\gamma$. We leave the generation of the ERG and subsequent elimination of vanishing markings to derive the CTMC as an exercise for the reader.

Example 12.14   *$M/E_k/1$ Queue*

We study a variation of the $M/M/1$ queue where we introduce a k-stage Erlang service time distribution. A GSPN model of this queue is shown in Figure 12.13.

When the transition "arr" fires, a token is deposited in place "buffer" indicating a new job arrival. When there is no job in service, indicated by no tokens in place "service" and place "phcompl," the immediate transition "sch" will be enabled and upon firing will remove one token from place "buffer" and deposit k tokens at once in place "service." This is to indicate that k phases each with rate $k\mu$ are to be executed in sequence for a single job to complete. Each time the transition "phases" fires, one phase of job execution finishes. Once k tokens accumulate in place "phcompl," job execution has finished and transition "completion" will fire, clearing the place "phcompl" of all tokens. Before completion, the inhibitor arc from place "phcompl" to transition "sch" prevents a new customer proceeding to service, but once place "phcompl" is cleared, transition "sch" is enabled if there are jobs (tokens) waiting in place "buffer."

Problems

12.7   Develop a GSPN reliability model of n repairable components with imperfect coverage, thus generalizing the model of Example 12.11. Generate the extended reachability graph by hand, eliminate vanishing markings and thence produce the underlying CTMC with an absorbing state.

12.8   For the GSPN model of the $M/M/1/K$ queue with failure, repair, and imperfect coverage shown in Figure 12.2, generate the ERG, eliminate vanishing markings and thus produce the CTMC.

12.9   Solve (in closed form) the steady-state availability of the two-component system by considering the ERG of Figure 12.12(b) as an SMP – that is, by using the preservation method. Compare the results obtained by solving the CTMC model of Figure 12.12(c).

12.10   Generate the reachability graph by hand for the GSPN model of the $M/E_k/1$ queue (Figure 12.13), eliminate vanishing markings and thus produce the underlying irreducible CTMC.

12.3   Stochastic Reward Nets

Although GSPNs provide a useful high-level language for evaluating large systems, representation of the intricate behavior of such systems often leads to large and complex structures in the GSPN. To alleviate some of these problems, several structural extensions to Petri nets were introduced in the literature [5, 18, 19] to increase the modeling power of PNs. We refer, in particular, to the stochastic reward net (SRN) model defined in [18]. The SRN formalism extends the GSPN formalism by including guard functions, very general marking dependency, variable cardinality arcs and the superposition of a reward structure. From an SRN, a Markov reward model is thus generated, facilitating the specification and computation of important reward-based metrics. Furthermore, quite often the model structure also becomes much simpler due to the availability of the reward specification at the net level [20] and due to guard functions and variable cardinality arcs. Stochastic reward net models developed in several papers illustrate the power of SRNs [21–31]. To emphasize the importance of reward definition at the net level, we will consider this topic separately in Section 12.4.

Guard Function

A (Boolean) guard function $\mathscr{G}(\cdot)$ can be associated with a transition. Whenever a transition satisfies all the input and inhibitor conditions in a marking M, i.e., the input places contain a sufficient number of tokens to enable the transition and the number of tokens in the inhibitor places is less than the multiplicity of the corresponding inhibitor arc, the guard function is evaluated. The transition is considered disabled unless the guard function $\mathscr{G}(M) = \text{TRUE}$. Guard functions are very useful in expressing complex interdependencies and simplifying the model structure [31–33]. Guard functions are also useful in implementing state truncation as a method of dealing with very large state spaces [34]. Guard functions are also known as *enabling functions* since they can modify the standard enabling rules by specifying conditions that must be satisfied for the transition to be enabled.

Variable Cardinality (or Multiplicity) Arc

The cardinality (or multiplicity) of an arc in an SRN can be expressed as a function of the SRN marking. This facility is useful in simplifying the structure of the PN. It is very useful especially if an input place needs to be flushed. This facility can be used with input, output and inhibitor arcs [31–33].

General Marking Dependency

The rates and probabilities of transitions can also be defined as general functions of the marking of an SRN. A simple example where this was found to be very useful is in the multiple server queue of Example 12.10. For other examples see [31].

Example 12.15   *GSPN Reliability Model of a Series-Parallel System*
Generalized stochastic Petri nets have a modeling power superior to any non-state-space

model and are equal in modeling power to CTMC [2, 6]. Hence, automated algorithms can be used to convert a dependability model from a less powerful language into a GSPN [20, 35]. As an example, we return to the series-parallel system modeled as an RBD in Example 4.11, and whose FT is given in Figure 6.11. Using the algorithmic approach of transforming the FT to a GSPN given in [20], we derive the corresponding GSPN, shown in Figure 12.14. In this model, the timed transitions represent the failure behavior of the five components. The immediate transitions and the corresponding places are used to encode the conditions under which the system fails.

The TE of the FT represents system failure, and was derived in Eq. (6.6) as

$$\text{TE} = ((\overline{A} \wedge \overline{B}) \vee \overline{C}) \wedge (\overline{D} \vee \overline{E}).$$

The presence of a token in the place Sys_{dn} represents the TE, i.e., the condition indicating that the system is down.

Example 12.16 *SRN Reliability Model of a Series-Parallel System*
In the GSPN model of Example 12.15, we noted that the tree-like structure of places and immediate transitions was used to determine conditions under which the system fails, thus depositing a token in place Sys_{dn} in the GSPN of Figure 12.14. Once the system is down, as indicated by a token in place Sys_{dn}, we wished to disable all the (timed) failure transitions – this is done by explicit inhibitor arcs in the GSPN model. Second, we could then compute the system failure probability by asking for the probability of a token in place Sys_{dn}.

In an SRN we have the possibility of guard functions, so we can disable the failure transitions based on a marking-dependent guard function without needing an explicit Sys_{dn} place that necessitates a tree of immediate transitions and corresponding places. Furthermore, due to the possibility of reward definition at the net level, we can ask for the computation of the system failure probability without having an explicit place Sys_{dn}.

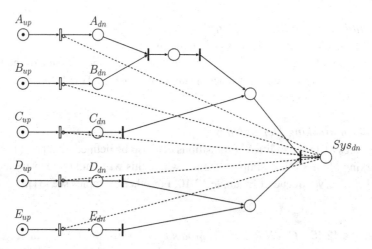

Figure 12.14   Generalized stochastic Petri net model of the RBD from Example 4.11.

Table 12.1 Boolean functions for the SRN in Figure 12.15.

Gate	Boolean function
G_3	$(\#(A_{dn}) == 1) \wedge (\#(B_{dn}) == 1)$
G_2	$(\#(C_{dn}) == 1) \vee G_3$
G_4	$(\#(D_{dn}) == 1) \vee (\#(E_{dn}) == 1)$
G_1	$G_2 \wedge G_4$

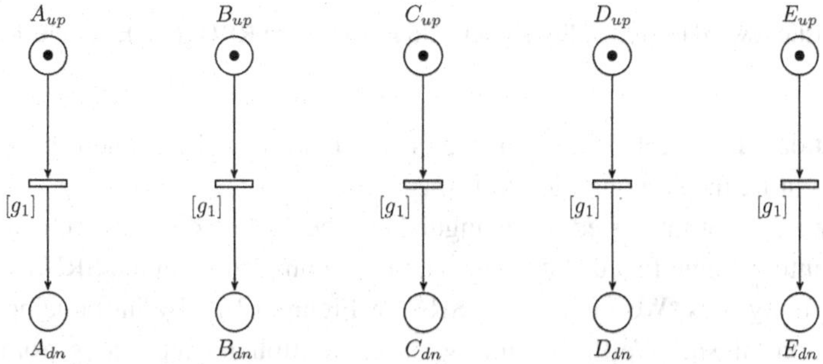

Figure 12.15 Stochastic reward net model of the RBD from Example 4.11.

We now present the SRN model of the system that is obtained through the FT-to-SRN conversion algorithm given in [20]. The SRN model is shown in Figure 12.15.

The simplicity of the structure of the SRN model compared to the GSPN model presented in Figure 12.14 is immediately apparent to the reader. The complexity of the system failure conditions are now captured by the Boolean functions presented in Table 12.1. The Boolean functions capture the essence of the FT model of the system.

When the system is down, the failure transitions should all be disabled since no failures can occur in this situation. This can be accomplished by assigning the guard function $[g_1] =!G_1$ to all the failure transitions of the SRN. Without this restriction, the state space of the model will contain 32 (= 2^5) states. With the guard function on the failure transitions, no new markings are generated once the system is down. We note that there is a shortcut for specifying such global guard functions using the function called *halting condition* in the SPNP software package [5, 16].

Example 12.17 *SRN Reliability Model with Repair of a Series-Parallel System (continued)*
We now modify the previous series-parallel system of Figure 12.15. In this modification, shown in Figure 12.16, we assume that each component is independently repairable upon failure. Furthermore, we assume that if the system fails, then it can no longer be repaired. Thus, when the system reaches a failed state, all repairs are halted. This can be accomplished in the SRN by associating the guard function $[g_1]$ with all the transitions. The repair transitions get disabled when $[g_1]$ evaluates to FALSE. Thus, with the guard function on the transitions, some of the markings become absorbing markings, i.e., no

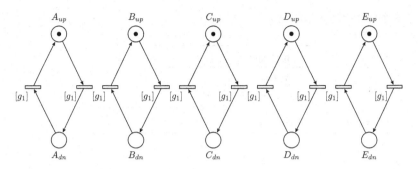

Figure 12.16   Stochastic reward net reliability model with repair of the RBD from Example 4.11.

transitions are enabled in them. Of the 31 markings, 14 of them become absorbing, representing the system failed states.

If we are not interested in distinguishing the 14 failed states, we can collapse all of them into a single failed state. This can be accomplished in the SRN by using variable multiplicity arcs. We modify the SRN in Figure 12.16 by introducing an immediate transition "flush." We then define variable multiplicity input arcs from all the places to this transition, with multiplicity equal to the number of tokens in the corresponding input place. Variable multiplicity is indicated on an arc by putting a zigzag sign on it. Furthermore, the immediate transition has a guard function $[g_2] = G_1$ associated with it. This immediate transition is thus enabled upon system failure and accomplishes the *flushing* of all the places of their tokens, thus resulting in a single marking representing the system failure condition. The modified SRN is shown in Figure 12.17. With this modification, the *truncated* state space now consists of only 18 markings, including one absorbing marking that represents the system failed state.

Example 12.18   *SRN Availability Model of a Redundant Repairable Fluid Level Controller*

We return to Example 9.10, but now we are prepared to model the entire system consisting of four subsystems that share a repair person. Recall that drawing the composite monolithic CTMC for the system as a whole will be unwieldy, and hence we resort to the higher-level specification afforded by SRNs. The overall SRN model generating the state space of the system as a whole for Example 9.10 is shown in Figure 12.18 [36].

The SRNs modeling the CTMCs of the individual subsystems are shown from right to left in order of their repair priority, and the correspondences with the subsystems modeled in Figure 9.14 are indicated at the top of the figure. The additional guard functions and the marking-dependent rate function for the failure transition P_F needed to completely specify the model are reported in the same figure.

Example 12.19   *A Redundant Repairable Fluid Level Controller with Travel Time*

When the travel time of the repair person is considered, as in Example 9.11, the SRN generating the monolithic CTMC for the model is shown in Figure 12.19, where we

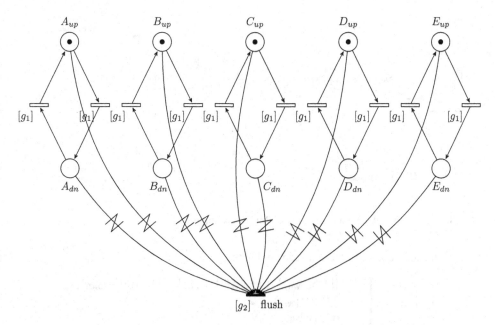

Figure 12.17 Reduced SRN reliability model with repair of the RBD from Example 4.11.

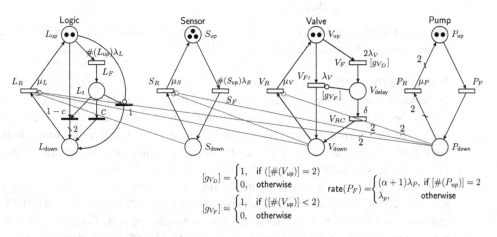

$$[g_{V_D}] = \begin{cases} 1, & \text{if } ([\#(V_{up})] = 2) \\ 0, & \text{otherwise} \end{cases}$$

$$[g_{V_F}] = \begin{cases} 1, & \text{if } ([\#(V_{up})] < 2) \\ 0, & \text{otherwise} \end{cases}$$

$$\text{rate}(P_F) = \begin{cases} (\alpha+1)\lambda_P, & \text{if } [\#(P_{up})] = 2 \\ \lambda_P, & \text{otherwise} \end{cases}$$

Figure 12.18 Stochastic reward net generating the overall CTMC for Example 9.10.

have used the SRN in Figure 12.18 with an extra subnet to indicate the on-site presence or absence of the single repair person [36].

The transition t_{arr} is enabled when any of the subsystems is needing a component repair, and thence the repair person arrives with exponentially distributed time delay with rate μ_t. The immediate transition t_{depart} is enabled when the repairs for all the subsystems are completed.

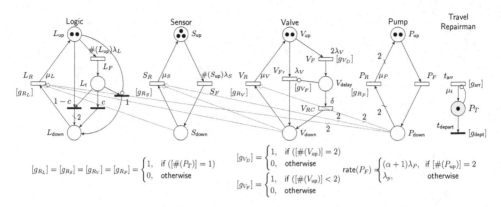

Figure 12.19 Stochastic reward net generating the CTMC of Example 9.11.

Hence, the guard functions can be defined as:

$$[g_{arr}] = \begin{cases} 1 & ([\#(S_{down})] \geq 1) \parallel ([\#(L_{down})] \geq 1) \parallel ([\#(V_{down})] \geq 1) \\ & \parallel ([\#(P_{down})] \geq 2), \\ 0 & \text{otherwise;} \end{cases}$$

$$[g_{depart}] = \begin{cases} 1 & ([\#(S_{down})] = 0) \;\&\; ([\#(L_{down})] = 0) \;\&\; ([\#(V_{down})] = 0) \\ & \&\; ([\#(P_{down})] = 0), \\ 0 & \text{otherwise.} \end{cases}$$

The monolithic CTMC can be automatically generated from the SRN to provide a value for the steady-state availability: $A = 0.997\,431\,402\,583$.

Example 12.20 *MMPP/M/1/K Queue*

In elementary queuing theory, the assumption of a Poisson arrival process is often made and has been rightly questioned by practitioners both on the grounds of exponentially distributed inter-arrival times and independence of successive inter-arrival times. Both these assumptions can be lifted at once by using the notion of a Markov-modulated Poisson process (MMPP) as an arrival process [37].

We have chosen a simple two-state MMPP as an arrival source. This is then a two-state CTMC drawn as an SRN, with two places labeled "MMPP1" and "MMPP2" and two timed transitions with rates α and β, shown in the top part of Figure 12.20. The bottom part of the figure is very similar to that of an $M/M/1/K$ queue (without failures), except that the firing rate of the arrival transition "arr" is now marking dependent and is specified as $\#(MMPP1)\lambda_1 + \#(MMPP2)\lambda_2$. In other words, the arrival rate at the queuing system is λ_1 when the MMPP is in state 1 (a token in place "MMPP1") and it is equal to λ_2 when the MMPP is in state 2 (a token in place "MMPP2").

Example 12.21 *SRN Model of a Multi-Server Retrial Queue*

Gharbi and Dutheillet [38] presented a model of a multi-server retrial system with K customers and m identical and parallel servers. They presented GSPN models of the system considering different failure rates for idle and in-service servers. We now present

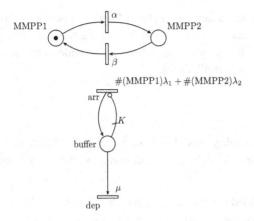

Figure 12.20   Stochastic reward net model of the *MMPP/M/1/K* queue.

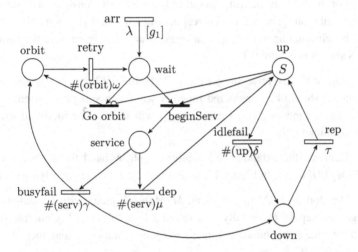

Figure 12.21   Stochastic reward net model of the multi-server retrial queue [38].

an SRN model of the system in Figure 12.21. The model assumes that when a customer arrives (a token deposited in place "wait"), if there are idle servers (indicated by tokens in place "up") then the customer proceeds to be served and completes service upon firing of the transition "dep." On the other hand, if a customer in place "wait" finds no idle server (indicated by the inhibitor arc from place "up" to the immediate transition "Go orbit") then the customer goes into an orbit and retries after a delay with mean $1/\omega$. Furthermore, a server currently serving a customer may fail at the rate γ, which is different from the idle server failure rate of δ. The repair rate of the server is τ and is marking independent, indicating shared repair. This model is a generalization of the *M/M/m/K* queue with retrials. It is interesting to note the structural similarities of the SRN model in Figure 12.21 with the SRN model in Figure 12.1.

In this SRN, we define a guard function $[g_1] = ((\#(\text{wait}) + \#(\text{service}) + \#(\text{orbit})) < K)$ associated with the arrival transition "arr." Thus, customer arrivals are disabled when the total number of customers in the system equals K. As is evident, the customers could be in the "wait," "service" or "orbit" places.

Problems

12.11 For the GSPN reliability model of Example 12.15, shown in Figure 12.14, construct the ERG by hand and subsequently eliminate vanishing markings to obtain the CTMC.

12.12 Generate the extended reachability graph by hand for the SRN reliability model of Example 12.17, first for the version without the transition "flush" (Figure 12.16) and then with the transition "flush" (Figure 12.17). In each case, eliminate vanishing markings to obtain the underlying CTMC.

12.13 For the SRN availability model of Figure 12.18, consider each subsystem SRN model in isolation, generate the corresponding ERG and subsequently the underlying CTMC by eliminating vanishing markings, if any. Compare with the four subsystem CTMCs shown in Figure 9.14.

12.14 Repeat the exercise suggested in Problem 12.13 except that the repair person travel subnet should be considered together with each subsystem subnet. Needless to say, the guard functions $[g_{\text{arr}}]$ and $[g_{\text{depart}}]$ will have to be modified to refer to the respective subnet only.

12.15 Generate the extended reachability graph by hand for the SRN model of the $MMPP/M/1/K$ queue in Figure 12.20 and hence derive the underlying CTMC.

12.16 Develop an SRN model of an $M/M/m/K$ queue with breakdown, imperfect coverage and repair, essentially generalizing Example 12.13. Generate the extended reachability graph by hand, eliminate vanishing markings and thence generate the underlying CTMC.

12.17 Generate the extended reachability graph by hand for the retrial queue SRN model of Example 12.21, eliminate vanishing markings and thence generate the underlying CTMC. Use $m = 1$ and $K = 3$.

12.18 Modify the SPN model (Figure 12.1) of Example 12.1 so that the server failure rate is γ_i when it is idle and γ_b when busy.

12.4   Computing the Dependability and Performance Measures

The primary goal of modeling the dynamic system behavior through an SRN is that it facilitates the definition and computation of a large number of performance and dependability metrics related to the system [39, 40]. Given that the analysis of an SPN (GSPN, SRN) involves the construction of the underlying reachability graph, the stochastic behavior of an SPN can be determined by calculating the occurrence probabilities over the states of the reachability set $\mathscr{R}(M_1)$ (and correspondingly for the underlying CTMC). It is clearly desirable that the output measures be defined at the net level, and the numerical computation carried out automatically by solving the underlying CTMC and then computing various performance and dependability measures. Several interesting metrics can be defined and associated with net-level entities like the places and transitions of the SRN. For example, we could talk in terms of the probability of a place being non-empty, the number of times a transition fires in a unit time or the amount of time spent by the SRN in any particular marking. Given that we have computed the steady-state probability vector π, the transient state probability vector $\pi(t)$, the cumulative probability vector $b(t)$ or the vector τ of mean times spent in each of the transient states before absorption of the CTMC associated with the tangible reachability set of the GSPN/SRN, the desired measure is often a weighted average of one of the above vectors.

We have already seen in Section 9.4 how Markov reward models provide a general framework to specify various performance and dependability measures of interest. Given that an SPN provides a compact formalism to specify the system behavior and automatically generate the underlying Markov chain, it behooves us to consider a concise formalism to specify the reward-based method of computing interesting measures. This is exactly where stochastic reward nets (SRNs) [4, 5] exhibit the full potential of a reward-based formalism.

In an SRN, the reward rate definitions are specified at the net level as functions of net primitives like the number of tokens in a place or the rate of a transition. In the rest of this section, #(p) represents the number of tokens in place p, *enabled*(t) is a Boolean function which returns a 1 if the transition t is enabled, and *rate*(t) returns the rate of a timed transition t, all in a marking M. For each marking of the SRN, the reward rate function is evaluated and the result is assigned as the reward rate for that marking. The underlying Markov model is now extended into an MRM, thus permitting the computation of various performance and dependability measures using the reward-based formalism. Thus, from the net-level reward function, a reward rate r_i is automatically computed for every tangible marking i. Suppose X represents the random variable corresponding to the reward rate in steady state, then the expected steady-state reward rate $E[X]$, see Eq. (9.55), can be computed as

$$E[X] = \sum_i r_i \pi_i, \tag{12.1}$$

where π_i is the steady-state probability of tangible marking i (recall that the probability of a vanishing marking is zero). We can not only perform steady-state analysis of SRNs but also transient and cumulative transient analysis. Thus, the expected value of the reward rate as a function of time, Eq. (9.54),

$$E[X(t)] = \sum_i r_i \pi_i(t), \tag{12.2}$$

and the expected value of the accumulated reward, Eq. (9.58),

$$E[Y(t)] = \sum_i r_i b_i(t) = \sum_i r_i \int_0^t \pi_i(u) du, \qquad (12.3)$$

can also be computed.

It must be noted that the definition of reward rates is orthogonal to the analysis type that is used. Thus, with the same reward definition we can compute the steady-state expected reward rate as well as instantaneous reward rate at time t, expected accumulated reward, and expected time-averaged reward over the interval $(0, t]$. Furthermore, more than one reward function can be (and usually is) defined for a given SRN. We can also compute the expected accumulated reward up to absorption in the case that there are absorbing markings:

$$E[Y(\infty)] = \sum_i r_i \tau_i. \qquad (12.4)$$

Furthermore, derivatives of all the above measures with respect to any or all input parameters can also be computed [4, 5, 41].

Example 12.22 *Queue with Breakdown and Repair*

Returning again to Example 12.1 and the SRN shown in Figure 12.1, let us examine how several interesting measures can now be specified using the reward-based formalism for the SRN. Table 12.2 lists several queue-level measures and the corresponding SRN-level reward rate functions that will enable the computation of the specified measures. Using the reward function in the first row of Table 12.2 and specifying steady-state solution to SPNP, we can obtain the steady-state unavailability – an application of Eq. (12.1). We can also ask SPNP to compute the derivative of the steady-state unavailability with respect to various input parameters such as λ, μ, γ, and τ. With the same reward specification, we can direct SPNP to carry out a transient analysis and hence compute the unavailability at time t – an application of Eq. (12.2) – or its derivatives with respect to the input parameters. We can direct SPNP to carry out a cumulative transient analysis and hence compute the expected downtime in the interval $(0, t]$ – an application of Eq. (12.3) – or its derivatives with respect to the input parameters. The reward function in the second row of Table 12.2 will enable us to compute the expected number of jobs in the system at steady state or at time t, as well as their derivatives. The reward function in the third row of Table 12.2 can be used to compute the steady-state or transient probability of an incoming job being rejected. A cumulative transient analysis using the reward function in the fourth row of Table 12.2 will yield the expected number of jobs rejected in the interval $(0, t]$. Cumulative transient analysis with the reward specification in the last row of Table 12.2 can be used to compute the expected number of repairs needed in the interval $(0, t]$. In order to compute the mean response time in the steady state, we first compute the mean number in the system (using the reward rate in the second row of the table), then compute the steady-state throughput (using the reward function in the sixth row of the table), and then divide the former by the latter.

Table 12.2 SRN-level reward rate specification of measures for the $M/M/1/K$ queue with breakdown.

Measure	Reward rate assignment for marking i
Server unavailability	$r_i = \begin{cases} 1 & \text{if } (\#(\text{down}) == 1) \\ 0 & \text{otherwise} \end{cases}$
Expected no. of jobs in system	$r_i = \#(\text{buffer})$
Prob. of job rejection	$r_i = \begin{cases} 1 & \text{if } (\#(\text{buffer}) == K) \\ 0 & \text{otherwise} \end{cases}$
Rate of job rejection	$r_i = \begin{cases} \lambda & \text{if } (\#(\text{buffer}) == K) \\ 0 & \text{otherwise} \end{cases}$
Server utilization	$r_i = \begin{cases} 1 & \text{if } ((\#(\text{buffer}) > 0) \;\&\&\; (\#(\text{up}) == 1)) \\ 0 & \text{otherwise} \end{cases}$
Throughput	$r_i = rate(\text{dep})$
Rate of repairs needed	$r_i = rate(\text{rep})$

Example 12.23 *Queue with Breakdown, Repair and Imperfect Coverage*
For Example 12.2, all the reward functions given in Table 12.2 can be used in the same way, but in addition we can use the reward function $r_i = (1 - c) \times enabled(\text{dep})$ to compute the probability of a job in service being dropped due to a failure. We can also use the reward function $r_i = \gamma(1 - c) \times enabled(\text{dep})$ to compute the rate at which a job in service is dropped due to a failure. With a cumulative transient analysis, with the latter reward definition, we can also compute the expected number of jobs in service dropped (due to a server failure) in the interval $(0, t]$.

Example 12.24 *Two-Component Reliability Model with Imperfect Coverage (continued)*
Returning to Example 12.11, different reward assignments are shown in Table 12.3. With the reward assignment in the first row of Table 12.3, we can compute the system reliability at time t using the transient analysis option (Eq. (12.2) is applied) or the mean time to system failure using the mean time to absorption (MTTA) option (Eq. (12.4) is applied). With the reward assignment in the second row of Table 12.3, we can compute the expected system capacity at time t using the transient analysis option (Eq. (12.2) is applied) or the mean capacity delivered until system failure using the MTTA option (Eq. (12.4) is applied). With the reward assignment in the third row of Table 12.3, we can compute the expected number of failures covered in the interval $(0, t]$ using the cumulative transient analysis option (Eq. (12.3) is applied) or the mean number of covered failures until system failure using the MTTA option (Eq. (12.4) is applied). With the reward assignment in the last row of Table 12.3, we can compute the expected number of repairs carried out in the interval $(0, t]$ using the cumulative transient analysis option (Eq. (12.3) is applied) or the mean number of repairs carried out until system failure using the MTTA option (Eq. (12.4) is applied). The derivatives of these measures can also be computed.

Example 12.25 *Two-Component Availability Model with Imperfect Coverage (continued)*
For the two-component availability model of Example 12.12, we utilize the same reward assignments shown in Table 12.3. With the reward assignment in the first row

Table 12.3 SRN-level reward rates for the two-component reliability model with imperfect coverage.

Measure	Reward rate assignment for marking i
System up	$r_i = \begin{cases} 1 & \text{if } (\#(\text{down}) == 0) \\ 0 & \text{otherwise} \end{cases}$
Capacity	$r_i = \#(\text{up})$
No. of covered failures	$r_i = c \times rate(\text{fail})$
Rate of needed repairs	$r_i = rate(\text{rep})$

Table 12.4 SRN-level reward rates for the $M/E_k/1$ queue of Example 12.14.

Measure	Reward rate assignment for marking i
Server utilization	$r_i = \begin{cases} 1 & \text{if } (\#(\text{service}) > 0) \\ 0 & \text{otherwise} \end{cases}$
Expected no. of jobs in system	$r_i = \begin{cases} \#(\text{buffer}) + 1 & \text{if } (\#(\text{service}) + \#(\text{phcompl}) \geq 1) \\ 0 & \text{otherwise} \end{cases}$
Expected response time	$r_i = \begin{cases} \dfrac{\#(\text{buffer})+1}{\lambda} & \text{if } (\#(\text{service}) + \#(\text{phcompl}) \geq 1) \\ 0 & \text{otherwise} \end{cases}$

of Table 12.3, we can compute the system availability at time t using the transient analysis option (Eq. (12.2) is applied) or the system availability in steady state using the steady-state analysis option (Eq. (12.1) is applied). With the reward assignment in the second row of Table 12.3, we can compute the expected system capacity at time t using the transient analysis option (Eq. (12.2) is applied). With the reward assignment in the third row of Table 12.3, we can compute the expected number of covered failures in the interval $(0, t]$ using the cumulative transient analysis option (Eq. (12.3) is applied). With the reward assignment in the last row of Table 12.3, we can compute the expected number of repairs carried out in the interval $(0, t]$ using the cumulative transient analysis option (Eq. (12.3) is applied). The derivatives of these measures can also be computed.

Example 12.26 $M/E_k/1$ *Queue (continued)*
For the $M/E_k/1$ queue of Example 12.14 we can specify various measures assigning different reward rates as in Table 12.4.

Example 12.27   *SRN Model of a Series-Parallel System*
Returning again to the SRN model of the series-parallel system presented in Example 12.16, we can now make use of the Boolean function $bool(G_1)$ to specify the reward rate r_i assignment to the marking i of the SRN in order to compute the system reliability, via Eq. (12.2), or system mean time to failure, Eq. (12.4), as follows:

$$r_i = \begin{cases} 1 & \text{if } (bool(G_1) == 0), \\ 0 & \text{otherwise.} \end{cases}$$

Table 12.5 SRN-level reward rate specification of measures for the $MMPP/M/1/K$ queue.

Measure	Reward rate assignment for marking i
Expected no. of jobs in system	$r_i = \#(\text{buffer})$
Rate of job rejection	$r_i = \begin{cases} \lambda_1 & \text{if } ((\#(\text{buffer}) == K) \;\&\&\; (\#(\text{MMPP1}) == 1)) \\ \lambda_2 & \text{if } ((\#(\text{buffer}) == K) \;\&\&\; (\#(\text{MMPP2}) == 1)) \\ 0 & \text{otherwise} \end{cases}$
Average arrival rate	$r_i = \begin{cases} \lambda_1 & \text{if } (\#(\text{MMPP1}) == 1) \\ \lambda_2 & \text{if } (\#(\text{MMPP2}) == 1) \end{cases}$

Notice that with the same reward assignment, same measures can be computed for the series-parallel system reliability models with repair shown in the SRN of Figure 12.16, as well as the SRN in Figure 12.17.

Example 12.28   *A Redundant Repairable Fluid Level Controller*
For the SRN in Example 12.18, the system is available when there are at least two tokens in place P_{up} for the sensor block, and at least one token each in places P_{up} for both the logic and pump blocks, and at least one token in place P_{up} and no token in place P_{delay} for the valve block. Thus, the assignment of the binary reward rate to marking i, which defines the up states, is the following:

$$r_i = \begin{cases} 1 & \text{if } (([\#(S_{up})] \geq 2) \;\&\&\; ([\#(L_{up})] \geq 1) \;\&\&\; ([\#(P_{up})] \geq 1) \\ & \quad \&\&\; ([\#(V_{up})] \geq 1 \;\&\&\; [\#(V_{delay})] == 0)), \\ 0 & \text{otherwise.} \end{cases} \quad (12.5)$$

Using the same parameter values as Example 9.10, the steady-state availability of this system obtained using SPNP is $A_{Sys} = 0.997\,485\,390\,685$. With the same reward assignment, the transient availability, expected interval availability and their derivatives can also be computed.

Example 12.29   *MMPP/M/1/K Queue*
We return to the $MMPP/M/1/K$ queue from Example 12.20 to define reward rates for various measures as shown in Table 12.5. The first row of the table will enable us to compute the expected number of jobs in the system, while the third row of the table will yield the average arrival rate. In order to compute the blocking probability for an arriving customer, we need to be more careful since the PASTA theorem is no longer applicable [42]. We first compute the rate at which incoming jobs are rejected using the reward function in the second row of the table and then divide by the average arrival rate computed using the third row of the table.

Problems

12.19   Numerically solve the steady-state system availability of the repairable fluid level controller with travel time SRN model of Example 12.19 using the SHARPE or the SPNP package.

12.20   Numerically solve the $MMPP/M/1/K$ SRN model of Example 12.20 using the SHARPE or the SPNP package. Compute various interesting measures via the reward rate assignments in Table 12.5. Besides steady-state measures, also compute transient measures.

12.21   Numerically solve the retrial SRN model of Example 12.21 using the SHARPE or the SPNP package. Compute various interesting measures via the reward rate assignments.

12.5   Further Reading

The journey into the world of Petri nets begins with the doctoral thesis of C. A. Petri in 1962 [1], where the formalism was first presented. A comprehensive survey of the formalism can be found in [43, 44]. Several textbooks on the subject are also available – see [2], which contains an extended annotated bibliography, and [45–47]. Early work in timed PNs with deterministic timing can be found in [8–11]. Stochastic Petri nets were initially proposed in two doctoral theses [12, 13]. Extensions to cover the case of generally distributed transition firing times have been considered in a number of papers [12, 14, 15, 48–50]. Many authors [3, 48, 51] have recognized that the use of SPNs for modeling real systems involves the presence of very brief or *fast* transitions, whose durations are short, or even negligible, with respect to the timescale of the problem. Different techniques have been proposed to tackle this problem. The generalization of SPNs to GSPNs introduced by Ajmone Marsan *et al.* [3] was widely adopted. Ciardo *et al.* [5, 18] introduced several structural extensions to Petri nets which increase their modeling power. These include guard functions, general marking dependency, variable cardinality arcs, and priorities, resulting in the SRN modeling formalism. The key addition in SRNs was the definition of net-level reward functions. Some of these structural constructs were also used in stochastic activity networks [19, 52] and GSPNs [53]. Extensions to non-Markovian nets such as Markov regenerative stochastic Petri nets [54–56] and jointly discrete and continuous state spaces via fluid stochastic Petri nets [57–62] are also available.

In this chapter we have considered only PNs with *black* tokens, i.e., PNs in which tokens are indistinguishable. There is a vast literature in which tokens are differentiated by means of some attribute attached to them. These classes of PN are generally called colored PNs [63] or high-level PNs [64]. The world of PNs with an extensive bibliography is managed and updated at www.informatik.uni-hamburg.de/TGI/PetriNets/.

References

[1] C. Petri, "Kommunikation mit automaten," Doctoral Thesis, University of Bonn, 1962, (Available in English as: *Communication with Automata*, Technical Report RADC-TR-65-377, Rome Air Development Center, Griffiss NY, 1966).

[2] J. Peterson, *Petri Net Theory and the Modeling of Systems*. Prentice Hall, 1981.

[3] M. Ajmone Marsan, G. Balbo, and G. Conte, "A class of generalized stochastic Petri nets for the performance evaluation of multiprocessor systems," *ACM Transactions on Computer Systems*, vol. 2, pp. 93–122, 1984.

[4] G. Ciardo, J. Muppala, and K. Trivedi, "On the solution of GSPN reward models," *Performance Evaluation*, vol. 12, pp. 237–253, 1991.

[5] G. Ciardo, J. Muppala, and K. S. Trivedi, "SPNP: Stochastic Petri net package," in *Proc. Third Int. Workshop on Petri Nets and Performance Models*, 1989, pp. 142–151.

[6] G. Ciardo, "Toward a definition of modeling power for stochastic Petri net models," in *Proc. Int. Workshop on Petri Nets and Performance Models*. IEEE Computer Society Press no. 796, 1987, pp. 54–62.

[7] A. Bobbio, "System modelling with Petri nets," in *System Reliability Assessment*, eds. A. Colombo and A. S. de Bustamante. Kluwer Academic P.G., 1990, pp. 103–143.

[8] P. Merlin and D. Faber, "Recoverability of communication protocols: Implication of a theoretical study," *IEEE Transactions on Communication*, vol. COM-24, pp. 1036–1043, 1976.

[9] J. Sifakis, "Use of Petri nets for performance evaluation," in *Measuring, Modelling and Evaluating Computer Systems*, eds. H. Beilner and E. Gelenbe. North Holland, 1977, pp. 75–93.

[10] C. Ramamoorthy and G. Ho, "Performance evaluation of asynchronous concurrent systems using Petri nets," *IEEE Transactions on Software Engineering*, vol. SE-6, pp. 440–449, 1980.

[11] W. Zuberek, "Timed Petri nets and preliminary performance evaluation," in *Proc. 7th Ann. Symp. on Computer Architecture*, 1980, pp. 88–96.

[12] S. Natkin, "Les reseaux de Petri stochastiques et leur application a l'evaluation des systemes informatiques," Thèse de Docteur Ingegneur, CNAM, Paris, 1980.

[13] M. Molloy, "On the integration of delay and throughput measures in distributed processing models," Phd Thesis, UCLA, 1981.

[14] G. Florin and S. Natkin, "Les reseaux de Petri stochastiques," *Technique et Science Informatique*, vol. 4, pp. 143–160, 1985.

[15] M. Ajmone Marsan, G. Balbo, A. Bobbio, G. Chiola, G. Conte, and A. Cumani, "On Petri nets with stochastic timing," in *Proc. Int. Workshop on Timed Petri Nets*. IEEE Computer Society Press no. 674, 1985, pp. 80–87.

[16] C. Hirel, B. Tuffin, and K. S. Trivedi, "SPNP: Stochastic Petri nets. Version 6," in *Int. Conf. on Computer Performance Evaluation: Modelling Techniques and Tools (TOOLS 2000)*, eds. B. Haverkort and H. Bohnenkamp, LNCS 1786, Springer Verlag, 2000, pp. 354–357.

[17] R. Sahner, K. Trivedi, and A. Puliafito, *Performance and Reliability Analysis of Computer Systems: An Example-Based Approach Using the SHARPE Software Package*. Kluwer Academic Publishers, 1996.

[18] G. Ciardo, A. Blakemore, P. F. Chimento, J. K. Muppala, and K. Trivedi, "Automated generation and analysis of Markov reward models using stochastic reward nets," in *Linear Algebra, Markov Chains, and Queueing Models*, eds. C. D. Meyer and R. J. Plemmons. Springer, 1993, vol. 48, pp. 145–191.

[19] J. Couvillon, R. Freire, R. Johnson, W. Obal, M. Qureshi, M. Rai, W. Sanders, and J. Tvedt, "Performability modeling with UltraSAN," *IEEE Software*, vol. 8, pp. 69–80, Sep. 1991.

[20] M. Malhotra and K. Trivedi, "Dependability modeling using Petri nets," *IEEE Transactions on Reliability*, vol. R-44, pp. 428–440, 1995.

[21] Y. Cao, H. Sun, and K. S. Trivedi, "Performance analysis of reservation media-access protocol with access and serving queues under bursty traffic in GPRS/EGPRS," *IEEE Transactions on Vehicular Technology*, vol. 52, no. 6, pp. 1627–1641, 2003.

[22] G. Ciardo, J. K. Muppala, and K. S. Trivedi, "Analyzing concurrent and fault-tolerant software using stochastic reward nets," *Journal of Parallel and Distributed Computing*, vol. 15, no. 3, pp. 255–269, 1992.

[23] R. Ghosh, F. Longo, F. Frattini, S. Russo, and K. S. Trivedi, "Scalable analytics for IaaS cloud availability," *IEEE Transactions on Cloud Computing*, vol. 2, no. 1, pp. 57–70, 2014.

[24] O. C. Ibe and K. S. Trivedi, "Stochastic Petri net models of polling systems," *IEEE Journal on Selected Areas in Communications*, vol. 8, no. 9, pp. 1649–1657, 1990.

[25] O. C. Ibe, H. Choi, and K. S. Trivedi, "Performance evaluation of client–server systems," *IEEE Transactions on Parallel and Distributed Systems*, vol. 4, no. 11, pp. 1217–1229, 1993.

[26] J. K. Muppala, K. S. Trivedi, V. Mainkar, and V. G. Kulkarni, "Numerical computation of response time distributions using stochastic reward nets," *Annals of Operations Research*, vol. 48, pp. 155–184, 1994.

[27] H. Sun, X. Zang, and K. S. Trivedi, "A stochastic reward net model for performance analysis of prioritized DQDB MAN," *Computer Communications*, vol. 22, no. 9, pp. 858–870, 1999.

[28] H. Sun, X. Zang, and K. S. Trivedi, "Performance of broadcast and unknown server (BUS) in ATM LAN emulation," *IEEE/ACM Transactions on Networking*, vol. 9, no. 3, pp. 361–372, 2001.

[29] L. A. Tomek, J. K. Muppala, and K. S. Trivedi, "Modeling correlation in software recovery blocks," *IEEE Transactions on Software Engineering*, vol. 19, no. 11, pp. 1071–1086, 1993.

[30] C. Wang, D. Logothetis, K. S. Trivedi, and Y. Viniotis, "Transient behavior of ATM networks under overloads," in *Proc. 15th Ann. Joint Conf. of the IEEE Computer and Communications Societies, Networking the Next Generation*, 1996, pp. 978–985.

[31] D. Wang, W. Xie, and K. S. Trivedi, "Performability analysis of clustered systems with rejuvenation under varying workload," *Performance Evaluation*, vol. 64, no. 3, pp. 247–265, 2007.

[32] J. K. Muppala, S. P. Woolet, and K. S. Trivedi, "Real-time systems performance in the presence of failures," *IEEE Computer*, vol. 24, no. 5, pp. 37–47, 1991.

[33] D. Wang, B. B. Madan, and K. S. Trivedi, "Security analysis of SITAR intrusion tolerance system," in *Proc. 2003 ACM workshop on Survivable and Self-Regenerative Systems: in assoc. with 10th ACM Conf. on Computer and Communications Security*, 2003, pp. 23–32.

[34] J. K. Muppala, A. Sathaye, R. Howe, and K. S. Trivedi, "Dependability modeling of a heterogeneous VAX-cluster system using stochastic reward nets," in *Hardware and Software Fault Tolerance in Parallel Computing Systems*, ed. D. R. Avresky. Horwood, 1992, pp. 33–59.

[35] D. Codetta-Raiteri, "The conversion of dynamic fault trees to stochastic Petri nets, as a case of graph transformation," *Electronic Notes on Theoretical Computer Science*, vol. 127, pp. 45–60, 2005.

[36] H. Sukhwani, A. Bobbio, and K. Trivedi, "Largeness avoidance in availability modeling using hierarchical and fixed-point iterative techniques," *International Journal of Performability Engineering*, vol. 11, no. 4, pp. 305–319, 2015.

[37] W. Fischer and K. Meier-Hellstern, "The Markov-modulated Poisson process (MMPP) cookbook," *Performance Evaluation*, vol. 18, no. 2, pp. 149–171, 1993.

[38] N. Gharbi and C. Dutheillet, "An algorithmic approach for analysis of finite-source retrial systems with unreliable servers," *Computers & Mathematics with Applications*, vol. 62, no. 6, pp. 2535–2546, 2011.

[39] M. Ajmone Marsan, A. Bobbio, G. Conte, and A. Cumani, "Performance analysis of degradable multiprocessor systems using generalized stochastic Petri nets," *IEEE Computer Society Newsletters*, vol. 6, SI-1, pp. 47–54, 1984.

[40] J. B. Dugan, A. Bobbio, G. Ciardo, and K. Trivedi, "The design of a unified package for the solution of stochastic Petri net models," in *Proc. Int. Workshop on Timed Petri Nets*. IEEE Comp Soc Press no. 674, 1985, pp. 6–13.

[41] J. K. Muppala and K. S. Trivedi, "GSPN models: Sensitivity analysis and applications," in *Proc. 28th Ann. Southeast Regional Conf.*, Greenville, South Carolina, USA, April 18–20, 1990, pp. 25–33.

[42] R. Wolff, "Poisson arrivals see time averages," *Operations Research*, vol. 30, no. 2, pp. 223–231, 1982.

[43] J. Peterson, "Petri nets," *Computing Surveys*, vol. 9, pp. 223–252, 1977.

[44] T. Agerwala, "Putting Petri nets to work," *IEEE Computer*, pp. 85–94, Dec. 1979.

[45] W. Reisig, *Petri Nets: An Introduction*. Springer-Verlag, 1982.

[46] G. Brams, *Réseaux de Petri: Théorie et pratique*. Masson, 1983 (in French).

[47] M. Silva, *Las Redes de Petri en la Automatica y la Informatica*. AC, 1985.

[48] J. B. Dugan, K. Trivedi, R. Geist, and V. Nicola, "Extended stochastic Petri nets: Applications and analysis," in *Proc. PERFORMANCE '84*, Paris, 1984.

[49] P. Haas and G. Shedler, "Regenerative stochastic Petri nets," *Performance Evaluation*, vol. 6, pp. 189–204, 1986.

[50] M. Ajmone Marsan, G. Balbo, A. Bobbio, G. Chiola, G. Conte, and A. Cumani, "The effect of execution policies on the semantics and analysis of stochastic Petri nets," *IEEE Transactions on Software Engineering*, vol. SE-15, pp. 832–846, 1989.

[51] A. Bobbio, A. Cumani, and R. Del Bello, "Reduced Markovian representation of stochastic Petri net models," *Systems Science*, vol. 10, pp. 5–23, 1984.

[52] W. Sanders and J. Meyer, "Reduced base model construction methods for stochastic activity networks," *IEEE Journal on Selected Areas in Communications*, vol. 9, no. 1, pp. 25–36, Jan. 1991.

[53] M. Ajmone Marsan, G. Balbo, G. Conte, S. Donatelli, and G. Franceschinis, *Modelling with Generalized Stochastic Petri Nets*. Wiley Series in Parallel Computing, 1995.

[54] A. Bobbio, A. Puliafito, M. Telek, and K. Trivedi, "Recent developments in non-Markovian stochastic Petri nets," *Journal of Systems Circuits and Computers*, vol. 8, no. 1, pp. 119–158, Feb. 1998.

[55] H. Choi, V. Kulkarni, and K. Trivedi, "Markov regenerative stochastic Petri nets," *Performance Evaluation*, vol. 20, pp. 337–357, 1994.

[56] R. German, *Performance Analysis of Communication Systems: Modeling with Non-Markovian Stochastic Petri Nets*. John Wiley & Sons, 2000.

[57] K. S. Trivedi and V. Kulkarni, "FSPNs: Fluid stochastic Petri nets," in *Proc. 14th Int. Conf. on Applications and Theory of Petri Nets*, ed. M. A. Marsan. Springer-Verlag, 1993, vol. 691, pp. 24–31.

[58] G. Horton, V. Kulkarni, D. Nicol, and K. S. Trivedi, "Fluid stochastic Petri nets: Theory, application, and solution techniques," *European Journal of Operations Research*, vol. 105, no. 1, pp. 184–201, Feb. 1998.

[59] G. Ciardo, D. M. Nicol, and K. S. Trivedi, "Discrete-event simulation of fluid stochastic Petri nets," *IEEE Transactions on Software Engineering*, vol. 25, no. 2, pp. 207–217, 1999.

[60] M. Gribaudo, M. Sereno, A. Horváth, and A. Bobbio, "Fluid stochastic Petri nets augmented with flush-out arcs: Modelling and analysis," *Discrete Event Dynamic Systems*, vol. 11 (1/2), pp. 97–117, Jan. 2001.

[61] M. Gribaudo, A. Horváth, A. Bobbio, E. Tronci, E. Ciancamerla, and M. Minichino, "Fluid Petri nets and hybrid model-checking: A comparative case study," *Reliability Engineering and System Safety*, vol. 81, pp. 239–257, 2003.

[62] A. Bobbio, S. Garg, M. Gribaudo, A.Horváth, M. Sereno, and M. Telek, "Compositional fluid stochastic Petri net model for operational software system performance," in *Int. Workshop on Software Aging and Rejuvenation WOSAR-08*. IEEE Computer Society, 2008, pp. 1–6.

[63] K. Jensen and L. Kristensen, *Coloured Petri Nets: Modelling and Validation of Concurrent Systems*. Springer, 2009.

[64] K. Jensen and G. Rozenberg, *High-Level Petri Nets: Theory and Application*. Springer Verlag, 1991.

Part IV

State-Space Models with Non-Exponential Distributions

In Part III we considered time-homogeneous continuous-time Markov chains, where the models satisfy the Markov property and the transition rates are time independent. We have also noted that the sojourn times (or holding times) in states of a homogeneous CTMC are exponentially distributed. In Part IV, we relax the time independence assumption.

If the system parameters are dependent on global time, but not the sojourn time in any given state, then the system behavior can still be represented as a Markov chain that satisfies the Markov property but not the homogeneity property. As a consequence, the infinitesimal generator matrix will have one or more rates that are time dependent. Here, the time dependence will be on the global time, that is, the time origin will be the beginning of system operation at $t = 0$. The resulting models are referred to as non-homogeneous CTMCs. We will also use the notation t_0 as the time epoch of the beginning of CTMC operation.

In contrast, a semi-Markov process permits the state sojourn time distributions to be non-exponential and hence have time-dependent rates, but the time in an SMP is measured from the time of entry into the state and hence an SMP allows local time dependence. In an SMP, the instants of entrance in any new state are regeneration points for the process. Relaxing this hypothesis, under certain conditions (Section 14.5), originates the so-called Markov regenerative processes (MRGPs).

In the analysis of physical systems, sometimes the hypotheses underlying the NHCTMC formalism (dependence on the global time) or the SMP formalism (all the state transitions must be regenerative) are not satisfied. In these cases the NHCTMC or SMP will not be the right model. A possible alternative is to represent (or to approximate) non-exponential distributions by means of PH distributions, see Eq. (10.45), so that the model with non-exponential distributions can be converted into an expanded homogeneous CTMC model. Thus, PH expansion, considered in Chapter 15, can be seen as a more general method than either NHCTMC or SMP.

Part IV will cover the following techniques:

- non-homogeneous CTMCs (in Chapter 13);
- semi-Markov and Markov regenerative models (in Chapter 14);
- phase-type expansion (in Chapter 15).

A comparison of the modeling power of the different model types considered in Part IV is discussed in Section 15.6.

13 Non-Homogeneous Continuous-Time Markov Chains

13.1 Introduction

Non-homogeneous continuous-time Markov chains satisfy the Markov property, but do not satisfy the homogeneity requirements with respect to time. In these models, the transition rates are dependent on the global time, so that the sojourn time distributions do not have the memoryless property. We note here the distinction between the Markov property and the memoryless property. The former is a property of some stochastic processes (e.g., DTMC, CTMC, Markov process) while the latter is the property of some distributions (e.g., the exponential and the geometric). To distinguish in this chapter between homogeneous and non-homogeneous CTMCs, we refer to the first type as HCTMCs and to the latter as NHCTMCs.

13.1.1 Kolmogorov Differential Equation

The transient behavior of an NHCTMC is defined by the system of Kolmogorov ODEs [1]:

$$\frac{d\boldsymbol{\pi}(t)}{dt} = \boldsymbol{\pi}(t)\boldsymbol{Q}(t), \tag{13.1}$$

with the initial probability vector $\boldsymbol{\pi}(t_0)$ at the beginning of the NHCTMC operation at time t_0, subject to the normalization condition (for an n-state NHCTMC):

$$\sum_{i=1}^{n} \pi_i(t) = 1.$$

Equation (13.1) is very similar to Eq. (9.19) for an HCTMC with one difference: the generator matrix entries are now assumed to be time dependent. If $\boldsymbol{Q}(t)$ is integrable, a solution to the above equation exists and is of the form

$$\boldsymbol{\pi}(t) = \boldsymbol{\pi}(t_0)\boldsymbol{P}(t_0,t),$$

where the entries of $\boldsymbol{P}(t_0,t)$ are $p_{ij}(t_0,t)$, that is, the probability that the Markov chain is in state j at time t given that it started in state i at time t_0. We may be tempted to write the solution as $\boldsymbol{P}(t_0,t) = e^{\int_{t_0}^{t} \boldsymbol{Q}(\tau)d\tau}$, analogous to the general solution of an HCTMC, but this does not hold in general unless $\boldsymbol{Q}(t)$ and its time integral $\int_{t_0}^{t} \boldsymbol{Q}(\tau)d\tau$ commute for all t [2].

First consider the special case of an NHCTMC in which the matrix $Q(t)$ can be factored so that $Q(t) = g(t)\,W$, with matrix W having all its entries independent of time. In this case the solution to the NHCTMC can be written as:

$$\boldsymbol{\pi}(t) = \boldsymbol{\pi}(t_0)\,e^{\left(\int_{t_0}^t g(\tau)d\tau\right)W} = \boldsymbol{\pi}(t_0)\,e^{Wg^*t}, \tag{13.2}$$

where $g^* = (\int_{t_0}^t g(\tau)d\tau)/t$ and Wg^* is the generator matrix of an HCTMC. Thus, by solving an "equivalent" HCTMC, we can obtain the solution of the original NHCTMC in this special case. We refer to this method of solution of an NHCTMC as the *equivalent-HCTMC method*. In the more general case where such a factorization is not feasible, $P(t_0, t)$ can be solved using the Peano–Baker series [3] as:

$$P(t_0, t) = I + \int_{t_0}^t Q(\tau_1)d\tau_1 + \int_{t_0}^t Q(\tau_1) \int_{t_0}^{\tau_1} Q(\tau_2)d\tau_2 d\tau_1 +$$
$$\int_{t_0}^t Q(\tau_1) \int_{t_0}^{\tau_1} Q(\tau_2) \ldots \int_{t_0}^{\tau_{k-1}} Q(\tau_k)d\tau_k d\tau_{k-1} \cdots d\tau_2 d\tau_1 + \cdots$$

However, the computation of $P(t_0, t)$ for a time-varying generator matrix using the above series is rather difficult. As a consequence, several alternative methods have been proposed for the transient analysis of NHCTMCs such as the convolution integration approach, the ODE approach, uniformization, and the piecewise constant approximation (PCA) approach. Yet another approach is based on using a phase-type expansion of the non-exponential distributions.

For the special case of an acyclic NHCTMC, the convolution integration approach can be recommended [1, 4] (for a comparison, see Section 10.1.1):

$$p_{ij}(t_0, t) = \delta_{ij} e^{-\int_{t_0}^t q_{ii}(\tau)d\tau} + \int_{t_0}^t \sum_k p_{ik}(t_0, x) q_{kj}(x) e^{-\int_x^t q_{jj}(\tau)d\tau} dx,$$

where δ_{ij} is the Kronecker delta function defined by $\delta_{ij} = 1$ if $i = j$ and 0 otherwise. The corresponding equation for the unconditional state probability is:

$$\pi_i(t) = \pi_i(t_0) e^{-\int_{t_0}^t q_{ii}(\tau)d\tau} + \int_{t_0}^t \sum_{k \neq i} \pi_k(t_0, x) q_{ki}(x) e^{-\int_x^t q_{ii}(\tau)d\tau} dx. \tag{13.3}$$

We will illustrate the convolution integration approach in Section 13.2 by means of an example, and the PCA approach in Section 13.3, while the ODE method and uniformization will be briefly described in Section 13.6. PH expansion will be covered in Chapter 15.

13.2   Illustrative Examples

Solution of NHCTMCs using the convolution integrals approach (13.3) is illustrated through the following examples.

Example 13.1 *Duplex Processor in Critical Applications*

Duplex processors are often used in control systems for train control and other such safety-critical systems. A possible model of such a system is given in Figure 13.1(a). This model incorporates several features of real systems. First, the time to failure distribution is not restricted to being exponential and is allowed to be general, resulting in a time-dependent (or age-dependent) failure rate function. Second, imperfect failure coverage is considered. The system operation is assumed to begin in state 2, where both the components are energized and correctly functioning. Upon the failure of one of these components, the system is able to cope with it with the coverage probability c_2 resulting in a transition to the operational state 1. With the complementary probability, $1 - c_2$, the failure is not covered leading to an unsafe system state UF. From state 1, a component failure can lead to the safe shutdown state SF (with probability c_1) or to the unsafe state UF (with probability $(1 - c_1)$). Both SF and UF are absorbing states. Note that if this was a reliability model we need not have separated the failure states UF and SF, but with this being a safety model we separate the down state into two: the safe shutdown state SF and unsafe state UF. With the assumption that the two components are in hot standby mode, the age of both the components is measured from the beginning of system operation, i.e., with the global clock. Hence we have an NHCTMC.

Assuming that the time to failure of both the components is Weibull distributed with shape parameter β and scale parameter η, Eq. (3.22), we solve for the transient state probabilities in this case using the convolution integration approach. The probabilities at time t of the transient states 1 and 2 are:

$$\pi_2(t) = e^{-2\int_0^t \lambda(x)dx}$$

$$= e^{-2\int_0^t \eta\beta(\eta x)^{\beta-1}dx}$$

$$= e^{-2(\eta t)^\beta},$$

$$\pi_1(t) = \int_0^t \pi_2(x)2\lambda(x)c_2 e^{-\int_x^t \lambda(y)dy}dx$$

$$= \int_0^t e^{-2(\eta x)^\beta}2\eta\beta(\eta x)^{\beta-1}c_2 e^{-(\eta t)^\beta+(\eta x)^\beta}dx$$

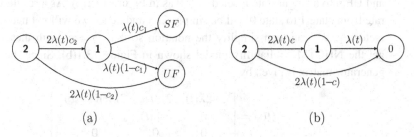

(a)   (b)

Figure 13.1   Duplex processor: (a) safety-critical system model; (b) reliability model.

$$= 2\eta\beta c_2 e^{-(\eta t)^\beta} \int_0^t e^{-(\eta x)^\beta}(\eta x)^{\beta-1}dx$$

$$= 2c_2 e^{-(\eta t)^\beta} \int_0^t e^{-(\eta x)^\beta}d((\eta x)^\beta)$$

$$= 2c_2 e^{-(\eta t)^\beta}(1 - e^{-(\eta t)^\beta})$$

$$= 2c_2(e^{-(\eta t)^\beta} - e^{-2(\eta t)^\beta}).$$

The probability of the fail-safe state SF at time t is given by:

$$\pi_{SF}(t) = \int_0^t \pi_1(x)\lambda(x)c_1 dx$$

$$= \int_0^t 2\eta\beta c_2(e^{-(\eta x)^\beta} - e^{-2(\eta x)^\beta})(\eta x)^{\beta-1}c_1 dx$$

$$= 2c_1 c_2 \int_0^t (e^{-(\eta x)^\beta} - e^{-2(\eta x)^\beta})d((\eta x)^\beta)$$

$$= c_1 c_2(1 - 2e^{-(\eta t)^\beta} + e^{-2(\eta t)^\beta}).$$

From the above, we get the eventual absorption probability to the safe state as $\pi_{SF}(\infty) = c_1 c_2$. Finally, the probability of the unsafe state UF at time t is given by:

$$\pi_{UF}(t) = \int_0^t \pi_1(x)\lambda(x)(1-c_1)dx + \int_0^t \pi_2(x)2\lambda(x)(1-c_2)dx$$

$$= \int_0^t 2c_2(e^{-(\eta x)^\beta} - e^{-2(\eta x)^\beta})(\eta\beta)(\eta x)^{\beta-1}(1-c_1)dx$$

$$+ \int_0^t e^{-2(\eta t)^\beta}2\eta\beta(\eta x)^{\beta-1}(1-c_2)dx$$

$$= c_2(1-c_1)(1 - 2e^{-(\eta t)^\beta} + e^{-2(\eta t)^\beta}) + (1-c_2)(1 - e^{-2(\eta t)^\beta}).$$

From the above, we get the eventual absorption probability to the unsafe state as $\pi_{UF}(\infty) = c_2(1-c_1) + (1-c_2) = 1 - c_1 c_2$.

Example 13.2   *Reliability of the Duplex Processor*
Now suppose that we are only interested in computing the system reliability (and unreliability) in the system of Example 13.1. Then we may merge the two states SF and UF into a single state labeled F or 0 as in Figure 13.1(b). As a result, the transition rate from state 1 to state 0 will be simply $\lambda(t)$ and hence we will not need the coverage factor c_1. We can then simplify the notation so that $c_2 = c$. With these changes, we get the NHCTMC reliability model shown in Figure 13.1(b). Since the infinitesimal generator matrix is given by

$$Q(t) = \begin{bmatrix} -2\lambda(t) & 2\lambda(t)c & 2\lambda(t)(1-c) \\ 0 & -\lambda(t) & \lambda(t) \\ 0 & 0 & 0 \end{bmatrix},$$

it can be factored into $\lambda(t)W$ where W is an HCTMC generator matrix. We can then use the equivalent-HCTMC solution method of Eq. (13.2). We leave the actual solution using this method as an exercise.

Example 13.3   *Reliability of a Multi-Voltage High-Speed Train [5]*
We return to the model of the multi-voltage propulsion system designed for the Italian high-speed railway system [5] studied earlier in Example 4.14 as an RBD and in Example 10.21 as an HCTMC. In [5], the train is divided into several modules where each module can be modeled as a simple RBD as shown in Figure 4.10.

Here we consider only one module. Later, in Chapter 16, we will consider multiple modules. In Example 10.21 we showed that a single module can be modeled as a three-state Markov reward model. Although in the original paper constant failure rates are assumed, here we generalize and allow time-dependent failure rates. The resulting three-state NHCTMC is shown in Figure 13.2, where $\lambda(t)$ is the sum of the failure rates of the transformer, the filters, the inverter and the motors, while $\gamma(t)$ is the failure rate of an individual four-quadrant converter.

We first derive the state probability expressions for the three states using the convolution integration approach as in Example 13.1:

$$\pi_2(t) = e^{-\int_0^t (\lambda(x)+2\gamma(x))dx}$$

$$= e^{-\int_0^t \lambda(x)dx} \cdot e^{-\int_0^t 2\gamma(x)dx} \tag{13.4}$$

$$= e^{-(\Lambda(t)+2\Gamma(t))},$$

where

$$\Lambda(t) = \int_0^t \lambda(x)dx, \qquad \Gamma(t) = \int_0^t \gamma(x)dx, \tag{13.5}$$

$$\pi_1(t) = \int_0^t \pi_2(x) \cdot 2\gamma(x) \cdot e^{-\int_x^t (\lambda(y)+\gamma(y))dy} dx$$

$$= \int_0^t e^{-(\Lambda(x)+2\Gamma(x))} \cdot 2\gamma(x) \cdot e^{-\int_x^t (\lambda(y)+\gamma(y))dy} dx$$

$$= 2\int_0^t e^{-\int_0^t (\lambda(y)+\gamma(y))dy} \cdot \gamma(x) \cdot e^{-\Gamma(x)} dx \tag{13.6}$$

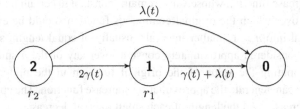

Figure 13.2   The propulsion module as a three-state model with time-dependent rates.

$$= 2e^{-(\Lambda(t)+\Gamma(t))} \cdot (1 - e^{-\Gamma(t)})$$

$$= 2e^{-(\Lambda(t)+\Gamma(t))} - 2e^{-(\Lambda(t)+2\Gamma(t))},$$

$$\pi_0(t) = 1 - \pi_2(t) - \pi_1(t). \tag{13.7}$$

From the state probability expressions, we can write down the expected power available at time t as an example of expected reward rate at time t:

$$E[X(t)] = \sum_{i=0}^{2} r_i \pi_i(t) = r_2 \pi_2(t) + r_1 \pi_1(t),$$

and the expected energy delivered in the interval $(0, t]$ as an example of expected accumulated reward:

$$E[Y(t)] = \sum_{i=0}^{2} \int_0^t r_i \pi_i(t)\, dt = r_2 \int_0^t \pi_2(t)\, dt + r_1 \int_0^t \pi_1(t)\, dt.$$

Problems

13.1 For the duplex processor system in Example 13.1, compute the derivatives of $\pi_{UF}(t)$ with respect to the coverage parameters c_2 and c_1, and thus find which of them has more of an effect on the result.

13.2 Solve the three-state reliability model of Example 13.2, first using the equivalent-HCTMC method and then by the convolution integration method, and thus cross-check the two results. Next, derive an expression for the MTTF of the system as well as the MTIF, defined as the mean time improvement factor, i.e., the ratio of system MTTF and component MTTF.

13.3 For the multi-voltage train in Example 13.3, study the effect of the distributional assumption by plotting the two measures $E[X(t)]$ and $E[Y(t)]$ as functions of t for different values of the Weibull shape parameter for both $\lambda(t)$ and $\gamma(t)$, varying the shape parameter in such a way as to keep the component MTTF invariant.

13.3   Piecewise Constant Approximation

Piecewise constant approximation is based on approximating a continuous time-variant function by a staircase function whose value remains constant in certain intervals. Given an interval $(0, t]$ over which the original continuous function should be evaluated, we divide the interval into $n > 1$ smaller intervals, usually of equal length, such that the function can be considered approximately constant over any of the n small intervals. In this way, computing the value of the original function in the midpoint of each small interval, we can generate the approximating staircase function. The approximation improves as n increases and the length of each small interval decreases.

Example 13.4 *Duplex Processor with Repair [6]*

As a variation of the three-state NHCTMC example of Figure 13.1(b), consider adding a repair transition from state 1 back to state 2 with a time-independent rate μ, as shown in Figure 13.3 [6].

Unlike the case of no repair as in the previous example, Figure 13.1(b), where the model was an NHCTMC, when repair from state 1 back to state 2 is introduced (Figure 13.3), we need to make two assumptions for the model to remain an NHCTMC:

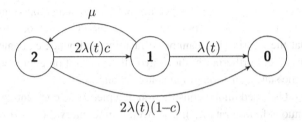

Figure 13.3 Non-homogeneous continuous-time Markov chain model of a duplex system with repair.

- repair time is negligible compared to the time to failure;
- repair is *minimal*, i.e., the repaired processor is in a condition equal to the one just before its failure (*as bad as old*).

The infinitesimal generator matrix of this approximate NHCTMC is:

$$Q(t) = \begin{bmatrix} -2\lambda(t) & 2\lambda(t)c & 2\lambda(t)(1-c) \\ \mu & -(\lambda(t)+\mu) & \lambda(t) \\ 0 & 0 & 0 \end{bmatrix}.$$

Notice that an alternative repair strategy, called *maximal* repair (or restoring the failed component to an as-good-as-new state), will create multiple age clocks and hence will not be captured by the single global clock of an NHCTMC. For the case of maximal repair, we will need to use the PH expansion method, as will be shown in Chapter 15.

The failure rate $\lambda(t)$ is assumed to be the hazard rate of a two-parameter Weibull distribution:

$$\lambda(t) = \eta\beta(\eta t)^{\beta-1}. \tag{13.8}$$

Table 13.1 shows the values of the parameters of the system where η and β are chosen to obtain a Weibull distribution with an increasing failure rate.

With the numerical values of Table 13.1, we get MTTR $= \frac{1}{\mu} = 0.0083$ yr and MTTF $= 1/\eta\Gamma(1+\frac{1}{\beta}) = 0.903$ yr.

With minimal repair, and assuming that the MTTR $= \frac{1}{\mu} \ll \beta\Gamma(\frac{1}{\beta}+1)$ (MTTF of the Weibull distribution), i.e., the MTTR is much smaller than the MTTF, the model is an approximate NHCTMC. Such a model requires only a global clock to describe

Table 13.1 Weibull failure
rate parameter values.

Parameter	Value
β	2.1
$1/\eta$	1.02 yr
μ	$120\,\mathrm{yr}^{-1}$
c	0.9
t_0	0

all the time-dependent transition rates, since in every state each component is as old as the system. To compute the reliability of this model, the PCA method of solution can be used. Note that the matrix $Q(t)$ cannot be factored and hence equivalent-HCTMC solution is not possible. Furthermore, the convolution integration approach will not be easy to use since the state transition graph is not acyclic.

The PCA method is based on the construction of a piecewise constant approximation of the time-continuous failure rate $\lambda(t)$. The overall time interval $(t_0 = 0, t_1]$ is divided into $n+1$ shorter intervals of length δ:

$$t \in (0, t_1] \rightarrow t \in (i\delta, (i+1)\delta], \ i = 0, 1, \ldots, n,$$

wherein the function is assumed to have the constant value $\lambda(i\delta)$:

$$\lambda(t) = \begin{cases} \lambda(\frac{\delta}{2}) & 0 < t \leq \delta \\ \lambda(\frac{\delta}{2} + \delta) & \delta < t \leq 2\delta \\ \ \vdots & \ \vdots \\ \lambda(\frac{\delta}{2} + i\delta) & i\delta < t \leq (i+1)\delta, \ i = 0, 1, \ldots, n \\ \ \vdots & \ \vdots \end{cases}$$

Figure 13.4 shows both the Weibull failure rate of Eq. (13.8) and its piecewise constant approximation with $\delta = 0.5$.

As a consequence of the failure rate approximation, the infinitesimal generator matrix $Q(t)$ also makes changes at discrete epochs:

$$Q(t) = \begin{cases} Q(\frac{\delta}{2}) & 0 \leq t < \delta \\ Q(\frac{\delta}{2} + \delta) & \delta < t \leq 2\delta \\ \ \vdots & \ \vdots \\ Q(\frac{\delta}{2} + i\delta) & i\delta < t \leq (i+1)\delta, \ i = 0, 1, \ldots, n \\ \ \vdots & \ \vdots \end{cases}$$

and hence the system is an HCTMC in each interval $(i\delta, (i+1)\delta], \ i = 0, 1, \ldots, n$. The transient state probabilities are computed successively in each subinterval by assigning the final state probability vector of the previous subinterval as the initial state probability

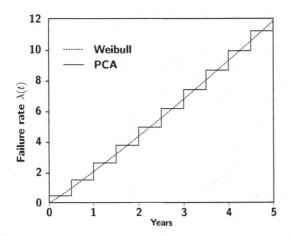

Figure 13.4 Weibull failure rate and piecewise constant approximation with $\delta = 0.5$.

vector of the next interval:

$$
\pi(t) = \begin{cases}
\pi(0)e^{Q(\frac{\delta}{2})(t-0)} & 0 < t \le \delta \\
\pi(\delta)e^{Q(\frac{\delta}{2}+\delta)(t-\delta)} & \delta < t \le 2\delta \\
\vdots & \vdots \\
\pi(i\delta)e^{Q(\frac{\delta}{2}+i\delta)(t-i\delta)} & i\delta < t \le (i+1)\delta, \ i = 0, 1, \dots, n \\
\vdots & \vdots
\end{cases}
$$

where

$$
\pi(\delta) = \pi(0)e^{Q(\frac{\delta}{2})(\delta-0)}
$$

$$
\pi(2\delta) = \pi(\delta)e^{Q(\frac{\delta}{2}+\delta)(2\delta-\delta)}
$$

$$
\vdots
$$

$$
\pi(i\delta) = \pi((i-1)\delta)e^{Q(\frac{\delta}{2}(i-1)+\delta)\delta}
$$

$$
\vdots
$$

The reliability for this NHCTMC can be computed using SHARPE [6, 7]. Figure 13.5 shows the reliability of the system for two different values of δ. We note that the two approximations are very similar.

Other measures besides system reliability can also be computed. For instance, the equivalent system failure rate $h(t)$ (as opposed to component failure rate) can be computed by extending Eq. (9.77) to the non-homogeneous case, as:

$$
h(t) = \frac{\lambda(t)\pi_1(t) + 2\lambda(t)(1-c)\pi_2(t)}{\pi_1(t) + \pi_2(t)}. \tag{13.9}
$$

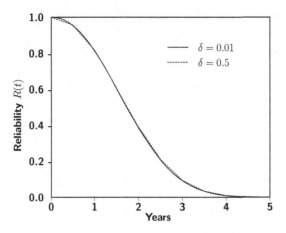

Figure 13.5 Reliability as a function of time for the system of Figure 13.3.

Note that, from the definition of hazard rate, Eq. (13.9) can also be interpreted as:

$$h(t) = \frac{\frac{d\pi_0(t)}{dt}}{1 - \pi_0(t)}. \qquad (13.10)$$

13.4 Queuing Examples

In this section we consider a variation of some of the queuing models introduced in Chapter 11 where we allow some random variables in the model to have non-exponential distributions.

Example 13.5 *M/M/m Queue for Dynamic Memory Allocation [8]*
In this example we consider a variation of the $M/M/m$ queue of Section 11.4 where an arriving job will request a certain amount of memory dynamically. This could also be any other similar type of resource being allocated dynamically. Most jobs will release the acquired resource once they finish using it but occasionally this may not occur. This implies that the resource is "lost" or depleted as far as the operating system (OS) is concerned. This phenomenon of resource leak/depletion is ubiquitous in almost all computer operating systems and has been called *software aging* [9]. The most well-known form of software aging is memory leaks within the dynamic storage allocation module of the OS. We shall study the time to resource exhaustion due to such leaks by means of an NHCTMC. Let M be the total amount of memory initially available. Assume that the resource requests form a Poisson process with rate λ, and the amount of resource for a request is a random variable X with probability density function $g(x)$. The system fails if the requested amount is more than the remaining available resource. We first define the variables necessary for modeling the depletion caused by resource leakage. Each incoming request can cause the system to transit to

the sink state (failure) if the amount of requested resource is more than the current available amount of the resource in the system pool, otherwise the request is queued up waiting for service.

Let T be the time to resource exhaustion or failure, i.e., the time to reach the sink state. The allocated amount of the resource is held for a random period of time, which depends on the processing or service rate, and determines the resource release rate. When the holding time per request is exponentially distributed with rate μ, the release rate in state k is $\mu_k = k\mu$.

The total amount of resource requested over k requests is

$$S_k = \sum_{i=0}^{k} X,$$

whose density and Cdf are, respectively:

$$g^{[k]}(x) \quad \text{and} \quad G^{[k]}(x) = \int_0^t g^{[k]}(u)\,du,$$

where $g^{[k]}(x)$ is the k-fold convolution of $g(x)$.

Let $\ell(t)$ be the current amount of leaked memory at time t; the resource available at time t is then $M - \ell(t)$, where $\ell(t)$ will likely be an increasing function of the time. Denote by $\zeta[k, \ell(t)]$ the conditional probability of system failure or crash in state k upon the arrival of a new request:

$$\zeta[k, \ell(t)] = P\{S_{k+1} > (M - \ell(t)) | S_k \le (M - \ell(t))\}$$

$$= \frac{G^{[k]}(M - \ell(t)) - G^{[k+1]}(M - \ell(t))}{G^{[k]}(M - \ell(t))}$$

$$= 1 - \frac{G^{[k+1]}(M - \ell(t))}{G^{[k]}(M - \ell(t))}.$$

In the leak-free case ($\ell(t) = 0$), the model is a homogeneous CTMC since all transition rates are then independent of the global time variable t.

In the presence of a leak, the CTMC of Figure 13.6 is non-homogeneous because $\zeta[k, \ell(t)]$ is a function of the global time variable t. This is because once the CTMC reaches the sink state, the system will have failed and will be rebooted. Thus t in the above is the time since the last reboot. We assume that $\ell(t)$ is any increasing function

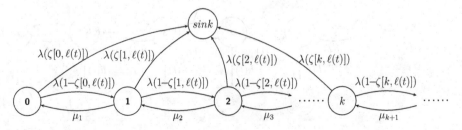

Figure 13.6 Model of dynamic memory allocation with leak present.

of time and differentiable except at a finite number of points. For the NHCTMC model of Figure 13.6, with generator matrix $Q(t)$, the transient state probability vector may be obtained by solving the Kolmogorov equation

$$\frac{d\pi(t)}{dt} = \pi(t)Q(t) \qquad \text{with initial condition} \qquad \pi_0 = 1$$

and $\quad \pi_{\text{sink}}(t) = 1 - \sum_k \pi_k(t)$.

The instantaneous system failure rate can be written, see also (13.10), as:

$$h(t) = \frac{\frac{d\pi_{\text{sink}}(t)}{dt}}{1 - \pi_{\text{sink}}(t)} = \frac{\sum_k \lambda \zeta[k, \ell(t)]\pi_k(t)}{\sum_k \pi_k(t)},$$

and it increases monotonically as the leaked resource accumulates [8]. Given a specific function $g(x)$ such as the uniform distribution, we can use the PCA method to numerically solve for $h(t)$. This can be said to be an analytic demonstration of the phenomenon of software aging.

Example 13.6   *Effect of Aging on Program Execution [10]*

We return to the $M/M/1/K$ queue with server breakdown, but unlike the case considered in Section 11.4 we will allow the server failure rate to be age dependent, as depicted in Figure 13.7. We further allow the service rate of the server to degrade with age. This example is adapted from [10]. We modify the behavior of an $M/M/1/K$ queue to capture two deleterious effects of software aging: performance degradation and crash/hang failures.

In Example 13.5, the server was the dynamic memory allocation module, and the memory allocator itself was the agent that produced aging effects due to memory leaks. By contrast, in the present example the server is the processor or a software server that is affected by software aging. We assume that requests for processing arrive according to a Poisson process of rate λ. The service rate μ will have a time-independent value in the absence of aging. However, software aging leads to slow performance degradation and hence μ will be a decreasing function $\mu(t)$ of time since the last reboot. In addition, each state of the system under consideration is subject to a crash/hang failure with rate $h(t)$ that is an increasing function of time since the last reboot. We could, for instance, use as an input here the function $h(t)$ produced as an output in the previous example

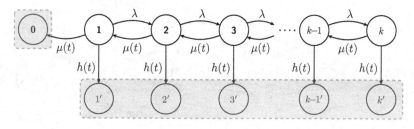

Figure 13.7   Two effects of software aging on program execution.

and given by Eq. (13.9). In any case, the NHCTMC can be solved using the PCA method.

Problems

13.4 Using the Bernoulli pmf for $g(j)$ in Example 13.5, use SHARPE and the PCA method to compute $h(t)$.

13.5 Use the $h(t)$ computed from Problem 13.4 to solve for the job rejection probability via the NHCTMC of Example 13.6. Use SHARPE and the PCA method.

13.5   Reliability Growth Examples

We test a system and we observe the time at which failures occur. The successive failure instants form a stochastic process, represented in Figure 13.8, where the labels in the states indicate the number of observed failures.

We consider a generalization of the homogeneous Poisson process to the case where the failure rate λ is allowed to be time (age) dependent. The time dependence of the failure intensity function $\lambda(t)$ is measured using a global clock. With this assumption, the process represented by the state diagram of Figure 13.8 becomes a non-homogeneous Poisson process (NHPP).

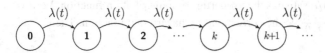

Figure 13.8   Non-homogeneous Poisson process.

Let $N(t)$ be the number of detected failures in the interval $(0, t]$ in the NHPP of Figure 13.8. Its pmf, $P\{N(t) = k\} = \pi_k(t)$, can be derived using the convolution integration method as:

$$\pi_k(t) = e^{-m(t)} \frac{[m(t)]^k}{k!} \qquad k \geq 0, \tag{13.11}$$

where the mean value function $m(t) = E[N(t)]$ is the expected number of failures detected by time t, and its derivative $\lambda(t) = \frac{dm(t)}{dt}$ is known as the failure intensity function. Notice that the label on the arc emanating from state $k \geq 0$ is $\lambda(t)$, where t is measured with the global clock and not the local clock (that is, the time of entry into state k). Hence this is an NHCTMC (and not an SMP).

Example 13.7   *The Duane Model [11]*
The Duane model has been used to capture reliability growth in many applications. In this model the NHPP failure intensity function $\lambda(t) = \eta\beta(\eta t)^{\beta-1}$ is used. Then

$m(t) = (\eta t)^{\beta}$. This model is also known as the power-law model and the AMSAA (Army materials systems analysis activity) model. We note that the time to first failure (and occupancy or holding time of state 0) is Weibull distributed since $\lambda(t)$ is the hazard rate of a Weibull distribution. However, the sojourn time of any state $k > 0$ will not be Weibull distributed. This can be verified as follows. Let T_k denote the occupancy (or sojourn) time in state k of the NHPP. Then first show that the conditional pdf of T_1 given T_0 is [12]

$$f_{T_1 | T_0}(t|u) = e^{-(m(t+u) - m(u))} \lambda(t+u),$$

and hence

$$f_{T_1}(t) = \int_0^{\infty} \lambda(t+u) \lambda(u) e^{-m(t+u)} du.$$

Example 13.8   *Software Reliability Growth Models*
Software testing for bugs and following removal of bugs is expected to enhance the reliability of software systems. Software reliability growth models (SRGMs) are used to quantify and assess the extent of reliability growth achieved by the testing and debugging activities. Assume that the number of failures $N(t)$ occurring during the time interval $(0, t]$ of testing has the Poisson pmf given by Eq. (13.11). In a so-called finite-failure SRGM, the mean value function $m(t)$ reaches a limit, say $\lim_{t \to \infty} m(t) = a$. Then $m(t)/a$ satisfies the properties of a distribution function. Let $F(t) = m(t)/a$ and let $h(t) = \frac{dF(t)/dt}{1 - F(t)}$ be the corresponding hazard rate function. Then

$$h(t) = \frac{\lambda(t)/a}{1 - m(t)/a} \quad \text{and} \quad \lambda(t) = h(t)[a - m(t)].$$

Hence, $h(t)$ can be interpreted as the failure occurrence rate per fault, and $a - m(t)$ as the average number of remaining faults at time t, while a is the expected number of faults to be found after an infinite amount of testing. Different finite-failure NHPP SRGMs can be obtained by specifying different functions $F(t)$ (or equivalently, $h(t)$). These are listed in Table 13.2 [13].

Table 13.2 Finite-failure NHPP SRGMs.

		$F(t)$	$h(t)$	$m(t)$	$\lambda(t)$
Goel–Okumoto	[14]	$1 - e^{-bt}$	b	$a(1 - e^{-bt})$	abe^{-bt}
Gen. Goel–Okumoto	[15]	$1 - e^{-bt^c}$	bct^{c-1}	$a(1 - e^{-bt^c})$	$abce^{-bt^c}t^{c-1}$
S-shaped	[16]	$1 - (1+gt)e^{-gt}$	$\frac{g^2 t}{1+gt}$	$a[1 - (1+gt)e^{-gt}]$	$ag^2 t e^{-gt}$
Log-logistic	[17]	$\frac{(\lambda t)^{\kappa}}{1+(\lambda t)^{\kappa}}$	$\frac{\lambda \kappa (\lambda t)^{\kappa-1}}{1+(\lambda t)^{\kappa}}$	$a\frac{(\lambda t)^{\kappa}}{1+(\lambda t)^{\kappa}}$	$a\frac{\lambda \kappa (\lambda t)^{\kappa-1}}{[1+(\lambda t)^{\kappa}]^2}$

Examples of infinite-failure SRGMs based on NHPPs include the Musa–Okumoto logarithmic model [18, 19] with

$$\lambda(t) = \frac{\gamma}{\gamma\theta t + 1} \quad \text{and} \quad m(t) = \frac{\ln(\gamma\theta t + 1)}{\theta},$$

so that $\lim_{t\to\infty} m(t) = \infty$. Note that the distribution of time between failures for a finite-failure NHPP model is defective [20]. Note also that SRGMs are not restricted to use in the testing phase but can be and have been used during the operational phase [21]. The SRGM above assumes that the time to fix a bug is zero. Gokhale *et al.* [17] extended the NHPP model to allow for non-zero time for the fault removal.

13.6 Numerical Solution Methods for NHCTMCs

Transient analysis of non-homogeneous CTMCs consists of solving the ODE IVP of Eq. (13.1) [22]. The transient solution methods for NHCTMCs are based on the same criteria as those for HCTMCs described in Section 10.4.4. The obvious difference is that in the HCTMC the infinitesimal generator Q is time independent, while in the NHCTMC $Q(t)$ is time dependent. In the following sections we discuss two methods for NHCTMC numerical transient analysis: the *ODE method* [23–25], and the *uniformization method* [26–28]. Numerical computation of the moments of accumulated reward in an NHCTMC reward model is discussed in [29].

13.6.1 Ordinary Differential Equation Method

We refer to Section 10.4.4 for a general description of ODE methods. Recall that we seek a solution on a discrete set of points $\{t_i | i = 0, 1, \ldots, n\}$ where $h = t_{i+1} - t_i$ is the step length, and we distinguish between non-stiff and stiff methods.

Most of the conventional ODE methods perform acceptably for transient analysis of non-stiff Markov models [25, 30], i.e., in cases where $\max_i\{|q_{ii}|\}t$ is not large. One of the most commonly used methods for the non-stiff case is the fourth-order Runge–Kutta [23, 24], which is an explicit single-step method.

However, for stiff Markov chains explicit methods have enormous difficulty solving the corresponding ODE. The problems stem from the fact that the solutions of stiff systems contain rapidly decaying transient terms [31].

Since explicit methods do not satisfy the so-called A-stability property, small step sizes must be used over the entire period of integration to avoid large numerical errors. Because transient solvers are often used to model a system from startup to steady state, explicit methods are time-consuming and often yield an incorrect steady-state solution. Moreover, for extremely stiff problems, even A-stability may be inadequate. To overcome this problem, [25] proposed the use of the TR-BDF2 method, which is an implicit two-step method with so-called L-stability. The adaptation of the TR-BDF2 method to NHCTMCs is as follows. The trapezoid rule applied to the interval $(t_i, t_i + \gamma h_i]$ is:

$$\pi_{i+\gamma}\left(I - \frac{\gamma h_i}{2}Q(t_i + \gamma h_i)\right) = \pi_i\left(I + \frac{\gamma h_i}{2}Q(t_i)\right). \qquad (13.12)$$

After computing $\pi_{i+\gamma}$, we use the *second-order backward difference formula* (BDF2) to step from $t_i + \gamma h_i$ to t_{i+1}; this step requires the solution of the linear system

$$\pi_{i+1}[(2-\gamma)I - (1-\gamma)h_i Q(t_{i+1})] = \gamma^{-1}\pi_{i+\gamma} - \gamma^{-1}(1-\gamma)^2\pi_i.$$

The local truncation error of the TR-BDF2 method is $O(h^3)$. To better control the error and provide a more efficient solution, the step size h_i can be varied from step to step. If too much error is introduced, the step is repeated with smaller h. After a successful step, h may be increased.

Another interesting technique for dealing with stiffness in NHCTMC analysis is proposed in [32] as an extension of the implicit Runge–Kutta method. Again, we refer to Section 10.4.4 for a general description of the method. In the present case, a third-order L-stable method with time step of constant size h is:

$$\pi_{i+1}(I - 2/3\,hQ(t_{i+1}) + 1/6h^2Q(t_{i+1})^2) = \pi_i(I + 1/3\,hQ(t_i)). \qquad (13.13)$$

This method involves only squaring the generator matrix and it is reasonable to expect that fill-in will not be very much. For further details of the solution methods for this equation we refer the reader to Section 10.4.4. Reference [32] provides a comprehensive overview of stiffness-tolerant methods for transient analysis of Markov chains.

13.6.2   Uniformization Method

Equation (13.1) can be solved again by adapting the *uniformization* method of Section 10.4.3 to the non-homogeneous case. For an NHCTMC with n states and generator matrix $Q(t) = [q_{ij}(t)]$, assume that there exists a $q < \infty$ such that, for all $t < \infty$, $q = \max_j |q_{jj}(t)|$. We define the Poisson process $\{N(t), t \geq 0\}$ and the embedded DTMC with one-step transition probability matrix

$$Q^*(t) = I + \frac{Q(t)}{q}$$

in the same way as for the homogeneous CTMC. The only difference is that now the transition probability matrix of the embedded DTMC $Q^*(t)$ is time dependent.

Given the definitions above, the uniformization series for an NHCTMC is given in [26], for all $0 \leq t \leq \infty$, as:

$$\pi(t) = \pi(0)\sum_{k=0}^{\infty}\frac{(qt)^k}{k!}e^{-qt}\cdot\underset{(t_1\leq t_2\leq\cdots\leq t_k)}{\int_0^t\int_0^t\cdots\int_0^t}Q^*(t_1)Q^*(t_2)\cdots Q^*(t_k)\,d\overline{H}(t_1,t_2,\ldots,t_k)$$

$$= \pi(0)\cdot\hat{U}(t), \qquad (13.14)$$

where $d\overline{H}(t_1,t_2,\ldots,t_k)$ is the joint density of the order statistics $t_1 \leq t_2 \leq \cdots \leq t_k$ of a k-dimensional uniform distribution at $(0,t] \times \cdots \times (0,t] \subset \mathbb{R}^k$, and

$Q^*(t_1)Q^*(t_2)\cdots Q^*(t_k)$ is the standard matrix product of the DTMC transition probability matrices at times $t_1 \le t_2 \le \cdots \le t_k$.

The main complication in computing the non-homogeneous expression in (13.14) is the fact that a continuum of transition matrices is required. To overcome this complication, in [26] a finite-grid approximation is used. Let $0 < h < q^{-1}$ be the step size and denote by $n_i = \lfloor t_i h^{-1} \rfloor$ $(i = 1, 2, \dots)$ the grid points. The discrete-time approximation of the uniformization series, for arbitrary n and $t = nh$, is as follows:

$$U(t) = U(nh) \tag{13.15}$$

$$= \sum_{k=0}^{\infty} \frac{(qt)^k}{k!} e^{-qt} \cdot \left[\sum_{\substack{0 \\ n_1 \le n_2 \le \cdots \le n_k}}^{n-1} \sum_{0}^{n-1} \cdots \sum_{0}^{n-1} Q(n_1 h) Q(n_2 h) \cdots Q(n_k h) \overline{H}(n_1, n_2, \dots, n_k) \right],$$

where $\overline{H}(n_1, n_2, \dots, n_k)$ is the pmf of the order statistics $n_1 \le n_2 \le \cdots \le n_k$ of a k-dimensional uniform distribution over $\{0, 1, \dots, n-1\}^k$. With the assumption that the transition rates are Lipschitz continuous for all t, i.e., there exists a positive constant ℓ such that

$$\|Q(t + \Delta t) - Q(t)\| \le \Delta t \cdot \ell,$$

it can be shown that

$$\|U(t) - \hat{U}(t)\| \le ht \cdot \ell,$$

where $\hat{U}(t)$ is the true value given in Eq. (13.14). Given a truncated series of $U(t)$ in (13.15), $\pi(t)$ can be computed using $\pi(0) U(t)$.

13.7   Further Reading

Besides those references cited in the text, other useful papers on the topic of NHCTMC include [33–38]. Though not covered in this book, there is literature on non-homogeneous discrete-time Markov chains. Applications of cyclic non-homogeneous discrete-time Markov models can be found in intermittent renewable energy production systems [39–41] such as wind, tidal, and solar power systems. These systems are not continuously available to provide power; however, they are generally functioning in different operational configurations and on a predictable cyclic basis (i.e., solar panels during daytime, wind turbines during particular seasons).

References

[1] K. Trivedi, *Probability and Statistics with Reliability, Queueing and Computer Science Applications*, 2nd edn. John Wiley & Sons, 2001.

[2] A. Rindos, S. Woolet, I. Viniotis, and K. Trivedi, "Exact methods for the transient analysis of nonhomogeneous continuous time Markov chains," in *Computations with Markov Chains.* Springer, 1995, pp. 121–133.

[3] T. E. Fortmann and K. L. Hitz, *An Introduction to Linear Control Systems (Control and System Theory).* CRC Press, 1977.

[4] W. Feller, *An Introduction to Probability Theory and its Applications*, Vols. I and II. John Wiley & Sons, 1968.

[5] G. Cosulich, P. Firpo, and S. Savio, "Power electronics reliability impact on service dependability for railway systems: A real case study," in *Proc. IEEE Int. Symp. on Industrial Electronics, ISIE '96*, vol. 2, Jun. 1996, pp. 996–1001.

[6] F. Frattini, A. Bovenzi, J. Alonso, and K. S. Trivedi, "Reliability indices," in *Wiley Encyclopedia of Operations Research and Management Science.* John Wiley & Sons, 2013, pp. 1–21.

[7] R. Sahner, K. Trivedi, and A. Puliafito, *Performance and Reliability Analysis of Computer Systems: An Example-based Approach Using the SHARPE Software Package.* Kluwer Academic Publishers, 1996.

[8] Y. Bao, X. Sun, and K. S. Trivedi, "A workload-based analysis of software aging, and rejuvenation," *IEEE Transactions on Reliability*, vol. 54, no. 3, pp. 541–548, 2005.

[9] Y. Huang, C. M. R. Kintala, N. Kolettis, and N. D. Fulton, "Software rejuvenation: Analysis, module and applications," in *Proc. Int. Symp. on Fault-Tolerant Computing (FTCS)*, 1995, pp. 381–390.

[10] S. Garg, A. Puliafito, M. Telek, and T. Trivedi, "Analysis of preventive maintenance in transactions based software systems," *IEEE Transactions on Computers*, vol. 47, no. 1, pp. 96–107, Jan. 1998.

[11] J. T. Duane, "Learning curve approach to reliability monitoring," *IEEE Transactions on Aerospace*, vol. 2, no. 2, pp. 563–566, 1964.

[12] E. Parzen, *Stochastic Processes.* Holden Day, 1962.

[13] S. Ramani, S. S. Gokhale, and K. S. Trivedi, "SREPT: Software Reliability Estimation and Prediction Tool," *Performance Evaluation*, vol. 39, no. 1–4, pp. 37–60, 2000.

[14] A. Goel and K. Okumoto, "Time-dependent error-detection rate model for software reliability and other performance measures," *IEEE Transactions on Reliability*, vol. R-28, no. 3, pp. 206–211, Aug. 1979.

[15] A. Goel, "Software reliability models: Assumptions, limitations, and applicability," *IEEE Transactions on Software Engineering*, vol. SE-11, no. 12, pp. 1411–1423, Dec. 1985.

[16] S. Yamada, M. Ohba, and S. Osaki, "S-shaped reliability growth modeling for software error detection," *IEEE Transactions on Reliability*, vol. R-32, no. 5, pp. 475–484, Dec. 1983.

[17] S. Gokhale, M. Lyu, and K. Trivedi, "Analysis of software fault removal policies using a non-homogeneous continuous time Markov chain," *Software Quality Journal*, vol. 12, no. 3, pp. 211–230, 2004.

[18] J. D. Musa and K. Okumoto, "A logarithmic Poisson execution time model for software reliability measurement," in *Proc. 7th Int. Conf. on Software Engineering.* IEEE Press, 1984, pp. 230–238.

[19] J. Musa, *Software Reliability Engineering: More Reliable Software Faster and Cheaper*, 2nd edn. Print on demand from http://johnmusa.com/book.htm, 2004.

[20] M. M. Grottke and K. S. Trivedi, "On a method for mending time to failure distributions," in *Proc. Int. Conf. on Dependable Systems and Networks, DSN 2005*, Jun. 2005, pp. 560–569.

[21] J. Alonso, M. Grottke, A. Nikora, and K. S. Trivedi, "The nature of the times to flight software failure during space missions," in *Proc. IEEE Int. Symp. on Software Reliability Engineering (ISSRE)*, 2012.

[22] J. D. Lambert, *Computational Methods in Ordinary Differential Equations*. John Wiley & Sons, 1973.

[23] L. F. Shampine, "Stiffness and nonstiff differential equation solvers, II: Detecting stiffness with Runge–Kutta methods," *ACM Transactions in Mathematical Software*, vol. 3, no. 1, pp. 44–53, 1977.

[24] E. Hairer, S. P. Nørsett, and G. Wanner, *Solving Ordinary Differential Equations I: Nonstiff Problems*, 2nd edn. Springer, 1993, vol. 8.

[25] A. Reibman and K. Trivedi, "Numerical transient analysis of Markov models," *Computers and Operations Research*, vol. 15, pp. 19-36, 1988.

[26] N. M. van Dijk, "Uniformization for nonhomogeneous Markov chains," *Operations Research Letters 12*, pp. 283–291, 1992.

[27] M. Malhotra, J. K. Muppala, and K. S. Trivedi, "Stiffness-tolerant methods for transient analysis of stiff Markov chains," *Microelectronics and Reliability*, vol. 34, pp. 1825–1841, 1994.

[28] K. W. A.van Moorsel, "Numerical solution of non-homogeneous Markov processes through uniformization," in *Proc. 12th European Simulation Multiconference on Simulation: Past, Present and Future*, 1998, pp. 710–717.

[29] M. Telek, A. Horváth, and G. Horváth, "Analysis of inhomogeneous Markov reward models," *Linear Algebra and its Applications*, vol. 386, pp. 383–405, 2004.

[30] W. Grassmann, "Transient solution in Markovian queueing systems," *Computers and Operations Research*, vol. 4, pp. 47–56, 1977.

[31] E. Hairer and G. Wanner, *Solving Ordinary Differential Equations II: Stiff and Differential Algebraic Problems*. Springer, 1991.

[32] M. Malhotra, J. K. Muppala, and K. S. Trivedi, "Stiffness-tolerant methods for transient analysis of stiff Markov chains," *Journal of Microelectronics and Reliability*, vol. 34, pp. 1825–1841, 1994.

[33] R. Geist, M. Smotherman, K. S. Trivedi, and J. B. Dugan, "The reliability of life-critical computer systems," *Acta Informatica*, vol. 23, no. 6, pp. 621–642, 1986.

[34] R. Geist, M. Smotherman, K. S. Trivedi, and J. B. Dugan, "The use of Weibull fault processes in modeling fault tolerant systems," *AIAA Journal of Guidance and Control*, vol. 11, no. 1, pp. 91–93, 1988.

[35] M. Smotherman and K. Zemoudeh, "A non-homogeneous Markov model for phased-mission reliability analysis," *IEEE Transactions on Reliability*, vol. 38, no. 5, pp. 585–590, 1989.

[36] K. Trivedi and R. Geist, *A Tutorial on the CARE III Approach to Reliability Modeling*, NASA contractor report 3488. National Aeronautics and Space Administration, Scientific and Technical Information Branch, 1981.

[37] K. Trivedi and R. Geist, "Decomposition in reliability analysis of fault-tolerant systems," *IEEE Transactions on Reliability*, vol. R-32, no. 5, pp. 463–468, Dec. 1983.

[38] Y. Li, E. Zio, and Y. Lin, *Methods of Solutions of Inhomogeneous Continuous Time Markov Chains for Degradation Process Modeling*. John Wiley & Sons, Ltd, 2013, pp. 3–16.

[39] A. Platis, N. Limnios, and M. L. Du, "Asymptotic availability of systems modeled by cyclic non-homogeneous Markov chains [substation reliability]," in *Proc. Ann. Symp. on Reliability and Maintainability*, Jan 1997, pp. 293–297.

[40] A. Platis, N. Limnios, and M. L. Du, "Dependability analysis of systems modeled by non-homogeneous Markov chains," *Reliability Engineering & System Safety*, vol. 61, no. 3, pp. 235–249, 1998.

[41] T. Scholz, V. Lopes, and A. Estanqueiro, "A cyclic time-dependent Markov process to model daily patterns in wind turbine power production," *Energy*, vol. 67, pp. 557–568, 2014.

14 Semi-Markov and Markov Regenerative Models

Recall that for a stochastic process to be called Markov, the Markov property (see Section 9.1) should be satisfied at all time instants t. Let us suppose that we can relax this restriction so that the Markov property is now required to be satisfied only at the time of entry into a state. The resulting stochastic process is known as a semi-Markov process (SMP) [1, 2]. If we consider the sequence of states of such a stochastic process at the instances of time when the state transitions occur (the transition epochs), the sequence will enjoy the Markov property and hence form a discrete-time Markov chain. This sequence is known as the embedded DTMC of the original SMP. A complete description of an SMP requires the definition of the transition probability matrix of the embedded DTMC and the Cdf of the continuous random variables assigned as sojourn times in each state. The strict requirement that at all transition epochs the Markov property be satisfied sometimes does not hold. Relaxing this requirement, under some conditions, may result in a Markov regenerative process (MRGP), as will be discussed in Section 14.5. After a brief definition of SMP in Section 14.1, the steady-state solution for an irreducible SMP is examined in Section 14.2. SMPs with single and multiple absorbing states are considered in Section 14.3 by computing the mean, variance, and probability of absorption. The transient analysis of an SMP is dealt with in Section 14.4, and Section 14.5 gives a brief account of Markov regenerative processes.

14.1 Introduction

Let $\{Z(t)|t \geq 0\}$ be the SMP under consideration. Let $t_0 = 0, t_1, t_2, \ldots, t_n, \ldots$ be the time instances at which $Z(t)$ undergoes a state transition. The sequence of states $\{X_n = Z(t_n)|n \geq 0\}$ is a DTMC. We will assume that the DTMC satisfies the homogeneity property and hence it is fully characterized by its one-step transition probability matrix (TPM) $P = [p_{ij}]$, where p_{ij} is the probability that the next jump of the DTMC is to state j given that the process is in state i. Let $H_i(t)$ be the sojourn time distribution in state i. If $H_i(t)$ is assumed to have an exponential distribution for all i, then the above SMP is simply an alternative characterization of a homogeneous CTMC. We will in general allow $H_i(t)$ to be generally distributed. In the rest of this chapter, those states that are drawn as squares in the figures represent states with non-exponential sojourn time distributions, and the labels on the transitions are the Cdfs of the transition times.

14.2   Steady-State Solution

In order to obtain the steady-state solution of an SMP, we first proceed to obtain the steady-state probability vector v of the embedded DTMC by solving the linear system of equations:

$$v = vP \quad \text{subject to} \quad ve^{\mathrm{T}} = 1, \tag{14.1}$$

where v_i is the steady-state probability of the embedded DTMC for state i and e is the vector with all entries equal to 1. We have assumed that the DTMC is irreducible and aperiodic, thus ensuring that it has a unique steady-state solution. Intuitively, $v_i N$ is the expected number of visits to state i out of the total of N observed transitions of the DTMC. Thus, v_i does not take into account the sojourn time spent in each state. Let $h_i = \int_0^\infty (1 - H_i(t))dt$ be the mean sojourn time in state i. Then the steady-state probability π_i for the SMP state i is given by [3]:

$$\pi_i = \frac{v_i h_i}{\sum_j v_j h_j}. \tag{14.2}$$

Example 14.1   *Two-State SMP*

Consider a two-state SMP modeling the failure/repair behavior of a single component. Let $F(t)$ and $G(t)$ respectively denote the distributions of time to failure and time to repair. Figure 14.1 shows this SMP graphically; it can be seen as a generalization of the two-state CTMC of Example 9.2, and as an alternative characterization of the alternating renewal process of Section 3.5.5. Note also that the repair of the component is considered to be maximal so that the component is returned to an as-good-as-new state upon completion of repair.

Following the steps described earlier, first we construct the TPM of the embedded DTMC:

$$P = \begin{array}{c} \\ U \\ D \end{array} \begin{array}{cc} U & D \\ \left[\begin{array}{cc} 0 & 1 \\ 1 & 0 \end{array} \right]. \end{array}$$

In this case, the DTMC is periodic, and thus limiting state probabilities do not exist as the DTMC alternates between two states. However, the stationary probability vector can

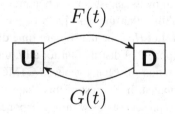

Figure 14.1   Two-state SMP availability model.

still be computed from

$$v = vP, \qquad v_U + v_D = 1$$

to obtain $v_U = v_D = \frac{1}{2}$.

Furthermore,

$$h_U = \int_0^\infty (1 - F(t))dt = \text{MTTF},$$

$$\text{and} \quad h_D = \int_0^\infty (1 - G(t))dt = \text{MTTR}.$$

Now, using Eq. (14.2) we obtain

$$\pi_U = \frac{\frac{1}{2}\text{MTTF}}{\frac{1}{2}\text{MTTF} + \frac{1}{2}\text{MTTR}}$$

$$= \frac{\text{MTTF}}{\text{MTTF} + \text{MTTR}} = \left[1 + \frac{\text{MTTR}}{\text{MTTF}}\right]^{-1}, \qquad (14.3)$$

which also gives the steady-state availability of the component – see Eq. (3.106).

Quite often in practice, state D will not be considered a down state unless the sojourn time in the state exceeds a threshold, say τ. Noting that the probability of an individual sojourn in state D not exceeding the threshold is given by $G(\tau)$, by assigning this value as a reward rate to state D and a reward rate 1 to state U, we get the steady-state availability as a function of the threshold τ as:

$$A(\tau) = \pi_U + G(\tau)\pi_D. \qquad (14.4)$$

Example 14.2 *Single Component Subject to Different Types of Failures [4]*
Next, we consider a single component subject to different types of failures and corresponding repairs (see Figure 14.2).

We assume that TTF and TTR for failure type i ($i = 1, 2, \ldots, n$) are respectively distributed as $F_i(t)$ and $G_i(t)$. We assume that only one type of failure can occur at a time, and that during the recovery/repair from a failure, another failure cannot

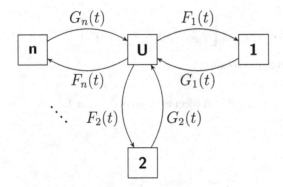

Figure 14.2   State diagram of a system with n outage types [4].

occur. Failures occur independently. Each repair completion brings the system to an as-good-as-new state. The age clock of each type of failure is thus reset upon the completion of a repair. With these assumptions, the underlying stochastic process is an SMP with $n+1$ states $\{U, 1, \ldots, n\}$. This model was used by Girtler [5] for the stability of a sea-going ship propulsion system, and the analysis shown below is from [4].

To derive the entries of the TPM of the embedded DTMC, we observe that exit from state U is determined by the minimum of the transition times emerging from state U (Section 3.3), hence

$$p_{i,U} = 1,$$

$$p_{U,i} = \int_0^\infty \prod_{j=1, j \neq i}^n (1 - F_j(x)) \, dF_i(x) \quad \text{with} \quad \sum_{i=1}^n p_{U,i} = 1.$$

Furthermore,

$$H_U(t) = 1 - \prod_{j=1}^n (1 - F_j(t)),$$

$$H_i(t) = G_i(t),$$

so that the mean sojourn times turn out to be

$$h_U = \int_0^\infty \prod_{j=1}^n (1 - F_j(t)) dt = \text{MTTF},$$

$$\text{and} \quad h_i = \int_0^\infty (1 - G_i(t)) dt = \text{MTTR}_i \quad (i = 1, 2, \ldots, n).$$

The TPM of the embedded DTMC can be seen to be

$$
P = \begin{array}{c} \\ U \\ 1 \\ \vdots \\ n \end{array}
\begin{array}{cccc}
U & 1 & \cdots & n \\
\left[\begin{array}{cccc}
0 & p_{U,1} & \cdots & p_{U,n} \\
p_{1,U} & 0 & \cdots & 0 \\
\vdots & \vdots & \ddots & \vdots \\
p_{n,U} & 0 & \cdots & 0
\end{array} \right]
\end{array}.
$$

It is then easy to see that the solution to

$$v = vP \quad \text{subject to} \quad v_U + \sum_{j=1}^n v_j = 1$$

is given by

$$v_U = \frac{1}{2} \quad \text{and} \quad v_i = \frac{p_{U,i}}{2}.$$

The steady-state availability becomes

$$A = \pi_U = \frac{h_U}{h_U + \sum_{j=1}^{n} p_{U,j} h_j} = \left[1 + \sum_{j=1}^{n} p_{U,j} \frac{h_j}{h_U} \right]^{-1}$$

$$= \left[1 + \sum_{j=1}^{n} p_{U,j} \frac{\text{MTTR}_j}{\text{MTTF}} \right]^{-1}. \qquad (14.5)$$

By substituting $n = 1$ in the above expression, we get Eq. (14.3).

Example 14.3 *Single Component Subject to Planned and Unplanned Outages [4]* We specialize Example 14.2 to the case of planned and unplanned outages, denoted by states PM and D respectively in Figure 14.3. The planned outage may be for system upgrades, configuration changes, maintenance, and so on. In practical situations, the planned downtime may be larger than the unplanned downtime. Assume that the TTF and TTR for planned outages have respective distributions $F(t)$ and $G(t)$. The TTF and TTR for unplanned outages are assumed to follow exponential distributions with respective rates λ and μ.

When we specialize the formula in (14.5) to the SMP of Figure 14.3, we get:

$$A = \left[1 + p_{U,D} \frac{h_D}{h_U} + p_{U,PM} \frac{h_{PM}}{h_U} \right]^{-1}, \qquad (14.6)$$

where

$$h_D = \int_0^\infty e^{-\mu t} dt, \qquad\qquad p_{U,D} = \int_0^\infty (1 - F(t)) \lambda e^{-\lambda t} dt,$$

$$h_{PM} = \int_0^\infty (1 - G(t)) dt, \qquad p_{U,PM} = \int_0^\infty e^{-\lambda t} dF(t),$$

$$h_U = \int_0^\infty (1 - F(t)) e^{-\lambda t} dt.$$

By denoting $\alpha(\lambda) = \int_0^\infty e^{-\lambda x} dF(x)$, $\theta(\lambda) = \frac{\alpha(\lambda)}{1 - \alpha(\lambda)}$, and $\frac{1}{\mu_2} = \int_0^\infty (1 - G(t)) dt$, we can further write $h_U = (1 - \alpha(\lambda)) 1/\lambda$. Noticing that $p_{U,D} = \lambda h_U$, we finally obtain from (14.6):

$$A = \left[1 + \frac{\lambda}{\mu} + \theta(\lambda) \frac{\lambda}{\mu_2} \right]^{-1}.$$

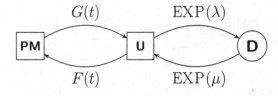

Figure 14.3   State diagram of a system with planned and unplanned outages [4].

From (14.6) we can see that the ratio of the downtime due to planned outage to the downtime due to failure is given by $\frac{\mu}{\mu_2}\theta(\lambda)$. In the case that the time to trigger an update is a constant equal to t_0, $F(t) = u(t - t_0)$ and the availability becomes

$$A = \left[1 + \frac{\lambda}{\mu} + \frac{\lambda}{\mu_2}\frac{e^{-\lambda t_0}}{1 - e^{-\lambda t_0}}\right]^{-1}.$$

The ratio of the downtime due to planned outage to the downtime due to failure is $\frac{\mu}{\mu_2}\frac{e^{-\lambda t_0}}{1 - e^{-\lambda t_0}}$.

Example 14.4   *Time-Based Preventive Maintenance [6]*
We now specialize the general model of Example 14.2 to the case of time-based preventive maintenance (PM) [6]. We will now assume that the TTF and TTR for unplanned outages have respective general distributions $F(t)$ and $G(t)$. The planned outage now will be for preventive maintenance which will be carried out after a deterministic duration t_0, and hence the corresponding distribution is the unit step function $u(t - t_0)$. For the time to carry out preventive maintenance, we will need only its mean, which we assume to be given by h_{PM}. Without loss of generality, we assume this distribution to be $u(t - h_{PM})$ (see Figure 14.4). We will also let the mean time to carry out reactive repair after a failure be given by

$$h = \int_0^\infty (1 - G(t))dt.$$

It follows that

$$p_{U,D} = P\{\text{failure occurs before the PM trigger}\} = F(t_0),$$
$$p_{U,PM} = 1 - F(t_0),$$
$$p_{PM,U} = 1 = p_{D,U},$$
$$h_U(t_0) = \int_0^{t_0}(1 - F(t))dt,$$
and $h_D = h.$

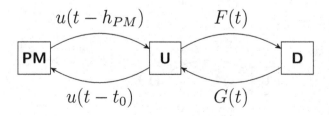

Figure 14.4   State diagram of time-based preventive maintenance model [6].

Hence the steady-state availability is given by

$$A = \pi_U = \frac{h_U(t_0)}{h_U(t_0) + p_{U,PM}h_{PM} + p_{U,D}h}$$

$$= \frac{h_U(t_0)}{h_U(t_0) + (1 - F(t_0))h_{PM} + F(t_0)h}. \tag{14.7}$$

Now, given the nature of the distribution function $F(t)$, we need to derive the formula for $h_U(t_0)$ and then we can compute the steady-state availability. Note that $F(t)$ should be an IFR distribution in order for PM to yield positive results. If we have the data of past inter-failure times, we could fit the data to some known distributions and then derive $h_U(t_0)$ and thence the optimal preventive maintenance schedule. We will go through the steps assuming $F(t)$ to be hypoexponential HYPO(λ_1, λ_2), Weibull WEIB(α, β), $\beta > 1$ and lognormal LN(μ, σ^2) as examples of three such IFR distributions. We note that Barlow and Campo [7] have developed a non-parametric technique, known as "total time on test transform," that avoids the step of fitting the data to a distribution and directly works with the sequence of inter-failure times to derive the optimal PM schedule. In Chapter 16 on hierarchical models, Example 16.1 will consider the possibility that the failure time distribution $F(t)$ is derived from a lower-level model.

Two-Stage Hypoexponentially Distributed TTF (Section 3.2.5)

$$F(t) = 1 - \frac{\lambda_2}{\lambda_2 - \lambda_1}e^{-\lambda_1 t} + \frac{\lambda_1}{\lambda_2 - \lambda_1}e^{-\lambda_2 t}. \tag{14.8}$$

Then,

$$h_U = \int_0^{t_0} (1 - F(t))dt$$

$$= \frac{\lambda_1}{\lambda_2(\lambda_2 - \lambda_1)}e^{-\lambda_2 t_0} - \frac{\lambda_2}{\lambda_1(\lambda_2 - \lambda_1)}e^{-\lambda_1 t_0} + \frac{\lambda_1 + \lambda_2}{\lambda_1 \lambda_2}. \tag{14.9}$$

Substituting Eq. (14.9) in (14.7), we get the final expression for the steady-state availability.

Weibull-Distributed TTF (Section 3.2.3)
Note that we use here the Cdf expression given in (3.27).

$$F(t) = 1 - \exp\left(-\alpha t^\beta\right), \tag{14.10}$$

and $$h_U = \int_0^{t_0} (1 - F(t))dt = \int_0^{t_0} e^{-\alpha t^\beta} dt$$

$$= \frac{1}{\beta \alpha^{1/\beta}} \int_0^{\alpha t_0^\beta} e^{-u} u^{\frac{1}{\beta} - 1} du \tag{14.11}$$

$$= \frac{\alpha^{-1/\beta}}{\beta} \Gamma\left(\frac{1}{\beta}\right) G\left(\alpha t_0^\beta, \frac{1}{\beta}\right),$$

where $G(x,a) = \frac{1}{\Gamma(a)} \int_0^x e^{-u} u^{a-1} du$ is the incomplete Gamma function.

Thus, substituting Eq. (14.11) in (14.7), we get the required expression for the steady-state availability.

Lognormal-Distributed TTF (Section 3.2.4)

Zhao *et al.* have shown in [8] that the best fit of experimental failure data in software aging experiments is provided by a lognormal distribution. The corresponding density and distribution functions are given by (3.32) and (3.33):

$$f(t) = \frac{1}{\sigma t \sqrt{2\pi}} e^{\frac{-(\ln(t)-\mu)^2}{2\sigma^2}},$$

$$F(t) = \frac{1}{2}\left[1 + \mathrm{erf}\left(\frac{\ln(t)-\mu}{\sigma\sqrt{2}}\right)\right].$$

Further analysis is similar to the earlier case of a Weibull-distributed TTF:

$$h_U = \int_0^{t_0} (1 - F(t)) dt$$

$$= \int_0^{t_0} \left(1 - \frac{1}{2}\left[1 + \mathrm{erf}\left(\frac{\ln(t)-\mu}{\sigma\sqrt{2}}\right)\right]\right) dt$$

$$= \int_0^{t_0} \left(\frac{1}{2}\left[1 - \mathrm{erf}\left(\frac{\ln(t)-\mu}{\sigma\sqrt{2}}\right)\right]\right) dt$$

$$= \int_0^{t_0} \left(\frac{1}{2}\left[\mathrm{erfc}\left(\frac{\ln(t)-\mu}{\sigma\sqrt{2}}\right)\right]\right) dt$$

$$= \frac{1}{2}\int_0^{t_0} \mathrm{erfc}\left(\frac{\ln(t)-\mu}{\sigma\sqrt{2}}\right) dt,$$

where erf is the error function and erfc is the complementary error function. Combining this result with the one-step TPM of the embedded DTMC, we get an expression for the steady-state availability (14.7):

$$p_{U,PM} = 1 - F(t_0) = \frac{1}{2}\mathrm{erfc}\left(\frac{\ln(t_0)-\mu}{\sigma\sqrt{2}}\right),$$

$$p_{U,D} = F(t_0) = \frac{1}{2}\mathrm{erf}\left(\frac{\ln(t_0)-\mu}{\sigma\sqrt{2}}\right).$$

Substituting in (14.7),

$$A_{\mathrm{lognorm}} = \frac{h_U}{h_U + \frac{1}{2}\left[\mathrm{erfc}(K) h_{PM} + \mathrm{erf}(K) h\right]},$$

where K is given by

$$K = \frac{\ln(t_0)-\mu}{\sigma\sqrt{2}}.$$

The final expression for steady-state availability is:

$$A_{\text{lognorm}} = \frac{\int_0^{t_0} \text{erfc}\left(\frac{\ln(t)-\mu}{\sigma\sqrt{2}}\right) dt}{\int_0^{t_0} \text{erfc}\left(\frac{\ln(t)-\mu}{\sigma\sqrt{2}}\right) dt + \left[\text{erfc}\left(\frac{\ln(t_0)-\mu}{\sigma\sqrt{2}}\right) h_{\text{PM}} + \text{erf}\left(\frac{\ln(t_0)-\mu}{\sigma\sqrt{2}}\right) h\right]}.$$

Example 14.5 *Optimal Preventive Maintenance Interval*
We wish to determine the optimal value of the preventive maintenance trigger interval t_0 for the general case first, and then for the special case of a WEIB(α, β) TTF distribution. We can show how to compute the optimal PM trigger interval t_0 by taking the derivative of the expression for A (14.7) with respect to t_0 and equating it to 0:

$$\frac{\partial A}{\partial t_0} = \frac{h_U'(t_0)[h_U(t_0) + h_{\text{PM}}(1 - F(t_0)) + hF(t_0)] - h_U(t_0)[h_U'(t_0) - h_{\text{PM}}F'(t_0) + hF'(t_0)]}{[h_U(t_0) + h_{\text{PM}}(1 - F(t_0)) + hF(t_0)]^2}$$

$$= 0.$$

Equating the numerator of the previous equation to 0 and expanding,

$$h_U(t_0)h_U'(t_0) + h_{\text{PM}}h_U'(t_0) - h_{\text{PM}}h_U'(t_0)F(t_0) + hh_U'(t_0)F(t_0)$$
$$= h_U(t_0)h_U'(t_0) - h_{\text{PM}}h_U(t_0)F'(t_0) + hh_U(t_0)F'(t_0),$$

and then

$$h_{\text{PM}}h_U'(t_0) + (h - h_{\text{PM}})F(t_0)h_U'(t_0) = (h - h_{\text{PM}})F'(t_0)h_U(t_0). \tag{14.12}$$

As an example, for the failure distribution $F(t)$ being Weibull (3.27),

$$F'(t_0) = f(t_0) = \alpha\beta(t_0)^{\beta-1}e^{-\alpha t_0^\beta}$$

and

$$h_U'(t_0) = \frac{\partial}{\partial t_0}\int_0^{t_0} e^{-\alpha t^\beta} dt = e^{-\alpha t_0^\beta}.$$

Substituting the expressions for $F'(t_0)$ and $h_0'(t_0)$ in (14.12),

$$h_{\text{PM}}e^{-\alpha t_0^\beta} + (h - h_{\text{PM}})(1 - e^{-\alpha t_0^\beta})(e^{-\alpha t_0^\beta}) = (h - h_{\text{PM}})\alpha\beta(t_0)^{\beta-1}e^{-\alpha t_0^\beta}h_U(t_0).$$

Thus, the general non-linear equation for the optimal t_0 becomes

$$(h + (h_{\text{PM}} - h)e^{-\alpha(t_0)^\beta}) + \beta\alpha(h_{\text{PM}} - h)(t_0)^{\beta-1}h_U(t_0) = 0, \tag{14.13}$$

similar to the expression provided in [8]. Clearly, there is no closed-form expression, hence a numerical solution is sought. Evaluating Eqs. (14.10), (14.11), and (14.13) iteratively, the optimal t_0 can be derived as shown in the routine written in R [9] below, using numerical values for the parameters from [8].

```
h_pm <- 60;
h <- 300;
alpha <- 1.811923e-12 # in /second
beta <- 17.3682;

weib_3state_solver <- function(t0) {
  x <- (alpha)*(t0^beta)
```

```
y <- (alpha ^ (-1/beta))
h0 <- (y / beta)*gamma(1/beta)*pgamma(x, 1/beta)
(h + (h_pm-h)*exp(-alpha*((t0)^beta))) +
    h0*beta*alpha*(h_pm-h)*(t0^(beta-1))
}

soln <- uniroot(weib_3state_solver, c(1, 100000000));

Results: t0* = 6162601 sec (1711.83 hrs), A(t0*)= 0.9999897
```

Example 14.6   *Two Repairable Components with Imperfect Coverage (continued)*
We again return to the two-component system with imperfect coverage that we
modeled using a GSPN in Example 12.12. Looking back at the ERG presented in
Figure 12.12(b), we realize that this ERG could easily be viewed as an SMP, as
presented in Figure 14.5. This SMP consists of two kinds of states, those in which
the sojourn time is exponentially distributed (the tangible states), and those in which
the sojourn time is zero (the vanishing states). Here, our attempt is to show the solution
of the ERG with *preservation* [10, 11], whereby the vanishing states are not eliminated
to construct the underlying CTMC. Instead, the vanishing states are preserved, and the
resulting ERG can be solved as an SMP.

The corresponding TPM of the embedded DTMC is given as follows:

$$
\bordermatrix{
 & 200 & 110 & 101 & 011 & 002 \cr
200 & 0 & 1 & 0 & 0 & 0 \cr
110 & 0 & 0 & c & 0 & 1-c \cr
\boldsymbol{P}= \quad 101 & \frac{\mu}{\lambda+\mu} & 0 & 0 & \frac{\lambda}{\lambda+\mu} & 0 \cr
011 & 0 & 0 & 0 & 0 & 1 \cr
002 & 0 & 0 & 1 & 0 & 0 \cr
}.
$$

Solving the steady-state equations in (14.1) for the embedded DTMC, we get

$$
v_{200} = v_{110} = \frac{\mu}{(4-c)\mu+3\lambda}, \qquad v_{101} = \frac{\lambda+\mu}{(4-c)\mu+3\lambda},
$$

$$
v_{011} = \frac{\lambda}{(4-c)\mu+3\lambda}, \qquad v_{002} = \frac{(1-c)\mu+\lambda}{(4-c)\mu+3\lambda}.
$$

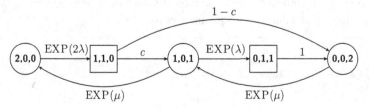

Figure 14.5   Two-component availability model with imperfect coverage: the corresponding ERG viewed as
an SMP.

Noting the mean sojourn times in the SMP states: $h_{2,0,0} = 1/(2\lambda)$, $h_{1,0,1} = 1/(\lambda + \mu)$, $h_{0,0,2} = 1/(\mu)$, $h_{1,1,0} = h_{0,1,1} = 0$, we can derive the steady-state probabilities of the SMP states using Eq. (14.2). The CTMC model for this example was considered in Example 9.5, where the steady-state availability was computed. The result computed using the SMP model above should match the results derived earlier. We leave this as a problem for the reader.

Example 14.7 *Availability of an Uninterruptible Power Supply [12]*
We now consider the uninterruptible power supply (UPS) model from [12, 13]. Devices that need an uninterruptible power supply often use rechargeable batteries as a backup for the normal AC power supply. Upon failure of the main power supply the battery supplies power to the device. However, batteries drain after the stored charge is exhausted. Assume that this charge will last for a fixed amount of time, say L. Thus, if the AC power supply is restored before the battery is fully discharged, the system will experience an uninterrupted power supply. This situation can be modeled by the three-state SMP shown in Figure 14.6.

From Figure 14.6, the one-step TPM is:

$$p_{B,U} = \int_0^L \mu e^{-\mu t} dt = 1 - e^{-\mu L}, \qquad p_{B,D} = e^{-\mu L},$$
$$p_{U,B} = 1, \qquad\qquad\qquad\qquad p_{D,U} = 1;$$

$$
P = \begin{array}{c} \\ U \\ B \\ D \end{array}
\begin{array}{c} \begin{array}{ccc} U & B & D \end{array} \\
\left[\begin{array}{ccc}
0 & 1 & 0 \\
1 - e^{-\mu L} & 0 & e^{-\mu L} \\
1 & 0 & 0
\end{array} \right].
\end{array}
$$

Solving for the steady-state probabilities of the embedded DTMC using Eqs. (14.1), we get

$$v_U = v_B(1 - e^{-\mu L}) + v_D, \qquad v_B = v_U, \qquad v_D = v_B e^{-\mu L},$$
$$v_U = \frac{1}{2 + e^{-\mu L}}, \qquad\qquad v_B = \frac{1}{2 + e^{-\mu L}}, \qquad v_D = \frac{e^{-\mu L}}{2 + e^{-\mu L}}.$$

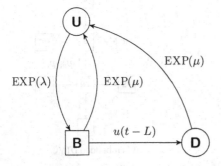

Figure 14.6   State diagram of the UPS availability model [12].

The state sojourn times and SMP steady-state probabilities, from Eq. (14.2), become:

$$H_B(t) = \begin{cases} 1 - e^{-\mu t} & t < L, \\ 1 & t \geq L, \end{cases}$$

$$h_U = \frac{1}{\lambda}, \qquad h_B = \int_0^L e^{-\mu t} dt = \frac{1 - e^{-\mu L}}{\mu}, \qquad h_D = \frac{1}{\mu},$$

$$\pi_U = \frac{\mu}{\lambda + \mu}, \qquad \pi_B = \frac{\lambda(1 - e^{-\mu L})}{\lambda + \mu}, \qquad \pi_D = \frac{\lambda e^{-\mu L}}{\lambda + \mu}.$$

Finally, the steady-state availability of the UPS is:

$$A = \pi_U + \pi_B = \frac{\mu}{\lambda + \mu} + \frac{\lambda}{\lambda + \mu}(1 - e^{-\mu L}).$$

Note that an alternative way of modeling this problem is by the two-state SMP of Example 14.1, by applying the idea that state D is a down state if the sojourn time in that state is longer than L. Hence the above equation is what we get by substituting $G(L) = 1 - e^{-\mu L}$ in Eq. (14.4).

Example 14.8   *Security Attributes of a Computer System [14]*

Our next example relates to quantifying the security attributes of a computer system or network and is adapted from [14]. In order to analyze the security attributes of an intrusion-tolerant system, we need to consider the actions undertaken by an attacker as well as the system's response to an attack. A generic state transition behavior of an intrusion-tolerant system is sketched in Figure 14.7(a), with the states identified in Table 14.1. We will assume that the system satisfies the assumption of an SMP. The embedded DTMC of this SMP is shown in Figure 14.7(b). The dashed lines represent transitions for recovering the full services after an attack and returning to the good state by manual intervention. A complete description of this SMP model requires knowledge

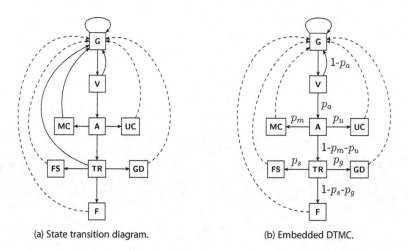

(a) State transition diagram.   (b) Embedded DTMC.

Figure 14.7   State transition diagram (a) and embedded DTMC (b) for an intrusion-tolerant system [14].

Table 14.1 State description for intrusion-tolerant computer system.

Label	Description
G	Good state
V	Vulnerable state
A	Active attack state
MC	Masked compromised state
UC	Undetected compromised state
TR	Triage state
FS	Fail-secure state
GD	Graceful degradation state
F	Failed state

of various parameters, such as the mean sojourn time in each state and the transition probabilities indicated in Figure 14.7(b).

The TPM of this DTMC is:

$$
\boldsymbol{P} = \begin{array}{c} \\ G \\ V \\ A \\ MC \\ UC \\ TR \\ FS \\ GD \\ F \end{array}
\begin{array}{c} \begin{array}{ccccccccc} G & V & A & MC & UC & \ \ TR & \ FS & GD & \ \ F \end{array} \\
\left[\begin{array}{ccccccccc}
0 & 1 & 0 & 0 & 0 & 0 & 0 & 0 & 0 \\
1-p_a & 0 & p_a & 0 & 0 & 0 & 0 & 0 & 0 \\
0 & 0 & 0 & p_m & p_u & 1-p_m-p_u & 0 & 0 & 0 \\
1 & 0 & 0 & 0 & 0 & 0 & 0 & 0 & 0 \\
1 & 0 & 0 & 0 & 0 & 0 & 0 & 0 & 0 \\
0 & 0 & 0 & 0 & 0 & 0 & p_s & p_g & 1-p_s-p_g \\
1 & 0 & 0 & 0 & 0 & 0 & 0 & 0 & 0 \\
1 & 0 & 0 & 0 & 0 & 0 & 0 & 0 & 0 \\
1 & 0 & 0 & 0 & 0 & 0 & 0 & 0 & 0
\end{array} \right]
\end{array}
$$

$$(14.14)$$

Solving for the steady-state probabilities of the DTMC, we get:

$$v_G = v_V(1-p_a) + v_{MC} + v_{UC} + v_{FS} + v_{GD} + v_F, \quad v_V = v_G, \quad v_A = v_V p_a = v_G p_a,$$

$$v_{MC} = v_A p_m, \quad v_{UC} = v_A p_u, \quad v_{TR} = v_A(1-p_m-p_u) = v_G p_a(1-p_m-p_u),$$

$$v_{FS} = v_{TR} p_s, \quad v_{GD} = v_{TR} p_g, \quad v_F = v_{TR}(1-p_s-p_g).$$

From the above equations, along with the normalization condition

$$\sum_i v_i = 1, \ i \in \{G,V,A, MC, UC, TR, FS, GD, F\},$$

we get

$$v_G = \frac{1}{2 + p_a(3 - p_m - p_u)}.$$

Once the mean sojourn times in all states are given, we can derive expressions for the SMP steady-state probabilities by applying Eq. (14.2). From these, we can get the steady-state system availability after recognizing that states F, FS, and UC are system down states:

$$A = 1 - (\pi_{FS} + \pi_F + \pi_{UC}).$$

To compute other security attributes such as the confidentiality and integrity, we need to consider specific attack scenarios. For several types of attacks identified in [14], the states UC and F will mean a loss of confidentiality, while state FS will not. Hence, the steady-state confidentiality is given by:

$$C = 1 - (\pi_F + \pi_{UC}).$$

Example 14.9 *Computer System Workload Model Reflecting Software Aging [15]*
Software aging is a phenomenon that leads to a depletion of OS resources such as the size of the dynamically allocated memory pool [16]. Recall the leakage function $\ell(t)$ considered in Example 13.5. Suppose that the actual size of the memory leaked at time t is measured and then an attempt is made to predict the time until exhaustion of all memory. Several different statistical approaches have been used for this purpose, including Sen's slope estimate [17]. In [17], the prediction of time to resource exhaustion was based only on elapsed time. However, the workload of a computer system has a natural and an important impact on the rate at which system resources deplete. In [15], the system workload is characterized by a number of measured variables (such as cpuContextSwitch, sysCall, pageOut, and pageIn) pertaining to CPU activity and file system I/O. These variables were measured at 10 min intervals. The actual measured values are then points in a five-dimensional space – four of the dimensions being the parameters mentioned above, with the fifth being the time at which they are measured. Thousands of points in this five-dimensional space are reduced to eight clusters or states through the well-known k-means clustering algorithm [18].

The next step, after the state clusters are identified, is to build a state–transition model for the system workload. This is done by determining the transition probabilities from one state to another. The transition probability p_{ij} from a state i to a state j was estimated from the actual measured data as:

$$p_{ij} = \frac{\text{observed no. of transitions from state } i \text{ to state } j}{\text{total observed no. of transitions from state } i}.$$

State transition probabilities were estimated from the observed data for the eight workload states, and the resulting 8×8 transition probability matrix $\mathbf{P} = [p_{ij}]_{8 \times 8}$ is

Table 14.2 Fitted sojourn time distributions in the eight workload states.

State	Sojourn time distribution, $H(t)$	Distribution type
1	$1 - 1.6029e^{-0.9t} + 0.6029e^{-2.39t}$	Hypoexponential
2	$1 - 0.9995e^{-0.446t} - 0.0005e^{-0.007t}$	Hyperexponential
3	$1 - 0.9952e^{-0.327t} - 0.0048e^{-0.0175t}$	Hyperexponential
4	$1 - 0.8414e^{-0.3275t} - 0.1586e^{-0.03825t}$	Hyperexponential
5	$1 - 1.426e^{-0.56t} + 0.4259e^{-1.875t}$	Hypoexponential
6	$1 - 0.807e^{-0.551t} - 0.193e^{-0.037t}$	Hyperexponential
7	$1 - 2.866e^{-1.302t} + 1.865e^{-2t}$	Hypoexponential
8	$1 - 0.9883e^{-0.2655t} - 0.0117e^{-0.0271t}$	Hyperexponential

shown below:

$$P = \begin{bmatrix} 0.000 & 0.155 & 0.224 & 0.129 & 0.259 & 0.034 & 0.165 & 0.034 \\ 0.071 & 0.000 & 0.136 & 0.140 & 0.316 & 0.026 & 0.307 & 0.004 \\ 0.122 & 0.226 & 0.000 & 0.096 & 0.426 & 0.000 & 0.113 & 0.017 \\ 0.147 & 0.363 & 0.059 & 0.000 & 0.098 & 0.216 & 0.088 & 0.029 \\ 0.033 & 0.068 & 0.037 & 0.011 & 0.000 & 0.004 & 0.847 & 0.000 \\ 0.070 & 0.163 & 0.023 & 0.535 & 0.116 & 0.000 & 0.023 & 0.070 \\ 0.022 & 0.049 & 0.003 & 0.003 & 0.920 & 0.003 & 0.000 & 0.000 \\ 0.307 & 0.077 & 0.154 & 0.231 & 0.077 & 0.154 & 0.000 & 0.000 \end{bmatrix}. \quad (14.15)$$

Using Eq. (14.1), the steady-state probability vector of the embedded DTMC was computed given the TPM matrix (14.15).

Then, the actual measured data was used to fit the sojourn time distribution vector $H(t) = [H_i(t)]$ to either a two-stage hyperexponential or a two-stage hypoexponential (Section 3.2.5) distribution function. As shown in Table 14.2, the sojourn times in workload states 1, 5, and 7 were fitted to hypoexponential distributions, while the sojourn times in all other states were fitted to hyperexponential distributions. The mean sojourn time h_i in state i was then calculated for each state. The steady-state probability vector $\pi = [\pi_1, \pi_2, \ldots, \pi_8]$ of the SMP was then computed using Eq. (14.2).

The computed steady-state probabilities from the SMP model were then compared with the observed values and found to be close enough to consider the SMP model a good fit. Once a validated SMP model of system workload is obtained, the next step is to assign a reward rate to each state corresponding to the rate of resource depletion in each state. From the measured data it was found that the rate of depletion did depend heavily on the workload state. From the SMP reward model, the expected time to resource depletion can be computed. This is an interesting example [15] where the structure (states and state transitions) as well as all the parameters of the model were estimated from measured data.

Example 14.10 *Two-Level Software Rejuvenation [19]*
Software aging is quite commonly observed, as has been pointed out in [15, 17]. It can be counteracted by a proactive technique called "software rejuvenation" that cleans

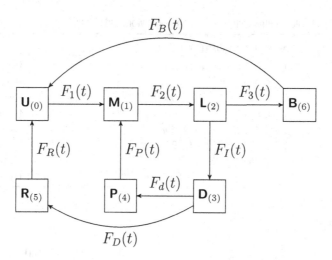

Figure 14.8   Semi-Markov process model for two-level software rejuvenation.

the environment in which the software is executing. This cleaning can take place at several levels of granularity [20, 21]. We focus here on software rejuvenation with actions offered at two levels of granularity. We study the semi-Markov model presented in Figure 14.8 for a software system with two-level rejuvenation actions. In Figure 14.8, the index (i) attached to the label of each state indicates the cardinal number of the state in the TPM.

The states are listed in Table 14.3 and have the following meanings:

- U is the highly efficient and highly robust software execution phase.
- M is the medium-efficient software execution phase.
- L is the low-efficient execution phase or alert phase. A rejuvenation action is needed or a system crash may happen.
- D is the state where it is determined which level of rejuvenation is appropriate.
- P is the partial rejuvenation state (level 1 rejuvenation).
- R is the full rejuvenation state (level 2 rejuvenation).
- B is the state where the system is recovering from a crash failure.

All the distributions $F_i(t)$ are assumed to be non-exponential. In practice, the time to trigger rejuvenation is typically a fixed duration, i.e., its Cdf has the form $F_I(t) = u(t - \tau)$, where $u(t)$ is the unit step function and τ is the time to trigger rejuvenation.

The non-zero entries of matrix P are:

$$p_{0,1} = 1,$$
$$p_{1,2} = 1,$$
$$p_{2,3} = 1 - F_3(\tau), \qquad p_{2,6} = F_3(\tau),$$
$$p_{3,4} = \int_0^\infty (1 - F_D(t))dF_d(t), \quad p_{3,5} = 1 - p_{3,4},$$
$$p_{4,1} = 1,$$
$$p_{5,0} = 1,$$
$$p_{6,0} = 1.$$

Table 14.3 State definition of the model of Figure 14.8.

No.	State	Description	Available?
0	U	Highly efficient phase	Yes
1	M	Medium-efficient phase	Yes
2	L	Low-efficient phase	Yes
3	D	Taken offline for rejuvenation	No
4	P	Partial rejuvenation	No
5	R	Full rejuvenation	No
6	B	Recovering from a crash failure	No

Solving the equations

$$v = vP \quad \text{subject to} \quad ve^{\mathrm{T}} = 1,$$

we obtain the steady-state probabilities of the embedded Markov chain for the SMP as

$$v = \frac{1}{D(\tau,p)}(1-p(1-F(\tau)),1,1,(1-F(\tau)),p(1-F(\tau)),(1-F(\tau))$$

$$-p(1-F(\tau)),F_3(\tau)),$$

in which $D(\tau,p) = 5 - p + pF_3(\tau) - F_3(\tau)$ and $p = p_{3,4}$. The expected sojourn times in the states of the SMP can be computed as:

$$h_0 = \int_0^\infty (1-F_1(t))dt,$$

$$h_1 = \int_0^\infty (1-F_2(t))dt,$$

$$h_2 = \int_0^\infty (1-F_1(t))(1-F_3(t))dt = \int_0^\tau (1-F_3(t))dt,$$

$$h_3 = \int_0^\infty (1-F_d(t))(1-F_D(t))dt,$$

$$h_4 = \int_0^\infty (1-F_P(t))dt,$$

$$h_5 = \int_0^\infty (1-F_R(t))dt,$$

$$h_6 = \int_0^\infty (1-F_B(t))dt.$$

The steady-state probability of each SMP state is:

$$\pi_i = \frac{v_i h_i}{\sum_{j=0}^6 v_j h_j}.$$

Finally, the steady-state availability of the system is:

$$A = \pi_0 + \pi_1 + \pi_2 = \frac{S(\tau,p)}{S(\tau,p) + V(\tau,p)},$$

where

$$S(\tau,p) = (1 - p + pF_3(\tau))h_0 + h_1 + \int_0^{\tau} (1 - F_3(t))dt,$$

$$V(\tau,p) = (p - pF_3(\tau))h_4 + (1 - p - F_3(\tau) + pF_3(\tau))h_5 + F_3(\tau)h_6.$$

Example 14.11   *Protocol in Vehicular Ad Hoc Networks [22]*

Our next example considers the DSRC (dedicated short-range communication) protocol used for safety messages in vehicular ad hoc networks [22]. Broadcasting safety messages is one of the fundamental services in DSRC that is being standardized as IEEE 802.11p. The SMP model capturing channel contention and backoff behavior is shown in Figure 14.9.

This SMP characterizes the packet transmission by a tagged vehicle. The vehicle is in the "idle" state if its queue is empty. Once a packet is ready to be transmitted, the vehicle first senses for channel activity for a duration known as DIFS (distributed inter-frame space). If the channel is detected to be idle during this period (associated probability $1 - q$), the vehicle transits to transmitting state XMT, implying that the packet is being transmitted. In the case that the channel is sensed to be busy during the DIFS period the vehicle will back off, with the backoff counter set randomly in the range $[0, W - 1]$ where W is the backoff window size. The backoff counter is decremented by 1 each time the channel is detected idle during a time slot of duration σ (this happens with probability $1 - p$); this action corresponds in the SMP to a transition from state $W - i$ to $W - i - 1$.

If another vehicle is transmitting and hence the channel is sensed busy during the backoff time slot σ (i.e., another vehicle is transmitting a packet), the backoff counter of the tagged vehicle is suspended and deferred for the duration of packet transmission, a period of length T. The corresponding transition in the SMP occurs from state $W - i$ to D_{W-i-1} with probability p. Once the backoff counter reaches the value 0, the packet will be transmitted – the SMP transits from state 0 to state XMT. Upon transmission, if the queue of the tagged vehicle is empty (with probability $1 - \rho$) the SMP transits to the idle state. Otherwise, if there are packets in the queue (with probability ρ) the vehicle

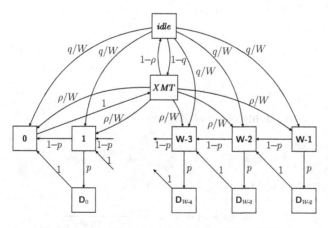

Figure 14.9   Semi-Markov process model for 802.11p broadcasts [22].

will sense the channel for a DIFS period and randomly chooses a backoff counter. From the state diagram and corresponding TPM, we can obtain the stationary probability vector of the embedded DTMC:

$$
\begin{cases}
v_j = (W - j)v_{W-1}, & j = 0, 1, \ldots, W - 1, \\
v_{D_j} = (W - j - 1)pv_{W-1}, & j = 0, 1, \ldots, W - 2, \\
v_{XMT} = \dfrac{W}{\rho + q(1 - \rho)}v_{W-1}, & \\
v_{idle} = \dfrac{(1 - \rho)W}{\rho + q(1 - \rho)}v_{W-1}. &
\end{cases}
$$

Finally, using the normalization condition, we get:

$$
v_{W-1} = \frac{2[\rho + q(1 - \rho)]}{[W + 1 + p(W - 1)][\rho + q(1 - \rho)]W + 2(2 - \rho)W}.
$$

As explained in [22], the mean sojourn times in the states of the SMP are given by:

$$
h_j =
\begin{cases}
\sigma & j = 0, 1, \ldots, W - 1, \\
T & j = D_0, D_1, \ldots, D_{W-2}, \\
T & j = XMT, \\
\frac{1}{\lambda} + DIFS & j = idle.
\end{cases}
$$

From the above, the steady-state probabilities of the SMP states can be computed using Eq. (14.2).

Problems

14.1   In Example 14.4 it was stated that the TTF distribution $F(t)$ should be IFR for the preventive maintenance to be effective. Explain why. What happens if $F(t)$ is exponential or DFR?

14.2   The steady-state availability of a repairable two-component system with imperfect coverage was analyzed in Example 14.6 as an SMP. Verify that the solution of the SMP coincides with the solution of the CTMC of Example 9.5.

14.3   For Example 14.7, derive expressions for the derivatives of the system availability A with respect to MTTF $= 1/\lambda$, MTTR $= 1/\mu$ and the battery constant L.

14.4   For the security SMP model of Example 14.8, derive expressions for the derivatives of the steady-state availability with respect to all the probability parameters of the embedded DTMC. Repeat for all the mean state sojourn times.

14.5   For the software aging model in Example 14.9, complete the computation of the steady-state probability vector of the embedded DTMC, and then the steady-state probability vector of the SMP. Next, using the reward rate assignment from the original paper [15], compute the expected steady-state reward rate and hence the approximate expected time to resource exhaustion. For those who want to derive the exact time to resource exhaustion, compute the distribution of accumulated reward $P\{Y(t) > x\}$. This is quite a bit more difficult [23, 24].

14.6   Consider a two-component parallel redundant system with generally distributed TTF and TTR. Assume that the component failure events and repair events are mutually independent; in other words, the two components can be modeled independently. Compare the steady-state availability of this system possessing redundancy with a system without redundancy but with optimal PM, as in Example 14.5. Carry out the comparison assuming a Weibull TTF distribution, with the following values: shape parameter $\alpha = 2,3$; scale parameter λ. Determine the value of λ such that MTTF = 1000 hr. Assume the reactive repair time is $h = $ MTTR = 1 hr. Carry out the comparison parametrically with respect to the ratio h_{PM}/h ranging from 0.000 01 to 1 (varying by factors of 10).

14.7   For the two-level rejuvenation model of Example 14.10, derive optimal values of the two parameters p and τ so as to maximize system availability.

14.3   Semi-Markov Processes with Absorbing States

Next we consider an SMP with one absorbing state. Thus the TPM of the embedded DTMC can now be seen to be in the following partitioned form:

$$P = \left[\begin{array}{c|c} P_u & c^T \\ \hline 0 & 1 \end{array}\right] \tag{14.16}$$

In the above, P_u is the $(n-1 \times n-1)$ partition of matrix P over the transient states and is a sub-stochastic matrix (row sum less than 1); c^T is a column vector grouping the transition probabilities from any transient state to the absorbing state. Note that the absorbing state needs a self-loop with probability 1 in order to make the TPM a stochastic matrix (so that all row sums are equal to 1).

14.3.1   Mean Time to Absorption

It can be shown that the expected number of visits V_j to state j until absorption is given by [25–27]

$$V_j = \alpha_j + \sum_{i=1}^{n-1} V_i p_{ij} \qquad j = 1, 2, \ldots, n,$$

where α_j is the initial probability of state j and state n is the absorbing state. The above equation can be written in vector–matrix form:

$$V = \alpha + V P_u, \tag{14.17}$$

where $\alpha = [\alpha_j]$ is the initial probability vector and $V = [V_j]$ the vector of the expected number of visits in each state. Once the average number of visits to state j is computed, the mean time to absorption is easily written as

$$\text{MTTA} = \sum_{j=1}^{n-1} V_j h_j,$$

where h_j is the mean sojourn time in state j.

Example 14.12 *Mean Time to Absorption in Example 14.4*

Returning to the periodic preventive maintenance of Example 14.4, but now disallowing (reactive) repair from the system failure state D, we wish to compute the mean time to absorption (see Figure 14.10).

Recalling the TPM for this case:

$$P = \begin{array}{c} \\ \text{U} \\ \text{PM} \\ \text{D} \end{array} \begin{array}{c} \begin{array}{ccc} \text{U} & \text{PM} & \text{D} \end{array} \\ \left[\begin{array}{cc|c} 0 & 1 - F(t_0) & F(t_0) \\ 1 & 0 & 0 \\ 0 & 0 & 1 \end{array} \right] \end{array}.$$

Assuming, as initial probability, $\alpha_U = 1$, we compute

$$V_U = 1 + V_{PM}, \qquad V_{PM} = [1 - F(t_0)] V_U.$$

Hence,

$$V_U = \frac{1}{F(t_0)}, \qquad V_{PM} = \frac{1 - F(t_0)}{F(t_0)}.$$

Then,

$$\text{MTTA} = \frac{h_U}{F(t_0)} + \frac{1 - F(t_0)}{F(t_0)} h_{PM} = \frac{h_U + h_{PM}}{F(t_0)} - h_{PM}.$$

Notice that during the time to absorption the system is down for PM multiple times. So if we attach reward rate 1 to state U and 0 to state PM, we can compute the expected accumulated reward to absorption as expected capacity to failure (MCTF):

$$\text{MCTF} = \frac{h_U}{F(t_0)} = \frac{\int_0^{t_0} (1 - F(t)) dt}{F(t_0)}.$$

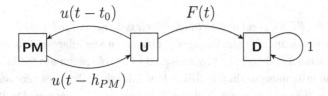

Figure 14.10 Absorbing state version of the time-based preventive maintenance model [6].

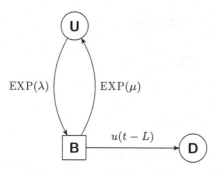

Figure 14.11   Absorbing state version of the UPS model.

Example 14.13   *Mean Time to Absorption in Example 14.7*
Returning to the UPS system in Example 14.7, by dropping the repair transition, we
have the modified SMP model with absorbing state as shown in Figure 14.11.
    The TPM is given by

$$P = \begin{array}{c} \\ U \\ B \\ D \end{array} \begin{array}{ccc} U & B & D \\ \left[\begin{array}{cc|c} 0 & 1 & 0 \\ 1 - e^{-\mu L} & 0 & e^{-\mu L} \\ 0 & 0 & 1 \end{array} \right] \end{array}.$$

Then, assuming as initial probability $\alpha_U = 1$, we have

$$V_U = 1 + V_B(1 - e^{-\mu L}),$$
$$V_B = V_U.$$

Hence, $V_U = 1 + V_U(1 - e^{-\mu L})$ or $V_U = e^{\mu L}$. Thus the MTTA, which in this case is also
MTTF, is given by:

$$\text{MTTF} = \frac{V_U}{\lambda} + V_B \left(\frac{1 - e^{-\mu L}}{\mu} \right)$$
$$= \frac{e^{+\mu L}}{\lambda} + \frac{e^{+\mu L}}{\mu} - \frac{1}{\mu}$$
$$= e^{+\mu L} \left(\frac{1}{\lambda} + \frac{1}{\mu} \right) - \frac{1}{\mu}.$$

Note that if $L = 0$, MTTF reduces to $1/\lambda$ since as soon as the power supply is down,
the system goes down without a battery backup.

Example 14.14   *RF Channel in a Cellular Wireless Network [28]*
Consider a model of an RF (radio frequency) channel in a cellular wireless network. The
quality of a channel is determined by its signal to interference ratio (SIR). As pointed
out in [29], an instantaneous drop in SIR below a threshold does not necessarily lead to
the occurrence of an outage event. A channel outage event is determined by the duration
of time that the SIR stays below a threshold. To reflect the impact of this time duration,

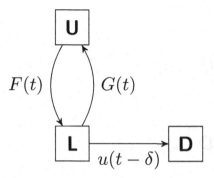

Figure 14.12 Semi-Markov process model of the minimum duration outage in an RF channel [28].

an RF channel outage event is said to occur when the SIR stays below a threshold for a duration longer than the "minimum duration," δ. This leads to the SMP model shown in Figure 14.12 [28]. The SMP model presented here is more general than that in [28] since we allow all the sojourn times to be generally distributed.

The TPM of the embedded DTMC is given by

$$P = \begin{array}{c} \\ U \\ L \\ D \end{array} \begin{array}{c} U \quad\quad L \quad\; D \\ \left[\begin{array}{cc|c} 0 & 1 & 0 \\ G(\delta) & 0 & 1-G(\delta) \\ 0 & 0 & 1 \end{array} \right] \end{array}.$$

The mean sojourn times are given by:

$$h_U = \int_0^\infty (1 - F(t))dt, \qquad h_L = \int_0^\delta (1 - G(t))dt.$$

Assuming, as initial probability, $\alpha_U = 1$, we obtain the average number of visits:

$$V_U = 1 + V_L G(\delta), \qquad V_L = V_U.$$

Hence, $V_U = 1 + V_U G(\delta)$ and $V_U = \frac{1}{1-G(\delta)}$. Thus, the MTTA, which is also the MTTF in this case, is given by:

$$\begin{aligned} \text{MTTF} &= \frac{h_U}{1 - G(\delta)} + \frac{h_L}{1 - G(\delta)} \\ &= \frac{\int_0^\infty (1 - F(t))dt}{1 - G(\delta)} + \frac{\int_0^\delta (1 - G(t))dt}{1 - G(\delta)}. \end{aligned}$$

We can also now study the steady-state availability of the channel by allowing recovery from state D, as in Figure 14.13.

We assume that once the channel has failed, the recovery activity, if any, that was started prior to failure needs to be restarted. Now substitute the MTTF from above, and

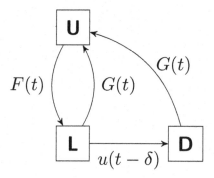

Figure 14.13   Semi-Markov process model of the minimum duration outage with recovery in an RF channel [28].

$\text{MTTR} = \int_0^\infty (1 - G(t))dt$ to get:

$$A = \frac{\text{MTTF}}{\text{MTTF} + \text{MTTR}} = \left(1 + \frac{\text{MTTR}}{\text{MTTF}}\right)^{-1}.$$

Example 14.15   *Recovery Process for Software Failures Caused by Mandelbugs [30]*
For a definition and classification of software failures, and in particular of Mandelbugs, the reader can refer to [31].

In this example, we depict the failure recovery behavior of an IT system during its operational phase by means of the flow chart shown in Figure 14.14. In this flow chart, the detection phase is followed by a diagnosis phase, during which the problem is investigated, and then by one or more recovery actions that are selected by diagnosis. Depending on the type of bug that caused the failure (restart-maskable Mandelbug, reboot-maskable Mandelbug, reconfiguration-maskable Mandelbug, Bohrbug or other type of Mandelbug), the recovery process follows one of four different branches of the flow chart in Figure 14.14. For a discussion of the software fault classification employed here, see [31].

Before the recovery actions, there is a diagnosis phase, in which either developers or automated tools select a recovery action among: restart, reconfiguration, reboot or fix. Then, depending on the bug type that triggered the failure, the action can successfully bring the system to a correct state, or the recovery action may not be successful and the system may still exhibit a failure upon re-execution. In the latter case, we assume that another recovery action is attempted, by performing another action with both a higher duration and a higher likelihood of avoiding a failure (e.g., if a restart fails, then a reboot is attempted; rebooting takes more time, but recovers from failures due to both restart-maskable and reboot-maskable Mandelbugs).

The proposed flow chart can be adopted to derive a semi-Markov model for the time to recovery from a failure, which is depicted in Figure 14.15. Recovery actions can have a generally distributed duration, with the only assumption being that their mean values should be finite. The model represents the distribution of the time to recovery from a software failure for a generic IT system. We use this model to derive a closed-form

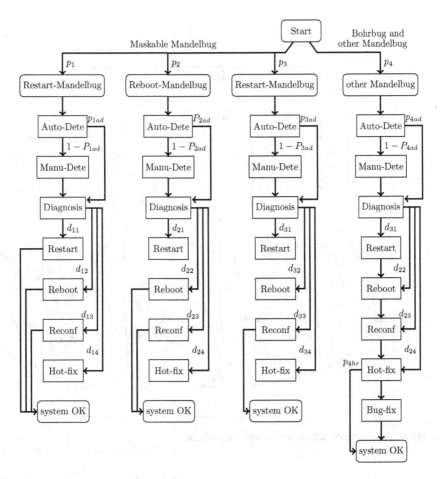

Figure 14.14 Flow chart of recovery after a failure.

expression of the MTTR, which can also be used to compute the steady-state availability of a system by means of the equation:

$$A = \frac{\text{MTTF}}{\text{MTTF} + \text{MTTR}} = \left(1 + \frac{\text{MTTR}}{\text{MTTF}}\right)^{-1}.$$

From the time to recovery SMP model shown in Figure 14.15, we obtain the following closed-form solution for the MTTR:

$$\begin{aligned}
\text{MTTR} = {}& p_1[E[D_{1ad}] + (1 - p_{1ad})E[D_{1md}] + E[D_{1dg}] + d_{11}E[D_{1rs}] \\
& + d_{12}E[D_{1rb}] + d_{13}E[D_{1rc}] + d_{14}E[D_{1hf}]] \\
& + p_2[E[D_{2ad}] + (1 - p_{2ad})E[D_{2md}] + E[D_{2dg}] + d_{21}(E[D_{2rs}] + E[D_{2rb}]) \\
& + d_{22}E[D_{2rb}] + d_{23}E[D_{2rc}] + d_{24}E[D_{2hf}]]
\end{aligned}$$

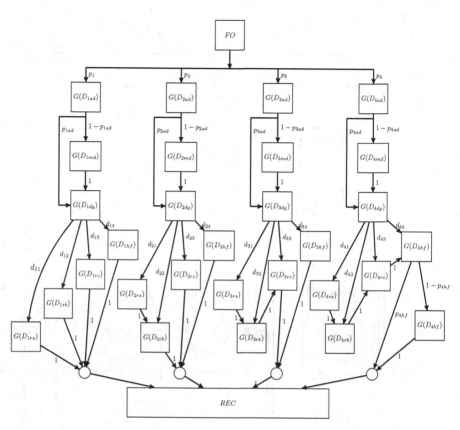

Figure 14.15   Time to recover SMP after a software failure.

$$+ p_3[E[D_{3ad}] + (1 - p_{3ad})E[D_{3md}] + E[D_{3dg}] + d_{31}(E[D_{3rs}] + E[D_{3rb}] + E[D_{3rc}])$$

$$+ d_{32}(E[D_{3rb}] + E[D_{3rc}]) + d_{33}E[D_{3rc}] + d_{34}E[D_{3hf}]]$$

$$+ p_4[E[D_{4ad}] + (1 - p_{4ad})E[D_{4md}] + E[D_{4dg}]$$

$$+ d_{41}(E[D_{4rs}] + E[D_{4rb}] + E[D_{4rc}] + E[D_{4hf}]) + d_{42}(E[D_{4rb}] + E[D_{4rc}] + E[D_{4hf}])$$

$$+ d_{43}(E[D_{4rc}] + E[D_{4hf}]) + d_{44}E[D_{4hf}] + (1 - p_{4hf}E[D_{4bf}]].$$

Problems

14.8   In Example 14.6, make state $(0,0,2)$ absorbing, thus turning the original availability model into a reliability model (with repair). Now derive an expression for the system MTTF. Also, derive expressions for the derivatives of the MTTF with respect to the three parameters λ, μ, and c. Now suppose that the overall cost of the system can be written as: $2K_1(\lambda)^{-\alpha} + K_2(\mu)^{\beta} + K_3(1 - c)^{-\gamma}$, for some positive constants K_1, K_2, K_3, α, β, and γ. Derive expressions for the values of λ, μ, and c that will maximize the system MTTF.

14.9 Consider the preventive maintenance in Example 14.12 and derive equations to determine the optimal value of the PM trigger interval t_0 in Figure 14.10 that will maximize the MCTF. Assume that the TTF is Weibull distributed.

14.3.2 Variance of the Time to Absorption

The TPM in partitioned form given in Eq. (14.16) can also be used to compute the variance of the time to absorption [25, 32]. First, note that the k-step transition probability matrix P^k has the form

$$P^k = \left[\begin{array}{c|c} P_u^k & c^T \\ \hline 0 & 1 \end{array}\right].$$

Entry p_{ij}^k of matrix P_u^k denotes the probability of arriving in (transient) state j after exactly k steps starting from (transient) state i. It can be shown that $\sum_{k=0}^{\ell} P_u^k$ converges as ℓ approaches infinity [27]. This implies that the inverse matrix $M = (I - P_u)^{-1}$, called the fundamental matrix, exists and is given by

$$M = (I - P_u)^{-1} = I + P_u + P_u^2 + \cdots = \sum_{k=0}^{\infty} P_u^k. \tag{14.18}$$

Let θ_{ij}^2 denote the variance of the number of visits to state j starting from state i. Then, in matrix form, we have [25, 32]:

$$\theta^2 = M(2M_d - I) - M_{sq}, \tag{14.19}$$

where M_d represents a diagonal matrix obtained from matrix M by zeroing out all its off-diagonal entries, and M_{sq} denotes the Hadamard product [33] of M with itself (i.e., $M_{sq\,ij} = M_{ij} \cdot M_{ij}$).

Once the variance of the number of visits to state j is computed, the variance of the time to absorption is written as

$$\text{VTTA} = \sum_{j=1}^{n-1} (\sigma_j \cdot V_j + h_j^2 \cdot \theta_{1j}), \tag{14.20}$$

where h_j and σ_j are the mean and variance of the sojourn time in state j, and θ_{1j} is the $(1, j)$ entry of matrix θ.

Example 14.16 *Variance of Time to Absorption for Composite Web Services*
Composition of multiple web services is a convenient way of defining new services within a business process. By combining existing services using a high-level language such as the Web Services Business Process Execution Language (WS-BPEL), commonly known as BPEL (Business Process Execution Language) [34, 35], service providers can quickly develop new services. When deploying these services, service providers need to study the mean and variance of a composite web service's completion time.

Suppose a composite service is constructed from four services, i.e., initial service (S_0), hotel searching service (S_1), supermarket searching service (S_2) and reply service (S_3). The BPEL process as illustrated in Figure 14.16 can be treated as a semi-Markov process, as shown in Figure 14.17. In the figure, C and F are the absorbing states, representing successful completion and failure, respectively. Here, p_1 and p_2 are the branching probabilities from the initial service (S_0) to the hotel searching service (S_1) and supermarket searching service (S_2), respectively. The hotel searching service (S_1), supermarket searching service (S_2) and reply service (S_3) may fail, and have the success probabilities q_1, q_2, and q_3, respectively.

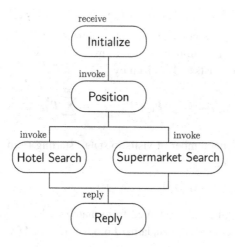

Figure 14.16   A composite web service in BPEL.

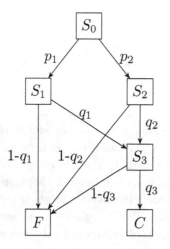

Figure 14.17   Composite web service SMP model.

Suppose we have obtained the following numerical values: $p_1 = 0.4$, $p_2 = 0.6$, $q_1 = 0.9$, $q_2 = 0.8$, $q_3 = 0.8$; then the TPM is

$$P = \begin{array}{c} \\ S_0 \\ S_1 \\ S_2 \\ S_3 \\ F \\ C \end{array} \begin{array}{cccccc} S_0 & S_1 & S_2 & S_3 & F & C \\ \left[\begin{array}{cccccc} 0 & 0.4 & 0.6 & 0 & 0 & 0 \\ 0 & 0 & 0 & 0.9 & 0.1 & 0 \\ 0 & 0 & 0 & 0.8 & 0.2 & 0 \\ 0 & 0 & 0 & 0 & 0.2 & 0.8 \\ 0 & 0 & 0 & 0 & 1 & 0 \\ 0 & 0 & 0 & 0 & 0 & 1 \end{array} \right] \end{array}.$$

Then, by Eq. (14.18) we can get

$$M = \begin{bmatrix} 1 & 0.4 & 0.6 & 0.84 \\ 0 & 1 & 0 & 0.9 \\ 0 & 0 & 1 & 0.8 \\ 0 & 0 & 0 & 1 \end{bmatrix}.$$

Further, we can calculate

$$M_{\text{d}} = \begin{bmatrix} 1 & 0 & 0 & 0 \\ 0 & 1 & 0 & 0 \\ 0 & 0 & 1 & 0 \\ 0 & 0 & 0 & 1 \end{bmatrix} \quad \text{and} \quad M_{\text{sq}} = \begin{bmatrix} 1 & 0.16 & 0.36 & 0.7056 \\ 0 & 1 & 0 & 0.81 \\ 0 & 0 & 1 & 0.64 \\ 0 & 0 & 0 & 1 \end{bmatrix}.$$

By Eq. (14.19), we can get

$$\theta^2 = \begin{bmatrix} 0 & 0.24 & 0.24 & 0.1344 \\ 0 & 0 & 0 & 0.09 \\ 0 & 0 & 0 & 0.16 \\ 0 & 0 & 0 & 0 \end{bmatrix}.$$

Thus, finally,

$$\text{VTTA} = \sum_{j=0}^{3} \left(\sigma_j \cdot V_j + h_j^2 \cdot \theta_{1i}^2 \right) \tag{14.21}$$

$$= \sigma_0 + (0.4 \cdot \sigma_1 + 0.24 \cdot h_1^2) + (0.6 \cdot \sigma_2 + 0.24 \cdot h_2^2) + (0.84 \cdot \sigma_3 + 0.1344 \cdot h_3^2),$$

where h_j and σ_j are the mean and variance of the sojourn time in state j ($j = 0, 1, 2, 3$).

For other examples showing the computation of the variance of the time to absorption, see [36–38].

Problems

14.10 In Example 14.6, make state $(0,0,2)$ absorbing, thus turning the original availability model into a reliability model (with repair). Now derive an expression for the variance of the system TTF.

14.3.3   Probability of Absorption in SMPs with Multiple Absorbing States

Next we consider SMPs with multiple absorbing states. Assume that there are m absorbing states $(1 \leq m < n)$ in an n-state SMP. Then the TPM can be partitioned so that P_u is an $(n - m \times n - m)$ sub-stochastic matrix, C is a rectangular $(n - m \times m)$ matrix and I is an $(m \times m)$ identity matrix:

$$P = \left[\begin{array}{c|c} P_u & C \\ \hline 0 & I \end{array} \right],$$

Medhi [39] shows that

$$B = (I - P_u)^{-1} C, \tag{14.22}$$

where matrix $B = [b_{ij}]$ is defined so that b_{ij} is the probability of being absorbed in state j given that the DTMC started in state i. The matrix $M = (I - P_u)^{-1}$, see (14.18), is known as the fundamental matrix of the embedded DTMC.

Example 14.17   *Security Attributes of a Computer System (continued) [14]*

We return to the security quantification problem of Example 14.8. We identify those states that are security compromised states and make them absorbing states. Subsequently, the MTTA in this case can be christened the mean time to security failure (MTTSF). For example, for a Sun web server bulletin board vulnerability (Bugtraq ID 1600), states UC, FS, GD, and F will form the set of absorbing states (and, for the MTTSF analysis, can be merged into a single absorbing state). Making these states absorbing states in the matrix (14.14), and applying Eq. (14.17), assuming $\alpha_G = 1$, we finally find:

$$V_G = \frac{1}{p_a(1 - p_m)} = V_V, \quad V_A = \frac{1}{1 - p_m}, \quad V_{MC} = \frac{p_m}{1 - p_m}, \quad V_{TR} = \frac{1 - p_m - p_u}{1 - p_m},$$

$$\text{and} \quad \text{MTTSF} = \frac{h_G p_a^{-1} + h_V p_a^{-1} + h_A + h_{MC} p_m + h_{TR}(1 - p_m - p_u)}{1 - p_m}.$$

Furthermore, after separating each of the absorbing states, using Eq. (14.22), we can find the absorption probabilities as:

$$b_{G,F} = \frac{(1 - p_s - p_g)(1 - p_m - p_u)}{1 - p_m},$$

$$b_{G,FS} = \frac{p_s(1 - p_m - p_u)}{1 - p_m},$$

$$b_{G,GD} = \frac{p_g(1 - p_m - p_u)}{1 - p_m},$$

$$b_{G,UC} = \frac{p_u}{1 - p_m}.$$

Problems

14.11 For the composite web service in Example 14.16, derive expressions for the probabilities of absorption into states C and F.

14.4   Transient Solution

We denote by $V(t) = [V_{ij}(t)]$ the matrix of the transient solution for the conditional transition probability, where each entry $V_{ij}(t)$ is defined as:

$$V_{ij}(t) = P\{Z(t) = j \mid Z(0) = i\}.$$

To arrive at the solution equation, define the sequence of time points at which $\{Z(t) \mid t \geq 0\}$ makes state transitions as

$$T_0 = 0, T_1, T_2, \ldots, T_n, T_{n+1}, \ldots, \quad \text{and} \quad X_n = Z(T_n).$$

Then define

$$k_{ij}(t) = P\{X_{n+1} = j, T_{n+1} - T_n \leq t \mid X_n = i\}$$

as the conditional probability that, given that the process has entered state i at time T_n, the next transition occurs at time t toward state j. We assume that the $k_{ij}(t)$ satisfy the homogeneity property so that they do not depend on n. We collect all the $k_{ij}(t)$ into a matrix, called the kernel matrix of the SMP:

$$K(t) = [k_{ij}(t)].$$

Based on our earlier characterization, it can be seen that

$$k_{ij}(t) = p_{ij} H_i(t), \tag{14.23}$$

or, alternatively,

$$p_{ij} = \lim_{t \to \infty} k_{ij}(t), \quad P = \lim_{t \to \infty} K(t), \quad H_i(t) = \sum_j k_{ij}(t), \tag{14.24}$$

where $H_i(t)$ is the distribution of the sojourn time in state i. The transient solution for the conditional probabilities $V_{ij}(t)$ can be shown to satisfy the equations [40, 41]:

$$V_{ij}(t) = [1 - H_i(t)]\delta_{ij} + \sum_k \int_0^t V_{kj}(t - \tau) \, dk_{ik}(\tau), \tag{14.25}$$

where δ_{ij} is the Kronecker delta, equal to 1 if $i = j$ and 0 otherwise. The coupled set of Volterra equations of the second kind (14.25) is a *Markov renewal equation* [41, 42] and can be solved in the LST domain:

$$V^{\sim}(s) = E^{\sim}(s) + K^{\sim}(s) V^{\sim}(s)$$
$$= [I - K^{\sim}(s)]^{-1} E^{\sim}(s), \tag{14.26}$$

where $E^{\sim}(s) = \int_0^\infty e^{-st} dE(t)$ is a diagonal matrix with entries $E_{ii}(t) = 1 - H_i(t)$, and $K^{\sim}(s) = \int_0^\infty e^{-st} dK(t)$.

After solving Eq. (14.26) and taking the inverse LS transform of $V^{\sim}(s)$, the unconditional state probabilities $\pi(t) = [\pi_i(t)]$ become

$$\pi(t) = \pi(0)V(t), \tag{14.27}$$

where $\pi(0)$ is the initial state probability vector.

Example 14.18   *Two-Level Rejuvenation Model*

Transient analysis of the two-level rejuvenation model discussed in Example 14.10 can be carried out by determining the kernel matrix $K(t)$ as follows:

$$K(t) = \begin{bmatrix} 0 & k_{0,1}(t) & 0 & 0 & 0 & 0 & 0 \\ 0 & 0 & k_{1,2}(t) & 0 & 0 & 0 & 0 \\ 0 & 0 & 0 & k_{2,3}(t) & 0 & 0 & k_{2,6}(t) \\ 0 & 0 & 0 & 0 & k_{3,4}(t) & k_{3,5}(t) & 0 \\ 0 & k_{4,1}(t) & 0 & 0 & 0 & 0 & 0 \\ k_{5,0}(t) & 0 & 0 & 0 & 0 & 0 & 0 \\ k_{6,0}(t) & 0 & 0 & 0 & 0 & 0 & 0 \end{bmatrix},$$

where the non-zero elements of $K(t)$ could be derived as:

$$k_{0,1}(t) = F_1(t), \qquad\qquad k_{1,2}(t) = F_2(t),$$

$$k_{2,3}(t) = \int_0^t (1 - F_3(u))dF_I(u), \quad k_{2,6}(t) = \int_0^t (1 - F_I(u))dF_3(u),$$

$$\tag{14.28}$$

$$k_{3,4}(t) = \int_0^t (1 - F_D(u))dF_d(u), \quad k_{3,5}(t) = \int_0^t (1 - F_d(u))dF_D(u),$$

$$k_{4,1}(t) = F_P(t), \quad k_{5,0}(t) = F_R(t), \quad k_{6,0}(t) = F_B(t).$$

Comparing Eq. (14.28) with the results in Example 14.10, it is easy to verify that Eq. (14.24) holds.

Example 14.19   *Transient Analysis of the UPS Model*

Returning to the UPS model of Figure 14.6, we now carry out its transient analysis [12]. First, we develop the kernel matrix:

$$K(t) = \begin{bmatrix} 0 & k_{UB}(t) & 0 \\ k_{BU}(t) & 0 & k_{BD}(t) \\ k_{DU}(t) & 0 & 0 \end{bmatrix},$$

where

$$k_{UB}(t) = P\{\text{power supply fails before or at time } t\},$$

$$k_{BU}(t) = P\{\text{repair of power supply completed before or at time } t \text{ and } t < L\},$$

$$k_{BD}(t) = P\{\text{repair of power supply not completed at time } t \text{ and } t \geq L\},$$

$$k_{DU}(t) = P\{\text{power supply is repaired before or at time } t\},$$

and $u(t-L)$ is the unit step function at L. From the above definitions, we can determine the entries in the kernel matrix both in the time domain and in the LST domain:

$$k_{UB}(t) = 1 - e^{-\lambda t}, \qquad\qquad \tilde{k}_{UB}(s) = \frac{\lambda}{s+\lambda},$$

$$k_{BU}(t) = \begin{cases} 1 - e^{-\mu t} & t < L, \\ 1 - e^{-\mu L} & t \geq L, \end{cases} \qquad \tilde{k}_{BU}(s) = (1 - e^{-(s+\mu)L})\left(\frac{\mu}{s+\mu}\right),$$

$$k_{BD}(t) = \begin{cases} 0 & t < L, \\ e^{-\mu L} & t \geq L, \end{cases} \qquad \tilde{k}_{BD}(s) = e^{-(s+\mu)L},$$

$$k_{DU}(t) = 1 - e^{-\mu t}, \qquad\qquad \tilde{k}_{DU}(s) = \frac{\mu}{s+\mu}.$$

Further,

$$E_{ii}(t) = 1 - H_i(t) = 1 - \sum_j k_{ij}(t), \quad i,j = U, B, D.$$

Hence, the individual entries for the diagonal matrix $E(t)$ are, in the time and LST domains:

$$E_{UU}(t) = e^{-\lambda t}, \qquad\qquad \tilde{E}_{UU}(s) = \frac{s}{s+\lambda},$$

$$E_{BB}(t) = \begin{cases} e^{-\mu t} & t < L, \\ 0 & t \geq L, \end{cases} \qquad \tilde{E}_{BB}(s) = \frac{1 - e^{-(s+\mu)L}}{s+\mu},$$

$$E_{DD}(t) = e^{-\mu t}, \qquad\qquad \tilde{E}_{DD}(s) = \frac{s}{s+\mu}.$$

Assuming as initial probability $\pi_U(0) = 1$, the availability can be expressed as $A(t) = 1 - \pi_D(t) = 1 - V_{UD}(t)$. Transient analysis based on Eq. (14.26) is difficult to use since inverting $\tilde{V}(s)$ to get $V(t)$ is itself a difficult problem that usually precludes closed-form solutions. However, in this particular case a closed-form solution has been derived in [12], to get:

$$A(t) = 1 - \pi_D(t)$$
$$= 1 - \frac{\lambda}{\lambda+\mu}[e^{-\mu L} - e^{-(\lambda+\mu)t+\lambda L}]u(t-L). \qquad (14.29)$$

The steady-state availability A can be determined either by taking the limit of Eq. (14.29) as $t \to \infty$ or by simply considering that

$$V(\infty) = \lim_{t\to\infty} V(t) = \lim_{s\to 0} \tilde{V}(s)$$

and determining A directly from $\tilde{V}(s)$ without the inverse Laplace transform:

$$A = 1 - \frac{\lambda}{\lambda+\mu}e^{-\mu L}.$$

Assuming as numerical values $L = 48\,\text{hr}$, $\lambda = 1/480\,\text{hr}^{-1}$, and $\mu = 1/24\,\text{hr}^{-1}$, the availability $A(t)$ resulting from (14.29) is plotted as a function of time in Figure 14.18.

Example 14.20 *RF Channel in a Cellular Wireless Network* [28]
We return to the RF channel with recovery from the down state D as represented

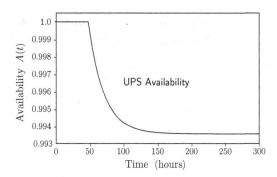

Figure 14.18   Availability of the UPS system.

in Figure 14.13. If we particularize the failure and repair distributions to be exponential, i.e.,

$$F(t) = 1 - e^{-\lambda t} \quad \text{and} \quad G(t) = 1 - e^{-\mu t},$$

the above model reduces to the model of Example 14.19, and the transient solution follows the same pattern. Hence the system availability is given by Eq. (14.29) where, with reference to Figure 14.12, the delay to get the down state is $L = \delta$.

Problems

14.12   In the UPS of Example 14.19, derive an expression for the transient availability assuming that the TTF is two-stage Erlang distributed and the TTR is two-stage hyperexponentially distributed.

14.13   Complete the steps in the solution of the transient analysis of the two-level rejuvenation model discussed in Example 14.18. You will need to carry out a numerical inversion of the LST via Mathematica or by writing your own routine.

14.5   Markov Regenerative Process

Consider a system with two components, each with the time to failure distribution EXP(λ), that share a single repair person whose repair time follows a general distribution $G(t)$. The components are utilized in a parallel redundant mode so that the system is up if at least one component is up. The state transition diagram for this system is given in Figure 14.19, where the label in state i indicates the number of components in the repair station.

The stochastic process underlying the state transition graph of Figure 14.19 is not an SMP as might be mistakenly surmised. Suppose that the system made a transition from state 0 to state 1 upon the failure of the first component. Then there is a competition between the failure of the second component and the repair of the first. If the failure of the second component occurs before the repair on the first is completed then the system

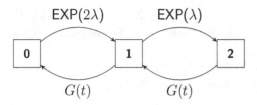

Figure 14.19 State transition graph of a two-component system with shared repair and general repair time distribution.

transits to state 2. Notice that the memory of the remaining time of the repair activity will be non-zero so that the entry into state 2 does not have the Markov property. In this model, the instants of entrance into states 0 and 1 are regeneration points for the process, but not the instants of entrance into state 2. Note that the process in Figure 14.19 may as well be interpreted as an $M/G/1/2/2$ queue.

This example motivates the need to explore more complex stochastic models in which not all the state transition epochs lead to regeneration points as would be the case in an SMP. To this end we introduce the class of *Markov regenerative processes* (MRGPs).

A Markov regenerative process is a generalization of many stochastic processes, including SMPs [3, 41]. In a CTMC, any time epoch t satisfies the Markov property, see (9.1); in an SMP, the Markov property is satisfied at all the time instances at which the process undergoes a state transition; in an MRGP the Markov property holds only when the process enters a subset of specific states called *regeneration states*.

Even though the MRGP $\{Z(t), t \geq 0\}$ may not have the Markov property in general, there are time instances $T_0, T_1, \ldots, T_n, \ldots$ such that the states of the process at those time points $(Y_0, Y_1, \ldots, Y_n, \ldots,$ respectively) satisfy the Markov property. More formally, $\forall i_0, i_1, \ldots, i_n \in \Omega$,

$$P\{Y_n = i_n \mid Y_{n-1} = i_{n-1}, \ldots, Y_1 = i_1, Y_0 = i_0\} = P\{Y_n = i_n \mid Y_{n-1} = i_{n-1}\}.$$

Therefore it does not matter which states the process $\{Z(t), t \geq 0\}$ has visited before reaching Y_n at time T_n. The state $Z(T_n)$ is the only information needed for the future development of $Z(T_n + t), t \geq 0$. The embedded time points $\{T_n, n \geq 0\}$ are the *regeneration time points* (RTPs): the process behaves in the same way each time it passes a regeneration time point:

$$P\{Z(T_n + t), t \geq 0 \mid Z(T_n) = i_n\} = P\{Z(t), t \geq 0 \mid Z(0) = i_n\}.$$

The evolution of the process between two consecutive RTPs is called the *subordinated process*.

14.5.1 Transient and Steady-State Analysis of an MRGP

The two matrices that must be defined for the transient analysis of an MRGP are commonly referred to as the *global kernel* and *local kernel* [3, 41]. The global kernel is a matrix $K(t) = [K_{ij}(t)]$ that describes the occurrence of the next RTP:

$$K_{ij}(t) = P\{Y_1 = j, T_1 \leq t \mid Y_0 = i\},$$

where Y_1 is the right continuous state hit by $Z(t)$ at the next RTP. The local kernel describes the behavior of the system inside a subordinated process, that is, between two consecutive RTPs.

The local kernel is a matrix $E(t) = [E_{ij}(t)]$ that gives the state transition probabilities within a regeneration interval, before the next RTP occurs:

$$E_{ij}(t) = P\{Z(t) = j, T_1 > t \mid Y_0 = i\}.$$

In the particular case in which the marking process $Z(t)$ is a semi-Markov process, any state change leads to the next RTP.

Let $V(t) = [V_{ij}(t)]$ define the transition probability matrix over $(0, t]$, i.e.:

$$V_{ij}(t) = P\{Z(t) = j \mid Z(0) = Y_0 = i\}, \qquad i, j \in \Omega. \tag{14.30}$$

Based on the global and the local kernels, the transient analysis can be carried out in the time domain by solving the following generalized Markov renewal equation [3, 41]:

$$V_{ij}(t) = E_{ij}(t) + \sum_{\Omega} \int_0^t dK_{ik}(y)\, V_{kj}(t - y). \tag{14.31}$$

In matrix form, (14.31) becomes:

$$V(t) = E(t) + K * V(t),$$

where $*$ is the convolution integral symbol. In the transform domain:

$$V^{\sim}(s) = \left[I - K^{\sim}(s)\right]^{-1} E^{\sim}(s), \tag{14.32}$$

where the superscript $^{\sim}$ indicates the Laplace–Stieltjes transform and s the complex transform variable of t, i.e., $F^{\sim}(s) = \int_0^\infty e^{-st} dF(t)$.

In order to use the above equations, we need to specify $K(t) = [K_{ij}(t)]$ and $E(t) = [E_{ij}(t)]$ matrices for $Z(t)$, and their LSTs. A time-domain solution for the transition probability matrix $V(t)$ can be obtained by numerically integrating Eq. (14.31). Alternatively, starting from the LST equation (14.32) a combination of symbolic and numeric computation is needed to obtain measures in the time domain [43].

For the purpose of the steady-state analysis of an MRGP, the following two matrices $\alpha = [\alpha_{ij}]$ and $\phi = [\phi_{ij}]$ should be evaluated. The two matrices are defined as:

$$\alpha_{ij} = \int_{t=0}^\infty E_{ij}(t)\, dt = \lim_{s \to 0} \frac{1}{s} E_{ij}^{\sim}(s), \tag{14.33}$$

$$\phi_{ij} = \lim_{t \to \infty} K_{ij}(t) = \lim_{s \to 0} K_{ij}^{\sim}(s).$$

α_{ij} is the expected time that the subordinated process starting from state i spends in state j. Matrix $\phi = [\phi_{ij}]$ is the one-step transition probability matrix of the DTMC embedded at the RTPs, and hence ϕ_{ij} gives the probability that the subordinated process starting from state i is followed by a regeneration interval starting from state j. We assume, here, that all the subordinated processes have a finite mean sojourn time, so that the measures in (14.33) exist and are finite. The evaluation of the measures in (14.33) is dependent on the nature of each subordinated process.

Once the matrices $\alpha = [\alpha_{ij}]$ and $\phi = [\phi_{ij}]$ have been generated from the model definition, the steady-state analysis of an MRGP requires the following three steps:

Step 1: Compute

$$\alpha_i = \sum_{j \in \Omega} \alpha_{ij};$$

α_i is the expected duration of the subordinated process starting from state i, before the next RTP.

Step 2: Evaluate the state probability vector $v = [v_i]$, whose elements are the unique solution of:

$$v = v\phi, \qquad \sum_{\Omega} v_i = 1. \tag{14.34}$$

Given that (14.34) exists and has a unique solution, v is the stationary state probability vector of the DTMC embedded at the RTPs.

Step 3: The steady-state probabilities of the MRGP are given by:

$$\pi_j = \lim_{t \to \infty} P\{Z(t) = j\} = \frac{\displaystyle\sum_{k \in \Omega} v_k \alpha_{kj}}{\displaystyle\sum_{k \in \Omega} v_k \alpha_k}. \tag{14.35}$$

Example 14.21 *The M/D/1/2/2 Queue [44]*

We illustrate the analysis methods of an MRGP through the example of an $M/D/1/2/2$ queuing system. This example reformulates the example of Figure 14.19 assuming as general distribution the deterministic of duration d, i.e., $G(t) = u(t - d)$. The $M/D/1/2/2$ queuing model represents two customers in the system, each of which submits a job at exponentially distributed intervals of rate λ, and one server with deterministic service times of duration d. The size of the queues for arriving jobs is two. Figure 14.20 shows the state space of the system.

In state 0, both costumers are thinking and the arrival rate of a job to the queue is 2λ. In state 1, two activities are in competition: the arrival of a second job at rate λ and the service of the job in the queue with a deterministic time d. In state 2, two jobs are

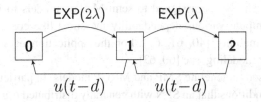

Figure 14.20 State space of the $M/D/1/2/2$ queue.

in the queue and the only action is the completion of the one already in service with a deterministic time d.

States 0 and 1 are regeneration states. We want to compute the state probability vector $\boldsymbol{\pi}(t) = [\pi_j(t)], j = 0, 1, 2$, at time t assuming that the system starts from state 0 with probability 1, so that $\pi_j(t) = V_{0j}(t)$.

We derive the global kernel $\boldsymbol{K}(t) = [K_{ij}(t)]$ $(i, j = 0, 1, 2)$ of this model row by row:

$$K_{00}(t) = 0, \qquad\qquad K_{01}(t) = 1 - e^{-2\lambda t}, \qquad\qquad K_{02}(t) = 0,$$

$$K_{10}(t) = \begin{cases} 0 & t < d, \\ e^{-\lambda d} & t \geq d, \end{cases} \qquad K_{11}(t) = \begin{cases} 0 & t < d, \\ 1 - e^{-\lambda d} & t \geq d, \end{cases} \qquad K_{12}(t) = 0,$$

$$K_{20}(t) = 0, \qquad\qquad K_{21}(t) = \begin{cases} 0 & t < d, \\ 1 & t \geq d, \end{cases} \qquad\qquad K_{22}(t) = 0.$$

Then we derive the local kernel $\boldsymbol{E}(t) = [E_{ij}(t)]$ $(i, j = 0, 1, 2)$ row by row:

$$E_{00}(t) = e^{-2\lambda t}, \qquad E_{01}(t) = 0, \qquad\qquad E_{02}(t) = 0,$$

$$E_{10}(t) = 0, \qquad E_{11}(t) = \begin{cases} e^{-\lambda t} & t < d, \\ 0 & t \geq d, \end{cases} \qquad E_{12}(t) = \begin{cases} 1 - e^{-\lambda t} & t < d, \\ 0 & t \geq d, \end{cases}$$

$$E_{20}(t) = 0, \qquad E_{21}(t) = 0, \qquad\qquad E_{22}(t) = \begin{cases} 1 & t < d, \\ 0 & t \geq d. \end{cases}$$

From the definition of the local and global kernels, the LSTs of $\boldsymbol{K}(t)$ and $\boldsymbol{E}(t)$ can be evaluated and then Eq. (14.32) can be applied. The solution is obtained by numerically inverting the LST using Jagerman's method [45] as adapted by Chimento and Trivedi [46]. A numerical solution for the present example is reported in [44], and an approximate solution using PH expansion in [47]. Analysis of related models but with general distributions is demonstrated in [42, 43, 48]. In [49], extensions to $M/D/1/m$ and $M/G/1/m$ queuing systems are provided and various numerical techniques are compared. For MRGP applications to computer networking problems, see [50–54]. For dependability applications of MRGP, see [53–57], and for applications to queues see [58].

14.6   Further Reading

Further studies on semi-Markov processes can be found in classical books such as [59] and [60]. Reward rates can also be assigned to semi-Markov models to increase the modeling power and compute various dependability measures like completion time, sojourn time and performability [40, 61, 62]. For the application of semi-Markov methods to dependability modeling, see [60, 63].

Petri nets have been used to generate SMP and MRGP models. In particular, various papers have discussed conditions that an SPN with generally distributed transition firing times should satisfy to generate an MRGP, and how the global and local kernels can be automatically derived from the SPN specification [43, 48, 51, 64–67].

References

[1] R. Pyke, "Markov renewal processes: Definitions and preliminary properties," *Annals of Mathematical Statistics*, vol. 32, pp. 1231–1242, 1961.

[2] R. Pyke, "Markov renewal processes with finitely many states," *Annals of Mathematical Statistics*, vol. 32, pp. 1243–1259, 1961.

[3] E. Cinlar, *Introduction to Stochastic Processes*. Prentice-Hall, 1975.

[4] Y. Cao, H. Sun, K. Trivedi, and J. Han, "System availability with non-exponentially distributed outages," *IEEE Transactions on Reliability*, vol. 51, no. 2, pp. 193–198, Jun. 2002.

[5] J. Girtler, "The semi-Markov model of the process of appearance of sea-going ship propulsion system ability and inability states in application to determining the reliablity of these systems," *Polish Maritime Research*, vol. 20, no. 4, pp. 18–24, Dec. 2013.

[6] D. Chen and K. Trivedi, "Analysis of periodic preventive maintenance with general system failure distribution," in *Proc. Pacific Rim Int.Symp. on Dependable Computing (PRDC)*, 2001, pp. 103–107.

[7] R. Barlow and R. Campo, "Total time on test processes and application to failure data analysis," Research report No. ORC 75-8, University of California, 1975.

[8] J. Zhao, Y. Wang, G. Ning, K. S. Trivedi, R. Matias, and K. Cai, "A comprehensive approach to optimal software rejuvenation," *Performance Evaluation*, vol. 70, no. 11, pp. 917–933, 2013.

[9] R Core Team, *R: A Language and Environment for Statistical Computing*, R Foundation for Statistical Computing, Vienna, Austria, 2012. [Online]. Available: www.R-project.org/

[10] G. Ciardo, J. Muppala, and K. Trivedi, "On the solution of GSPN reward models," *Performance Evaluation*, vol. 12, pp. 237–253, 1991.

[11] C. Hirel, B. Tuffin, and K. S. Trivedi, "SPNP: Stochastic Petri Nets. Version 6," in *Proc. Int. Conf. on Computer Performance Evaluation: Modelling Techniques and Tools (TOOLS 2000)*, eds. B. Haverkort and H. Bohnenkamp, LNCS 1786, Springer Verlag, 2000, pp. 354–357.

[12] L. Yin, R. Fricks, and K. Trivedi, "Application of semi-Markov process and CTMC to evaluation of UPS system availability," in *Proc. Reliability and Maintainability Symp.*, 2002, pp. 584–591.

[13] A. Pievatolo and I. Valadè, "{UPS} reliability analysis with non-exponential duration distribution," *Reliability Engineering and System Safety*, vol. 81, no. 2, pp. 183–189, 2003.

[14] B. B. Madan, K. Goševa-Popstojanova, K. Vaidyanathan, and K. S. Trivedi, "A method for modeling and quantifying the security attributes of intrusion tolerant systems," *Performance Evaluation*, vol. 56, pp. 167–186, 2004.

[15] K. Vaidyanathan and K. Trivedi, "A comprehensive model for software rejuvenation," *IEEE Transactions on Dependable and Secure Computing*, vol. 2, no. 2, pp. 124–137, Apr. 2005.

[16] Y. Huang, C. M. R. Kintala, N. Kolettis, and N. D. Fulton, "Software rejuvenation: Analysis, module and applications," in *Proc. Int. Symp. on Fault-Tolerant Computing (FTCS)*, 1995, pp. 381–390.

[17] S. Garg, A. van Moorsel, K. Vaidyanathan, and K. Trived, "A methodology for detection and estimation of software aging," in *Proc. 9th Intl. Symp. on Software Reliability Engineering*, 1998, pp. 282–292.

[18] J. Hartigan, *Clustering Algorithms*. John Wiley & Sons, 1975.

[19] W. Xie, Y. Hong, and K. Trivedi, "Analysis of a two-level software rejuvenation policy," *Reliability Engineering and System Safety*, vol. 87, no. 1, pp. 13–22, 2005.

[20] A. Bobbio, M. Sereno, and C. Anglano, "Fine-grained software degradation models for optimal rejuvenation policies," *Performance Evaluation*, vol. 46, pp. 45–62, Sep. 2001.

[21] J. Alonso, R. Matias, E. Vicente, A. Maria, and K. S. Trivedi, "A comparative experimental study of software rejuvenation overhead," *Performance Evaluation*, vol. 70, no. 3, pp. 231–250, 2013.

[22] X. Yin, X. Ma, and K. Trivedi, "An interacting stochastic models approach for the performance evaluation of DSRC vehicular safety communication," *IEEE Transactions on Computers*, vol. 62, no. 5, pp. 873–885, May 2013.

[23] B. R. Haverkort, R. Marie, G. Rubino, and K. S. Trivedi (eds.), *Performability Modelling: Techniques and Tools*. John Wiley & Sons, 2001.

[24] R. Smith, K. Trivedi, and A. Ramesh, "Performability analysis: Measures, an algorithm and a case study," *IEEE Transactions on Computers*, vol. C-37, pp. 406–417, 1988.

[25] E. P. C. Kao, "A note on the first two moments of times in transient states in a semi-Markov process," *Journal of Applied Probability*, vol. 11, no. 1, pp. pp. 193–198, 1974.

[26] J. G. Kemeny and J. L. Snell, *Finite Markov Chains: With a New Appendix "Generalization of a Fundamental Matrix."* Springer, 12 1983.

[27] K. Trivedi, *Probability and Statistics with Reliability, Queueing and Computer Science Applications*, 2nd edn. John Wiley & Sons, 2001.

[28] Y. Ma, J. Han, and K. Trivedi, "Transient analysis of minimum duration outage for RF channel in cellular systems," in *IEEE Vehicular Technology Conference*, vol. 2, Jul. 1999, pp. 1698–1702.

[29] N. B. Mandayam, P. C. Chen, and J. Holtzman, "Minimum duration outage for cellular systems: A level crossing analysis," in *Proc. IEEE 46th Vehicular Technology Conf., 1996. Mobile Technology for the Human Race*, vol. 2, Apr. 1996, pp. 879–883.

[30] M. Grottke, D. S. Kim, R. K. Mansharamani, M. K. Nambiar, R. Natella, and K. S. Trivedi, "Recovery from software failures caused by Mandelbugs," *IEEE Transactions on Reliability*, vol. 65, no. 1, pp. 70–87, 2016.

[31] M. Grottke and K. S. Trivedi, "Fighting bugs: Remove, retry, replicate, and rejuvenate," *IEEE Computer*, vol. 40, no. 2, pp. 107–109, 2007.

[32] U. Bhat and G. K. Miller, *Elements of Applied Stochastic Processes*, 3rd edn. John Wiley & Sons, 2002.

[33] C. Davis, "The norm of the Schur product operation," *Numerische Mathematik*, vol. 4, no. 1, pp. 343–344, 1962.

[34] B. Margolis, *SOA for the Business Developer: Concepts, BPEL, and SCA*. MC Press, LLC, 2007.

[35] L. Gönczy, S. Chiaradonna, F. D. Giandomenico, A. Pataricza, A. Bondavalli, and T. Bartha, *Dependability Evaluation of Web Service-Based Processes*. Springer, 2006, pp. 166–180.

[36] S. Gokhale and K. S. Trivedi, "Reliability prediction and sensitivity analysis based on software architecture," in *Proc. 13th Int, Symp. on Software Reliability Engineering (ISSRE)*, 2002, pp. 64–78.

[37] S. Gokhale, "Quantifying the variance in application reliability," in *Proc. 10th IEEE Pacific Rim Int. Symp. on Dependable Computing (PRDC 2004)*, 2004, pp. 113–121.

[38] Z. Zheng, K. Trivedi, K. Qiu, and R. Xia, "Semi-Markov models of composite web services for their performance, reliability and bottlenecks," *IEEE Transactions on Services Computing*, vol. 6, no. 1, 2016.

[39] J. Medhi, *Stochastic Processes*. New Age International, 1994.

[40] G. Ciardo, R. Marie, B. Sericola, and K. Trivedi, "Performability analysis using semi-Markov reward processes," *IEEE Transactions on Computers*, vol. 39, no. 10, pp. 1252–1264, 1990.

[41] V. G. Kulkarni, *Modeling and Analysis of Stochastic Systems*. Chapman & Hall, 1995.

[42] R. Fricks, M. Telek, A. Puliafito, and K. Trivedi, "Markov renewal theory applied to performability evaluation," in *Modeling and Simulation of Advanced Computer Systems: Applications and Systems*, ser. State-of-the Art in Performance Modeling and Simulation, eds. K. Bagchi and G. Zobrist. Gordon and Breach Publishers, 1996, ch. 11, pp. 193–236.

[43] A. Bobbio, A. Puliafito, and M. Telek, "A modeling framework to implement preemption policies in non-Markovian SPN," *IEEE Transactions Software Engineering*, vol. 26, pp. 36–54, 2000.

[44] H. Choi, V. Kulkarni, and K. Trivedi, "Transient analysis of deterministic and stochastic Petri nets," in *Application and Theory of Petri Nets 1993*, ser. Lecture Notes in Computer Science, ed. M. Ajmone Marsan, vol. 691. Springer, 1993, pp. 166–185.

[45] D. Jagerman, "An inversion technique for the Laplace transform," *The Bell System Technical Journal*, vol. 61, pp. 1995–2002, Oct. 1982.

[46] P. Chimento and K. Trivedi, "The completion time of programs on processors subject to failure and repair," *IEEE Transactions on Computers*, vol. 42, pp. 1184–1194, 1993.

[47] A. Bobbio and M. Telek, "Computational restrictions for SPN with generally distributed transition times," in *First European Dependable Computing Conf. (EDCC-1)*, LNCS 852, eds. D. H. K. Echtle and D. Powell. Springer Verlag, 1994, pp. 131–148.

[48] H. Choi, V. Kulkarni, and K. Trivedi, "Markov regenerative stochastic Petri nets," *Performance Evaluation*, vol. 20, pp. 337–357, 1994.

[49] R. German, D. Logothetis, and K. Trivedi, "Transient analysis of Markov regenerative stochastic Petri nets: A comparison of approaches," in *Proc. 6th Int. Conf. on Petri Nets and Performance Models, PNPM95*. IEEE Computer Society, 1995, pp. 103–112.

[50] S. Dharmaraja, K. S. Trivedi, and D. Logothetis, "Performance modelling of wireless networks with generally distributed hand-off interarrival times," *Computer Communications*, vol. 26, pp. 1747–1755, 2003.

[51] R. German, *Performance Analysis of Communication Systems: Modeling with Non-Markovian Stochastic Petri Nets*. John Wiley & Sons, 2000.

[52] D. Logothetis and K. S. Trivedi, "Transient analysis of the leaky bucket rate control scheme under poisson and ON–OFF sources," in *Proc. IEEE INFOCOM '94, 13th Annual Joint Conf. of the IEEE Computer and Communications Societies*, 1994, pp. 490–497.

[53] D. Logothetis and K. S. Trivedi, "The effect of detection and restoration times on error recovery in communication networks," *Journal of Network System Management*, vol. 5, no. 2, pp. 173–195, 1997.

[54] W. Xie, H. Sun, Y. Cao, and K. S. Trivedi, "Modeling of user perceived webserver availability," in *Proc. IEEE Int. Conf. on Communications*, 2003, pp. 1796–1800.

[55] D. Logothetis and K. S. Trivedi, "Time-dependent behavior of redundant systems with deterministic repair," in *Proc. 2nd Int. Workshop on the Numerical Solution of Markov Chains*, ed. W. J. Stewart. Kluwer Acedemic Publishers, 1995.

[56] K. Vaidyanathan, S. Dharmaraja, and K. S. Trivedi, "Analysis of inspection-based preventive maintenance in operational software systems," in *Proc. 21st Symp. on Reliable Distributed Systems*, 2002, pp. 286–295.

[57] D. Wang, W. Xie, and K. S. Trivedi, "Performability analysis of clustered systems with rejuvenation under varying workload," *Performance Evaluation*, vol. 64, no. 3, pp. 247–265, 2007.

[58] D. Logothetis, V. Mainkar, and K. S. Trivedi, "Transient analysis of non-Markovian queues via Markov regenerative processes," in *Probability Models and Statistics: A. J. Medhi Festschrift*, eds. A. C. Borthakur and H. Choudhury. New Age International Limited, 1996, pp. 109–131.

[59] R. Howard, *Dynamic Probabilistic Systems, Volume II: Semi-Markov and Decision Processes*. John Wiley & Sons, 1971.

[60] N. Limnios and G. Oprisan, *Semi-Markov Processes and Reliability*. Birkhäuser Engineering, 2001.

[61] A. Bobbio and L. Roberti, "Distribution of the minimal completion time of parallel tasks in multi-reward semi-Markov models," *Performance Evaluation*, vol. 14, pp. 239–256, 1992.

[62] G. Rubino and B. Sericola, "Sojourn times in semi-Markov reward processes: Application to fault-tolerant systems modeling," *Reliability Engineering and System Safety*, vol. 41, no. 1, pp. 1–4, 1993.

[63] M. Carravetta, R. D. Dominicis, and M. D'Esposito, "Semi-Markov processes in modelling and reliability: A new approach," *Applied Mathematical Modelling*, vol. 5, no. 4, pp. 269–274, 1981.

[64] A. Bobbio and M. Telek, "Markov regenerative SPN with non-overlapping activity cycles," in *Int. Computer Performance and Dependability Symp., IPDS95*. IEEE Computer Society Press, 1995, pp. 124–133.

[65] R. German, A. van Moorsel, M. Qureshi, and W. Sanders, "Expected impulse rewards in Markov regenerative stochastic Petri nets," in *Application and Theory of Petri Nets*, Lecture Notes in Computer Science, vol. 1091, eds. J. Billington and W. Reisig. Springer, 1996, pp. 172–191.

[66] L. Malhis and W. Sanders, "An efficient two-stage iterative method for the steady-state analysis of Markov regenerative stochastic Petri net models," *Performance Evaluation*, vol. 27–28, pp. 583–601, 1996.

[67] A. Bobbio, A. Puliafito, M. Telek, and K. Trivedi, "Recent developments in non-Markovian stochastic Petri nets," *Journal of Systems Circuits and Computers*, vol. 8, no. 1, pp. 119–158, Feb. 1998.

15 Phase-Type Expansion

Recall that the Erlang and hypoexponential distributions can be synthesized by summing independent, exponentially distributed random variables. The hyperexponential distribution is a mixture of exponential distributions and can be visualized as a parallel combination of stages of exponentials. We also observed in Chapter 8 that the time to failure random variable of many systems is found to follow such a distribution. Even when imperfect coverage of failures is considered, the time to failure distribution will have both series and parallel stages of exponentials. This idea of stringing together stages or phases of exponentials has been generalized in several ways and has been used extensively in system modeling. After a brief overview of the origin of distributions with exponential stages, we define the class of *phase-type* (PH) distributions in Section 15.1. In Section 15.2, the main properties of the PH class are discussed. Section 15.3 illustrates the use of PH distributions in system modeling. Applications to the computation of completion times are elaborated in Section 15.4 and to queuing models in Section 15.5. Section 15.6 illustrates the potentialities and restrictions of the non-exponential model types considered in Part IV.

15.1   Introduction

A general class of distributions with rational Laplace transform was studied in two seminal papers by Cox in 1955 [1, 2]. A Coxian distribution is one whose LST is a rational function (a fraction whose numerator and denominator are polynomial functions of the Laplace variable s) such that the degree of the numerator is less than the degree of the denominator. A Coxian distribution can be graphically represented as a sequence of exponential stages (possibly with complex transition rates) connected by weighted arcs and with a final absorbing state. This class of distributions, also known as exponomial or exponential polynomial (EP) [3], is very general and contains the exponential, the Erlang, the hypoexponential and the hyperexponential distributions. The class EP is closed under polynomial addition, multiplication, differentiation, integration and under random variable operations like maximum, minimum, convolution and probabilistic mixture [3]. A non-negative random variable with an EP distribution can be expressed as

$$F(t) = \sum_{j=1}^{n} a_j t^{k_j} e^{\lambda_j t}, \tag{15.1}$$

where k_j is a non-negative integer and a_j and λ_j are real or complex numbers. This class of distributions allows a mass at origin (that is, $F(0) > 0$) as well as a mass at infinity (that is, $1 - F(\infty) > 0$). The tool SHARPE utilizes EP distributions in the form of Eq. (15.1). Note, however, that the functions expressed by (15.1) are not bounded to be Cdfs, and no simple relations among the parameters are known to establish whether a function of the form (15.1) is a Cdf or not.

If we consider a set of stages of exponentials connected together by means of weighted edges, with weights being probabilities, and append an absorbing state then we have an SMP. Since each of the state sojourn time distributions is exponential, we can transform such a graph into a homogeneous CTMC. Based on this idea, a subclass of Coxian distributions was formally defined in [4] and named "PH distributions." Examples of PH distributions were introduced in Section 3.2.5, while a formal definition of the PH class was given in Section 10.2.1. The class of PH distributions can approximate as closely as desired any Cdf as the number of states of the corresponding CTMC grows to infinity [5]. Hence, any discrete-state stochastic process with non-exponential distributions can be approximated by a CTMC over an expanded state space by replacing the non-exponential distributions by a proper PH approximation. Furthermore, excellent methods are now available to fit known probability distributions or empirical samples of data to PH distributions [6–9].

The time to absorption of a homogeneous CTMC with one absorbing state is called a *phase-type* (PH) random variable and its distribution a *phase-type* (PH) distribution. Due to its great power and flexibility, the phase-type PH approach [4] is commonly applied in the analysis of a large class of stochastic processes such as SMPs, NHCTMCs, MRGPs, and beyond.

We call the procedure of approximating a non-Markovian process by means of an expanded CTMC its *Markovization*, wherein a non-exponential distribution is replaced by its PH approximation. Note that the definition of PH allows a mass at origin but not a mass at infinity. However, by allowing multiple absorbing states, we can define PH distributions with a mass at infinity. The class PH is a proper subset of the class EP (or Coxian) since it has a rational LST. We note also that replacing the exponential with a geometric distribution, we can get a discrete PH distribution [10] (which will not be dealt with in this book).

15.1.1   Definition

Recalling the definition in Section 10.2.1, suppose we have a CTMC with n transient states and state $n + 1$ as an absorbing state with infinitesimal generator Q and initial state probability vector $\pi(0) = (\pi_u(0), \pi_{n+1}(0))$. To avoid trivialities, we assume in the following that the initial probability in the absorbing state is equal to zero: $\pi_{n+1}(0) = 0$ (otherwise there will be a mass at origin). The infinitesimal generator Q is an $(n+1) \times (n+1)$ matrix that can be partitioned in the following form:

$$Q = \left[\begin{array}{c|c} Q_u & a^T \\ \hline 0 & 0 \end{array} \right], \tag{15.2}$$

where Q_u is the $n \times n$ partition of Q over the transient states, $a^T = -Q_u e^T$ is a column vector grouping the transition rates from any transient state to the absorbing state, and e^T is a column vector with all the entries equal to 1. The time to reach the absorbing state starting from time $t = 0$ (time to absorption) T_a is called a PH random variable of *order n* with *representation* $(Q_u, \pi_u(0))$ [4, 11]. Its Cdf, see Eq. (10.45), is given by:

$$F_a(t) = P\{T_a \le t\} = P\{Z(t) = a\} = \pi_a(t)$$
$$= 1 - \pi_u(0) e^{Q_u t} e^T. \tag{15.3}$$

In the Laplace transform domain, a PH distribution and its density are given, from Section 10.2.1, by:

$$F_a^*(s) = \frac{1}{s} \pi_u(0)(sI - Q_u)^{-1} a^T,$$
$$f_a^*(s) = sF_a^*(s) = \pi_u(0)(sI - Q_u)^{-1} a^T. \tag{15.4}$$

A PH distribution of order n and with representation $(Q_u, \pi_u(0))$ has, in general, $n^2 + n = n(n+1)$ free parameters and it does not have a unique generating CTMC [12–14]; many CTMCs with different infinitesimal generators may provide the same time to absorption distribution. A common and useful subclass of the PH family is the subclass known as *acyclic* PH (APH). An APH is characterized by a state transition graph with no cycles and hence its states can always be ordered in such a way that the infinitesimal generator matrix is upper triangular. Hence an APH of order n has, in general, $n(n+1)/2 + n = n(n+3)/2$ free parameters. Several examples of APHs have already been introduced in Section 3.2.5, like the hyperexponential, the hypoexponential, and the Erlang. A distinctive feature of APHs that is not shared by the general cyclic PH family [15] is that they have a unique minimal representation known as its canonical form [12, 16]. We say that an APH of order n in canonical form has a minimal representation since the number of free parameters is $2n+1$ (but the transition rates should be ordered in a decreasing sequence) and corresponds to the degrees of freedom of the distribution and cannot be reduced any further.

The theorems in [16] show that any APH of order n may be converted into a minimal canonical representation of order n.

Example 15.1 *Canonical APH of Order Two [17]*
A general APH of order 2 (2-APH) is shown in Figure 15.1; its infinitesimal generator

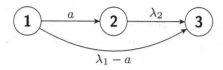

Figure 15.1   State space of a two-state APH.

Q is given by:

$$Q = \left[\begin{array}{cc|c} -\lambda_1 & a & \lambda_1 - a \\ 0 & -\lambda_2 & \lambda_2 \\ \hline 0 & 0 & 0 \end{array}\right].$$

With the additional constraint $\lambda_1 \geq \lambda_2$ and $\pi_1(0) = 1$ (initial probability mass concentrated in state 1), the 2-APH of Figure 15.1 is canonical [16], i.e., can represent any APH of order 2 with any combination of parameters and initial probability vectors. If $\lambda_1 > \lambda_2$, the Cdf and the expected value assume the form:

$$F(t) = 1 - \frac{\lambda_1 - \lambda_2 - a}{\lambda_1 - \lambda_2} e^{-\lambda_1 t} - \frac{a}{\lambda_1 - \lambda_2} e^{-\lambda_2 t},$$

(15.5)

$$E[X] = \frac{\lambda_2 + a}{\lambda_1 \lambda_2}.$$

Note that if $a = 0$ or $\lambda_1 - a = \lambda_2$, Eq. (15.5) reduces to an exponential Cdf with parameters λ_1 or λ_2, respectively.

If $\lambda_1 = \lambda_2 = \lambda$, the Cdf becomes:

$$F(t) = 1 - (1 + at) e^{-\lambda t},$$

which becomes a two-stage Erlang with the further condition $a = \lambda$.

If $\lambda_1 - a < \lambda_2$, the Cdf of Eq. (15.5) is IFR, while if the inequality is reversed, the Cdf becomes DFR. To show the different behaviors, the Cdf, density, and hazard rate are shown in Figure 15.2 for the three sets of parameters reported in Table 15.1.

Table 15.1 Parameter values for the 2-APH of Figure 15.2.

Curve	λ_1	a	λ_2
1	2	2	2
2	1	0	1
3	4	1.5	0.5

Example 15.2   *Tire Subsystem with One Spare*

We return to Example 8.1 where the tire subsystem of a car with one spare was considered in the framework of dynamic redundancy of a warm standby system with imperfect coverage. We derived the reliability of the system in Eq. (8.6). The system

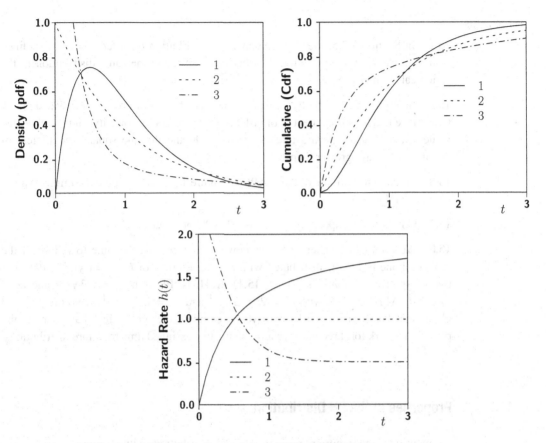

Figure 15.2 The density, Cdf, and hazard rate of the 2-APH of Table 15.1.

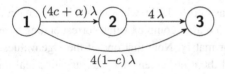

Figure 15.3 State space of the two-state APH representing the tire subsystem.

time to failure is seen to be an APH random variable of order two. The distribution can then be expressed in the canonical form as in Figure 15.1. Remembering that for a canonical 2-APH we should have $\lambda_1 \geq \lambda_2$, from Eq. (8.6) we see that the mapping is obtained by replacing, in Figure 15.1,

$$\lambda_1 = (4+\alpha)\lambda \quad \text{and} \quad \lambda_2 = 4\lambda.$$

Comparing Eq. (8.6) with Eq. (15.5), we obtain $a = (4c + \alpha)\lambda$. The canonical representation of Eq. (8.6) is finally depicted in Figure 15.3.

Problems

15.1 In Section 3.2.5, the hyperexponential distribution of order $n = 2$ was defined as a mixture of two exponential distributions. Represent the same distribution in the canonical form (15.5).

15.2 In Section 3.2.5, the hyperexponential distribution of order $n = 2$ was defined. Generalize to a hyperexponential of order n. Find the Cdf, the density, and the expected value. Show that the hazard rate is always DFR and that the squared coefficient of variation is always greater than 1.

15.3 Find the hazard rate of the 2-APH of Figure 15.1 and compute its limiting values for $t = 0$ and $t \to \infty$.

15.4 Compute the expected value of the 2-APH of Figure 15.1 when $\lambda_1 = \lambda_2$.

15.5 We return to Problem 3.8, but now we assume that the time to failure of the mobile phone is 2-APH distributed with a mean lifetime of $E[X] = 6$ yr. We consider two alternative distributions (Figure 15.1). APH$_1$ is IFR with $\lambda_1 = 1/3\,\mathrm{yr}^{-1}$ and $\lambda_2 = 1/3\,\mathrm{yr}^{-1}$. APH$_2$ is DFR with $\lambda_1' = 26/12\,\mathrm{yr}^{-1}$ and $\lambda_2' = 1/12\,\mathrm{yr}^{-1}$. If you buy a used phone that has been in use for two years, compute the probability that your mobile phone will work for three more years with the two assigned time to failure distributions.

15.2   Properties of the PH Distribution

We mention some important properties of PH distributions without proof.

1. A PH distribution, being a subclass of the EP distribution family, can always be expressed as a sum of terms as in Eq. (15.1), where $k_j \geq 0$ is an integer, a_j is a real or a complex constant and λ_j is a real or complex value corresponding to an eigenvalue of the infinitesimal generator matrix. Note that one of the eigenvalues of a CTMC generator matrix is 0 and all the other eigenvalues are negative real numbers or, if complex, their real parts are always negative.
2. In the case of an APH distribution, the eigenvalues are given by the diagonal entries of the infinitesimal generator and hence are all non-positive real numbers.
3. As $t \to \infty$, a PH distribution assumes the behavior of a single exponential with parameter equal to the eigenvalue with minimal absolute value (the minimal diagonal entry in absolute value for an APH distribution). Hence the hazard rate of a PH distribution is asymptotically constant and equal to the eigenvalue with minimal absolute value.
4. The least variable PH distribution of order n is the Erlang [18], and its squared coefficient of variation is $cv^2 = 1/n$.
5. PH distributions are closed under various basic operations arising in reliability theory: finite mixtures, finite convolutions, minimum, maximum, formation of coherent systems of independent components [19, 20]. Further, they are closed under computation of completion times when the work requirement is also PH [21, 22].

6. PH distributions of order n can approximate as closely as desired any continuous distribution function (including the deterministic) as $n \to \infty$. A simple intuitive proof based on the infinite replication of mixtures of Erlangs is given in [5]. Cumani [16] showed that the same property holds for the APH subclass, for which a unique minimal canonical form exists (as in Example 15.1).

Property 4, which links the minimum coefficient of variation to the order of the distribution, may force the use of a very large number of stages when the distribution to approximate is very sharp or even deterministic. However, in dependability problems it is rather unusual to be faced with sharp time-to-failure or time-to-repair distributions, and good fits may often be achieved with few states.

Various fitting techniques are proposed in the literature, mainly based on subclasses of PH distributions. Fitting mixtures of Erlangs with moment-matching techniques is proposed in [23–26]. The maximum likelihood estimation criterion is used in conjunction with canonical APH in [27–29] and the related tool, called PhFit, is available from [30]. Various fitting procedures for the complete class of PH distributions, based on the EM algorithm of Dempster *et al.* [31], have been elaborated in [6, 8, 9, 32], and related tools are available. Various examples of fitting commonly used distributions in dependability are in [28], and from experimental data in [17, 33].

Example 15.3 *Fitting a Weibull Distribution*
Using the tool PhFit [30], a Weibull distribution with shape parameter $\beta = 1.5$ and scale parameter $\eta = 0.1$ that has an expected value $E[X] = 9.03$ and a variance $\sigma^2 = 37.6$ is fitted with a canonical APH of order 2, 3, and 4, respectively. The results are shown in Figure 15.4.

To show the quality of the result, the expected value, the variance and the coefficient of variation of the fitted APH distributions are given in Table 15.2 together with the maximum deviation of the APH Cdf from the original Weibull Cdf.

It should be stressed, however, that by Properties 1 and 3 above, the hazard rate of a PH distribution has a finite value for $t = 0$ and tends to a constant value as $t \to \infty$. It turns out that, with a finite number of stages, a PH distribution cannot fit the behavior of the hazard rate of a Weibull distribution; in fact, with shape parameter $\beta < 1$ the Weibull hazard rate tends to ∞ for $t \to 0$, and with shape parameter $\beta > 1$ tends to

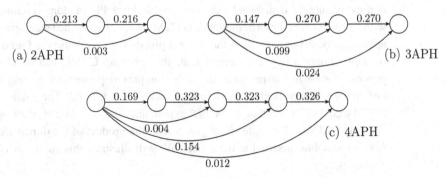

Figure 15.4 Best-fit approximation of a Weibull distribution with $\beta = 1.5$, $\eta = 0.1$: (a) 2APH, (b) 3APH, (c) 4APH

Table 15.2 Comparison of the $W(1.5, 0.1)$ with the n-APH distributions of Figure 15.4.

Distribution	$E[X]$	σ^2	cv^2	max dev
Weibull	9.03	37.57	0.46	—
2-APH	9.17	42.7	0.50	0.0104
3-APH	9.07	39.4	0.48	0.0036
4-APH	9.03	37.6	0.46	6.3×10^{-4}

∞ for $t \to \infty$ (Section 3.2.3). The problem of fitting a PH to long-tail distributions is addressed in [34].

In [35], the authors fit Weibull Cdfs by resorting to hypoexponential in the case of IFR and hyperexponential in the case of DFR. The use of the canonical representation for fitting non-exponential distributions or empirical data samples is more efficient, since the only parameter to fix is the order n of the APH. All the APHs of order n are included in the representation.

15.3   Phase-Type Distributions in System Modeling

Models with non-exponential distributions were considered in the framework of NHCTMC in Chapter 13 and in the framework of SMP (and MRGP) in Chapter 14. Both techniques have modeling power limitations in the dependability context. Under the umbrella of NHCTMC, although the rates are allowed to be time dependent, they can only be function of a unique global clock and thus these models are not able to properly capture repair activities and hence availability models [36], as was discussed in Chapter 13. Semi-Markov processes, on the contrary, require that at each change of state all the activities restart from scratch. In a parallel system when the first component fails, the second one forgets the time already elapsed and its time to failure restarts with the same Cdf as a new one.

An alternative approach very often used in practice is the Markovization of a model with non-exponential distributions using the phase-type expansion (also referred to as the *device of stages*) [24, 36–39]. The method consists in representing a non-exponentially distributed random variable by a PH random variable, so that the whole process becomes a homogeneous CTMC over an expanded state space. The major advantage of PH expansion is that once a proper PH distribution is found to represent or approximate a non-exponential Cdf, the resulting CTMC can be algorithmically generated even in the presence of fairly complex dependencies among components, and numerically solved using standard solvers for CTMC. The main drawback of the Markovization technique is the exponential growth in the state space of the expanded CTMC. The state space grows with the product of the dimensions of the PH random variables inserted in the model. We will illustrate this approach via a series of examples.

Example 15.4 *Comparison of Non-Exponential Models*
A component labeled A has a PH-distributed time to failure represented by the two-state APH of Figure 15.1 (and thus including any possible behavior, increasing, constant or decreasing, of the failure rate). The Cdf $F_A(t)$ is given in Eq. (15.5), and the time-dependent failure rate can be computed with the usual formula (see Table 3.1):

$$h_A(t) = \frac{dF_A(t)}{dt} \frac{1}{1 - F_A(t)}. \tag{15.6}$$

The non-repairable single component can be modeled as a CTMC, Figure 15.5(a1), with states 1 and 2 as up states and state 3 as a down state. An alternative equivalent model is the NHCTMC of Figure 15.5(a2), where the failure time Cdf $F_A(t)$ has the expression given in (15.5). These two representations for the component reliability model are completely equivalent. Next we consider an availability model wherein we assume that the component is repairable with a constant repair rate equal to μ. Different repair strategies can be considered. The minimal repair strategy [40] is said to be used when the component after repair is as bad as old, and a maximal repair or replacement strategy is said to be used [36] when the component after repair is as good as new.

The availability model of Figure 15.5(b1) captures a replacement policy wherein the TTF has a PH distribution represented by the CTMC of Figure 15.5(a1). The availability model of Figure 15.5(b2) can be interpreted as an SMP; in this case, a replacement policy is assumed and the two models of Figures 15.5(b1) and 15.5(b2) are equivalent. On the other hand, if the model of Figure 15.5(b2) is interpreted as an NHCTMC, we need to ponder for a moment. Recall that the global clock continues its journey through all failure and repair events. Thus, after a repair event the failed component is *worse than old* since the global clock is incremented by the repair time with respect to the previous epoch of failure. If the repair time is negligible with respect to the failure time,

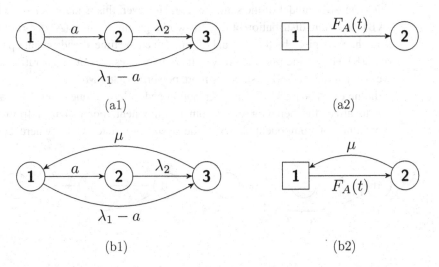

Figure 15.5 A single PH component.

as is usually the case, the above NHCTMC approximates the minimal repair policy, that is, as bad as old repair.

The model of Figure 15.5(b2) with non-exponential failure time distribution provides different results if interpreted as an SMP or as an NHCTMC. To pinpoint the difference, we derive the steady-state availability expressions for the two cases. MTTF $= E[X_a]$, given by (15.5), is the expected value of the 2-APH distribution of Figure 15.5(a1). The steady-state availability of the repairable component of Figure 15.5(b1), or of Figure 15.5(b2) interpreted as an SMP, is then:

$$A_{b_1} = \frac{\text{MTTF}}{\text{MTTF} + \text{MTTR}} = \frac{E[X_a]}{E[X_a] + 1/\mu}.$$

On the other hand, if the model of Figure 15.5(b2) is interpreted as an NHCTMC, the failure time distribution changes after each repair, and the model becomes regenerative as $t \to \infty$ due to the behavior of $h_A(t)$ (see Property 3 above). Since λ_2 is the smallest diagonal element in absolute value, by assumption, the steady-state availability in this case is:

$$A_{b_2} = \frac{1/h_A(\infty)}{1/h_A(\infty) + 1/\mu} = \frac{1/\lambda_2}{1/\lambda_2 + 1/\mu}.$$

Example 15.5 *A Two-Component Redundant System with PH Time to Failure Distributions*

A redundant system is composed of two components A and B having failure time Cdfs $F_A(t)$ and $F_B(t)$ represented by the 2-APH of Figure 15.6(a) with parameters (a, α_1, α_2) and Figure 15.6(b) with parameters (b, β_1, β_2), respectively. States A1 and A2 (resp. B1 and B2) are up states and state A3 (B3) is the down state.

The CTMC reliability model of the two-component parallel system is shown in Figure 15.7(a), where the only down state is A_3B_3. Consider the model shown in Figure 15.7(b) without using the method of stages. We may be tempted to think that this is an SMP model of the same system. However, this is incorrect since in state $\overline{A}B$ ($A\overline{B}$) the failure distribution of the still-working component B (A) does not restart from scratch, but is given by the residual lifetime given the time already spent up to the failure of A (B). Hence, the times of entry to these two states are not regeneration epochs; thus we see that the method of stages is more powerful than SMP.

In the expanded CTMC, the degradation and failure of one component are assumed not to affect the behavior of the other component. For instance, from state A_2B_1 the failure of component A leads the system to state A_3B_1, where component B

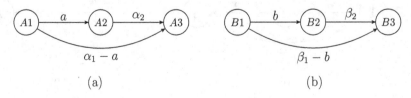

(a)   (b)

Figure 15.6   Failure time distributions of components A and B.

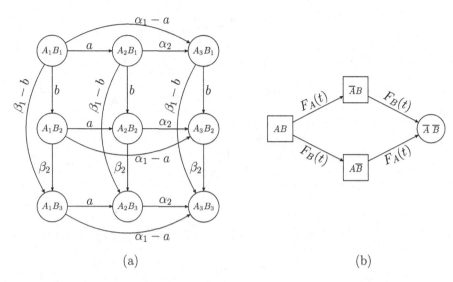

Figure 15.7   A redundant system of two PH units.

maintains its phase. We note that it is easy to introduce these dependencies in the model if we so choose. We further note that using the method of stages we obtain a homogeneous CTMC for which solution algorithms are well known and implemented in many software packages. The only drawback is the increased number of states in the resulting CTMC. For instance, if the PH distribution of A has $n_A + 1$ states and the PH distribution of B has $n_B + 1$ states, the CTMC modeling the two-component system will have $(n_A + 1) \times (n_B + 1)$ states. We leave the solution of the CTMC in Figure 15.7(a) as an exercise.

Example 15.6   *A Two-Component Redundant System with Common Cause Failure and Repair [36]*
Example 15.5 is now made more realistic by including two common features encountered in practical dependability problems: common cause failure and repair. Components A and B have TTF distributions $F_A(t)$ and $F_B(t)$, respectively. The two components fail independently with probability $(1 - q)$, while with probability q the failure of one of the two components causes the failure of the second one due to a common cause mechanism. The common cause TTF distribution is denoted by $F_{AB}(t)$ and it is reasonable to assume that $F_{AB}(t)$ is the distribution of the minimum between TTF_A and TTF_B. Further, a more complex repair policy is applied. When a single component fails, it can be repaired independently with constant repair rate μ_A or μ_B, respectively, but if both components fail, either due to successive failures of the two components or due to the common cause failure, a single repair action takes place from state $(\overline{A}\,\overline{B})$ to state AB with rate μ_R (Figure 15.8).

The model of Figure 15.8(a) is neither an SMP (states $(\overline{A}B)$ and $(A\overline{B})$ are not regenerative) nor an NHCTMC, since state AB is regenerative. If the components A and B have a PH-distributed TTF, the CTMC derived from the phase-type expansion can be

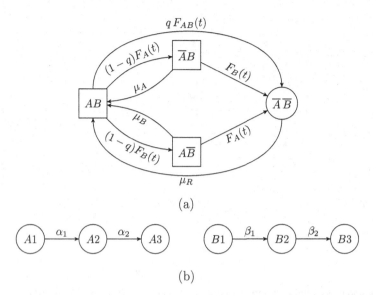

(a)

(b)

Figure 15.8   A redundant system with common cause failure and repair and the PH representation of A and B.

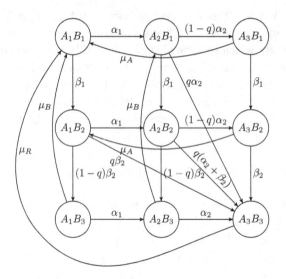

Figure 15.9   Phase-type expansion of the system of Figure 15.8.

correctly generated. To simplify the notation we assume that both the components have the two-stage hypoexponential TTF distribution shown in Figure 15.8(b).

The expanded CTMC is given in Figure 15.9, where the states are labeled with the name of the component and its current phase.

In the parallel configuration, A_3B_3 is the only **down** state. Any change of phase in one component does not affect the phase of the other one, thus preserving its working, degraded or failed condition. If a component fails (transition from phase 2 to phase 3), its failure is propagated with probability q to the other component (common cause failure) and the system jumps to the failure state A_3B_3. Hence, for instance, from state

A_2B_1, the failure of A leads the system to state A_3B_1 with rate $(1-q)\alpha_2$ and to state A_3B_3 with rate $q\alpha_2$. When a single component fails, it is restored to an as-good-as-new condition without affecting the possible degraded state of the other component. For instance, from state A_2B_3 the repair of component B occurs with rate μ_B and leads the system to state A_2B_1, where component B is as good as new and component A remains in its degraded state. When both components have failed (state A_3B_3), the system is down and both components are replaced with a single repair action represented by the transition to state A_1B_1 with rate μ_R.

Problems

15.6 Solve the CTMC of Figure 15.7(a) for its time-dependent state probabilities, starting in state A_1B_1 with probability 1, using the convolution integration approach. Verify the obtained solution by solving the model in semi-symbolic form using the SHARPE software package.

15.7 Develop an SRN model for Example 15.5 and solve numerically for the system reliability using either the SHARPE or the SPNP software package.

15.8 Solve the CTMC of Figure 15.9 for its steady-state probabilities and hence the steady-state system availability. Verify your closed-form solution by carrying out a numerical solution via the SHARPE software package.

15.9 Develop an SRN model for Example 15.6 and solve numerically for the system availability using either the SHARPE or the SPNP software package.

15.10 Return to the UPS availability model of Example 14.7 and approximate the deterministic duration L by a k-stage Erlang random variable. Numerically compute the system availability as a function of k from the resulting homogeneous CTMC, and compare the results with those obtained using exact SMP equations. You may solve the CTMC using SHARPE or a similar software package.

15.11 Return to the UPS availability model of Example 14.7 and approximate the deterministic duration L by a k-stage Erlang random variable. Develop a GSPN or an SRN model and thence compute the system availability as a function of k. Compare the numerical results with those obtained using exact SMP equations and with those obtained using the direct solution of the CTMC as in the problem above. You may use SHARPE, SPNP or a similar software package.

15.4   Task Completion Time under a PH Work Requirement

Consider a task running on a multi-state server that requires τ units of work to be completed. Assume that the task arrives at time $t=0$. The *task completion time*, as defined in Section 3.7, is a random variable $T(\tau)$ with Cdf $F_T(t,\tau)$. Equations for $F_T(t,\tau)$ are derived in the transform domain in [41] when the multi-state server is a semi-Markov reward process where the reward rate attached to an SMP state is the server speed in that state. Furthermore, upon change of server state, different execution policies including preemptive resume (prs – the work already done is resumed at each state transition) or preemptive repeat (prt – the work already done is lost at each state transition and the execution of the task restarts from scratch in the new state) were allowed.

If the work requirement τ is a PH-distributed random variable and the server is modeled as a CTMC reward process, the task completion time is a PH random variable under any mixture of prs and prt execution policies. This closure property for PH distributions was proved in [21, 42].

Example 15.7 *Probability of Completing a Scheduled Production [21]*
A production plant consists of two statistically identical machine tools working in parallel. Each machine tool has a productivity of r pieces per unit time, fails with a constant failure rate λ, is repaired with a constant repair rate μ and a single repair crew is available. Common mode failures are considered by allowing the two machines to fail simultaneously with rate λ'. The MRM of the server is shown in Figure 15.10, where the label inside each state represents the number of working machines.

In state 2 the production capacity (reward rate) is $r_2 = 2r$, in state 1 it is $r_1 = r$, and in state 0, $r_0 = 0$. We assume that the production of a single piece takes an exponentially distributed amount of time with parameter α. We wish to compute the probability that the scheduled production of m pieces will be completed by a given time t. By the assumptions made, the work requirement τ for producing m pieces is an Erlang-distributed random variable with m stages and parameter α. When a machine fails, the execution policy may be either prs or prt. Under the prs policy, when a machine fails the work is interrupted but is not lost; the work is restarted on the machine still in operation, or when the failed machine is repaired. Under the prt policy, on the other hand, when a machine fails the piece being worked on is lost and its production should be restarted.

Consider the case in which all the server states are prs. The expanded CTMC that includes the events of piece completion and the failure and repair of the machines is shown in Figure 15.11(a). The first component inside each state indicates the server state, while the second component represents the number of already completed pieces. In server state 2 the time to complete a single piece is an exponential random variable with parameter $r_2\alpha$, while in server state 1 the time to complete a single piece is $\text{EXP}(r_1\alpha)$. The distribution $F_T(t, \tau)$ of the time to complete m pieces can be calculated

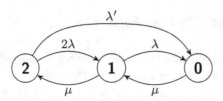

Figure 15.10   Server model of a two-machine parallel system.

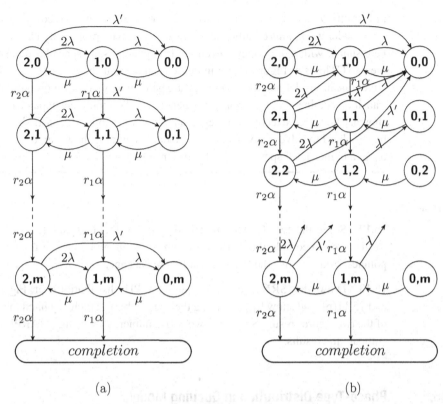

Figure 15.11 Expanded Markov chains for the server model of Figure 15.10, when all the states are of prs (a) or prt (b), and the work requirement is Erlang.

as the distribution of the time to reach the absorbing state labeled "completion" (a numerical solution can be obtained from any CTMC solver such as SHARPE).

If all the server states are prt, the corresponding expanded CTMC whose time to absorption distribution provides the completion time distribution $F_T(t, \tau)$ is shown in Figure 15.11(b). To correctly represent the prt policy, each time a failure transition occurs the piece count should be decreased by one (or two in the case of common mode failures). So, from state $(2, i)$ the completion of piece i with rate $r_2\alpha$ leads the system to state $(2, i+1)$, while the failure of one machine leads the system to state $(1, i-1)$ with rate 2λ. Similarly, from state $(1, i)$ the completion of piece i with rate $r_1\alpha$ leads the system to state $(1, i+1)$, while the failure of the machine leads the system to state $(0, i-1)$ with rate λ. When a common mode failure occurs from state $(2, i)$, the system transits to state $(0, i-2)$ since, under the prt policy, two pieces are lost. Once again, $F_T(t, \tau)$ can be computed numerically by using a software package such as SHARPE.

In a more realistic setting, we assume that a failure in a machine tool can sometimes have unrecoverable effects on the whole production, and sometimes simply interrupt the work without causing any defect in the piece. In the first case the machine failure is modeled by a prt policy (the work must be restarted), while in the second case the failure transition is of prs type. The complete model specification requires the assignment of the probability p that a machine failure causes a permanent and unrecoverable defect in the piece under work, thus entailing a prt policy, while with probability $(1-p)$ the work can be resumed with a prs policy. Given p, the expanded state space becomes a mixture of the two CTMCs of Figure 15.11, in the sense that from each state (i,j), with probability p the transitions defined by a prt policy are generated, while with probability $(1-p)$ the transitions defined by a prs policy are generated. The complete expanded CTMC is not shown. The mixed execution policy entails a very complex stochastic behavior that can be very hard to solve analytically. The model behavior is, however, easier to capture by using a higher-level specification, for instance by using an SRN.

Problems

15.12   Solve the two CTMC models of Figures 15.11(a) and 15.11(b) numerically for the time to completion distribution and its mean, and compare the two preemption policies. You may use the SHARPE software package.

15.13   Draw the SRN model for the two PH distributions of Figures 15.11(a) and 15.11(b), and show by generating their reachability graphs by hand that the CTMCs of the two figures result. Solve the two SRNs numerically using SHARPE or SPNP and compare the results.

15.5   Phase-Type Distribution in Queuing Models

One way the PH distribution family is used in the context of queuing models is to generalize the service time and/or inter-arrival time distribution to be PH, or to approximate a general distribution by a PH distribution. Thus, for example, a $PH/PH/m/K$ queue can be modeled as a homogeneous CTMC [4, 43, 44]. More complex arrival processes like the Markov arrival process, the Markov-modulated Poisson process and the batch Markov arrival process, with PH service-time distributions can be represented as expanded CTMCs [45, 46].

We will soon consider the example of an $M/ER_2/1$ queue. The second way the PH distribution arises is when we consider the distribution of response time in certain queuing systems. For example, we showed that the steady-state response time distribution in an $M/M/1$ FCFS queue is exponential. This can be extended to an $M/M/m$ queue and to open networks of such queues [47–49]. We will see an example of this soon.

Example 15.8   *Approximating MRGP Using PH Cdf: The M/G/1/2/2 Queue*
We return to Example 14.21, where the $M/G/1/2/2$ queue was solved as an MRGP. However, if the service time distribution is PH, or is approximated by a PH distribution, the model can be converted into an expanded CTMC. Further, different service preemption policies or even mixtures of preemption policies can be accommodated.

In the case of Example 14.21, the service distribution was assumed to be deterministic. In this case the best approximating PH is known to be the Erlang (with the number of stages tending to infinity), since the Erlang is the distribution with the minimum coefficient of variation among PH distributions with the same order. Discussion and results for this example are in [50].

Example 15.9 *Queuing Model of the Apache HTTP Server [51]*
The Apache HTTP server is structured as a pool of workers (either threads or processes, depending on the specific software release), as shown in Figure 15.12. Requests enter the Apache HTTP server at the accept queue according to a Poisson process of rate λ, where they wait until a worker is available.

Since the protocol limits the size of the worker pool, in the queuing model the total number of requests is limited to K, and the service discipline is FCFS. The service time is assumed to have a two-stage Erlang (ER_2) distribution with parameter 2μ and expected value equal to $1/\mu$. The queuing model for the system is $M/ER_2/1/K$, and the underlying CTMC is shown in Figure 15.13. The state index denotes the number of $EXP(2\mu)$ phases to be completed by the server.

The steady-state balance equations are:

$$-\lambda\pi_0 + 2\mu\pi_1 = 0,$$

$$-(\lambda + 2\mu)\pi_1 + 2\mu\pi_2 = 0,$$

$$-(\lambda + 2\mu)\pi_i + 2\mu\pi_{i+1} + \lambda\pi_{i-2} = 0, \quad 2 \le i \le 2K - 2,$$

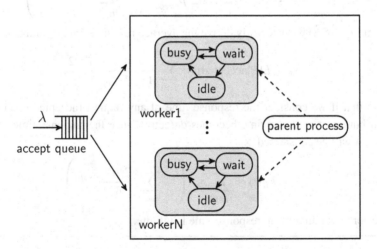

Figure 15.12 Architecture of the Apache HTTP server.

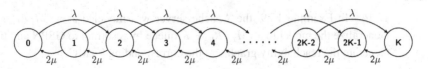

Figure 15.13 Continuous-time Markov chain for the $M/ER_2/1/K$ queuing model.

$$-2\mu\pi_{2K-1}+2\mu\pi_{2K}+\lambda\pi_{2K-3}=0,$$
$$-2\mu\pi_{2K}+\lambda\pi_{2K-2}=0, \tag{15.7}$$
$$\sum_{i=0}^{2K}\pi_i=1.$$

The system of equations in (15.7) is solved in [51] using the method of generating functions. An incoming request is rejected when the system is full, implying $2K$ or $2K-1$ phases in the system. The probability of rejection, P_{reject}, is therefore given by

$$P_{\text{reject}}=\pi_{2K}+\pi_{2K-1}.$$

Example 15.10 *Response Time Distribution in $M/ER_2/1/K$ Queue*
In this example, we first derive the cumulative distribution function of the random variable R, the response time of an accepted request in the $M/ER_2/1/K$ FCFS queue in steady state. As in the case of the $M/M/1$ queue, we use the tagged job approach for this purpose. The probability that an arriving request sees $i \le 2K-2$ phases in the system, given that it is accepted, is given by $\pi_i/(1-P_{\text{reject}})$. For such a request, the response time is the sum of the completion times of $i+2$ phases. As these completion times are independent and are each exponentially distributed with rate 2μ, the conditional response time follows an $(i+2)$-stage Erlang distribution with rate 2μ. Therefore, the Cdf of the response time R of an accepted request is a PH distribution given by:

$$F_{R|\text{accepted}}(t)=\sum_{i=0}^{2K-2}\frac{\pi_i}{1-P_{\text{reject}}}\cdot\left(1-\sum_{j=0}^{i+1}\frac{(2\mu t)^j}{j!}e^{-2\mu t}\right). \tag{15.8}$$

From Eq. (15.8) we can easily derive the average response time formula:

$$E[R|\text{accepted}]=\sum_{i=0}^{2K-2}\frac{\pi_i}{1-P_{\text{reject}}}\cdot\frac{i+2}{2\mu}.$$

Note that if we consider the response time of any request (accepted or not), then the distribution of response time becomes defective since the response time of a rejected request will be considered ∞:

$$F_R(t)=\sum_{i=0}^{2K-2}\pi_i\cdot\left(1-\sum_{j=0}^{i+1}\frac{(2\mu t)^j}{j!}e^{-2\mu t}\right). \tag{15.9}$$

The corresponding mean response time is infinite.

Problems

15.14 In Example 15.9, the Apache server was modeled as an $M/ER_2/1/K$ queue. Generalize the model to an $M/ER_n/1/K$ queue by resorting to GSPN and then to SRN. Generate the reachability graph and the corresponding CTMC.

15.15 Draw the SRN model for an $M/ER_n/1/K$ queue with server failure and repair.

15.6   Comparison of the Modeling Power of Different Model Types

The examples of this chapter have shown that the various model types introduced in the previous chapters have restrictions, and each one can cope with different features of a system. The flow chart of Figure 15.14 is intended to clarify the potentialities of the different model types and to guide the modeler in selecting the right model type with respect to the particular system under analysis and its characteristics in terms of events and distributions.

If all the events are characterized by exponential distributions – constant hazard rate (CHR) – and statistical dependencies in the failure or repair processes of the components are to be taken into account, the appropriate model type is the homogeneous CTMC considered in Chapters 9 and 10. This is by far the most utilized technique in dependability since the governing equations are easily derived from the model

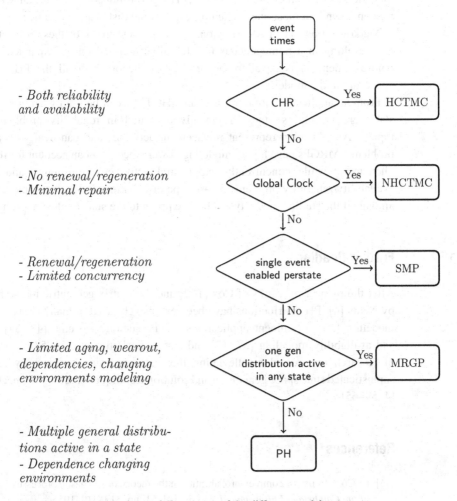

Figure 15.14   Flow chart comparing the modeling power of the different model types.

description, and general, powerful numerical solution techniques are known and implemented in many software packages (Sections 9.8 and 10.4).

When failure and/or repair events have non-exponential distributions, several alternatives can be considered. If all the event times are measured from a single global clock then the use of the NHCTMC of Chapter 13 is appropriate. In this framework, repair times are assumed to be very small in comparison with failure times and a repair is assumed to restore the system to an as-bad-as-old condition. This condition is sometimes referred to as minimal repair [36, 40]. A repair action modeled by means of a PH cannot represent an as-bad-as-old repair, but the repair restores the component to a previous stage. In this context a minimal repair can be interpreted as a transition that recovers the component just to the previous stage [52].

If repair is regenerative and at any change of state the clocks of all the transitions are reset, the appropriate model type is the SMP of Chapter 14. If one or more events with exponentially distributed times occur concurrently with a generally distributed event time, then the MRGP framework can be used. The governing equations for an MRGP can be written down if at most one single non-exponential distribution is active in any state.

Markovization via PH distributions can overcome many of these restrictions while still resulting in a homogeneous CTMC, albeit defined on an expanded state space whose dimension is equal to the product of the order of all the PH distributions appearing in the model.

From an application point of view, an NHCTMC can be used when no regeneration states are present, so they are mainly exploited in reliability problems (without repair). An SMP can represent replacement activities and can deal with availability problems. MRGPs have higher modeling power since they can account for regenerative phenomena while concurrently exponential activities are running and can cover availability/reliability problems. Phase-type distributions have greater modeling power among all the studied model types, but are prone to the state explosion problem.

15.7   Further Reading

After the pioneering works of Cox [1, 2] and the matrix geometric methods proposed by Neuts [4], PH distributions have been extensively used in many areas of stochastic modeling [53]. Prominent applications are in queuing systems [43, 54], reliability and availability modeling [37, 55] and stochastic Petri nets [56–58]. A current area of research is directed at alleviating the state explosion problem by resorting to sophisticated state space generation and solution techniques based on Kronecker algebra [4, 59–65].

References

[1] D. Cox, "A use of complex probabilities in the theory of stochastic processes," *Proceedings of the Cambridge Philosophical Society*, vol. 51, pp. 313–319, 1955.

[2] D. Cox, "The analysis of non-Markovian stochastic processes by the inclusion of supplementary variables," *Proceedings of the Cambridge Philosophical Society*, vol. 51, pp. 433–440, 1955.

[3] R. Sahner, K. Trivedi, and A. Puliafito, *Performance and Reliability Analysis of Computer Systems: An Example-based Approach Using the SHARPE Software Package*. Kluwer Academic Publishers, 1996.

[4] M. Neuts, *Matrix Geometric Solutions in Stochastic Models*. Johns Hopkins University Press, 1981.

[5] D. Cox and H. Miller, *The Theory of Stochastic Processes*. Chapman and Hall, 1965.

[6] S. Asmussen, O. Nerman, and M. Olsson, "Fitting phase-type distributions via the EM algorithm," *Scandinavian Journal of Statistics*, vol. 23, no. 4, pp. 419–441, 1996.

[7] A. Horváth and M. Telek, "PhFit: A general phase-type fitting tool," in *Computer Performance Evaluation: Modelling Techniques and Tools*, Lecture Notes in Computer Science, vol. 2324, eds. T. Field, P. Harrison, J. Bradley, and U. Harder. Springer, 2002, pp. 82–91.

[8] H. Okamura, T. Dohi, and K. S. Trivedi, "A refined EM algorithm for PH distributions," *Performance Evaluation*, vol. 68, no. 10, pp. 938–954, 2011.

[9] G. Horváth and M. Telek, "BuTools 2: a rich toolbox for Markovian performance evaluation," in *Valuetools 2016 - Conf. on Performance Evaluation Methodologies and Tools*, 2016.

[10] A. Bobbio, A. Horváth, M. Scarpa, and M. Telek, "Acyclic discrete phase type distributions: Properties and a parameter estimation algorithm," *Performance Evaluation*, vol. 54, no. 1, pp. 1–32, 2003.

[11] C. O'Cinneide, "Characterization of phase-type distributions," *Stochastic Models*, vol. 6, no. 1, pp. 1–57, 1990.

[12] C. O'Cinneide, "On non-uniqueness of representations of phase-type distributions," *Stochastic Models*, vol. 5, pp. 247–259, 1989.

[13] C. Commault and J. Chemla, "On dual and minimal phase-type representations," *Stochastic Models*, vol. 9, pp. 421–434, 1993.

[14] T. Osogami and M. Harchol-Balter, "Closed form solutions for mapping general distributions to quasi-minimal PH distributions," *Performance Evaluation*, vol. 63, no. 6, pp. 524–552, 2006.

[15] G. Horváth and M. Telek, "On the canonical representation of phase-type distributions," *Performance Evaluation*, vol. 66, no. 8, pp. 396–409, 2009.

[16] A. Cumani, "On the canonical representation of homogeneous Markov processes modelling failure-time distributions," *Microelectronics and Reliability*, vol. 22, pp. 583–602, 1982.

[17] A. Bobbio and A. Cumani, "Markov models: A new class of distributions for the analysis of lifetime data samples," in *Proc. 4th EUREDATA Conf.*. Paper 10.3/1–13, 1983.

[18] D. Aldous and L. Shepp, "The least variable phase type distribution is Erlang," *Stochastic Models*, vol. 3, pp. 467–473, 1987.

[19] D. Assaf and B. Levikson, "Closure of phase type distributions under operations arising in reliability theory," *The Annals of Probability*, vol. 10, pp. 265–269, 1982.

[20] R. Maier and C. O'Cinneide, "A closure characterisation of phase-type distributions," *Journal of Applied Probability*, vol. 29, no. 1, pp. 92–103, 1992.

[21] A. Bobbio and K. S. Trivedi, "Computation of the distribution of the completion time when the work requirement is a PH random variable," *Stochastic Models*, vol. 6, pp. 133–149, 1990.

[22] A. Bobbio and L. Roberti, "Distribution of the minimal completion time of parallel tasks in multi-reward semi-Markov models," *Performance Evaluation*, vol. 14, pp. 239–256, 1992.

[23] W. Bux and U. Herzog, "The phase concept: Approximation of measured data and performance analysis," in *Computer Performance*, eds. K. Chandy and M. Reiser. North-Holland, 1977, pp. 23–38.

[24] C. Singh, R. Billinton, and S. Lee, "The method of stages for non-Markovian models," *IEEE Transactions on Reliability*, vol. R-26, pp. 135–137, 1977.

[25] M. Johnson and M. Taaffe, "Matching moments to phase distributions: Mixtures of Erlang distribution of common order," *Stochastic Models*, vol. 5, pp. 711–743, 1989.

[26] L. Schmickler, "MEDA: Mixed Erlang distributions as phase-type representations of empirical distribution functions," *Stochastic Models*, vol. 8, pp. 131–156, 1992.

[27] A. Bobbio and A. Cumani, "ML estimation of the parameters of a PH distribution in triangular canonical form," in *Computer Performance Evaluation*, eds. G. Balbo and G. Serazzi. Elsevier Science Publishers, 1992, pp. 33–46.

[28] A. Bobbio and M. Telek, "A benchmark for PH estimation algorithms: Results for acyclic-PH," *Stochastic Models*, vol. 10, pp. 661–677, 1994.

[29] A. Bobbio, A. Horváth, and M. Telek, "PhFit: A general phase-type fitting tool," in *Proc. Int. Conf. on Dependable Systems and Networks*, 2002, p. 543.

[30] M. Telek, "A tool for fitting distributions of phase type." [Online]. Available: http://webspn.hit.bme.hu/~telek/tools.htm

[31] A. Dempster, N. Laird, and D. Rubin, "Maximum likelihood from incomplete data via the EM algorithm," *Journal of the Royal Statistical Society*, vol. 39, pp. 1–38, 1977.

[32] A. Thumler, P. Buchholz, and M. Telek, "A novel approach for phase-type fitting with the EM algorithm," *IEEE Transactions on Dependable and Secure Computing*, vol. 3, pp. 245–258, 2006.

[33] M. Faddy, "Phase-type distributions for failure times," *Mathematical and Computer Modelling*, vol. 22, no. 10–12, pp. 63–70, 1995.

[34] A. Horváth and M. Telek, "Approximating heavy tailed behavior with phase type distributions," in *Proc. 3rd Int. Conf. on Matrix-Analytic Methods in Stochastic Models, MAM3*. Notable Publications Inc, 2000, pp. 191–214.

[35] Y. Lu, A. Miller, R. Hoffmann, and C. Johnson, "Towards the automated verification of Weibull distributions for system failure rates," in *Proc. Int. Workshop on Formal Methods for Industrial Critical Systems and Automated Verification of Critical Systems (FMICS-AVoCS 2016)*, 2016.

[36] S. Distefano, F. Longo, and K. S. Trivedi, "Investigating dynamic reliability and availability through state-space models," *Computers and Mathematics with Applications*, vol. 64, no. 12, pp. 3701–3716, 2012.

[37] M. Neuts and K. Meier, "On the use of phase type distributions in reliability modelling of systems with two components," *OR Spektrum*, vol. 2, pp. 227–234, 1981.

[38] A. Bobbio and A. Cumani, "A Markov approach to wear-out modelling," *Microelectronics and Reliability*, vol. 23, pp. 113–119, 1983.

[39] D. Montoro-Cazorla and R. Pérez-Ocón, "A deteriorating two-system with two repair modes and sojourn times phase-type distributed," *Reliability Engineering and System Safety*, vol. 91, no. 1, pp. 1–9, 2006.

[40] U. Jensen, "Stochastic models of reliability and maintenance: An overview," in *Reliability and Maintenance of Complex Systems*, ed. S. Özekici. NATO ASI Series, Springer, 1996, ch. 1, pp. 3–37.

[41] V. Kulkarni, V. Nicola, and K. Trivedi, "The completion time of a job on a multi-mode system," *Advances in Applied Probability*, vol. 19, pp. 932–954, 1987.

[42] M. Neuts, "Two further closure properties of PH-distributions," *Asia-Pacific Journal of Operational Research*, vol. 9, pp. 459–477, 1992.

[43] V. Ramaswami, "Algorithms for the multi-server queue with phase type service," *Stochastic Models*, vol. 1, no. 3, pp. 339–417, 1985.

[44] B. Sericola, F. Guillemin, and J. Boyer, "Sojourn times in the M/PH/1 processor sharing queue," *Queueing Systems*, vol. 50, no. 1, pp. 109–130, 2005.

[45] S. Asmussen, *Markov Additive Models*. Springer, 2003, pp. 302–339.

[46] S. Chakravarthy, *Markovian Arrival Processes*. John Wiley & Sons, 2010.

[47] D. Gross and C. Harris, *Fundamentals of Queuing theory*. John Wiley & Sons, 1985.

[48] S. Woolet, "Performance analysis of computer networks," PhD Thesis, Department of Computer Science, Duke University, 1993.

[49] K. Trivedi, *Probability and Statistics with Reliability, Queueing and Computer Science Applications*, 2nd edn. John Wiley & Sons, 2001.

[50] A. Bobbio and M. Telek, "Non-exponential stochastic Petri nets: An overview of methods and techniques," *Computer Systems: Science and Engineering*, vol. 13, no. 6, pp. 339–351, 1998.

[51] J. Zhao, K. Trivedi, M. Grottke, J. Alonso, and Y. Wang, "Ensuring the performance of Apache HTTP server affected by aging," *IEEE Transactions on Dependable and Secure Computing*, vol. 11, no. 2, pp. 130–141, Mar. 2014.

[52] D. Chen and K. S. Trivedi, "Closed-form analytical results for condition-based maintenance," *Reliability Engineering and System Safety*, vol. 76, no. 1, pp. 43–51, 2002.

[53] G. Latouche and V. Ramaswami, *Introduction to Matrix Analytic Methods in Stochastic Modeling*. Society for Industrial and Applied Mathematics, 1999.

[54] B. V. Houdt, "A phase-type representation for the queue length distribution of a semi-Markovian queue," in *Proc. 7th Int. Conf. on Quantitative Evaluation of Systems*, Sep 2010, pp. 49–58.

[55] J. E. Ruiz-Castro, R. Pérez-Ocón, and G. Fernández-Villodre, "Modelling a reliability system governed by discrete phase-type distributions," *Reliability Engineering and System Safety*, vol. 93, no. 11, pp. 1650–1657, 2008.

[56] A. Cumani, "ESP: A package for the evaluation of stochastic Petri nets with phase-type distributed transition times," in *Proc. Int. Workshop on Timed Petri Nets*. IEEE Computer Society Press no. 674, 1985, pp. 144–151.

[57] P. Chen, S. Bruell, and G. Balbo, "Alternative methods for incorporating non-exponential distributions into stochastic timed Petri nets," in *Proc. Int. Workshop on Petri Nets and Performance Models (PNPM89)*. IEEE Computer Society, 1989, pp. 187–197.

[58] S. Haddad, P. Moreaux, and G. Chiola, "Efficient handling of phase-type distributions in generalized stochastic Petri nets," in *Proc. 18th Int. Conf. on Application and Theory of Petri Nets, ICATPN'97*, LNCS 1248, eds. P. Azéma and G. Balbo. Springer Verlag, 1997, pp. 175–194.

[59] J. Brewer, "Kronecker products and matrix calculus in system theory," *IEEE Transactions on Circuits and Systems*, vol. CAS-25, pp. 772–781, 1978.

[60] M. Davio, "Kronecker products and shuffle algebra," *IEEE Transactions on Computers*, vol. C-30, pp. 116–125, 1981.

[61] S. Donatelli, S. Haddad, and P. Moreaux, "Structured characterization of the Markov chain of a phase type SPN," in *Proc. 10th Int. Conf. on Modelling Techniques and Tools for Computer Performance Evaluation*. Springer, 1998, pp. 243–254.

[62] M. Scarpa and A. Bobbio, "Kronecker representation of stochastic Petri nets with discrete PH distributions," in *Int. Computer Performance and Dependability Symp. (IPDS98)*. IEEE Computer Society Press, 1998, pp. 52–61.

[63] G. Ciardo and A. Miner, "A data structure for the efficient Kronecker solution of GSPNs," in *Proc. 8th Int. Conf. on Petri Nets and Performance Models (PNPM99)*. IEEE Computer Society, 1999, pp. 22–31.

[64] P. Buchholz, G. Ciardo, S. Donatelli and P. Kemper, "Complexity of memory-efficient Kronecker operations with applications to the solution of Markov models," *INFORMS Journal on Computing*, vol. 12, no. 3, pp. 203–222, 2000.

[65] P. Buchholz and P. Kemper, "Kronecker based matrix representations for large Markov chains," in *Validation of Stochastic Systems*, LNCS 2925, eds. M. S. B. Haverkort, H. Hermanns. Springer Verlag, 2004, pp. 256–295.

Part V

Multi-Level Models

The previous chapters introduced several modeling formalisms that can be used to represent the dependability behavior of systems. So far, the implicit assumption has been that the entire system model is developed using a single formalism. While these approaches are quite useful, the complexity of real-life systems makes it infeasible for a single formalism to adequately capture the entire system behavior. Each formalism has its own advantages and drawbacks, thus limiting the range of system behavior that a formalism can accurately represent in the model. This is especially true with non-state-space-based approaches, which have limited capability to incorporate interdependencies and dynamic changes in behavior. While state-space-based formalisms are highly capable in this regard, they suffer from the state space explosion problem. We now explore the possibility of combining several model types in a multi-level model by selecting an appropriate formalism to represent component- or subsystem-level behavior and then composing the submodel results together to come up with the overall measures of the system behavior. Indeed, such multi-level models that combine the capabilities of different formalisms are commonly used to solve real-world problems.

Multi-level models are sometimes used for mere specification, so that an underlying monolithic model is nevertheless generated, stored, and solved. We use multi-level modeling here not only for specification but also for solution, so that a monolithic model is never generated, stored or solved. Each individual model is solved and the appropriate part of the solution is passed on to other submodels as input parameters, possibly after some processing. Such a transfer of quantities leads us to construct an import graph indicating communication paths between submodels. If the import graph is acyclic then we call the overall model *hierarchical*. Otherwise, cycles in the import graph result in a fixed-point problem that needs an iterative solution via successive substitution. In Chapter 16 we will study hierarchical models and in Chapter 17 we will study fixed-point iterative models. Real-life case studies will be explored in Part VI.

16 Hierarchical Models

In this chapter, we explore the possibility of combining several model types in a hierarchical manner where the exchange of quantities across submodels is acyclic in nature. We first give the motivation for hierarchical modeling in Section 16.1. In Section 16.2, we describe our notion of hierarchical modeling and its formal specification. In the subsequent sections we present several examples to illustrate the power of this approach. These examples differ in the kinds of parameters passed among the submodels (probabilities, expected values, variances, distributions, expected reward rates, etc.) and the solution techniques (non-iterative hierarchical solution, two-level performability solution, behavioral decomposition, state aggregation, etc.) employed. The set of examples is divided into availability models (Section 16.3), reliability models (Section 16.4), dynamic fault tree models (Section 16.5), performance models (Section 16.6), performability models (Section 16.7), and survivability models (Section 16.8).

16.1 Introduction

Dependability and performance models of complex systems suffer from two afflictions: model *largeness* and model *stiffness*. We briefly alluded to these earlier in the book. Model largeness arises due to the complex interactions among the large number of components that need to be represented in the model, while stiffness is primarily a result of extreme disparity in various time constants related to different aspects of system behavior. Approaches to address these two drawbacks can be broadly classified into two categories: *avoidance* and *tolerance*. This chapter is devoted to examining several such techniques.

Largeness avoidance techniques aim to avoid the generation and solution of large models by exploiting the independence in the failure and repair behavior of components. Non-state-space methods such as reliability block diagrams, reliability graphs, and fault trees that we covered in Part II can be broadly classified as largeness avoidance techniques. However, these methods cannot easily capture various dependencies such as repair dependency, reliability models with different repair policies, dynamic redundancy, coverage, and common cause failures. Among performance model types, product-form queuing networks and series-parallel directed acyclic graphs could be used for compact system representation and efficient solution [1, 2] and hence avoid largeness. But queuing networks in practice do not often satisfy product-form constraints [1], while series-parallel task graphs do not capture resource contention [2]. For capturing dependencies, non-product-form behavior or contention for resources amid concurrency, we need to resort to state-space methods.

Continuous-time (or discrete-time) Markov chains and semi-Markov models are the most commonly used state-space-based methods for dependability and performance modeling. Unfortunately, state-based methods suffer from state space explosion, i.e., an extremely large state space is required for the accurate modeling of real systems. Generation of large Markov models is in itself a complex task. To alleviate the problem of model specification and generation, higher-level paradigms such as stochastic Petri nets [3, 4], stochastic automata networks [5], stochastic activity networks [6], and stochastic process algebras [7] provide a more compact representation of a model and can be automatically converted into the underlying Markov chain [4, 7]. These approaches come under the purview of largeness tolerance methods. Such higher-level formalisms, however, do not alleviate the problems of generating, storing, and solving the large model. To overcome the problem of largeness in model solution, several techniques have been proposed that decompose the Markov chain generator matrix and yield approximate solutions [8, 9]. However, the problem of large model generation and storage still persists. Kronecker-algebra-based methods have been used to some extent to efficiently store and solve large Markov models [10–12]. Model decomposition approaches that work at the model or system level rather than at the matrix level are thus needed. Examples of such techniques include the flow-equivalent server approximation introduced by Chandy *et al.* [13], the behavioral decomposition technique used in the software package HARP [14], the approach used by Balbo *et al.* [15], the hierarchical (or multi-level) model solution employed in the software package SHARPE [2] and the use of fixed-point iterative solutions [16–18]. In these approaches, the overall model is never generated/stored/solved, but instead smaller models are generated whose solution is combined (or rolled up) to yield the overall model solution. Such multi-level modeling often leads to stiffness avoidance as well as largeness avoidance, since it is often possible to separate in the different levels rates differing by orders of magnitude.

16.2   Hierarchical Modeling

The hierarchical (or multi-level) model solution used in [2] and the fixed-point iterative solution methods presented in [16–18] are examples of *hierarchical composition* as opposed to decomposition. We consider hierarchical composition in its most general sense. Our view of hierarchical modeling encompasses a wide variety of modeling and solution techniques including reward-based performability analysis [19–21] (decomposition of the overall model into dependability and performance models for both specification and solution for combined performance and dependability measures), and non-iterative hierarchical (hierarchical composition or decomposition) models considered in [2, 22–25].

The overall system model consists of one or more submodels of possibly different types (referred to as hybrid hierarchical modeling) that interact with each other. The model solution is obtained by first solving the submodels and thereafter combining the submodel solutions in order to obtain the overall system model solution.

Malhotra and Trivedi [26] presented a formal method of expressing hierarchy with many examples. A general hierarchical modeling environment is provided by the software package SHARPE [2]. SHARPE allows hierarchy among models of different types (such as fault trees, RBDs, reliability graphs, CTMCs, SMPs, MRGPs, stochastic reward nets, product-form queuing networks, etc.). In contrast to several other approaches that utilize hierarchy only for model generation and specification, our approach uses hierarchy for model solution as well. In the SHARPE framework, the hierarchy among various submodels is implicit. The order of solution of the submodels is specified in the textual input file (or in the GUI) and syntactic constructs for parameter passing between the submodels are provided. This chapter briefly describes an approach for the formal specification of hierarchy both in model construction and solution. Our approach brings hierarchy in model solution to the fore and provides a better understanding of the overall system model to the user.

16.2.1 Import Graph

We consider the overall system model as an import graph whose nodes are the set $\{M_1, M_2, \ldots, M_k\}$ of submodels. For instance, the import graph for Example 16.1 is shown in Figure 16.1. These submodels can encompass the entire gamut of model types, including both non-state-space- and state-space-based formalisms. In general, models solved using discrete event simulation could also be added in the above list, but in this book we do not discuss such a possibility further. Similarly, data obtained from measurements on real systems in production and/or in a laboratory environment followed by statistical analysis can be included but will not be considered further here. At the level of the import graph, we treat each submodel as a black box that has some inputs and outputs. Specialized solution techniques are used within each black box to compute the outputs from its inputs. Thus each submodel can simply be considered a mapping from its inputs to its outputs. The outputs could be expressed in fully symbolic, semi-symbolic or numeric form depending on a combination of the model type, the measure computed and the options selected by the modeler. The same holds for the inputs to any black box. The actual values of the inputs may be computed by other submodels, may be computed from several submodel outputs or may be supplied directly by the modeler. There could be many possible outputs from an individual submodel.

Consider, for example, a homogeneous CTMC with a single absorbing state. Suppose it models the reliability of a (sub)system. The set of inputs to this model consists of the initial state probabilities for the CTMC and the entries of the generator matrix. Optionally, reward rates for the states may be specified as inputs as well. These input parameters can be directly supplied by the user or derived from outputs of other submodels. The possible outputs that can be computed from this model are mean time to absorption (which in this case will be the (sub)system mean time to failure), (sub)system reliability or unreliability at time t, individual probabilities of being in different states at some time t, or in general any reward-based measure. The user can specify that one or more of these outputs be computed. Likewise, the user can also specify the solution method depending upon the type of solution desired and the available software package. For example, a semi-symbolic method may yield the reliability of a system as a closed-form expression in t (the time variable), whereas a numerical method will yield a numerical value of the reliability at a given instant of time t.

The solution of the overall system model is computed by considering the interactions among various submodels. By interactions we mean that parameters are passed between a pair of connected submodels. These parameters could be a numerical value (of mean, variance, transition rates, coverage probabilities, reward rates, steady-state availability, expected reward rate in the steady state, etc.), a closed-form expression (e.g., time-dependent coverage probability), or a semi-symbolic distribution function (e.g., (un)reliability at time t, instantaneous (un)availability at time t, distribution of accumulated reward to absorption, etc.). If output parameters are passed from submodel M_i as inputs to submodel M_j, then we write $M_i \succ M_j$. The relation "$\succ$" defines an order of execution among the submodels $M_1, M_2, \ldots, M_k$. This order is represented by a directed graph G in which each node represents a submodel and there is an edge directed from node i to j if $M_i \succ M_j$. Such a graph has been named an *import graph* [27]. Let the transitive closure of relation $\succ$ be denoted by $\succ\succ$.

To compute the solution of the overall model, two cases need to be considered:

- Import graph G is acyclic. This implies that the relation $\succ$ defines a partial order among various submodels and provides one or more linear orderings by which each submodel may be solved. If $M_i \succ\succ M_j$, then M_i must be solved before M_j since M_j receives input from the outputs of M_i. If neither $M_i \succ\succ M_j$ nor $M_j \succ\succ M_i$ hold, then M_i and M_j can be solved in any order since the solution of one does not affect the solution of the other. We carry out a topological sort [28] to linearize the solution sequence of submodels. Examples of this type are considered in the current chapter. As we will point out, a hierarchical model solution need not always result in an approximation, but it often does.
- Import graph G is non-acyclic. This implies that there exists at least one pair of submodels M_i and M_j such that $M_i \succ\succ M_j$ and $M_j \succ\succ M_i$. This case presents a problem regarding solution of the overall model. We must resort to an iterative solution in this case. Examples involving such import graphs with cycles will be dealt with in Chapter 17.

16.3   Availability Models

In this section we first discuss preventive maintenance for optimizing availability in the software context. The second example in this section is that of the fluid valve system level controller that has been considered in previous chapters, but now hierarchical modeling is illustrated through this example.

Example 16.1 *Time-Based Preventive Maintenance*

We return to the three-state SMP for optimizing availability (see Example 14.4 and Figure 14.4). Earlier we assumed that the time to failure distribution $F(t)$ is either given or is obtained by fitting it to real data. Now we compute $F(t)$ from a lower-level model. First consider the lower-level model from Example 13.6, where an NHCTMC is used to capture two effects of software aging: performance degradation by means of server slowdown as a function of the time since the last reboot (proactive or reactive), and a server failure rate that is increasing with time as measured from the last reboot (proactive or reactive). As can be seen from Eq. (14.7), we need to compute the distribution function $F(t_0)$ of the time to reach the failure state and the integral $h_U(t_0) = \int_0^{t_0} (1 - F(t))dt$. Both these can be computed by using piecewise constant approximation as discussed in Section 13.3. The import graph for this case is shown in Figure 16.1. In this case the import graph is acyclic and has two submodels: the top-level model M_S is the three-state SMP and the bottom-level model M_N is the NHCTMC shown in Figure 13.7. The quantities being passed up the hierarchy are $F(t_0)$ and $h_U(t_0) = \int_0^{t_0} (1 - F(t))dt$. We leave the computations as a homework exercise, where use of the SHARPE software package is recommended. For further details see [29].

As a second case, we use the same top-level model but the bottom-level model is now the NHCTMC of Example 13.5 and Figure 13.6. Once again, the two quantities $F(t_0)$ and the integral $h_U(t_0) = \int_0^{t_0} (1 - F(t))dt$ can be computed from the NHCTMC using the PCA method. In this case the import graph is also hierarchical and has two submodels: the top level is the three-state SMP and the bottom level is the NHCTMC shown in Figure 13.6. The quantities being passed up the hierarchy are the above two values. The import graph of Figure 16.1 applies in this case as well. We leave the computations as a homework exercise, where use of the SHARPE software package is recommended. For further details see [30].

Next we consider the same three-state SMP as the top-level model, but the seven-state SMP reward model from Example 14.9 is used as the lower-level model that supplies the needed parameters to the upper level. Both the structure (states and state transitions) and parameters (transition probabilities and the sojourn time distributions) of the seven-state SMP are derived from collected real data. For further details, see [31].

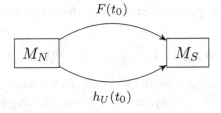

Figure 16.1 Import graph for the preventive maintenance example.

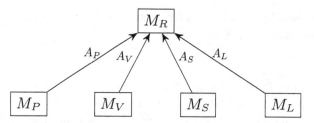

Figure 16.2   Import graph for fluid level controller with independent repair.

In all three cases above, the hierarchy does not introduce any approximation *per se*, but numerical solution errors may still be incurred.

Example 16.2   *A Repairable Fluid Level Controller*

We return to fluid level controller of Example 4.1, but with the addition of independent repair of components. Each component is modeled by a two-state CTMC. At the top level we have the RBD shown in Figure 4.2. We have four lower-level CTMC submodels (M_P, M_V, M_S, and M_L) that supply steady-state availability values to the top-level RBD (M_R) of Figure 4.2. Notice that the effect of solving the top-level RBD in this case is simply multiplying the availabilities of the four subsystems in order to obtain the overall system availability; hence, there may be a tendency to omit this submodel altogether. The simple import graph with the five submodels is shown in Figure 16.2.

If the assumption of independent repair does hold then there is no approximation involved in this hierarchical solution. The system steady-state availability expression is then easily written:

$$A_{\text{indep. repair}} = \frac{\mu_P}{\lambda_P + \mu_P} \frac{\mu_V}{\lambda_V + \mu_V} \frac{\mu_S}{\lambda_S + \mu_S} \frac{\mu_L}{\lambda_L + \mu_L}. \tag{16.1}$$

In a similar way, an expression for the transient availability can also be written. An important point to note is that by using a hierarchical model solution, we not only avoid largeness (and stiffness) but often the submodels become simple enough that we can derive a closed-form solution, as in this case.

As pointed out, at times a hierarchical model solution does not imply an approximate solution, but at other times it does. An approximate hierarchical solution can be sought by appealing to the concept of *nearly independent subsystems* developed in [27, 32]. A system is said to be composed of nearly independent subsystems if the probability of a state of the overall system can be approximated by the product of probabilities of the states of individual subsystems:

$$P\{S = s_1, s_2, \ldots, s_n\} \approx P\{S_1 = s_1\} P\{S_2 = s_2\} \cdots P\{S_n = s_n\},$$

where $P\{S = s_1, s_2, \ldots, s_n\}$ is the probability of a state of the overall system, and where subsystem i is in state $S_i = s_i$. A system with dependent subsystems can be treated as nearly independent if each individual subsystem model can be modified to (approximately) account for the effect of the dependence. The following example illustrates the use of this concept.

Example 16.3 *Non-redundant Fluid Level Controller with Shared Repair: Approximate Solution*

In Example 9.9, the steady-state availability of this system was computed by generating and solving a monolithic CTMC model. In Example 16.2, we used a hierarchical approach for deriving the system availability but invoked independence of repair across subsystems. Since the shared repair facility introduces dependence across subsystems, we will now use a hierarchical approach by invoking near independence across subsystems.

As in Example 9.9, we assume shared repair with preemptive repair priority order $P \succ V \succ S \succ L$. Since P has the highest repair priority, its repair can proceed without any dependence on other subsystems. The CTMC availability model of this subsystem can thus be solved on its own. The repair of the other subsystems with lower priority needs to be cognizant of any repair needs of higher-priority subsystems. We can account for this dependence by reducing the repair rates of the subsystems with lower priority. The CTMC M_P for subsystem P is solved first, and the probability Q_P that subsystem P has an ongoing repair is calculated. Thus the probability that the repair person is idle is $q_P = (1 - Q_P)$. Subsystem V can be repaired only if the repair person is not busy with subsystem P, which can be accounted for by reducing the repair rate of subsystem V by the factor q_P:

$$\mu'_V = \mu_V q_P = \mu_V (1 - Q_P).$$

Applying this idea sequentially, in order to start repair of components in subsystem M_i, the repair person must be idle on all the subsystems M_j, $j \le (i - 1)$. With this modification, the four CTMCs (each with two states) are solved in sequence (P, V, S, and L) and then the subsystem availabilities are supplied to the series RBD of Figure 4.2. The acyclic dependency between the five submodels in this case is represented by the import graph of Figure 16.3. Compare the import graphs of this example having shared repair with that of Figure 16.2 with independent repair. The steady-state availability expression for this case is easily written in closed form:

$$A_{\text{shared repair}} = \frac{\mu_P}{\lambda_P + \mu_P} \frac{\mu'_V}{\lambda_V + \mu'_V} \frac{\mu'_S}{\lambda_S + \mu'_S} \frac{\mu'_L}{\lambda_L + \mu'_L}. \tag{16.2}$$

The updated expressions for the repair rates for each subsystem are reported in the fourth column of Table 16.1, and the numerical values of the approximate steady-state availabilities of the subsystems are compared with the independent case [33].

The results for the steady-state system availability are compared in Table 16.2. The approximation technique provides fairly good results, with match up to seven digits.

It is clear that the hierarchy in this case gives an approximation. One way to improve the approximation is to consider transient values of parameters such as $q_P(t)$. As well as needing a transient analysis of all the submodels, this will mean that all lower-priority

Table 16.1 Nearly independent approximation for shared repair.

Subsystem	A_i	Q_i	μ_i'	$A_{\text{indep. repair}}$	$A_{\text{shared repair}}$
P	π_1	π_0	μ_P	0.996 810 207	0.996 810 207
V	π_1	π_0	$\mu_V(1 - Q_P)$	0.999 200 639	0.999 198 084
S	π_1	π_0	$\mu_S(1 - Q_P)(1 - Q_V)$	0.998 402 556	0.998 396 168
L	π_1	π_0	$\mu_L(1 - Q_P)(1 - Q_V)(1 - Q_S)$	0.999 000 999	0.998 995 392

Table 16.2 Comparing approximate and exact system availabilities.

Indep. repair	Shared repair (exact)	Shared repair (approx)
0.995 754 485	0.993 414 494 000	0.993 414 420

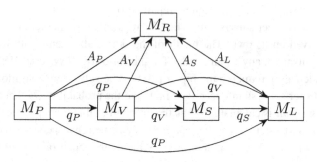

Figure 16.3 Import graph for fluid level controller with shared repair.

CTMC models will become non-homogeneous CTMCs. We leave this extension as an exercise for the reader.

Example 16.4 *Redundant Fluid Level Controller with Independent Repair*

The redundant configuration for each component of this system was introduced in Example 9.10, where the subsystems were assumed to be independent. In this case, individual CTMC availability submodels of the four subsystems were presented in Example 9.10. The steady-state availability results computed individually from each submodel will be supplied to the top-level RBD. The import graph of Figure 16.2 from the non-redundant case with independent repair still applies, though all the submodels in this case are different and correspond to the CTMC submodels of Example 9.10.

Example 16.5 *Redundant Fluid Level Controller with Shared Repair: Approximate Solution*

The dependence due to the shared repair person in the redundant fluid level controller was investigated in Example 12.18, where the monolithic state space was generated by means of an SRN model. By adopting the same approximation idea as the previous example, we can derive a hierarchical approximate solution with the same import graph as Figure 16.3, though all the submodels in this case are different and correspond to the

Table 16.3 Nearly independent approximation for shared repair of system with redundancy.

Subsystem	A_i	Q_i	μ'_i	$A_{indep.\,repair}$	$A_{shared\,repair}$
P	$\pi_{11} + \pi_{01} + \pi_{10}$	π_0	μ_P	0.997 907 835	0.997 907 835
V	$\pi_2 + \pi_1$	$\pi_1 + \pi_0$	$\mu_V(1 - Q_P)$	0.999 799 082	0.999 799 077
S	$\pi_3 + \pi_2$	$\pi_2 + \pi_1 + \pi_0$	$\mu_S(1 - Q_P)$ $(1 - Q_V)$	0.999 984 689	0.999 984 576
L	$\pi_2 + \pi_1$	$\pi_1 + \pi_0$	$\mu_L(1 - Q_P)$ $(1 - Q_V)(1 - Q_S)$	0.999 798 444	0.999 796 706

Table 16.4 Comparing availabilities of different system configurations and repair scenarios.

Configuration	Indep. repair	Shared repair (exact)	Shared repair (approx)
Redundancy	0.997 490 970	0.997 485 391	0.997 489 119

CTMC availability submodels of Example 9.10. The result for the independent case and the approximate solution for each subsystem are given in Table 16.3.

The results for the system availability are compared in Table 16.4. The shared repair with preemptive priority reduces the steady-state system availability with respect to the independent case, and the hierarchical solution still provides a fairly good approximation. Once again, the approximation can be improved by computing and passing transient values of the parameters.

Problems

16.1 Carry out the computational details for the three versions of Example 16.1. You may find it convenient to use a software package such as SHARPE for this purpose.

16.2 For Example 16.2, derive expressions for the parametric sensitivity of the steady-state system availability to all the parameters. Using the numerical values we have been using in this running example, determine the availability bottleneck.

16.3 Reproduce the numerical results of Example 16.3. You may find it convenient to use a software package such as SHARPE for this purpose.

16.4 For Example 16.3, derive expressions for the parametric sensitivity of the steady-state system availability to all the parameters. Using the numerical values we have been using in this running example, determine the availability bottleneck. Compare the results with the independent repair case considered in Problem 16.2.

16.5 Extend Example 16.3 by computing the transient availabilities of each submodel and hence the transient availability of the system as a whole. Note that you still compute and pass the steady-state values of parameters such as q_P from one submodel to another. You may find it convenient to use a software package such as SHARPE for this purpose.

16.6 Extend Example 16.3 by computing the transient probabilities of each submodel and passing such values of parameters from one submodel to another. Thus, for example

$q_P(t) = 1 - Q_P(t)$ will be computed and passed to submodel V. Each of the submodels V, S, and L will become non-homogeneous CTMCs. To solve these CTMCs you can use the PCA method of solution discussed in Chapter 13. You may find it convenient to use a software package such as SHARPE for this purpose.

16.7 Reproduce the numerical results of Example 16.5. You may find it convenient to use a software package such as SHARPE for this purpose.

16.8 For Example 16.5, derive expressions for the parametric sensitivity of the steady-state system availability to all the parameters. In order to obtain the derivatives, you may use the chain rule from calculus [34]. Using the numerical values we have been using in this running example, determine the availability bottleneck.

16.9 Extend Example 16.5 by computing the transient availabilities of each submodel and hence the transient availability of the system as a whole. Note that you still compute and pass the steady-state values of parameters such as q_P from one submodel to another. You may find it convenient to use a software package such as SHARPE for this purpose.

16.4   Reliability Models

The first example in this section shows that the FT modularization described in Section 6.5 is an instance of a multi-level model. The second example considers interval reliability computation for systems modeled by CTMCs, while the third example illustrates the reliability model of a RAID system drawn from [35]. The fourth example in this section discusses phased mission system reliability. The fifth and sixth examples deal with so-called behavioral decomposition [36].

Example 16.6 *Fault Tree Modularization*

The modularization procedure in an FT, illustrated in Section 6.5, is an example of hierarchical modeling. In Example 6.26 we showed that the subtree emerging from node G_4 is a module since its terminal events do not occur elsewhere in the tree. The subtree module, depicted in Figure 16.4(a), is an FT with repeated events and can be seen as submodel M_{G_4} in a hierarchical structure. When we replace the submodel M_{G_4} in the original FT of Figure 6.14, we get the top-level model M_{FT} which is the FT with repeated events represented in Figure 16.4(b).

The import graph for this example is shown in Figure 16.5, where submodel M_{G_4} passes to the top-level model M_{FT} the TE probability of the corresponding subtree.

Example 16.7 *Interval Reliability for Systems with Redundancy*

We introduced the notion of interval reliability in Section 3.6, and then used it in Section 4.1.4 for systems without redundancy. Now we wish to develop expressions for interval reliability for systems with redundancy and modeled by CTMCs. Consider again the three-state CTMC availability model in Example 9.4 of two identical components sharing a repair person as shown in Figure 9.7. The CTMC reliability model

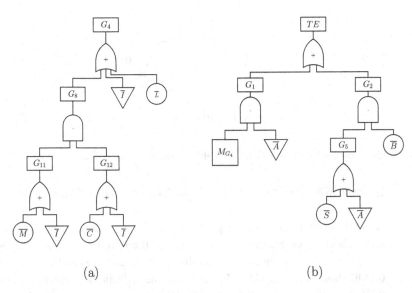

(a)   (b)

Figure 16.4   Modularization of the FT of Figure 6.14: (a) Subtree module M_{G_4}; (b) top-level model M_{FT}.

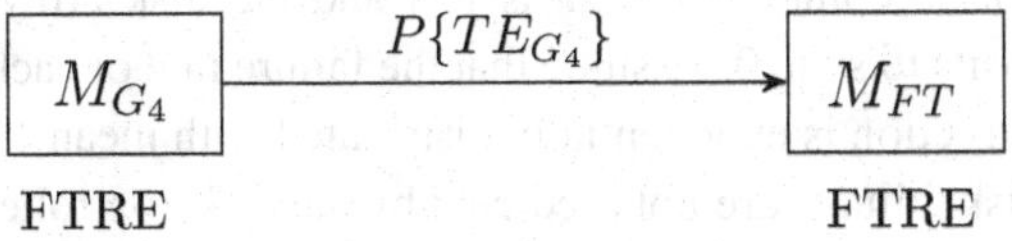

Figure 16.5   Import graph for the FT of Figure 16.4.

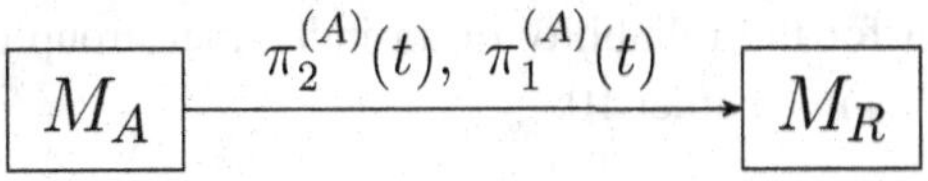

Figure 16.6   Import graph for the interval reliability model.

of this system was considered in Example 10.12 and in Figure 10.9. In order to derive an expression for the interval reliability of this system, we first solve the availability model of Figure 9.7 (say model M_A) for its transient state probabilities, now labeled $\pi_2^{(A)}(t)$, $\pi_1^{(A)}(t)$, and $\pi_0^{(A)}(t)$. Then we solve the reliability model of Figure 10.12 (call it M_R) for the conditional reliabilities $R_2(\tau)$, $R_1(\tau)$, and $R_0(\tau) = 0$, where $R_i(\tau)$ is the reliability of the system at time τ given that it starts in state i. Then the interval reliability expression for the two identical component parallel redundant system is given by:

$$IR(t,\tau) = \pi_2^{(A)}(t)R_2(\tau) + \pi_1^{(A)}(t)R_1(\tau). \tag{16.3}$$

The import graph in this case consists of two models, M_A and M_R, as shown in Figure 16.6. The first model M_A computes the two transient probabilities $\pi_2^{(A)}(t)$, $\pi_1^{(A)}(t)$ and passes them on to the upper-level model M_R. This approach can be easily generalized to any pair of CTMC availability and reliability models of a given system.

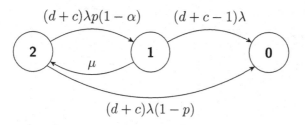

Figure 16.7   Continuous-time Markov chain reliability model for an individual RAID group.

Example 16.8   *Reliability Model of Redundant Array of Inexpensive Disks (RAID) System [35]*
A RAID system consists of a group of n disks wherein each group has d data disks and c check disks. Specifically, we consider the RAID 2 architecture that can correct one error within a group. The reliability model of an individual group is the three-state CTMC shown in Figure 16.7. In state 2, all the disks in the group are operational. Upon the failure of one of the disks, the CTMC transits to state 1 where data reconstruction is initiated while the disk array keeps functioning. If any other disk in the group fails before the reconstruction is completed, data is lost and the disk array is considered failed as the CTMC transits to state 0. Assume that the failure rate of each disk is λ, and the time for disk reconstruction is exponentially distributed with mean $1/\mu$. A fraction, $(1-\alpha)$, of individual disk failures are not predicted by the disk controller and will lead the group to degraded state 1. A small fraction, $1-p$, of individual disk failures will escape the fault tolerance provided by RAID and hence will lead the group to transit to state 0. Thus results the three-state CTMC reliability model shown in Figure 16.7.

A closed-form expression for the reliability of an individual group can be derived using the techniques discussed in Chapter 10:

$$R_i(t) = a_1 e^{-\beta_1 t} + a_2 e^{-\beta_2 t}, \tag{16.4}$$

where

$$\beta_1 = \frac{-(((d+c)(2-p\alpha)-1)\lambda + \mu) + \sqrt{x}}{2},$$

$$\beta_2 = \frac{-(((d+c)(2-p\alpha)-1)\lambda + \mu) - \sqrt{x}}{2},$$

$$a_1 = \frac{\beta_1 + \mu + (d+c-1)\lambda + (d+c)\lambda p(1-\alpha)}{\beta_1 - \beta_2},$$

$$a_2 = \frac{\beta_2 + \mu + (d+c-1)\lambda + (d+c)\lambda p(1-\alpha)}{\beta_2 - \beta_1},$$

$$x = ((d+c)((d+c)p^2\alpha^2 - 2p\alpha) + 1)\lambda^2 + \mu^2$$
$$+ 2((d+c)p(2-\alpha) - 1)\lambda\mu.$$

With the independence assumption across groups, the overall system reliability is modeled by the reliability block diagram shown in Figure 16.8. Hence the overall disk

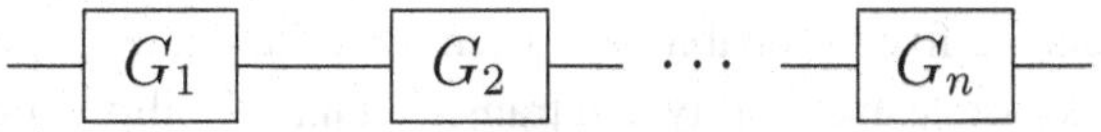

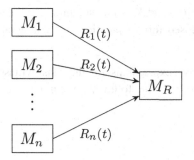

Figure 16.8   Reliability block diagram for RAID system.

Figure 16.9   Import graph for the RAID model.

array reliability $R_{da}(t)$ can be written as:

$$R_{da}(t) = \prod_{i=1}^{n} R_i(t).$$

By integrating under the reliability curve of Eq. (16.4), we obtain the mean time to data loss (MTTDL) for an individual group:

$$
\begin{aligned}
\text{MTTDL}_i &= \int_0^\infty R_i(t)dt = a_1/\beta_1 + a_2/\beta_2 \\
&= \frac{\mu + ((d+c)(1+p(1-\alpha)) - 1)\lambda}{(d+c)\lambda(\mu(1-p) + (d+c-1)(1-\alpha)\lambda)}.
\end{aligned}
\tag{16.5}
$$

The overall MTTDL for the disk array is then obtained as:

$$\text{MTTDL}_{da} = \int_0^\infty R_{da}(t)dt = \int_0^\infty [a_1 e^{-\beta_1 t} + a_2 e^{-\beta_2 t}]^n dt = \sum_{j=0}^{n} \frac{\binom{n}{j} a_1^j a_2^{n-j}}{j\beta_1 + (n-j)\beta_2}. \tag{16.6}$$

The import graph for the RAID example is shown in Figure 16.9, and is acyclic. In this example, the fully symbolic expression of individual group reliability is passed to the upper-level reliability block diagram. Thus there is no error introduced by the hierarchical model and its solution. For further details of this model, refer to [35]. For an MRGP model of RAID, see [37]. For several other models of RAID reliability see [38–40].

Problems

16.10   First solve the original fault tree of Figure 6.14, and then the two-level hierarchy in Figures 16.4(a), 16.4(b) and 16.5. Use SHARPE or a similar software package. First assume that all inputs are fixed probabilities. Then use failure rates for each leaf node, and then use the steady-state availability and transient unavailability for each leaf node.

16.11 Derive closed-form expressions for $\pi_2^{(A)}(t)$ and $\pi_1^{(A)}(t)$ by solving the availability model of Figure 9.7 assuming the system starts in state 2 at time 0. Next, derive closed-form expressions for $R_2(\tau)$ and $R_1(\tau)$ by solving the reliability model of Figure 10.12. Thence write down a closed-form expression for the interval reliability of this system using Eq. (16.3).

16.12 For the RAID example, derive expressions for the sensitivities (derivatives) of the system reliability with respect to the following input parameters: λ, μ, p and α.

16.4.1   Phased Mission Systems

We consider the so-called phased mission system (PMS) models, also known as mission-based systems. Many systems in the nuclear, aerospace, chemical, geothermal and electronic industries are required to operate in distinct phases. For an aircraft, the three phases of a flight mission are takeoff, cruise, and landing [41]. For NASA's Mars Rover mission, for instance, the phases include vehicle launch, cruise, approach, entry, descent, landing on Mars and rover egress. System configuration, parameters and operational requirements may vary from phase to phase, depending on the application. There have been several approaches for the reliability analysis of such PMSs. One group of methods uses non-state-space solution techniques as in [42–44], while others use state-space techniques including HCTMC [45, 46], NHCTMC [47], and deterministic SPN [48].

Example 16.9   *A Three-Phase System*
We use a simple CTMC reliability model to illustrate PMSs in this example. The same example is solved using fault-tree-based non-state-space methods in [43, 44]. Our system consists of three independent components A, B, and C, with phase-independent failure rates. However, the conditions under which the system fails change from phase to phase. In Phase 1, the three components form a parallel redundant configuration, Figure 16.10(a); in Phase 2, A is in series with the parallel combination of B and C, Figure 16.10(b); in Phase 3, the three components are used in a series configuration, Figure 16.10(c).

The CTMC models for the three phases have different numbers of states. In these CTMC models, states labeled F are failure states while all the other states are up states. The three submodels are solved in sequence. First, the submodel of Phase 1 is solved for its mission time t_1 with initial state (111). Denote by $\pi_{ijk}^{(1)}(t)$, $0 \le t \le t_1$ the state probabilities of the Phase 1 CTMC at time t. In order to solve the CTMC model of Phase 2, we need to map the final state probabilities from Phase 1, $\pi_{ijk}^{(1)}(t_1)$, as initial state probabilities $\pi_{ijk}^{(2)}(0)$ for Phase 2.

But since not all the states of the CTMC model of Phase 1 are explicitly present in the CTMC model of Phase 2, we need to carefully map these probabilities. For the

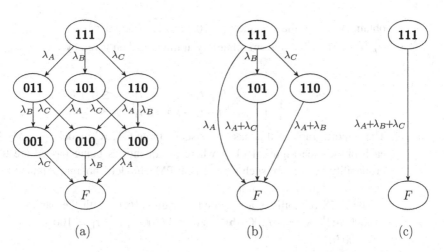

Figure 16.10 Continuous-time Markov chain models for the three phases of a PMS.

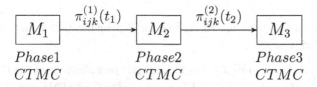

Figure 16.11 Import graph for the three-phase PMS model.

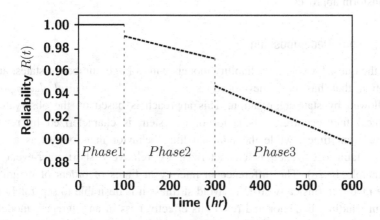

Figure 16.12 Reliability comparison for different phases of the PMS.

up states $\{(111),(101),(110)\}$ that are common to the two CTMC models, we have $\pi_{ijk}^{(2)}(0) = \pi_{ijk}^{(1)}(t_1)$. The remaining state probabilities are simply fed into the failure state since if the system is in any of these states at the end of Phase 1 they will be immediately declared as failure states at the beginning of Phase 2: $\pi_F^{(2)}(0) = \pi_{011}^{(1)}(t_1) + \pi_{001}^{(1)}(t_1) + \pi_{010}^{(1)}(t_1) + \pi_{100}^{(1)}(t_1) + \pi_F^{(1)}(t_1)$. With this initial state probability vector, we solve the Phase 2 CTMC model as $\pi_{ijk}^{(2)}(t)$, $0 \le t \le t_2$. The initial probabilities for Phase 3 are obtained in a similar manner: $\pi_{111}^{(3)}(0) = \pi_{111}^{(2)}(t_2)$ and $\pi_F^{(3)}(0) = \pi_{101}^{(2)}(t_2) + \pi_{110}^{(2)}(t_2) + \pi_F^{(2)}(t_2)$. The PMS system reliability at time t is then given by:

$$R_{PMS}(t) = \begin{cases} 1 - \pi_F^{(1)}(t) & \text{if } 0 \le t \le t_1, \\ 1 - \pi_F^{(2)}(t - t_1) & \text{if } t_1 \le t \le t_1 + t_2, \\ 1 - \pi_F^{(3)}(t - t_1 - t_2) & \text{if } t_1 + t_2 \le t \le t_1 + t_2 + t_3. \end{cases} \qquad (16.7)$$

The import graph in this case consists of three CTMC models $\{M_1, M_2, M_3\}$, one for each phase (see Figure 16.11), where the quantities being passed are transient state probability vectors. In such hierarchical PMS models, no approximation is introduced due to the hierarchical solution.

The PMS reliabilities are plotted in Figure 16.12 with the parameter values $\lambda_A = 1/1000$ **hr**$^{-1}$, $\lambda_B = 1/2000$ **hr**$^{-1}$, $\lambda_C = 1/3000$ **hr**$^{-1}$, $t_1 = 100$ **hr**, $t_2 = 300$ **hr**, and $t_3 = 600$ **hr**.

Problems

16.13   With reference to Figure 16.12, note and explain the discontinuities at the phase change boundaries.

16.14   Derive a closed-form solution for the state probabilities for the CTMCs of all three phases and hence derive a closed-form expression for the PMS system reliability of Example 16.9. First use the convolution integration approach and then use the Laplace transform approach.

16.4.2   Behavioral Decomposition

In the quest for solving reliability models with a large number of states, an important notion that has been developed is the so-called *behavioral decomposition* [36] followed by state aggregation. This approach is based on the observation that the typical fault-occurrence behavior of a system is characterized by relatively long inter-event times, while the fault-handling behavior that follows as a consequence of a fault occurrence is characterized by relatively short inter-event times (see Example 10.28). The difference in inter-event times of orders of magnitude makes the model intrinsically stiff. To avoid stiffness it is desirable to separately analyze the fault-handling behavior and reflect its effectiveness in an aggregate model by one or more parameters or functions. This decomposition, based on widely differing time constants, has been called *behavioral decomposition*. Models of the fault-handling processes are either semi-Markov or simulative in nature, thus removing the usual restrictions of exponential holding times within the coverage model. The aggregate fault-occurrence model is allowed to be a non-homogeneous Markov chain in [36], thus allowing the times to failure to possess Weibull-like distributions. We will assume that times to failure are exponentially distributed for the sake of simplicity. The reader may refer to the original papers to see the use of non-exponential distributions.

Example 16.10 *Hierarchical Reliability Model Based on Behavioral Decomposition* *[14]*

We now consider a hierarchical reliability model of a computer system consisting of three processors and two shared memories communicating over a bus. The system is operational if at least one processor can communicate with at least one of the memories via an operational bus. In the current example we develop a two-level model, while in the next example we will recast the same modeling problem as a three-level hierarchy.

Figure 16.13 shows the CTMC reliability model (assuming perfect coverage for now) of the system; each state is labeled with three indices (i,j,k), where i is the number of working processors, j the number of memories, and k the bus. The failure rate of each processor is λ, that of each memory is α, and that of the bus is β. The failure states due to the exhaustion of resources are separated: F1 indicates exhaustion of the processor cluster, F2 exhaustion of the memories, and F3 the failure of the bus. In the context of the HARP reliability modeling package, this CTMC is the *fault-occurrence model* (FOM) [14]. In HARP, the FOM can be specified as a CTMC, as in this case, or as a fault tree from which HARP will automatically generate such a CTMC. Furthermore, Weibull time to failure distributions are allowed, leading the FOM to be a non-homogeneous CTMC. Upon occurrence of an individual component failure a set of procedures and actions are executed by the system that are abstracted here under the aegis of an FEHM. Many different types of FEHM are available in HARP, and different FEHMs can be used in the same model if needed. We reuse the simple CTMC FEHM considered in Example 10.28, and use the same FEHM all through this model. Including the appropriate FEHM at each fault occurrence in the CTMC of Figure 16.13 generates the model of Figure 16.14.

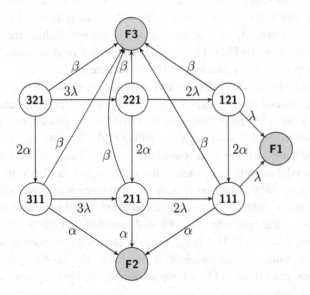

Figure 16.13 Fault-occurrence CTMC of the multiprocessor system.

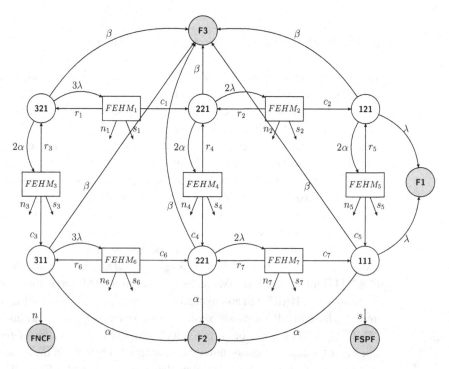

Figure 16.14 Overall reliability model of the multiprocessor system with FEHM included.

We assume here that each of the FEHMs is a homogeneous CTMC. The overall model is then a CTMC. But this CTMC is not only large but also stiff as there are time constants that are several orders of magnitude apart. Behavioral decomposition exploits the widely differing time constants in FEHM and FOM and consists in solving each FEHM separately for its exit probabilities and replacing them by branch points. Once an FEHM is entered after the occurrence of a component failure, there are at least three eventual outcomes: the FEHM is able to detect the failure to be a transient one and recover the system to the original state, the FEHM recognizes the failure as a permanent one and is able to recover the system to a degraded state, or the FEHM is not able to cope with the failure and the system is brought down to the coverage failure state FSPF.

But if we now consider the execution of an FEHM in the context of a real system, the occurrence of a second fault during the handling of the first one must be considered, since the second fault can cause an immediate system failure. In the context of the overall CTMC model as well, from each of the states within an FEHM transitions due to further component failures must be considered. So far we have ignored this possibility since the time spent in an FEHM is negligible compared with the time between failures. In order to improve the approximation, we will now consider the effect of a second component failure while an FEHM is handling the previous one. Such near-coincident faults do have a significant effect on model results for highly reliable systems. We thus have four eventual exits from an FEHM, the additional one being system failure due to a near-coincident fault.

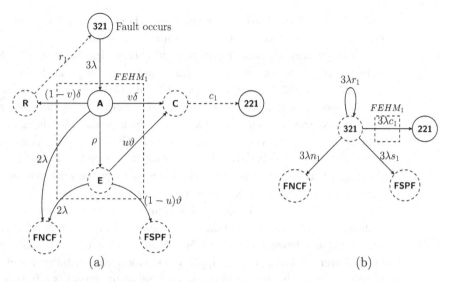

Figure 16.15 Reducing the NCF-FEHM to a branch point.

To be more concrete, with reference to Figure 16.14, we consider explicitly the FEHM$_1$ inserted between state (321) and state (221) that is entered with rate 3λ from state (321). The concrete version of the FEHM that we choose in this illustration is the CTMC shown in Figure 16.15(a) that is adapted from Figure 10.33. We see in this figure that during the handling of the fault in states A and E, two processors are still working and they may fail at the rate 2λ leading to the system failure state denoted by FNCF (failure due to near-coincident fault). We have assumed here that a near-coincident fault in the same subsystem brings the system down, while a near-coincident fault in another subsystem does not. Different combinations of this assumption are allowed in HARP. The CTMC of Figure 16.15(a) has two transient states (A and E) and four absorbing states (C, R, FSPF, FNCF). State R refers to a transient restoration state, state C refers to recovery to a degraded state and state FSPF refers to system failure due to a single fault that could not be handled by the FEHM. Let us term an FEHM extended with near-coincident faults an NCF-FEHM. Because of the different values of the near-coincident fault rates, the details of each NCF-FEHM are different. The probability of ultimate absorption in one of the four states starting from state A with probability 1 can be derived from Eq. (10.86) to give:

$$c_1 = \frac{(2\lambda + \vartheta)v\delta + u\rho\vartheta}{(2\lambda + \delta + \rho)(2\lambda + \rho)},$$

$$r_1 = \frac{(1-v)\delta}{2\lambda + \delta + \rho},$$

$$s_1 = \frac{(1-u)\rho}{(2\lambda + \delta + \rho)(2\lambda + \rho)},$$

$$n_1 = \frac{2\lambda(2\lambda + \vartheta) + 2\lambda\rho}{(2\lambda + \delta + \rho)(2\lambda + \rho)}.$$

(16.8)

Replacing the NCF-FEHM by an instantaneous branch point [8] with four branches, the *imperfect coverage* model of Figure 16.15(b) is obtained.

In a similar way, we can compute the exit probabilities from any of the NCF-FEHMs of Figure 16.14 and we can replace them with the resulting branch points to obtain the final, reduced CTMC reliability model of Figure 16.16. In order to reduce the clutter, we have shown several arcs emanating from different states that do not reach their destinations. Arcs with a multiplier s_i will terminate on the absorbing state FSPF. Those with the multiplier n_i will terminate on state FNCF. This CTMC is much smaller and is not stiff, and hence behavioral decomposition at once helps largeness and stiffness avoidance. Furthermore, since the FEHM is solved separately, it can be any other model type such as an SMP, an SRN or even more complex, needing a simulative solution [49]. Thus we have a two-level hierarchical reliability model in which the top-level submodel is shown in Figure 16.16 and at the bottom level we have a a set of NCF-FEHM models such as the one shown in Figure 16.15(a). Values such as c_1, r_1, s_1, and n_1 are computed by the lower-level models and supplied to the top-level reliability model. Drawing the import graph for this two-level hierarchical reliability model is left as an exercise for the reader. Nevertheless, this two-level reliability model is an approximation; the error incurred by this approximation was analyzed in [50].

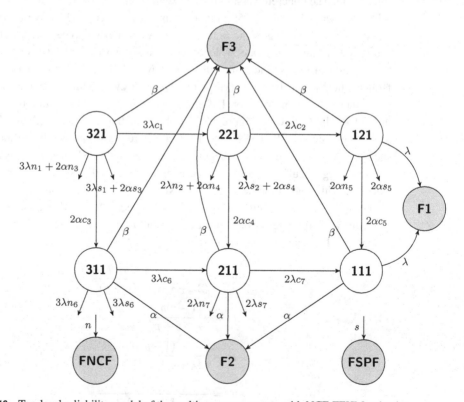

Figure 16.16   Top-level reliability model of the multiprocessor system with NCF-FEHM reduction.

Example 16.11 *A Three-Level Hierarchical Reliability Model Based on Behavioral Decomposition*

The previous example showed the use of a two-level hierarchical model where the lower-level NCF-FEHM was providing the inputs to the higher-level FOM model. The native FEHM model was discussed earlier, in Example 10.28, where in Figure 10.32 the FEHM was presented in isolation and the three possible exits were considered. Subsequently, in Figure 10.33, the effect of the near-coincident fault was considered. The resulting NCF-FEHM is then reduced to a branch point, as shown in Figure 10.34. This approach was used in the previous example, where the branch point was again illustrated in Figure 16.15. A consequence of using this approach is that each of the NCF-FEHM models in Figure 16.14 is now dependent on the number of remaining functioning components, be it the processors or the memories.

A closer examination of the FEHM models in Figures 10.32 and 10.33 reveals clearly that the former by itself is independent of the number of remaining functional components. The dependence is induced because of the near-coincident faults. This suggests yet another level of decomposition of the NCF-FEHM model. The model in Figure 10.32 becomes the lowest FEHM submodel. This submodel M_1 captures all the subsequent actions after the occurrence of a single fault exclusive of the near-coincident fault such as that in Figure 10.32. Note that this model needs to be solved only once since it is independent of the number of remaining functional components. This model M_1, in Figure 16.17(a), is solved to yield three defective distribution functions of the time to be absorbed into the respective absorbing states:

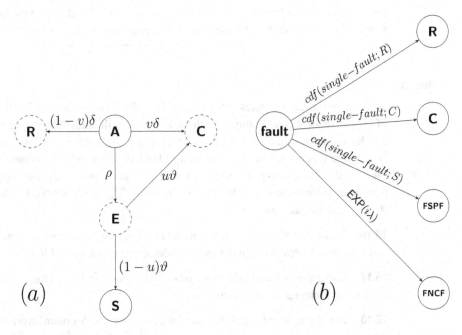

(a) (b)

Figure 16.17 (a) Native FEHM for submodel M_1 (see Figure 10.32); (b) Submodel M_2.

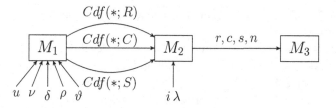

Figure 16.18 Overall import graph for the three-level hierarchical reliability model.

Cdf(single-fault;R), Cdf(single-fault;C), and Cdf(single-fault;S). These Cdfs are equal to the respective probabilities of being in the absorbing state at time t. The next-level submodel M_2, termed the near-coincident fault competition (NCFC) submodel, shown in Figure 16.17(b), captures the competition between the resolution of the first fault and the occurrence of a second component failure. This model captures the number of remaining functional components, and hence has to be repeatedly solved for different values of i. Furthermore, this is now a semi-Markov model.

The outputs from M_1 are three distribution functions that are input to M_2. The outputs from M_2 are the branching probabilities, r, c, s, and n. These branching probabilities become inputs to the FOM submodel M_3, which can then be solved to yield the reliability metrics. The overall import graph for the model is captured in Figure 16.18. Submodel M_1 has five input parameters, u, v, δ, ϑ, and ρ. There are three outputs from this submodel that represent Cdf(single-fault;R), Cdf(single-fault;C), and Cdf(single-fault;S). Submodel M_2 receives the inputs Cdf(single-fault;R), Cdf(single-fault;C) and Cdf(single-fault;S) from M_1 and an external input $i\lambda$ representing the near-coincident fault rate. M_2 has four outputs, representing the probabilities that submodel M_2 reaches one of the four absorbing states. Submodel M_3 is solved for the system reliability, which is represented by the output of this submodel.

Problems

16.15 First insert the states of the CTMC NCF-FEHMs in the overall reliability model of Figure 16.14. In our example, each NCF-FEHM will amount to adding two states (A and E) and corresponding transitions, including those for the near-coincident failures. Now solve this model numerically using SHARPE. Compute the system reliability at time t and the eventual probabilities of being absorbed in various absorbing states. For parameter values, use $\lambda = 1/50000$, $\alpha = 1/100000$, $\beta = 1/200000$, $\delta = 3600$, $\rho = 60$, $\vartheta = 60$, $v = 0.4$, and $u = 0.8$.

16.16 Solve the reduced reliability model of Figure 16.16 numerically using SHARPE and compare the results with the exact model in the previous problem.

16.17 Solve the reduced reliability model of Figure 16.16 in closed form using the convolution integration approach.

16.18 For the reduced reliability model of Figure 16.16 obtain expressions for the derivatives of the system reliability with respect to all the parameters of the FOM and the FEHM.

16.19 Next, extend the reduced model to make the failure rates λ, α, and β functions of the global time t, thus making the CTMC non-homogeneous. Now solve for the state probabilities and system reliability in closed form using the convolution integration approach.

16.20 Derive expressions for the three Cdfs for the FEHM of Figure 16.17(a) as submodel M_1 and then derive the four absorption probabilities from submodel M_2 of Figure 16.17(b). Compare these four probabilities with those derived directly in Example 16.10, Eq. (16.8). Observe that there is no additional error introduced by this extra level of decomposition.

16.21 Numerically solve the three-level hierarchical model of Example 16.11, and thus learn how to specify such hierarchical models in SHARPE.

16.5   Dynamic Fault Tree

Traditional fault trees, considered in Chapter 6, are well accepted models in many areas of dependability, safety, and security [51, 52] since they are simple to manipulate and are supported by powerful software tools for qualitative and quantitative analysis. However, traditional (static) FTs (SFTs) suffer from the main limitation that the basic components must be assumed to be statistically independent. Dynamic fault trees are a hierarchical extension of standard FTs proposed by Dugan *et al.* in [53] with the aim of combining the simplicity of the graphical structure of standard FTs with a more powerful state-space modeling technique able to incorporate selected forms of functional or temporal dependencies that often arise in fault-tolerant systems.

In particular, Dugan *et al.* in [53] introduce four dynamic gates called the priority AND (PAND) gate, the sequence enforcing gate (SEQ), the functional dependency gate (FDEP) and the warm spare gate (WSP). Since the original proposal by Dugan *et al.* in [53], a plethora of DFT variants have been proposed in the literature. These variants are systematically examined in [54], showing that they are not always semantically equivalent. In most of the DFT formalisms, only two dynamic gates, PAND, and WSP, are considered as primary gates since the FDEP reduces to a Boolean OR [52, 55], and the sequence enforcing gate to a particular form of WSP.

The dynamic gates are inserted in the FT as traditional Boolean gates, but their probabilistic output requires a CTMC to be solved first.

The PAND gate in Figure 16.19(a) reaches a failure condition (that is, its output becomes true) if and only if all of its input components have failed in a preassigned order (from left to right in the graphical notation). If B fails before A, we reach an operational state; if A fails before B we reach the failure state whose probability is the output probability of the PAND gate. The gate can thus be equivalently represented by the CTMC of Figure 16.19(b), assuming the respective failure rates of the components A and B as λ_A and λ_B. Solution of the CTMC yields the failure probability at time t for the PAND gate:

$$\pi_f(t) = \frac{\lambda_A}{\lambda_A + \lambda_B} + \frac{\lambda_B}{\lambda_A + \lambda_B} e^{-(\lambda_A + \lambda_B)t} - e^{-\lambda_B t}. \tag{16.9}$$

The PAND gate of Figure 16.19 can be generalized to a gate with n inputs in which the output occurs only if the inputs fail in a predefined order (for instance from left to right in the graphical representation).

In the warm standby configuration, introduced in Section 8.2 and Example 10.10, a spare can fail even while it is dormant, but its failure rate is reduced by a dormancy factor. The dependency introduced by the dormancy factor cannot be properly accounted for in a standard FT. In the DFT setting, the WSP gate is represented as in Figure 16.20(a), where A is the primary component that can be replaced, upon failure, by one or more spares, ordered from S_1 to S_n. The WSP gate fails if its primary component fails and all of its spares have failed or are unavailable (a spare may become unavailable if it is shared and being used by another WSP gate).

Figure 16.20(b) reports the corresponding CTMC in the case in which the failure rate of the main component is λ_P, the number of spares is equal to 2, with equal failure rate λ_S and dormancy factor α. The output probability of the WSP is the probability of the state "fail" in the CTMC.

Dynamic fault tree gates are typically solved by automated conversion to the equivalent Markov model [53]. However, when dynamic gates are nested the direct

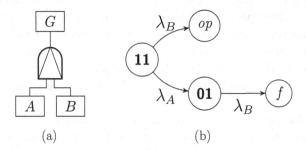

(a)   (b)

Figure 16.19   Priority AND gate. (a) PAND gate symbol; (b) CTMC representation.

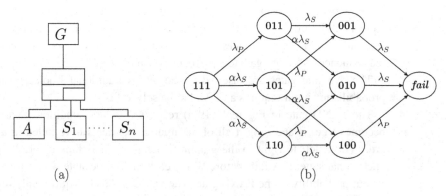

(a)   (b)

Figure 16.20   Warm spare gate. (a) WSP gate symbol; (b) CTMC representation with two equal spares.

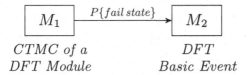

Figure 16.21 Import graph for a DFT.

automated conversion to a CTMC is not easy and may encounter the state explosion problem. Two techniques together have been used to help mitigate these difficulties: *modularization* and use of a higher-level language.

Modularization for a standard FT was illustrated in Section 6.5 as a procedure to identify independent subtrees that can be converted and solved separately [56–59]. Modularization provides a hierarchical model structure, as illustrated in Example 16.6. The same idea can now be extended to DFTs. A module of a DFT may be classified as static or dynamic. Static modules may contain repeated basic events but not dynamic gates, and can be analyzed by means of the techniques discussed in Chapter 6. Dynamic modules contain dynamic gates and require a state-space generation/solution. A module is minimal if it does not contain any other module; minimal modules are those to be isolated, and replaced in the original SFT or DFT by a single basic event to which the module TE probability is assigned. A linear-time algorithm for the identification and isolation of modules is given in [56].

The other approach is conversion of minimal dynamic modules into a GSPN [60, 61] or an SRN. We note here that because of the possibility of defining reward rates, the use of SRNs might be especially useful in reducing the size of the net. This conversion enjoys the same benefits as illustrated in Chapter 12, and makes easier the representation of nested dynamic gates from which the state space is automatically generated and solved using SPN solvers such as SPNP and SHARPE.

The import graph for the DFT solution is shown in Figure 16.21.

An alternative way is to use a system-level decomposition followed by hierarchical model composition as advocated and illustrated in this chapter. As is shown here, we can mix and match many different formalisms in this manner so as to use the flexibility and power of more complex state-space formalisms with exponentially and non-exponentially distributed transition times.

Example 16.12 *A Multiprocessor System [60]*
A multiprocessor system is composed of three compute nodes C_1, C_2, and C_3 working in a parallel redundant configuration, each one composed of one processor and one memory. The storage subsystem is formed by two hard disks D_1 and D_2, where D_1 is the primary disk and D_2 is the backup copy in a warm standby mode. All the components are connected to the system bus B (Figure 16.22).

For the task of updating the secondary disk, the main compute node C_1 needs to be up. The system is up as long as at least one compute node, an updated disk and the bus are up. The DFT model for the system reliability is shown in Figure 16.23. Note that

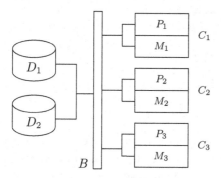

Figure 16.22 A multiprocessor system.

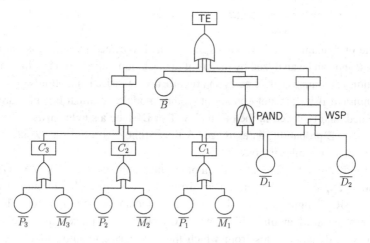

Figure 16.23 Dynamic fault tree model of the multiprocessor system of Figure 16.22.

D_1 and D_2 are inputs to a WSP gate, and C_1 is an input to a PAND gate whose other input is D_1. The PAND propagates the failure only if the primary disk D_1 fails after C_1 leaving the backup copy not updated.

To solve the DFT, the first step is to locate independent modules and solve each module hierarchically, as by the import graph of Figure 16.21. To solve a dynamic module, constant failure rates must be assigned to all the components belonging to the module, in order to convert the module into a CTMC. In the DFT of Figure 16.23, all the basic events, excluding only $\bar{B}$, belong to a single dynamic module which must be converted into the corresponding CTMC. The direct conversion is not easy. A complete solution in which the gates of the DFT are progressively converted into a GSPN is given in [60]. From the GSPN the failure probability of the dynamic module is obtained from any of the many SPN software packages, and then the module is replaced in the original DFT by a single basic event to which the module probability is assigned. When all the dynamic modules are solved and replaced, the DFT is reduced to a standard FT to which the solution techniques of Chapter 6 are applied.

Recently, the solution of DFTs has also taken two different directions, an *algebraic* approach and an approach based on *Bayesian belief networks*. The algebraic approach in [55] proposes to define events as temporal binary variables, and to construct a new algebra by adding to the usual Boolean operators new temporal operators able to represent sequences of events. With this new algebra, it is possible to derive an algebraic expression of the structure function of the TE as a function of the basic events in a DFT. The basic temporal operator is the *precedence operator*, indicated by $X \lhd Y$, which acts on two temporal binary variables X and Y and has the following definition:

$$\begin{cases} X \lhd Y = \text{true} & \text{event } X \text{ occurs before event } Y, \\ X \lhd Y = \text{false} & \text{otherwise.} \end{cases}$$

A similar algebraic approach has been investigated in [62], leading to the definition of an extension of BDDs called a *sequential* BDD or SBDD [63].

The second approach to mention is the conversion of a DFT into a dynamic Bayesian network (DBN), adapting and extending the procedure for the conversion of a static FT into a Bayesian network (BN) already presented in Section 6.10 [64–66]. In the DBN model, time is represented as a sequence of time slices separated by a constant small interval. When the Markov assumption holds, as in the present case of a DFT, it is sufficient to represent two consecutive time slices and the network is fully specified if it is provided with:

- the prior probabilities for the root variables at time $t = 0$;
- the intra-slice conditional dependency model, together with the corresponding conditional probability tables;
- the inter-slice conditional dependency model and probability tables, which make explicit the temporal probabilistic dependencies between variables. In particular, a variable at time $t + \Delta t$ may depend not only on its value at time t, but also on the values of other variables in the previous time slice.

The motivation for adopting a DBN instead of the corresponding CTMC is that, by decomposing the state of a complex system into its constituent variables, the network is able to take advantage of sparseness in the temporal probability model [67]. The conditional independence assumption enables a compact representation, avoiding the complexity of specifying and using a global state model. The two models, DBN and DFT, are not exactly equivalent, since DFT uses CTMC where transitions occur in a continuous fashion, while DBN uses the factorized representation of a DTMC. As a consequence, while in a discrete-time model there is a non-null probability for the coincidence of events in the interval between two time slices, the same is no longer true in a continuous-time model.

Rules for converting the basic DFT gates into a DBN are provided in [68], and a tool (with examples) implementing a modular algorithm for automatically translating a DFT into the corresponding DBN is given in [65].

Resorting to the DBN formalism has the advantage of exploiting all the modeling capabilities of graphical probabilistic models [66]: multi-valued variables (instead of binary events), local dependencies among components (instead of the classical statistical-independence assumption) and noisy interaction among component behaviors (instead of deterministic interaction). In addition, a general inference mechanism (combining prediction as well as diagnosis) can be performed naturally on a DBN, while they are not easily implemented in standard DFT analysis, especially if experimental evidence is gathered during the life of the system. A complex case study where the parameters of the BN can be updated online from experimental observations is presented in [69] related to the models of an FDIR module in autonomous spacecraft.

Problems

16.22 Verify Eq. (16.9) by solving the CTMC of Figure 16.19(b) using the convolution integration approach. Then obtain expressions for the derivatives of the PAND failure probability with respect to the failure rates λ_A and λ_B.

16.23 Generalize Eq. (16.9) to the case of an n-input PAND gate with the sequence of input events: $A_1, A_2, \ldots, A_n$.

16.24 Derive the transient state probability expressions for the CTMC of Figure 16.20(b), and thus obtain the failure probability expression for the WSP gate with two equal spares. Use either the Laplace transform approach or the convolution integration approach. Next, derive expressions for the derivatives of the WSP failure probability with respect to the three input parameters, λ_P, λ_S, and α.

16.25 Develop a GSPN reliability model for the multiprocessor system in Example 16.12; solve the model numerically using SHARPE.

16.26 Develop an SRN reliability model for the multiprocessor system in Example 16.12. Make sure that you use guard functions and reward assignments to keep the model as simple as possible, and solve the model numerically using SHARPE or SPNP.

16.27 Develop a two-level hierarchical reliability model for the multiprocessor system of Example 16.12. The top-level model is a fault tree with one input being $\overline{B}$ and the other input being the failure probability of a simplified SRN model of the DFT excluding the event of bus failure $\overline{B}$. Solve the model numerically using SHARPE.

16.28 Simplify the GSPN or the SRN models in the above problems by observing that within a compute node, the processor and memory are in series and since each of them has a time-independent failure rate, the failure rate of each compute node is simply the sum of the two. Hence each compute node can be directly represented by a timed transition in the Petri nets without having the processor and memory failure events explicitly represented.

16.6 Performance Models

The example in this section develops a model for the response time distribution of the Common Object Request Broker Architecture (CORBA) notification service [70]. This model is hierarchical, with two submodels. The lower-level model is the $M/M/1/k$

queue that we studied in Section 11.5. The upper-level model here is a CTMC with two absorbing states. The initial probability vector of the upper-level model is obtained from the steady-state probability vector of the lower-level model. Many other examples of hierarchical performance models are available in the literature.

Example 16.13 *Modeling Notification Services in Real-Time Distributed Systems*
The Object Management Group's CORBA notification service [70] is a messaging service specification for distributed systems. It specifies configurable events (messages) and connection persistence, and standards-based QoS and administrative properties that include event delivery and discard policies. In this example, we briefly describe performance models for the notification services that were developed in [71]. The QoS properties relevant to this example include:

OrderPolicy: this specifies the event delivery order to be Any, FIFO (first in first out), Priority or Deadline.

MaxEventsPerConsumer: this specifies the maximum number of events that may be queued in the master queue for a consumer at any time. A value of 0 implies that there is no limit.

DiscardPolicy: this specifies the order in which events are discarded when overflow conditions arise in queues. This could be Any, FIFO, LIFO (last in first out), Priority or Deadline order.

Any practical notification service deployment is likely to limit maximum queue lengths and to limit resource exhaustion. When *MaxEventsPerConsumer* is set to some value k, further incoming events will result in the discarding of events from within the proxy supplier queue (PSQ), the queue between a supplier and consumers, if the number of events queued for that consumer is already k. The event to be discarded is picked according to that consumer's *DiscardPolicy* setting. We now develop performance models to analyze the effect of these discards.

We describe a performance model for a notification service that is configured with both delivery and discard orders set to FIFO. Models for other configurations can be found in [71]. We use two homogeneous CTMCs for this purpose. We assume that the input of events forms a Poisson process with rate λ, and that a delivery operation to a consumer completes at a constant rate μ. Thus the PSQ is an $M/M/1/k$ queue with *MaxEventsPerConsumer* = k. Therefore, if we let the number of events in the PSQ denote the state of the queue, the steady-state probabilities of the PSQ with arrival rate λ and delivery rate μ are given by the equations in Section 11.5. The state probabilities $\pi_j^{(L)}$ are not affected by whether or not any discards occur within the queue, so long as the discarded event is replaced by another event (as is the case with notification service discards), and an event undergoing service is not discarded or preempted. We use the superscript L for these state probabilities to indicate the lower-level model that is the $M/M/1/k$ queue. Let us denote the lower-level model by M_1. In the upper-level CTMC model (denoted by M_2), we use a "tagged event" approach to find the distribution of time to delivery and discard for a given event. This involves observing the behavior of an event (the "tagged" event) that enters a PSQ that is at steady state until it leaves the queue. By the well-known PASTA property [72], each tagged event that arrives at a queue at steady state sees a queue with the same steady-state queue length pmf given by the equations in Section 11.5. Thus, the steady-state probability vector of the lower-level model will be supplied to the upper-level CTMC and will be used to determine its initial state probability vector. The corresponding import graph is shown in Figure 16.24.

An event can be removed from its PSQ after delivery to the consumer or when it is discarded from the queue in accordance with the *DiscardPolicy* QoS property when an overflow condition occurs. Let the random variable H denote the time that a newly arrived event spends in the system before being removed. The CTMC model for the FIFO discard policy is shown in Figure 16.25.

In the case of FIFO discard, the tagged event will be discarded if an overflow condition occurs when it is the oldest waiting event in the queue. Hence it is necessary to keep track of the events that arrive behind the tagged event in addition to those in front of it. Hence, each state in Figure 16.25 is a 2-tuple (i,j). Here, j denotes the number of events in front of the tagged event and i denotes the total number of events in the queue, including the tagged event. When the event at the head of the queue gets delivered, the

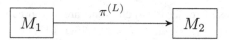

Figure 16.24   Import graph for the notification service model.

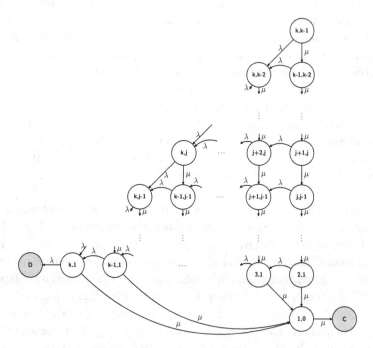

Figure 16.25   Continuous-time Markov chain when *OrderPolicy* = FifoOrder, *DiscardPolicy* = FifoOrder.

tagged event moves closer to the head, indicated by a transition from state (i,j) to state $(i-1,j-1)$ at the rate μ. In any of the states labeled (k,j) (with $j \in \{2,\ldots,k-1\}$), if an arrival occurs (at the rate λ) before an event in front gets delivered (at the rate μ), the event at the head gets discarded, resulting in a transition to state $(k,j-1)$. If a new event arrives when the CTMC is in state $(k,1)$, the tagged event is discarded (transition to state D occurs) since it is the one at the head of the queue. State $(1,0)$ indicates that the tagged event has been taken up for delivery. Transition to state C denotes the delivery completion of the event to the consumer.

Before we can solve the CTMC of Figure 16.25 for various time to absorption distributions, we need to specify the initial state probability vector. Based on the PASTA property, we can assert that: $\pi_{j,j-1}(0) = \pi_{j-1}^{(L)}$ for $j = 1,2,\ldots,k-1$ and $\pi_{k,k-1}(0) = \pi_{k-1}^{(L)} + \pi_{k}^{(L)}$. The distribution of the time to delivery $(F_C(t))$ and time to discard $(F_D(t))$ can now be computed from the transient solution of the two-dimensional CTMC of Figure 16.25 as the probability of being in these respective states at time t, that is, $F_C(t) = \pi_C(t)$ and $F_D(t) = \pi_D(t)$. Note that the holding time distribution (time to absorption distribution of the CTMC) for the tagged event in the queue, by the theorem of total probability, is given by:

$$F_H(t) = F_C(t) + F_D(t) \tag{16.10}$$
$$= \pi_C F_{H|C}(t) + \pi_D F_{H|D}(t),$$

where $F_{H|C}(t)$ is the conditional holding time distribution for the tagged event given that it gets delivered successfully, and $F_{H|D}(t)$ is the conditional holding time distribution for the tagged event given that it gets discarded. π_C and π_D are the eventual probabilities that the event gets delivered successfully and gets discarded, respectively.

Since D and C are multiple absorbing states (Section 10.2.4), the distribution of the time to delivery or the response time, $F_C(t) = \pi_C F_{H|C}(t)$, is a defective distribution [73] with the defect at $t = \infty$ equal to the probability of discard, π_D. Figure 16.26 plots $F_C(t)$ versus time, assuming the following numerical values: $k = 25$, $\lambda = 800\,\text{s}^{-1}$, $\mu = 430\,\text{s}^{-1}$ [71].

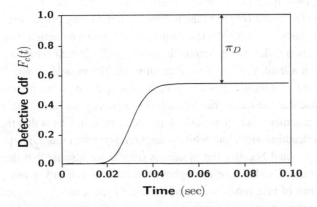

Figure 16.26   Defective nature of the $F_C(t)$ distribution.

Similarly, the distribution of the time to discard, $F_D(t) = \pi_D F_{H|D}(t)$, is a defective distribution with the defect at $t = \infty$ equal to the probability of successful delivery, π_C. The probability of deadline violations (p_{DV}) at steady state is given by:

$$p_{DV} = [1 - F_{H|C}(d)], \tag{16.11}$$

where $F_{H|C}(t)$ is the conditional holding time distribution of a successfully delivered message, and d is the deadline associated with each message. Note that $F_{H|C}(t) = \frac{F_C(t)}{F_C(\infty)} = \frac{F_C(t)}{\pi_C}$, where $F_C(t)$ is the defective response time distribution.

Problems

16.29 Assuming $k = 2$, derive closed-form expressions for all the measures in Example 16.13.

16.30 Compute the defective response time distribution shown in Figure 16.26 using SHARPE or a similar package, using the following numerical values: $k = 25$, $\mu = 430$, $\lambda = 200$.

16.7   Performability Models

This section considers performability [19] models of systems. We consider a reward-based performability model [74], i.e., performance and dependability models are separated. Such models represent a different kind of hierarchy. The performance models are solved to obtain performance measures. These measures are used as reward rates that are assigned to various states of the dependability model (assuming that the dependability model is a state-space type). The dependability model is then solved to obtain overall performability measures.

Example 16.14 *Missile Control System*

Consider the CTMC model of a missile control system illustrated in Figure 16.27. Here, the state labels 0, 1, and 2 represent the number of functioning processors [75]. Missile launch is to be guaranteed within a short interval after target detection. For a successful missile launch, two tasks (target acquisition and missile launch setup) must complete execution within a hard, pre-established deadline d. The missile system is considered functional as long as at least one of the processors is operational and no hard deadline is violated. Assume that the failure rate of each processor is γ and the repair rate is τ. The fault coverage parameter is c. State DF (dynamic failure) indicates that the system has failed due to a deadline violation, while state RP represents failure due to the arriving task set being rejected because the system is down. We assume that the system will remain in the same CTMC state throughout the execution of any given task set. The inter-arrival times of task requests are independent exponentially distributed with rate λ. Assume that the arrival rate λ is low enough that the system is stable even with a single processor, and that there is hardly any queuing.

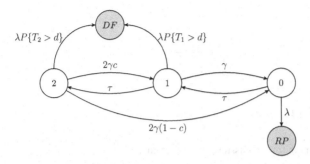

Figure 16.27 Hard deadline CTMC model.

The completion time of the task set must be considered under all the different CTMC states. Assume that the two task execution times are exponentially distributed with means $1/\mu_1$, and $1/\mu_2$, respectively. The two tasks are executed in parallel when both processors are operational. We denote by T_2 the task set completion time when both processors are active and T_1 the task set completion time when only one processor is active.

Hence, the task set completion time, T_2, will be the maximum of the two tasks' completion times and its distribution function is the product of the individual task distribution functions assuming independent task execution (see Section 3.3.2). The no deadline violation probability in this case is:

$$F_2(d) = P\{T_2 \leq d\} = 1 - e^{-\mu_1 d} - e^{-\mu_2 d} + e^{-(\mu_1 + \mu_2)d}. \tag{16.12}$$

With a single available processor, the tasks must be executed sequentially and, consequently, the task set completion time, T_1, is hypoexponentially distributed with parameters μ_1 and μ_2, and two-stage Erlang if $\mu_1 = \mu_2$ (Section 3.2.5). Hence the no deadline violation probability in this case is:

$$F_1(d) = P\{T_1 \leq d\} = \begin{cases} 1 - \dfrac{\mu_2}{\mu_2 - \mu_{t1}} e^{-\mu_1 d} + \dfrac{\mu_1}{\mu_2 - \mu_1} e^{-\mu_2 d} & \text{if } \mu_1 \neq \mu_2, \\ 1 - e^{-\mu_1 d} - \mu_1 d e^{-\mu_1 d} & \text{if } \mu_1 = \mu_2. \end{cases} \tag{16.13}$$

The two lower-level submodels M_1 – Eq. (16.12) – and M_1' – Eq. (16.13) – supply the two deadline violation probabilities (or complementary distribution functions) in fully symbolic form to the upper-level CTMC (Figure 16.27) from which the system failure probability at time t due to hardware failures (reaching state 0), dynamic (or deadline violation) failure probability (reaching state DF) or task rejection failure (reaching state RP) probability can be computed as shown below. The acyclic import graph is shown in Figure 16.28.

Define:

$$q_1 = P\{T_1 > d\},$$

$$q_2 = P\{T_2 > d\}. \tag{16.14}$$

Assume that the initial state of the upper-level CTMC (Figure 16.27) model is state 2; that is, $\pi_2(0) = 1$, $\pi_0(0) = \pi_1(0) = 0$. Then the system of differential equations

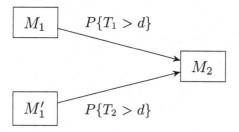

Figure 16.28 Acyclic import graph for the missile control system.

becomes:

$$\frac{d\pi_2(t)}{dt} = \tau\pi_1(t) - (2\gamma c + 2\gamma(1-c) + \lambda q_2)\pi_2(t),$$

$$\frac{d\pi_1(t)}{dt} = 2\gamma c\pi_2(t) + \tau\pi_0(t) - (\tau + \gamma + \lambda q_1)\pi_1(t),$$

$$\frac{d\pi_0(t)}{dt} = 2\gamma(1-c)\pi_2(t) + \gamma\pi_1(t) - (\tau + \lambda)\pi_0(t), \qquad (16.15)$$

$$\frac{d\pi_{DF}(t)}{dt} = \lambda q_1\pi_1(t) + \lambda q_2\pi_2(t),$$

$$\frac{d\pi_{RP}(t)}{dt} = \lambda\pi_0(t).$$

Using the technique of Laplace transform, we can reduce the above system of ODEs to a set of linear algebraic equations:

$$s\pi_2^*(s) - 1 = \tau\pi_1^*(s) - (2\gamma + \lambda q_2)\pi_2^*(t),$$

$$s\pi_1^*(s) = 2\gamma c\pi_2^*(s) + \tau\pi_0^*(s) - (\tau + \gamma + \lambda q_1)\pi_1^*(s),$$

$$s\pi_0^*(s) = 2\gamma(1-c)\pi_2^*(s) + \gamma\pi_1^*(s) - (\tau + \lambda)\pi_0^*(s), \qquad (16.16)$$

$$s\pi_{DF}^*(s) = \lambda q_1\pi_1^*(s) + \lambda q_2\pi_2^*(s),$$

$$s\pi_{RP}^*(s) = \lambda\pi_0^*(s).$$

From the above system of equations, we can get Laplace domain answers:

$$\pi_2^*(s) = \frac{\tau\pi_1^*(s) + 1}{s + 2\gamma + \lambda q_2},$$

$$\pi_1^*(s) = \frac{\tau(s + 2\gamma + \lambda q_2)\pi_0^*(s) + 2\gamma c}{(s + \tau + \gamma + \lambda q_1)(s + 2\gamma + \lambda q_2) - 2\gamma c\tau},$$

$$\pi_0^*(s) = \frac{(2\gamma(1-c)\tau + \gamma(s + 2\gamma + \lambda q_2))\bar{\pi}_1(s) + 2\gamma(1-c)}{(s + \tau + \lambda)(s + 2\gamma + \lambda q_2)}, \qquad (16.17)$$

$$s\pi_{DF}^*(s) = \lambda q_1\pi_1^*(s) + \lambda q_2\pi_2^*(s),$$

$$s\pi_{RP}^*(s) = \lambda\pi_0^*(s).$$

We leave the transform inversion as an exercise. We can also get absorption probabilities using the final value theorem of Laplace transforms:

$$\pi_{DF}(\infty) = \lim_{s->0} s\pi_{DF}^*(s),$$

$$\pi_{RP}(\infty) = \lim_{s->0} s\pi_{RP}^*(s). \tag{16.18}$$

Dynamic failure is sometimes referred to as performance failure and it is one way in which the performance and reliability of a system are interdependent. Another aspect of such a dependence is that the failure rate γ could be considered a function of the utilization of the processor [76]. Yet another type of interdependence is the fact that the service time of a job can be affected by server failure while the job is being executed. Such considerations have been studied in detail in several papers [77, 78].

Example 16.15 *Messaging Services in Distributed Systems*
This example is adapted from [79]. There are several settings of the notification service that are relevant to this example. We consider two of them: *best effort* (BE) and *persistent connections*.

> *EventReliability:* When set to BestEffort, the messaging service does not provide delivery guarantees. All messages inside the messaging service will be lost if the messaging service process crashes. When set to Persistent, the messaging service maintains the internal queues and state information in persistent storage.
>
> *ConnectionReliability:* When set to BestEffort, connections between clients and the messaging service are lost when any of them crashes and is then restarted. Also, if a consumer crashes, the messaging service will discard all events queued for that consumer that have not been delivered yet. When set to Persistent, the messaging service keeps retrying connections, ignoring temporary failure indications, in the hope that the clients will come back up. See [80] for details of the *EventReliability* and *ConnectionReliability* settings.

In this hierarchical model, the upper level is the system availability model (denoted by M_A) in which the supplier, messaging service and the consumer process can fail and get repaired. We assume that each of these three entities have detailed lower-level availability models (not shown here) to produce the steady-state availabilities of the supplier, messaging service and the consumer, denoted by A_S, A_M, and A_C, respectively. When steady-state availability measures are required, it is convenient to reduce the models to two-state availability models [73] and work with equivalent failure (γ_{eq}) and repair rates (τ_{eq}) (Section 9.5.6). The two-state equivalent models also simplify the derivation of closed-form solutions using a hierarchical approach. We use the superscripts (S), (M), and (C) when referring to the failure and repair rates of the supplier, messaging service and consumer, respectively.

The system availability model (M_A) is an eight-state CTMC representing all combinations in which the supplier, consumer and messaging service are up or have failed and are being repaired. Assuming independence, the steady-state probabilities of these states are:

$$\pi_{1,1,1} = A_S A_M A_C, \qquad\qquad \pi_{0,1,1} = (1 - A_S) A_M A_C,$$
$$\pi_{1,0,1} = A_S (1 - A_M) A_C, \qquad \pi_{1,1,0} = A_S A_M (1 - A_C),$$
$$\pi_{0,1,0} = (1 - A_S) A_M (1 - A_C), \qquad \pi_{1,0,0} = A_S (1 - A_M)(1 - A_C),$$
$$\pi_{0,0,1} = (1 - A_S)(1 - A_M) A_C, \qquad \pi_{0,0,0} = (1 - A_S)(1 - A_M)(1 - A_C).$$

Next, we discuss the lower-level pure performance model (denoted by M_P) that will supply reward rates to the availability model M_A. Since failures are rare, the queue within the messaging service can be assumed to be in its steady state. Let $E[N] = \sum_{i=0}^{K} i \pi_i^{(P)}$ denote the expected number of messages in the queue at steady state, where $\pi_i^{(P)}$ is the steady-state probability that the queue has i messages. $\pi_k^{(P)}$ is the probability of the queue being full [73]. For the performance models in this section, we assume that message arrivals constitute a Poisson process with parameter λ and that the message consumption times are exponentially distributed with rate μ. Thus the queue is an $M/M/1/k$ queue, where k is the maximum number of messages that can be held in the PSQ. The import graph for this model is shown in Figure 16.29.

We now proceed to develop the performability model by assigning reward rates computed from M_P to the states of the availability model M_A. First we derive the reward rate (actually they are loss rates in this case) equations without considering deadline violations. We only consider the best effort QoS setting. For the other cases, the reader is referred to the original paper [79]. It is clear that in the fully up state, the only loss occurs when an incoming message finds the buffer full. But when either the messaging service or the consumer crashes, there is an impulse reward (actually loss) for all the jobs, that is, $E[N]$. When converted to equivalent reward rate values, we get respective reward rates: $\gamma_{eq}^{(M)} E[N]$ and $\gamma_{eq}^{(C)} E[N]$. Hence the reward rate attached to the fully up state is:

$$r_{1,1,1} = \lambda \pi_k^{(P)} + \gamma_{eq}^{(M)} E[N] + \gamma_{eq}^{(C)} E[N].$$

In every other state of M_A, all arriving messages will imply a message loss and hence the reward rate for all the remaining states is λ. Thus the expected loss rate in the steady state for the best effort setting is:

$$ELR_{BE} = A_S A_M A_C (\lambda \pi_k^{(P)} + \gamma_{eq}^{(M)} E[N] + \gamma_{eq}^{(C)} E[N]) + (1 - A_S A_M A_C)\lambda. \qquad (16.19)$$

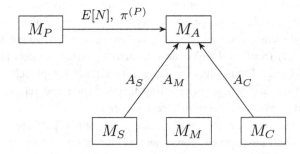

Figure 16.29 Import graph for the messaging service model.

When messages have deadlines associated with them, in addition to the losses accounted for above, losses due to deadline violations also need to be considered. Note that response time distributions, rather than mean response times, are required to compute the probability of deadline violations.

In order to evaluate the losses due to deadline violations, we need performance models such as those of Example 16.13.

In this example, for notational simplicity, we shall use $F_R(t) = \frac{F_C(t)}{F_C(\infty)} = \frac{F_C(t)}{1-\pi_D}$ to denote the *conditional* response time distribution of a message given that it is successfully delivered (a non-defective distribution). The expected loss rate due to deadline violations (DV) at steady state, to be added to that in Eq. (16.19), can be written as:

$$ELR_{DV} = A_S A_M A_C (1 - F_R(d))\lambda(1 - \pi_k^{(P)}), \tag{16.20}$$

where $\lambda(1 - \pi_k^{(P)}) = \mu(1 - \pi_0^{(P)})$ is the throughput at the consumer at steady state, and d is the deadline associated with each message.

Example 16.16 *Multi-Voltage Propulsion System of a High-Speed Train*

We now revisit the multi-voltage propulsion system of a high-speed train; in Example 4.14, we presented a block diagram with three blocks in parallel. Later, in Example 10.21, we developed a three-state CTMC model of an individual block. Then, in Example 13.3, we generalized the three-state CTMC to allow for time-dependent failure rates for individual components.

Now we consider n blocks in parallel (a generalization of the three-block case considered by the original authors). To build up a model for the whole system, we assume statistical independence among the n parallel modules.

Recall that from the NHCTMC (Example 13.3), we have closed-form expressions for the three state probabilities $\pi_0(t)$, $\pi_2(t)$, and $\pi_1(t)$. Considering this three-state NHCTMC as the lower-level model (call it M_N), we use a simple multi-state fault tree as the top level (call it M_F) with a single AND gate and n inputs, as depicted in Figure 16.30 (with $n = 3$). Each of the inputs is fed from the NHCTMC model discussed above and hence is a three-state input. Let us indicate with n_0, n_1, and n_2 the number of subsystems in states 0, 1, and 2, respectively. Clearly, the possible system states are characterized by the following combination of indices: $\{n_0, n_1, n_2 \mid n_0, n_1, n_2 \geq 0; n_0 + n_1 + n_2 = n\}$. The probability of each of the states is easily written down:

$$\pi_{n_0,n_1,n_2} = \binom{n}{n_0, n_1, n_2}[\pi_2(t)]^{n_2} \cdot [\pi_1(t)]^{n_1} \cdot [\pi_0(t)]^{n_0}. \tag{16.21}$$

From this, the expected power supplied in kW at time t is computed as:

$$E[X(t)] =$$

$$\sum_{n_2=0}^{n}\sum_{n_1=0}^{n-n_2}(r_2 n_2 + r_1 n_1)\binom{n}{n-n_1-n_2, n_1, n_2}[\pi_2(t)]^{n_2} \cdot [\pi_1(t)]^{n_1} \cdot [\pi_0(t)]^{n-n_1-n_2}. \tag{16.22}$$

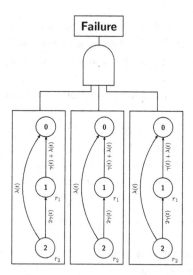

Figure 16.30   Top-level model, M_F, as an FT.

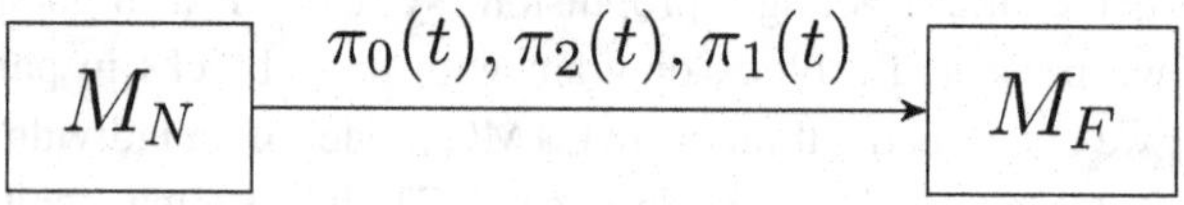

$$M_N \xrightarrow{\pi_0(t),\, \pi_2(t),\, \pi_1(t)} M_F$$

Figure 16.31   Import graph for the multi-voltage propulsion system of a high-speed train.

The expected total energy supplied in kW h during the interval $(0, t]$ can also be written as:

$$E[Y(t)] = \sum_{n_2=0}^{n} \sum_{n_1=0}^{n-n_2} (r_2 n_2 + r_1 n_1) \binom{n}{n - n_1 - n_2, n_1, n_2}$$

$$\cdot \int_0^t ([\pi_2(x)]^{n_2} \cdot [\pi_1(x)]^{n_1} \cdot [\pi_0(x)]^{n - n_1 - n_2}) dx. \tag{16.23}$$

In this example, the model is hierarchical and the quantities passed from the lower-level model to the upper-level model are time-dependent state probabilities. The import graph of this system is shown in Figure 16.31.

Example 16.17   *Multi-Voltage Propulsion System of a High-Speed Train: Sensitivity and Uncertainty Propagation [81]*
For Example 16.16, we derive here the parametric sensitivity and the epistemic uncertainty propagation through the analytic closed-form approach described in Section 3.4. We consider in this example the MTTF as the output measure [81], and we restrict our analysis to time-independent transition rates in the lower-level model

Table 16.5 Ranking based on scaled sensitivities of MTTF.

ϑ	SS_ϑ (MTTF)
λ_I	-4.16997×10^{-1}
λ_M	-3.51155×10^{-1}
γ	-2.03316×10^{-1}
λ_T	-2.41419×10^{-2}
λ_F	-4.38944×10^{-3}

(Figure 10.18). From the FT of Figure 16.30, with $n = 3$, we obtain:

$$R_{sys} = 1 - [1 - 2e^{-(\gamma+\lambda)t} + e^{-(2\gamma+\lambda)t}]^3,$$

$$\text{MTTF} = \int_0^\infty R_{sys}(t)dt$$

$$= \frac{8}{3(\gamma+\lambda)} - \frac{29}{6(2\gamma+\lambda)} + \frac{12}{(3\gamma+2\lambda)} - \frac{12}{(4\gamma+3\lambda)} + \frac{6}{(5\gamma+3\lambda)}.$$

With the numerical values of Table 4.5, we obtain the result MTTF $= 21\,662.9$ hr.

The dimensionless scaled sensitivity is computed from Eq. (9.79):

$$SS_\vartheta(\text{MTTF}) = \frac{\partial \text{MTTF}}{\partial \vartheta} \cdot \frac{\vartheta}{\text{MTTF}}, \qquad \vartheta = \{\lambda_T, \lambda_F, \lambda_I, \lambda_M, \gamma\}. \tag{16.24}$$

Table 16.5 reports the numerical values of the scaled sensitivity of the MTTF sorted by importance. Note that all the values are negative, meaning that if the value of the parameter ϑ increases, the MTTF decreases. The most influential component in affecting the MTTF is the inverter, which is the component with the largest failure rate value.

To analytically evaluate the effect of the epistemic uncertainty in the input parameters on the uncertainty of the output measure, the MTTF can be viewed as a random function of the input parameters $(\Lambda_T, \Lambda_I, \Lambda_F, \Lambda_M, \Gamma)$ that are now assumed to be random variables. The approach of Section 3.4 consists in computing the unconditional expected MTTF through the following equation:

$$E[\text{MTTF}] = \int \cdots \int \text{MTTF}(\bullet) f(\bullet) d\lambda_T \, d\lambda_I \, d\lambda_F \, d\lambda_M \, d\gamma, \tag{16.25}$$

where the input parameters are set to $\Lambda_T = \lambda_T$, $\Lambda_I = \lambda_I$, $\Lambda_F = \lambda_F$, $\Lambda_M = \lambda_M$, $\Gamma = \gamma$, and $f(\bullet)$ is the joint epistemic density $f_{\Lambda_T, \Lambda_I, \Lambda_F, \Lambda_M, \Gamma}(\lambda_T, \lambda_I, \lambda_F, \lambda_M, \gamma)$ of the input parameters.

Following Section 3.4, each parameter is an exponentially distributed random variable of rate Θ with prior density $f_\Theta(\vartheta)$. We observe a sample of k i.i.d. realizations of Θ with total testing time equal to S. The conditional density of Θ given $S = s$ is given by Eq. (3.70), in the form of a k-stage Erlang distribution with parameter s:

$$f_{\Theta|S}(\vartheta|s) = \frac{\vartheta^{k-1} s^k e^{-\vartheta s}}{(k-1)!}. \tag{16.26}$$

The maximum likelihood estimate for the rate ϑ in Eq. (16.26) is $\hat{\vartheta} = k/s$, from which we obtain $s = k/\hat{\vartheta}$. In this setting, the unconditional expected MTTF is computed from:

$$E[\mathrm{MTTF}] = \int_0^\infty \int_0^\infty \int_0^\infty \int_0^\infty \int_0^\infty \mathrm{MTTF}(\bullet) \cdot \prod_\vartheta \frac{\vartheta^{k_\vartheta-1} s_\vartheta^{k_\vartheta} e^{-\vartheta s_\vartheta}}{(k_\vartheta - 1)!} \, d\vartheta, \quad (16.27)$$

where $\vartheta = \{\lambda_T, \lambda_I, \lambda_F, \lambda_M, \gamma\}$ and k_ϑ is the number of samples for the parameter ϑ. s_ϑ is the total time of test for the parameter ϑ and is given by $s_\vartheta = k/\hat{\vartheta}$, where for $\hat{\vartheta}$ we assume the corresponding value in Table 4.5. In Eq. (16.27) the joint pdf $f(\bullet)$ can be computed as a product of the marginal pdfs since we have assumed independence in the model input parameters.

The uncertainty introduced in the MTTF by the single parameter ϑ, while all the other parameters are assumed to have the constant value of Table 4.5, is given by:

$$E[\mathrm{MTTF}] = \int_0^\infty \mathrm{MTTF}(\bullet) \frac{\vartheta^{k_\vartheta-1} s_\vartheta^{k_\vartheta} e^{-\vartheta s_\vartheta}}{(k_\vartheta - 1)!} d\vartheta. \quad (16.28)$$

The results are depicted in Figure 16.32(a), where each curve represents $E[\mathrm{MTTF}]$ of the system computed through Eq. (16.28) for the considered parameter. For $k \to \infty$, $E[\mathrm{MTTF}]$ tends to the MTTF computed with the parameter values of Table 4.5.

The variance of the MTTF is $\mathrm{Var}[\mathrm{MTTF}] = E[\mathrm{MTTF}^2] - (E[\mathrm{MTTF}])^2$. The variance of the MTTF when a single parameter ϑ is assumed to be affected by uncertainty is reported in Figure 16.32(b), as a function of the number of samples k. In this case, the variance tends to zero as $k \to \infty$.

The closed-form approach to the uncertainty propagation requires the computation of a multiple integral for each parameter and may become impractical for large and complex models. From Table 16.5 and Figure 16.32 it is possible to observe that for the parameters with a larger absolute value of the sensitivity, a larger number of samples is needed to reach the same tight confidence interval. The results obtained indicate that a relationship between parametric sensitivity and uncertainty propagation exists. Since parametric sensitivity evaluation is easier and quicker to carry out than uncertainty propagation through integral computation, the results suggest using

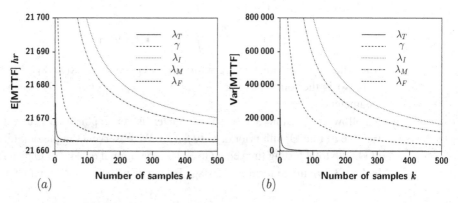

Figure 16.32 Expected value **(a)** and variance **(b)** of MTTF as functions of the number of samples k, with epistemic uncertainty in a single parameter at a time.

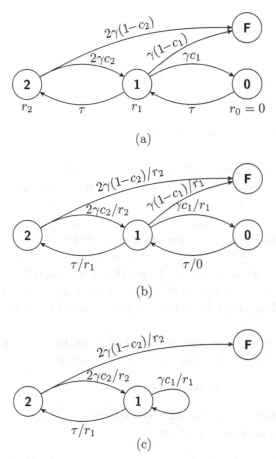

Figure 16.33 Upper-level reliability models for the $M/M/2/k$ queue with server failures. (a) Time domain; (b) reward domain; (c) reward domain after elimination of the zero-reward state.

parametric sensitivity analysis to identify the parameters that need a tighter confidence interval to get more accurate output measures.

Example 16.18 *M/M/2/k Queue with Server Failure and Single Repair Person*
Next we consider an $M/M/2/k$ queue with server failures and single repair person. The arrival rate of jobs is λ, the service rate is μ, the failure rate of each server is γ, and the repair rate is τ. First, consider the pure reliability model of the system shown in Figure 16.33(a). When both servers are up (state 2), a failure of one of the two servers is not covered with probability $1 - c_2$ and leads to system failure state F, an absorbing state. A covered failure will lead to state 1 from where a server repair (with rate τ) leads back to state 2, and a second server failure before the completion of repair on the first server leads to state 0, in the case of a covered failure (with probability c_1), or to state F (with probability $1 - c_1$). In state 0, the server can be repaired at rate τ and the system will return to state 1.

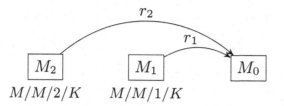

Figure 16.34 Import graph for the $M/M/2/k$ queue with server failures.

We are interested in computing the expected number of jobs completed and the expected number of jobs rejected in the interval $(0,t]$. We are also interested in computing distribution functions of these two measures until absorption. We first compute the throughput of the $M/M/2/k$ queue and attach it as the reward rate r_2 to state 2 of the reliability model. Similarly, we attach the throughput of an $M/M/1/k$ queue as reward rate r_1 to state 1 of the reliability model. Reward rate 0 is assigned to both the states 0 and F. The import graph in Figure 16.34 shows the relationship between the three models: the $M/M/2/k$ queue labeled M_2, the $M/M/1/k$ queue labeled M_1 and the reliability model labeled M_0. The import graph is acyclic, like all the others in this chapter.

Once the transient state probabilities of the reliability model are computed, we can write the throughput at time t, an example of expected reward rate at time t, as:

$$E[T(t)] = r_2 \pi_2(t) + r_1 \pi_1(t),$$

and the expected number of jobs completed in the interval $(0,t]$, an example of expected accumulated reward in the interval $(0,t]$, as:

$$E[C(t)] = \int_0^t [r_2 \pi_2(x) + r_1 \pi_1(x)] dx.$$

In order to compute the distribution of the number of jobs completed until absorption, we apply the transformation due to Beaudry [82], already utilized in Example 10.21. We divide the exit rates from each state by the reward rate of the state as shown in Figure 16.33(b). But, as we see, there is a non-absorbing state with a zero reward rate giving rise to a transition rate that is infinite. In order to deal with this situation, we use the approach in [83] and eliminate the instantaneous state 0 from the CTMC, leading to Figure 16.33(c). Now, computing the time to absorption distribution of this reduced CTMC will give the distribution of accumulated reward for the CTMC of Figure 16.33(a).

Problems

16.31 For the missile control system in Example 16.14, first complete the Laplace domain solution of the equations in (16.17) and then symbolically invert to get a time domain result of all state probabilities. Thence obtain expressions for the eventual probabilities of reaching the states DF and RP.

16.32 For the missile control example, obtain expressions for the derivatives of $\pi_{DF}(t)$ as functions of all the input parameters of the model.

16.33 For the top-level CTMC of the missile control example, derive a closed-form expression for the mean time to absorption to any of the absorbing states.

16.34 For the multi-voltage train example, specialize the equations for the expected power supplied at time t and the expected energy supplied in the interval $(0, t]$ when the three-state CTMC for each block is a homogeneous one. In this case, the integral in the latter should be possible to carry out in a closed form.

16.35 For the $M/M/2/k$ queue in Example 16.18, write down explicit expressions for r_2 and r_1 as functions of the arrival rate λ, the service rate μ and the system size k. Next, write down expressions for the derivatives of the expected accumulated reward in the interval $(0, t]$ as functions of the parameters $\lambda, \mu, \gamma, \tau, c_2$, and c_1.

16.36 For the $M/M/2/k$ queue in Example 16.18, develop a monolithic CTMC model either directly by hand or via an SRN model. Solve the monolithic model numerically for the measures discussed in that example and compare with those obtained from the hierarchical model.

16.8 Survivability Models

In this section, we present a general survivability quantification approach that is applicable to a wide range of system architectures, applications, failure/recovery behaviors and metrics. The following definition of survivability was proposed by the T1A1.2 network survivability performance working group [84]. By this definition, survivability shows the time-varying system behavior after a failure occurs. We use their definition as the basis for the example in this section. Here, failure may be a hardware/software failure, a security failure or a natural disaster.

> *Survivability:* Suppose a measure of interest M has the value m_0 just before a failure occurs. The survivability behavior can be depicted by the following attributes: m_a is the value of M just after the failure occurs, m_u is the maximum difference between the value of M and m_a after the failure, m_r is the restored value of M after some time t_r and t_R is the time for the system to restore the value of m_0.

Recall that dependability primarily concerns itself with hardware/software failures and security primarily concerns itself with attacks. By contrast, survivability deals with both these types of failures as well as natural disasters. Further, survivability is concerned with the transient behavior of the system (performance) just after the occurrence of a failure (attack or natural disaster) until the system stabilizes (gets repaired or recovers). Survivability can thus be seen as a generalization of system recovery since it considers system performance during the recovery stage as well as simply the time to recovery. Another concept closely related to survivability is that of resilience. We leave the discussion of resilience to the literature [85–87].

Example 16.19 *Survivability of a Telecommunication Switching System*
Returning to Example 11.7, where we considered the performability modeling of a telecommunication switching system, we now extend the example to consider the survivability behavior [88].

We consider the blocking probability $P_{bk}(t)$ as the measure of interest M in our example. Since the definition of survivability given earlier explicitly indicates the time-dependent characteristics of the system behavior, transient solutions are needed in this case. The failure that we inject just before considering the recovery from that failure could well be a massive failure of a large number of trunks at once. But for the sake of simplicity, we only consider the situation of a single failure. Furthermore, we assume that no additional failures will occur before the first one has been recovered from. For our purposes, states indicating no failure or one failure (the first two rows in Figure 11.21) are included and all other irrelevant states are truncated from the composite model. Now, to inject a single trunk failure in the model we force all the failure transitions from the first row to the second row since we want to study the system behavior given that a failure has occurred. These transitions are marked with dashed arcs suggesting that these transitions are forced to begin with. The truncated composite model is shown in Figure 16.35.

In fact, we are interested in the transient analysis of system recovery after the first failure has occurred while disallowing any further failures. Thus the CTMC model whose transient behavior is the object of the survivability study is shown in Figure 16.36 and is denoted by M_S. The initial probability values $\pi_{i,j}(0)$ for the states of the survivability model M_S are the probability values of the system states just after the occurrence of the failure. We set $\pi_{n,j}(0) = 0$ since the failure has already occurred. The value of $\pi_{n-1,j}(0)$ is determined by two factors. One is the system behavior before the failure, which can be computed from the steady-state probability $\pi_{n,j}^{(P)}$ in the pure performance model shown in Figure 16.37, denoted by M_P. The steady-state probability vector of this model has already been derived in Eq. (11.17). The other factor is the set of failure transitions from the first row (n,j) to the second row $(n-1,j)$ in Figure 16.35. Since state $(n-1,j)$ can be reached from either state (n,j) with transition rate $(n-j)\gamma$ or

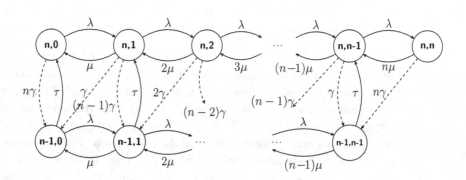

Figure 16.35   T1A1.2 Markov model: Initial probability assignment for the survivability model.

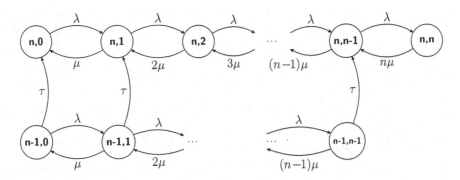

Figure 16.36 Continuous-time Markov chain model M_S to compute the transient behavior after the injected failure.

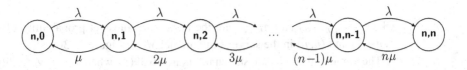

Figure 16.37 Continuous-time Markov chain model M_P to compute the steady state before the failure.

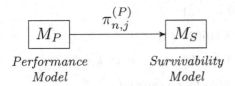

Performance   Survivability
Model   Model

Figure 16.38 The import graph of the survivability model.

state $(n, j+1)$ with transition rate $(j+1)\gamma$, upon the firing of the corresponding failure transitions, $\pi_{n-1,j}(0)$ is determined as follows:

$$\pi_{n-1,j}(0) = \frac{n-j}{n}\pi_{n,j}^{(P)} + \frac{j+1}{n}\pi_{n,j+1}^{(P)}. \tag{16.29}$$

Figure 16.35 suggests the above equation for initial probability assignment.

The import graph for the survivability model is shown in Figure 16.38.

Note that the initial probabilities $\pi_{n-1,j}(0)$ do not depend on the failure rate γ, and γ does not appear in the transition rates of the survivability model M_S. Therefore, the survivability measure $P_{bk}(t)$ is independent of the failure rate γ. This is important since for attacks and natural disasters, the rate of occurrence may not be easily obtained.

For the measure of interest P_{bk}, $m_0 = \pi_{n,n}^{(P)}$ is the blocking probability in normal operation with no failure occurrence. The transient blocking probability $P_{bk}(t)$ is given by:

$$P_{bk}(t) = \pi_{n-1,n-1}(t) + \pi_{n,n}(t), \tag{16.30}$$

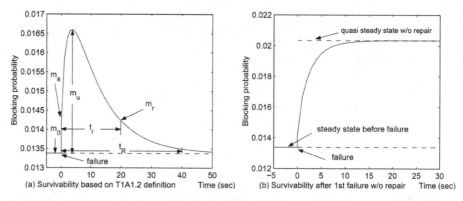

Figure 16.39   Survivability results. (a) Survivability after first failure; (b) Survivability after first failure without repair.

where $\pi_{n-1,n-1}(t)$ and $\pi_{n,n}(t)$ are the respective transient probabilities of states $(n-1, n-1)$ and (n, n) in the survivability model M_S of Figure 16.36.

The numerical results are shown in Figure 16.39(a), where the survivability attributes m_a, m_u, m_r and t_R can be obtained from the $P_{bk}(t)$ curve as

$$m_0 = 1.337\,604 \times 10^{-2},$$

$$m_a = 1.417\,861 \times 10^{-2},$$

$$m_u = 3.216\,020 \times 10^{-3},$$

$$m_r = 1.422\,506 \times 10^{-2} \text{ when } t_r = 20\,\text{s},$$

$$t_R = 39\,\text{s} \quad (1\% \text{ relative error}).$$

From the earlier quantifications of performance, availability, performability, and now of survivability, we can see that all the measures of interest can be derived from the Markov models developed in Example 11.7. The quantification procedure shows how the performance, availability, and combined performance, and availability models are constructed and solved in different circumstances.

More importantly, both availability and performability depend on the failure rate γ but survivability does not. Since it is usually difficult to have agreement on the value of γ, especially in the context of attacks and natural disasters, those availability and performability measures tend to be controversial and unconvincing. The survivability definition in Section 16.8 seems appropriate for telecommunications systems. Note that although the T1A1.2 working group gives this definition, the method of its quantification via modeling has only recently been discussed [88–91].

More Survivability Measures
While the survivability definition given earlier has addressed several important aspects of the survivability behavior, the quantification study is not necessarily limited to current definitions. More characteristics can be revealed by either new measures of interest or the same measure under different circumstances.

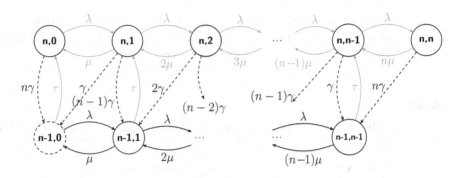

Figure 16.40 Truncated composite model without repair.

For the same survivability measure $P_{bk}(t)$, we now study it in a different situation. Rather than allowing repair right after the failure, suppose we are interested in the system behavior before any repair action is taken. Consequently, the truncated composite model is shown in Figure 16.40 where the circles and arcs in gray represent the removed states and transitions. The dotted arcs are also forced, as they indicate that instantaneous transitions have taken place and their rates are used to compute the initial probability assignment $(\pi_{i,j}(0))$ of the truncated composite model. For $i = n - 1$, $\pi_{n-1,j}(0)$ is computed according to Eq. (16.29). For $i < n - 1$, $\pi_{i,j}(0) = 0$ for the rest of states.

The transient blocking probability can be computed as before: $P_{bk}(t) = \pi_{n-1,n-1}(t)$. The numerical result after the first failure is shown in Figure 16.39(b). As expected, $P_{bk}(t)$ increases over time and reaches a quasi-steady-state value before repair.

Another measure that might be of interest is ELF (excess loss due to failures) [92]. Although the maximum blocking probability and the relaxation time t_R (duration needed for P_{bk} to reach its steady state) after the first failure can be easily read from Figure 16.39, it may be desirable to use a single number to reflect the total loss due to failure. One such measure is the ELF, which consists of two parts. One part, denoted by N_d, represents the dropping of ongoing calls (if any) carried by the failed trunk. From the pure performance model (see Example 11.1), the probability of carrying an ongoing call on the failed trunk in state j is j/n, which means

$$N_d = \frac{j}{n} \pi_j^{(P)}.$$ (16.31)

The other part, denoted by N_b, represents the extra blocked calls due to a temporarily higher than normal $P_{bk}(t)$. In other words, N_b is defined as the integral of the difference between the transient $P_{bk}(t)$ and the steady-state P_{bk} over the relaxation time t_R. Therefore, N_b can be computed as

$$N_b = \int_0^{t_R} (P_{bk}(t) - P_{bk})\lambda\, dt.$$ (16.32)

Finally, ELF is the sum of N_d and N_b. The numerical results are $N_d = 0.6557$, $N_b = 0.2457$, and ELF $= N_d + N_b = 0.9014$. Note that the new survivability measures introduced in this section are also independent of the failure rate γ.

Problems

16.37   Verify the numerical results given in Example 16.19 using SHARPE or a similar package.

16.38   Redo the above example using stochastic reward nets.

16.39   Redo the above example so that two trunk failures occur together initially.

16.9   Further Reading

There is a rich literature on hierarchical models. For hierarchical availability models, see [93–97]. For more examples of hierarchical reliability models, see [2, 14, 22, 98]. For hierarchical performance models, see [13, 15, 99, 100]. For hierarchical performability models, see [2, 31, 74, 83, 101–106]. For further examples of survivability models, see [89, 91, 107–110].

References

[1] G. Bolch, S. Greiner, H. de Meer, and K. S. Trivedi, *Queueing Networks and Markov Chains: Modeling and Performance Evaluation with Computer Science Applications*, 2nd edn. Wiley-Interscience, Apr. 2006.

[2] R. Sahner, K. Trivedi and A. Puliafito, *Performance and Reliability Analysis of Computer Systems: An Example-Based Approach Using the SHARPE Software Package*. Kluwer Academic Publishers, 1996.

[3] M. Ajmone Marsan, G. Balbo, and G. Conte, "A class of generalized stochastic Petri nets for the performance evaluation of multiprocessor systems," *ACM Transactions on Computer Systems*, vol. 2, pp. 93–122, 1984.

[4] G. Ciardo, A. Blakemore, P. F. Chimento, J. K. Muppala, and K. Trivedi, "Automated generation and analysis of Markov reward models using stochastic reward nets," in *Linear Algebra, Markov Chains, and Queueing Models*, eds. C. D. Meyer and R. J. Plemmons. Springer, 1993, vol. 48, pp. 145–191.

[5] B. Plateau and K. Atif, "Stochastic automata network for modeling parallel systems," *IEEE Transactions on Software Engineering*, vol. 17, pp. 1093–1108, 1991.

[6] W. Sanders and J. Meyer, "Reduced base model construction methods for stochastic activity networks," *IEEE Journal on Selected Areas in Communications*, vol. 9, no. 1, pp. 25–36, Jan. 1991.

[7] J. Hillston, *A Compositional Approach to Performance Modelling*. Cambridge University Press, 2005, vol. 12.

[8] A. Bobbio and K. S. Trivedi, "An aggregation technique for the transient analysis of stiff Markov chains," *IEEE Transactions on Computers*, vol. C-35, pp. 803–814, 1986.

[9] P. J. Courtois, *Decomposability: Queueing and Computer System Applications*, ACM monograph series. Academic Press, 1977.

[10] P. Buchholz and P. Kemper, "Kronecker based matrix representations for large Markov models," in *Validation of Stochastic Systems: A Guide to Current Research*, 2004, pp. 256–295.

[11] A. Benoit, B. Plateau, and W. J. Stewart, "Memory-efficient Kronecker algorithms with applications to the modelling of parallel systems," *Future Generation Computer Systems*, vol. 22, no. 7, pp. 838–847, 2006.

[12] Y. Bao, I. Bozkurt, T. Dayarl, X. Sun, and K. Trivedi, "Decompositional analysis of Kronecker structured Markov chains." *Electronic Transactions on Numerical Analysis*, vol. 31, pp. 271–294, 2008 [electronic only].

[13] K. M. Chandy, U. Herzog, and L. Woo, "Parametric analysis of queuing networks," *IBM Journal of Research and Development*, vol. 19, no. 1, pp. 36–42, Jan. 1975.

[14] S. J. Bavuso, J. Bechta Dugan, K. Trivedi, E. M. Rothmann, and W. E. Smith, "Analysis of typical fault-tolerant architectures using HARP," *IEEE Transactions on Reliability*, vol. R-36, no. 2, pp. 176–185, Jun. 1987.

[15] G. Balbo, S. Bruell, and S. Ghanta, "Combining queueing network and generalized stochastic Petri nets for the solution of complex models of system behavior," *IEEE Transactions on Computers*, vol. 37, pp. 1251–1268, 1988.

[16] R. Ghosh, F. Longo, V. K. Naik, and K. S. Trivedi, "Modeling and performance analysis of large-scale IaaS clouds," *Future Generation Computer Systems*, vol. 29, no. 5, pp. 1216–1234, 2013.

[17] L. Lei, Y. Zhang, X. S. Shen, C. Lin, and Z. Zhong, "Performance analysis of device-to-device communications with dynamic interference using stochastic Petri nets," *IEEE Transactions on Wireless Communications*, vol. 12, no. 12, pp. 6121–6141, Dec. 2013.

[18] X. Yin, X. Ma, and K. Trivedi, "An interacting stochastic models approach for the performance evaluation of DSRC vehicular safety communication," *IEEE Transactions on Computers*, vol. 62, no. 5, pp. 873–885, May 2013.

[19] J. F. Meyer, "On evaluating the performability of degradable computing systems," *IEEE Transactions on Computers*, vol. 29, no. 8, pp. 720–731, Aug. 1980.

[20] K. Trivedi, J. Muppala, S. Woolet, and B. R. Haverkort, "Composite performance and dependability analysis," *Performance Evaluation*, vol. 14, no. 3-4, pp. 197–216, Feb. 1992.

[21] Y. Ma, J. Han, and K. Trivedi, "Composite performance and availability analysis of wireless communication networks," *IEEE Transactions on Vehicular Technology*, vol. 50, no. 5, pp. 1216–1223, Sep. 2001.

[22] J. Blake and K. Trivedi, "Multistage interconnection network reliability," *IEEE Transactions on Computers*, vol. C-38, pp. 1600–1604, 1989.

[23] K. Trivedi, R. Vasireddy, D. Trindade, S. Nathan, and R. Castro, "Modeling high availability systems," in *Proc. IEEE Pacific Rim Int. Symp. on Dependable Computing (PRDC)*, 2006.

[24] K. S. Trivedi, D. Wang, J. Hunt, A. Rindos, W. E. Smith, and B. Vashaw, "Availability modeling of SIP protocol on IBM© WebSphere©," in *Proc. Pacific Rim Int. Symp. on Dependable Computing (PRDC)*, 2008, pp. 323–330.

[25] W. E. Smith, K. S. Trivedi, L. Tomek, and J. Ackaret, "Availability analysis of blade server systems," *IBM Systems Journal*, vol. 47, no. 4, pp. 621–640, 2008.

[26] M. Malhotra and K. Trivedi, "A methodology for formal expression of hierarchy in model solution," in *Proc. 5th Int. Workshop on Petri Nets and Performance Models*, Oct 1993, pp. 258–267.

[27] G. Ciardo and K. Trivedi, "A decomposition approach for stochastic reward net models," *Performance Evaluation*, vol. 18, pp. 37–59, 1993.

[28] T. H. Cormen, C. E. Leiserson, R. L. Rivest, and C. Stein, *Introduction to Algorithms*, 3rd edn. The MIT Press, 2009.

[29] S. Garg, A. Puliafito, M. Telek, and T. Trivedi, "Analysis of preventive maintenance in transactions based software systems," *IEEE Transactions on Computers,*, vol. 47, no. 1, pp. 96–107, Jan. 1998.

[30] Y. Bao, X. Sun, and K. S. Trivedi, "A workload-based analysis of software aging, and rejuvenation," *IEEE Transactions on Reliability*, vol. 54, no. 3, pp. 541–548, 2005.

[31] K. Vaidyanathan and K. Trivedi, "A comprehensive model for software rejuvenation," *IEEE Transactions on Dependable and Secure Computing*, vol. 2, no. 2, pp. 124–137, Apr. 2005.

[32] L. Tomek and K. Trivedi, "Fixed point iteration in availability modeling," in *Proc. 5th Int. GI/ITG/GMA Conference on Fault-Tolerant Computing Systems, Tests, Diagnosis, Fault Treatment*. Springer-Verlag, 1991, pp. 229–240.

[33] H. Sukhwani, A. Bobbio, and K. Trivedi, "Largeness avoidance in availability modeling using hierarchical and fixed-point iterative techniques," *International Journal of Performability Engineering*, vol. 11, no. 4, pp. 305–319, 2015.

[34] R. Matos, J. Araujo, D. Oliveira, P. Maciel, and K. Trivedi, "Sensitivity analysis of a hierarchical model of mobile cloud computing," *Simulation Modelling Practice and Theory*, vol. 50, pp. 151–164, 2015.

[35] M. Malhotra and K. S. Trivedi, "Reliability analysis of redundant arrays of inexpensive disks," *Journal of Parallel and Distributed Computing*, vol. 17, pp. 146–151, 1993.

[36] K. Trivedi and R. Geist, "Decomposition in reliability analysis of fault-tolerant systems," *IEEE Transactions on Reliability*, vol. R-32, no. 5, pp. 463–468, Dec. 1983.

[37] F. Machida, R. Xia, and K. S. Trivedi, "Performability modeling for raid storage systems by Markov regenerative process," *IEEE Transactions on Dependable and Secure Computing*, vol. 67, 2016.

[38] J. G. Elerath and J. Schindler, "Beyond MTTDL: A closed-form RAID-6 reliability equation," *ACM Transactions on Storage (TOS)*, vol. 10, no. 2, p. 7, 2014.

[39] I. Iliadis and V. Venkatesan, "Rebuttal to 'Beyond MTTDL: A closed-form RAID-6 reliability equation,'" *ACM Transactions on Storage (TOS)*, vol. 11, no. 2, pp. 9:1–9:10, Mar. 2015.

[40] A. Thomasian and Y. Tang, "Performance, reliability, and performability of a hybrid RAID array and a comparison with traditional RAID1 arrays," *Cluster Computing*, vol. 15, no. 3, pp. 239–253, 2012.

[41] K. Hjelmgren, S. Svensson, and O. Hannius, "Reliability analysis of a single-engine aircraft FADEC," in *Proc. Ann. Reliability and Maintainability Symposium*, 1998, pp. 401–407.

[42] L. Xing, "Reliability evaluation of phased-mission systems with imperfect fault coverage and common-cause failures," *IEEE Transactions on Reliability*, vol. 56, no. 1, pp. 58–68, 2007.

[43] Y. Ma and K. Trivedi, "An algorithm for reliability analysis of phased-mission systems," *Reliability Engineering and System Safety*, vol. 66, pp. 157–170, 1999.

[44] X. Zang, H. Sun, and K. Trivedi, "A BDD-based algorithm for reliability analysis of phased-mission systems," *IEEE Transactions on Reliability*, vol. 48, pp. 50–60, 1999.

[45] J. McGough, A. Reibman, and K. Trivedi, "Markov reliability models for digital flight control systems," *AIAA Journal of Guidance, Control, and Dynamics*, vol. 12, no. 2, pp. 209–219, 1989.

[46] B. Çekyay and S. Özekici, "Performance measures for systems with Markovian missions and aging," *IEEE Transactions on Reliability*, vol. 61, no. 3, pp. 769–778, 2012.

[47] M. Smotherman and K. Zemoudeh, "A non-homogeneous Markov model for phased-mission reliability analysis," *IEEE Transactions on Reliability*, vol. 38, no. 5, pp. 585–590, 1989.

[48] I. Mura, A. Bondavalli, X. Zang, and K. S. Trivedi, "Dependability modeling and evaluation of phased mission systems: A DSPN approach," in *IEEE DCCA-7, IFIP Int. Conf. on Dependable Computing for Critical Applications*, San Jose, CA, USA, Jan. 6–8 1999, pp. 319–337.

[49] J. Bechta Dugan, and K. Trivedi, "Coverage modeling for dependability analysis of fault-tolerant systems," *IEEE Transactions on Computers*, vol. 38, no. 6, pp. 775–787, Jun. 1989.

[50] J. McGough, M. Smotherman, and K. Trivedi, "The conservativeness of reliability estimates based on instantaneous coverage," *IEEE Transactions on Computers*, vol. C-34, pp. 602–609, 1985.

[51] W. Lee, D. Grosh, F. Tillman, and C. Lie, "Fault tree analysis, methods and applications: A review," *IEEE Transactions on Reliability*, vol. R-34, pp. 194–203, 1985.

[52] M. Stamatelatos and W. Vesely, *Fault Tree Handbook with Aerospace Applications*. NASA Office of Safety and Mission Assurance, 2002, vol. 1.1.

[53] J. B. Dugan, S. Bavuso, and M. Boyd, "Dynamic fault-tree models for fault-tolerant computer systems," *IEEE Transactions on Reliability*, vol. 41, pp. 363–377, 1992.

[54] S. Junges, D. Guck, J. Katoen, and M.Stoelinga, "Uncovering dynamic fault trees," in *Proc. Int. Conf. on Dependable Systems and Networks*, Jun. 2016.

[55] G. Merle, J. Roussel, J. Lesage, and A. Bobbio, "Probabilistic algebraic analysis of fault trees with priority dynamic gates and repeated events," *IEEE Transactions on Reliability*, vol. 59, no. 1, pp. 250–261, 2010.

[56] Y. Dutuit and A. Rauzy, "A linear-time algorithm to find modules of fault tree," *IEEE Transactions on Reliability*, vol. 45, pp. 422–425, 1996.

[57] R. Gulati and J. Dugan, "A modular approach for analyzing static and dynamic fault-trees," in *Proc/ IEEE Ann. Reliability and Maintainability Symposium*, 1997, pp. 57–63.

[58] J. B. Dugan, K. Sullivan, and D. Coppit, "Developing a low-cost high-quality software tool for dynamic fault-tree analysis," *IEEE Transactions on Reliability*, vol. 49, no. 1, pp. 49–59, 2000.

[59] A. Bobbio and D. Codetta-Raiteri, "Parametric fault-trees with dynamic gates and repair boxes," in *Proc. Ann. Reliability and Maintainability Symp.*. IEEE, Jan. 2004, pp. 459–465.

[60] D. Codetta-Raiteri, "The conversion of dynamic fault trees to stochastic Petri nets, as a case of graph transformation," *Electronic Notes on Theoretical Computer Science*, vol. 127, pp. 45–60, 2005.

[61] D. Codetta-Raiteri, "Integrating several formalisms in order to increase fault trees' modeling power," *Reliability Engineering and System Safety*, vol. 96, no. 5, pp. 534–544, 2011.

[62] L. Xing, O. Tannous, and J. B. Dugan, "Reliability analysis of nonrepairable cold-standby systems using sequential binary decision diagrams," *IEEE Transactions on Systems, Man and Cybernetics, Part A: Systems and Humans*, vol. 42, no. 3, pp. 715–726, May 2012.

[63] D. Ge, D. Li, Q. Chou, R. Zhang, and Y. Yang, "Quantification of highly coupled dynamic fault tree using IRVPM and SBDD," *Quality and Reliability Engineering International*, 2014.

[64] H. Boudali and J. B. Dugan, "A new Bayesian network approach to solve dynamic fault trees," in *Proc. Ann. Reliability and Maintainability Symposium*, Jan. 2005, pp. 451–456.

[65] S. Montani, L. Portinale, A. Bobbio, and D. Codetta-Raiteri, "Radyban: A tool for reliability analysis of dynamic fault trees through conversion into dynamic Bayesian networks," *Reliability Engineering and System Safety*, vol. 93, pp. 922–932, 2008.

[66] L. Portinale and D. Codetta-Raiteri, *Modeling and Analysis of Dependable Systems: A Probabilistic Graphical Model Perspective*. World Scientific, 2015.

[67] K. Murphy, "Dynamic Bayesian networks: Representation, inference and learning," Ph.d. Thesis, University of California, Berkeley, 2002.

[68] S. Montani, L. Portinale, and A. Bobbio, "Dynamic Bayesian networks for modeling advanced fault tree features in dependability analysis," in *Proc. 16th European Conf. on Safety and Reliability, Leiden, The Netherlands, AA Balkema*, 2005, pp. 1415–1422.

[69] A. Bobbio, D. Codetta-Raiteri, L. Portinale, A. Guiotto, and Y. Yushtein, "A unified modelling and operational framework for fault detection, identification, and recovery in autonomous spacecrafts," in *Theory and Application of Multi-Formalism Modeling*, eds. M. Gribaudo and M. Iacono. IGI-Global, 2014, ch. 11, pp. 239–258.

[70] M. Aleksy, M. Schader, and A. Schnell, "Design and implementation of a bridge between CORBA's notification service and the Java message service," in *Proc. 36th Ann. Int. Conf. on System Sciences*, 2003.

[71] S. Ramani, K. Trivedi, and B. Dasarathy, "Performance analysis of the CORBA notification service," in *Proc. 20th Symp. on Reliable Distributed Systems*, 2001, pp. 227–236.

[72] R. Wolff, "Poisson arrivals see time averages," *Operations Research*, vol. 30, no. 2, pp. 223–231, 1982.

[73] K. Trivedi, *Probability and with Reliability, Queueing and Computer Science Applications*, 2nd edn. John Wiley & Sons, 2001.

[74] K. Trivedi, J. Muppala, S. Woolet, and B. Haverkort, "Composite performance and dependability analysis," *Performance Evaluation*, vol. 14, pp. 197–215, 1992.

[75] S. Woolet, "Performance analysis of computer networks," Ph.d. Thesis, Department of Computer Science, Duke University, 1993.

[76] B. Iyer, L. Donatiello, and P. Heidelberger, "Analysis of performability for stochastic models of fault-tolerant systems," *IEEE Transactions on Computers*, vol. C-35, pp. 902–907, 1986.

[77] P. Chimento and K. Trivedi, "The completion time of programs on processors subject to failure and repair," *IEEE Transactions on Computers*, vol. 42, pp. 1184–1194, 1993.

[78] V. Kulkarni, V. Nicola, and K. Trivedi, "The completion time of a job on a multi-mode system," *Advances in Applied Probability*, vol. 19, pp. 932–954, 1987.

[79] S. Ramani, K. Goseva-Popstojanova, and K. S. Trivedi, "A framework for performability modeling of messaging services in distributed systems," in *Proc. 8th IEEE Int. Conf. on Engineering of Complex Computer Systems*, Dec. 2002, pp. 25–34.

[80] S. Ramani, K. S. Trivedi, and B. Dasarathy, "Reliable messaging using the CORBA notification service," in *3rd Int. Symp. on Distributed Objects and Applications*, 2001, pp. 229–238.

[81] R.Pinciroli, K. Trivedi, and A. Bobbio, "Parametric sensitivity and uncertainty propagation in dependability models," in *Proc. 10th Int. Conf. on Performance Evaluation Methodologies and Tools (Valuetools 2016)*, 2016.

[82] M. Beaudry, "Performance-related reliability measures for computing systems," *IEEE Transactions on Computers*, vol. C-27, pp. 540–547, 1978.

[83] G. Ciardo, R. Marie, B. Sericola, and K. Trivedi, "Performability analysis using semi-Markov reward processes," *IEEE Transactions on Computers*, vol. 39, no. 10, pp. 1252–1264, 1990.

[84] T1A1.2 Working Group on Network Survivability Performance, *Technical Report on Enhanced Network Survivability Performance*. Standards Committee T1 Telecommunications, Feb. 2001.

[85] J. Laprie, "From dependability to resilience," in *Proc. Dependable Systems and Networks*, 2008.

[86] L. Simoncini, "Resilient computing: An engineering discipline," in *Proc. IEEE Int. Symp. on Parallel Distributed Processing*, May 2009.

[87] R. Ghosh, D. Kim, and K. Trivedi, "System resiliency quantification using non-state-space and state-space analytic models," *Reliability Engineering and System Safety*, vol. 116, pp. 109–125, 2013.

[88] Y. Liu and K. S. Trivedi, "A general framework for network survivability quantification," *Proc. 12th GI/ITG Conf. on Measuring, Modelling and Evaluation of Computer and Communication Systems*, 2004, pp. 369–378.

[89] Y. Liu, V. Mendiratta, and K. Trivedi, "Survivability analysis of telephone access network," in *Proc. 15th Int. Symp. on Software Reliability Engineering*, 2004, pp. 367–378.

[90] Y. Liu and K. Trivedi, "Survivability quantification: The analytical modeling approach," *International Journal of Performability Engineering*, vol. 2, no. 1, p. 29, 2006.

[91] P. Heegaard and K. Trivedi, "Survivability quantification of communication services," in *Proc. 38th Ann. IEEE/IFIP Int. Conf. on Dependable Systems and Networks*, 2008, pp. 462–471.

[92] D. Chen, S. Garg, and K. Trivedi, "Network survivability performance evaluation: A quantitative approach with applications in wireless ad-hoc networks," in *Proc. 5th ACM Int. Workshop on Modeling Analysis and Simulation of Wireless and Mobile Systems*, ACM, 2002, pp. 61–68.

[93] O. C. Ibe, R. C. Howe, and K. S. Trivedi, "Approximate availability analysis of VAXcluster systems," *IEEE Transactions on Reliability*, vol. 38, no. 1, pp. 146–152, 1989.

[94] M. Lanus, Y. Liang, and K. Trivedi, "Hierarchical composition and aggregation of state-based availability and performability models," *IEEE Transactions on Reliability*, vol. 52, pp. 44–52, 2003.

[95] F. Lee and M. Marathe, *Beyond Redundancy: A Guide to Designing High-Availability Networks*. Cisco, 1999.

[96] R. de S. Matos, J. Araujo, D. Oliveira, P. R. M. Maciel, and K. S. Trivedi, "Sensitivity analysis of a hierarchical model of mobile cloud computing," *Simulation Modelling Practice and Theory*, vol. 50, pp. 151–164, 2015.

[97] D. Tang and K. Trivedi, "Hierarchical computation of interval availability and related metrics," in *Proc. 2004 Int. Conf. on Dependable Systems and Networks*, 2004, p. 693.

[98] J. Blake and K. Trivedi, "Reliability analysis of interconnection networks using hierarchical composition," *IEEE Transactions on Reliability*, vol. 38, no. 1, pp. 111–120, 1989.

[99] J. K. Muppala, K. S. Trivedi, V. Mainkar, and V. G. Kulkarni, "Numerical computation of response time distributions using stochastic reward nets," *Annals of Operational Research*, vol. 48, pp. 155–184, 1994.

[100] H. Qian, D. Medhi, and K. S. Trivedi, "A hierarchical model to evaluate quality of experience of online services hosted by cloud computing," in *Proc. 12th IFIP/IEEE Int. Symp. on Integrated Network Management*, 2011, pp. 105–112.

[101] J. T. Blake, A. L. Reibman, and K. S. Trivedi, "Sensitivity analysis of reliability and performability measures for multiprocessor systems," *SIGMETRICS Performance Evaluation Review*, vol. 16, no. 1, pp. 177–186, May 1988.

[102] S. Garg, Y. Huang, C. Kintala, S. Yajnik, and K. Trivedi, "Performance and reliability evaluation of passive replication schemes in application level fault tolerance," in *Proc. 29th Ann. Int. Symp. on Fault-Tolerant Computing*, 1999, pp. 322–329.

[103] R. Ghosh, K. S. Trivedi, V. Naik, and D. S. Kim, "End-to-end performability analysis for infrastructure-as-a-service cloud: An interacting stochastic models approach," in *Proc. IEEE Pacific Rim Int. Symp. on Dependable Computing*, Dec. 2010, pp. 125–132.

[104] B. R. Haverkort, R. Marie, G. Rubino, and K. S. Trivedi, eds., *Performability Modelling: Techniques and Tools*. John Wiley & Sons, 2001.

[105] W. Sanders and J. Meyer, "A unified approach for specifying measures of performance, dependability and performability," in *Dependable Computing for Critical Applications*, eds. A. Avižienis and J.-C. Laprie. Springer, 1991, vol. 4, pp. 215–237.

[106] K. S. Trivedi, S. Ramani, and R. M. Fricks, "Recent advances in modeling response-time distributions in real-time systems," *Proceedings of the IEEE*, vol. 91, no. 7, pp. 1023–1037, 2003.

[107] A. Avritzer, S. Suresh, D. S. Menasché, R. M. M. Leão, E. de Souza e Silva, M. C. Diniz, K. S. Trivedi, L. Happe, and A. Koziolek, "Survivability models for the assessment of smart grid distribution automation network designs," in *Proc. 4th ACM/SPEC Int. Conf. on Performance Engineering*. ACM, 2013, pp. 241–252.

[108] A. Koziolek, A. Avritzer, S. Suresh, D. S. Menasché, M. C. Diniz, E. de Souza e Silva, R. M. M. Leão, K. S. Trivedi, and L. Happe, "Assessing survivability to support power grid investment decisions," *Reliability Engineering and System Safety*, vol. 155, pp. 30–43, 2016.

[109] A. Koziolek, A. Avritzer, S. Suresh, D. Sadoc Menasche, K. Trivedi, and L. Happe, "Design of distribution automation networks using survivability modeling and power flow equations," in *Proc. IEEE 24th Int. Symp. on Software Reliability Engineering*, Nov. 2013, pp. 41–50.

[110] M. Beccuti, S. Chiaradonna, F. D. Giandomenico, S. Donatelli, G. Dondossola, and G. Franceschinis, "Quantification of dependencies between electrical and information infrastructures," *International Journal of Critical Infrastructure Protection*, vol. 5, no. 1, pp. 14–27, 2012.

17 Fixed-Point Iteration

In Chapter 16 we examined the development of hierarchical models of systems. The import graph as a formalism to represent the model hierarchy was explained. In the previous chapter we primarily considered examples where the import graph was acyclic. In cases where the import graph is non-acyclic, the solution of the model requires repeated solution of the submodels in an iterative manner. This will be examined further in this chapter.

The rest of this chapter is organized as follows. We give a general introduction to fixed-point iteration in Section 17.1. In the subsequent sections we present several examples to illustrate the power of this approach. The set of examples is divided into availability (Section 17.2), reliability (Section 17.3), and performance models (Section 17.4).

17.1 Introduction

Hierarchical models where the import graph is acyclic are straightforward from a solution perspective. We carry out a topological sort of the import graph and then we are able to solve the models in such a way that the required information from other models is always available. But when the import graph has cycles, overall model solution poses some interesting issues. We need to resort to an iterative approach in this case. A typical fixed-point equation can be expressed in general as follows:

$$x = G(x), \tag{17.1}$$

where $x = (x_1, \ldots, x_n)$ is the vector of iteration variables. A standard approach to finding a solution to the above fixed-point equation is to use successive substitution, where we start with an initial guess x_0 and then iterate the equation as follows:

$$x_n = G(x_{n-1}).$$

The iteration terminates when the difference between two successive iterates is below a threshold value. The following subsections highlight some aspects of the iterative solution approach.

17.1.1   Revisiting the Import Graph

We have already seen that if parameters are passed from submodel M_i to submodel M_j, then we write $M_i \succ M_j$. This yields the *import graph* G [1, 2]. Let the transitive closure of relation $\succ$ be denoted by $\succ\succ$. If G is non-acyclic, this implies that there exists at least one pair of submodels M_i and M_j such that $M_i \succ\succ M_j$ and $M_j \succ\succ M_i$. A typical approach to solving such an iteration is to break the cyclicity among the submodels. One of the submodels, say M_i, is chosen and some starting values (initial guesses) for the input parameters of this submodel are supplied. Using these starting values, the submodel is solved and its results are used to solve the rest of the submodels in the cycle. The outputs of these submodels determine the next input parameters for submodel M_j. The cycle is repeated, and it is said to constitute an *iteration*. Iterations are continued until convergence is achieved. Such a solution method based on successive substitution and iteration has been used successfully by many researchers [3–10].

Existence, Uniqueness, and Convergence in Fixed-Point Iteration

The difficult issues with the use of fixed-point iteration for the solution of multi-level models is the theoretical proof of the existence, uniqueness, and convergence of the fixed-point iteration equations [11]. Under certain conditions it is possible to establish sufficient conditions for the existence of the solution. The uniqueness of the solution is more difficult to prove [12]. Other questions regarding the rate of convergence and the additional error introduced by the iteration need further research.

Based on the following theorem, proof of existence has been shown by many authors [4–6, 8, 9, 12, 13].

Theorem 17.1 *Brouwer's Fixed-Point Theorem [14]*
Let $G : S \subset \mathbb{R}^N \to \mathbb{R}^N$ be continuous on the compact, convex set S, and suppose that $G(S) \subseteq S$, where $G(S)$ stands for $\cup_{x \in S}\{G(x)\}$. Then, G has a fixed point in S.

17.2   Availability Models

The fixed-point iteration methodology is illustrated through the example of a fluid level controller that has been considered in the previous chapters.

Example 17.1 *A Redundant Repairable Fluid Level Controller with Travel Time: Approximate Solution*
In Example 9.11 we considered various levels of redundancy of the fluid level controller units and explicit travel time associated with the repair person. The generation of the state space was considered in Example 12.19, where we showed the corresponding SRN model.

Now we derive an approximate solution by considering the subsystems as *nearly independent*. Compared to Figure 9.15 of Example 9.11 and Example 16.3, the effect

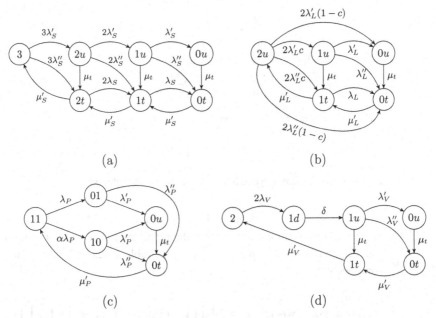

Figure 17.1   Continuous-time Markov chains of the individual units including shared repair and travel time: (a) sensor; (b) control logic; (c) pump; (d) valve.

of sharing the travel time is more complex, as shown in the individual CTMCs of Figure 17.1.

The states are labeled with a letter t at the end to indicate that the repair person is on site repairing a component, while a letter u is used to indicate that the repair person is not on site. When a component failure occurs in subsystem M_i, there are two situations to consider:

- If the failure does not trigger a repair action or the failure occurs in states labeled with a t, the repair person is on site and actually repairing a component in subsystem M_i; hence, the failure transition under consideration will emanate at the same rate as in Figure 9.15.
- If the failure occurs in any other state, the failure transition is split into two transitions. If the repair person is not on site for any of the subsystems, they must be notified and hence a transition occurs to a state marked with a u. The failure rates in this case are denoted by λ'. If the repair person is already on site in some subsystem $M_j \neq M_i$ then they do not need to be notified. The failure rates in this case are denoted by λ'' and the failure transition is directed to a state labeled with a t.

For each subsystem k ($k = $ P, V, S, L), the probability that the repair person is on site repairing subsystem k is given by

$$r^{(k)} = \sum_{i \in *t} \pi_i^{(k)},$$

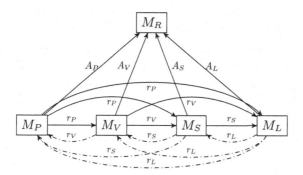

Figure 17.2   Import graph showing a cyclic interaction between the submodels.

where the symbol $\in *t$ denotes the subset of states with label t. The subset of states $*u$ of the subsystem k indicates that the repair person is not working on subsystem k. The failure transitions emanating from repairable states with label u are split into two components:

$$\lambda = \lambda' + \lambda'' \quad \text{where} \quad \lambda' = \lambda \prod_{j \neq k} \left(1 - r^{(j)} \right) \quad \text{and} \quad \lambda'' = \lambda \left[1 - \prod_{j \neq k} \left(1 - r^{(j)} \right) \right].$$

(17.2)

The term $\prod_{j \neq k}(1 - r^{(j)})$ in Eq. (17.2) is the probability that the repair person is not on site repairing any subsystem, and hence the subsystem k must notify the repair person and account for the travel time. Thus, for evaluating the solution for each submodel, there is a cyclic dependency with the solution from the other submodels, as shown in the import graph in Figure 17.2. Here, the solid arrows indicate the transfer of parameter values in the current iteration, and the dotted arrows indicate the parameter values from the previous iteration. Using Brouwer's fixed-point theorem, the existence of a solution to fixed-point iteration is proven in [5, 6, 8, 9, 11, 13].

We use the method of successive substitutions to solve for the fixed point. In our experience, a small number of iterations is usually enough for convergence.

The model for the four subsystems has cyclic dependencies for the failure rates, as provided in Eq. (17.2), and acyclic dependencies for the repair rates (as given in Table 17.1). The analysis (including the fixed-point iteration) is performed using SHARPE, and the results are reported in Table 17.1.

The system steady-state availability is $A_{Sys} = 0.997\,255\,235$. From the two rightmost columns in Table 17.1, it is interesting to note that the availability for each subsystem improves in the presence of shared repair with respect to the independent case.

The results for the system availability are compared in Table 17.2. This was unlike the case in Table 16.1 in Example 16.3, where shared repair slightly reduced the availability, especially for lower-repair-priority subsystems. The reason is that when the repair person arrives on site to repair a subsystem, they would take care of the repairs for failures in other subsystems that are already **down** or whose failure occurs while the

Table 17.1 Nearly independent approximation for shared repair with travel time.

Subsys.	Up state A_{Sys}	Repair state q	Updated repair rate μ'	Indep. repair availability $A_{indep. repair}$	Shared repair availability $A_{shared repair}$
P	$\pi_{11} + \pi_{01} + \pi_{10}$	π_{00t}	μ_P	0.997 7773	0.997 7784
V	$\pi_2 + \pi_{1n} + \pi_{1t}$	$\pi_{1t} + \pi_{0t}$	$\mu_V(1 - Q_P)$	0.999 7989	0.999 7989
S	$\pi_3 + \pi_{2n} + \pi_{2t}$	$\pi_{2t} + \pi_{1t} + \pi_{0t}$	$\mu_S(1 - Q_P)$ $(1 - Q_V)$	0.999 9799	0.999 9798
L	$\pi_2 + \pi_{1n} + \pi_{1t}$	$\pi_{1t} + \pi_{0t}$	$\mu_L(1 - Q_P)$ $(1 - Q_V)(1 - Q_S)$	0.999 6976	0.999 6967

Table 17.2 Comparison of availabilities of different system configurations and repair scenarios.

Configuration	Indep. repair	Shared repair (exact)	Shared repair (approx)	# digits match
$\mu_t = 0.1$	0.997 255 093	0.997 431 402 583	0.997 255 235	3

first repair is in process, so that the travel time is "paid for" only once for possible multiple repair actions.

Problems

17.1 With the numerical values given in Table 9.1, numerically verify the solution to Example 17.1 with SHARPE or a similar tool.

17.2 Redo Example 17.1 using stochastic reward nets instead of directly using CTMCs. Numerically solve the model using SHARPE or SPNP.

17.3   Reliability Models

In the literature, we find a large number of case studies on the use of fixed-point iteration in availability and in performance models. However, the application of this approach to reliability and performability models is infrequent. The following example, an extension of the fluid level controller problem that we have presented in this book from several angles, examines the use of fixed-point iteration in a reliability model.

Example 17.2 *A Redundant Repairable Fluid Level Controller Reliability Model with Travel Time: Approximate Solution*
We return to Example 17.1, but consider reliability instead of availability. We start by making all failure states absorbing in each CTMC submodel of Figure 17.1. The other key difference is that since we will need to consider transient analysis of all the

submodels, the variable $r^{(k)}(t)$ becomes time dependent:

$$r^{(k)}(t) = \sum_{i \in *t} \pi_i^{(k)}(t).$$

Thence,

$$\lambda'(t) = \lambda \prod_{j \neq k} \left(1 - r^{(j)}(t)\right) \quad \text{and} \quad \lambda''(t) = \lambda \left[1 - \prod_{j \neq k} \left(1 - r^{(j)}(t)\right)\right]. \tag{17.3}$$

As a result, all the submodels become NHCTMCs, even though all event times are exponentially distributed. Note that the monolithic model in this case is still an HCTMC but because the parameters to be transferred are time dependent and these are part of the transition rates of the receiving submodel, we have a resultant NHCTMC. Recall from Chapter 13 that the solution of each NHCTMC submodel can be carried out by using PCA coupled with the HCTMC solution method. Such iterative solutions of individual CTMCs will need to be intertwined with the fixed-point iteration among the four submodels.

Problems

17.3 Numerically carry out the solution to Example 17.2 using SHARPE or a similar tool.

17.4 Redo Example 17.2 using stochastic reward nets instead of directly using CTMCs. Numerically solve the model using SHARPE.

17.5 Numerically compare the results of a monolithic reliability model (which will be an HCTMC that can be either generated by hand or automatically starting from an SRN) with those obtained by the fixed-point iterative method above.

17.4   Performance Models

In the example given in this section, we discuss a fixed-point iterative performance model of a wireless handoff protocol in a cellular wireless system.

Example 17.3   *Wireless Handoff Performance Model*
This example, adapted from [15], models the performance of a single cell in a wireless cellular network. New calls originate in the cell at the rate of λ_n, and handoff calls from adjacent cells arrive at the rate λ_h. Sometimes a handoff call to a new base station system (BSS) suffers from poor signal quality. In the model, we assume that a handoff call receives good (poor) signal quality with probability α $(1 - \alpha)$, respectively. Furthermore, the dwell time of a call in a BSS (both new and a good handoff) is assumed to be exponentially distributed with rate μ. This rate is the sum of the rate of call

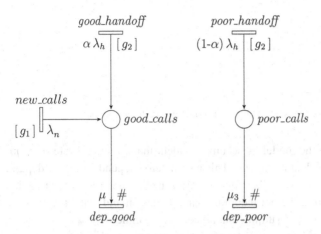

Figure 17.3   Stochastic reward net model, M_{SRN}, of wireless handoff in a cellular network.

completion μ_1 and handoff out rate μ_2 to neighboring cells. A poor handoff call will dwell in a BSS for an exponentially distributed time with rate μ_3. We further assume that c is the number of channels, and $g < c$ is the number of guard channels reserved for handoff calls only. Hence, any new incoming call will be blocked when the number of busy channels exceeds $(c - g)$. The authors in [15] developed models for several algorithms that use a combination of guard channels (GC) and mobile-assisted handoff (MAHO) options. Poor-quality handoff calls are dealt with by either dropping (Drop) or rehanding off (ReHo) to another cell. The three combinations considered in the paper include GC+Drop, GC+ReHo, and GC+MAHO.

We now present in Figure 17.3 a generic SRN model, M_{SRN}, for the wireless handoff system dealing with both good and poor-quality handoff calls. In the SRN, the meaning of the transitions are self-explanatory, but we now clarify them further. Transition new_-calls represents the arrival of new calls to the cell, while transitions good_handoff and poor_handoff respectively represent a good (poor) call handoff. Transitions dep_good and dep_poor represent the departure of good calls and poor calls, respectively. Places good_calls (poor_calls) respectively keep track of the current good and poor-quality calls in the BSS.

In this SRN, we define a guard function $[g_1] = ((\#(\text{good\_calls}) + \#(\text{poor\_calls})) < c - g)$ associated with the arrival transition new_calls. Thus, the arrival of new calls is disabled when the total number of calls in the system equals $(c - g)$. As is evident, the customers could be in the good_calls or poor_calls places, so that the total number of ongoing calls in the cell is given by the sum of tokens in places good_calls and poor_-calls. Similarly, we define a guard function $[g_2] = ((\#(\text{good\_calls}) + \#(\text{poor\_calls})) < c)$ associated with the handoff transitions good_handoff and poor_handoff, respectively. Thus, the arrival of handoff calls is disabled when the total number of calls in the system equals c.

The SRN model in its entirety represents the model for the GC+ReHo option, while completely disabling the transition poor_handoff yields the model for the GC+Drop option. In the latter case, allowing $\alpha \to 1$ yields the model for the GC+MAHO option.

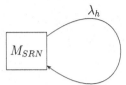

Figure 17.4   Import graph for the wireless handoff model in a cellular network.

Examining the model, it becomes evident that the arrival rate of handoff calls λ_h is not an independent parameter. This parameter is dependent on the departure of handoff calls from the other cells. This sets up a fixed-point problem, in that, in a multicell cellular network, the net incoming handoff rate should equal the net outgoing handoff rate from any cell, assuming that all cells are symmetric.

This relationship in the SRN model can be expressed as follows:

$$\lambda_h = \frac{\mu_2}{\mu} Throughput(\text{dep\_good}) + Throughput(\text{dep\_poor}).$$

In the above equation, the throughputs of the transitions can be computed by assigning a reward rate equal to the rates of the respective transitions and computing the expected steady-state reward. The import graph in this case will have a single node with a self-loop, as shown in Figure 17.4. Obtaining a closed-form solution for the model is not feasible. However, we can set up the fixed-point iterative solution using either SHARPE or SPNP.

Example 17.4 *Performance Analysis of Device-to-Device Communications with Dynamic Interference [7]*
This example, derived from [7], studies the performance of device-to-device (D2D) communications with dynamic interference. In particular, the authors of [7] analyze the performance of frequency reuse among D2D links with dynamic data arrival setting. A D2D link consists of a source D2D user equipment (UE) transmitting to its destination D2D UE. This transmission could experience potential interference from other overlapping D2D links. In their model, each source D2D UE maintains a queue with finite capacity to buffer the dynamically arriving data. An interfering link is potential, since it only exists when the queue of the source D2D UE associated with its transmitter is non-empty. The transmission rate of each link is determined by its own signal to interference and noise ratio, which varies with time due to the fast fading effects of its own wireless channel, and also the fast fading effects and changing-off status of its potential interfering link. Moreover, the transmission rates of different links vary asynchronously over time. The system can be modeled as an interfering queuing system where there are multiple queues, each one serviced by a single server, whose service rate is determined by the transmission rate of the corresponding link.

Since the corresponding state-space-based analysis suffers from the state space explosion problem, the authors separate the model into two submodels, and use the model decomposition and iteration procedure derived from [1] to simplify the analysis.

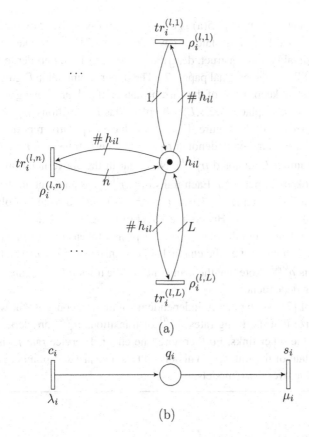

Figure 17.5   Stochastic reward net model of D2D communications with dynamic interference.

The first submodel keeps track of the channel state, and the second submodel keeps track of the queuing behavior. The authors use a discrete stochastic Petri net (DSPN) model [16] to represent the queuing behavior, with the service times being deterministic. Here, we present an SRN model of the system with the simplifying assumption that the service times are exponentially distributed with rate μ_i, which varies according to the state of the specific link. The DSPN can be approximated by either Erlangian phases (see Chapter 15) or by doing a transient analysis of the subordinated CTMC for the duration of the deterministic transition [16–18], if required. Yet another alternative is to carry out a discrete-event simulative solution of the DSPN via the SPNP package [19].

The system has K D2D links, and each link is represented by the two SRN submodels as shown in Figure 17.5, where i denotes a generic link.

The SRN submodel of Figure 17.5(b) models the queuing behavior of a generic link i. Place q_i keeps track of the number of messages at link i. The firing time of transition c_i is exponentially distributed with rate λ_i. When it fires, one packet arrives at the queue place q_i. Similarly, transition s_i represents the service time for packets at the rate μ_i. When it fires, one packet leaves the queue place q_i.

The channel state in Figure 17.5(a) is tracked by an L-state Markov model in which the system can move from any state to any other state. The SRN takes advantage of the variable cardinality arc construct, described in Section 12.3, in order to simplify the structure of the SPN in the original paper [7]. The upper submodel in Figure 17.5(a) has a single place h_{il} that keeps track of the current state of the channel using an appropriate number of tokens in the place $(1,\ldots,L)$. The place has L transitions, $tr_i^{(l,n)}, 1 \leq l, n \leq L$ that are connected to it. In Figure 17.5(a) we have explicitly reported only three transitions; transition number 1 (denoted $tr_i^{(l,1)}$), a generic transition n (denoted $tr_i^{(l,n)}$) and transition number L (denoted $tr_i^{(l,L)}$). Note that in the above notation $l = \#(h_{il})$ is the number of tokens in place h_{il}. Each transition $tr_i^{(l,n)}$ has an input arc from place h_{il} with variable cardinality equal to $\#(h_{il})$, and an output arc directed to place h_{il} with variable cardinality equal to n. Hence, the effect of firing transition $tr_i^{(l,n)}$ is to remove all the $l = \#(h_{il})$ tokens from the place h_{il} and deposit n tokens in the same place. When $tr_i^{(l,n)}$ fires, the token count, i.e., the channel state, transits from l to n. The firing rate of transition $tr_i^{(l,n)}$ is $\rho_i^{(l,n)}$. Note that the firing rates of the transitions are functions of l and thus are marking dependent.

The authors of [7] use the nearly independent submodels concept that we examined earlier. We realize that the firing rates $\rho_i^{(l,n)}$ of transitions $tr_i^{(l,n)}$ are dependent on the queue states of the other links. Furthermore, the channel service rate μ_i is dependent on the current state of the link $\#h_{il}$. This introduces a clear fixed-point equation that is solved by iterating over these models.

Problems

17.6 Numerically carry out the solution to Example 17.3 using SHARPE, SPNP or a similar tool.

17.7 Numerically solve Example 17.4 in two different ways:

(a) using our SRN model shown in Example 17.4 with parameter values from [7];
(b) after modifying the SRN to expand the service transition s_i into k Erlang stages (vary k over $1, 5, 10$).

17.5   Further Reading

For further examples of fixed-point iteration in performance models, see [3, 6, 11, 15, 20–24]. Further examples of fixed-point iteration in availability modeling can be found in [9, 11].

References

[1] G. Ciardo and K. Trivedi, "A decomposition approach for stochastic reward net models," *Performance Evaluation*, vol. 18, pp. 37–59, 1993.

[2] M. Malhotra and K. Trivedi, "A methodology for formal expression of hierarchy in model solution," in *Proc. 5th Int. Workshop on Petri Nets and Performance Models*, Oct. 1993, pp. 258–267.

[3] J. S. Baras, V. Tabatabaee, G. Papageorgiou, and N. Rentz, "Modelling and optimization for multi-hop wireless networks using fixed point and automatic differentiation," in *Proc. 6th Int. Symp. on Modeling and Optimization in Mobile, Ad Hoc, and Wireless Networks*, Apr. 2008, pp. 295–300.

[4] X. Ma, X. Yin, and K. Trivedi, "On the reliability of safety applications in VANETs," *International Journal of Performability Engineering*, vol. 8, no. 2, pp. 115–130, Mar. 2012.

[5] X. Yin, X. Ma, and K. Trivedi, "An interacting stochastic models approach for the performance evaluation of DSRC vehicular safety communication," *IEEE Transactions on Computers*, vol. 62, no. 5, pp. 873–885, May 2013.

[6] R. Ghosh, F. Longo, V. K. Naik, and K. S. Trivedi, "Modeling and performance analysis of large scale IaaS clouds," *Future Generation Computer Systems*, vol. 29, no. 5, pp. 1216–1234, 2013.

[7] L. Lei, Y. Zhang, X. S. Shen, C. Lin, and Z. Zhong, "Performance analysis of device-to-device communications with dynamic interference using stochastic Petri nets," *IEEE Transactions on Wireless Communications*, vol. 12, no. 12, pp. 6121–6141, Dec. 2013.

[8] R. Ghosh, F. Longo, R. Xia, V. K. Naik, and K. S. Trivedi, "Stochastic model driven capacity planning for an infrastructure-as-a-service cloud," *IEEE Transactions on Services Computing*, vol. 7, no. 4, pp. 667–680, 2014.

[9] R. Ghosh, F. Longo, F. Frattini, S. Russo, and K. S. Trivedi, "Scalable analytics for IaaS cloud availability," *IEEE Transactions on Cloud Computing*, vol. 2, no. 1, pp. 57–70, 2014.

[10] G. Casale, M. Tribastone, and P. G. Harrison, "Blending randomness in closed queueing network models," *Performance Evaluation*, vol. 82, pp. 15–38, 2014.

[11] V. Mainkar and K. S. Trivedi, "Sufficient conditions for existence of a fixed point in stochastic reward net-based iterative models," *IEEE Transactions on Software Engineering*, vol. 22, no. 9, pp. 640–653, Sep. 1996.

[12] G. Haring, R. Marie, R. Puigjaner, and K. Trivedi, "Loss formulas and their application to optimization for cellular networks," *IEEE Transactions on Vehicular Technology*, pp. 654–673, 2001.

[13] L. Tomek and K. Trivedi, "Fixed point iteration in availability modeling," in *Proc. 5th Int. GI/ITG/GMA Conf. on Fault-Tolerant Computing Systems, Tests, Diagnosis, Fault Treatment*. Springer-Verlag, 1991, pp. 229–240.

[14] J. M. Ortega and W. C. Rheinboldt, *Iterative Solution of Nonlinear Equations in Several Variables*. Academic Press, NY, 1970.

[15] B. B. Madan, S. Dharmaraja, and K. S. Trivedi, "Combined guard channel and mobile-assisted handoff for cellular networks," *IEEE Transactions on Vehicular Technology*, vol. 57, no. 1, pp. 502–510, Jan. 2008.

[16] C. Lindemann, *Performance Modelling with Deterministic and Stochastic Petri Nets*. Wiley-Interscience, 1998.

[17] I. Mura, A. Bondavalli, X. Zang, and K. S. Trivedi, "Dependability modeling and evaluation of phased mission systems: A DSPN approach," in *Proc. IEEE IFIP Int. Conf. on Dependable Computing for Critical Applications*, San Jose, CA, USA, Jan. 6–8 1999, pp. 319–337.

[18] D. Wang, W. Xie, and K. S. Trivedi, "Performability analysis of clustered systems with rejuvenation under varying workload," *Performance Evaluation*, vol. 64, no. 3, pp. 247–265, 2007.

[19] C. Hirel, B. Tuffin, and K. S. Trivedi, "SPNP: Stochastic Petri Nets. Version 6," in *Proc. Int. Conf. on Computer Performance Evaluation: Modelling Techniques and Tools*, eds. B. Haverkort and H. Bohnenkamp, LNCS 1786, Springer Verlag, 2000, pp. 354–357.

[20] P. Heidelberger and K. S. Trivedi, "Queueing network models for parallel processing with asynchronous tasks," *IEEE Transactions on Computers*, vol. 31, no. 11, pp. 1099–1109, 1982.

[21] P. Heidelberger and K. S. Trivedi, "Analytic queueing models for programs with internal concurrency," *IEEE Transactions on Computers*, vol. 32, no. 1, pp. 73–82, 1983.

[22] M. Liu and J. S. Baras, "Fixed point approximation for multirate multihop loss networks with state-dependent routing," *IEEE/ACM Transactions on Networking*, vol. 12, no. 2, pp. 361–374, Apr. 2004.

[23] X. Ma and K. S. Trivedi, "Reliability and performance of general two-dimensional broadcast wireless network," *Performance Evaluation*, vol. 95, pp. 41–59, 2016.

[24] X. Yin, X. Ma, and K. S. Trivedi, "MAC and application level performance evaluation of beacon message dissemination in DSRC safety communication," *Performance Evaluation*, vol. 71, pp. 1–24, 2014.

Part VI

Case Studies

The previous chapters presented several modeling formalisms, both non-state-space and state-space based. Subsequently, in the last two chapters we examined the multi-level approach via a combination of formalisms to enable modeling of large systems. Throughout the chapters, several examples were presented to illustrate the applications of the concepts. The success of these approaches is best illustrated through the evaluation of real-life models of large-scale systems.

18 Modeling Real-Life Systems

In this chapter, we present several large-scale models to illustrate the modeling prowess of the formalisms presented earlier and to illustrate the ability to tackle modeling of real-life systems. These case studies are grouped into availability (Section 18.1), reliability (Section 18.2) and combined performance and reliability models (Section 18.3).

18.1 Availability Models

This section presents four real-world system availability models: a Cisco GSR 12000 router, a high-availability platform from Sun Microsystems, an IBM BladeCenter, and a SIP implementation on WebSphere. These models are hierarchical with acyclic import graphs. In all these real-world models, steady-state (un)availability is computed from the lower-level model and is assigned to an equivalent component at the higher level. We could easily compute the transient unavailability from the lower-level model and propagate it up the hierarchy so as to compute the transient unavailability for the system as a whole. We leave this as an exercise. Furthermore, we can also compute the derivatives of the system steady-state availability with respect to all input parameters – we also leave this as an exercise.

Example 18.1 *Availability Modeling of the Cisco GSR 12000 Router [1]*
We return to the Cisco GSR 12000 Router example for which the top-level reliability block diagram was discussed in Example 4.7. Each of the blocks in that RBD will now be modeled by a homogeneous CTMC whose output will be fed into the RBD, thus giving rise to an example of hierarchical composition. The Cisco 12000 series follows a distributed routing architecture, shown in Figure 18.1. The routing function is performed in the gigabit route processors (GRPs). The GRP is responsible for running the protocols and computing the routing tables from the network topology, and for system control and administrative functions. The packet forwarding functions are performed by individual line cards (LCs). An LC maintains a routing table updated from the GRPs. Each LC independently performs lookup for a destination address from the routing table. All LCs are connected to the switch fabric through point-to-point serial lines built on a passive backplane. The switch fabric includes switch fabric cards (SFCs) and clock

Table 18.1 Configuration of Cisco 12000 GSR.

Module	Redundancy	Outages
LC-in	No	Hardware, Software
LC-out	No	Hardware, Software
CSC	2	Hardware
SFC	3	Hardware
GRP	1 active and 1 standby	Hardware
Chassis	No	Hardware
IOS	1 active and 1 standby	Software

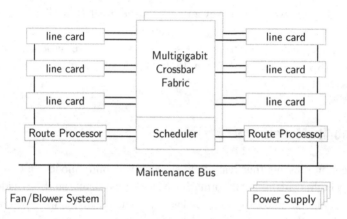

Figure 18.1   Architecture of Cisco GSR 12000 Series.

and scheduler cards (CSCs). SFCs provide multigigabit switching capacity through the multibus aggregation technique while CSCs handle requests from LCs, issue grants to access the fabric and provide the reference clock to maintain synchronization.

The high availability features in the Cisco GSR 12000 include the following:

- GRP redundancy: To avoid a single point of failure, standby redundancy is employed with one primary and one standby unit.
- CSC/SFC redundancy: The GSR 12000 requires at least one CSC. An additional CSC is recommended for reliability and performance. In general, the SFCs offer standby redundancy, that is, for every n SFCs, there will be one standby SFC.
- Power supply/cooling system redundancy: A GSR 12000 requires a minimum of two power supplies to form a redundant scheme. The same type of redundancy is used in the cooling system.
- Online insertion and removal: Online insertion and removal is supported for all cards (LC, CSC, SFC, GRP, etc.).

We assume that (i) any upgrade will cause an outage of the whole system, and (ii) the power supply is considered as part of the chassis. Table 18.1 shows the configuration of the Cisco 12000 GSR, which is considered to be available when at least four out of five CSC/SFC modules are functional.

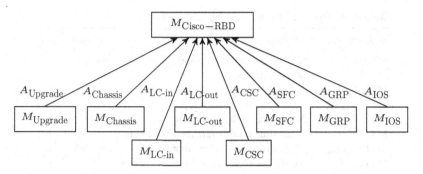

Figure 18.2   Import graph for the Cisco GSR 12000 router case study.

We use a two-level hierarchical model for the Cisco router system steady-state availability. At the higher level, we use the RBD shown in Figure 4.7, in which each block corresponds to a submodel of a module or event (upgrade). At the lower level, we use SMPs or CTMCs to model each module or event. The import graph is reported in Figure 18.2.

In the original report [1], all the submodels and the overall hierarchical model were solved numerically using SHARPE. Here we solve all the models by hand to obtain symbolic closed-form expressions for their steady-state availabilities. Furthermore, we generalize several submodels from CTMC to SMP. The parameters used in the Cisco case study are summarized in Table 18.2; they are used for illustration only and not to be taken as the true values for the real system.

The router submodels can be grouped into two-state, three-state, and multi-state cases as follows. A non-redundant module (such as Chassis or Upgrade) with only one kind of outage/recovery is modeled as the two-state SMP shown in Figure 14.1. The steady-state availability expression for this SMP is given in Eq. (14.3). From the first two rows of Table 18.2, we get MTBU $= 4320$ hr and MTTU $= 20$ min. Therefore, $A_{\text{Upgrade}} = 0.999\,922\,845$. Likewise for the Chassis module, where MTTF $= 396\,510$ hr and MTTR $= 4$ hr give $A_{\text{Chassis}} = 0.999\,989\,912$.

A non-redundant module with two kinds of outage/recovery (such as LC, having both hardware and software failures and repairs) is a special case of the SMP model shown in Figure 14.2 with the number of failure types $n = 2$. For LC-in (rows 3 to 6) we get MTTSF $= 18\,000$ hr, MTTSR $= 20$ min, MTTHF $= 111\,050$ hr, and MTTHR $= 2$ hr, and hence $A_{\text{LC-in}} = 0.999\,963\,472$. Likewise, for LC-out (rows 7 to 10), where MTTSF $= 18\,000$ hr, MTTSR $= 20$ min, MTTHF $= 103\,402$ hr, and MTTHR $= 2$ hr, we get $A_{\text{LC-out}} = 0.999\,962\,141$.

A redundant module, with n active units and one spare unit (such as the GRP and the IOS) and with only one type of outage/recovery, is modeled as the CTMC shown in Figure 18.3, where imperfect coverage is considered [2, 3]. In state 1, all hardware components are working properly. If one of the active units fails and the failure is undetected, the CTMC transits to state 13 with rate $n\lambda_3(1 - c_3)$, where λ_3 is the failure rate of the individual unit and c_3 is the coverage factor for the active unit. In state 13, repair occurs at the rate of μ_4 and the CTMC returns to state 1. While the CTMC is

Table 18.2   Assumed parameters for the Cisco 12000 GSR.

Symbol	Meaning	Value	Unit
MTBU	Mean time between upgrade	4 320	hr
MTTU	Mean time to upgrade	20	min
$\text{MTTSF}_{\text{LC-in}}$	Mean time to software failure of LC-in	18 000	hr
$\text{MTTSR}_{\text{LC-in}}$	Mean time to software repair of LC-in	20	min
$\text{MTTHF}_{\text{LC-in}}$	Mean time to hardware failure of LC-in	111 050	hr
$\text{MTTHR}_{\text{LC-in}}$	Mean time to hardware repair of LC-in	2	hr
$\text{MTTSF}_{\text{LC-out}}$	Mean time to software failure of LC-out	18 000	hr
$\text{MTTSR}_{\text{LC-out}}$	Mean time to software repair of LC-out	20	min
$\text{MTTHF}_{\text{LC-out}}$	Mean time to hardware failure of LC-out	103 402	hr
$\text{MTTHR}_{\text{LC-out}}$	Mean time to hardware repair of LC-out	2	hr
MTTF_{CSC}	Mean time to failure of CSC	272 584	hr
MTTR_{CSC}	Mean time to repair of CSC	2	hr
MTTF_{SFC}	Mean time to failure of SFC	422 115	hr
MTTR_{SFC}	Mean time to repair of SFC	2	hr
MTTF_{GRP}	Mean time to failure of GRP	108 304	hr
MTTR_{GRP}	Mean time to repair of GRP	2	hr
MTTS_{GRP}	Mean time to switchover of GRP	20	min
$C_{\text{A-GRP}}$	Coverage of GRP active failure	0.9	—
$C_{\text{S-GRP}}$	Coverage of GRP standby failure	0.6	—
$\text{MTTF}_{\text{Chassis}}$	Mean time to failure of Chassis	396 510	hr
$\text{MTTR}_{\text{Chassis}}$	Mean time to repair of Chassis	4	hr
MTTF_{IOS}	Mean time to failure of IOS	18 000	hr
MTTR_{IOS}	Mean time to repair of IOS	20	min
MTTS_{IOS}	Mean time to switchover of IOS	20	min
$C_{\text{A-IOS}}$	Coverage of IOS active failure	0.9999	—
$C_{\text{S-IOS}}$	Coverage of IOS standby failure	0.9999	—

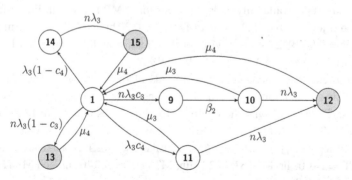

Figure 18.3   Continuous-time Markov chain for n-out-of-$n+1$ redundancy.

in state 1, if failure of the active unit is detected, the CTMC goes to state 9 with rate $n\lambda_3 c_3$. From state 9, a switchover occurs from active to standby at a rate of β_2 and the CTMC moves to state 10. In state 10, either repair of non-service-affecting failures occurs at the rate of μ_3 and the model returns to state 1, or a second unit fails with rate $n\lambda_3$ before the repair is completed and the CTMC goes to state 12. In state 1, the standby unit can fail. If such a failure is detected, the CTMC moves to state 11 with rate $\lambda_3 c_4$, where c_4 is the coverage factor for standby unit failures. If such a failure is undetected, the CTMC moves to state 14, with rate $\lambda_3(1 - c_4)$. In state 11, the standby unit can be repaired with rate μ_3 and the model returns to state 1. If the second unit fails during the repair in state 11, the model goes to state 12, with rate $n\lambda_3$. In state 12, repair of service-affecting failures occurs at the rate of μ_4 and the CTMC returns to state 1. From state 14, an attempt to switch in a standby unit with latent failure occurs with rate $n\lambda_3$, and the model goes to state 15 where repair of service-affecting failures occurs at the rate of μ_4 and the CTMC returns to state 1. States 12, 13, and 15 are down states, while all the remaining states are up states. Writing and solving the steady-state balance equations, we first obtain the state probability expressions:

$$\pi_1 = \frac{n\beta_2\mu_4\,(\mu_3 + n\lambda_3)}{E}, \qquad \pi_9 = \frac{n^2 c_3\lambda_3\mu_4\,(\mu_3 + n\lambda_3)}{E},$$

$$\pi_{10} = \frac{n^2\beta_2 c_3\lambda_3\mu_4}{E}, \qquad \pi_{11} = \frac{n\beta_2 c_4\lambda_3\mu_4}{E},$$

$$\pi_{12} = \frac{n^2\beta_2\lambda_3^2\,(c_4 + nc_3)}{E}, \qquad \pi_{13} = \frac{n^2\beta_2\lambda_3\,(1 - c_3)\,(\mu_3 + n\lambda_3)}{E},$$

$$\pi_{14} = \frac{\beta_2\mu_4\,(1 - c_4)\,(\mu_3 + n\lambda_3)}{E}, \qquad \pi_{15} = \frac{n\beta_2\lambda_3\,(1 - c_4)\,(\mu_3 + n\lambda_3)}{E}, \tag{18.1}$$

where

$$E = \beta_2\mu_3\mu_4 - \beta_2 c_4\mu_3\mu_4 + n^3 c_3\lambda_3^2\mu_4 - n^2\beta_2 c_3\lambda_3\mu_3 + n^2\beta_2 c_3\lambda_3\mu_4$$
$$+ n^2 c_3\lambda_3\mu_3\mu_4 - n\beta_2 c_4\lambda_3\mu_3 + n^3\beta_2\lambda_3^2 + n^2\beta_2\lambda_3\mu_3 + n^2\beta_2\lambda_3\mu_4 + n^2\beta_2\lambda_3^2$$
$$+ n\beta_2\lambda_3\mu_3 + n\beta_2\lambda_3\mu_4 + n\beta_2\mu_3\mu_4. \tag{18.2}$$

Then we obtain the steady-state availability expression as:

$$A_{n+1} = 1 - \frac{\beta_2 n\left(-c_4\lambda_3\mu_3 - c_3\lambda_3\mu_3 n + \lambda_3\mu_3 + \lambda_3^2 n^2 + \lambda_3\mu_3 n + \lambda_3^2 n\right)}{E}. \tag{18.3}$$

Using (18.3) for the GRP and the values in Table 18.2 (rows 15 to 17), with $n = 1$, MTTF $= 108\,304$ hr, MTTR $= 2$ hr, MTTS $= 20$ min, $c_3 = 0.9$, and $c_4 = 0.6$, we obtain $A_{\text{GRP}} = 0.999\,993\,405$.

Likewise for IOS (rows 22 to 24), where MTTF $= 18\,000$ hr, MTTR $= 20$ min, MTTS $= 20$ min, $c_3 = 0.9999$, and $c_4 = 0.9999$, we get $A_{\text{IOS}} = 0.999\,993\,405$.

The CSC/SFC module is modeled by the CTMC shown in Figure 18.4. In this figure, a pair of numbers *i-j* represents a state in which *i* CSCs and *j* SFCs are up. State 2-3 represents the initial configuration with all the CSC/SFC cards up. These five modules operate as a four-out-of-five system, so any state with less than four operational units is a down state. The failure and repair rates of a CSC are denoted by λ_1 and μ_1, respectively; the failure and repair rates of an SFC are denoted by λ_2 and μ_2, respectively. Solving the

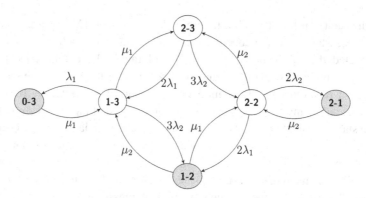

Figure 18.4   CTMC for CSC-SFC module

steady-state balance equations of this CTMC, we obtain the steady-state probabilities:

$$\pi_{2\text{-}3} = \frac{1}{G}, \qquad \pi_{2\text{-}2} = \frac{3\lambda_2}{\mu_2 G}, \qquad \pi_{1\text{-}3} = \frac{2\lambda_1}{\mu_1 G},$$

$$\pi_{0\text{-}3} = \frac{2\lambda_1^2}{\mu_1^2 G}, \qquad \pi_{2\text{-}1} = \frac{6\lambda_2^2}{\mu_2^2 G}, \qquad \pi_{1\text{-}2} = \frac{6\lambda_1 \lambda_2}{\mu_1 \mu_2 G},$$

(18.4)

where

$$G = 1 + \frac{3\lambda_2}{\mu_2} + \frac{6\lambda_2^2}{\mu_2^2} + \frac{2\lambda_1}{\mu_1} + \frac{2\lambda_1^2}{\mu_1^2} + \frac{6\lambda_1 \lambda_2}{\mu_1 \mu_2}.$$

(18.5)

Based on a four-out-of-five logic, states 0-3, 2-1 and 1-2 are down states and hence the steady-state availability is:

$$A_{\text{CSC/SFC}} = \pi_{1\text{-}3} + \pi_{2\text{-}3} + \pi_{2\text{-}2} = \frac{2\lambda_1 \mu_2 + \mu_1 \mu_2 + 3\mu_1 \lambda_2}{\mu_1 \mu_2 G}.$$

(18.6)

Given the parameter values from Table 18.2, $\text{MTTF}_{\text{CSC}} = 272\,584$ hr, $\text{MTTR}_{\text{CSC}} = 2$ hr, $\text{MTTF}_{\text{SFC}} = 422\,115$ hr, and $\text{MTTR}_{\text{SFC}} = 2$ hr, we have $A_{\text{CSC/SFC}} = 1.0$. Finally, from the top-level RBD of Figure 4.4, and assuming independence across subsystems, we get the overall Cisco 12000 GSR steady-state availability:

$$A_{\text{router}} = A_{\text{Upgrade}} * A_{\text{LC-in}} * A_{\text{LC-out}} * A_{\text{CSC/SFC}} * A_{\text{GRP}} * A_{\text{Chassis}} * A_{\text{IOS}}$$

$$= 0.999\,825\,19.$$

We divide overall router failures into three types: hardware failures, software failures and upgrades. The overall downtime is composed of the individual downtimes caused by each type of failure. We also observe that upgrades cause major downtime in the system. Table 18.3 shows the overall downtime due to different failure types.

Example 18.2   *Availability Modeling of High-Availability Platform from Sun Microsystems [4]*

Next we consider the availability model of a carrier-grade high-availability server

Table 18.3 Downtime breakdown (in minutes per year) of GSR.

System	Hardware	Software	Upgrade	Total
Cisco 12000 GSR	28	23	41	92

platform from Sun Microsystems, whose RBD was examined in Example 4.24. The hardware configuration of the platform includes multiple independent drawers interconnected by dual-redundant SCSI chains or Ethernet as discussed in Example 4.16. The modeling here focuses on the availability of each drawer, which consists of multiple components with possible redundancy:

- Power supply subsystem
- Fan subsystem
- Alarm card
- CPU
- Ethernet
- PCI backplane
- SCSI interconnects and disk drives connected via SCSI
- Satellite cards.

A hierarchical composition is adopted to compute the system availability. Only the steady-state availability is computed here, but transient and expected interval availability can also be computed. At the top level is the RBD of Figure 4.24, in which each block either corresponds to a submodel of a subsystem or is an individual component with a specified failure rate and repair rate. The RBD for the SCSI subsystem is shown in Figure 4.12. The second level in the hierarchy contains the Markov chains for the subsystems, except for the satellite card subsystem. The CTMC submodel for the fan subsystem was already presented in Example 9.16 (see Figure 9.21) together with the numerical results. In the last level of the hierarchy the satellite card subsystem is described by an RBD and contains two Markov chain models, SatNet, and SatCard. The output of the submodels in the form of steady-state availability values are passed as inputs to the higher levels, and eventually to the top-level RBD model in order to compute the overall steady-state system availability. This structure can be described by the import graph in Figure 18.5, which is similar to the one of Figure 18.2.

Power Supply Subsystem
The dual-redundant power supply (PS) subsystem is modeled as the six-state CTMC shown in Figure 18.6.

In state 2 the two units of the subsystem are up. If either one suffers a failure that is covered, the CTMC enters state 1. If the first fault is latent, then the CTMC enters state 1p; another fault while in this state will lead the CTMC to state 0, bringing the power supply down. One of the two units is up in state 1, and upon successful repair of the failed unit the CTMC returns to state 2. If the repair is unsuccessful, the CTMC enters state "irep." Upon repair completion from irep, the subsystem returns to state 2. Failure of the second unit before the completion of repair of the first failed unit leads the

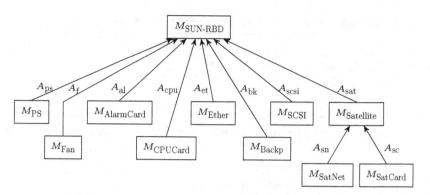

Figure 18.5 Import graph for the Sun Microsystems case study.

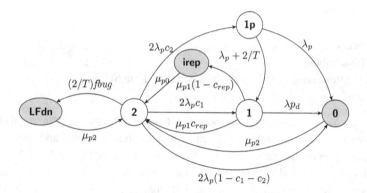

Figure 18.6 Continuous-time Markov chain for the power supply subsystem, M_{PS}.

CTMC to state 0. State "LFdn" is used to allow for diagnostics to bring the subsystem down. State 1p is used to indicate that a latent fault has occurred, which will eventually lead to failure of the power supply subsystem. Transition from state 1p to state 1 can occur if a non-latent fault occurred in the subsystem with a latent fault or if diagnostics detected a latent fault. The power supply system is up as long as it is in states 2, 1p or 1. The failure and repair rates of the power supply subsystem are given in Table 18.4. The parameter values are for illustrative purpose only, and should not be construed as actual values of the real system.

Alarm Card Subsystem

The alarm card subsystem has two kinds of failure/recovery (as in Example 9.7), including hardware failure/repair and software failure/reboot. The CTMC for this case is shown in Figure 18.7. State UP is the only up state; F_h and F_s are the hardware and software failure states, respectively.

Assuming a hardware failure rate of λ_{ACh}, a hardware repair rate of μ_{AC}, a software failure rate of λ_{ACs} and a software repair (i.e., reboot) rate of β_{AC}, the closed-form

Table 18.4 Failure and repair rates for the power system submodel of the Sun Microsystems case study.

Symbol	Meaning	Value	Unit
λ_P	PS failure rate	4.0662×10^{-6}	hr^{-1}
λ_{Pd}	PS failure rate in degraded state	8.1324×10^{-6}	hr^{-1}
c_1	Coverage for first fault	0.998	—
c_2	Probability of latent fault	0.001	—
T	Diagnostics period	10 000	hr
$fbug$	Probability of diagnostics bringing system down	0.0001	—
μ_{p0}	Repair rate when person on site	2.4	hr^{-1}
μ_{p1}	Repair rate for one unit for PS	0.226 41	hr^{-1}
μ_{p2}	Repair rate for two units for PS	0.226 41	hr^{-1}
c_{rep}	Probability of successful repair	0.99	—

Table 18.5 Failure and repair rates of the alarm card submodel of the Sun Microsystems case study.

Symbol	Meaning	Value	Unit
λ_{ACh}	Hardware failure rate for Alarm	7.1627×10^{-9}	hr^{-1}
λ_{ACs}	Software failure rate for Alarm	0.000 0001	hr^{-1}
μ_{AC}	Repair rate of Alarm	0.24	hr^{-1}
β_{AC}	Reboot rate of CMC	720	hr^{-1}

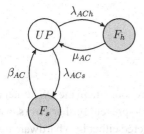

Figure 18.7 Continuous-time Markov chain for the alarm card subsystem, $M_{\text{AlarmCard}}$.

solution for the steady-state availability of this card is written, see Eq. (9.49), as:

$$A_{AC} = \frac{1}{1 + \lambda_{ACh}/\mu_{AC} + \lambda_{ACs}/\beta_{AC}}. \tag{18.7}$$

Illustrative parameter values are reported in Table 18.5.

CPU Card Subsystem

The CPU card subsystem is modeled by the five-state CTMC shown in Figure 18.8. It can experience failures due to hardware or software (e.g., the operating system).

The failure types are distinguished by the type of recovery action that follows. A cold reboot may be required to deal with hardware transients and some fraction of OS failures. While in state UP, a permanent hardware failure leads the system to state H; upon repair from that state it returns to state UP. State Sp represents a panic reboot state

Table 18.6 Failure and recovery rates of the CPU card subsystem submodel.

Symbol	Meaning	Value	Unit
λ_{CPh}	Hardware failure rate	6.6771×10^{-6}	hr^{-1}
μ_{CPh}	Hardware repair rate	0.23077	hr^{-1}
λ_{CPs}	Software failure rate	0.000002	hr^{-1}
λ_{t}	Transient failure rate	$1/8760$	hr^{-1}
μ_{trans}	Database rebuild rate	0.125	hr^{-1}
β_{c}	Cold reboot rate	6	hr^{-1}
β_{p}	Panic reboot rate	12	hr^{-1}
q_{p}	Probability of successful panic reboot	0.95	—
q_{c}	Probability of successful cold reboot	0.99	—
p_{p}	Probability of occurrence of panic reboot	0.5	—
p_{c}	Probability of occurrence of cold reboot	0.5	—

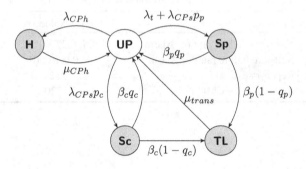

Figure 18.8   Markov chain for the CPU card subsystem, M_{CPUCard}.

of the subsystem. A panic reboot is a software reboot where the OS kernel or application decides to reboot because of a severe situation to prevent further errors, which could get more serious. Transition from UP to Sp is triggered either by a hardware transient failure or by a type of software failure that may lead to the panic reboot with a probability of p_{p}. The system enters state Sc from state UP when a cold reboot is required. Reboots can be successful, leading the system to return to state UP. Reboots of the system may not be successful (state TL), triggering the need for a database rebuild with rate μ_{trans}.

Transition rates for the CTMC of Figure 18.8 are given in Table 18.6 with illustrative values. The CPU card subsystem is up as long as the Markov chain is in state UP.

Ethernet Subsystem

The Ethernet subsystem is modeled as the four-state CTMC shown in Figure 18.9. This CTMC has one type of failure and two types of recovery. In state 2, both the ports of the subsystem are up. Failure of one of the ports, with a total rate of $2\lambda_{\text{E}}$, leads the subsystem to state 1. The second port can fail in state 1 before the repair of the first failed port with rate λ_{E}, leading the subsystem to enter state 0. The subsystem may undergo repair of the failed port, which proceeds in two stages. First is the logistic time with mean value denoted by *logt*. This explains the transition from state 1 to state

Table 18.7 Failure and repair rates of the Ethernet submodel of the Sun Microsystems case study.

Symbol	Meaning	Value	Unit
λ_E	Ethernet failure rate	1.79492×10^{-7}	hr^{-1}
μ_S	Repair rate for two ports (Ethernet)	0.23529	hr^{-1}
$logt$	Logistic time	4	hr

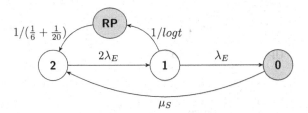

Figure 18.9 Markov chain for the Ethernet subsystem, M_{Ether}.

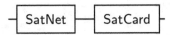

Figure 18.10 Reliability block diagram for the satellite card subsystem, $M_{Satellite}$.

RP. State RP represents the subsystem under repair. The Ethernet subsystem will enter state 2 from state 0 through the repair of the two failed ports with the rate μ_S. The Ethernet subsystem is considered up as long as it is in states 2 or 1.

Transition rates for the CTMC of Figure 18.9 are given in Table 18.7 with illustrative values.

Backplane, SCSI, and Drive Subsystems

The PCI backplane block is a non-redundant module with only one kind of failure/recovery, which is modeled as a single component in the block diagram (which is a two-state CTMC) with failure rate $\lambda_{BP} = 0.00000112$ and repair rate $\mu_{BP} = 0.22641$; this provides a component availability of $A_{BP} = 0.999995053$.

The SCSI block has already been considered in Example 4.16, and the related illustrative failure and repair rate values reported in Table 4.6.

Satellite Card Subsystem

The four-out-of-seven satellite block in Figure 4.24 is modeled by the next-level reliability block diagram shown in Figure 18.10. Each of the SatNet and SatCard blocks in Figure 18.10 is modeled as a CTMC.

The SatNet block is modeled using the same CTMC model as for the Ethernet subsystem of Figure 18.9, so $M_{SatNet} = M_{Ethernet}$. SatCard is a non-redundant module modeled by the five-state CTMC shown in Figure 18.11. This model of a satellite card has the same structure as that of the CPU card, except for the different labels for the transition rates, whose values are reported in Table 18.11.

Table 18.8 Failure and recovery rates of the SatCard subsystem submodel.

Symbol	Meaning	Value	Unit
λ_{Sth}	Hardware failure rate	$6.635\,93 \times 10^{-6}$	hr^{-1}
μ_{Sth}	Hardware repair rate	$0.237\,15$	hr^{-1}
λ_{Sts}	Software failure rate	$0.000\,02$	hr^{-1}

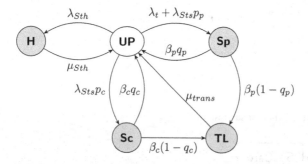

Figure 18.11 Markov chain for the SatCard submodel, M_{SatCard}.

Results
The steady-state availability values for every subsystem, and for the system as a whole, have already been reported in Table 4.7 associated with the top-level RBD of Figure 4.24. Further details can be found in [4].

Example 18.3 *Availability Modeling of the IBM BladeCenter [5]*
We return to the IBM BladeCenter example already considered in Example 6.33 and in Section 9.6. In contrast to the previous two examples of this chapter, the top-level model here is a fault tree rather than an RBD because of the need to incorporate repeated events. Recall the top-level giant fault tree model, named $M_{\mathrm{Blade\text{-}FT}}$, shown in Figure 6.23. We make a small change in this fault tree so that event Sw_i for software failure of blade Bl_i is no longer a basic event but a square block that is expanded into the CTMC submodel M_{Sw} already developed in Example 9.23. The CTMC submodels of Md, C, P_1, P_2, CPU, M, and D were examined in detail in Examples 9.19–9.27, respectively. Recall that these submodels were named M_{Md}, M_{C}, M_{P}, M_{P}, M_{CPU}, M_{M}, and M_{D}.

The event $\overline{B}_i$ indicating the failure of the base of blade server i in the FT is modeled by a simple two-state CTMC with a specified MTTF and MTTR, and hence is not shown explicitly. Similarly, for events corresponding to Ethernet and fiber channel switch failures and corresponding ports, we use a two-state CTMC with given values of MTTF and MTTR.

The import graph, shown in Figure 18.12, is very similar to the ones reported in Figures 18.2 and 18.5 except that the top-level model is an FT instead of an RBD.

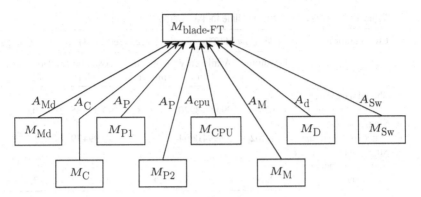

Figure 18.12   Import graph for the IBM BladeCenter case study.

The quantities computed from the CTMCs are steady-state availabilities, A_{Md}, A_{C}, A_{P}, A_{cpu}, A_{M}, A_{D}, and A_{Sw}, which are passed up the hierarchy as in the previous examples. Note that M_{P} needs to be solved only once as both power domains have the same availability, A_{P}.

Closed-form expressions for each of the submodel availabilities are given in Examples 9.19–9.23. SHARPE can produce a closed-form availability expression for the top-level FT, but the length of the expression and corresponding storage space is so enormous that it is necessary to resort to numerical solution. Note that across subsystems we have assumed independence. For further details, the reader is encouraged to refer to [5]. If dependence among subsystems is necessary (for instance to allow for shared repair across subsystems), it is possible to resort to fixed-point iteration as in the previous chapter.

Utilizing the low and high MTTF values reported in Table 6.8, as well as the predicted MTTF for each field replaceable unit, the availability and expected annual downtime for a single blade are examined. In Table 18.9, the shared hardware composed of the midplane Md, power supplies P_1 in power domain 1 and cooling system C (see the top OR gate in Figure 6.23) is examined. For all three MTTF values, the shared hardware is responsible for less than 1 min of annual downtime. This is well below the $5\,\text{min}\,\text{yr}^{-1}$ limit for a five-nines availability solution. Secondly, the single blade server along with the network switches and blade server software is examined. Line 2 of Table 18.9 shows that a blade server expects more than 5 min but less than 50 min of annual downtime. Thus, the blade server hardware is a four-nines availability solution. Finally, the combined values are shown in line 3 and the total blade server solution is expected to provide availability greater than 99.99%.

In Table 18.10, the downtime contributions in $\text{min}\,\text{yr}^{-1}$ of the hardware components are itemized along with the contribution from the software. Here, the software MTTF is identical for all three cases of hardware MTTF considered. The portion of the hardware and software downtime contributed by the service response time is also determined. For predicted hardware MTTF values, service response represents about 35% of the total downtime and represents a significant availability bottleneck that can only be addressed with changes to the service delivery process.

Table 18.9   Availability and downtime for hardware and software.

Configuration	Low MTTF		Predicted MTTF		High MTTF	
	Downtime (min yr^{-1})	Availability (%)	Downtime (min yr^{-1})	Availability (%)	Downtime (min yr^{-1})	Availability (%)
Shared hardware (excl. switches)	0.909	99.9998270	0.782	99.9998512	0.674	99.9998717
Single blade + network switches	31.502	99.9940064	27.367	99.9947931	23.222	99.9955817
Single blade + shared + network switches	32.411	99.9936134	28.149	99.9946443	23.896	99.9954534

Table 18.10   Component contributions to single blade downtime.

Hardware Component	Downtime (min yr^{-1})		
	Low hardware MTTF	Predicted hardware MTTF	High hardware MTTF
Blade CPU	0.34690	0.29700	0.25507
Blade memory	1.81770	1.52985	1.32197
Blade disk drive	7.88400	5.24879	2.10240
Blade base + NICs + network switches	7.16677	6.00447	5.25585
Chassis midplane	0.86087	0.74105	0.63885
Chassis power subsystems	0.04839	0.04116	0.03563
Chassis cooling	0.00000	0.00000	0.00000
Total hardware component downtime	18.12463	13.86232	9.60977
Software downtime	14.28711	14.28711	14.28711
Total downtime	32.41174	28.14943	23.89688
Minutes of total downtime attributed to service response time	13.15197	9.98276	6.73376

The non-redundant electronics on the blade server along with disk drives are the largest contributors of hardware-related downtime. While high-reliability disk drives might appear to be the answer to improved downtime, high MTTF values for drives may be more a matter of operational and environmental conditions than any really significant differences in the hardware. So, other options should be considered. The downtime associated with the disk drive can be virtually eliminated using a RAID-1 configuration if a scheduled maintenance window can be used for disk replacement.

Spare blades are often considered as a means to achieve high availability. Table 18.11 gives the probability of exactly k operational out of 14 total blade servers. Either 14 or 13 operational blades is the most likely scenario, due to 0 or 1 blade server failures. There is only a very small probability of more than two simultaneous blade server

Table 18.11 Probabilities of exactly k operational blades (including software) in a chassis with 14 blades.

Number of required blade servers	Number of hot-standby spare blade servers	Availability (%)	Probability of exactly k operational blade servers	Downtime $(\min \mathrm{yr}^{-1})$
14	0	99.926 9727	0.999 269 727	383.83
13	1	99.999 8186	0.000 728 459	0.95
12	2	99.999 8433	0.000 016 573	0.82
11–7	3–7	99.999 8433	0	0.82
6	8	99.999 8511	0.000 000 078	0.78
5–1	9–13	99.999 8511	0	0.78

failures; hence, BladeCenter availability does not improve with more than two spare blade servers.

Example 18.4 *Availability Model of SIP on IBM WebSphere [6]*

In the IBM SIP availability model, the main emphasis is on software failures and recovery from these failures, since the hardware submodels are simply adopted from the previous example of the IBM BladeCenter. Demand for highly available platforms has grown in response to the increasing popularity of the IETF (Internet Engineering Task Force) [7] standard on SIP for converged services combining data, voice, and video. Analysis of IBM's SIP platform combining BladeCenter hardware and WebSphere Application Server software was described in [6] using hierarchical composition combining fault trees and CTMCs to model the overall system availability.

Figure 18.13 shows the architecture for one SIP application server cluster. A cluster is composed of two replicated BladeCenters, each containing four blade servers. Fault tolerance with the goal of high availability is achieved by a combination of hardware and software redundancy. One BladeCenter would suffice from the performance perspective but two BladeCenters (each having four blade servers) are used to provide hardware fault tolerance. One proxy server would fulfill the performance requirements but two identical copies are used to provide software fault tolerance. Similarly, six WebSphere applications servers (AS) would be enough for the required performance but 12 identical copies of the AS are used to provide software fault tolerance. The session information in each AS is replicated in a peer AS on the other BladeCenter. These two ASs constitute a replication domain. There are six replication domains numbered 1 through 6, and six blades (each with two ASs) labeled A through F.

To develop a model, first the overall system failure is decomposed into subsystem (or component) failures, as shown in Figure 18.14. This failure classification informs the subsequent model logic.

Due to the size of the underlying state space, a monolithic CTMC would be intractable. Exploiting the independence across subsystems, a two-level hierarchical model is employed wherein the top-level model is an FT with repeated events, and

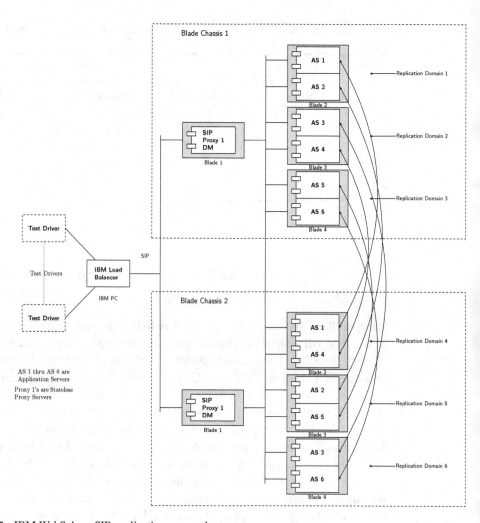

Figure 18.13 IBM WebSphere SIP application server cluster.

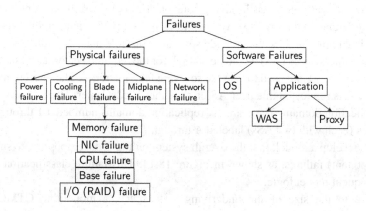

Figure 18.14 Component or subsystem failures in the SIP system.

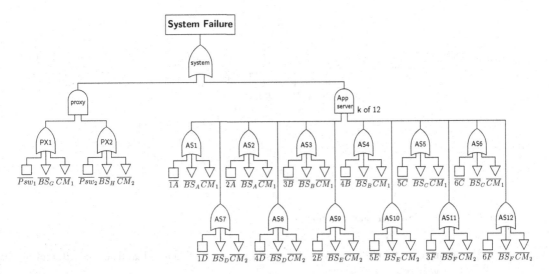

Figure 18.15   Fault tree model for SIP system.

bottom-level models are CTMCs. We start with the FT shown in Figure 18.15, which includes both hardware and software failures.

The labels in the leaf nodes in the FT of Figure 18.15 have the following meanings:

- $\overline{ij}$ ($i = 1,\ldots,6$, $j = A,\ldots,F$) represents software failure of the application server in domain i on blade X.
- $\overline{Psw_i}$ ($i = 1,2$) represents software failure of proxy server i.
- $\overline{BS_j}$ ($j = A,\ldots,F$) represents hardware failure of blade server j.
- $\overline{CM_i}$ ($i = 1,2$) represents common hardware failure on chassis i.

As usual in FT notation, a leaf node drawn as a circle indicates a basic event, a square means that there is a submodel defined and an inverted triangle means a repeated event.

Proxy server PX_1 will not be usable in the case when the proxy software is down or the blade server BS_G is down or the chassis common hardware CM_1 is down. Similarly for the proxy server PX_2 and for all the applications servers. The system as a whole is down if no proxy server can be accessed or k or more application servers cannot be accessed.

Each $\overline{BS_j}$ ($j = A,\ldots,F$) in Figure 18.15 is replaced by the FT on the right-hand side of Figure 18.16. Similarly, all the $\overline{CM_1}$ and $\overline{CM_2}$ occurrences in Figure 18.15 are replaced by the FT on the left-hand side of Figure 18.16. While doing this replacement, the two Ethernet switch components (esw_1 and esw_2) are shared by all blades on Chassis 1 while the blade servers on Chassis 2 share two additional Ethernet switch components (esw_3 and esw_4).

Chassis $\overline{CM_i}$ fails if the corresponding midplane fails or its cooling subsystem fails or its power supply fails. A blade server $\overline{BS_j}$ fails if its base or its CPU subsystem or its memory subsystem or its RAID subsystem or its OS fails or its network link fails. For the network link, each blade has two network interface cards (nic_1 and nic_2) that

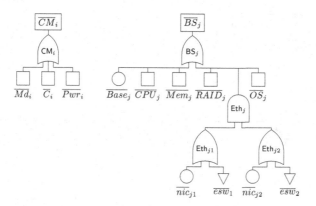

Figure 18.16   Middle-level SIP fault tree models.

connect to two Ethernet switches (esw_1 and esw_2) that are shared across different blade servers.

The giant FT thus constructed is the top-level FT model $M_{SIP\text{-}FT}$. Thus the middle-level fault trees are separately specified to ease the construction of the giant fault tree and not for hierarchical solution. For the cooling and power supply subsystem, we utilize the CTMC submodels M_C and M_P that we developed in the previous example of the IBM BladeCenter. We now discuss six distinct CTMCs that are used to model the availability of the blade chassis midplane, blade CPU, blade memory, hard disk, application or proxy server software, and the operating system. Of these, the OS model (M_{OS}) was already discussed in Example 9.17 and was shown in Figure 9.22. The remaining CTMC submodels are discussed next.

Midplane CTMC

The midplane CTMC model, M_{Md}, is shown in Figure 18.17. The midplane is working properly in states UP and U1, where UP refers to the state where both redundant communication paths are functioning properly. This is distinct from U1 where only one of the two paths is operational. Faults occur in the midplane at the rate λ_{mp}, and a fraction of possible failures denoted as common-mode failures will bypass U1 and bring the midplane system down. This is modeled here with the direct transition from state UP to the down state DN at rate λ_{mp} modulated by the common-mode factor c_{mp}. From either the DN or U1 states, a repair person will arrive at rate α_{sp}, transitioning the system to the offline repair state RP. A complete repair occurs at rate μ_{mp}, returning the system to the UP state.

CPU CTMC

The blade CPU CTMC, M_{CPU}, shown in Figure 18.18 models a two-processor system that is operational when both processors are up. In state UP two processors are up so the subsystem is up. The subsystem is down in state D1 where one processor is down, wherefrom a call for repair occurs at rate α_{sp}, and in state RP a repair is ongoing. Assume that λ_{cpu} and μ_{cpu} are the respective failure and repair rates of the CPU.

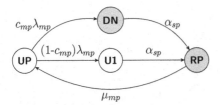

Figure 18.17 Continuous-time Markov chain availability model, M_{Md}, for the chassis midplane.

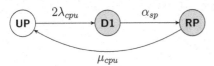

Figure 18.18 Continuous-time Markov chain availability model, M_{CPU}, for the blade CPU subsystem.

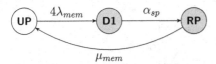

Figure 18.19 Continuous-time Markov chain availability model, M_{Mem}, for the node memory subsystem.

Memory CTMC

The memory subsystem model is a variation of the CPU submodel to describe two banks of two memory DIMMs for a total of four discrete memory modules, as shown in Figure 18.19, and is named M_{Mem}. The respective failure and repair rates of a memory module are λ_{mem} and μ_{mem}. The subsystem is considered down if any module in either bank has failed.

RAID CTMC

The SIP storage consists of two hard disks in a RAID 1 configuration. Figure 18.20 depicts the CTMC availability model, M_{RAID}, of the RAID system on each blade. When the RAID controller chip detects the first drive failure, the system will continue to function by loading all data from the one disk that is up (state U1 in the model). As before, summoning a repair person takes a mean time of $1/\alpha_{sp}$; however, since the device is still in operation there is a possibility of another disk failure before the repair person arrives, resulting in the down state DN. If the repair person arrives before the second failure, the system transitions to the RP state, where the failed hard drive is replaced at rate μ_{hd}. After replacement, the first disk will be copied onto the second in state CP at rate χ_{hd}. If copying is successful, the system returns to normal operation in state UP. If a second failure occurs during copying (at rate λ_{hd}), or if the repair person arrives to find the system in state DN, the subsystem requires a complete replacement with fresh preloaded drives in state DW. Models can penalize reaching this unfavorable subsystem state by reward rate assignment, or simply by a slower μ_{2hd} rate as is seen in this submodel.

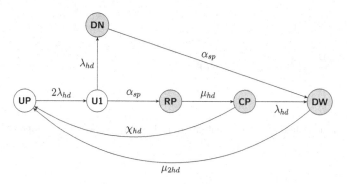

Figure 18.20 RAID 1 CTMC availability model, M_{RAID}.

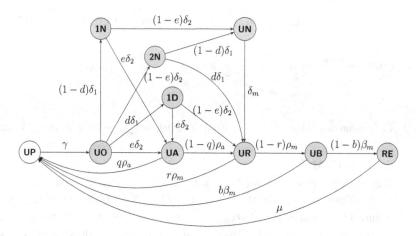

Figure 18.21 Continuous-time Markov chain availability model, M_{AS} (and M_{Proxy}), for AS and proxy.

Application Server CTMC

Figure 18.21 shows the CTMC availability model of the AS software, named M_{AS} and M_{Proxy}. The model does not include the states and transitions characterizing failover or switchover behavior (to its backup in the same replication domain) because they do not affect the availability of individual software servers. However, failover plays an important role in reducing the SIP message loss and thus is included in the models that compute service-oriented measures (to be covered in a later section of this chapter). The description of all the states is provided in Table 18.12. When a failure occurs in an application server, the server transitions from state UP to state UO. In state UO, failure detection is attempted by two automated detectors: WLM (workload manager) and NA (node agent). The detection probabilities of WLM and NA are assumed to be d and e, respectively. If the WLM detects the failure first, the server enters state 1D, in which case failover is carried out while the node agent is trying to detect the failure. Assume that the node agent will not detect the failure before failover is completed. With probability e the failure will be detected by NA; the server then enters state UA, and subsequently states UR, UB, and RE in turn for automatic process restart,

Table 18.12 States and their meanings in the software server CTMC availability model.

State	Meaning
UP	Software server is up
UO	Server is in an undetected failure state
1D	Failure is detected by WLM (IP Sprayer), NA has not yet detected failure
1N	WLM (IP Sprayer) is unable to detect the failure
2N	Node agent (NA) is unable to detect the failure
UN	Neither WLM nor NA is able to detect the failure
UA	NA detected the failure; auto-process restart ongoing
UR	Manual process restart on the failed server
UB	Manual reboot of the blade assigned to the failed server
RE	Manual repair of the failed server

manual process restart, manual reboot and manual repair. The use of increasingly complex recovery actions in this manner is known as *escalated levels of recovery*. With probability $(1-d)(1-e)$, neither WLM nor NA detects the failure and the server enters state UN, in which the failure is manually detected. After manual detection, the server enters states UR, UB, and RE for manual process restart, manual node reboot, and manual repair. The use of restart or reboot as a means of recovery from failure and the use of identical software copies to failover imply that the major cause of AS (and proxy) software failures is assumed to be Mandelbugs [8] rather than traditional Bohrbugs. Accordingly, the small fractions of failures treated by escalated repair actions are presumably due to residual Bohrbugs. The model for the SIP proxy is identical, except that in place of WLM an IP sprayer is the second automated detector besides NA.

Operating System CTMC

The operating system CTMC submodel, M_{OS}, was already discussed in Example 9.17 (see Figure 9.22), and the steady-state availability was computed given the parameter numerical values reported in Example 9.17.

Model Solution

Each bottom-level CTMC model is solved for their respective steady-state availabilities in closed form or numerically, and propagated upwards to the top-level fault tree, M_{SIP-FT}. Therefore, system solutions are achieved without constructing, storing or solving a single (inefficient or impossible) monolithic CTMC model. The import graph for this hierarchical availability model is shown in Figure 18.22.

The parameter values for the models in Figures 18.17–18.21 are given in Tables 18.13–18.17. Since the real parameter values are confidential, we have used arbitrary (but reasonable) values.

The set of model input parameters can be divided into five groups: hardware component failure rates (or equivalently MTTFs), Table 18.13; software component MTTFs, Table 18.14; the mean times to carry out various recovery steps, Table 18.15; the coverage probabilities of various recovery steps, Table 18.16; and mean times

Table 18.13 SIP system: Hardware component MTTFs.

Parameter	Meaning	Value
$1/\lambda_{mp}$	Mean time to midplane failure	10^6 hr
$1/\lambda_c$	Mean time to blower failure	10^6 hr
$1/\lambda_{ps}$	Mean time to power module failure	10^6 hr
$1/\lambda_{cpu}$	Mean time to processor failure	10^6 hr
$1/\lambda_{base}$	Mean time to base failure	10^6 hr
$1/\lambda_{swh}$	Mean time to Ethernet switch failure	10^6 hr
$1/\lambda_{nic}$	Mean time to NIC failure	10^6 hr
$1/\lambda_{mem}$	Mean time to memory DIMM failure	10^6 hr
$1/\lambda_{hd}$	Mean time to hard disk failure	10^6 hr

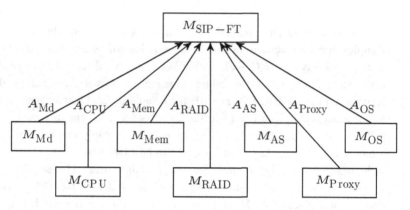

Figure 18.22 Import graph for IBM SIP on WebSphere.

associated with the repair actions, Table 18.17. The hardware component MTTFs are available in most companies in the form of tables [5].

Software component MTTFs are not readily available in most companies, and experiments are necessary to estimate these values. These experiments are hard to conduct because of the time they take, due to the difficulty of attributing the cause of a failure and the difficulty of ensuring the representativeness of the workload. The use of accelerated life testing is recommended [9, 10]. The values we assume for our example here are given in Table 18.14. The third group of input parameters to the above models, consisting of the mean times to detect, failover, restart, and reboot, are estimated from data collected during fault/error injection experiments. This was indeed done in the present example by an IBM team. The assumed values are given in Table 18.15. The fourth group of model input parameters are various coverage probabilities that also need to be estimated by means of fault/error injection experiments [11, 12]. The values we assumed are listed in Table 18.16. The fifth group of model input parameters relate to various repair actions. The assumed values for this case study are listed in Table 18.17.

Using the default values listed in these tables, the predicted system steady-state unavailability is 2.2×10^{-6}. The contribution of AS and SIP/proxy to the unavailability

Table 18.14 SIP system: Software MTTFs.

Parameter	Meaning	Value
$1/\gamma$	Mean time to AS and proxy server failure	1000 hr
$1/\lambda_{OS}$	Mean time to OS failure	4000 hr

Table 18.15 SIP system: Mean times for various recovery actions.

Parameter	Meaning	Value
$1/\delta_{OS}$	Mean time to detect OS failure	1 hr
$1/\beta_{OS}$	Mean time for node OS reboot	10 min
$1/\chi_{hd}$	Mean time to copy disk data	10 min
$1/\delta_1$	Mean time for WLM failure detection	2 s
$1/\delta_2$	Mean time for node agent failure detection	2 s
$1/\delta_m$	Mean time for manual AS failure detection	10 min
$1/\phi$	Mean time for failover	1 s
$1/\rho_a$	Mean time for automatic process restart	10 s
$1/\rho_m$	Mean time for manual process restart	60 s
$1/\beta_m$	Mean time for manual node reboot	10 min

Table 18.16 SIP system: Various coverage parameters.

Parameter	Meaning	Value
c_{mp}	Probability of midplane common mode failure	0.001
c_{ps}	Coverage factor for power module failure	0.99
b_{OS}	Coverage factor for OS reboot	0.9
c	Coverage factor for AS/proxy failover	0.9
d	Coverage factor for WLM detection	0.9
e	Coverage factor for node agent detection	0.9
q	Coverage factor for automatic process restart	0.9
r	Coverage factor for manual process restart	0.9
b	Coverage factor for manual node restart	0.9

Table 18.17 SIP system: Mean times for various repair actions.

Parameter	Meaning	Value
$1/\alpha_{sp}$	Mean time to repair person arrival	2 hr
$1/\mu_{mp}$	Mean time to repair midplane	1 hr
$1/\mu_c$	Mean time to repair blower	1 hr
$1/\mu_{2c}$	Mean time to repair two blowers	1.5 hr
$1/\mu_{ps}$	Mean time to repair a power module	1 hr
$1/\mu_{2ps}$	Mean time to repair two power modules	1.5 hr
$1/\mu_{cpu}$	Mean time to repair a processor	1 hr
$1/\mu_{base}$	Mean time to repair base	1 hr
$1/\mu_{OS}$	Mean time to repair OS	1 hr
$1/\mu_{swh}$	Mean time to repair Ethernet switch	1 hr
$1/\mu_{nic}$	Mean time to repair NIC	1 hr
$1/\mu_{mem}$	Mean time to repair a memory bank	1 hr
$1/\mu_{hd}$	Mean time to repair hard disk	1 hr
$1/\mu_{2hd}$	Mean time to repair two hard disks	1.5 hr

Table 18.18 SIP system: Downtime by various causes.

k	OS	Hardware	Proxy	App server	Total downtime
1	1155	73.65	0.000 36	165.7	1394
2	1155	73.65	0.000 36	0.024	1228
3	1.13	1.13	0.000 36	0.000 0005	2.7350
4	1.13	1.13	0.000 36	0	2.413
5	0.07	1.13	0.000 36	0	1.219
6	0.07	1.13	0.000 36	0	1.218
7	0.07	0.000 38	0.000 36	0	0.093

is minuscule: 2.7×10^{-9}, indicating the effectiveness of escalated levels of recovery. Varying the number of spare AS instances (that is, $k - 1$), we computed the total downtime, OS-only downtime, hardware-only downtime, proxy-only downtime and application-server-only downtime; the numbers are shown in Table 18.18. As seen from the table, the OS-only and hardware-only downtimes for $k = 2i - 1$ ($i = 1, 2, 3$) are the same as those for $k = 2i$. This happens because an OS failure or a hardware failure will bring down an even number (at least two) of the application servers. When k is small ($k = 1, 2$), OS-only downtime is dominant because the OS has a higher failure rate; as k increases to 5 and 6, there are more redundant blade servers that host the application servers, and we need three or more OS failures to bring down the system. The application-server-only downtime decreases rapidly as k increases because of more redundant application servers in the system. For further details, see [6].

Problems

18.1 For the Cisco router in Example 18.1, first re-derive all the submodel steady-state availability expressions by hand to verify those given in the book. Then derive expressions for the derivatives of each submodel steady-state availability with respect to all its input parameters. Finally, derive expressions for the derivatives of the overall router availability with respect to all its input parameters.

18.2 For the Sun Microsystems availability model in Example 18.2, derive closed-form expressions for the steady-state availability of each submodel described by a CTMC. Thence write down a closed-form expression for the overall system steady-state availability.

18.3 For the Sun Microsystems availability model in Example 18.2, carry out epistemic (parametric) uncertainty propagation using a sampling-based method as discussed in Section 3.4 [13].

18.4 Solve the IBM BladeCenter availability model discussed in Example 18.3 and thus verify the results displayed in Tables 18.9–18.11. You may find it convenient to use the SHARPE software package.

18.2   Reliability Models

This section presents a single complex hierarchical reliability model derived from [14].

Example 18.5 *Reliability Model of the NASA Remote Exploration and Experimentation System*

We next consider the reliability analysis of the NASA Remote Exploration and Experimentation (REE) system [14]. The system is a set of processing elements (also referred to as PEs or as nodes) connected via a Myrinet [15] interconnect. Each PE has a commercial off-the-shelf (COTS) operating system derived from UNIX. The REE system architecture is shown in Figure 18.23. Hardware, software, time, and data redundancy are employed to achieve fault tolerance to component faults. Fault detection and recovery are engineered by means of software-implemented fault tolerance (SIFT) and by a system executive (SE) [16]. The SE does local error detection and failure recovery, while the system control processor (SCP) is responsible for fault detection and recovery at the system level. The SCP uses periodic PE health status messages or heartbeats from the SE on each PE to determine their status. Processing elements can be individually reset or restarted by either the SCP or by the SE.

A service request to the system is dispatched to multiple PEs for concurrent processing by a fault-tolerant scheduler. Unlike the IBM SIP example, traditional design diversity [18, 19] is used in the software components deployed on different PEs to provide software fault tolerance. The results from these PEs are then compared by the SE to detect and tolerate faults. Transient and intermittent faults can be detected and recovered from, by SIFT, SE or SCP. However, as the REE system operates without human intervention, components cannot be replaced after permanent failures.

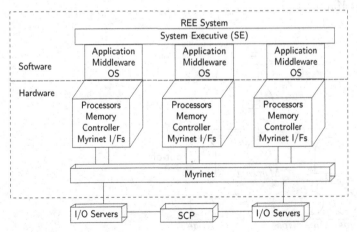

Figure 18.23   Remote Exploration and Experimentation system architecture block diagram (adapted from [17]).

Hence, permanent failures require the system to be reconfigured to a degraded mode of operation. For a detailed flow chart of the recovery actions see [17].

Component failures can be transient or permanent in nature. The majority of transient faults are masked due to the use of error-correcting codes and redundant hardware. Unmasked transient failures may freeze the processors, making the system unresponsive, or generate erroneous outputs. The heartbeat messages are relied upon by the SE to detect unresponsive or frozen nodes.

Erroneous outputs are detected with the help of acceptance tests by SIFT [19]. Upon detection of either an unresponsive node or erroneous output, the node is rebooted after transferring its tasks to other spare nodes (the tasks are restarted from the last checkpoint). In addition, the SCP selectively reboots several nodes as a preventive action, according to their uptime and current health status. This preventive action, which is called *software rejuvenation*, helps in getting rid of latent faults and effectively tolerates "soft" or aging-related faults such as memory leaks [20].

Transient and intermittent faults are assumed not to affect the reliability of the system after recovery actions, since the failed nodes will eventually return to use and since the recovery delay can be assumed to be negligible. If a permanent fault has occurred, the faulty node will not reboot successfully. In such a case, a notification is sent to the SE, the faulty node is removed from service, and the system reconfigured with the remaining working nodes. Subsequently, if the number of working nodes is smaller than the minimum requirement (two in the testbed), the whole system fails. We will only consider permanent faults in the current exposition.

The reliability model of the REE system that we describe here is a three-level hierarchical model adopted from [13]. The top-level model is the FT shown in Figure 18.24 and named M_{FT}. The node subsystem is configured as a k-out-of-n system of such nodes (with $k = 2$ in the prototype implementation).

The I/O subsystem (ios) also has redundancy in its configuration and is implemented as a parallel system with several redundant nodes. In the middle level, the failure behavior of individual nodes is captured using the RBD shown in Figure 18.25 and named M_{RBD}.

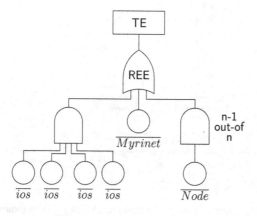

Figure 18.24   Top-level FT model (M_{FT}) of the REE system (adapted from [13, 17]).

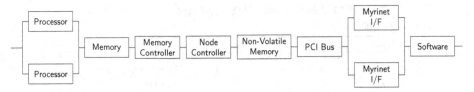

Figure 18.25   Middle-level RBD model (M_{RBD}) of the REE system (adapted from [13, 17]).

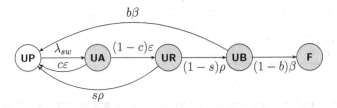

Figure 18.26   Low-level CTMC model (M_{CTMC}) of the REE system (adapted from [13, 17]).

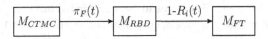

Figure 18.27   Import graph of the REE system (adapted from [13, 17]).

Each node uses redundancy in the form of spare processor chips and spare Myrinet interfaces to increase its reliability. All the nodes (compute and I/O) are assumed to exhibit the same stochastic behavior regarding fault occurrences and recovery, and the faults are mutually independent. The reliability (unreliability) of each individual node $(1 - R_{Node}(t))$ and $(1 - R_{ios}(t))$ is obtained by solving the middle-level RBD model, M_{RBD}, and is supplied as inputs to the top-level FT, M_{FT}. At the lowest level, the CTMC shown in Figure 18.26 captures the failure behavior of the software subsystem and is referred to as M_{CTMC}.

Most software failures in the nodes are expected to be recovered by retry, process restart or node reboot. The software CTMC model M_{CTMC} captures escalated levels of recovery actions. The software is considered to have failed only if the software failures cannot be recovered from, even after node reboot. Software failures that cannot be recovered from even after retry, restart or reboot have been identified as Bohrbugs, while those that can be recovered from by use of these actions have been called Mandelbugs [8]. The unreliability computed from the CTMC M_{CTMC}, at the lowest level, provides one of the inputs to the RBD model M_{RBD}. In this case, the software fails when the CTMC reaches State F and hence $\pi_F(t)$ is computed from M_{CTMC} and passed on to M_{RBD}. The import graph of the three-level hierarchical model is shown in Figure 18.27.

Example 18.6   *Epistemic Uncertainty Propagation for the NASA Remote Exploration and Experimentation System*
In the case study of Example 18.5, besides the computation of reliability, we also

Table 18.19 Point estimates of REE model input parameters.

Parameter	Value	Parameter	Value
λ_{pro}	0.01	ε	31 536 000
λ_{mem}	0.03	ρ	3 153 600
λ_{bmc}	0.01	β	52 560
λ_{ndc}	0.01	c	0.9
λ_{nvm}	0.01	r	0.9
λ_{bus}	0.02	b	0.9
λ_{nif}	0.03		
λ_{net}	0.0006		
λ_{sw}	8.76		

demonstrate epistemic parametric uncertainty propagation through the hierarchical reliability model. Unlike our earlier approach where the propagation was done analytically by the unconditioning integral as in Section 3.4, we will now be using a Monte Carlo sampling procedure. The times to failure of individual components in all the models are assumed to have an exponential distribution. Table 18.19 summarizes the point estimates of the parameters of the time to failure distributions of various components in the system and the coverage probabilities of various recovery actions in the CTMC of Figure 18.26. For the uncertainty in the model input parameters, we choose the half-width of the 95% two-sided confidence interval of the rate parameters to be ranging from 15% to 33% of the point estimate. For the coverage parameters, we choose the lower limit of the 95% upper one-sided confidence interval to be ranging from 80% to 90% of the point estimate.

The steps in uncertainty propagation through the reliability model of the REE system are as follows (see Section 3.4):

(a) Determining Epistemic Distributions of Model Input Parameters

We first determine the number of observations that would have been used by someone who computed the point estimate and 95% confidence interval, for each model input parameter, as explained in [13]. The number of observations calculated in this way for each model input parameter is summarized in Table 18.20. Since the times to failure of the components in the system have been assumed to follow exponential distributions, the epistemic distributions of the parameters as random variables are Erlang with parameters as summarized in Table 18.20. The parameters of the beta epistemic distributions (of the coverage parameters as random variables) are also given in the table.

(b) Determining Numbers of Samples from Epistemic Distributions

We compute the number of observations (e.g., the number of failures observed or number of faults injected) that would have been used to compute the confidence interval of the input parameter by inverting the relation between the width of the confidence interval, the number of observations and the point estimate of the input

Table 18.20 Epistemic distributions of parameters for the REE system.

Parameter	Num. observations	Epistemic distribution
λ_{pro}	117	Erlang($117, 8.57 \times 10^{-5}$)
λ_{mem}	167	Erlang($167, 1.79 \times 10^{-4}$)
λ_{bmc}	92	Erlang($92, 1.08 \times 10^{-4}$)
λ_{ndc}	117	Erlang($117, 8.54 \times 10^{-5}$)
λ_{nvm}	103	Erlang($103, 9.71 \times 10^{-5}$)
λ_{bus}	146	Erlang($146, 1.37 \times 10^{-4}$)
λ_{nif}	208	Erlang($208, 1.44 \times 10^{-4}$)
λ_{net}	19	Erlang($19, 4.21 \times 10^{-5}$)
λ_{sw}	65	Erlang($65, 7.42$)
β	81	Erlang($81, 1.54 \times 10^{-3}$)
b	Faults injected: 57	beta($52, 5$)
	Faults detected: 51	

parameter. We consider determining the number of observations when the aleatory distribution for time to failure follows an exponential distribution and the aleatory distribution of successfully detecting or recovering from failures follows a Bernoulli distribution.

Exponential Distribution: Number of Observations from Two-Sided Confidence Interval: When the time to failure of a component is exponentially distributed, the point estimate of the rate parameter $\hat{\lambda}$ of the exponential distribution is given by $\hat{\lambda} = r/\psi_r$, where r is the number of observed failures and ψ_r is the value of the accumulated life on test random variable Ψ_r. Making use of the upper and lower limits of the $100(1 - \alpha)\%$ two-sided confidence interval of λ [21, 22], the half-width of the confidence interval of λ, given by d, can be determined as:

$$d = \frac{1}{4\psi_r} \left\{ \chi_{2r,\alpha/2}^2 - \chi_{2r,1-\alpha/2}^2 \right\}, \tag{18.8}$$

where $\chi_{2r,1-\alpha/2}^2$ is the critical value of the chi-square distribution with $2r$ degrees of freedom. From Eq. (18.8), the number of observations (number of observed failures) that would have resulted in a given half-width d at a confidence coefficient $(1 - \alpha)$ can be computed as:

$$r = \left\lceil \frac{\hat{\lambda}}{4d} \left\{ \chi_{2r,\alpha/2}^2 - \chi_{2r,1-\alpha/2}^2 \right\} \right\rceil. \tag{18.9}$$

Starting with an initial assumption (using the normal approximation to the chi-square distribution) for the value of r, the fixed-point Eq. (18.9) is solved by successive substitution to converge at a value of r. Similar reasoning can be used to compute r when an upper or lower one-sided confidence interval is provided.

Bernoulli Distribution: Number of Observations from One-Sided Confidence Interval: The point estimate of the coverage probability is given by $\hat{c} = \psi_r/r$, where ψ_r is the value of the random variable Ψ_r denoting the number of faults/errors detected and recovered, and r is the total number of faults/errors injected. The lower limit of the upper one-sided $100(1 - \alpha)\%$ confidence interval for the coverage probability is given by [22]:

$$c_{\text{L}} = 1 - \frac{\chi^2_{2(r-\psi_r+1),\alpha}}{2r}. \tag{18.10}$$

Inverting Eq. (18.10), the number of injections that would have resulted in this lower limit of the confidence interval, c_{L}, at a confidence coefficient $(1 - \alpha)$ can be obtained as:

$$r = \left\lceil \frac{\chi^2_{2(r(1-\hat{c})+1),\alpha}}{2(1 - c_{\text{L}})} \right\rceil. \tag{18.11}$$

Equation (18.11) is solved by successive substitutions to obtain the value of r by using an initial approximation for r. The initial value is calculated using the normal approximation of the chi-square distribution.

The number of samples that we utilize in our Monte Carlo procedure from the epistemic distribution of each input parameter is determined to be $n = 208$, the largest of all the numbers of observations of the input parameters.

(c) Sampling Procedure

Subsequent to the determination of the epistemic distribution of each input parameter, samples or random deviates need to be drawn from these distributions. We employ the Latin hypercube sampling (LHS) [23] procedure in our uncertainty propagation method, which divides the entire probability space into equal intervals (the number of intervals is equal to the number of samples needed) and draws a random sample from each of the intervals. Thus it is almost guaranteed to address the entire probability space evenly and to easily reach low probability / high impact areas of the epistemic distribution of input parameters. Once the samples are generated, random pairings without replacement are carried out between samples of different parameters, to ensure randomness in the sequence of the samples. It has been shown that for a given sample size, a statistic of model output obtained by the LHS sampling procedure will have a variance less than or equal to that obtained by a random sampling procedure [23, 24]. The actual samples of random numbers or the random deviates can be generated by any of a number of methods, such as the inverse transform method, rejection sampling, Box–Mueller transform or Johnson's translation [25].

A sample from the distribution of each of the k model input parameters will result in a vector of k parameter values. We use the LHS sampling procedure to draw a total number, n, of samples (sample size as determined in point (b) above) from each distribution. In the examples shown in this paper, for the rate parameters in the models we sample from the Erlang distribution. For the coverage factors in the models, we sample from the beta distribution defined in Table 18.20. To implement the LHS procedure, we generate random deviates from a uniform distribution in the intervals the probability space has been divided into, for LHS sampling. Using those random deviates, we obtain quantiles from the Erlang or beta distributions. We also use the function *RandomSample* to obtain random pairing without replacement between samples from different epistemic distributions.

(d) Solving the Analytic Model
For each vector of the $n = 208$ set of values of the input parameters, the three-level hierarchical model is solved using SHARPE.

(e) Summarizing the Output Values
The empirical Cdf of the reliability of the system output is constructed from the $n = 208$ values of the output obtained by solving the system reliability model using SHARPE. The two-sided 95% confidence interval of the reliability at $t = 5$ yr is computed to be $(0.949485, 0.981994)$. The small width of the 95% two-sided confidence interval (0.032509) suggests that the epistemic uncertainties in the model input parameters used in this example did not introduce significant uncertainty in the model output parameters. Higher epistemic uncertainty in the model input parameters (wider confidence intervals) may result in a higher uncertainty in the model output.

Problems

18.5 Derive a closed-form solution for the reliability of the REE system using the three-level hierarchical model of Example 18.5.

18.6 Reproduce the results of Example 18.6 using ordinary Monte Carlo sampling and then using LHS sampling and compare the results.

18.7 Obtain derivatives of the system reliability with respect to all the input parameters using the results of Problem 18.5.

18.8 Carry out numerical integration for the uncertainty propagation using the results of Problem 18.5.

18.3 Combined Performance and Reliability Models

The first case study in this section concerns the performance and reliability of a composite web service for a travel agent. We consider a sequence of one-level CTMC models with various options for failures and recovery. We use a composite model that incorporates both failures and performance [26]. In the second case study, we consider a hierarchical model for the DPM computation in the IBM SIP case study whose hierarchical availability model was presented in Example 18.4.

Example 18.7   *Performance and Reliability of Composite Web Services*

This example is adapted from [26]. Consider a business process that is a composition of multiple web services, called TravelAgent (see Figure 18.28).

By combining existing services using a high-level language such as BPEL, service providers can quickly develop new services. Figure 18.29 shows a concrete implementation of this process in BPEL. The process tries to make an airline reservation in a parallel manner: First, it looks up two different airlines for available routes and fares. When they both respond, it chooses one of the airlines based on some criterion such as fare, schedule, and so forth. Otherwise, when either one of the two airlines fails to respond, it chooses the other airline. When the flight is reserved, the process goes to the hotel reservation step. In the case that both airlines fail to respond, then the process gives up and aborts.

We assume throughout that the times to complete all the individual web services are exponentially distributed. Similarly, we assume that the overhead time to conduct a restart is also exponentially distributed. Thus, each of the models we develop is a homogeneous CTMC. We consider the following three cases.

Case 1: No Failures

We start with the simple case where none of the web services fails. In this case, the BPEL process in Figure 18.29 can be modeled by the CTMC shown in Figure 18.30(a). The parallel invocation in Figure 18.29 gets translated into four states. In the state labeled Airline selection (1/2), both activities are proceeding concurrently. After one of them finishes, only the other one is active. Finally, when both finish, we proceed to make the reservation. All the transitions are labeled with the reciprocal of the mean response time of the related activity.

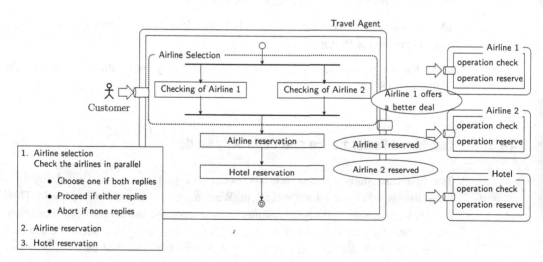

Figure 18.28   The TravelAgent process.

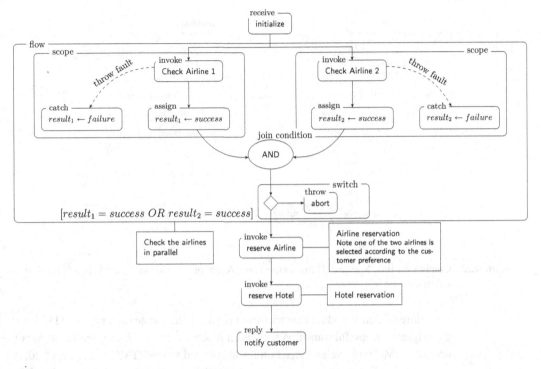

Figure 18.29   A composite web service in BPEL.

Denoting the response time random variable of service j by T_j, the overall response time as a random variable can easily be written as:

$$T_{\text{sys}} = T_{\text{i}} + \max\{T_{\text{a1}}, T_{\text{a2}}\} + T_{\text{ar}} + T_{\text{ht}} + T_{\text{cn}}. \tag{18.12}$$

Recalling that the expected value of a sum is the sum of the expected values, and recalling Eq. (4.30) regarding the expected value of the maximum of two independent exponentially dsitributed random variables, we have an expression for the overall mean response time:

$$mr_{\text{sys}} = mr_{\text{i}} + \left(mr_{\text{a1}} + mr_{\text{a2}} - \frac{1}{\dfrac{1}{mr_{\text{a1}}} + \dfrac{1}{mr_{\text{a2}}}} \right) + mr_{\text{ar}} + mr_{\text{ht}} + mr_{\text{cn}}. \tag{18.13}$$

In this case, in the absence of failures, the overall service reliability is clearly $R_{\text{sys}} = 1$.

Case 2: CTMC with Failures but No Recovery

Invocation of each web service within a BPEL process may fail. To take account of this possibility, we add a single failure state to the CTMC of Figure 18.30(b). In order to account for individual web service failures, we need to modify the transition rates of the (no failures) CTMC in the following manner. Suppose an individual web service j takes λ_j^{-1} on average and it has a probability of R_j for successful completion (i.e., $(1 - R_j)$

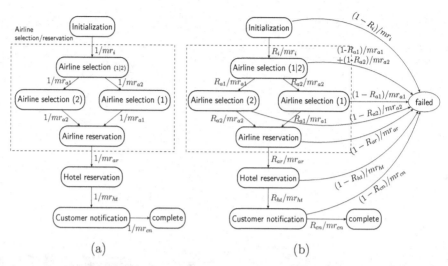

Figure 18.30   Continuous-time Markov chains for the TravelAgent process in two cases: (a) no failures; (b) failures but no recovery,

for failure). Then we add a new transition (to the failure state) at a rate $\lambda_j \cdot (1 - R_j)$ and the original successful transition now has a modified rate $\lambda_j \cdot R_j$. Figure 18.30(b) is the revised CTMC with web service failures introduced to the CTMC of Figure 18.30(a).

The system's overall mean response time (for completed requests) can be shown, using Eq. (10.74), to be:

$$mr_{\text{sys}} = mr_i + R_i \cdot \frac{1 + R_{a2} \cdot \dfrac{mr_{a1}}{mr_{a2}} + R_{a1} \cdot \dfrac{mr_{a2}}{mr_{a1}}}{\dfrac{1}{mr_{a1}} + \dfrac{1}{mr_{a2}}} + R_i \cdot R_{a1} \cdot R_{a2}$$

$$\cdot (mr_{ar} + R_{ar} \cdot (mr_{ht} + R_{ht} \cdot mr_{cn})).$$

In this case, with failures but no recovery [Figure 18.30(b)], the overall service reliability can easily be shown, using Eq. (10.72) for instance, to be:

$$R_{\text{sys}} = R_i \cdot R_{a1} \cdot R_{a2} \cdot R_{ar} \cdot R_{ht} \cdot R_{cn}. \tag{18.14}$$

Case 3: CTMC with Failures and Recovery

For high reliability, BPEL processes often specify recovery procedures via fault handlers that are invoked for restarting failed invocations.

Figure 18.31 shows the CTMC with failures and recovery via restarts. We have assumed that restart is successful with probability C_j, while it fails to recover with probability $1 - C_j$. We also allow for an overhead time for restart. Thus, for instance, upon the failure of the hotel invocation, a restart attempt is made with the mean overhead time of ot_{ht} and probability of success C_{ht}. We assume that there is no restart for the airline invocation. Further, if either one or both airline invocations is successful, we proceed further in the flow. Upon the failure of both invocations, we abort. In Figure 18.31, states R_{ini}, R_{Ar}, R_{Ht}, and R_{Cn} indicate that the corresponding activity

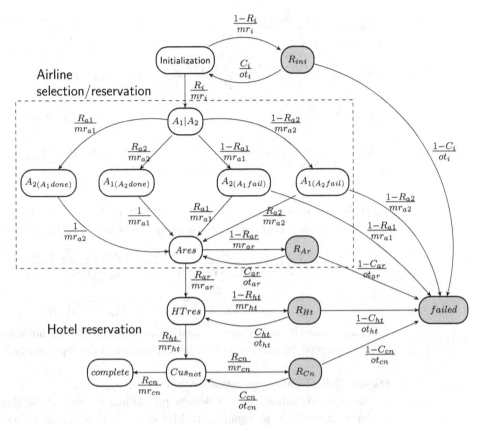

Figure 18.31 Continuous-time Markov chain model for the TravelAgent process in the case with restarts.

did not complete and a restart is attempted with mean overhead times ot_i, ot_{ar}, ot_{ht}, and ot_{cn}, respectively.

In this case, the system mean response time can be derived using Eq. (10.74) to be:

$$mr_{sys} = V_i \cdot mr_i + V_{1|2} \cdot \cfrac{1}{\cfrac{1}{mr_{a1}} + \cfrac{1}{mr_{a2}}} + V_{a2} \cdot mr_{a2} + V_{f1} \cdot mr_{a2} + V_{a1} \cdot mr_{a1} + V_{f2} \cdot mr_{a1}$$

$$+ V_{ar} \cdot mr_{ar} + V_{ht} \cdot mr_{ht} + V_{cn} \cdot mr_{cn} + V_{Rini} \cdot ot_i + V_{Rar} \cdot ot_{ar}$$

$$+ V_{Rht} \cdot ot_{ht} + V_{Rcn} \cdot ot_{cn}, \tag{18.15}$$

where the average numbers of visits to the states are:

$$V_i = \frac{1}{1 - C_i(1 - R_i)}, \qquad V_{1|2} = R_i \cdot V_i,$$

$$V_{a1} = \frac{R_{a2} \cdot V_{1|2}}{mr_{a2} \cdot \left(\dfrac{1}{mr_{a1}} + \dfrac{1}{mr_{a2}}\right)}, \qquad V_{a2} = \frac{R_{a1} \cdot V_{1|2}}{mr_{a1} \cdot \left(\dfrac{1}{mr_{a1}} + \dfrac{1}{mr_{a2}}\right)},$$

$$V_{F1} = \frac{(1 - R_{a1}) \cdot V_{1|2}}{mr_{a1} \cdot \left(\dfrac{1}{mr_{a1}} + \dfrac{1}{mr_{a2}}\right)}, \qquad V_{F2} = \frac{(1 - R_{a2}) \cdot V_{1|2}}{mr_{a2} \cdot \left(\dfrac{1}{mr_{a1}} + \dfrac{1}{mr_{a2}}\right)},$$

$$V_{ar} = \frac{V_{a1} + R_{a2} \cdot V_{F1} + V_{a2} + R_{a1} \cdot V_{F2}}{1 - C_{ar} \cdot (1 - R_{ar})}, \qquad V_{Rar} = (1 - R_{ar}) \cdot V_{ar},$$

$$V_{ht} = \frac{R_{ar} \cdot V_{ar}}{1 - C_{ht} \cdot (1 - R_{ht})}, \qquad V_{Rht} = (1 - R_{ht}) \cdot V_{ht},$$

$$V_{cn} = \frac{R_{ht} \cdot V_{ht}}{1 - C_{cn} \cdot (1 - R_{cn})}, \qquad V_{Rcn} = (1 - R_{cn}) \cdot V_{cn},$$

$$V_{Rini} = (1 - R_i) \cdot V_i.$$

Finally in this case with failures and restarts (CTMC of Figure 18.31), the overall service reliability in closed form can be shown, using, for instance, Eq. (10.72), to be:

$$R_{sys} = \frac{R_i}{(1 - C_i \cdot (1 - R_i))} \cdot \frac{(R_{a1} + R_{a2} - R_{a1} \cdot R_{a2}) \cdot R_{ar}}{(1 - C_{ar} \cdot (1 - R_{ar}))}$$
$$\cdot \frac{R_{ht}}{(1 - C_{ht} \cdot (1 - R_{ht}))} \cdot \frac{R_{cn}}{(1 - C_{cn} \cdot (1 - R_{cn}))}.$$

For the sensitivity and bottleneck analysis as well as parametrization and validation of the above models, see [26]. For further exploration of this topic, see [27].

Example 18.8 *DPM Computation for the IBM SIP System [28]*
We have already introduced the defects per million measure in Section 9.5.7, and we have shown how to compute DPM for a simple application server servicing SIP request messages in Examples 9.18 and 10.19. This case study is adapted from [28]. The DPM computation method is applied to IBM SIP SLEE (Service Logic Execution Environment) platform consisting of the WebSphere application server whose availability was analyzed in Example 18.4 and whose architecture is sketched in Figure 18.13. Our method takes into account software/hardware failures, different stages of recovery, different phases of call flow, retry attempts and the interactions between call flow and failure/recovery behavior. It can be regarded as a two-level hierarchical model of combined performance and dependability.

The SIP application installed on the application servers is termed a back-to-back user agent (B2BUA), which mediates SIP messages between two legs in VoIP call sessions. During the call session, the user agent client (UAC) and the user agent server (UAS) exchange SIP messages to establish, maintain, and terminate the call session. All SIP messages from one side have to go through the B2BUA in order to be forwarded to the other side. The call flow for the B2BUA is shown in Figure 18.32.

The UAC first sends an INVITE message to the UAS through the B2BUA proxy. The UAS replies to the UAC with a RINGING message, then pauses for 15 s before sending an OK message to the UAC indicating that the phone has been picked up. The UAC replies to the UAS with an ACK message and the call session is then set up. The UAC pauses for 45 s, assuming a phone conversation of 45 s, and then sends an INFO message to the UAS. The UAS sends back an OK message upon receiving the INFO

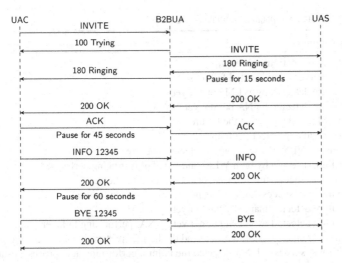

Figure 18.32   Call flow for B2BUA.

message from the UAC. The UAC pauses for another 60 s, then sends a BYE message to terminate the call session. The UAS replies with an OK message and the session is then terminated.

We consider three categories of failures in the IBM SIP SLEE cluster: replication domain failure, node reboot, and proxy server failure. We do not consider hardware failures [6] in this case study as their contribution to DPM is found to be negligible. For each failure type i, we calculate the failure frequency f_i and the average number of calls, n_i, lost per failure. Then the overall DPM is written as:

$$\text{DPM} = \left(\sum_{i \in \Phi} f_i n_i \right) \frac{10^6}{6\lambda}, \tag{18.16}$$

where $\Phi = \{\text{domains, reboots, proxies}\}$ is the set of failure modes, λ is the call arrival rate at a single replication domain and 6λ is the total call arrival rate in the SIP SLEE cluster. The quantity $f_i n_i$ is the call loss rate per unit time, and since the average number of incoming calls per unit time to the cluster is 6λ (recall the property of a Poisson process), $f_i n_i / 6\lambda$ gives the fraction of calls that are lost.

We now present a CTMC availability model for a single replication domain for computing the failure frequency f_i. Recall the CTMC availability model of an individual application server (AS) that we presented in Figure 18.21. The current model shown in Figure 18.33 takes two such single AS models and combines them, adding failover actions from one AS to the other upon its failure.

In state 2U, both application servers are functioning properly. In state UO, one application server has failed. The failure is detected by the workload manager (WLM) or the node agent (NA), or manually. A failover is triggered if the failure is detected by WLM, while the failed server process is automatically restarted if the failure is detected by NA. Manual recovery will occur if these automatic recoveries do not succeed. Manual recovery starts by manually restarting the failed server process; if that does not

Table 18.21 States in replication domain CTMC model.

State	Meaning
2U	Both applications servers up
UO	One server in undetected failure state
1D	Failure detected by WLM but not yet by NA
2D	Failure detected by NA but not yet by WLM
1N	WLM unable to detect the failure
2N	NA unable to detect the failure
UN	Neither WLM nor NA able to detect the failure
1D2D	Failure detected by both WLM and NA; failover being performed
FS	Failover successful, NA has not yet detected the failure
FN	Failover unsuccessful, NA has not yet detected the failure
MD	Failure detected manually but not yet by WLM
1D2N	Failure detected by WLM but not yet by NA; performing failover
UC	Failover unsuccessful, NA detected the failure; performing auto process restart
UA	Failover successful, NA detected the failure; performing auto process restart
UR	Failover successful, performing manual process restart
US	Failover unsuccessful, performing manual process restart
UB	Failover successful, manually reboot blade
UT	Failover unsuccessful, manually reboot blade
RE	Failover successful, performing manual repair
RP	Failover unsuccessful, performing manual repair

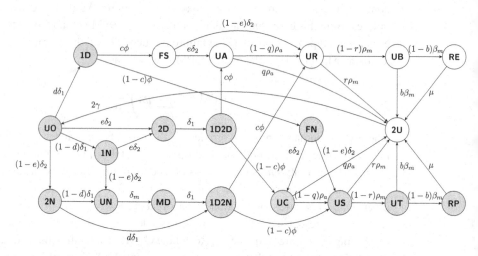

Figure 18.33 Continuous-time Markov chain availability model, $M_{\text{Rep-ss}}$, for a replication domain.

work, manual reboot of the node for the failed server is carried out. If still unsuccessful, manual repair is finally done. The states are described in Table 18.21.

Assuming that the successful detection probabilities by WLM and NA are d and e, respectively, then if the WLM detects the failure first, the CTMC enters state 1D, where a failover is performed, and in the meantime the NA is still trying to detect the failure. We assume that the NA will not detect the failure before failover is completed (which overestimates the DPM due to replication domain failures). Then, with probability c the failover is successful and the CTMC enters state FS, where the NA is still attempting to detect the failure. With probability e the failure is detected by NA, and the CTMC enters states UA, UR, UB, and RE, in sequence, for auto process restart, manual process restart, manual reboot and manual repair. With probability $1 - e$, NA is not able to detect the failure, and hence from state FS the CTMC enters the states UR, UB, and then RE. If the failover is unsuccessful in state 1D, the CTMC will go through states FN, UC, US, UT, and RP that correspond to NA detection, auto process restart, manual process restart, manual reboot and manual repair.

If NA detects the failure first, the CTMC enters state 2D. We assume that the WLM will subsequently always detect the failure if NA detects it first. Therefore, after the WLM detection delay the CTMC enters state 1D2D, wherein both failover and auto process restart are being performed. We assume that the auto process restart will not finish before the failover is completed (which also overestimates the DPM of the replication domain). Then from state 1D2D the CTMC will go through either the state sequence UA → UR → UB → RE or the sequence UC → US → UT → RP, depending on whether or not the failover is successful. While in state UO, if the WLM detection delay has elapsed but the failure is not detected by WLM, the model enters state 1N, where the failure is still being detected by NA. With probability e, NA detects the failure and the CTMC enters state 2D; with probability $1 - e$, the CTMC enters state UN, in which case neither WLM nor NA has been able to detect the failure. While in state UO, if the NA detection delay has elapsed but the failure is not detected by NA, the CTMC enters state 2N where the failure is still being detected by WLM. With probability d, WLM detects the failure and the CTMC enters state 1D2N; with probability $1 - d$, the WLM is not able to detect the failure and the CTMC enters state UN. In state UN, the failure is being manually detected. After manual detection the CTMC enters state MD, in which manual recovery is performed. We assume that the WLM can always detect the failure as soon as some recovery procedure is initiated. Therefore, the CTMC enters state 1D2N from MD after the WLM detection delay, and will enter states US → UT → RP in turn for manual process restart, node reboot and manual repair. Some of the parameters of the CTMC model were already presented earlier (in Tables 18.14–18.16). The remaining ones are shown in Table 18.22.

From the CTMC availability model, we can compute the failure frequency needed for the DPM computation. For a single replication domain, the failure rate is 2γ. Given its availability model in Figure 18.33, the failure frequency for an individual replication domain is $2\gamma \pi_{2U}$, where π_{2U} is the steady-state probability that the CTMC is in state 2U. The total failure frequency for the six replication domains is then

$$f_{\text{domains}} = 12\gamma \pi_{2U}. \tag{18.17}$$

The proxy CTMC availability model is the same as that of the replication domain except that the WLM is replaced by the IP sprayer. The input parameters of the proxy CTMC model are also the same as those for the replication domain model. Thus the failure

Table 18.22 Replication domain parameters.

Parameter	Meaning	Value
$1/\phi$	Mean time for failover	1 s
c	Coverage factor for failover	0.9
t_e	Call setup phase duration	15 s
t_1	Stable phase 1 duration	45 s
t_2	Stable phase 2 duration	60 s
rw_o	Retry window for non-INVITE messages	12 s
rw_i	Retry window for INVITE message	4 s
s_o	Number of retries for non-INVITE messages	4
s_i	Number of retries for INVITE message	3

frequency of the two proxies paired together is

$$f_{\text{proxies}} = 2\gamma\pi_{2U}. \tag{18.18}$$

Consider replication domain 1, that runs on two separate nodes which are shared by two other replication domains 2 and 4. When a failure occurs in domain 2 (or 4), a hardware reboot might be necessary to recover from the failure. With 50% probability the node being rebooted has replication domain 1 residing on it. Therefore the frequency of reboot failure in domain 1 is 0.5 times the sum of the reboot frequencies in domain 2 and domain 4. Since each domain has the same reboot frequency, the reboot failure frequency of domain 1 equals its own reboot frequency, computed as $\beta_m(\pi_{UB} + \pi_{UT})$, where π_{UB} and π_{UT} are the respective probabilities that the domain availability model (Figure 18.33) is in state UB and UT. The overall reboot failure frequency for the six replication domains is

$$f_{\text{reboots}} = 6\beta_m(\pi_{UB} + \pi_{UT}). \tag{18.19}$$

We next derive the formula for the mean number of lost calls per replication domain failure. We omit the derivation of the mean number of lost calls due to proxy or reboot failures – interested readers may consult [28] for these and other details. For this derivation, a call session is divided into four phases:

- *New call arrival:* starts from the initial INVITE message and ends after the B2BUA $\rightarrow$ UAC 180 Ringing message.
- *Setup phase:* starts with $t_e = 15$ s pause and ends after the B2BUA $\rightarrow$ UAS ACK message.
- *Stable phase 1:* starts with $t_1 = 45$ s pause and ends after B2BUA $\rightarrow$ UAC 200 OK message.
- *Stable phase 2:* starts with $t_2 = 60$ s pause and ends with the session.

The four phases in the call session are seen as the user behavior graph [29, 30] shown in Figure 18.34.

The following scenarios will result in call losses:

Figure 18.34 Graphical representation of a call session.

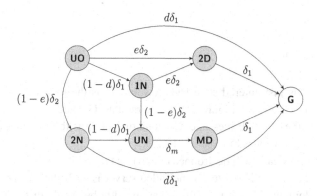

Figure 18.35 Loss model, $M_{\text{Rep-new}}$, for newly arriving calls.

- Newly arriving calls are dropped if the initial INVITE message is sent to a failed application server and this failure is not detected by the WLM within the rw_i seconds retry window. If the WLM fails to detect the failure, all newly arriving calls are dropped until the failed server is manually recovered.
- When an application server fails (indicated by the transition from state 2U to state UO), all calls in the setup phase in the failed server will be lost.
- All calls in stable phase 1 will be lost if the UAC → B2BUA INFO message was sent to a failed application server and the failover does not occur within the rw_o seconds retry window (either because the failover is unsuccessful or the time between the INFO message arrival and the failover is longer than rw_o seconds).
- All calls in stable phase 2 will be lost if the UAC → B2BUA BYE message was sent to a failed application server and the failover does not occur within the rw_o seconds retry window (either because the failover is unsuccessful or the time between the BYE message arrival and the failover is longer than rw_o seconds).

We use rw_i to denote the maximum retry window for the INVITE messages, and rw_o to denote the maximum retry window for all non-INVITE messages. We also use s_i for the maximum number of retries for an INVITE message during the rw_i seconds period, and s_o for the maximum number of retries for a non-INVITE message during the rw_o seconds period.

To compute the mean number of newly arriving calls lost per failure, we consider the CTMC model of Figure 18.35, which shows the state transitions after a replication domain failure has occurred. This CTMC is extracted from the replication domain availability model of Figure 18.33.

In Figure 18.35, G is an absorbing state obtained by merging the states where WLM has already detected the failure – that is, by merging states 1D, 1D2D, 1D2N, and 2U of Figure 18.33. Suppose a domain failure occurs at time 0 and thus the CTMC

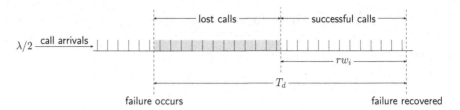

Figure 18.36 Lost newly arriving calls.

of Figure 18.35 has just entered state UO. Assume T_d is the time to absorption (i.e., the time until the CTMC of Figure 18.35 enters state G), and rw_i is the maximum retry window for the INVITE message. Computation of T_d is then a first-passage time problem like those considered in Section 10.2.5.

If $T_d < rw_i$, no new calls will be dropped due to this failure. If, on the other hand, $T_d \geq rw_i$, then since the call arrival rate for the failed server is $\lambda/2$ (the call arrival rate for one replication domain is λ and the calls are assumed to be evenly distributed between the two application servers in that domain, therefore for one server the call arrival rate is $\lambda/2$), the mean number of new calls dropped is then $(T_d - rw_i)\lambda/2$. This follows from the property of a Poisson arrival process (Section 3.5.3) where the expected number of arrivals in a duration of length t is the arrival rate multiplied by t. Figure 18.36 shows how the newly arriving calls are affected by T_d.

T_d is a random variable and its cumulative distribution function, $F_d(x)$, can be computed from the CTMC of Figure 18.35: $F_d(x) = \pi_G(x)$, where $\pi_G(x)$ is the transient probability that the model of Figure 18.35 is in state G at time x given that it started in state UO at time 0. Hence the mean number of newly arriving calls dropped due to a server failure is given by:

$$n_a = \int_{rw_i}^{\infty} (x - rw_i)\,\lambda/2\,dF_d(x) \tag{18.20}$$

$$= \int_{rw_i}^{\infty} (x - rw_i)\,\lambda/2\,d\pi_G(x)$$

$$= \left[\text{MTTA} - rw_i + \int_{0}^{rw_i} \pi_G(x)dx \right] \lambda/2, \tag{18.21}$$

where $\text{MTTA} = \int_{0}^{\infty} x\,d\pi_G(x)$ is the mean time to absorption in state G measured from the time to entry into state UO, and can be computed from the CTMC of Figure 18.35 as described in Section 10.2.2 or using a software package such as SHARPE. Similarly, the integral of the transient state probability above can also be computed using SHARPE [31].

Given the call arrival rate λ (calls per second) for the replication domain, the mean number of calls in the setup phase is λt_e, calls in stable phase 1 is λt_1, and those in stable phase 2 is λt_2. These expressions are based on the $M/G/\infty$ formula [32]. The parameters t_e, t_1, and t_2 are defined in Table 18.22.

The up states for different call types are shown in Table 18.23. The remaining states are down states for the corresponding call type. All setup calls will be lost when a failure

Table 18.23 Up states by call type.

Call type	Up states
Setup calls	2U
Stable calls	2U, FS, UA, UR, UB, RE
New calls	2U, 1D, 1D2D, FS, FN, UA, UR, UB, UC, RE, 1D2N

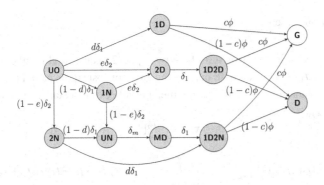

Figure 18.37 Loss model, $M_{\text{Rep-stable}}$, for stable calls.

occurs. Therefore the only up state for setup calls is 2U, which represents no failure in the system. New calls can be processed once WLM detects the failure. Therefore, besides 2U, the up states are those representing WLM detecting the failure. Stable calls can be processed after successful failover, and the up states in addition to 2U are those in which the failover has been successfully carried out.

The expected number of existing setup calls in the replication domain is λt_e. Half of these calls were being processed by the failed application server. Hence the expected number of setup calls that will be lost during one application server failure is given by:

$$n_e = \lambda t_e / 2. \tag{18.22}$$

When an application server fails, there are $\lambda t_1/2$ calls in the failed server that are in stable phase 1. The INFO requests sent by these calls will still be directed to the failed server before the failure is recovered. Because these $\lambda t_1/2$ calls arrive at the application server at different times, the time for them to issue the INFO message is also different. And because the call arrival rate is $\lambda/2$ for each server, we assume that the INFO request rate issued by these $\lambda t_1/2$ calls is $\lambda/2$.

Figure 18.37 shows the loss model for stable calls. In this model, G and D are two absorbing states, where G represents failover completing successfully, and D means failover is not performed or not successful. As stated earlier, if the time period between the INFO message being sent and the model's entry into state G is less than the maximum retry window, the call will not be lost, otherwise it will be dropped.

Suppose that T_d is the time period between the failure occurrence and the model entering state G ($T_d = \infty$ if the model never enters G), and rw_o is the maximum retry window for non-INVITE messages. If T_d is less than rw_o, no call will be lost; otherwise,

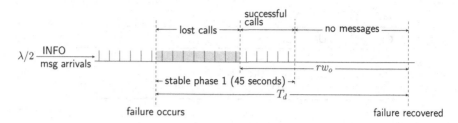

Figure 18.38 Lost stable calls in phase 1.

the mean number of lost calls is $\min(T_d - rw_0, t_1)\lambda/2$, as the total number of lost calls cannot exceed $\lambda t_1/2$. Figure 18.38 depicts the case for lost stable calls in phase 1.

From the explanation above we get the mean number of lost stable calls in phase 1 as

$$
\begin{aligned}
n_1 &= \int_{rw_0}^{t_1+rw_0} (x - rw_0)\lambda/2\, d\pi_G(x) + \lambda t_1/2[1 - \pi_G(t_1 + rw_0)] \\
&= \left[(x - rw_0)\pi_G(x)\big|_{rw_0}^{t_1+rw_0} - rw_0 \int_{rw_0}^{t_1+rw_0} \pi_G(x)dx + t_1 - t_1\pi_G(t_1 + rw_0) \right]\frac{\lambda}{2} \\
&= \left[t_1 - \int_{rw_0}^{t_1+rw_0} \pi_G(x)dx \right]\lambda/2.
\end{aligned}
\tag{18.23}
$$

Similarly, we can get the mean number of stable calls lost in phase 2 by replacing t_1 with t_2 in Eq. (18.23):

$$
n_2 = \left[t_2 - \int_{rw_0}^{t_2+rw_0} \pi_G(x)dx \right]\lambda/2.
\tag{18.24}
$$

From Eqs. (18.22), (18.21), (18.23), and (18.24), the total number of lost calls due to one replication domain failure is the sum of lost calls for each call type:

$$
n = n_e + n_a + n_1 + n_2.
\tag{18.25}
$$

Of the twelve application servers in the six replication domains, there is one core group coordinator. The WLM detection delay for a coordinator failure is longer (on the order of seconds) than for a non-coordinator failure. This small change of parameter δ_1 in Table 18.22 has little impact on failure frequency, which is on the order of one per thousands of hours. Therefore, we ignore the change when computing π_{2U} from Figure 18.33. However, it has a great impact on the mean number of lost calls per failure (n in Eq. (18.25), which will now be considered a function of δ), as calls might be lost if the failure cannot be recovered in seconds. Therefore, when applying the previous method, a longer δ_1 is used for coordinator failure.

Let the average number of lost calls computed from Eq. (18.25) be $n^{(c)}$ for a coordinator failure, and n for a non-coordinator failure. The former occurs with probability $1/12$ and the latter with probability $11/12$. Then the average number of lost calls per failure is

$$
n_{\text{domains}} = n^{(c)}/12 + 11n/12.
\tag{18.26}
$$

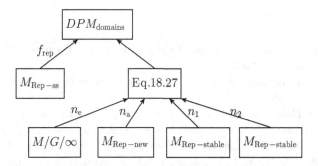

Figure 18.39 Import graph for the computation of the DPM caused by self-failures for replication domains.

Hence, the DPM caused by self-failures over six replication domains is

$$\text{DPM}_{\text{domains}} = \pi_{2U} 12\gamma \left(\frac{n^{(c)}}{12} + \frac{11n}{12} \right) \frac{10^6}{6\lambda}. \tag{18.27}$$

The import graph for the computation of the DPM due to domain failures is presented in Figure 18.39.

For the computation of the mean number of lost calls per reboot failure and due to proxy failures, the reader is referred to [28, 33]. Numerical values and the sensitivity of DPM to various parameters can also be found in the cited references.

Problems

18.9 Derive Eq. (18.13) for the overall mean response time of the CTMC of Case 1 shown in Figure 18.30(a) using Eq. (10.49).

18.10 Derive Eq. (18.15) for the overall mean response time of the CTMC of Case 3 shown in Figure 18.31 using Eq. (10.74).

18.11 Derive Eq. (18.7) for the overall reliability of the CTMC of Case 3 shown in Figure 18.31 using Eq. (10.72).

18.12 Obtain a closed-form solution for π_{2U} and for n_{domains}, and thus derive a closed-form expression for $\text{DPM}_{\text{domains}}$.

18.13 Derive Eq. (18.21) by integrating by parts Eq. (18.21).

18.4 Further Reading

For other case studies on reliability modeling, see [34–46]. For further case studies on availability modeling, see [47–58]. For more case studies on performance modeling, see [59–64], while for performability modeling, see [65–72]. For more case studies on survivability, see [73–75].

References

[1] K. S. Trivedi, "Availability analysis of Cisco GSR12000 and Juniper M20/M40," Duke University, Tech. Rep., 2000.

[2] F. Lee and M. Marathe, *Beyond Redundancy: A Guide to Designing High-Availability Networks*. Cisco, 1999.

[3] F. Lee, *Cisco High-Availability Initiatives and Related Design Parameters*. Cisco, 2000.

[4] K. Trivedi, R. Vasireddy, D. Trindade, S. Nathan, and R. Castro, "Modeling high availability systems," in *Proc. IEEE Pacific Rim Int. Symp. on Dependable Computing (PRDC)*, 2006.

[5] W. E. Smith, K. S. Trivedi, L. Tomek, and J. Ackaret, "Availability analysis of blade server systems," *IBM Systems Journal*, vol. 47, no. 4, pp. 621–640, 2008.

[6] K. S. Trivedi, D. Wang, J. Hunt, A. Rindos, W. E. Smith, and B. Vashaw, "Availability modeling of SIP protocol on IBM© WebSphere©," in *Proc. Pacific Rim Int. Symp. on Dependable Computing (PRDC)*, 2008, pp. 323–330.

[7] Internet Engineering Task Force. [Online]. Available: www.ietf.org/

[8] M. Grottke and K. S. Trivedi, "Fighting bugs: Remove, retry, replicate, and rejuvenate," *IEEE Computer*, vol. 40, no. 2, pp. 107–109, 2007.

[9] R. Matias, K. S. Trivedi, and P. R. M. Maciel, "Using accelerated life tests to estimate time to software aging failure," in *Proc. IEEE 21st Int. Symp. on Software Reliability Engineering*, 2010, pp. 211–219.

[10] J. Zhao, Y. Jin, K. S. Trivedi, R. M. Jr., and Y. Wang, "Software rejuvenation scheduling using accelerated life testing," *JETC*, vol. 10, no. 1, pp. 9:1–9:23, 2014.

[11] M. Hsueh, T. Tsai, and R. Iyer, "Fault injection techniques and tools," *IEEE Computer*, vol. 30, no. 4, pp. 75–82, Apr. 1997.

[12] R. K. Iyer, N. Nakka, W. Gu, and Z. Kalbarczyk, "Fault injection," in *Encyclopedia of Software Engineering*, Nov 2010, ch. 29, pp. 287–299.

[13] K. Mishra, K. S. Trivedi, and R. Some, "Uncertainty analysis of the remote exploration and experimentation system," *Journal of Spacecraft and Rockets*, vol. 49, pp. 1032–1042, 2012.

[14] J. H. Lala and J. T. Sims, "A dependability architecture framework for remote exploration and experimentation computers," *Fast Abstract, FTCS-29*, 1999.

[15] Myrinet. [Online]. Available: https://www.hpcresearch.nl/euroben/Overview/web11/myrinet.php

[16] J. Rohr, "Software-implemented fault tolerance for supercomputing in space," in *Proc. 28th Int. Symp. Fault-Tolerant Computing*. IEEE Computer Society, 1998.

[17] D. Chen, S. Dharmaraja, D. Chen, L. Li, K. S. Trivedi, R. R. Some, and A. P. Nikora, "Reliability and availability analysis for the JPL remote exploration and experimentation system," in *Proc. 2002 Int. Conf. on Dependable Systems and Networks*, 2002, pp. 337–344.

[18] A. Avizienis, "The methodology of *n*-version programming," in *Software Fault Tolerance*, ed. M. R. Lyu. John Wiley & Sons, 1994, ch. 2, pp. 23–46.

[19] B. Randell and J. Xu, "The evolution of the recovery block concept," in *Software Fault Tolerance*, ed. M. R. Lyu. John Wiley & Sons, 1994, ch. 1, pp. 1–22.

[20] Y. Huang, C. M. R. Kintala, N. Kolettis, and N. D. Fulton, "Software rejuvenation: Analysis, module and applications," in *Proc. Int. Symp. on Fault-Tolerant Computing (FTCS)*, 1995, pp. 381–390.

[21] D. Rasch, "Sample size determination for estimating the parameter of an exponential distribution," *Akademie der Landwirtacheftswissenschaften der DDR Forschungszentrurn für Tierproduktion*, vol. 19, pp. 521–528, 1977.

[22] K. Trivedi, *Probability and Statistics with Reliability, Queueing and Computer Science Applications*, 2nd edn. John Wiley & Sons, 2001.

[23] J. Helton and F. Davis, "Latin hypercube sampling and the propagation of uncertainty in analyses of complex systems," *Reliability Engineering and System Safety*, vol. 81, no. 1, pp. 23–69, 2003.

[24] M. D. McKay, R. J. Beckman, and W. J. Conover, "A comparison of three methods for selecting values of input variables in the analysis of output from a computer code," *Technometrics*, vol. 42, no. 1, pp. 55–61, Feb. 2000.

[25] M. E. Johnson, *Multivariate Statistical Simulation*. John Wiley & Sons, 1987.

[26] N. Sato and K. S. Trivedi, "Stochastic modeling of composite web services for closed-form analysis of their performance and reliability bottlenecks," in *Proc. 5th Int. Conf. on Service-Oriented Computing*, eds. B. J. Krämer, K.-J. Lin, and P. Narasimhan. Springer, 2007, pp. 107–118.

[27] Z. Zheng, K. Trivedi, K. Qiu, and R. Xia, "Semi-Markov models of composite web services for their performance, reliability and bottlenecks," *IEEE Transactions on Services Computing*, vol. 6, no. 1, 2016.

[28] K. S. Trivedi, D. Wang, and J. Hunt, "Computing the number of calls dropped due to failures," in *Proc. 21st IEEE Int. Symp. on Software Reliability Engineering*, IEEE, 2010, pp. 11–20.

[29] M. Calzarossa, R. A. Marie, and K. S. Trivedi, "System performance with user behavior graphs," *Performance Evaluation*, vol. 11, no. 3, pp. 155–164, 1990.

[30] D. Wang and K. S. Trivedi, "Modeling user-perceived reliability based on user behavior graphs," *International Journal of Reliability, Quality and Safety Engineering*, vol. 16, no. 4, pp. 303–329, 2009.

[31] R. Sahner, K. Trivedi, and A. Puliafito, *Performance and Reliability Analysis of Computer Systems: An Example-based Approach Using the SHARPE Software Package*. Kluwer Academic Publishers, 1996.

[32] G. Bolch, S. Greiner, H. de Meer, and K. S. Trivedi, *Queueing Networks and Markov Chains: Modeling and Performance Evaluation with Computer Science Applications*, 2nd edn. Wiley-Interscience, Apr. 2006.

[33] S. Mondal, X. Yin, J. Muppala, J. Alonso Lopez, and K. S. Trivedi, "Defects per million computation in service-oriented environments," *IEEE Transactions on Services Computing*, vol. 8, no. 1, pp. 32–46, Jan. 2015.

[34] Y. Bao, X. Sun, and K. S. Trivedi, "A workload-based analysis of software aging, and rejuvenation," *IEEE Transactions on Reliability*, vol. 54, no. 3, pp. 541–548, 2005.

[35] J. Blake and K. Trivedi, "Multistage interconnection network reliability," *IEEE Transactions on Computers*, vol. C-38, pp. 1600–1604, 1989.

[36] J. Blake and K. Trivedi, "Reliability analysis of interconnection networks using hierarchical composition," *IEEE Transactions on Reliability*, vol. 38, no. 1, pp. 111–120, 1989.

[37] S. J. Bavuso, J. Bechta Dugan, K. Trivedi, E. M. Rothmann, and W. E. Smith, "Analysis of typical fault-tolerant architectures using HARP," *IEEE Transactions on Reliability*, vol. R-36, no. 2, pp. 176–185, Jun. 1987.

[38] K. Hjelmgren, S. Svensson, and O. Hannius, "Reliability analysis of a single-engine aircraft FADEC," in *Proc. Ann. Reliability and Maintainability Symp.*, 1998, pp. 401–407.

[39] H. Kantz and K. Trivedi, "Reliability modeling of the MARS system: A case study in the use of different tools and techniques," in *Proc. 4th Int. Workshop on Petri Nets and Performance Models*, 1991, pp. 268–277.

[40] D. Logothetis and K. S. Trivedi, "Reliability analysis of the double counter-rotating ring with concentrator attachments," *IEEE/ACM Transactions on Networking*, vol. 2, no. 5, pp. 520–532, Oct. 1994.

[41] D. N. P. Murthy and W. Blishke, *Reliability: Modeling, Prediction, and Optimization*. John Wiley & Sons, 2000.

[42] A. Ramesh, D. Twigg, U. Sandadi, and T. Sharma, "Reliability analysis of systems with operation-time management," *IEEE Transactions on Reliability*, vol. 51, no. 1, pp. 39–48, 2002.

[43] A. V. Ramesh, D. W. Twigg, U. R. Sandadi, T. C. Sharma, K. S. Trivedi, and A. K. Somani, "An integrated reliability modeling environment," *Reliability Engineering and System Safety*, vol. 65, no. 1, pp. 65–75, 1999.

[44] W. E. Smith and K. S. Trivedi, "Dependability evaluation of a class of multi-loop topologies for local area networks," *IBM Journal of Research and Development*, vol. 33, no. 5, pp. 511–423, 1989.

[45] D. Codetta-Raiteri and L. Portinale, "Dynamic Bayesian networks for fault detection, identification, and recovery in autonomous spacecraft," *IEEE Transactions on Systems, Man, and Cybernetics: Systems*, vol. 45, no. 1, pp. 13–24, Jan. 2015.

[46] K. Tiassou, K. Kanoun, M. KaÃćniche, C. Seguin, and C. Papadopoulos, "Aircraft operational reliability: A model-based approach and a case study," *Reliability Engineering and System Safety*, vol. 120, pp. 163–176, 2013.

[47] S. Garg, Y. Huang, C. Kintala, S. Yajnik, and K. Trivedi, "Performance and reliability evaluation of passive replication schemes in application level fault tolerance," in *Proc. 29th Ann. Int. Symp. on Fault-Tolerant Computing (FTCS)*, 1999, pp. 322–329.

[48] R. Ghosh, F. Longo, F. Frattini, S. Russo, and K. S. Trivedi, "Scalable analytics for IaaS cloud availability," *IEEE Transactions on Cloud Computing*, vol. 2, no. 1, pp. 57–70, 2014.

[49] O. C. Ibe, R. C. Howe, and K. S. Trivedi, "Approximate availability analysis of VAXcluster systems," *IEEE Transactions on Reliability*, vol. 38, no. 1, pp. 146–152, 1989.

[50] O. Ibe, A. Sathaye, R. Howe, and K. Trivedi, "Stochastic Petri net modeling of VAXcluster system availability," in *Proc. 3rd Int. Workshop on Petri Nets and Performance Models, PNPM*, 1989, pp. 112–121.

[51] M. Lanus, Y. Liang, and K. Trivedi, "Hierarchical composition and aggregation of state-based availability and performability models," *IEEE Transactions on Reliability*, vol. 52, pp. 44–52, 2003.

[52] J. K. Muppala, A. Sathaye, R. Howe, and K. S. Trivedi, "Dependability modeling of a heterogeneous VAX-cluster system using stochastic reward nets," in *Hardware and Software Fault Tolerance in Parallel Computing Systems*, ed. D. R. Avresky. Ellis Horwood, 1992, pp. 33–59.

[53] D. Tang and K. Trivedi, "Hierarchical computation of interval availability and related metrics," in *Proc. 2004 Int. Conf. on Dependable Systems and Networks*, 2004, p. 693.

[54] K. Vaidyanathan and K. Trivedi, "A comprehensive model for software rejuvenation," *IEEE Transactions on Dependable and Secure Computing*, vol. 2, no. 2, pp. 124–137, Apr. 2005.

[55] K. Vaidyanathan, R. Harper, S. Hunter, and K. Trivedi, "Analysis and implementation of software rejuvenation in cluster systems," in *Proc. Joint Int. Conf. on Measurements and Modeling of Computer Systems, SIGMETRICS/Performance*, 2001, pp. 62–71.

[56] A. Bobbio, G. Bonanni, E. Ciancamerla, R. Clemente, A. Iacomini, M. Minichino, A. Scarlatti, R. Terruggia, and E. Zendri, "Unavailability of critical SCADA communication

links interconnecting a power grid and a telco network," *Reliability Engineering and System Safety*, vol. 95, pp. 1345–1357, 2010.

[57] J. Zhao, Y. Wang, G. Ning, K. S. Trivedi, R. Matias, and K. Cai, "A comprehensive approach to optimal software rejuvenation," *Performance Evaluation*, vol. 70, no. 11, pp. 917–933, 2013.

[58] A., Avritzer, F. D. Giandomenico, A. Remke, and M. Riedl, *Assessing Dependability and Resilience in Critical Infrastructures: Challenges and Opportunities*. Springer, 2012, pp. 41–63.

[59] Y. Cao, H. Sun, and K. S. Trivedi, "Performance analysis of reservation media-access protocol with access and serving queues under bursty traffic in GPRS/EGPRS," *IEEE Transactions on Vehicular Technology*, vol. 52, no. 6, pp. 1627–1641, 2003.

[60] R. Ghosh, F. Longo, V. K. Naik, and K. S. Trivedi, "Modeling and performance analysis of large scale IaaS clouds," *Future Generation Computer Systems*, vol. 29, no. 5, pp. 1216–1234, 2013.

[61] R. Ghosh, F. Longo, R. Xia, V. K. Naik, and K. S. Trivedi, "Stochastic model driven capacity planning for an infrastructure-as-a-service cloud," *IEEE Transactions on Services Computing*, vol. 7, no. 4, pp. 667–680, 2014.

[62] J. K. Muppala, K. S. Trivedi, V. Mainkar, and V. G. Kulkarni, "Numerical computation of response time distributions using stochastic reward nets," *Annals of Operations Research*, vol. 48, pp. 155–184, 1994.

[63] K. S. Trivedi, S. Ramani, and R. M. Fricks, "Recent advances in modeling response-time distributions in real-time systems," *Proceedings of the IEEE*, vol. 91, no. 7, pp. 1023–1037, 2003.

[64] X. Yin, X. Ma, and K. Trivedi, "An interacting stochastic models approach for the performance evaluation of DSRC vehicular safety communication," *IEEE Transactions on Computers*, vol. 62, no. 5, pp. 873–885, May 2013.

[65] S. Garg, A. Puliafito, M. Telek, and T. Trivedi, "Analysis of preventive maintenance in transactions based software systems," *IEEE Transactions on Computers*, vol. 47, no. 1, pp. 96–107, Jan. 1998.

[66] R. Ghosh, K. S. Trivedi, V. Naik, and D. S. Kim, "End-to-end performability analysis for infrastructure-as-a-service cloud: An interacting stochastic models approach," in *Proc. IEEE Pacific Rim Int. Symp. on Dependable Computing (PRDC)*, Dec 2010, pp. 125–132.

[67] M. Hsueh, R. K. Iyer, and K. S. Trivedi, "Performability modeling based on real data: A case study," *IEEE Transactions on Computers*, vol. 37, no. 4, pp. 478–484, 1988.

[68] F. Machida, R. Xia, and K. S. Trivedi, "Performability modeling for RAID storage systems by Markov regenerative process," *IEEE Transactions on Dependable and Secure Computing*, vol. 67, 2016.

[69] W. Sanders and J. Meyer, "Reduced base model construction methods for stochastic activity networks," *IEEE Journal on Selected Areas in Communications*, vol. 9, no. 1, pp. 25–36, Jan. 1991.

[70] R. Smith, K. Trivedi, and A. Ramesh, "Performability analysis: Measures, an algorithm and a case study," *IEEE Transactions on Computers*, vol. C-37, pp. 406–417, 1988.

[71] L. Gönczy, S. Chiaradonna, F. D. Giandomenico, A. Pataricza, A. Bondavalli, and T. Bartha, *Dependability Evaluation of Web Service-Based Processes*. Springer, 2006, pp. 166–180.

[72] S. Wang, X. Cui, J. Shi, M. Tomovic, and Z. Jiao, "Modeling of reliability and performance assessment of a dissimilar redundancy actuation system with failure monitoring," *Chinese Journal of Aeronautics*, vol. 29, no. 3, pp. 799–813, 2016.

[73] P. Heegaard and K. Trivedi, "Survivability quantification of communication services," in *Proc. 38th Ann. IEEE/IFIP Int. Conf. on Dependable Systems and Networks, DSN*, 2008, pp. 462–471.

[74] A. Koziolek, A. Avritzer, S. Suresh, D. S. Menasché, M. C. Diniz, E. de Souza e Silva, R. M. M. Leão, K. S. Trivedi, and L. Happe, "Assessing survivability to support power grid investment decisions," *Reliability Engineering and System Safety*, vol. 155, pp. 30–43, 2016.

[75] Y. Liu, V. Mendiratta, and K. Trivedi, "Survivability analysis of telephone access network," in *Proc. 15th Int. Symp. on Software Reliability Engineering (ISSRE)*, 2004, pp. 367–378.

Author Index

Abdallah, H., 416
Abraham, J., 150, 155, 156
Ackaret, J., 8, 22, 72, 237, 239, 342, 343, 345, 347, 348, 578, 656, 657, 666
Adams, R., 7
Agerwala, T., 482
Aggarwal, K., 184
Agrawal, A., 150, 163, 196
Ajmone Marsan, M., 461, 463–465, 476, 482, 578
Akers, J., 184, 247, 248
Albert, R., 196
Aldous, D., 61, 376, 556
Aleksy, M., 604, 605
Alonso Lopez, J., 9, 16, 23, 33, 341, 381, 497, 503, 567
Amari, S., 22, 152, 184, 222, 247, 248, 292
Amer, H., 22, 292
Andrews, J., 222, 226, 227, 230, 265
Apostolakis, G., 265
Araujo, J., 586, 624
Arlat, J., 394
Arnold, T. F., 22
Asmussen, S., 552, 557, 566
Assaf, D., 376, 556
Atif, K., 578
Avi-Itzhak, B., 443
Avizienis, A., 3–7, 11, 35, 74, 137, 243, 292, 363, 394, 407
Avogadri, S., 404
Avritzer, A., 397, 400, 433, 624, 689
Axelsson, O., 418
Ayyub, B., 213

Babar, J., 189
Balakrishnan, N., 61
Balan, A., 150
Balbo, G., 461, 463–465, 482, 570, 578
Ball, M., 150, 192, 193, 196
Bank, R., 417
Bao, Y., 498–500, 578, 581, 689
Barabasi, A., 196
Baras, J. S., 632, 640
Barbetta, P., 17

Barlow, R., 36, 44, 56, 80, 93, 94, 99, 150, 163, 196, 201, 203, 217, 271, 274, 515
Bartha, T., 535, 689
Bauer, E., 7
Bavuso, S. J., 22, 250, 292, 295, 406, 578, 593, 599, 600, 689
Beaudry, M., 353, 387, 618
Beauregard, M. R., 21
Beccuti, M., 192–194, 624
Beckman, R. J., 674
Beeson, S., 265
Beichelt, F., 193
Bellmore, M., 160
Bello, G., 404
Ben-Haim, H., 121
Benoit, A., 578
Beounes, C., 394
Berénguer, C., 156
Bernardi, S., 202
Bhat, U., 535
Bhave, P., 161
Billinton, R., 557
Birnbaum, Z., 110
Birolini, A., 36
Blake, J., 33, 127, 129, 130, 350, 578, 624, 689
Blakemore, A., 353, 469, 482, 578
Blishke, W., 689
Bobbio, A., 8, 10, 26, 33, 59, 61, 99, 110, 118, 119, 125, 138, 150, 170, 172–176, 179, 181, 184, 191–194, 240, 245, 250, 253, 255, 325–327, 331, 384, 385, 400, 409, 444, 445, 458, 461, 476, 482, 544, 546, 552, 553, 556–558, 564, 567, 578, 583, 596, 599, 601, 603, 604, 689
Boccaletti, S., 196
Bolch, G., 353, 423, 439, 450, 578, 686
Bologna, S., 240
Bonanni, G. A., 170, 172, 173, 689
Bondavalli, A., 535, 590, 639, 689
Bondi, A., 433
Boudali, H., 250, 603
Bouissou, M., 243
Bouricius, W., 134, 138, 292
Bovenzi, A., 497

Box, G. E., 16
Boyd, M., 250, 599, 600
Boyer, J., 566
Bozkurt, I., 578
Brace, K., 152
Brameret, P. A., 33
Brams, G., 482
Brayton, R, 185
Brewer, J., 570
Brown, A., 92
Brown, D., 237
Bruell, S., 570, 578
Bryant, R., 150, 152, 153, 155
Bucci, G., 220, 221
Buchholz, P., 557, 570, 578
Bux, W., 557
Byres, J., 243

Cai, K., 17, 516, 517
Caldarelli, G., 196
Caldarola, L., 247
Campo, R., 515
Candea, G., 92
Cao, Y., 511, 513, 689
Carlson, C. S., 21
Carnevali, L., 220, 221
Carrasco, J. A., 416
Carravetta, M., 546
Carter, W., 134, 138, 292
Casale, G., 632
Castro, R., 123, 124, 137, 338, 339, 578, 650, 656
Catanzaro, M., 196
Çekyay, B., 590
Châtelet, E., 156
Chakravarthy, S., 566
Chandy, K. M., 578
Chao, M. T., 140
Chavez, M., 196
Chemla, J., 553
Chen, D., 5, 11, 59, 514, 529, 570, 623, 669–671
Chen, P., 570
Chen, P.-C., 530
Chiang, D., 140, 141
Chiaradonna, S., 535, 624, 689
Chimento, P., 10, 97, 353, 450, 469, 482, 546, 578, 611
Chiola, G., 461, 482, 570
Choi, H., 388, 390, 391, 482, 545, 546
Chou, Q., 603
Chu, T., 265
Cian, A. D., 243
Ciancamerla, E., 170, 172, 173, 192, 240, 250, 253, 255, 689
Ciarambino, I., 22, 202, 265

Ciardo, G., 25, 33, 185, 353, 458, 464, 469, 470, 476–478, 482, 539, 546, 570, 578, 580, 582, 618, 632, 639
Cinlar, E., 316, 510, 543, 544
Clark, G., 25
Clarke, E., 174
Clarotti, C., 264
Clemente, R., 170, 172, 173, 689
Codetta-Raiteri, D., 8, 22, 250, 251, 255, 262, 470, 601–604, 689
Coit, D. W., 25
Cojazzi, G., 265
Colbourn, C. J., 150, 196
Colombari, V., 404
Commault, C., 553
Conover, W. J., 674
Conte, G., 461, 463–465, 476, 482, 578
Contini, S., 201, 229, 264, 265
Cooper, G., 251
Coppit, D., 601
Cormen T., 580
Cosulich, G., 122, 146, 386, 493
Coughran, J., 417
Courtney, T., 25
Courtois, P. J., 33, 578
Couvillon, J., 469
Cox, D. R., 74, 305, 313, 314, 353, 419, 552, 557, 570
Cramer, H., 16
Credle, R., 237
Cristian, F., 5
Cruon, R., 160, 274
Cui, X., 689
Cullinane, M., 161
Cumani, A., 59, 461, 476, 482, 553, 554, 557, 558, 570

Dai, P., 155, 156
Dai, Y., 184, 191, 247, 248
Daly, D., 25
Dal Cin, M., 22
Daneshmand, M., 340
Dasarathy, B., 605, 607, 611
Davio, M., 570
Davis, C., 535
Davis, F., 674
Davis, L., 237
Dayar, T., 353, 578
de Meer, H., 22, 353, 423, 439, 450, 578, 686
de Souza e Silva, E., 33, 397, 400, 412, 624, 689
Deavours, D., 25
Del Bello, R., 482
Demichela, M., 265
Dempster, A., 557
Derisavi, S., 25
Derman, C., 140, 141

Der Kiureghian, A., 27
D'Esposito, M., 546
Deswarte, Y., 243
de LietoVollaro, A., 147
de LietoVollaro, R., 147
Dharmaraja, S., 439, 636, 637, 640, 669–671
Di Giandomenico, F., 535, 624, 689
Diniz, M. C., 397, 624, 689
Distefano, S., 558, 559, 561, 570
Ditlevsen, O., 27
Dohi, T., 16, 552
Dohmen, K., 155
Dominicis, R. D., 546
Donatelli, S., 570, 624
Donatiello, L., 611
Dondossola, G., 624
Dorogovtsev, S., 196
Doyle, J., 25
Drechsler, R, 185
Du, M. L., 505
Duane, J. T., 501
Dugan, J. B., 17, 22, 229, 250, 292, 295, 406, 476,
    482, 505, 578, 593, 596, 599–601, 603, 689
Dutheillet, C., 474
Dutuit, Y., 156, 222, 228–230, 265, 601

Egidi, L., 245
Elerath, J. G., 589
Elmakis, D., 121
Elsayed, E. A., 36
Estanqueiro, A., 505

Faber, D., 460, 482
Faddy, M., 557
Feller, W., 363, 490
Feng, C., 155, 156
Fernández-Villodre, G., 570
Fichtner, W., 417
Filho, P., 17
Finger, J. M., 31, 36
Firpo, P., 122, 146, 386, 493
Fischer, W., 474
Fishman, G. S., 196
Fleming, J. L., 62
Florin, G., 461, 482
Fortmann, T. E., 490
Fovino, I., 243
Fox, A., 92
Fox, B. L., 416
Franceschinis, G., 192–194, 624
Frank, P. M., 26
Franz, M., 243
Frattini, F., 497, 632, 634, 640, 689
Freire, R., 469
Fricks, R., 32, 34, 217, 384, 385, 517, 539, 540,
    624, 689

Fu, J. C., 140
Fujita, M., 174
Fulton, N. D., 11, 59, 498, 522, 670

Gail, H., 412
Garg, S., 5, 482, 500, 522, 581, 623, 624, 689
Gaver, D., 443, 450
Ge, D., 603
Ge, L., 213, 214
Geist, R., 292, 407, 482, 505, 592
German, R., 332, 482, 546
Gertsbakh, I. B., 196
Ghanta, S., 578
Gharbi, N., 474
Ghosh, R., 33, 336, 578, 619, 624, 632, 634, 640,
    689
Girtler, J., 512
Glynn, P. W., 416
Goševa-Popstojanova, K., 5, 520, 522, 538
Goel, A., 502, 503
Gokhale, S., 61, 502
Gönczy, L., 535, 689
Gopal, K., 184
Goyal, A., 33, 98, 99, 331, 353
Goševa-Popstojanova, K., 16, 611, 612
Grassmann, W., 412, 413, 415, 416, 503
Gray, J., 89
Green, D., 237
Gregg, D. M., 16
Greiner, S., 353, 423, 439, 450, 578, 686
Gribaudo, M., 482
Grosh, D. L., 201, 599
Gross, D., 566
Grosse, E., 417
Grottke, M., 5, 11, 16, 33, 349, 433, 503, 532, 567,
    665, 671
Grouchko, D., 160, 274
Gu, T., 174
Gu, W., 17
Guck, D., 599
Guillemin, F., 566
Guiotto, A., 8, 604
Gulati, R., 229, 601
Gupta, J., 184
Gupta, R., 161

Haas, P., 482
Hachtel, G. D., 174
Haddad, S., 570
Hairer, E., 503
Han, J., 511, 513, 530, 541, 578
Hannius, O., 8, 235, 590, 689
Happe, L., 397, 400, 624, 689
Hardy, G., 150, 163, 192, 193
Haring, G., 632
Harper, R. E., 689

Harris, C., 566
Harrison, P. G., 632
Hartigan, J. A., 522
Haverkort, B. R., 331, 353, 578, 608, 624
Heegaard, P., 622, 689
Heidelberger, P., 611, 640
Heidtmann, K., 155, 156
Heimann, D., 331, 450
Helton, J., 674
Henley, E., 10, 11, 201, 264
Herrmann, J., 192, 193
Herzog, U., 557, 578
Hillston, J., 578
Hirel, C., 25, 464, 639
Hitz, K. L., 490
Hixenbaugh, A., 201, 264
Hjelmgren, K., 8, 235, 590, 689
Ho, G., 460, 482
Hoath, P., 340
Hoffmann, G., 294
Hoffmann, R., 558
Holtzman, J., 530
Hong, Y., 523
Horton, G., 482
Horváth, A., 482, 503, 552, 557, 558
Horváth, G., 552, 553, 557
Houdt, B. V., 570
Howard, R., 329, 353, 546
Howe, R. C., 33, 34, 624, 689
Hsu, G. H., 397
Hsueh, M., 17, 292, 689
Huang, Y., 11, 59, 498, 522, 624, 670, 689
Hudson, J., 184
Hunt, J., 7, 9, 17, 25, 99, 339, 340, 578, 659, 668, 680, 681
Hunter, L., 94
Hunter, S. W., 689
Huslende, R., 450
Hwang, D. U., 196

Iacomini, A., 170, 172, 173, 689
Ibe, O., 34, 624, 689
Iliadis, I., 589
Imai, H., 163
Incalcaterra, P., 240
Iung, B., 250
Iversen, G., 16
Iyer, R. K., 17, 292, 611, 689

Jagerman, D., 546
Jane, C., 174, 193, 194
Jenkins, G. M., 16
Jensen, A., 415
Jensen, K., 482
Jensen, P., 160
Jensen, U., 559, 570

Jessep, D., 134, 138, 292
Jiao, Z., 689
Johnson, C., 340, 558
Johnson, M., 557
Johnson, M. E., 674
Johnson, R., 469
Junges, S., 599
Jürjens, J., 202

Kaaniche, M., 243, 689
Kai, Y., 247, 689
Kalbarczyk, Z., 17
Kam, T., 185
Kanoun, K., 394, 689
Kantam, R., 61
Kantz, H., 689
Kao, E. P. C., 528, 535
Kapur, K. C., 36, 184
Katoen, J. P., 599
Kaufmann, A., 160, 274
Kececioglu, D., 16
Kemeny, J. G., 305, 528
Kemper, P., 570, 578
Kendall, D., 427
Khurana, H., 245
Kim, D. S., 3, 26, 33, 243–246, 336, 350, 619, 624, 689
Kintala, C. M., 11, 59, 498, 522, 624, 670, 689
Kleinrock, L., 423, 450
Klir, G. J., 25
Kogan, Y., 340
Kolettis, N., 11, 59, 498, 522, 670
Kolowrocki, K., 184
Koutras, M. V., 140
Koziolek, A., 397, 400, 624, 689
Kristensen, L., 482
Kropp, C., 240
Kulkarni, V., 10, 96, 97, 305, 313, 314, 331, 353, 419, 450, 482, 539, 543–546, 564, 611, 624, 689
Kumamoto, H., 11, 201, 264
Kuo, S., 192

Lacey, P., 213
Laih, Y. W., 193, 194
Laird, N., 557
Lakatos, L., 423
Lala, J. H., 669
Lambert, J. D., 503
Lampka, K., 329, 332
Landers, T. L., 160
Landwehr, C., 3, 5–7, 11, 74, 243, 407
Langseth, H., 250
Lanus, M., 624, 689
Laprie, J.-C., 3–7, 11, 35, 74, 243, 394, 407, 619
Latora, V., 196

Latouche, G., 570
Lavenberg, S., 331, 353
Lawless, J. F., 16
Leão, R. M. M., 397, 400, 624, 689
Lee, F., 624, 647
Lee, S., 557
Lee, S. C., 16
Lee, W. S., 201, 599
Leemis, L., 16, 36
Leguesdron, P., 431
Lei, L., 578, 632, 638, 640
Leiserson, C. E., 580
Lesage, J. J., 599, 603
Levashenko, V., 184
Leveson, N. G., 11
Levikson, B., 376, 556
Levitin, G., 121, 184
Levy, Y., 340
Li, D., 603
Li, L., 669–671
Li, Y. F., 505
Liang, Y., 624, 689
Lie, C. H., 201, 599
Lieberman, G., 140, 141
Liew, S. C., 450
Limbourg, P., 25, 27, 33
Limnios, N., 150, 163, 192, 193, 505, 546
Lin, C., 578, 632, 638, 640
Lin, Y. H., 505
Lindemann, C., 363, 412, 413, 639
Lisnianski, A., 121, 184
Liu, M., 640
Liu, Y., 194, 324, 439, 620, 622, 689
Logothetis, D., 439, 546, 689
Longo, F., 558, 559, 561, 570, 578, 632, 634, 640, 689
Lopes, V., 505
Lu, K. W., 450
Lu, S., 192
Lu, Y., 558
Lucet, C., 150, 163, 192, 193
Luo, T., 155, 156
Lüttgen, G., 185
Lyu, M. R., 11, 502

Ma, X., 526, 578, 632, 634, 640, 689
Ma, Y., 530, 541, 578, 590
Ma, Z., 398
Machida, F., 26, 33, 350, 589, 689
Maciel, P., 26, 33, 350, 586, 624
Madan, B. B., 5, 520, 522, 538, 636, 637, 640
Mainkar, V., 33, 632, 634, 640, 689
Makam, S., 292
Malhis, L., 546

Malhotra, M., 28, 32, 202, 206, 253, 363, 412, 413, 418, 419, 470, 471, 503, 504, 579, 586, 588, 632
Malik, H., 61
Mancieri, L., 147
Mandayam, N. B., 530
Marathe, M., 624, 647
Marcot, B., 250
Margolis, B., 535
Maria, A., 23
Marie, R., 331, 363, 397, 400, 409, 411, 413, 416, 539, 546, 618, 624, 632
Marquez, D., 250
Marshall, F. M., 35, 255
Martin, B., 35
Masera, M., 243
Masuda, Y., 331
Matias, R., 17, 23, 516, 517, 689
Matos, R., 26, 33, 586, 624
Matuzas, V., 229, 264
Mayer, R., 556
Mayes, R. L., 27, 69
Mays, L., 161
McCluskey, E., 22, 292
McDermid, J., 22
McDermott, R. E., 21
McGeer, P. C., 174
McGough, J., 409, 590, 596
McKay, M. D., 674
McMillan, K., 174
Medhi, D., 3, 243, 624
Medhi, J., 538
Medina-Oliva, G., 250
Meier, K., 558, 570
Meier-Hellstern, K., 474
Menasche, D. S., 397, 400, 624, 689
Mendes, J., 196
Mendiratta, V., 622, 689
Meng, F., 232
Merle, G., 599, 603
Merlin, P., 460, 482
Merseguer, J., 202
Meyer, J., 184, 329, 332, 353, 446, 482, 578, 608, 624, 689
Mi, J., 56
Mikulak, R. J., 21
Miller, A., 558
Miller, D., 185, 243, 353
Miller, G. K., 535
Miller, H., 305, 419, 552, 557
Miner, A., 185, 189, 570
Minichino, M., 170, 172, 173, 192, 240, 250, 253, 255, 689
Mishra, K., 25, 27, 33, 69, 71, 296, 668, 670–672
Misra, K. B., 11
Misra, R., 292

Mittal, N., 331, 450
Moler, C., 414
Molloy, M., 460, 482
Mondal, S., 9, 341, 381
Montani, S., 250, 603
Montgomery, D., 16
Montoro-Cazorla, D., 558
Moreaux, P., 570
Moreno, Y., 196
Moslehl, A., 35, 255
Muntz, R. R., 33
Muppala, J., 9, 25, 27, 32, 33, 331, 341, 353, 381,
     412, 418, 464, 469, 477, 478, 482, 503, 504,
     578, 608, 624, 689
Mura, I., 590, 639
Murphy, K., 603
Murthy, D. N., 11, 689
Musa, J., 5, 503
Myers, A., 22

Nabli, H., 331
Naik, V., 33, 578, 624, 632, 634, 640, 689
Nail, M., 250
Nakka, N., 17
Nam, P., 250
Naor, P., 443
Nathan, S., 123, 124, 137, 578, 650, 656
Natkin, S., 460, 461, 482
Naylor, T. H., 31, 36
Nelson, W. B., 17
Nerman, O., 552, 557
Neuts, M., 376, 552, 553, 558, 564, 566, 570
Newman, M., 173, 196
Ng, Y., 363
Nguyen, D. G., 11
Nguyen, L., 194
Nicol, D., 243, 482
Nicola, V., 10, 96–99, 331, 450, 482, 564, 611
Nikora, A., 5, 16, 33, 503, 669–671
Ning, G., 17, 516, 517, 689
Niu, S., 140, 141
Norpoth, H., 16
Nørsett, S. P., 503

O'Cinneide, C., 553, 556
Obal, W., 469
Ohba, M., 502
Okamura, H., 552
Okumoto, K., 502, 503
Oliva, S., 265
Oliveira, D., 586, 624
Olsson, M., 552, 557
Oodan, A., 340
Oprisan, G., 546
Ortalo, R., 243
Ortega, J. M., 632

Osaki, S., 502
Ouyang, M., 19
Özekici, S., 590

Paez, T. L., 27, 69
Page, L., 150
Papadopoulos, C., 689
Papageorgiou, G., 632, 640
Papoulis, A., 44, 99, 419
Parhami, B., 4
Parzen, E., 501
Pataricza, A., 535, 689
Patterson, D., 92
Pearl, J., 250, 251, 262
Pecht, M., 36
Pellaumail, J., 431
Peng, R., 327
Pérez-Ocón, R., 570
Pérez-Ocón, R., 558
Perry, J., 150
Peterson, J., 18, 456, 470, 482
Petri, C., 455, 482
Petriu, D., 202
Piccinini, N., 22, 202, 265, 400
Pietrantuono, R., 11
Pietre-Cambacedes, L., 243
Pievatolo, A., 517
Plateau, B., 578
Platis, A., 505
Poole, D., 251
Portinale, L., 8, 22, 250, 251, 253, 255, 262, 263,
     603, 604, 689
Poucet, A., 201
Pourret, O., 250
Premoli, A., 26, 110, 118, 119, 125, 150
Proschan, F., 36, 44, 56, 80, 93, 99, 201, 203, 217,
     271, 274
Provan, J. S., 193, 196
Puigjaner, R., 632
Puliafito, A., 25, 143, 147, 194, 202, 206, 220–222,
     239, 292, 353, 413, 482, 497, 500, 539, 544,
     546, 552, 578, 579, 581, 686, 689
Pumfrey, D., 22
Puthenpura, S., 61
Pyke, R., 509

Qian, H., 624
Qian, X., 155, 156
Qui, K., 680
Qureshi, M., 332, 469, 546

Rai, M., 469
Rai, S., 156, 174, 181, 191
Ramamoorthy, C., 460, 482
Ramani, S., 605, 607, 611, 612, 624, 689

Ramaswami, V., 566, 570
Ramesh, A., 8, 33, 194, 221, 331, 689
Randell, B., 3, 5–7, 11, 74, 144, 243, 394, 407
Rao, G., 61
Rasch, D., 673
Rasmuson, D. M., 35, 255
Rauzy, A., 22, 33, 156, 168, 169, 222, 228–230, 265, 601
Reibman, A., 33, 317, 329, 350, 353, 363, 412, 415, 417, 418, 503, 590, 624
Reinsel, G. C., 16
Reisig, W., 482
Remke, A., 689
Renda, G., 265
Rentz, N., 632, 640
Rheinboldt, W. C., 632
Riedl, M., 689
Rindos, A., 17, 25, 99, 339, 340, 489, 578, 659, 668, 681
Rivest, R. L., 580
Roberti, L., 546, 556
Robertson, D., 237
Rohr, J., 669
Rose, D., 417
Rosenthal, A., 150
Ross, S., 99, 140, 141, 184
Rothmann, E. M., 22, 292, 295, 406, 578, 593, 689
Roussel, J. M., 33, 599, 603
Roy, A., 3, 243–246
Rozenberg, G., 482
Rubens de Matos, S., 350
Rubin, D., 557
Rubino, G., 196, 305, 353, 431, 624
Rudell, R., 152
Ruijters, E., 201, 202, 264
Ruiz-Castro, J. E., 570
Russo, S., 11, 632, 634, 640, 689

Saheban, F., 340
Sahner, R., 25, 143, 147, 194, 202, 206, 220–222, 239, 292, 353, 413, 497, 552, 578, 579, 686
Salata, F., 147
Sandadi, U., 8, 221, 689
Sanders, W., 25, 243, 245, 329, 332, 416, 469, 482, 546, 578, 624, 689
Sangiovanni-Vincentelli, A., 185
Sathaye, A., 33, 34, 689
Sato, N., 26, 62, 676, 680
Satyanarayana, A., 152
Savio, S., 122, 146, 386, 493
Savolaine, C., 340
Scarlatti, A., 170, 172, 173, 689
Scarpa, M., 552, 570
Schader, M., 604, 605
Schindler, J., 589
Schmickler, L., 557

Schneeweiss, W., 154, 201, 208, 274, 275
Schneider, F. B., 5
Schneider, P., 134, 138, 292
Schneier, B., 243
Schnell, A., 604, 605
Scholz, T., 505
Sebastio, S., 194, 195
Seguin, C., 689
Sekine, K., 163
Sereno, M., 482
Sericola, B., 305, 331, 353, 431, 539, 546, 566, 618
Shampine, L. F., 503
Shanthikumar, J., 331
Sharma, T., 8, 194, 221, 689
Shedler, G., 482
Shen, W. H., 193, 194
Shen, X., 578, 632, 638, 640
Shepp, L., 61, 376, 556
Shi, J., 689
Shooman, M., 36, 90, 126, 142, 147, 271
Shpungin, Y., 196
Shrestha, A., 184, 191, 247, 248
Siegle, M., 329, 332
Siewiorek, D. P., 5, 89
Sifakis, J., 460, 482
Silva, M., 482
Simon, C., 250
Simoncini, L., 619
Sims, J. T., 669
Singh, C., 557
Sinnamon, R., 222, 226, 227, 230
Smith, R., 329, 331, 353, 406, 417, 689
Smith, W. E., 8, 17, 22, 25, 72, 99, 237, 239, 292, 295, 339, 340, 342, 343, 345, 347, 348, 578, 593, 656, 657, 659, 666, 668, 681, 689
Smotherman, M., 409, 505, 590, 596
Snell, J. L., 305, 528
Soh, S., 174, 181, 191
Somani, A., 221, 689
Some, R., 668–672
Somenzi, F., 174
Spross, L., 193
Stamatelatos, M., 35, 201, 221, 264, 599
Stein, C., 580
Stewart, W. J., 314, 352, 353, 412, 419, 578
Stoelinga, M., 201, 202, 264, 599
Sucar, L., 250
Sukhwani, H., 325–327, 583
Sullivan, K., 601
Sumita, U., 331
Sun, H., 163, 184, 247, 248, 511, 513, 590, 689
Sun, X., 498–500, 578, 581, 689
Suresh, S., 397, 400, 624, 689
Svensson, S., 8, 235, 590, 689
Swarz, R. S., 5
Sweet, A., 54

Swiler, L. P., 27, 69
Szeid, L., 423

Taaffe, M., 557
Tabatabaee, V., 632, 640
Taha, A., 160
Takacs, L., 93
Tang, D., 22, 88, 624, 689
Tang, Y., 589
Tannous, O., 603
Tantawi, A., 98, 99
Tarapore, P., 340
Telek, M., 10, 61, 423, 482, 500, 503, 539, 544, 546,
    552, 553, 557, 558, 567, 581, 689
Ternau, T., 237
Terruggia, R., 170, 172–176, 179, 181, 184,
    191–194, 245, 689
Thomasian, A., 589
Thumler, A., 557
Tiassou, K., 689
Tillman, F. A., 201, 599
Tobias, P. A., 16, 36
Tomek, L., 8, 22, 72, 237, 239, 327, 342, 343,
    345, 347, 348, 578, 582, 632, 634, 656, 657,
    666
Tomovic, M., 689
Torasso, P., 263
Torres-Toledano, J., 250
Tortorella, M., 11
Traldi, L., 150
Tribastone, M., 632
Trindade, D., 16, 36, 123, 124, 137, 338, 339, 578,
    650, 656
Trivedi, K., 3, 5, 7–11, 16, 17, 22, 23, 25–28, 32–34,
    42, 48, 52, 53, 57, 59, 61, 62, 69–72, 84, 88,
    96–99, 109, 113, 115, 123, 124, 127, 129, 130,
    137, 143, 147, 150, 155, 156, 163, 184, 194,
    195, 202, 206, 217, 220–222, 230, 237, 239,
    243–248, 253, 265, 292, 295, 296, 305, 312,
    317, 324–327, 329, 331, 336, 338–343, 345,
    347–350, 353, 363, 364, 381, 384, 385, 388,
    390, 391, 397, 400, 406, 407, 409, 412, 413,
    415, 417–419, 423, 428, 430, 433, 437, 439,
    444, 445, 447, 450, 464, 469–471, 476–478,
    482, 489, 490, 497–500, 502–505, 511, 513,
    514, 516, 517, 520, 522, 523, 526, 528–530,
    532, 535, 538–541, 545, 546, 552, 556, 558,
    559, 561, 564, 566, 567, 570, 578–583, 586,
    588–590, 592, 593, 596, 605, 607–609, 611,
    612, 618–620, 622–624, 632, 634, 636, 637,
    639, 640, 645, 647, 650, 656, 657, 659, 665,
    666, 668–674, 676, 680, 681, 686, 689
Tronci, E., 240
Tsai, T., 17, 292
Tuffin, B., 25, 464, 639
Tung, Y., 161

Tvedt, J., 469
Twigg, D., 8, 194, 221, 689

Vaidyanathan, K., 5, 8, 9, 11, 520, 522, 538, 581,
    689
Valadè, I., 517
van Asseldonk, M., 213, 214
van Galen, M., 213, 214
Van Loan, C. F., 414
van Moorsel, A., 332, 416, 503, 546
Van Dijk, M. N., 503, 504
Vashaw, B., 17, 25, 99, 339, 340, 578, 659, 668, 681
Vasireddy, R., 123, 124, 137, 338, 339, 578, 650,
    656
Vaurio, J., 220, 221
Veeraraghavan, M., 150, 156, 230, 247
Venkatesan, V., 589
Verna, A., 400
Vesely, W., 35, 201, 221, 264, 599
Vicario, E., 220, 221
Vicente, E., 23
Villa, T., 185
Viniotis, I., 489

Wadia, A., 134, 138, 292
Wang, D., 7, 9, 17, 25, 32, 99, 115, 184, 194, 195,
    247, 248, 265, 339, 340, 578, 639, 659, 668,
    680, 681
Wang, S., 689
Wang, W., 388, 390, 391
Wang, Y., 17, 516, 517, 567, 689
Wanner, G., 503
Ward, K., 340
Weber, P., 250
Webster, P. G., 25
Weyuker, E., 433
Wolff, R., 432, 606
Woo, L., 578
Wood, A., 247
Wood, R. K., 152, 163
Woolet, S. P., 331, 353, 489, 566, 578, 608
Wright, R., 35

Xia, R., 589, 632, 634, 680, 689
Xie, W., 523, 639
Xing, J., 155, 156
Xing, L., 152, 184, 191, 222, 247, 248, 327, 590,
    603
Xu, J., 144, 394
Xu, Z., 174

Yajnik, S., 624, 689
Yamada, S., 502
Yan, L., 160
Yang, J., 327
Yang, Y., 174, 603

Yardley, T., 245
Yeh, F., 192
Yin, L., 517, 540
Yin, X., 9, 194, 195, 341, 381, 526, 578, 632, 634, 640, 689
Yongtae, P., 213
Youngjung, G., 213
Yuan, J., 174
Yuan, X. M., 397
Yushtein, Y., 8, 604

Zaitseva, E., 184
Zang, X., 163, 184, 247, 248, 590, 639

Zemoudeh, K., 505, 590
Zendri, E., 170, 172, 173, 689
Zhai, Q., 327
Zhang, N., 251
Zhang, R., 603
Zhang, Y., 578, 632, 638, 640
Zhao, J., 17, 516, 517, 567, 689
Zhao, X., 174
Zheng, Z., 680
Zhong, Z., 578, 632, 638, 640
Zio, E., 10, 505
Zonouz, S., 245
Zuberek, W., 460, 482

Subject Index

accelerated life-testing, 17
ACT, *see* attack and countermeasure tree
ADT, *see* attack and defense tree
aggregation, 444
agropark appraisal, 213
aircraft network
  upper and lower bounds, 194
all-terminal reliability, 152
alternating renewal process, 81
  exponential case, 85
  limiting behavior, 83
analytical solution, 18
  closed form, 18
  numerical, 19
AND gate, 205, 253
AndMin, 179, 187
AndSum, 176, 187
Apache HTTP server, 567
APH, *see* distribution, acyclic phase-type
AT, *see* attack tree
atomic exploit, 243
attack and countermeasure tree, 243, 245
  propagation rules, 244
attack and defense tree, 243
attack cost, 244
attack tree, 243
availability, 3, 6, 74, 280, 305
  models, 305
availability analysis, 81
  state transition diagram, 87
availability model, 645
availability taxonomy, 89

balance equation, 425
base repeater, 123, 213, 215, 227
basic event, 203, 221, 243
  probability, 219
Bayesian belief network, *see* Bayesian network
Bayesian network, 22, 250
  definition, 250
  k-out-of-n, 253
BDD, *see* binary decision diagram
Beaudry's method, 387

behavioral decomposition, 592
benchmark network
  weighted, 181
Bernoulli distribution, 62
binary decision diagram, 150, 155, 222, 224, 226, 247, 248
  basic properties, 152
  ordered, 153
  probability computation, 154
  reduced, 153
binary decision tree, 227
binary event, 203
binary probabilistic networks, 151
binary probabilistic weighted networks, 173
Birnbaum importance index, 110, 230, 261
birth–death process, 424
  steady state, 426
blade server system, 237
BN, *see* Bayesian network
Bohrbug, 5, 8, 349, 671
Boolean algebra
  useful properties, 265
bridge network, 155
  BDD, 157
  capacity weight, 180
  cost weight, 177
  factoring, 162
  graph-visiting algorithm, 163
  inclusion–exclusion, 156
  mincut, 160
  multi-state version, 189
  multi-valued version, 191
  SDP, 156
Brouwer's fixed-point theorem, 632

canonical disjunctive normal form, 275
CCF, *see* common cause failure
Cdf, *see* cumulative distribution function
CDNF, *see* canonical disjunctive normal form
CFR, *see* constant failure rate
Chapman–Kolmogorov equation, 306, 307
Cisco GSR 12000 router, 113, 645
CK, *see* Chapman–Kolmogorov equation

coefficient of variation, 44
coincident fault, 34
cold standby, 286, 296
common cause failure, 35, 255, 372
comparison of different model types, 569
comparison of non-exponential models, 559
completion time, 96
    constrained repair, 98
component redundancy, 126, 211
composite web service, 535, 676
computer system
    interval reliability, 114
conditional expected absorption time, 392
conditional independence, 251
conditional probability table, 251, 253
consecutive k-out-of-n, 140
    circular, 140
    linear, 140
constant failure rate, 44
continuous-time Markov chain, 305
    absorbing state, 313, 374
    balance equation, 315
    classification of states, 313
    dependability models, 317
    expected downtime, 318
    expected number of failures, 373
    expected state occupancy, 315
    expected time to absorption, 376
    expected uptime, 318
    fault/error-handling model, 406
    holding time, 312
    homogeneous, 308
    instantaneous availability, 318
    integral equation, 363
    interval availability, 318
    irreducible, 313
    Laplace transform, 364
    mean time to failure, 397
    mean time to repair, 397
    moments of time to absorption, 381
    multiple absorbing state, 388
    non-homogeneous, *see* non-homogeneous
        continuous-time Markov chain
    non-irreducible, 357
    non-null recurrent state, 313
    null recurrent state, 313
    ODE-based methods, 416
        hybrid, 419
        implicit Runge–Kutta, 418
        TR-BDF2, 417
    parametric sensitivity analysis, 351
    performance, 424
    positive recurrent state, 313
    power method, 352
    recurrent state, 313
    reliability models, 357

reward rate, 358
reward structure, 330
self-loop, 405
series expansion, 413
single absorbing state, 375
sojourn time, 312
solution
    steady-state, 314
standby configuration, 367
state
    absorbing, 357
    recurrent, 357
    transient, 357
state probability, 306
    equation, 309
stationary distribution, 313
steady-state analysis, 352
steady-state availability, 318
steady-state probability vector, 314
steady-state properties, 314
structure state, 330
successive over-relaxation, 353
symbolic solution, 412
time-averaged expected state occupancy, 316
transient solution methods, 412
transient state, 313
transition probability, 306
transition rates, 308
uniformization, 409, 415
countermeasure, 243
coverage, 292, 295
coverage factor, 22
CPT, *see* conditional probability table
criticality importance index, 231
CTMC, *see* continuous-time Markov chain
cumulative distribution function, 41
cumulative downtime distribution, 93
cybersecurity, 243, 245
cyclic queuing system, 442

decision power, 18
decomposition, 446
decreasing failure rate, 44
deductive reasoning, 201
defective distribution, 359
defects per million, 7, 340, 381
    application server, 341
    IBM SIP system, 680
    sensitivity analysis, 350
dependability, 3
    evaluation, 15
    measures, 4
    metrics, 4, 41
    service-oriented, 7, 9
    system-oriented, 7
    task-oriented, 9

depth-first search, 212
design specifications, 10
device-to-device communication, 638
DFR, *see* decreasing failure rate
DFS, *see* depth-first search
DFT, *see* dynamic fault tree
different failure causes, 391
directed networks, 151
discrete-time Markov chain
    embedded, 510
disjunctive normal form, 154, 208, 275
distribution
    acyclic phase-type, 553
        canonical form, 553
        eigenvalue, 556
        order two, 553
    Coxian, 552
    defective, 64
    Erlang, 552
    gamma, 61
    hyperexponential, 55, 552
    hypoexponential, 58, 292, 515, 552
    log-logistic, 61
    lognormal, 54, 516
    mass at origin, 63
    normal, 52
        standard, 52
    phase-type, 54, 376, 552
        asymptotic hazard rate, 556
        closure properties, 556
        eigenvalue, 556
        fitting algorithms, 557
        fitting long-tail distribution, 558
        fitting Weibull distribution, 557
        minimum coefficient of variation, 556
        properties, 556
        queuing models, 566
        system modeling, 558
        task completion time, 563
    Weibull, 50, 109, 491, 495, 515
DNF, *see* disjunctive normal form
dormancy factor, 289, 367
dormant condition, 220
drop policy, 96
Duane model, 501
duplex processor, 396, 491, 492
    with repair, 495
dynamic fault tree, 599
dynamic redundancy, 286

early-life failure, 45
embedded DTMC, 509
ENF, *see* expected number of failures
EP, *see* exponential polynomial
epistemic uncertainty propagation, 69, 295
    cold standby, 296

hot standby, 299
    multi-voltage high-speed train, 614
    warm standby, 298
equivalent failure rate, 89
equivalent repair rate, 89
Erlang distribution, 59, 288
Erlang random variable, 287
expected first-passage time, 397
expected interval availability, 88
expected number of failures, 264
expected reliability, 71
expected value, 43
exponential distribution, 46
    normalized, 48
    shifted, 50
    typical behavior, 47
exponential polynomial, 552
exponomial, 552

factoring, 162, 226
fail-danger, 64, 133
fail-safe, 64, 133
failure mode and effect analysis, 21
failure mode effects and criticality analysis, 21
failure rate, 44
    bathtub curve, 45
fault detection, isolation, and repair, 22
fault management, 286
fault tree, 201
    analysis, 201
    binary, 203
    case study, 234
    coherent, 203
    graph-visiting algorithm, 212
    k-out-of-*n*, 207
    logical expression, 208
    mincut, 208
    multi-state, 203, 247
    non-coherent, 203, 265
    OR and AND gate, 206
    qualitative analysis, 208
    quantitative analysis, 219
        non-repeated events, 221
    repeated events, 206, 213, 226, 229
    two-level representation, 209
fault-occurrence model, 593
fault/error handling, 22
fault/error-handling mechanism, 292, 295
fault/error-handling model, 406, 593
FDIR, *see* fault detection, isolation, and repair
FEH, *see* fault/error handling
FEHM, *see* fault/error-handling model
field measurement, 16
fire-fighting station, 400
first-passage time, 397
five-component system, 282

fixed-point
   availability models, 632
   equation, 631
   import graph, 632
   iteration, 631
   performance models, 636
   reliability models, 635
fluid level controller, 108, 111
   approximate solution, 583, 632, 635
   hierarchical solution, 582
   redundancy, 120, 121, 127, 584
   Weibull distribution, 109
FMEA, *see* failure mode and effect analysis
FMECA, *see* failure mode effects and criticality
      analysis
FOM, *see* fault-occurrence model
FT, *see* fault tree
FTA, *see* fault tree, analysis
functional dependence, 34
fundamental renewal equation, 76

GARR network, 169
   weighted, 182
Goel–Okumoto model, 503
graph-visiting algorithm, 152, 163, 212

hazard operability analysis, 21, 22
hazard rate, 44
HAZOP, *see* hazard operability analysis
hearing aid, 287
hierarchical composition, 578
hierarchical model, 577
   availability, 577, 580
   behavioral decomposition, 592
   dynamic fault tree, 577, 599
   performability, 577, 608
   performance, 577, 604
   phased mission systems, 590
   reliability, 577, 586
      behavioral decomposition, 593, 597
   survivability, 577, 619
homogeneity property, 509
homogeneous Markov chain, 307
hot standby, 291, 299
human error, 34
hypoexponential random variable, 288

IBM blade server system, 342
   cooling, 343
   memory, 345
   midplane, 342
   power domain, 343
   processor, 345
   RAID, 347
   software, 348

IBM BladeCenter, 656
IBM WebSphere model, 339, 659
   application server, 664
   CPU, 662
   memory, 663
   midplane, 662
   operating system, 665
   RAID, 663
identical repairable components, 319
   common cause failures, 322
   expected cost, 331
   imperfect coverage, 321
IFR, *see* increasing failure rate
impact cost, 244
imperfect coverage, 292, 380, 454
import graph, 579
   acyclic, 580
   non-acyclic, 580
importance analysis, 110, 118, 125
importance measures, 229
improving dependability, 99
impulse reward, 329
inclusion–exclusion formula, 155
increasing failure rate, 44
inference
   diagnostic, 261, 262
   predictive, 261
infinitesimal generator matrix, 308, 358
INH, *see* inhibit gate
inhibit gate, 264
interval reliability, 94, 114, 586, 590

k-out-of-n, 291, 292, 299
k-out-of-n gate, 253
K-terminal reliability, 152
Kolmogorov differential equation, 309, 489
Kronecker delta function, 490

labeled directed graph, 272
Laplace transform, 311, 553
largeness, 577
   avoidance, 577
   tolerance, 577, 578
Little's law, 430
load dependence, 34
log-logistic model, 503
logic gates, 203

$M/ER_2/1/K$
   response time distribution, 568
maintainability, 3, 6, 73
maintenance, 22
majority voting system
   exponential case, 136
   k-out-of-n, 133, 134, 137, 278

majority voting system (*cont.*)
    consecutive, 140
    non-identical, 143
    two groups, 144
    two-out-of-three, 134, 138, 139, 278, 282
Mandelbug, 5, 8, 349, 532, 671
marginal prior probability, 251
Markov model, 578
Markov property, 305, 509
Markov regenerative model, 509
Markov regenerative process, 509, 542, 543,
    552
    global kernel, 543
    local kernel, 543
    $M/D/1/2/2$ queue, 545
    $M/PH/1/2/2$ queue, 566
    steady-state analysis, 543, 544
    subordinated process, 543
    transient analysis, 543, 544
Markov reward model, 329, 330
    availability measures, 333
    equivalent failure and repair rate, 337
    expected accumulated reward, 331
    expected downtime, 333
    expected instantaneous reward rate, 330
    expected number of failures, 335
    expected number of repairs, 335
    expected number of transitions, 334
    expected number of visits, 334
    expected uptime, 333
    impulse reward, 332
    reward rate, 330
Markov-modulated Poisson process, 474, 481
Markovization, 552, 558
maximal repair, 495
maximum of random variables, 68, 117
    exponential, 68
MCS, *see* minimal cut set
MDD, *see* multi-valued decision diagram
mean downtime, 86
mean time to failure, 43, 280
mean uptime, 86
measurement, 15
measurement-based evaluation, 16
memory leak, 498
memoryless property, 47, 305
MFT, *see* fault tree, multi-state
mincut
    analysis, 152, 159
    definition, 159
    from BDD, 167
    order, 159
minimal cut set, 208, 244
    analysis, 209
    order, 210
minimal repair, 495, 559

minimum of random variables, 66, 107
    exponential, 67
minpath, 154
    analysis, 152
    connectivity, 154
    disconnectivity, 154
    from BDD, 167
    order, 154
    two-terminal reliability, 155
minterm, 275
model, 15
    abstraction, 24
    decomposition, 23
    error, 32
    idealization, 24
    monolithic, 24
    multi-level, 24
    validation, 25, 31
    verification, 25, 29
model-based evaluation, 17
modeling formalism, 27
    multi-level, 28
    non-state-space, 27
    power hierarchy, 28
    state-space, 28
modeling power, 18
modeling process, 20
modified renewal process, 81
modularization, 228, 586, 601
MRGP, *see* Markov regenerative process
MRM, *see* Markov reward model
MTBDD, *see* multi-terminal binary decision
    diagram
MTTF, *see* mean time to failure
multi-server retrial queue, 474
multi-server retrial system, 474
multi-state component, 247, 257, 272
multi-state event, 203
multi-state networks, 183
    weighted, 185
multi-state variable, 257
multi-terminal binary decision diagram, 174
    definition, 174
multi-valued decision diagram, 185, 247–249
multi-voltage high-speed train, 122, 146, 386, 493,
    613
multiprocessor system, 253, 259, 262
mutually exclusive, 271

n-identical component system, 440
NASA remote exploration system, 669
    uncertainty propagation, 671
near-coincident fault, 295
network connectivity, 152
network reliability, 150, 201
    algorithm limitation, 192

upper and lower bounds, 193
networks and graphs, 151
NHCTMC, *see* non-homogeneous continuous-time
     Markov chain
NHPP, *see* non-homogeneous Poisson process
noisy gates, 256
     noisy-AND, 257
     noisy-OR, 257
non-homogeneous continuous-time Markov chain,
     308, 489, 552, 558, 559
     convolution integral approach, 490
     equivalent-HCTMC method, 490
     numerical solution, 503
     ODE method, 490, 503
     piecewise constant approximation, 490, 494
     Runge–Kutta method, 504
     uniformization, 490, 504
non-homogeneous Poisson process, 501
non-repairable unit, 41
non-self-revealing failure, 220
non-series-parallel system, 142, 282
     factoring, 141
non-shared (independent) repair, 319

ODE IVP, *see* ordinary differential equation initial
     value problem
one repairable component
     availability, 311
operational computerized model, 24
optimal preventive maintenance, 517
OR gate, 203, 253
ordinary differential equation initial value problem,
     310
OrMin, 176, 187
OrSum, 179, 188

parallel redundancy, 115
parallel system
     availability, 119
     exponential case, 118
     reliability, 116
parameter uncertainty, 25
parametric sensitivity analysis, 349
partial parallel redundancy, 121
parts count method, 111
PCA, *see* piecewise constant approximation
pdf, *see* probability density function
Peano–Baker series, 490
performability, 446
performability model, 675
performance, 604
Petri net, 453, 455
     arc
          inhibitor, 458
          input, 456
          multiplicity, 458

output, 456
     variable cardinality, 469
dependability measures, 476
enabling function, 469
execution sequence, 457
extended reachability graph, 464
firing time, 461
guard function, 469
marking, 456
marking dependency, 469
place, 456
priority level, 460
random switch, 464
reachability graph, 457
reachability set, 457
reward-based measures, 476
state
     tangible, 464
     vanishing, 464
stochastic, 460
     generalized, 463
timed, 460
token, 456
transition, 456
     enabled, 456
     immediate, 463
     timed, 463
two repairable components, 456–459, 462, 466,
     467, 479
PH, *see* distribution, phase-type
phase-type random variable, 552
phase-type representation, 553
phased mission systems, 590
piecewise constant approximation, 494,
     500
Poisson distribution, 63
Poisson probability mass function, 78
Poisson process, 77
     buffer design, 79
posterior probability, 251
power network reliability, 170
power smart grid, 397
predictive dependability assessment, 10
preemptive repeat policy, 96
preemptive resume policy, 96
preventive maintenance, 11, 529,
     581
probabilistic approach, 33
probabilistic gate, 255
probability density function, 43
product law of reliability, 106
prototype measurement, 17

qualitative analysis, 202
quantitative analysis, 202
quantitative evaluation, 15

queue
  breakdown, 442, 453, 454, 459, 462, 467, 478, 479
  Erlang B formula, 439
  Kendall notation, 427
  $M/E_k/1$, 468, 480
  $M/M/\infty$, 435
  $M/M/1$, 428
    closed, 441
    response time distribution, 431
  $M/M/1/K$, 435
  $M/M/m$, 433, 498
  $M/M/m/m$, 438
  $MMPP/M/1/K$, 474
  single, 427
  state aggregation, 444

race model, 313
RAID
  reliability model, 588
random variable
  moments, 43
recovery block, 394
recovery process, 532
redundancy, 115
  cold standby, 368
  hot standby, 368
  warm standby, 368
redundant system with spare, 411
regeneration time point, 543
regular network, 158
  graph-visiting algorithm, 166
rejuvenation, 540
reliability, 3, 6, 42, 279
  independent components, 280
reliability block diagram, 105, 201, 292
reliability growth, 501
reliability model, 669
renal disease model, 384
renewal density, 76
renewal equation, 77
renewal function, 75
renewal process, 74
repair
  as bad as old, 495, 559
  as good as new, 495, 559
repair policy, 325
  deferred, 328
  non-preemptive priority, 325
  preemptive priority, 325
repairable fluid level controller, 325
  redundant, 326, 472
    travel time, 327
  travel time, 472
repairable systems
  independent components, 284

repairable unit, 41, 72
replacement, 559
retrial queue, 474
return on investment, 245
reward rate, 329, 400
RF channel, 541
risk assessment, 10
ROI, *see* return on investment
root variable, 251

(s,t)-terminal reliability, 152
S-shaped model, 503
safety, 3, 240, 599
safety analysis, 240
safety-critical, 64, 67, 133
SCADA, 245
scaled sensitivity
  availability, 92
SDP, *see* sum of disjoint products
security, 243, 245, 599
security computer system, 520, 538
security modeling, 243, 245
security quantification, 243, 245
self-revealing failure, 219
semi-Markov model, 509, 578
semi-Markov process, 509, 552, 558, 559
  absorbing state, 528
  different types of failures, 511
  mean time to absorption, 528
  multiple absorbing states, 538
  planned and unplanned outages, 513
  probability of absorption, 538
  RF channel, 530
  sojourn time, 509
  steady-state solution, 510
  time-based preventive maintenance, 514
  transient solution, 539
  two-state model, 510
  variance of time to absorption, 535
SEN, *see* shuffle-exchange network
sensitivity
  availability, 91
sensitivity analysis, 26, 110, 118, 125
  multi-voltage high-speed train, 614
sequentially dependent failure, 259
series system, 105
  availability, 112
  exponential case, 107
  generally distributed TTF, 109
  interval reliability, 114
  reliability, 106
series-parallel system, 119
Sf, *see* survival function
Shannon decomposition, 152, 154, 163, 174, 217, 279

shared repair, 320
ship propulsion system, 512
shuffle-exchange network, 127
    plus, 129
simulative solution, 19
single component, 358
    mass at origin, 359
single unit, 41
skydiver, 294
SMP, *see* semi-Markov process
software aging, 8, 498, 500, 522
software dependability, 8
software rejuvenation, 523
software reliability growth model, 502
    finite failure, 502
    infinite failure, 503
SRGM, *see* software reliability growth model
standby redundancy, 286
    cold, 286
    hot, 286
    warm, 286
state enumeration, 271
state probability vector, 307
state space, 271
state vector, 271
static redundancy, 286
statistical dependence, 34
steady-state availability, 86
stiffness, 577
stochastic reward nets, 469
structural importance, 217
structure function, 271, 274
    Boolean expression, 274, 275
success tree, 216
sum of disjoint products, 155, 222, 224,
    226, 227
sum of random variables, 286
Sun Microsystems, 123, 137, 650
    alarm card, 652
    backplane, 655
    CPU, 653
    Ethernet, 654
    power supply, 651
    satellite card, 655
    SCSI, 655
survivability, 397, 619
survival function, 42
system failure rate, 497, 500
system redundancy, 126, 211

task-oriented measures, 96
TE, *see* top event
telecommunication switching system, 324, 336,
    362, 439
terminal event, 203
test measurement, 17

three-component system, 273, 281
    series-parallel configuration, 276
time-dependent infinitesimal generator,
    489
tire subsystem, 554
TMR, *see* triple modular redundancy
top event, 201, 202
TPM, *see* transition probability matrix
transient availability, 86
transition probability matrix, 307, 509
transition rate matrix, 358
triple modular redundancy, 134, 136, 137
truth table, 274
turbine control system, 240
two failure modes, 359
two-component system, 272, 360, 364
    common cause failure, 372, 397
    identical, 361
    imperfect coverage, 380
    mean time to failure, 377
    parallel configuration, 276, 281
    phase-type distribution, 560
    phase-type distribution and CCF, 561
    repairable, 284, 370, 518
    series configuration, 276, 281
    series-parallel, 366
    standby configuration, 367
    with repair
        mean time to failure, 379
two-out-of-three PLC system, 234
two-terminal reliability, 152

unavailability on demand, 370
uncertainty, 27
    propagation, 27
undirected networks, 151
uninterruptible power supply, 519, 530, 540
unreliability, 42
UPS, *see* uninterruptible power supply
useful life region, 45

variance, 44
vehicular network, 526
Vesely–Fussell importance index, 232,
    261, 262

warm standby, 289, 292, 298
    safety analysis, 391
    with repair
        safety analysis, 393
warm standby system
    mean time to failure, 378
WAT, *see* weighted attack tree
wearout failure, 46
web service request, 114

weight
    capacity, 174, 178
    cost, 174, 175
weighted attack tree, 244

wireless handoff performance model,
    636
wireless sensor network, 140, 141
WSN, *see* wireless sensor network

Printed in the United States
By Bookmasters

Printed in the United States
By Bookmasters